APATITES AND THEIR SYNTHETIC ANALOGUES - SYNTHESIS, STRUCTURE, PROPERTIES AND APPLICATIONS

Edited by **Petr Ptáček**

Apatites and their Synthetic Analogues - Synthesis, Structure, Properties and Applications
http://dx.doi.org/10.5772/59882
Edited by Petr Ptáček

Contributors

Petr Ptáček

Published by InTech

Janeza Trdine 9, 51000 Rijeka, Croatia

Publishing Process Manager Danijela Duric

Technical Editor InTech DTP team

Cover InTech Design team

First published April, 2016

Printed in Croatia

Additional hard copies can be obtained from orders@intechopen.com

Apatites and their Synthetic Analogues - Synthesis, Structure, Properties and Applications , Edited by Petr Ptáček

p. cm.

Print ISBN 978-953-51-2265-4
Online ISBN 978-953-51-2266-1

For EU product safety concerns:
IN TECH d.o.o., Prolaz Marije Krucifikse Kozulić 3, 51000 Rijeka, Croatia,
info@intechopen.com or visit our website at intechopen.com.

Meet the author

Petr Ptáček was born in 1978 in Kutná Hora, Czech Republic. He received his PhD in materials science and technology from Brno University of Technology, Faculty of Chemistry in 2005. Since 2005 up to now he has been working at the Institute of Materials Chemistry and at the Materials Research Centre of Brno University of Technology as an Assistant Professor and a Senior Researcher, respectively. Current teaching and research activities include chemistry, materials structure and thermodynamics, heat treatment of silicates, ceramics and advanced ceramics, refractory materials, special inorganic cements, geopolymers, colloidal chemistry and heterogeneous kinetics. He was awarded several times for his out-standing academic performance in the fields mentioned above. Ptáček has published over 100 scientific articles, including full-length articles, book chapters and contributions to international conferences; he registered 15 national patents and participated in the development of 5 applied industrial technologies.

Contents

Preface

This book deals with the synthesis, structure, properties and applications of "sensu lato" apatite and its synthetic analogues. The book is divided into 10 chapters. In the beginning, the minerals from the supergroup of apatite and their synthetic analogues are presented, including their structure, properties, natural occurrence and utilization. Furthermore, analytical techniques used for the analysis, investigation of properties and the structure of apatites are described. Since the structure of apatites is open to the extensive substitution on cationic as well as anionic sites and tends to form solid solutions as well, immense numbers of compounds with the structure of apatite type were prepared and described in current literature. From this point of view, the book provides a basic survey over different types of these compounds, but it is far from exhaustive enumeration and description of all possible combinations. The phosphate rocks, their classification, geological role, mining and beneficiation of phosphate ore are also described. The book continues with a brief description of the utilization of phosphate ore for the production of elemental phosphorus, phosphoric acid and fertilizers. The last two chapters are dedicated to the utilization of phosphorus-bearing compounds and materials, including phosphate ceramics, glass and glass-ceramics, materials for tissue engineering, solid fuel cells, sensors, catalysts, etc.

The overall intent of this book is to provide a comprehensive insight into the structure, properties and utilization of compounds of apatite type and related phosphates. In this respect, the words used by JAFFE (please refer to **Chapters 4** and **9**) are appropriate:

The apatite group of minerals has been studied from the point of view of the physiologist, biochemist, or soil scientist, as well as from that of the geologist or mineralogist. Although all these people have had the same problems, they have not always been sufficiently acquainted with each other´s work.

On behalf of all authors, I hope that this publication holds your interest. If you, as readers, will find published information useful for your work, it will be the nicest reward for us.

Petr Ptáček
Brno University of Technology,
Czech Republic

Special Thanks

I wish to thank all the authors working with appetites and related or derived compounds for their inestimable contribution to the book. Moreover, I would like to express my appreciation to HALINA SZKLORZOVÁ for language and grammar corrections.

Acknowledgments

The book has been supported by the project: Materials Research Centre at FCH BUT – Sustainability and Development, REG LO1211, with financial support from the National Program for Sustainability I (Ministry of Education, Youth and Sports of the Czech Republic).

Chapter 1

Introduction to Apatites

Petr Ptáček

Additional information is available at the end of the chapter

http://dx.doi.org/10.5772/62208

Abstract

Apatite is the generic name, which was first introduced by German geologist A.G. Werner. These minerals and their synthetic analogs represent a major class of ionic compounds and the most common crystalline form of calcium phosphates, which are of interest of many industrial branches and scientific disciplines. Since, apatite (fluorapatite) is the most abundant phosphate mineral, apatite bearing phosphate rocks represents an important source of inorganic phosphorus. First chapter of this book introduces the basic concepts of nomenclature, composition, classification, crystal structure, mineralogy and properties of minerals from the supergroup of apatite. Furthermore, the minerals from the group of apatite and polysomatic apatites are described. Since, the most of the topics mentioned in this chapter will be developed in the following chapters, the key concepts provided in this chapter are important to understood before proceeding further.

Keywords: Apatite, Group of Apatite, Polysomatic Apatites, Fluorapatite, Hydroxylapatite, Chlorapatite, Vanadinite

The minerals[1] [1],[2],[3],[4],[5] from the apatite group[2] [6] are classified as hexagonal or pseudo-hexagonal monoclinic anhydrous phosphates containing hydroxyl or halogen of the generic formula[3]:

[1]**Minerals** are individual components comprising rocks formed by geological processes classified according to their crystal structure and chemical composition. The total number of minerals accepted by mineralogical community is about 4000. **Mineraloids** are mineral-like phases including synthetic materials, human-treated substances, and some biological materials, which do not fulfill the criteria for the definition of mineral species [2]. **Anthropogenic substances** are not considered as minerals. If such substances are identical to minerals, they can be referred as the "synthetic equivalents" of given mineral. If the synthetic substance has a simple formula, then the preference should be given to the use of a chemical formula instead of a mineral name. **Biogenic substances** can be accepted as minerals if geological processes were involved in the genesis of these compounds [1],[3],[4].

[2] The mineral classification system developed by German mineralogist K.H. STRUNZ.

[3] The variable formula should be written as: $(Ca,Sr,Pb,Y,Mn,Na)_5(PO_4,AsO_4,SO_4CO_3)_3(F,Cl,OH)$.

$$M_5(XO_4)_3Z_q$$

(**Dana classification** [7],[8]) or as phosphates, arsenates and vanadates with additional anions without water (**Strunz classification**[4] [8],[9]), where ***M*** = Ba, Ca, Ce, K, Na, Pb, Sr and Y; ***X*** = As, P,[5] and Si; and **Z** = F, Cl, O, OH and H_2O. Apatite minerals form a numerous and diverse group of minerals, while in addition a large number of synthetic compounds with the apatite type structure are known [10].

Apatite minerals can be formally derived from phosphoric acid[6] [11] (or H_3AsO_4 and H_3VO_4 for arsenates and vanadates, respectively). According to the Werner's coordination theory [12], [13], e.g. fluorapatite was considered as a compound formed by the substitution of calcium phosphate ($Ca_3(PO_4)_2$) into halide mineral fluorite (calcium fluoride, CaF_2 [14]). The coordination formula of fluorapatite can then be written as follows:

$$\left[Ca \begin{pmatrix} O & & PO_3 & Ca \\ & \rangle & Ca & \\ O & & PO_3 & Ca \end{pmatrix}_3 \right] F_2$$

The ratios of the mean sizes of ions "***M***" to "***X***" vary in the range from 1.89 to 4.43 for apatite compositions, but there are discontinuities between the ratios 2.50–2.60 and 3.25–3.35. These two gaps provide the structural base, which was used for the suggestion of classifying[7] apatites into three groups named after well-known mineral species occurring in each group [15],[16]:

1. Vanadinite–svabite group with the ***M***:***X*** ratio less than 2.5;
2. Apatite–mimetite group with the ***M***:***X*** ratio in the range from 2.60 to 3.25;
3. Pyromorphite group with the ***M***:***X*** ratio higher than 3.25.

The summary of some apatite species is listed in **Table 1**. The current nomenclature of minerals from apatite supergroup is described in Section **1.1.**

---

[4] A mineral group consists of two or more minerals with the same or essentially the same structure (i.e. isotypic structure belonging to one structural type) and composed of chemically similar elements (i.e. elements with similar crystal-chemical behavior). Crystal structures considered as being "essentially the same" can be denoted by the term *homeotypic*. The hierarchical scheme for the group nomenclature includes (1) mineral class, (2) mineral subclass, (3) mineral family, (4) mineral supergroup, (5) mineral group(s), and (6) mineral subgroup or mineral series [8].

[5] The phosphorus element was discovered by a Hamburg alchemist H. Brand. As was described in the book of G.E. Leibnitz (1646–1716) **Historia Invetions Phosphori**; phosphorus was extracted from "the spirit of urine" during the search for the philosopher´s stone. The name of phosphorus (in Latin means "morning star") was derived from Greek word "*phosphoris*," which meaning "*bringing light*" [35].

[6] In 3 moles of H_3PO_4, 8 H^+ ions were replaced by 4 Ca^{2+}, and the last H^+ was replaced by Ca-Z (Z = OH, F, Cl…).

[7] Despite the fact that this classification is out of fashion, there are some interesting structural consequences with general validity. Therefore, this classification was included to the introduction of this book.

http://dx.doi.org/10.5772/62208

The specific gravity of *sensu lato*[8] apatite ranges from 3.1 to 3.3 g·cm^{-3}. Apatites show basal and imperfect cleavage, conchoidal and uneven fracture and the hardness on the Mohs scale is 5. The color of streak is white and luster vitreous to subresinous. Apatite occurs usually in the shades of green to gray-green, also white, brown, yellow, bluish, or reddish, transparent to translucent and some specimens can be multicolored. The habit of apatite crystals ($Ca_5(PO_4)_3(F,Cl,OH)$) is usually prismatic, dipyramidal or tabular, and also massive compact or granular. Some varieties are phosphorescent when heated, and others become electric by friction. On the other hand, the morphology of apatite crystals is very complex, and there is large amount (~53) of described forms [17],[18].

Apatite occurs in a wide range of igneous and sedimentary rocks (**Chapter 7**) and deposits as isolated crystals in grains, usually as small as 1–2 mm. The largest known apatite deposit is in Kirovsk, Russia. The largest individual crystals were found in Renfrew, Ontario, Canada [19], [20].

Group of apatites					
I. The vanadite–svabites		II. The apatite–mimetites		III. The pyromorphites	
Composition	M:X	Composition	M:X	Composition	M:X
$La_8[GeO_4]_6$	1.89	$Ca_6Nd_4(SiO_4)_8$	2.62	$Pb_{10}(GeO_4)_2(PO_4)_4$	3.33
$Ca_{10}(CrO_4)_6(OH)_2$	1.94	$Ca_4Nd_6(SiO_4)_6(OH)_2$	2.65	$Pb_{10}(SiO_4)(GeO_4)(PO_4)_4$	3.44
$Pb_8Na_2(VO_4)_6$	2.13	$Ca_8Ca_2La_2(SiO_4)_6$	2.68	$Pb_{10}(PO_4)_4(SiO_4)_2$	3.49
$Ca_4Ce_6[GeO_4]_6Cl_2$	2.18	$Ca_8Ce_2(GeO_4)_2(PO_4)_4$	2.68	$Pb_8Na_2(SO_4)_2(PO_4)_2(SiO_4)_2$	3.50
$Ca_6La_4[GeO_4]_6$	2.18	$Pb_8Na_2(AsO_4)_6$	2.71	$Sr_{10}(PO_4)_6(OH,F)_2$	3.51
$Ca_4La_3Ce_3[GeO_4]_6(OH)_2$	2.20	$Pb_{10}(GeO_4)_2(AsO_4)_4$	2.71	$Ca_4Na_6(SO_4)_6F_2$	3.52
$Pb_{10}[VO_4]_6(F,ClBr,I)_2$	**2.21**	$Ca_{10}La_5Ce_5(SiO_4)_{12}(OH)_2$	2.71	$Pb_9Na(PO_4)_5(SiO_4)$	3.59
$Pb_8Tl_2[VO_4]_6$	2.22[a)]	$Ca_6La_4(SiO_4)_6$	2.72	$Pb_8Na_2(PO_4)_6$	3.62
$Ca_4La_6[GeO_4]_6(OH)_2$	2.24	$Ca_4La_8Ce_8(SiO_4)_6(OH)_2$	2.74	$Pb_8KNa(PO_4)_6$	3.73
$Pb_8K_2[VO_4]_6$	2.26	$Ca_5La_5(SiO_4)_6OH$	2.76	**$Pb_{10}(PO_4)_6(F,Cl,Br,OH)_2$**	**3.76**
$Sr_4Dd_6[SiO_4]_6(OH)_2$	2.26	$Ca_4La_6(SiO_4)_6(OH)_2$	2.80	$Pb_8Tl_2(PO_4)_6$	3.80[c)]
$Pb_{10}[GeO_4]_2(VO_4)_4$	2.30	$Ca_4La_6(SiO_4)_6F_2$	2.80	$Pb_8K_2(PO_4)_6$	3.85
$Sr_4a_6(GeO_4)_8(OH)_2$	2.34	$Pb_8KNa(ASO_4)_6$	2.80	$Pb_8Rb_2(PO_4)_6$	3.94
$Ca_{10}(AsO_4)_6(F,Cl)_2$	**2.34**	**$Pb_{10}(AsO_4)_6(F,Cl,Br,I)_2$**	**2.82**	$Pb_8Cs_2(PO_4)_6$	4.06
$Ca_4Ce_6(GeO_4)_4(SiO_4)_2Cl_2$	2.34	$Pb_8Tl_2(AsO_4)_6$	2.82[b)]	$Ba_{10}(PO_4)_6(F,OH)_2$	4.24
$La_8(SiO_4)_6$	2.36	$Tl_2La_8(SiO_4)_6(OH)_2$	2.85	$Pb_{10}(SiO_4)_2(BO_3)_4$	4.43

[8] Latin phrase (abbreviated as s.l.) used, which means "in the broad sense."

Group of apatites				
$Ca_6Ce_4(GeO_4)_4(PO_4)_2Cl_2$	2.41	$Ba_{10}(MnO_4)_6(OH)_2$	2.86	
$Ca_4Y_6(SiO_4)_6(OH)_2$	2.45	$Ba_{10}(CrO_4)_6(OH)_2$	2.86	
$Ba_2La_8(GeO_4)_6O_2$	2.45	$Pb_8K_2(AsO_4)_6$	2.89	
$Ba_3La_7(GeO_4)6O_{1.5}$	2.49	$Pb_{10}(SiO_4)_2(AsO_4)_4$	2.91	
$Ca_4Ce_6(GeO_4)_2(SiO_4)_4Cl_2$	2.49	$Sr_4La_6(SiO_4)_6(OH)_2$	2.93	
$Ba_{10}(VO_4)_6(OH)_2$	2.50	$Pb_8Rb_2(AsO_4)_6$	2.96	
$Pb_{10}(SiO_4)_2(VO_4)_4$	2.51	$Pb_{10}(SiO_4)_2(VO_4)_2(PO_4)_2$	3.02	
$Sr_{10}(MnO_4)_6(OH)_2$	2.52	$Ca_9Mg(PO_4)_6Cl_2$	3.02	
		$Ca_9Ni(PO_4)_6O$	3.03	
		$Ca_{10}(SiO_4)_8(SO_4)_3(OH)_3$	**3.03**	
		$Pb_{10}(SiO_4)(GeO_4)(PO_4)_2\ (AsO_4)_2$	3.05	
		$Ca_5Cd_5(PO_4)_6F_2$	3.09	
		$Ba_3La_7(SiO_4)_6(OH)_2$	3.11	
		$Ca_9Cd(PO_4)_6F_2$	3.11	
		$Ca_{10}(PO_4)_6(F,Cl,Br,OH)_2$	**3.12**	
		$Pb_8Bi_2(SiO_4)_4(PO_4)_2$	3.16	
		$Ca_9Sr(PO_4)_6O$	3.16	
		$Ca_9Pb(PO_4)_6(O,Cl_2)$	3.18	
		$Ca_9Ba(PO_4)_6(O,Cl_2)$	3.23	
a) 2.13–2.32		b) 2.70–2.95		c) 3.66–3.94

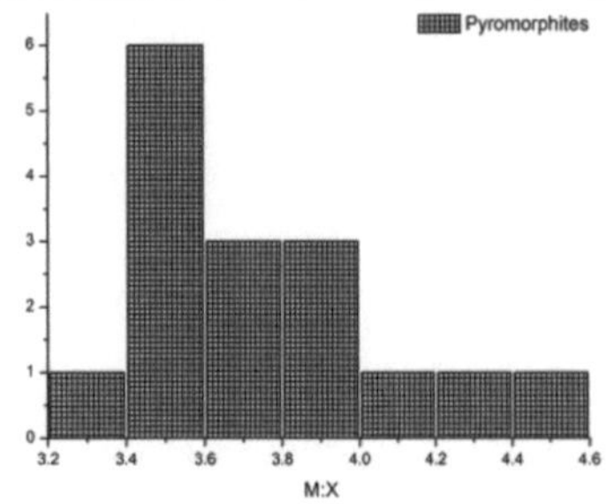

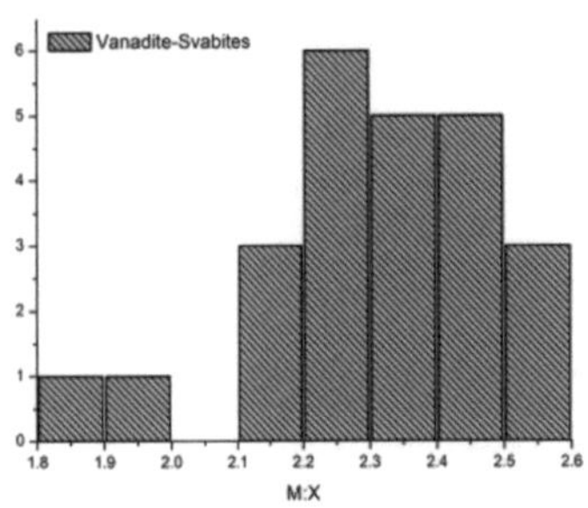

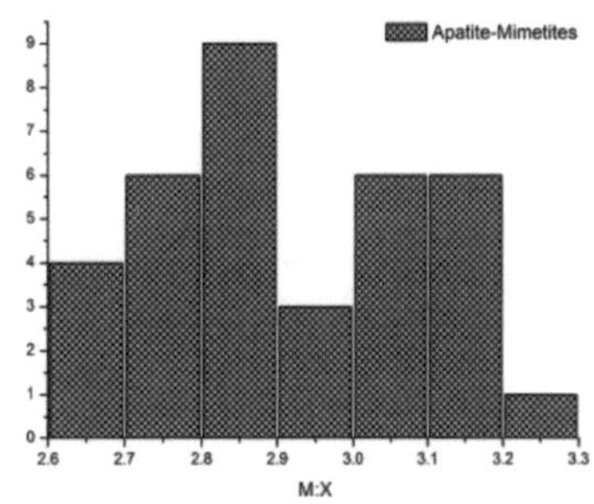

Table 1. The classification of synthetic (marked by bold) and natural apatites into three groups [15].

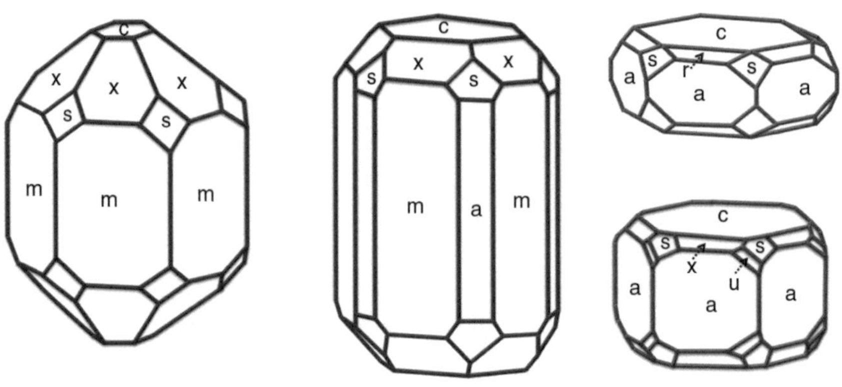

Fig. 1. Some of known forms of apatite crystals: c (0001), m (11$\bar{1}$0), a (11$\bar{2}$0), x (10$\bar{1}$1), s (11$\bar{2}$1), r (10r2) and u (2131).

http://dx.doi.org/10.5772/62208

Some examples of morphology of apatite crystals are shown in **Fig. 1**. The most abundant crystal faces of apatite possess the Miller–Bravais indices[9] [21],[22],[23],[24],[25] (0001), i.e. basis (or basal pinacoid[10]), (11$\bar{1}$0),[11] i.e. protoprism (or the prism of the first order), (10$\bar{1}$1) and (20$\bar{2}$1), i.e. the first-order dipyramids, but the faces such as (11$\bar{2}$0), i.e. deuteroprism (or the prism of the second order), (11$\bar{2}$1), i.e. dipyramid of the second order as well as (10$\bar{1}$2), i.e. the dipyramids of the first order, and (21$\bar{3}$1), i.e. dipyramid of the third order, are also common. Other faces such as (31$\bar{4}$1), (31$\bar{4}$2), (21$\bar{3}$0), (12$\bar{3}$2), etc., i.e. dipyramids of the third order, are also possible but rare (**Fig. 2**).

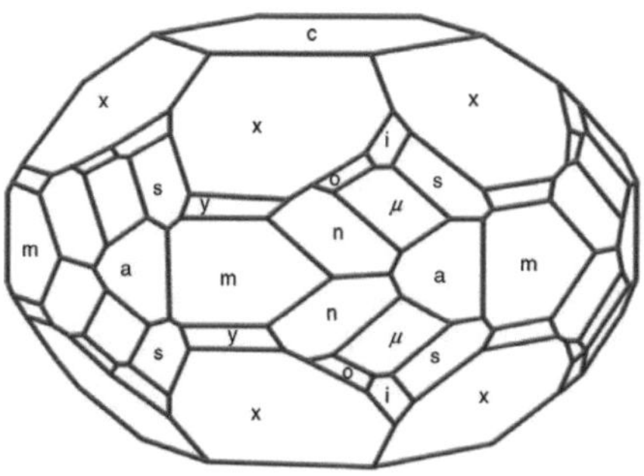

Fig. 2. Apatite from Gletsch (Switzerland): *y* (20$\bar{2}$1), *n* (31$\bar{4}$1), *o* (31$\bar{4}$2), *i* (12$\bar{3}$2) and μ (21$\bar{3}$1).

This means that the shapes pinacoid {0001}, hexagonal prisms {10$\bar{1}$0} and {11$\bar{2}$0}, dihexagonal prism {21$\bar{3}$0} and hexagonal dipyramid[12] {10$\bar{1}$1}, {10$\bar{1}$2}, {20$\bar{2}$1} and {11$\bar{2}$1} are possible to can be found on apatite crystals.

According to the crystal system, apatite minerals belong to the hexagonal[13] dipyramidal class (sometimes also termed as apatite type). The point group of apatite is 6/M (HERMAN–MAUGUIN SYMBOLS, H-M) or C_6h (SCHÖNFLIES SYMBOLS), and it is centrosymmetric [26],[27],[28], [29],[30], [31],[32]. The symmetry operation of this point group includes hexad perpendicular to mirror plane and the center of symmetry (**Fig. 3**).

---

[9] The orientation of planes in crystal was determined by Miller indices, i.e. three integers (hkl), which referee to the plane (or planes $hb_1 + kb_2 + lb_3$, where b_1, b_2 and b_3 are reciprocal lattice vectors). Negative integer, e.g. -i is written as $\bar{i}$. The group of planes equivalent to *(hkl)* is written as *{hkl}*. The square brackets *[hkl]* denote a direction on the basis of the direct lattice vectors instead of the reciprocal lattice and the set of equivalent planes ($ha_1 + ka_2 + la_3$, where a_1, a_2 and a_3 are direct lattice vectors) is written as <*hkl*>.

[10] Pinacoid {0001} consists of two opposite faces perpendicular to the 6-fold axis. It commonly occurs in combination with hexagonal prism and hexagonal or dihexagonal truncated pyramids [25]. The single face is termed as pedion.

[11] The hexagonal (and rhombohedral) crystal system uses four (Miller–Bravais) indices *(hkil)*, where *i* is termed as redundant index and $h + k + i = 0$, e.g. (20$\bar{2}$1) is equivalent to (201) and (110) to (11$\bar{2}$0).

[12] Hexagonal dipyramids $\{h0\bar{h}\,l\}$ and $\{hh(2\bar{h})l\}$ consist of 12 isosceles faces which intersect in a point along the vertical axis. Those two dipyramids differ only in their orientation with respect to three horizontal axes.

[13] In hexagonal crystal system $a = b \neq c$, $\alpha = \beta = 90°$, and $\gamma = 120°$. The hexagonal system is usually referred to four crystallographic axes designated a_1, a_2, a_3, and *c*. The *c*-axis is vertical and the three *a*-axes lie in horizontal plane (*c*-axis is perpendicular to this plane) with the angle of 120° between their positive ends [25].

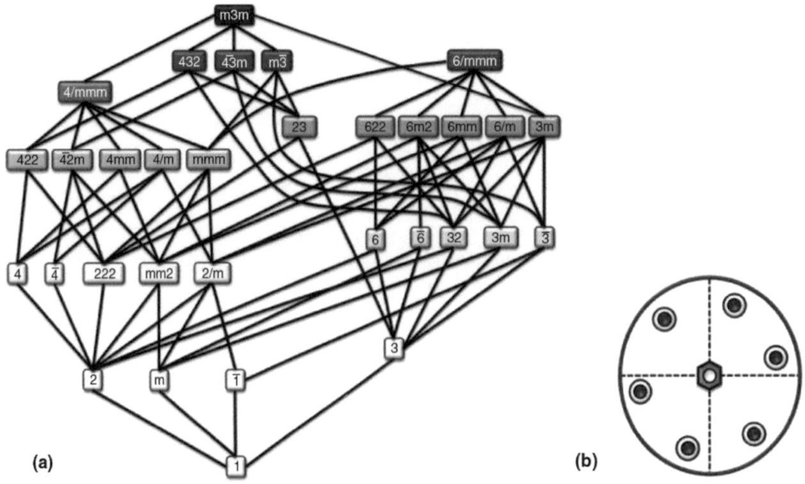

Fig. 3. Point group–subgroup relationship of point groups [31] (a) and stereogram of point group 6/M [36] (b) showing a sixfold rotation axis (hexad) plus the centre of symmetry and the mirror plane.

The projection of hexagonal crystal divided to twelve parts (dodecant) by the lateral axial planes is shown in **Fig. 4(a)**. The pole (×) of each plane requires five other points above as well as six other located on the same place below the projection plane (b). This leads to the formation of hexagonal dipyramid. The cross section provides hexagon (c) inclined to right or left. Therefore, the right or right-handed form (d) and the left or left-handed form (e) of dipyramids were recognized. One is formed from the poles marked by sign (×) and the second one is in the void dodecants of projections (b). The right form can be turned left by the rotation of 180° along one of the crystallographic axis ***a***.

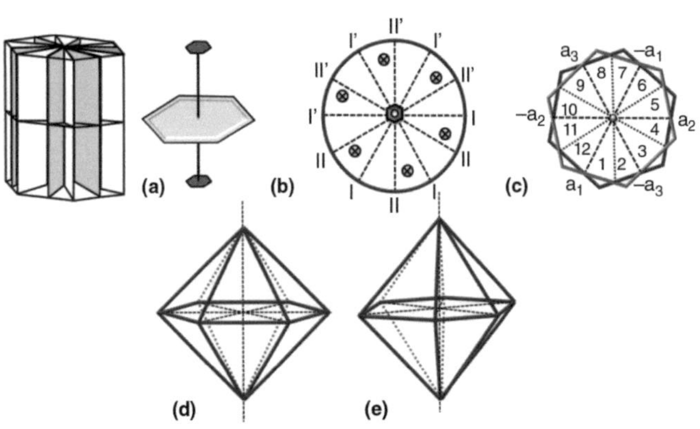

Fig. 4. Hexagonal dipyramidal crystal system: hexagonal lattice (a), stereographic projection (b), hexagon (c) and right (d) and left (e) dipyramids.

The crystal shaper related to the hexagonal–dipyramidal crystal system can be derived via extend or skip altering planes (the same up and down) according to **Table 2**.

1	2	3	4	5	6	7	8	9	10	11	12
1	2	3	4	5	6	7	8	9	10	11	12

Table 2. Derivation of shapes in hexagonal–dipyramidal crystal system.

The space group[14] of apatite is $P6_3/M$ (HERMAN–MAUGUIN SYMBOLS[15]), where P denotes primitive type of Bravais lattice,[14] 6_3 is sixfold (360°/6 = 60°) screw rotation axis parallel to (001) followed by a translation through a distance $3t/6$ (t is the magnitude of the shortest lattice vector along the axis), and the symbol /M refers to the mirror plane perpendicular to 6_3 screw axis (i.e. the mirror plane normal vector is parallel to the hexad[16] that coincides with glide plane[17] at z = ¼ for the monoclinic variant in the space group $P2_1/B$ (**Fig. 9**, the structure of hexagonal and monoclinic apatite was described in **Section 1.2**) [28],[29],[30], [31],[32],[38],[39].

The composition of natural apatites ($Ca_5(PO_4)_3(F,Cl,OH)$) exhibits large variations in the content of F, Cl and OH. Pure end members, e.g. hydroxylapatite ($Ca_5(PO_4)_3OH$), fluorapatite ($Ca_5(PO_4)_3F$) and chlorapatite ($Ca_5(PO_4)_3Cl$), are uncommon in nature, but binary and ternary compositions are widely reported in igneous, metamorphic and sedimentary rocks. Petrologists proposed that the variations in OH–F–Cl ratio in apatites or between biotite and apatite may be used as a geothermometer [33],[34],[35] and an indicator of volatile fugacity of halogens and water in magmatic and hydrothermal processes [33],[36]. The concentration of OH, F and Cl also directly correlates with the properties such as etching rates, annealing characteristics of U fission track in apatite (**Section 7.3.3**) and investigation of paleoenvironment and diagenesis (**Section 6.5**) [33].

Apatite minerals represent a major class of ionic compounds [37] of interest to many disciplines, including medical and biomaterial sciences (**Section 10.9**), geology (**Chapter 7**), cosmology ([40], **Section 7.3.4**), environmental (**Chapters 7** and **9**) and nuclear sciences (**Chapter 10**). Apatite also represents an important source of inorganic phosphorus for natural ecosystems and may favor the establishment of microbial communities able to exploit it [42].

1.1. Nomenclature of apatite minerals and apatite supergroup

The mineral apatite was first recognized by German geologist ABRAHAM GOTTLOB WERNER (1950–1817) and named in 1786 from Greek word *"apatao"* (απα'ταω, which means to mislead, to cheat, or to deceive because the mineral was often mistaken for other species, e.g. mineral beryl ($Be_3Al_2Si_6O_{18}$, space group P_6/MCC)) [43],[44],[45],[46].

[14] There are 14 Bravais lattices, 32 point groups (crystal classes), and 230 crystallographic space groups in three dimensions [28]. The space groups were independently described by E.S. FEDEROV (1853–1919, Russian), A.M. SCHÖNFLIES (1853–1928, German), and W. BARLOW (1845–1934, English) [32].

[15] The corresponding Schönflies symbol is C_{6h}^2. The number of group is 176.

[16] Hexad denotes the sixfold rotation axis. The 4-, 3-, 2-, and 1-fold rotation axes are termed as tetrad, triad, diad, and monad [27].

[17] Glide plane combines a reflection with a translation parallel to the plane.

Since 1856–1860, these minerals have been named fluorapatite, chlorapatite, and hydroxylapatite, depending on the dominant Z^- anion. The increasing number of new discovered species[18] resulted in a revision of the mineralogical nomenclature, which was initiated by the chairman of the IMA Commission on New Minerals, Nomenclature and Classification, E.A.J. Burke, and aimed at adopting, as far as possible, modified Levinson suffixes[19] (or Levinson modifier [47],[49]) instead of adjectival prefixes such as fluor-, chlor- and hydroxyl. The above-mentioned minerals were renamed to apatite–(CaF), apatite–(CaCl) and apatite–(CaOH).[20]

One of the rationales for that change was the benefit of having the names of these minerals appear consecutively in alphabetical listings and databases. These changes did not fully consider the structural complexity of minerals with the apatite structure [45],[49],[50]. The name of these minerals was currently changed back from apatite–CaF) apatite–(CaOH), apatite–(CaCl) to fluorapatite, hydroxylapatite and chlorapatite. Furthermore, the monoclinic variants[21] [51] fluorapatite—M, hydroxylapatite—M and chlorapatite—M are not considered to be distinct species [45].

The recently approved nomenclature scheme [49] could logically be extended to other minerals from the group of apatite, e.g. pyromorphite [52] should be named as apatite–(PbCl) or alforsite [53],[54] as apatite–(BaCl). It is also possible to include various tetrahedral cations (P, As, or V) into the extended suffix, e.g. apatite–(PbAsCl) instead of mimetite [55]. The results would be the mineral names, which are more similar to chemical formula [45]. The name mimetite–M is used for the polymorphic variant of mimetite. The mineral was previously known as clinomimetite and currently is not considered a distinct species [45].

The "apatite group" traditionally includes phosphate, arsenate and vanadate minerals. Other minerals belonging to different chemical classes, namely, silicates (e.g. britholite–(Ce) [56],[57] or britholite–(Y) [57],[58]), silicate-sulfates (e.g. hydroxylellestadite [45],[59], fluorellestadite [45],[60] and chlorellestadite[22] [45]) and sulfates (e.g. cesanite [61]) display the structural

[18] The concept of mineral species is defined mainly on the basis of its chemical composition and crystallographic properties. For example, hydroxylapatite and fluorapatite both crystallize in the hexagonal system, with the same space group and have similar unit-cell parameters. They are considered as the separate species because the relevant structural site is predominantly occupied by OH^- in hydroxylapatite and by F^- in fluorapatite [4].

[19] The nomenclature system based on chemical-symbol suffixes described by Levison [47] and originally applied only to rare-earth mineral species, which are defined to have the total atomic percentage of rare-earth elements and Y greater than any other element within a single set of crystal-structure sites, e.g. (~REE, Ca). A species name is related to a rare-earth mineral whenever the presence of rare-earth element distribution is determined. The chemical symbol for the predominant rare-earth element is appended, in parentheses, by means of a hyphen to the group name; this results in mineral species names such as monazite-(Ce), monazite-(La), and monazite-(Nd) [48]. If a rare-earth mineral appears together with considerable quantities of another rare-earth element which is unusual, or for any reason deserving the notice, two or more chemical symbols may be placed in the parentheses [47]. For example, a monazite–(Ce) with a considerable amount of samarium would be written as monazite–(Ce,Sm).

[20] Another examples are strontium apatite named as apatite-(SrOH) and clinohydroxylapatite named as apatite–(CaOH)–M.

[21] In essence, the polytypes are distinguished by alphanumerical symbols appended to the root name and joined to by a hyphen, for example, wollastonite-3*T* or graphite-2*H*. The numerical part of the symbol represents the layering periodicity and the alphabetical part, rendered in italic print represents the crystallographic system as follows: cubic (C), hexagonal (H), rhombohedral (R), trigonal (T), orthorhombic (O), monoclinic (M), and triclinic (A).

[22] The names of ellestadite–(OH), ellestadite–(F), and ellestadite–(Cl) minerals were changed back to hydroxylellestadite, fluorellestadite, and chlorellestadite, respectively [45],[49].

http://dx.doi.org/10.5772/62208

morphology of apatite. In accordance with the newly approved standardization of mineral group hierarchies, all of these minerals can be included in the broader apatite supergroup [45].

The valid IMA-accepted mineral species within the apatite supergroup can be divided into five groups [45]:

a. **Apatite group**: hexagonal and pseudo-hexagonal phosphates, arsenates and vanadates containing the same prevailing (species-defining) cations at both **M**(1) and **M**(2) sites.

b. **Hedyphane group**: hexagonal and pseudo-hexagonal phosphates, arsenates and sulfates containing different prevailing (species-defining) cations at **M**(1) and **M**(2) sites. Minerals from the group of hedyphane are described in **Section 2.1**.

c. **Belovite group**: hexagonal and trigonal phosphates with **M**(1) site split into the **M**(1) and **M**(1′) sites containing different prevailing (species-defining) cations. Minerals from the group of belovite are described in **Section 2.2**.

d. **Britholite group**: hexagonal and pseudo-hexagonal silicates, typically with partially ordered **M**(1) and **M**(2) cations. Minerals from the group of britholite are described in **Section 2.3**.

e. **Ellestadite group**: hexagonal and pseudo-hexagonal sulfato-silicates with ideal ratio $[SiO_4]^{4-}$:$[SO_4]^{2-}$ = 1:1. Minerals from the group of ellestadite are described in **Section 2.4**.

All of valid species within the apatite supergroup are listed in **Table 3**. There are also other minerals with the apatite structure [45]:

i. Borosilicates with REEs atoms strongly prevalent over calcium in M sites, including three valid minerals: tritomite–(Ce), $Ce_5(SiO_4,BO_4)_3(OH,O)$ [57], melanocerite-(Ce), $Ce_5(SiO_4,BO_4)_3(OH,O)$[23] [45],[57] and tritomite–(Y), $Y_5(SiO_4, BO_4)_3(O,OH,F)$ [62],[63]. These minerals could be formally included as REE silicates in the britholite subgroup.

ii. Carbonate–fluorapatite and carbonate–hydroxylapatite: Although the carbonate–fluorapatite names are extensively used in literature to denote the mineral portion of bones a teeth of vertebrates, their validity as distinct mineral species belonging to the group of apatite remains disputable. CO_3^{2-} anion is known to occur as subordinate component in the members of apatite group [64],[65], but both minerals mentioned above are discredited.

Group	Existing name (IMA list of minerals)	Approved name	End-member formula
Apatite group	Apatite–(CaF)	Fluorapatite	$Ca_5(PO_4)_3F$
	Apatite–(CaCl)	Chlorapatite[a]	$Ca_5(PO_4)_3Cl$
		Chlorapatite–*M*[b]	

[23] Melanocerite–Ce and tritomite–(Ce) are probably the same mineral, and tritomite-(Y) could be the Y-dominant analogue [45].

Group	Existing name (IMA list of minerals)	Approved name	End-member formula
	Apatite–(CaOH)	Hydroxylapatite[a]	$Ca_5(PO_4)_3OH$
	Apatite–(CaOH)–*M*	Hydroxylapatite–*M*[b]	
	Svabite	Svabite	$Ca_5(AsO_4)_3F$
	Turneaureite	Turneaureite	$Ca_5(AsO_4)_3Cl$
	Johnbaumite	Johnbaumite[a]	$Ca_5(AsO_4)_3OH$
	Fermorite	Johnbaumite–*M*[b]	
	2008-009[c]	Stronadelphite	$Sr_5(PO_4)_3F$
	Pyromorphyte	Pyromorphyte	$Pb_5(PO_4)_3Cl$
	Mimetite	Mimetite[a]	$Pb_5(AsO_4)_3Cl$
	Clinomimetite	Mimetite–*M*[b]	
	Alforsite	Alforsite	$Ba_5(PO_4)_3Cl$
	Vanadinite	Vanadinite	$Pb_5(VO_4)_3Cl$
Hedyphane group	Hedyphane	Hedyphane	$Ca_2Pb_3(AsO_4)_3Cl$
	–	**"Hydroxylhedyphane"**[d]	$Ca_2Pb_3(AsO_4)_3OH$
	Phosphohedyphane	Phosphohedyphane	$Ca_2Pb_3(PO_4)_3Cl$
	Phosphohedyphane–(F)	**Fluorphosphohedyphane**	$Ca_2Pb_3(PO_4)_3F$
	–	**"Hydroxylphosphohedy phane"**[d]	$Ca_2Pb_3(PO_4)_3OH$
	–	**New root name**[d]	$Ca_2Sr_3(PO_4)_3F$
	Morelandite	Morelandite	$\mathbf{Ca_2Ba_3(AsO_4)_3F,Cl}$
	–	**New root name**[d]	$Mn_2Ca_3(PO_4)_3Cl$
	Cesanite	Cesanite	$Ca_2Na_3(SO_4)_3OH$
	Caracolite	Caracolite	$Na_2(Pb_2Na)(SO_4)_3Cl$
	Aiolosite	Aiolosite	$Na_2(Na_2Bi)(SO_4)_3Cl$
Belovite group	Fluorcaphite	Fluorcaphite	$SrCaCa_3(PO_4)_3F$
	Apatite–(SrOH)	**Fluorstrophite**	$SrCaSr_3(PO_4)_3F$[e]
	Deloneite–(Ce)	**Deloneite**	$\mathbf{(Na_{0.5}REE_{0.25}Ca_{0.25})(Ca_{0.75}REE_{0.25})Sr_{1.5}(CaNa_{0.25}REE_{0.25})(PO_4)_3F_{0.5}(OH)_{0.5}}$
	Belovite–(Ce)	Belovite–(Ce)	$NaCeSr_3(PO_4)_3F$
	Belovite–(La)	Belovite–(La)	$NaLaSr_3(PO_4)_3F$
	Kuannersuite–(Ce)	Kuannersuite–(Ce)	$NaCeBa_3(PO_4)_3F_{0.5}Cl_{0.5}$
Britholite group	Britholite–(Ce)	Britholite–(Ce)	$(Ce,Ca)_5(SiO_4)_3OH$
	Britholite–(Y)	Britholite–(Y)	$(Y,Ca)_5(SiO_4)_3OH$

Group	Existing name (IMA list of minerals)	Approved name	End-member formula
	Fluorbritholite–(Ce)	Fluorbritholite–(Ce)	$(Ce,Ca)_5(SiO_4)_3F$
	2009-005[c]	**Fluorbritholite–(Y)**	$(Y,Ca)_5(SiO_4)_3F$
	Fluorcalciobritholite	Fluorcalciobritholite	$(Ca,REE)_5(SiO_4,PO_4)_3F$
	Melanocerite–(Ce)	–[f]	
	Tritomite–(Ce)	Tritomite–(Ce)	$Ce_5(SiO_4,BO_4)_3(OH,O)$
	Tritomite–(Y)	Tritomite–(Y)	$Y_5(SiO_4,BO_4)_3(O,OH,F)$
Ellestadite group	Ellestadite–(OH)	**Hydroxylellestadite**	$Ca_5(SiO_4)_{1.5}(SO_4)_{1.5}OH$
	Ellestadite–(F)	**Fluorellestadite**	$Ca_5(SiO_4)_{1.5}(SO_4)_{1.5}F$
	Ellestadite–(Cl)	–[g]	–
	Mattheddleite	Mattheddleite	$Pb_5(SiO_4)_{1.5}(SO_4)_{1.5}Cl$
	–	"Hydroxylmattheddleite"[d]	$Pb_5(SiO_4)_{1.5}(SO_4)_{1.5}OH$

[a] The suffix –H could be used to denote the hexagonal polymorph.
[b] The name of monoclinic polymorph that should no longer to be considered as distinct species.
[c] Mineral approved by the IMA CNMNC without a name.
[d] Potentially new mineral species.
[e] A mistake in the IMA list of minerals, please see Pasero et al [45] for further details. Since the mineral was initially considered to be the Sr-dominant analogue of fluorapatite with simplified formula $(Sr,Ca)_5(PO_4)_3(F,OH)$, it was named as strontium apatite. Later structural study determined the idealized formula of mineral as $SrCaSr_3(PO_4)_3F$. Recently, the mineral was renamed as apatite–(SrOH) [45],[49]. However, the name should have been changed to apatite–(SrF) given that fluorine is the dominant Z^- anion.
[f] Mineral to be potentially discredited (= tritomite–(Ce)).
[g] Mineral be discredited (a mineral with ideal end-member formula $Ca_5(SiO_4)_{1.5}(SO_4)_{1.5}Cl$ is assumed not to exist).

Table 3. Existing (IMA) approved names and end-member formulas for the minerals within the apatite supergroup. Approved changes are marked by bold and names in quotes are the most appropriate for potential new minerals.

The nomenclature of minerals from the apatite group is very confusing because many of the names are initially used to ill-defined varieties, which did not deserve specific status. There are many historical names,[24] which still appear in literature [10],[67],[68]:

- **Apatite**: is currently used as a generic name for apatite group of minerals (**Section 1.5**).
- **Dahllite**: is an obsolete name for carbonate–hydroxylapatite[25] (soluble in HCl with the evolution of CO_2) of the composition of $H_2Ca_{14}P_8C_2O_{39}$ or $2Ca_3P_2O_8{\cdot}CaCO_3{\cdot}½H_2O$. Dahllite was described by Brögger and Bäckström [69] as a mineral which occurs in crusts with fibrous structure on apatite, and also in a pure state as nodular masses and concretions. The mineral is pale yellowish white, greenish, or colorless, has white streak, the hardness on the Mohs scale is 5 and the density is 3.053 $g{\cdot}cm^{-3}$ [70].
- **Francolite**: is an obsolete name for carbonate–fluorapatite[23].

[24] These names are no longer accepted by IMA/CNMMN.
[25] Described deeply in **Section 2.6** and **Chapter 7**.

- **Staffelite:** is an obsolete name for a variety of carbonate–fluorapatite from Staffel, Germany, which form nodular-stalactitic aggregates and crusts, of the composition $3Ca_3(PO_4)_2{\cdot}CaF_2{\cdot}CaCO_3$ [71],[72].
- **Voelckerite**: the modern equivalent is oxyapatite ($3Ca_3(PO_4)_2{\cdot}CaO$) named according to agricultural chemist J.A. Voelcker [73], who proved the fact that apatite is often deficient in fluorine and chlorine (**Section 1.4**). Voelckerite is a white, subtranslucent mineral with imperfect cleavage and faint luster. The specific gravity is 3.06, and the hardness on the Mohs scale is about 5. In a thin basal section, a negative uniaxial interference figure was obtained [74].
- **Kurskite**: is an obsolete name for carbonate–fluorapatite[23] [75],[76].
- **Quercyite**: is an obsolete name for carbonate–hydroxylapatite applied to mixtures of amorphous collophanite (collophane) and crystalline dahllite, francolite, etc. Quercyite is often composed of alternating layers α-quercyite (optically negative) and β-quercyite (optically positive). The density of quercyite ranges from 2.83 to 2.87 $g{\cdot}cm^{-3}$ [70].
- **Wilkeite:** sicicatian strontian apatite [67]. Wilkeite is a former name (Eakle and Rogers [77] in 1914) for a variety of mineral fluorellestadite (**Section 2.4.1**).
- **Manganapatite:** (manganoan fluorapatite, Mn-bearing fluorapatite) was named by Siewert [78] in allusion to its chemical relationship to fluorapatite.
- **Lasurapatite:** sky blue variety, it occurs in crystals with lapis-lazuli at Bucharei in Siberia [79].
- **Moroxite**: greenish-blue variety of conchoidal apatite (the green varieties are named as asparagus stone). The name moroxite was given to this mineral by Karsten and is borrowed from the morochites of Pliny, according to author's statement: "*Est gemma, per se porrcea viridisque, trita autem candicans*" [81].
- **Foliated apatite**: according to the system of mineralogy of Jameson [81] in which minerals are arranged according to the natural history method, rhomboidal apatite (fist specie) is divided in three subspecies, including (1) foliated apatite, (2) conchoidal apatite and (3) phosphorite described below.

  > The most frequent color of foliated apatite is snow-white, yellowish-whiter, reddish-white and greenish-white. Several of these colors occur frequently in the same piece. It sometimes occurs also as massive and disseminated in distinct concretions, which are large and small angulo-granular, and sometimes thin and straight lamellar, generally crystallized. The crystals are small, very small and middle-sized and occur sometimes single, sometimes many irregularly superimposed on each other. It is brittle and easily fragile. Foliated apatite becomes electric by heating and by being rubbed with woolen cloth. On glowing coals, it emits a pale grass-green phosphoric light [79]

- **Conchoidal apatite or asparagus-stone**: a yellow-green variety of apatite. The name of asparagus stone has been given to this variety in allusion to asparagus green color, which it frequently exhibits. It was distinguished from the apatite partly by its crystalline form, and more particularly by tendency to phosphoresce on hot coals. It presents either the primitive or the perfect form, or it occurs with truncated lateral edges, but most frequently, the hexahedral prism is terminated by six-sided pyramids, the faces of which correspond with the sides of the prism and form with them the angle of 129°14′. These prisms are usually longer than in case of apatite. It sometimes occurs in small crystalline masses [80],[81]. This variety was former described also as a kind of schorl ($NaFe_3^{2+}Al_6(Si_6O_{18})(BO_3)_3(OH)_3OH$ [82],[83]), afterwards as a variety of beryl ($Be_3Al_2Si_6O_{18}$ [84][85]) [86].

> In Europe, asparagus stone occurs as a porous iron-shot limestone near Cape de Gate, in Murcia in Spain, in granite near Nantes and is basalt at Mont Ferrier, in France, imbedded in green talc, in the Zillertal in Salzburg, in beds of magmatic ironstone, along with sphere or rutilite (the name is used for almandine, rutile, titanate, or rutilized quartz), calcareous-spar, hornblende, quartz and augite, at Arendal in Norway. In America, it was found imbedded in granite at Baltimore, and in mica-slate in West Greenland [86].

- **Phosphorite**: the third subspecies in Jameson's [86] system of mineralogy. There are two kinds of phosphorites (described in **Section 7.1.1**): (1) common phosphorite and (2) earthy phosphorite.

- **Naurite**: a colloidal variety of apatite from Nauru Island, Pacific. The mineral was considered as a mineral species of carbonate–hydroxyl–fluorapatite from the group of collophane [87].

- **Eupychroite**: a fibrous mammillary variety of apatite (carbonate–fluorapatite) from Point Crown, Essex County, New York, United States [88].

- **Podolite**: carbonate–fluorapatite (soluble in HCl with the evolution of CO_2) of the composition of $Ca_{10}P_6CO_{27}$ or $3Ca_3P_2O_5{\cdot}CaCO_3$. The mineral was described by Tschirwinsky [89] and occurs as crystalline masses or in prismatic crystals on phosphorite, in spherulites of prismatic crystals. The color of the mineral is yellowish, and its density is 3.077 $g{\cdot}cm^{-3}$. A comparison of the properties and chemical composition of podolite and dahllite shows they are essentially identical [70].

- **Dehrnite**: a varietal name for sodian hydroxylapatite, i.e. apatite where a part of calcium is replaced by alkali metals and the content of Na > K. Originally, it was described from Dehrn, Germany, by Larsen and Shannon as a potassium sodium calcium phosphate that belongs to the apatite group [90],[91],[92].

- **Lewistonite**: a varietal name for alkali bearing [93][26] hydroxylapatite where a part of calcium is replaced by alkali metals and the content of K > Na. The mineral was described from Fairfield, Utah [90],[91],[92].

[26] Adjectival modifier that gives some information on the chemistry of the mineral, for example, "sodian," "sodium-rich," "sodium-bearing," or "sodium equivalent of" [93].

- **Saamite**: a rare earth element–strontian or calcium variety of fluorapatite of the composition of $(Ca,Sr,REE)_5(PO_4)_3(F,O)$ [40],[94]. According to VOLKOVA AND MELENTIEV [95], it is a saamite variety of apatite containing 5.58–11.42% SrO. Currently, the mineral name saamite refers to IMA approved mineral of the composition of $Ba\square Na_3Ti_2Nb(Si_2O_7)_2O_2(OH)_2\cdot 2H_2O$ from the Kirovskii mine, Mount Kukisvumchorr, Khibiny alkaline massif, Kola Peninsula, Russia [96].

- **Collophane**: a varietal name for massive cryptocrystalline carbonate-rich apatite[23] ($Ca_5(PO_4,CO_3,OH)_3(F,OH)$ [97]). It is often used when the specific phase of apatite cannot be identified. The deposits of collophane are often associated with fossilized bone or coprolites. The term is sometimes used also in the context of bone composition and structure [98], [99]. Most of the collophane occurs as pelletal phosphorite, but some occurs as a collophane mudstone and as cement in sedimentary breccia[27] [100],[101].

The list of calcium phosphate species accepted by Commission on New Minerals and Mineral Names (CNMMN)[28] of International Mineralogical Association (IMA) [67] is given in **Table 3.**

1.2. The crystal structure of apatite minerals

The stoichiometric end-member formula for phosphate-bearing apatites is $M_{10}(PO_4)_6Z_2$ (Z = 2) (reduced formula $M_5(PO_4)_3Z$ (Z = 2) can be used [45]), where ***M*** and ***Z*** represent divalent cations and monovalent anion, respectively. In the case of oxyapatites ($M_{10}(PO_4)_6O$), $2Z^-$ ions are replaced by O^{2-} [37]. Various representations of orthophosphate ion are introduced in **Fig. 5.**

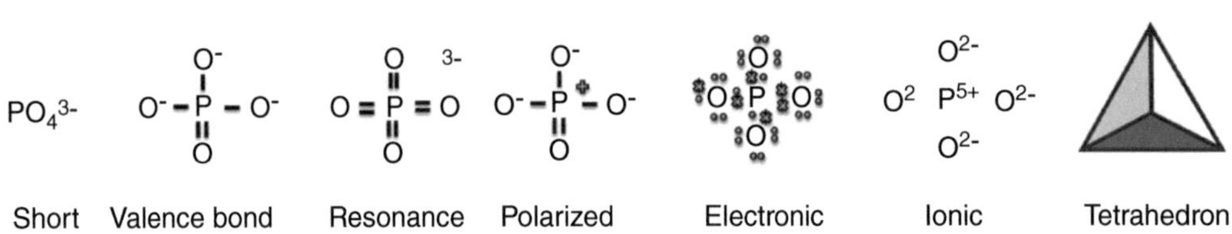

Fig. 5. Representation used for orthophosphate anion [106].

The PO_4 unit in the apatite structure can be partially substituted by AsO_4 (e.g. hedyphane ($Ca_2Pb_3(AsO_4)_3Cl$) [102] and minerals from the group of pyromorphite ($Pb_5(PO_4)_3Cl$ [17],[52], etc.), SiO_4, SO_4, VO_4 (vanadinite ($Pb_5(VO_4)_3Cl$) [103] and CO_3[29] [104],[105]. **Table 4** compares the properties of orthophosphate anion with other isoelectronic ortho-oxyanions [106].

[27] Breccias are among the most common features in ore deposits. They are associated with numerous types of ores, of either endogenic or supergenic origin, and in both subsurface and submarine environments. Breccia is a rock composed of angular fragments of preexisting rocks embedded in fine-grained matrix cement [101].

[28] The Commission on New Minerals, Nomenclature and Classification (CNMNC) of the International Mineralogical Association (IMA) was formed in July 2006 by a merger between the Commission on New Minerals and Mineral Names (CNMMN) and the Commission on Classification of Minerals, at the request of both commissions.

[29] Please see the discussion dedicated to the nomenclature of other minerals with the structure of apatite in **Section 1.1.**

Anion	AlO_4^{5-}	SiO_4^{4-}	PO_4^{3-}	SO_4^{2-}	ClO_4^{-}
Electron formula	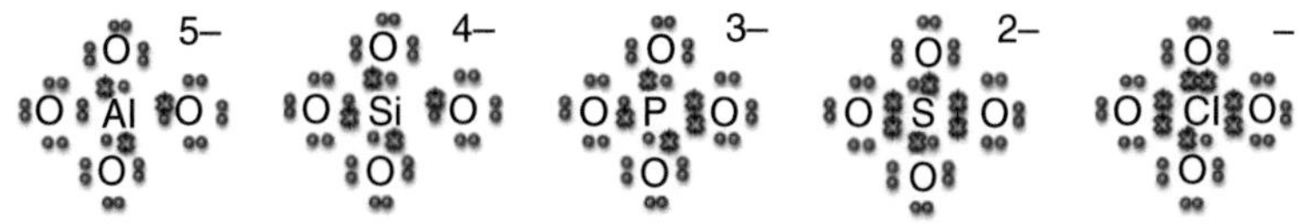				
Valence bonds	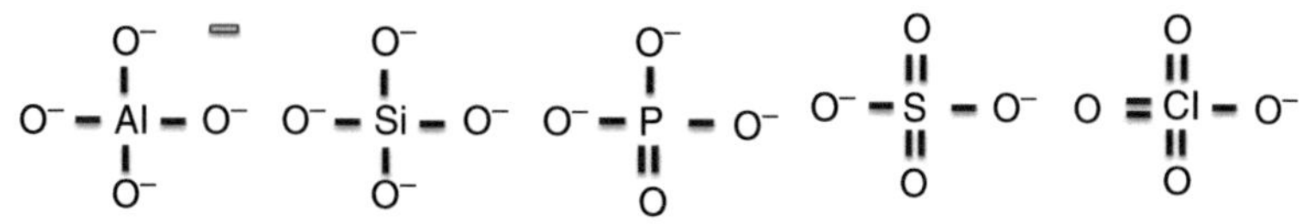				
Electrostatic formula	O^{2-} O^{2-} Al^{3+} O^{2-} O^{2-}	O^{2-} O^{2-} Si^{4+} O^{2-} O^{2-}	O^{2-} O^{2-} P^{5+} O^{2-} O^{2-}	O^{2-} O^{2-} S^{6+} O^{2-} O^{2-}	O^{2-} O^{2-} Cl^{7+} O^{2-} O^{2-}
Electrostatic strength	2/3	2/4	2/5	2/6	2/7
Observed length [Å]	1.77	1.62	1.54	1.49	1.46
Sum of covalent radii	1.92	1.83	1.76	1.70	1.65
Corrected sum	1.74	1.68	1.63	1.61	1.60
Ionic character of bond [%]	63	51	39	22	6
Acid strength (K_1)	10^{-1}	10^{-10}	10^{-2}	10^{3}	10^{8}
Sodium salt	Na_5AlO_4	Na_4SiO_4	Na_3PO_4	Na_2SO_4	$NaClO_4$
Density [$g{\cdot}cm^{-3}$]	2.57	–	2.54	2.66	2.50
Melting point [°C]	–	1018	1530	884	468
Solubility [g/100 g H_2O]	–	5.0	12.1	19.5	201.0

Table 4. The comparison of properties of isoelectric ortho-oxyanions [106].

The crystal structure of apatite (*sensu stricto* fluorapatite) was first independently solved in 1930 by MEHMEL [107] and NÁRAY-SZABÓ [108]. The archetype crystalline[30] structure of apatite is hexagonal with the space group $P6_3/M$ and the unit-cell parameters $\boldsymbol{a}$ = 9.3–9.6 and $\boldsymbol{c}$ = 6.7–6.9 Å. The above-mentioned generic crystal-chemical formula can be also written as follows [45], [109],[148]:

$$^{IX}M(1)_4\,^{VII}M(2)_6\left[^{IV}XO(1)O(2)O(3)_2\right]_6 Z_2;$$

[30] The term crystalline means atomic ordering on the scale that can produce an "idexable" (i.e. with Mille indices) diffraction pattern when the substance is traversed by a wave with suitable wavelength (X-ray, electrons, neutrons, etc.). However, some amorphous substances (e.g. georgeite, calciouranoite) were also accepted as minerals by the CNMMN.

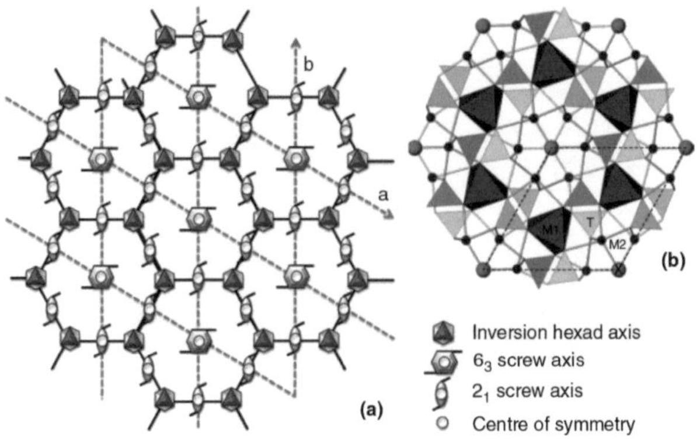

Fig. 6. Vertical symmetry elements of the space group $P6_3/M$ (a) [38] and the crystal structure of apatite as seen along *c*-axis (b) [45].

where left superscripts indicate ideal coordination numbers. **M** represents large cations, **X** represents metal or metalloids and $\mathbf{Z}^-$ represents anion (sometimes termed as column anion) [109].

Minerals from apatite group belong to the hexagonal–dipyramidal crystal system with the space group $P6_3/M$ (fluorapatite, hydroxylapatite and chlorapatite [45]), to trigonal–pyramidal system with the space group P3 (belovite–Ce [110]), also to trigonal–rhomobohedral system with the space group $P\bar{3}$ (belovite–La [45],[111]) and to hexagonal–pyramidal system with the space group $P6_3$ (fluorcaphite [45],[112],[113]).

Fluorapatite ($Ca_{10}(PO_4)_6F_2$, FA), chlorapatite ($Ca_{10}(PO_4)_6Cl_2$, ClA) and hydroxylapatite ($Ca_{10}(PO_4)_6OH_2$, HA) are the most important end members of apatite groups of minerals (**Table 7**). The hexagonal ($P6_3/M$) and monoclinic ($P2_1/B$) polymorphs[31] of apatite were described in literature [45],[104],[114].

The $P6_3/M$ space group **Fig. 6(a)** has three kinds of vertical symmetry elements [37]:

1. Sixfold screw axes passing through the corners of the unit cells. These symmetry elements are equivalent to a threefold rotation axis with a superimposed twofold screw axis.
2. Threefold rotation axes passing through 2/3, 1/3, 0 and 1/3, 2/3, 0.
3. Twofold screw axes passing through the midpoints of the cell edges and its center.

There are also mirror planes perpendicular to the *c*-axis and $z = 1/4$, $z = 3/4$ and numerous centers of symmetry.

---

[31] Polymorphic minerals are those that have essentially the same chemical composition but different crystal structures. The polymorphic forms of minerals are considered as different species if their structures are topologically different. For example, graphite and carbon are polymorphs of crystalline carbon; both have the same composition, but their structures are topologically different and therefore the minerals such as these are considered as separate species [4].

http://dx.doi.org/10.5772/62208

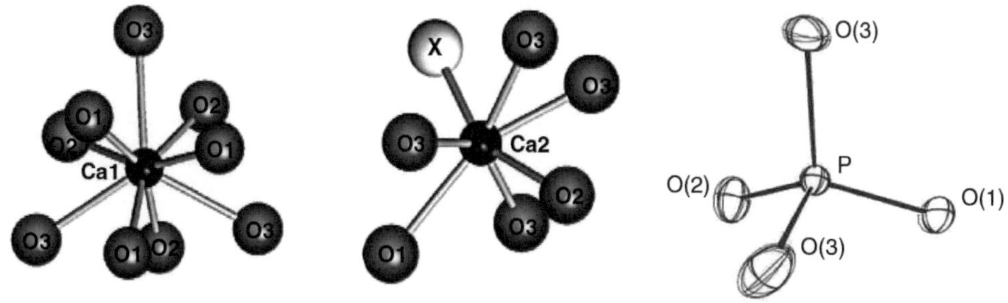

Fig. 7. The nearest neighbor of $Ca(1)O_9$ and $Ca(2)O_5Z(O)$ cationic polyhedral sites in the apatite structure (a) [115] and in the structure of $[PO_4]^{3-}$ tetrahedra (b) [33].

The $P6_3/M$ structure of calcium apatites **Fig. 6(b)** consists of isolated PO_4 (in general XO_4) tetrahedra centered at z = ¼ and ¾ are linked by **Ca(1)** (**M**(1)) in ninefold (6+3, 3 × O_1, 3 × O_2 and 3 × O_3 atoms) coordinated cation polyhedron (**M**$(1)O_9$) with a **Ch** site symmetry and **Ca(2)** (**M(2)**) in irregular sevenfold (6+1, O_1, O_2 and 4 × O_3 + **Z**) coordinated polyhedron (**M**$(2)O_5$**Z**(O) with **Cs** site symmetry (**Fig. 7**). The **M**$(1)O_9$ polyhedra share (0001) pinacoid[10] faces to form channel parallel to *c*-axis. In some cases, the **M(1)** sites are split into pairs of nonequivalent sites, which correspond to lowering of space group symmetry. The **M(2)** sites may be more irregular and the central cation may be considered to be eightfold (**M**$(2)O_5$**Z**) or ninefold (**M**$(2)O_5$**Z**(O)) coordinated. A prominent feature of the structure is large *c*-axis channels (apatitic channels [105] or anionic columns), which accommodate **Z** anion (**Fig. 8**). In other words, the **M**$(1)_4(XO_4)_6$ framework creates tunnels with the diameter adjusted to filling characteristics of **M**$(2)O_5$**Z**(O) components [33],[109],[115],[116].

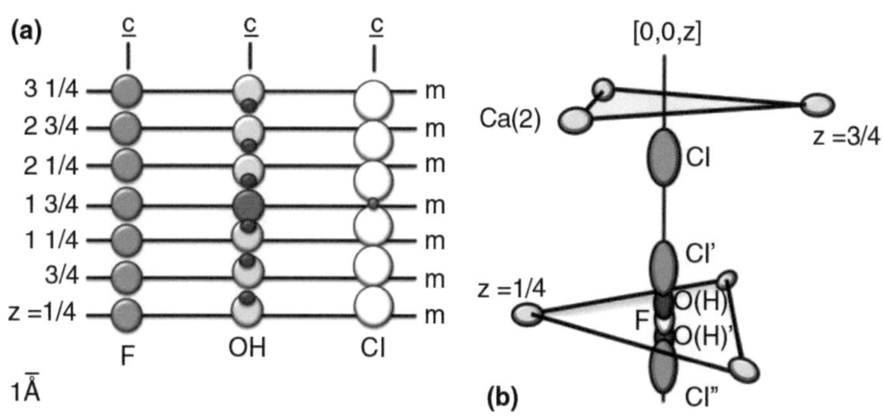

Fig. 8. Anion column in the hexagonal structure of fluor-, hydroxyl- and chlorapatite (a) [33] and depiction of triangle formed by Ca(2) atoms: Z′ and Z″, which denotes atoms disordered above and below the mirror plane (b) [116].

Adjacent Ca(l) and Ca(2) polyhedra are linked through oxygen atoms shared with PO_4 tetrahedra. The relationships among ionic sites and multiplicity and Wyckoff positions in all known space groups of apatite supergroup mineral are shown in **Table 5** [33],[38],[45], [66], [104],[117],[118].

Site	$P6_3/m$	$P6_3$	$P\bar{6}$	$P\bar{3}$	P3	$P2_1/m$	$P2_1$
M(1)	4f[32]	2 × 2b	2i, 2h	2 × 2d	2 × 1b, 2 × 1c	4f	2 × 2a
M(2)	6h	6c	3k, 3j	6g	2 × 3d	2a, 2 × 2e	3 × 2a
X	6h	6c	3k, 3j	6g	2 × 3d	3 × 2e	3 × 2a
O	2 × 6h, 12i	4 × 4c	2 × 3k, 2 × 3j, 2 × 6l	4 × 6g	8 × 3d	6 × 2e, 3 × 4f	12 × 2a
Z	2a or 2b or 4e (×0.5)	2a	1a, 1b or 2g	1a, 1b	2 × 1a	2a or 2e	2a

Table 5. Structure site multiplicities and Wyckoff positions for all known space groups of apatite supergroup minerals [45].

The anion column in fluorapatite **Fig.** **8**(F) shows fluoride anion located on the mirror plane at z = ¼ and ¾ in successive unit cells. Three successive hydroxyls in hydroxylapatite (column **OH**) are disordered 0.35 Å above the mirror planes and three successive hydroxyls are below the mirror planes, with the sense of disordering reversed by an impurity of fluoride anion (stippled circle at the position z = 1¾). In the column "**Cl**" of chlorapatite, there are three successive anions disordered about 1.2 Å above and three below the mirror planes. The vacancy (□) at z = 1¾ in chlorapatite must exist in order to reverse the sense of ordering, as F and OH species are prohibited [33],[116].

Each F is bonded to three Ca(2) atoms, which form a triangle within the mirror plane **Fig.** **8**(**b**) and **Fig.** **16**(**b**). Because of its larger size and longer Ca–Cl bond length as compared to Ca–OH and Ca–F, Cl anions in chlorapatite are displaced from the (0,0,¼) special position on the mirror plane to two equivalent half-occupied positions at (0,0,0.4323) and (0,0,0.0677). In chlorapatite, Cl anion is displaced so far from the mirror plane (1.2 Å) that a weak bond (0.09 valence units) forms between Ca(2) and the second Cl anion, Cl′, located one-half unit cell away along c (Ca(2)-Cl′ bond distance is 3.27 Å). Slight overbonding of Ca(2) because of this weak Ca(2)-Cl′ interaction is balanced through reduced bonding between Ca(2) and, O(l) in chlorapatite [33].

The species-forming substitutions at the **Z** anionic site are limited to the monovalent anions F^-, Cl^- and OH^-. This implies 50 negative charges per unit cell (i.e. 24 O^{2-} + $(F,Cl,OH)^-$) for all known apatite minerals. In addition, many studies of synthetic compounds with the apatite structure show that Z site is occupied by O^{2-}, which increases total negative charges, vacancies and water molecules. The M site can be occupied by Cd, Co, K and by almost all REE. The X site can be occupied by Be, Cr, Ge and Mn [45].

[32] Consists of Wyckoff letter (f) and multiplicity (4). The multiplicity of the Wyckoff position is equal to the number of equivalent points per unit cell.

http://dx.doi.org/10.5772/62208

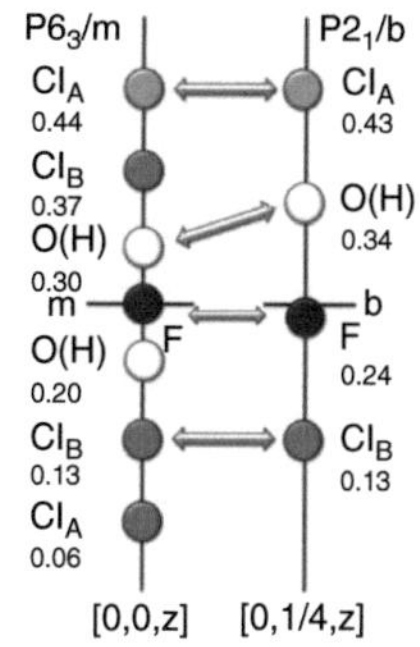

Fig. 9. Correlation between P6$_3$/m and P2$_1$/b sites in ternary apatites [116].

The difference between hexagonal and monoclinic polymorphs (e.g. apatite–(CaOH) and apatite–(CaOH)–M, see **Table 3**) lies in the position of **Z**$^-$ anions along large *c*-axis channels[33] (apatitic channels or anionic columns). This gives the rise (or not) to a mirror plane but does not correspond to a large ion rearrangement [37]. As the average ionic radii increases, i.e. the size of the tunnel ion increases through the series F > OH > Cl > Br, the metaprism untwists to accommodate larger anion [45],[119]. The correlation between the column-anion sites is introduced in **Fig. 9**.

Monoclinic symmetry of mineral with the space group **P2$_1$/b** named as clinohydroxylapatite [120] and subsequently renamed as apatite–(CaOH)–M [49] has typical axial setting of apatites resulting from the orientation ordering of OH$^-$ anions within [00z] anionic columns, with consequent doubling periodicity along [010] (*b*-axis). The following unit-cell parameters are given: $a = 9.445$ Å, $b = 18.853$ Å, $c = 6.8783$ Å and $\gamma = 120°$ [33],[45],[116].

The structure of minerals from the family of apatite allows to formulate seven requirements, including the following [109]:

1. Cation ordering in chemically complex members with minimal $M(1)_4(XO_4)_6$ framework distortion (metaprism twist angle <25°) that is accommodated in the P6$_3$, P$\bar{6}$ and P$\bar{3}$ maximal isomorphic subgroup of P6$_3$/M.

2. Intra- and inter-tunnel anion ordering that leads to P2$_1$/B varieties, possible modulation and extension of the (001) basal plane.

3. Framework topological tuning where the $M(2)Z_2$ tunnel contents are sufficiently small or substoichiometric so that the framework must constrict by increasing $M(1)O_6$ metaprism twist angle to >~25°, which is accompanied by a reduction to P2$_1$/M, P2$_1$, or P symmetry.

4. Framework hybrid intergrowth in which oxygen super- and substoichiometry leads to partial or complete replacement of BO_4 tetrahedra by BO_5 and BO_3 polyhedra, sometimes accompanied by a reduction in symmetry.

[33] The positions of noncolumn atoms in P2$_1$/B ternary apatite are similar to those in P6$_3$/M ternary apatite [116].

5. Polymorphic transformations initiated by the application of temperature/pressure that changes ionic sizes to drive the framework tuning.

6. Pseudomorfism whereby quite compositional adjustments lead to breaches in the critical limit of the metaprism twist angle and to the change in symmetry.

7. Polysomatism that arises by rotational twinning of $M_5X_3O_{18}Z_\delta$ modules in ordered and disordered sequences [109].

1.3 Polysomatic apatites

Less known polysomes of apatite are **ganomalite**[34] ($Pb_3Ca_2(SiO_4)(Si_2O_7)$ [121],[122],[123]) and **nasonite**[35] ($Ca_4Pb_6(Si_2O_7)_3Cl_2$ [124]) [109],[125],[126]. The concept of polysomatism was extensively developed by THOMPSON [127] and VEBLEN [128] for the crystallochemical analysis of rock-forming silicates and is a widely applied taxonomic principle for the description of condensed matter. Numerous polysome families include perovskite derivatives such as layered high T_c superconductors [129],[130],[131],[132], fluorite superstructures found in high-level nuclear ceramics [133] and β-alumina-hibonite ($CaAl_{12}O_{19}$) materials [134],[135],[136], which are encountered in superionic conductivity [109].

N	Crystallochemical formulae	Chemical formulae	Stacking sequence
2	$M_{10}(XO_4)_6Z_{2\delta}$	$M_{10}X_6O_{24}Z_{2\delta}$ 1)	$...\beta(\alpha\beta)\alpha...$
3	$M_{15}(X_2O_7)_3(XO_4)_3Z_{3\delta}$	$M_{15}X_9O_{33}Z_{3\delta}$ 2)	$...\beta(\alpha\alpha\beta)\alpha...$
4	$M_{20}(X_3O_{10})_3(XO_3)3Z_{4\delta}$	$M_{20}X_{12}O_{42}Z_{4\delta}$	$...\beta(\alpha\alpha\alpha\beta)\alpha...$
	$M_{20}(X_2O_7)_6Z_{4\delta}$	$M_{20}X_{12}O_{42}Z_{4\delta}$ 3)	$...\beta(\alpha\alpha\beta\beta)\alpha...$
5	$M_{25}(X_4O_{13})_3(XO_4)_3Z_{5\delta}$	$M_{25}X_{15}O_{51}Z_{5\delta}$	$...\beta(\alpha\alpha\alpha\alpha\beta)\alpha...$
	$M_{25}(X_3O_{10})_3(X_2O_7)_3Z_{5\delta}$		$...\beta(\alpha\alpha\alpha\beta\beta)\alpha...$
6	$M_{30}(X_5O_{15})_3(XO_4)_3Z_{6\delta}$	$M_{30}X_{18}O_{60}Z_{6\delta}$	$...\beta(\alpha\alpha\alpha\alpha\alpha\alpha)\alpha...$
	$M_{30}(X_4O_{13})_3(X_2O_7)_3Z_{6\delta}$		$...\beta(\alpha\alpha\alpha\alpha\beta\beta)\alpha...$
	$M_{30}(X_3O_{10})_6Z_{6\delta}$		$...\beta(\alpha\alpha\alpha\beta\beta\beta)\alpha...$
7	$M_{35}(X_6O_{19})_3(XO_4)_3Z_{7\delta}$	$M_{35}X_{21}O_{69}Z_{7\delta}$	$...\beta(\alpha\alpha\alpha\alpha\alpha\alpha\beta)\alpha...$
	$M_{35}(X_5O_{16})_3(X_2O_7)_3Z_{7\delta}$		$...\beta(\alpha\alpha\alpha\alpha\alpha\beta\beta)\alpha...$
	$M_{35}(X_4O_{13})_3(X_4O_{10})_3Z_{7\delta}$		$...\beta(\alpha\alpha\alpha\alpha\beta\beta\beta)\alpha...$
8	$M_{40}(X_7O_{22})_3(XO_4)_3Z_{8\delta}$	$M_{40}X_{24}O_{78}Z_{8\delta}$	$...\beta(\alpha\alpha\alpha\alpha\alpha\alpha\beta\beta)\alpha...$
	$M_{40}(X_5O_{16})_3(X_3O_{10})_3Z_{8\delta}$		$...\beta(\alpha\alpha\alpha\alpha\alpha\beta\beta\beta)\alpha...$
	$M_{40}(X_4O_{13})_6X_{6\delta}$		$...\beta(\alpha\alpha\alpha\alpha\beta\beta\beta\beta)\alpha...$

[34] Greek name reflecting luster appearance of mineral.

[35] Mineral is named according to American geologist LEWIS NASON.

http://dx.doi.org/10.5772/62208

N Crystallochemical formulae	Chemical formulae	Stacking sequence
∞ $M_5(X_3O_9)X_\delta$	$M_5X_3O_9Z_\delta$	$\ldots\alpha(\alpha)\alpha\ldots$
[1] Apatite	[2] Ganomalite	[3] Nasonite

Table 6. Staking sequence and compositions of polysomic apatites [109].

Certain complex structures are logically considered as intergrowths of chemically or topologically discrete modules. When the proportions of these components vary systematically, a polysomatic series is created. Certain complex structures are logically considered as intergrowths of chemically or topologically discrete modules. When the proportions of these components vary systematically a polysomatic series is created, the creation of which provides a basis for understanding the defects, the symmetry alternation and the trends in physical properties. The composition of polysomic family, which is formed by the condensation of *N*-modules of apatite (where *N* is the number of modules in the crystallographic repeat, **Table 6**) units of $M_5X_3O_{18}X_\delta$, can be described by the formula [109]:

$$M_{5N}X_{3N}O_{9N+6}X_N \ .$$

The apatite modules condense in a configuration to create B_nO_{3n+1}, where the values of $\boldsymbol{n}$ vary in range: $1 \le \boldsymbol{n} \le \infty$. For $N = 2$, the composition of polysome corresponds to the hydroxylapatite, e.g. $Ca_{10}(PO_4)_6(OH)_2$ if **M** = Ca, **X** = P and **Z** = OH.

Based on the value of parameter *N*, the following polysomes of apatite are recognized [109]:

1. Apatite: the value of $N = 2$ and the general composition corresponds to the formula $M_{10}X_6O_{24}Z_2$. Although, there are an infinite number of possible arrangements ($2 \le N \le \infty$), the ideal composition of apatite has the value of *N* from 2 to 8.

2. Ganomalite: the value of $N = 3$ and the general composition corresponds to the formula $M_{15}(X_2O_7)_3(XO_4)_3Z_3$.

3. Nasonite: the value of $N = 4$ and the general composition corresponds to the formula $M_{20}(X_2O_7)_6Z_4$.

The **Z** site could be vacant or partially occupied as required from the charge balance, and the **M:X** ratio must be maintained 5:3 (**Table 7**).

The apatite modules, while topologically identical, are often compositionally or symmetrically distinct and an infinite number of polysomes are feasible. The end members are the $N = 2$ polysome with all tetrahedra separated, and $N = \infty$ polysome, in which the hypothetical compound $M_5X_3O_9Z$ contains infinite, corner connected tetrahedral strings [109].

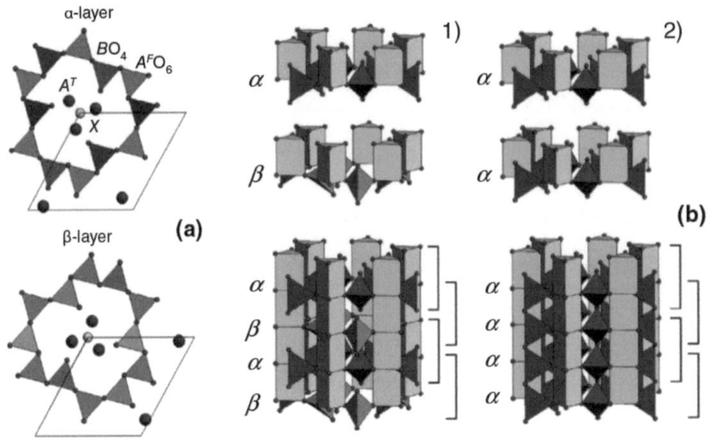

Fig. 10. Schematic representation of α and β, $M_5X_3O_{18}Z_2$ apatite modules (assuming a hexagonal basal plane), which are related by $[0001]_{hex}$ 60°rotation twinning. The principal idealization is that the M^FO_6 polyhedron is represented as a trigonal prism, but in real polysomes, twisting of the triangular faces through an angle φ creates a metaprism (a). Stacking of a and b modules show the construction of ...$\beta(\alpha\beta)\alpha$... apatite–2H $M_{10}(XO_4)_6Z_2$ (1) and the hypothetical structure ...$\alpha(\alpha)\alpha$... apatite–1H $M_5(X_3O_9)Z$ polysome end members (2). The coincident lattice where the condensation and elimination of oxygen take place is emphasized by brackets (M^T and Z ions are not included) (b) [109].

An idealized polysome module of apatite has the composition of $\mathbf{M(1)_2M(2)_3X_3O_{18}Z}$ and the thickness of ~3.5Å. These modules can occupy a hexagonal unit cell in two orientations, designated **α** and **β** layers **Fig. 10(a)**, which are rotated by 60° with respect to each other, with the condensation leading to the elimination of oxygen from coincident lattice position. The layers joint without the rotation create corner-connected B_nO_{3n+1} ($\boldsymbol{n}$ = ∞) tetrahedral strings,

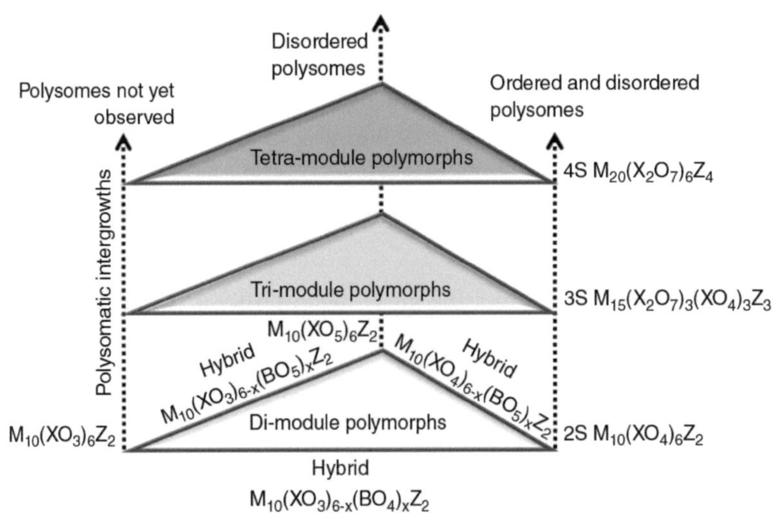

Fig. 11. Expanded apatite phase space containing all permutations of polymorphs, pseudomorphs, polysomes and hybrid structures, which may be feasible [109].

http://dx.doi.org/10.5772/62208

which can be broken through the introduction of a rotated layer. The case when the modules are placed directly upon one another in the sequence ...$\alpha(\alpha)\alpha$... (**Fig. 10(b)**), of the composition that corresponds to the hypothetical compounds $\mathbf{M}(1)_2\mathbf{M}(2)_3\mathbf{X}_3\mathrm{O}_9\mathbf{Z}$ with continuous chains of corner connected tetrahedral is created. Alternatively, if every module is rotated by 60° (rotationally twined) with respect to the order ...$\beta(\alpha\beta)\beta$..., six oxygens per layer pair are duplicated in the trigonal prisms and overall composition of polysome is $M(1)_4M(2)_6X_6O_{24}Z_2$. This motive can be found in the mineral mattheddleite ($Pb(1)_4Pb(2)_6[Si/SO_4]_6(Cl,OH)_2$) [109].

The prism of expanded apatite space (**Fig. 11**) enables to formulate new derivatives via the creation of $M(1)_4M(2)_6(XO_3/XO_4/XO_5)_6Z_2$ hybrids, which may display the polysomatic character [109].

1.4. Chemical composition of apatite supergroup minerals

The general formula of apatite compounds several times mentioned above ($Ca_{10}(PO_4)_6(F,Cl,O)_2$ or $3Ca_3(PO_4)_2{\cdot}Ca(F_2,Cl_2,O)$ was given by Voelcker [74] and Hoskyns-Abrahall [137]. Rammelsberg [138] assumed the existence of compound $Ca_{10}P_6O_{25}$ (equivalent to $3Ca_3(PO_4)_2{\cdot}CaO$) to explain the chemical composition of various apatites, although he thought the presence of this molecule was due to alteration. Groth [139] substituted hydroxyl for oxygen and gave the formula $(PO_4)_3(Cl,F,OH)Ca_5$. He was followed by Lacroix, Naumann-Zirkel and others [73]. Many minerals of the apatite group are deficient in combined fluorine and chlorine [73],[74],[140]. This deficiency was generally explained by assuming the compound $3Ca_3(PO_4)_2{\cdot}Ca(OH)_2$ [73].

Based on the results of chemical analysis, the calculation of apatite formula ($M_5(XO_4)_3Z$) can be determined according to the following criteria [45]:

a. Calculation on the basis of 13 total anions

b. Calculation on the basis of 8 total (M+X) cations

c. Calculation on the basis of 3X cations

In principle, criterion a is preferable to criteria b and c. In fact, the calculation based on any subset of all atoms does not affect the stoichiometric ration between them but automatically shifts the analytical error to the atoms not belonging to that subset. Criteria b and c would be best to use in cases in which structural vacancies are possible at some sites, but this does not seem to be the case for any apatite supergroup minerals [45].

Cations P^{5+}, As^{5+}, V^{5+}, Si^{4+} and S^{6+} can be assumed to be in tetrahedral coordination and assigned to *X* site in the formula of apatite. The total sum of these cation should be equal to 3 apfu[36] (in the single formula $M(1)_2M(2)_3(XO_4)_3Z$). All remaining cations enter **M**(1) and **M**(2) sites. The elucidation of partitioning between these two sites is almost impossible without an accurate evaluation of the electron density at each of them, which makes the structural study manda-

[36] The abbreviation for **a**toms **p**er **f**ormula **u**nit (**apfu**).

tory. Generally, **M**(1) sites (Wyckoff positions *4f* in $P6_3/M$ structure, **Table 5**) are occupied by smaller cations (in particular Ca) and **M**(2) sites (Wyckoff positions *6h*) accommodate larger cation[37] such as Ba^{2+} or Pb^{2+} [45].

1.5. Apatite group minerals

The apatite group includes minerals listed in **Table 3**. The most important are three minerals with ideal apatite formula $Ca_5(PO_4)_3F$, $Ca_5(PO_4)_3OH$ and $Ca_5(PO_4)_3Cl$ known as fluorapatite, hydroxylapatite and chlorapatite, respectively. They are recently renamed as apatite–(CaF), apatite–(CaOH) and apatite–(CaCl). The origin of those three distinct names to denote the F^-, OH^- and Cl^- variants and their distinction with respect to the "original apatite" *sensu lato* is uncertain [41],[45]. The composition and the cell dimension of major end members of apatite group minerals are listed in **Table 7**.

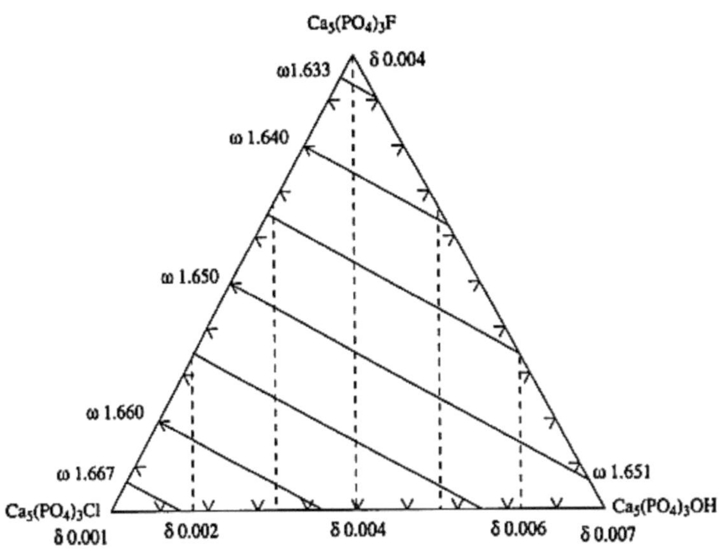

Fig. 12. Optical properties of fluoro-, hydroxyl- and chlorapatite series based on the values of end-member apatites [39].

Apatite is optically negative and normally uniaxial, although biaxial varieties with optic angle up to 20° are known. Carbonate bearing apatites (e.g. francolite), in particular way, have anomalous optics. Large variations in the composition within the apatite group make the accurate correlation of optical data difficult. The refractive index is the highest for chlorapatite (**Fig. 12**) and decreases by the substitutions of OH and F [39].

[37] The distribution of atoms at cationic sites can be affected also by Coulombic effect as was recognized for the structure of mineral aiolosite (**Section 2.1.5**) and cesanite (**Section 2.1.7**).

Mineral	Hydroxylapatite[38]	Fluorapatite	Chlorapatite	Carbonate–apatite
	Apatite–(CaOH)	**Apatite–(F)**	**Apatite–(Cl)**	–
Formula	$Ca_5(PO_4)_3OH$	$Ca_5(PO_4)_3F$	$Ca_5(PO_4)_3Cl$	$Ca_5[(PO_4)_{3-x}(CO_3,F)_x](F,OH,Cl)$
Abbreviation[39]	HAP, OHAp, HAp	FAP, FAp	CLAP	CHAP, CFAP, CCLAP
M [g mol^{-1}]	502.31	504.30	520.76	483.30[40]
Crystal system	Hexagonal–bipyramidal, space group $P6_3/M$			
a [Å]	9.42	9.39	9.60	9.36–9.32
c [Å]	6.87	6.88	6.78	6.89
a:c	1:0.73	1:0.73	1:0.71	1:0.73–0.74
V [Å^3]	527.59	526.03	540.59	520.97–529.06
Z	2			
CaO [%]	55.82	55.60	53.84	57.83 (CHAP) 57.60 (CFAP)
P_2O_5 [%]	42.39	42.22	40.89	36.66 (CHAP) 36.45 (CFAP)
X [%]	1.79	3.77	6.81	4.54(CO_2),1.86(H_2O) 4.52(CO_2),3.90(F)
Ca:P ratio	1.67 (5/3, ideal Ca:P ratio in the stoichiometric apatite[41]			
density[42] [g·cm^{-3}]	3.10–3.21/3.16	3.10–3.25/3.18	3.17–3.18/3.17	3.04 (CHAP) 3.12 (CFAP)
Hardness	5 (Mohs scale)			
Refractive index, $\omega - \varepsilon = \delta$	Negative uniaxial (-) 1.651 - 1.644 = 0.007	1.633 - 1.1.629 = 0.004	1.677 - 1.676 = 0.001	1.603 - 1.598 = 0.005 1.628 - 1.619 = 0.009

Table 7. Composition and properties of calcium apatite end-member $Ca_5(PO_4)_3Z$.

1.5.1. Fluorapatite (Apatite–CaF))

Fluorapatite ($Ca_{10}(PO_4)_6F_2$, F-rich apatite, FAP and FAp) [45],[141] is the most common member in the group of apatite that can be found mainly in igneous rocks (fluorapatite is a common accessory mineral in syenites [39] and metamorphic environments, for example, in carbona-

[38] In some works termed as hydroxylapatite.

[39] The symbol Ap, e.g. in HAp, reflects the abbreviation of mineral in rocks and ores, i.e. apatite = Ap.

[40] Depends on the type of carbonated apatite (**Section 2.6**), **Table 7** provides data for $Ca_5(PO_4,CO_3)_3OH$ (ideal composition of TYPE-B).

[41] Apatite is known to be often nonstoichiometric, especially on the surface [41].

[42] Measured/calculated density of mineral.

Fig. 13. Fluorapatite from (a) Cerro de Mercado Mine, Victoria de Durango, Mexico and (b) Sljudjanka, Bajkal, Russia.

tites and mica-pyroxenites [142]. The mineral crystallizes in the form of well-formed hexagonal crystals elongated on [0001], forms tabular to discoidal crystals flattened on {0001} or occurs as granular, globular to reniform, nodular and massive. The properties and chemical composition of fluorapatite mineral are listed in **Table 7**.

The color of fluorapatite mineral (**Fig. 13**) is pale green, green, light blue, yellow, purple, or white. Some varieties are colorless. Very brittle mineral gets broken to small fragments showing conchoidal fractures. Apatite shows poor cleavage on {0001} and $\{10\bar{1}1\}$. Fluorapatite is the most insoluble from the phosphate minerals [143],[144]. Apatite forms rare contact twins on $\{11\bar{2}1\}$ or $\{10\bar{1}3\}$. Some examples of fluorapatite crystals morphology are introduced in **Fig. 14**.

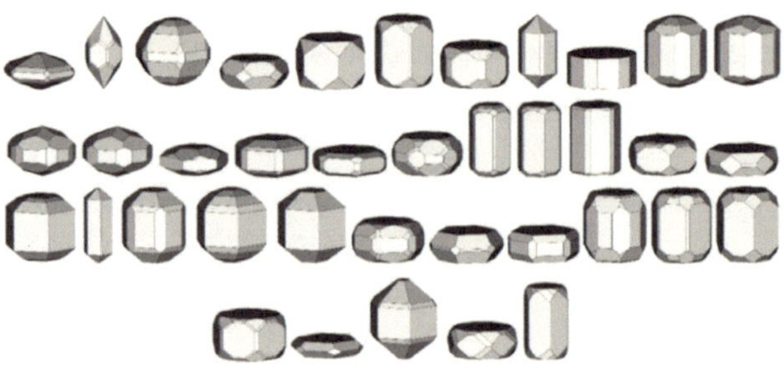

Fig. 14. Habit of fluorapatite crystals.

As was mentioned above (**Section 1.2**), fluorapatite was the first apatite with established structure [107],[108]. There is slight disagreement on the position of fluoride anion [145] Fluorapatite crystallizes in hexagonal dipyramidal crystal system with the parameters listed in **Table 7**. Atomic parameters including the number of atoms (*N*, where $\Sigma N = 42$) with equivalent site symmetry, their positions (coordinates *x*, *y* and *z*) in fractional coordinate system and equivalent isotropic temperature factor (*B*) for fluorapatite by Hughes et al [33] are listed in **Table 8**.

Atom	*N*	Site symmetry	*x*	*y*	*z*	*B* [Å²]
Ca(1)	4	C_{3h}	2/3	1/3	0.0010	0.91
Ca(2)	6	C_3	-0.00712	0.24227	¼	0.77
P	6	C_s	0.36895	0.39850	¼	0.57

Atom	*N*	Site symmetry	*x*	*y*	*z*	*B* [Å²]
O_1	6	C_s	0.4849	0.3273	¼	0.99
O_2	6	C_s	0.4667	0.5875	¼	1.19
O_3	12	C_s	0.2575	0.3421	0.0705	1.32
F	2	E	0	0	¼	1.93

Table 8. Positional parameters and equivalent isotropic factor for fluorapatite [33],[148].

The lengths of bonds in the fluorapatite structure are listed in **Table 9**.

Bond	Length [Å]	Bond	Length [Å]
P-O(1)	1.537	Ca(2)-F	2.3108
P-O(2)	1.539	Ca(2)-O(1)	2.701
P-O(3)	1.532	Ca(2)-O(2)	2.374
Ca(1)-O(1)	2.399	Ca(2)-O(3)	2.349
Ca(1)-O(2)	2.457	Ca(2)-O(3)	2.501
Ca(1)-O(3)	2.807	–	–

Table 9. Bond lengths in the structure of fluorapatite [33].

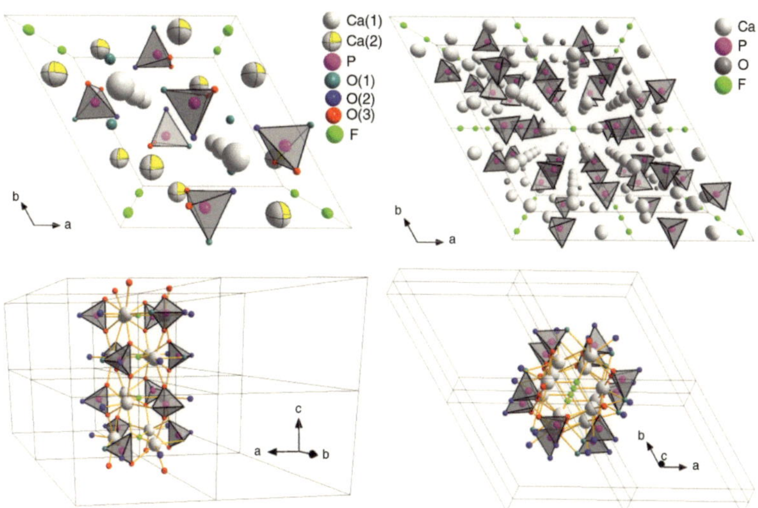

Fig. 15. Primitive unit cell of fluorapatite ($C_{10}(PO_4)_6F_2$) with atoms labeled according to symmetric type of element (a). The crystal structure of apatite (b, perspective view along the *c*-axis) and structure of columns (c, d).

The hexagonal crystal structure of fluorapatite of $P6_3/M$ space group is shown in **Fig. 15**. The atoms of Ca occupy two distinct sites [146], [147]:

i. $Ca(I)O_9$ polyhedra in sevenfold coordination

ii. $Ca(II)O_5Z(O)$ polyhedra in ninefold coordination

The length of Ca(2)-F bond is 2.311 Å (**Table 7**). As was described in **Section 1.2**, each fluoride anion is surrounded by three Ca atoms (**Fig. 16(a)**) at one level, and in addition, Ca-O columns are linked with PO_4^{3-} groups forming hexagonal networks. This arrangement gives very stable structure to fluorapatite while the structure of hydroxylapatite is lightly expanded and less stable to compare with fluorapatite. F anions lie in special position at the intersections of hexads with mirror planes at $z = ¼$ and ¾ (**Fig. 16(b)**) [39],[147],[148].

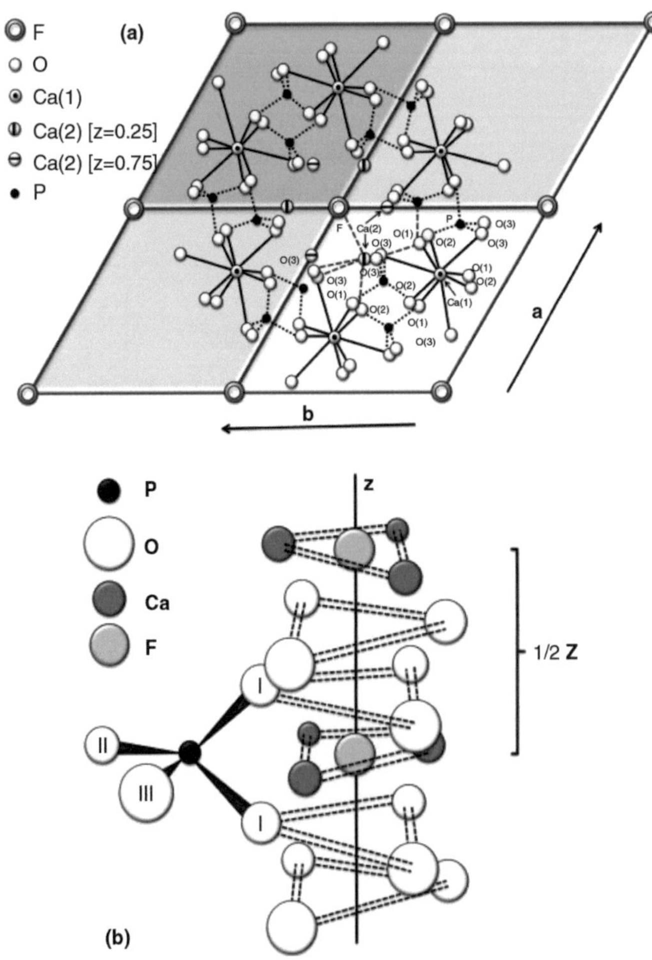

Fig. 16. Schematic depiction of a portion of four unit cells of the fluorapatite structure projected onto the (0001) plane. (a) Solid, dashed and dotted lines indicate bonds in different polyhedra. (b) The arrangement of Ca- and O-triangle on the screw axis (Z anion column) and attached PO_4^{3-} group [39].

http://dx.doi.org/10.5772/62208

The structure field of fluorapatite **Fig. 17** was investigated by KREIDLER and HUMMEL [149]. It is defined within the limits of radii of ions occupying the positions of phosphorus ($\boldsymbol{R}_p$) and calcium ($\boldsymbol{R}_c$) and is in the following range:

i. $0.29 \le \boldsymbol{R}_p \le 0.60$ Å

ii. $0.95 \le \boldsymbol{R}_c \le 1.35$ Å

The size limitations of ions occupying the calcium positions were less well defined because upper limit radius of ion was not found from the following reasons:

1. The only divalent cation larger than Ba^{2+} is Ra^{2+}.

2. Near the lower limit, apatites tended to have distorted structures, which no longer enabled hexagonal symmetry.

From practical proposes, the upper limit of $\boldsymbol{R}_c$ was set to 1.35 Å, which is slightly greater than ionic radius of barium. The lower limit of $\boldsymbol{R}_c$ was set to 0.95 Å so that both normal and distorted phases were included in the fluorapatite structure field.

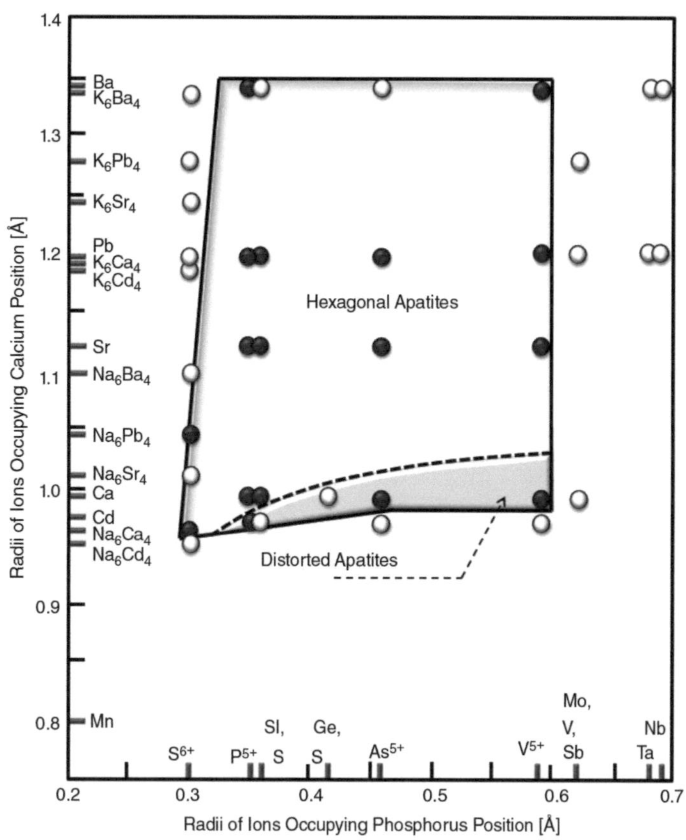

Fig. 17. The structure field of fluorapatite (Ahrens's radii): filled circles = apatite; open circles = no apatite [149].

The structure of monoclinic polymorph of fluorapatite is related to the space group $P2_1/B$ with the crystallographic parameters $a = 9.488$, $b = 18.963$, $c = 6.822$Å, $\beta = 119.97°$ and Z = 6 [145].

1.5.2. Hydroxylapatite (apatite–(CaOH))

Hydroxylapatite ($Ca_{10}(PO_4)_6(OH)_2$, HAP, pentacalcium monohydroxyorthophosphate) can be found mainly in igneous and metamorphic environments but also in biogenic deposits, e.g. in bone deposits [38],[44],[142]. Hydroxylapatite is very rare mineral. Wax yellow crystals up to 6 × 6 × 11 mm have been described from talc schist[43] from the Old Verde Antique serpentine quarry near Holly Springs, Cherokee County, Georgia [38]. The properties of hydroxylapatite are summarized in **Table 7**. The structure of hydroxylapatite and the comparison of sizes of ions is shown in **Fig. 18** and **Fig. 19**, respectively.

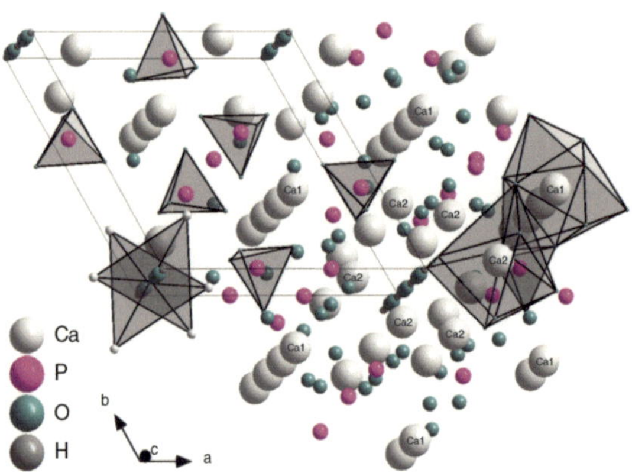

Fig. 18. The structure (view according to the *c*-axis) of hydroxylapatite showing the location of Ca(1) and Ca(2) sites.

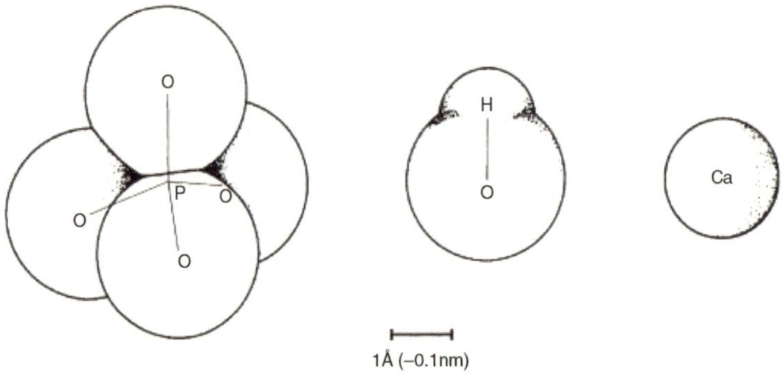

Fig. 19. The comparison of size of ions in the structure of hydroxylapatite [38].

[43] Medium-grade metamorphic rock occurred in almost infinite varieties, which was formed by the metamorphosis at high temperatures and pressure which leads to preferred orientation of flat (sheet-like) grains. The schist is medium to coarse grained.

http://dx.doi.org/10.5772/62208

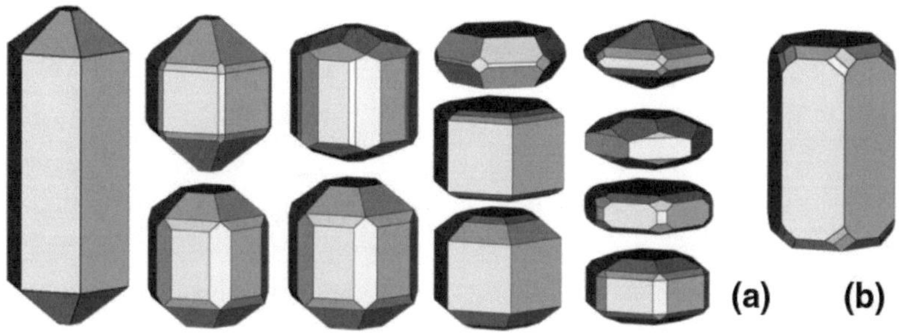

Fig. 20. Habit of hexagonal (a) and monoclinic (b) hydroxylapatite crystals.

The crystal habit and the structure of monoclinic polymorphs is show in **Fig. 20**. The monoclinic structure of hydroxylapatite (apatite–(CaOH)–M) appears to exist only in completely pure stoichiometric hydroxylapatite, and it is transformed to hexagonal form above about 250°C. The properties such as acid solubility and OH infusibility along the channels are related to the degree of disorder of OH positions. Hydroxyl anions lie in ordered positions in channels, whereas in hexagonal variety there is some disorder. Electrical properties are probably also dependent on exact channel position [106].

The atomic parameters (**Table 10**) and the length of bonds (**Table 11**) in the structure of hydroxylapatite were refined by Posner et al [150] from the structural data collected on the crystal of synthetic hydroxylapatite specimen. The results of natural sample of near-end member hydroxylapatite were provided by Hughes et al [33] and these data are also listed in **Table 10** and **Table 11**.

Hydroxylapatite is mainly used in the biomedical field for the preparation of bioceramics[44] [151],[152]. Since hydroxylapatite (HAp) is chemically similar to inorganic component of bone matrix and has excellent biocompatibility and surface active properties with living tissues, it has become one of the most important materials for artificial bone and bone regeneration [153], [154]. HA ceramics together with β-tricalcium phosphate have been the most extensively used substitution materials for artificial bone grafts for nearly three decades [153] (described in **Section 10.9**).

Atom	*N*	*x*	*y*	*z*	*B* [Å²]
Ca(1)	4	0.333/2/3	0.667/1/3	0.001/0.00144	0.666/0.929
Ca(2)	6	0.246/-0.00657	0.993/0.24706	0.250/1/4	0.328/0.859
P	6	0.400/0.36860	0.369/0.39866	0.250/1/4	0.192/0.62
O1	6	0.329/0.4850	0.484/0.3289	0.250/1/4	0.295/1.00

[44] Bioceramics can be divided into two large groups: bioinert and bioactive ceramics. The bioinert ceramics have almost no influence on surrounding living tissues like ZrO_2 and Al_2O_3. In contrast, the bioactive ceramics like calcium phosphates are able to bond with living tissues (**Section 10.9**).

Atom	*N*	*x*	*y*	*z*	*B* [Å²]
O2	6	0.589/0.4649	0.466/0.5871	0.250/1/4	0.496/1.25
O3	12	0.348/0.2580	0.259/0.3435	0.073/0.0703	0.632/1.57
OH		20.0/0.0	0.0/0.0	0.250/0.1979 O(H)	0.875/1.31
				–/0.04 H	–/3.3

Table 10. Atomic parameters for synthetic [150]/natural [33] hydroxylapatite: number of atoms per formula unit (*N*), positional parameters (*x*, *y* and *z*) and equivalent isotropic temperature factor (*B*).

Bond	Length [Å]	Bond	Length [Å]
P-O1	1.533/1.534	Ca(2)-OH	2.354/2.3851
P-O2	1.544/1.537	Ca(2)-O1	2.712/2.711
P-O3	1.514/1.529	Ca(2)-O2	2.356/2.353
Ca(1)-O1	2.416/2.404	Ca(2)-O3	2.367/2.343
Ca(1)-O2	2.449/2.452	Ca(2)-O3	2.511/2.509
Ca(1)-O3	2.802/2.802	–	–

Table 11. Length of bonds in the structure of synthetic [150]/natural [33] hydroxylapatite based on the parameters from **Table 10**.

The dehydroxylation of stoichiometric hydroxylapatite to oxyapatite takes place within the temperature range from 900°C to 1200°C [114]:

$$Ca_{10}\left(PO_4\right)_6\left(OH\right)_2 \xrightarrow{900-1000^\circ C} Ca_{10}\left(PO_4\right)_6 O + H_2O(g) \tag{1}$$

Further heating to temperatures higher than 1450°C leads to the thermal decomposition of oxyapatite into tricalcium phosphate (TCP, α-$Ca_3(PO_4)_2$) and tetracalcium phosphate (TTCP, $Ca_4P_2O_9$):

$$Ca_{10}\left(PO_4\right)_6 O \xrightarrow{T \geq 1450^\circ C} 2\ Ca_3\left(PO_4\right)_2 + Ca_4P_2O_9 \tag{2}$$

Calcium deficient hydroxylapatite decomposes at lower temperatures (at about 800°C) to stoichiometric hydroxylapatite and tricalcium phosphate according to their stoichiometry:

http://dx.doi.org/10.5772/62208

$$Ca_{10-x}(PO_4)_{6-x}(HPO_4)_x(OH)_{2-x} \xrightarrow{T\approx 800°C} (1-x)\ Ca_{10}(PO_4)_6(OH)_2 + 3x\ Ca_3(PO_4)_2 + x\ H_2O(g) \quad (3)$$

Another important option for utilization of hydroxylapatite is the preparation of porous high-temperature sorbent for carbon dioxide [155]. Apatite materials can be employed in discontinuous operations for removing CO_2 from gaseous streams in the form of structured monoliths or foams, with reduced pressure drops and enhanced refractory properties. High-temperature capture of carbon dioxide by hydroxylapatite proceeds via following reversible chemical reaction[45] [155]:

$$Ca_{10}(PO_4)_6(OH)_2 + CO_2(g) \leftrightarrow Ca_{10}(PO_4)_6CO_3 + H_2O(g) \quad (4)$$

A similar reaction is also possible with oxy-apatites:

$$Ca_{10}(PO_4)_6O + CO_2(g) \leftrightarrow Ca_{10}(PO_4)_6CO_3 \quad (5)$$

The highest CO_2 carrying capacity of HA macrogranules was detected at temperatures from 1000°C to 1100°C, achieving the values close to the theoretical limits. The changes in the HA microstructure induced during the thermal treatment (sintering) reduce the reactivity [155].

Since in the next decades the exploitation of fossil resources will continue and is expected to increase, rising the impact of fossil energy on the pollution and greenhouse effect, current technologies must be improved to become less harmful to the environment and more sustainable (zero emissions). The capture and the sequestration of CO_2 generated from the conversion of fossil fuels are being investigated as effective measures to reduce greenhouse gas emissions [156]. The apatite materials seem to be suitable sorbents for this purpose [155].

1.5.3. Chlorapatite (apatite–CaCl)

Calcium chlorapatite, as mineral, is relatively much less frequent than fluorapatite or hydroxylapatite and is formed primarily in flour-deficient environment [157]. The mineral crystallizes in the hexagonal system and the crystals are prismatic in habit, usually long, sometimes short and may have rounded ends or be terminated by pyramidal faces. Sometimes it occurs in granular massive to compact form [158]. The crystal habit and the structure of chlorapatite and monoclinic polymorphs chlorapatite–M are shown in **Fig. 21**. The crystallographic parameters and the properties are listed in **Table 7**.

---

[45] The preparation of carbonated apatites (**Section 4.6.1**) is based on the same principle as shows more general equation **Eq. 37**.

Chlorapatite mineral can be found as a mineral in calcium silicate marble, is an accessory mineral in layered mafic intrusions, occurs in veins such as "diabase," and replaces "triphylite" in some granite pegmatites. Such deposits are found in the USA; Quebec, Canada; Bushveld complex of Transvaal, South Africa; Angarth-Sud Tazenekht Plain of Morocco; Rajagarth, India; Kurokura, Japan; and Snarum, Norway [157].

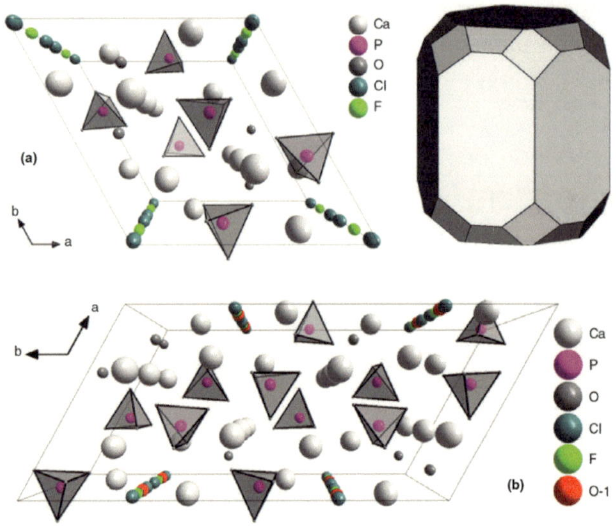

Fig. 21. Structure (perspective view according to the *c*-axis) and crystal habit of chlorapatite ([33], a) and chlorapatite–M ([116], b).

Atom	*N*	*x*	*y*	*z*	*B* [Å^2]
Ca(1)	4	2/3	1/3	0.0027	0.99
Ca(2)	6	0.00112	0.25763	1/4	1.14
P	6	0.37359	0.40581	1/4	0.77
O_1	6	0.4902	0.3403	1/4	1.34
O_2	6	0.4654	0.5908	1/4	1.47
O_3	12	0.2655	0.3522	0.0684	1.88
Cl	2	0	0	0.4323	2.68

Table 12. Positional parameters and equivalent isotropic factor for chlorapatite [33].

The atomic parameters and equivalent isotropic temperature factor for chlorapatite by Hughes et al [33] are listed in **Table 12**. The lengths of bonds in the chlorapatite structure are listed in **Table 13**.

Bond	Length [Å]	Bond	Length [Å]
P-O1	1.533	Ca(2)-F	2.759
P-O2	1.538	Ca(2)-O1	2.901
P-O3	1.524	Ca(2)-O2	2.306
Ca(1)-O1	2.407	Ca(2)-O3	2.331
Ca(1)-O2	2.448	Ca(2)-O3	2.254
Ca(1)-O3	2.793	–	–

Table 13. Bond lengths in the structure of chlorapatite [33].

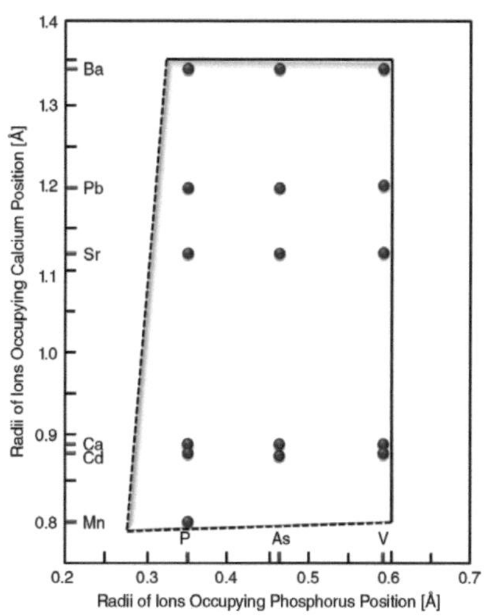

Fig. 22. The structure field of chlorapatite (Ahrens's radii) [149].

Together with fluorapatite, chlorapatite forms the solution[46] ($[Ca_5(PO_4)_3F]_{1-x}\cdot[Ca_5(PO_4)_3Cl]_x$, where parameter x varies in the range from 0 to 1) on whole range of composition [157].

The chlorapatite structural field (**Fig. 22**) was investigated by Kreidler and Hummel [149]. It has the boundaries of $0.19 \leq R_p \leq 0.60$ Å and $0.80 \leq R_c \leq 1.35$ Å differing from those of fluorapatite in two aspects:

1. Structurally distorted chlorapatite does not occur near the lower limit of R_c.
2. The chlorapatite structure could tolerate much smaller cations at the calcium sites than the fluorapatite structure.

Both of these differences are probably related to the difference in the position of halide ions in the fluor- and chlorapatite structure, but more detailed explanation of how the structural

[46] Activated by Sb^{3+} and Mn^{2+}, the resulting phosphor was used in fluorescent lamps until about 1990 when it was replaced by rare-earth activated alkaline earth aluminates [157].

difference enables chlorapatite to accept smaller cations without the distortion will not be attempted.

1.6. Other members of the apatite group

1.6.1. Svabite

Svabite is rare accessory mineral in calc-silicate skarns and arsenate analogue of fluorapatite [15] with the composition given by the formula $Ca_{10}(AsO_4)_6F_2$ (calcium fluorarsenate, CAAP). The mineral was named in 1981 by SJÖGREN [159] according to the Swedish chemist ANTON VON SWAB (1703–1768), who distilled zinc from calamine (smithsonite, $ZnCO_3$ [160]) [161]. Svabite has the same crystal habit as apatite, with rough hexagonal prisms, some of which show rounded pyramidal termination [162]. The structure of mineral svabite is shown in **Fig. 23**.

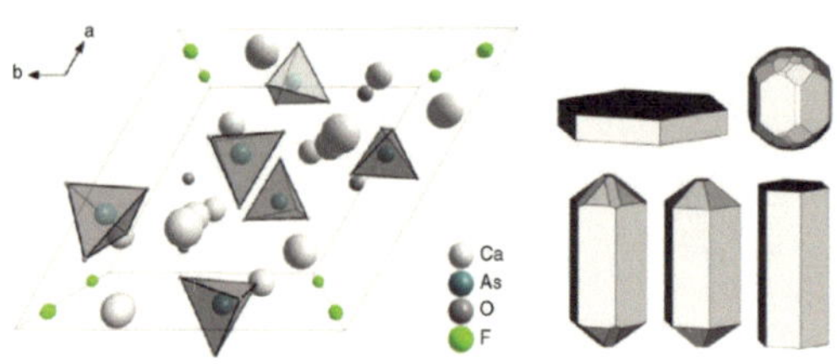

Fig. 23. The structure (perspective view along the *c*-axis) and examples of crystal habits of svabite.

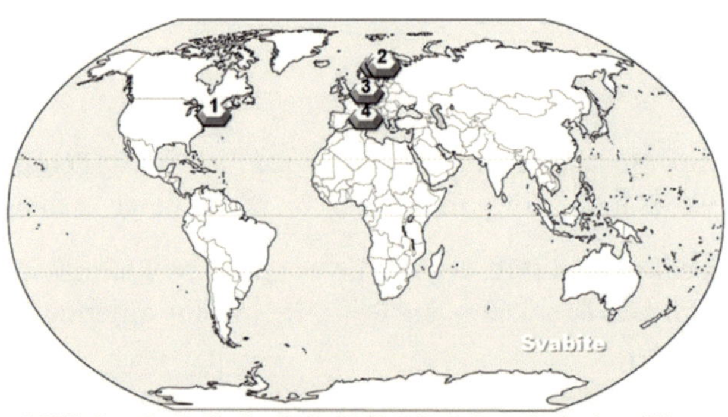

1) Mill Site Dump, Franklin Mine, Franklin, Franklin Mining District, Sussex Co., New Jersey, USA
2) Långban, Filipstad, Värmland, Sweden; Harstigen Mine, Pajsberg, Persberg district, Filipstad, Värmland, Sweden; Jakobsberg Mine, Jakobsberg ore field, Nordmark district, Filipstad, Värmland, Sweden
3) Clara Mine, Rankach valley, Oberwolfach, Wolfach, Black Forest, Baden-Württemberg, Germany
4) Sa Duchessa Mine, Oridda, Domusnovas, Carbonia-Iglesias Province, Sardinia, Italy

Fig. 24. Known localities for the mineral svabite.

It is brittle mineral that can be white, gray, gray-green, or colorless and transparent. The mineral is considered to be the intermediate between the apatite and the pyromorphite series. Svabite has the average density of 3.7 $g{\cdot}cm^{-3}$, and the hardness on the Mohs scale ranges from 4 to 5. It is hexagonal mineral belonging to the space group $P6_3/M$, $a = 9.75$ Å and $c = 6.92$ Å, $a{:}c = 1{:}0.71$, $V = 569.7$ Å^3, and **Z** = 2. The mineral occurs in the localities introduced in **Fig. 25**.

1.6.2. Turneaureite

Turneaureite ($Ca_5(AsO_4)_3Cl$, calcium chloroarsenate [7],[163]) is hexagonal mineral with the space group $P6_3/M$ with following crystallographic parameters $a = 9.81$ and $c = 6.868$ Å, $a{:}c = 1{:}0.700$, $V = 572.4$ Å^3 and Z = 2. The mineral is Cl analogue of svabite (**Section 1.6.1**) and OH analogue of johnbaumite (**Section 1.6.3**). The structure of turneaureite is shown in **Fig. 25**.

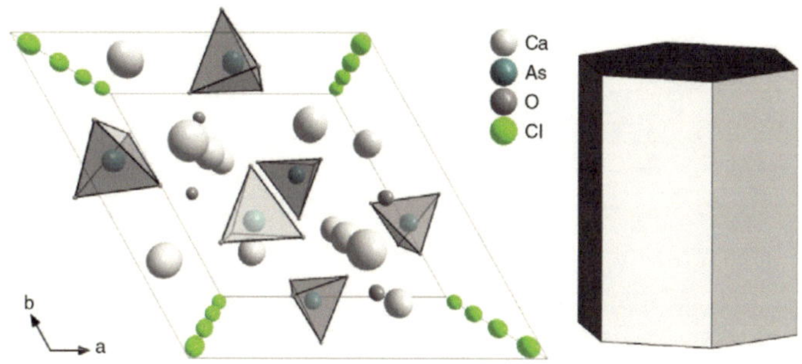

Fig. 25. The structure of turneaureite (perspective view along the *c*-axis) and examples of crystal habits of turneaureite.

The mineral occurs[47] in Franklin, New Jersey; Längban, Sweden (holotype[48]); and Balmat, New York [164]. The name honors Dr. Frederick Stewart Turneaure, Professor Emeritus at the University of Michigan, in recognition of his contributions to the geology and mineralogy of ore deposits. The mineral occurring occasionally as colorless, slightly turbid, prismatic (only the forms {1010} and {0001} are present) crystals up to 1.5 mm long. The mineral is colorless with a vitreous to slightly greasy luster. The hardness on the Mohs scale is 5. Measured and calculated density of mineral is 3.60 and 3.63 g·cm^{-3}, respectively [163], [164].

1.6.3. Johnbaumite

Johnbaumite ($Ca_5(AsO_4)_3OH$ [165],[166]) is a hexagonal mineral from the Franklin mine from Franklin, Sussex County, New Jersey, in 1944. The mineral is arsenate analogue of hydroxylapatite and hydroxyl analogue of svabite (**Section 1.6.1**). Johnbaumite was named in 1980 according to Baum [164]. The structure and crystal habits of johnbaumite is shown in **Fig. 26**.

[47] Turneaureite was found at three localities, but only the sample from Längban, Varmland, Sweden, provided single crystals of a size and quality adequate for the characterization of the specie [163].

[48] Holotype is the definition for single specimen (designated by the author) from which all the data from the original description were obtained. If the portion of such a specimen was send to other museum for preservation, the author shall designate this portion as "part of holotype." Other types of specimen are the cotype and the neotype. The cotype is a specimen used to obtain quantitative data for the original description (specimen examined only visually should not be considered a cotype). The neotype is a specimen chosen by author for a redefinition or reexamination of a species to represent the species when the holotype or cotypes cannot be found [5].

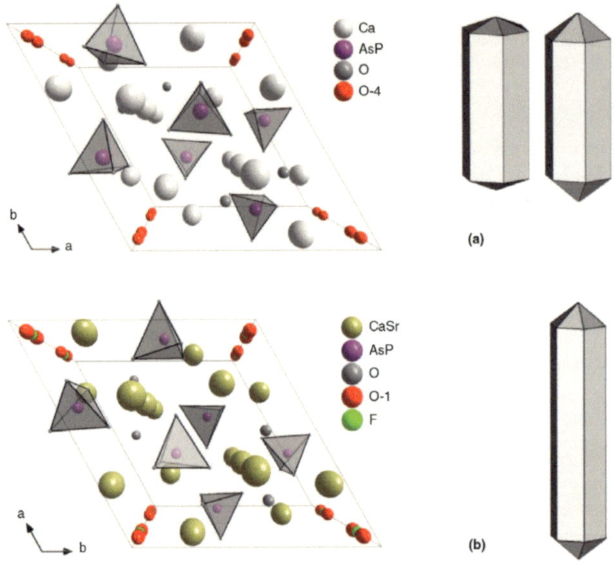

Fig. 26. The structure (perspective view along the *c*-axis) and the crystal habit of johnbaumite (a) and johnbaumite–M (b).

The space group of johnbaumite is $P6_3/M$ or $P6_3$ with $a = 9.70$ Å and $c = 6.93$ Å, $a{:}c = 1{:}0.714$, $\boldsymbol{V} = 564.96$ Å^3 and $\mathbf{Z} = 2$. Johnbaumite occurs as massive anhedral grayish-white granular material with individual grains of approximately 8 mm or less in diameter and is white to colorless with vitreous luster. Johnbaumite is colorless in thin section and has white streaks. The Mohs hardness is approximately 4½. The luster is vitreous on cleavage surfaces and slightly adamantine to greasy on fracture surfaces. The cleavage is distinct, parallel to {100}. Measured and calculated density of mineral is 3.68 and 3.73 g·cm^{-3}, respectively. Johnbaumite is fluorescent in short-wave UV radiation with a medium pinkish orange response color [165].

The monoclinic variant of mineral, which was previously named strontiumarsenapatit and fermorite ($3[(Ca,Sr)_3(P,As)_2O_8]\cdot Ca(OH,F)_2$ [74],[167], is now named johnbaumite–M and is not considered a distinct species [45]. The mineral crystallizes in the space group $P2_1/M$ with crystallographic parameters $a = 9.594$ Å, $b = 6.975$ Å and $c = 9.579$ Å; $\alpha = \gamma = 90°$ and $\beta = 119.97°$; $V = 556.341$ Å^3; and $\mathbf{Z} = 2$.

1.6.4. Pyromorphite

Pyromorphite (lead chlorophosphate, $Pb_5(PO_4)_3Cl$ [52],[168],[169],[170],[171],[172],[188], **Fig. 27**) was named by F.L. Hausmann in 1813. The name was derived from Greek words for "fire" and "form" to describe the recrystallization of a mineral from the melt. The structure and the example of shape of pyromorphite are shown in **Fig. 28**. Since the structures of vanadinite and pyromorphite are similar except for the tetrahedrally coordinated cations, it is described in **Section 1.6.8**.

http://dx.doi.org/10.5772/62208

Fig. 27. Pyromorphite: (a) Brandy Gill Mine, Great Britain and (b) Friedrichsegen Mine, Bad Ems, Nassau, Germany.

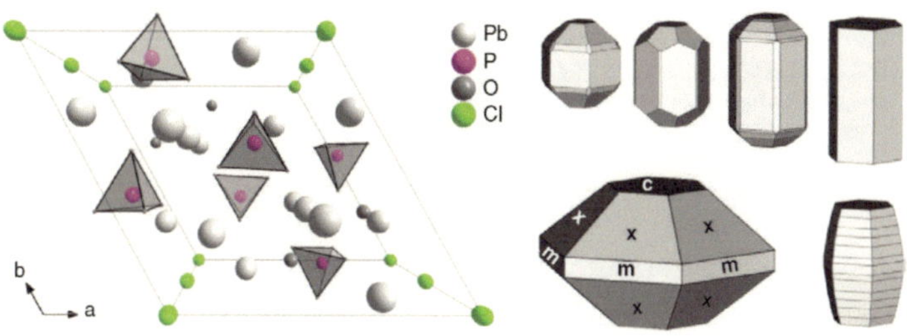

Fig. 28. The structure of pyromorphite mineral (perspective view according to the *c*-axis) (a) and the shape of crystal from Beaujeu (b): c (0001), m (10$\bar{1}$0) and x (10$\bar{1}$1).

Lead bearing minerals pyromorphite, mimetite (**Section 1.6.7**) and vanadinite (**Section 1.6.8**) occur independently or in great variety of isomorphous mixtures. All these species were prepared synthetically and pyromorphite is also known as a furnace product in slag [173], [174].

Pyromorphite is brittle hexagonal mineral that crystallizes in the space group **P6$_3$/M** with the crystallographic parameters $a = 9.9764$, $c = 7.3511$ Å, $a{:}c = 1{:}0.737$, $V = 634.14$ Å^3 and **Z** = 2. The mineral has the hardness from 3½ to 4 (Mohs scale) and shows the density in the range from 6.5 to 7.1 g·cm^{-3}. The luster is resinous and the color of mineral is commonly green, varying from grass green to the lighter and darker shades. It may be also pale brown. The secondary mineral formed together with limonite, cerussite ($PbCO_3$ [175]), hemimorphite ($Zn_4Si_2O_7(OH)_2{\cdot}H_2O$ [176],[177]) and smithsonite [160] in the upper oxidized portion of lea veins [178].

1.6.5. Stronadelphite

The mineral is named after the chemical element strontium and αδελφόσ, Greek word for "brother," as the full strontium analogue of fluorapatite, the most widespread member of the apatite supergroup. The mineral chemically close to stronadelphite, found in peralkaline pegmatite and Mt. Karnasurt, Lovozero alkaline complex, Kola Peninsula, Russia (**Fig. 29**), was first reported as "strontium-apatite."

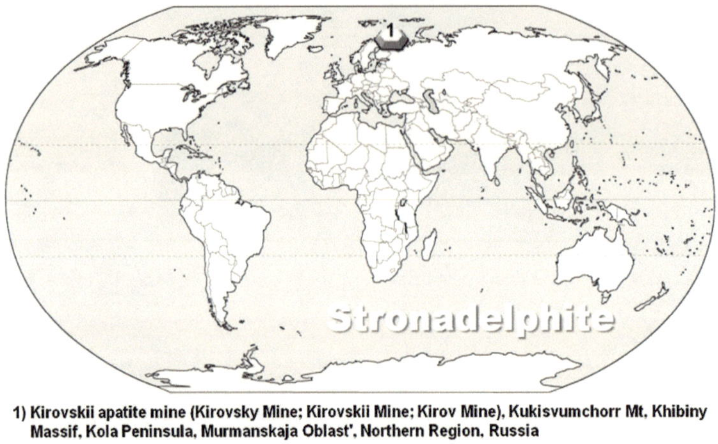

Fig. 29. Known localities for the mineral stronadelphite.

Stronadelphite ($Sr_5(PO_4)_3F$, strontium fluorophosphates, strontium apatite,[49] apatite–(SrF), SFAP [45],[179]) is a hexagonal mineral that crystallizes in the space group $P6_3/M$ and has the cell parameters $\boldsymbol{a}$ = 9.845 and $\boldsymbol{c}$ = 7.383 Å, $\boldsymbol{a}$:$\boldsymbol{c}$ = 1:0.75, $\boldsymbol{V}$ = 619.7 Å^3 and $\boldsymbol{Z}$ = 2. The brittle mineral is transparent and colorless with a pale greenish tint. There are no cleavages, and the hardness of mineral on the Mohs scale is 5 (apatite). Calculated and measured densities of the mineral are 3.98 and 3.92 g·cm^{-3}, respectively. The structure of mineral stronadelphite is shown in **Fig. 30**.

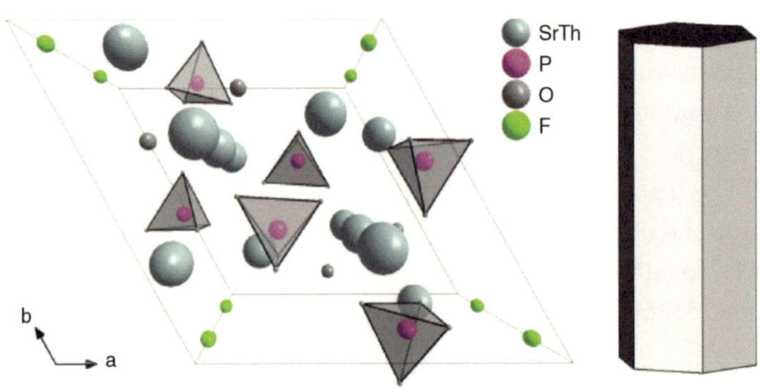

Fig. 30. Structure (view according to the *c*-axis) and crystal habit of stronadelphite mineral.

[49] Strontium apatite was also used as the old name for the mineral named apatite-(SrOH). The original name strontium–apatite was given before the structural study of the mineral and incorrectly reflects its relationship with apatite (please consult with **Table 3**). The name fluorostrophite (**Section 2.2.6**) was suggested by Passero et al [45].

http://dx.doi.org/10.5772/62208

1.6.6. Alforsite

Alforsite ($Ba_5(PO_4)_3Cl$, barium analogue of chlorapatite, barium chlorapatite, pentabarium tris[arsenate(V)] chloride [53],[54],[180]) is a colorless hexagonal mineral from the group of apatite named according to the geologist J.T. ALFORS. The mineral occurs in the localities introduced in **Fig. 31**.

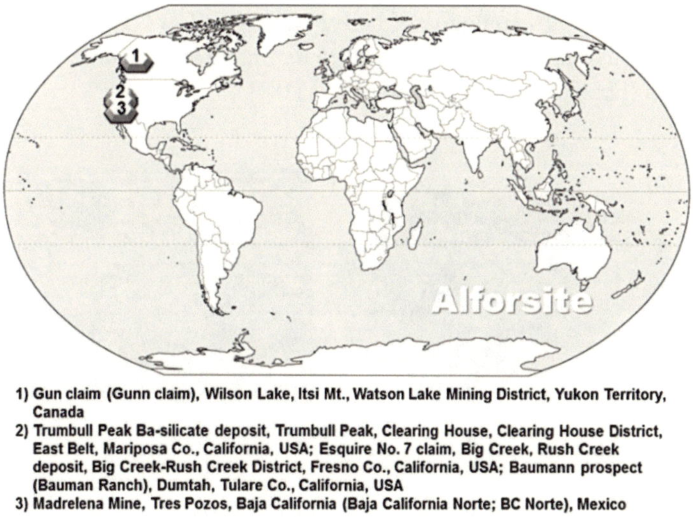

Fig. 31. Known localities for the mineral alforsite.

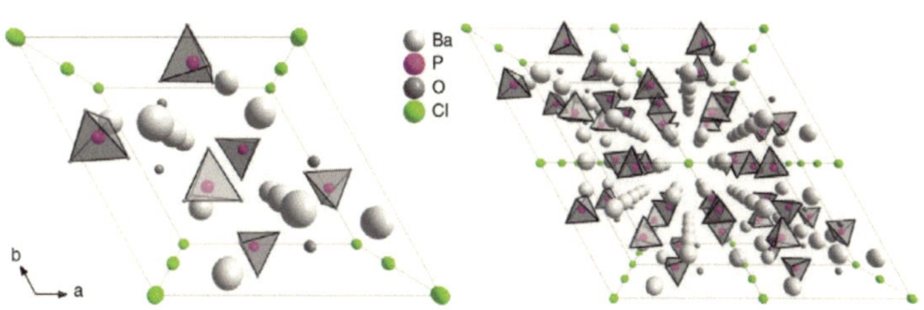

Fig. 32. The structure of mineral alforsite (perspective view along the *c*-axis).

Alforsite occurs as isolated small subhedral grains, generally less than 0.05 mm in diameter but rarely up to 0.2 mm. These colorless grains resemble typical fluorapatite, exhibiting low birefringence and high relief. This makes it difficult to distinguish alforsite from fluorapatite and from many of the associated high-relief barium minerals except by using the electron microprobe. Mineral is uniaxial and negative, with extremely low birefringence [53].

The structure of mineral alforsite is shown in **Fig. 32**. The space group of alforsite is $P6_3/M$ with the crystallographic parameters a = 10.25 Å, c = 7.64 Å, $a:c$ = 1:0.745, V = 700.77 Å^3 and Z = 2.

The mineral hardness on the Mohs scale is equal to 5. Calculated and measured density of mineral are 4.81 and 4.77 g·cm^{-3}, respectively [53],[54],[180].

1.6.7. Mimetite and clinomimetite

Mimetite (arsenopyromorphite, lead chloroarsenate, mimetite-H, $Pb_5(AsO_4)_3Cl$) [45],[55],[181],[182]), is the end member in the ternary system pyromorphite–vanadinite–mimetite. This is also the reason why the name of mineral is derived from Greek word *"mimethes,"* i.e. the imitator. Mimetite is also Pb_5 analogue of hedyphane (**Section 2.1**). Mimetite (**Fig. 33**) is an arsenate mineral; it usually forms as a secondary mineral in lead deposits through the oxidation of galena (PbS [183], Gn[50]) and arsenates [184].

Fig. 33. Mimetite (Příbram, Czech Republic).

Mimetite usually crystallizes in oxidized zones of lead deposits as small hexagonal prism with colors ranging from pale to bright yellow, orange, yellowish-brown, white, translucent, to opaque [184]. In accordance with other hexagonal apatite-group minerals, it crystallizes in the space group $P6_3/M$. The unit cell parameters are a = 10.46, c = 7.44 Å, a:c = 1:0.71 Z = 2 and V = 704.96 Å^3. The calculated density is 7.10 g·cm^{-3}. The hardness of the mineral on the Mohs scale is in the range from 3½ to 4. The structure and the shape of mimetite crystal is shown in **Fig. 34**.

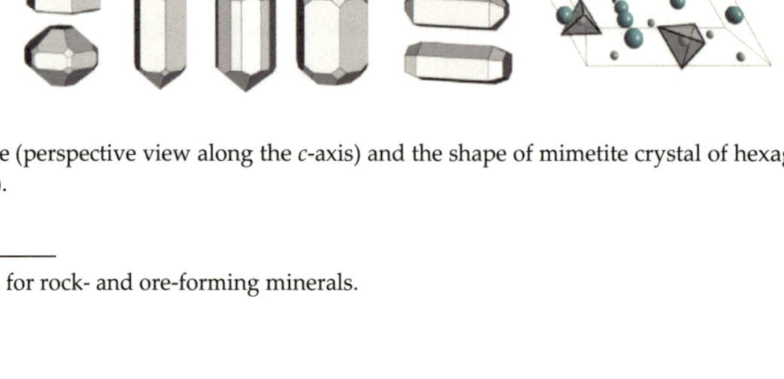

Fig. 34. The structure (perspective view along the c-axis) and the shape of mimetite crystal of hexagonal (a) and monoclinic polymorph (b).

[50] Symbol of mineral for rock- and ore-forming minerals.

The polymorphs of mimetite[51] are the monoclinic mimetite–M formerly known as clinomimetite mineral [185],[186] and mimetite-*2M* [182]. Monoclinic mimetite–M crystallizes in the space group $P2_1/B$ with the crystallographic parameters $\boldsymbol{a}$ = 10.189, $\boldsymbol{b}$ = 20.371 and $\boldsymbol{c}$ = 7.46 Å; angles $\boldsymbol{\alpha} = \boldsymbol{\beta} = 90°$ and $\boldsymbol{\gamma}$ = 119.88°; $\boldsymbol{V}$ = 1342.57 Å^3; and $\boldsymbol{Z}$ = 4. It is brittle mineral with the hardness of 4 (Mohs scale). The color of mimetite–M is pale greenish, yellow or white, and the mineral has white streak. Measured and calculated densities of the mineral are 7.36 and 7.37 g·cm^{-3}, respectively. Monoclinic polymorphs of mimetite-2M crystallize in the space group of $P2_1$ with the crystallographic parameters a = 20.422, b = 7.438 and c = 20.435 Å; ratio a:b:c = 2.746:1:2.747; angles $\boldsymbol{\alpha} = \boldsymbol{\beta} = 90°$ and $\boldsymbol{\gamma}$ = 119.95°; $\boldsymbol{V}$ = 2689.5 Å^3; and $\boldsymbol{Z}$ = 8.

1.6.8. Vanadinite

Vanadinite was named with regard to the content of vanadium[52] ($Pb_5(VO_4)_3Cl$ or $3Pb_3(VO_4)_2 \cdot PbCl_2$ [103],[187],[188],[189]). Since simple vanadates[53] incorporate other ions in their lattices, several series are known, including descloizite [190],[191]–mottramite [192],[193] and mounanaite[54] [194],[195]–krettnichite[55] [196],[197].

Fig. 35. Vanadinite (Mibladén, Morocco).

[51] Both monoclinic polymorphs were discredited from the IMA list of minerals in 2010.

[52] The most common vanadinite minerals are vanadate, descloizite ($PbZnVO_4(OH)$, orthorhombic [191]), and mottramite ($PbCuVO_4(OH)$, orthorhombic [193]) [204]. Vanadium was discovered in 1891 within lead vanadate ore by a Mexican mineralogist Andres Manuel del Rio (and named as erythronium), but it was mistaken as a form of chromium. The new element was recognized and named in 1830 by Swedish chemist Nils Gabriel Seftström. Vanadium was named after the Norse goddess Vanadis, who represented beauty and fertility [199],[205]. About 80% of vanadium is used for the production of special steel and alloys. Other applications include catalysts (e.g. in the production of sulfuric acid), pigments, and the manufacture of batteries [210]. Vanadium is used as alloy with a number of metals, e.g. ferro-vanadium (40–80% of vanadium), nickel-vanadium, alumino-vanadium, etc. [189].

[53] There are about 65 vanadium minerals, the most important sources of vanadium are titaniferous magnetites, carnotite (potassium uranyl vanadate used for the extraction of uranium, vanadium and radium), vanadinite, roscoelite ($K(V,Al,Mg)_2AlSi_3O_{10}(OH)_2$), patronite ($VS_4$), sulvanite ($Cu_3V^{2+}S_4$), uvanite ($U_2V^{5+}{}_6O_{21} \cdot 15H_2O$), bravoite, and davidite [210]. Vanadium is also found in clays, crude oil [189], and vanadium-rich variety of lignite (quisqueite).

[54] Monoclinic mineral (space group C2/M) of the composition $PbFe^{3+}{}_2(VO_4)_2(OH)_2$.

[55] Monoclinic mineral (space group C2/M) of the composition $PbMn^{2+}{}_2(VO_4)_2(OH)_2$.

Vanadinite mineral is the lead chlorovanadate (lead chloro orthovanadate) analogue of minerals mimetite (**Section 1.6.7**) and pyromorphite (**Section 1.6.4** [52]) and is considered as the end member in the ternary system pyromorphite–vanadinite–mimetite [198],[199]. The arseniferous variety of vanadinite was named as endlichite (arsenian vanadinite, $PbCl_2{\cdot}3Pb_3(V,As)_2O_8$ [200],[201],[202]).

Vanadinite is formed as the secondary product in oxidized zone of lead-bearing deposits[56] [203],[204],[205],[206] associated usually with galena, cerussite, or limonite [207]. It is known to form typically well-developed hexagonal prismatic crystals with smooth faces and sharp edges along [0001] [203]. It occurs frequently as acicular, hairlike, fibrous, rounded, globular, or hollow prismatic crystals. Synthetic vanadinite was first presented in 1957 by Durand [208].

The structure of hexagonal vanadinite belongs to the space group $P6_3/M$. The unit cell parameters are $a = 10.33$, $c = 7.34$ Å, $a{:}c = 1{:}\,0.71$, $\mathbf{Z} = 2$ and $V = 678.72$ Å^3. Calculated and measured average densities are 6.93 and 6.94 $g{\cdot}cm^{-3}$, respectively. It is brittle mineral of brown, brownish yellow, brown red, orange or yellow color, and some varieties can be colorless. The hardness of vanadinite on the Mohs scale varies in the range from 3½ to 4 [207],[209]. The structure of vanadinite and the example of crystal habit is shown in **Fig. 36**.

Vanadinite and pyromorphite (**Section 1.6.4**) possess similar structure where Pb(1) bonds to six oxygen atoms (3 × O(1) and 3 × O(2)) in the form of an approximate trigonal prism, with three longer bonds to oxygen atoms (3 × O(3)) through the prism faces. Adjacent $Pb(1)\text{-}O_9$ "prisms" share pinacoidal faces at the mirror planes (z = ¼ and ¾) to form $Pb(1)\text{-}O_9$ polyhedral chains parallel to the c-axis. Pb(2) lies in the mirror planes z = ¼ and ¾ and bonds to two oxygen atoms within the plane (O(1) and O(2)), four oxygen atoms (4 × O(3)) and two Cl atoms located on the hexad at 0,0,0 and 0,0,½ positions [188].

The major structural difference between vanadinite and pyromorphite occurs in XO_4 tetrahedra, which are occupied by V^{5+} (radius 0.59 Å) in vanadinite and by P^{5+} (0.35 Å) in pyromor-

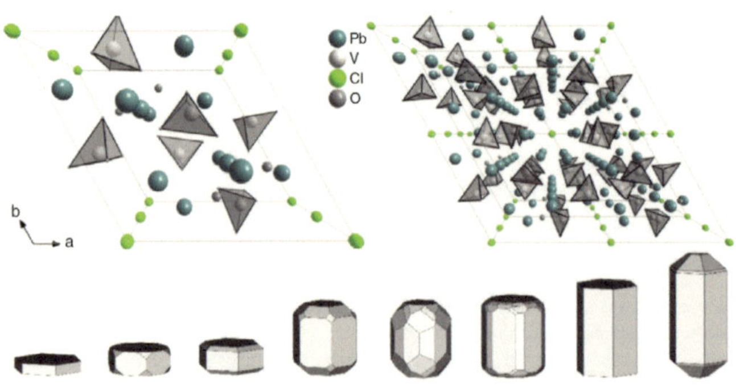

Fig. 36. The structure (perspective view along the c-axis) and the example of crystals habit of vanadinite.

[56] Almost all base-metal vanadate deposits occur in oxidized zones of the base-metal vein and replacement deposits. They also occur in other vanadium minerals in sediments. Vanadate deposits are largely restricted to tropical and temperature zones and to regions of dry climate [204],[206].

phite. The O-V-O and O-P-O angle in vanadate (VO_4^{3-}) and phosphate ion (**Section 1.2**) tetrahedron varies from 105.4 to 113.1° and from 107.1 to 111.8°, respectively. The most important structural features are the octahedral coordination of Pb(2) around Cl^- ions and the tetrahedral coordination of oxygen atoms around the vanadium atom. Each Cl^- ion is surrounded by six Pb(2) at the corner of a regular octahedron in which the Pb(2)-Cl distance is 3.17 Å and the shortest Pb(2)-Pb(2) distance (along an edge of the octahedron) is 4.48 Å [204], [210].

The production of vanadium metal from vanadinite ore consists of the following steps [188], [212]:

1. **Preparation of ammonium metavanadate** (NH_4VO_3) from powdered vanadinite ore, which is treated by concentrated HCl. This leads to the precipitation of $PbCl_2$ and to the solution of complex salt of $[V(OH)_4]Cl$ (**Eq. 6**). Ammonium metavanadate precipitates when $[V(OH)_4]Cl$ solution is boiled with NH_4Cl (**Eq. 7**).

$$Pb_5(VO_4)_3Cl + 12\ HCl \rightarrow 5\ PbCl_2 \downarrow + 3\left[V(OH)_4\right]Cl \tag{6}$$

$$\left[V(OH)_4\right]Cl + NH_4Cl \rightarrow NH_4VO_3 \downarrow + 2\ HCl + H_2O \tag{7}$$

2. **Conversion of ammonium metavanadate into V_2O_5** is reached by ignition. The process can be described by the following equation:

$$2\ NH_4VO_3 \rightarrow V_2O_5 + 2\ NH_3 + H_2O \tag{8}$$

3. **Reduction of $V_2O_5$to vanadium metal** via fluxing with Ca and $CaCl_2$ (**Eq. 9**) or by the aluminothermic process (**Eq. 10**):

$$V_2O_5 + 5\ Ca + 5\ CaCl_2 \rightarrow 2\ V + 5(CaO \cdot CaCl_2) \tag{9}$$

$$3\ V_2O_5 + 10\ Al \rightarrow 6\ V + 5\ Al_2O_3 \tag{10}$$

4. **Purification of vanadium metal** by *Van Arkel-de Boer method* [211],[216][57] in which the impure vanadium metal is heated with a limited amount of I_2 under vacuum.[58] Formed

[57] The method of reactive distillation of metal compounds (or also the chemical vapor transport reactions [214] developed by Dutch chemists Anton Eduard van Arkel and Jan Hendrik de Boer in 1925. The process can be applied if volatile iodides (from that they are also termed as the iodine process) of metal were formed and if the metal has higher melting point than the dissociation temperature of formed iodide. This technique has practical importance for Ti, Zr, Hf, Th, Cr, and V. The method which uses CO reactive gas instead of I_2 is known as the carbonyl or **Mond–Langer process** [213], e.g. $Ni(s) + 4\ CO(g) \leftrightarrow Ni(CO)_4(g)$.

[58] Chemical vapor transport (CVT) or vaporization reaction with iodine [214]. The investigation of the vaporization processes for several vanadium halide systems shown the existence of mixed halides $VX_{4-n}Y_n$ ($X = Cl, Br$; $Y = Br, I$) formed in reaction of Br_2 or I_2 with VX_2 or VX_3 solid phases at elevated temperature [215].

volatile iodide (VI_4, vanadium tetraiodide or vanadium(IV) iodide) decomposes at a higher temperature on the wolfram filament into pure vanadium and I_2, which becomes available to react with the impure vanadium, thus sustaining the process [189],[213],[214],[215],[216]:

$$\begin{aligned} &2\ \mathrm{V}(\mathrm{s,\ impure\ metal}) + 2\ \mathrm{I_2}(\mathrm{g}) \rightarrow \mathrm{VI_4}(\mathrm{g}) \\ &\xrightarrow{\mathrm{W-filament}} \mathrm{V}(\mathrm{s,\ pure}) + 2\ \mathrm{I_2}(\mathrm{g}) \end{aligned} \tag{11}$$

Author details

Petr Ptáček

Brno University of Technology, Czech Republic

References

[1] Nickel EH. The definition of a mineral. Canadian Mineralogist 1995;33 689–690.

[2] Nickel EH. Minerals: definition and classification. Reference module in earth systems and environmental sciences. Encyclopaedia of Geology 2005; 498–503 (current as of 25 March 2013).

[3] Nickel EH. Mineral names applied to synthetic substances. Canadian Mineralogist 1995;33 1335–1335.

[4] Nickel EH, Grice JD. The IMA commission on new minerals and mineral names: procedures and guidelines on mineral nomenclature. Canadian Mineralogist 1998;36 3–16.

[5] Dun PJ, Mandarino JA. Formal definitions of type mineral specimens. American Mineralogist 1987;72: 1269–1270.

[6] Clara M, Magalhães F, Williams PA. Apatite group minerals: solubility and environmental remediation. Thermodynamics, Solubility and Environmental Issues 2007;18 327–340.

[7] Richard V, Gaines H, Skinner C, Foord EE, Mason B, Rosenzweig A, King VT. Dana's New Mineralogy: The System of Mineralogy of James Dwight Dana and Edward Salisbury Dana. 8th ed., New York: John Wiley & Sons, 1997. ISBN 0-471-19310-0

[8] Mills SJ, Hatert F, Nickel EH, Ferraris G. The standardisation of mineral group hierarchies: application to recent nomenclature proposals. European Journal of Mineralogy 2009;21(5) 1073–1080.

[9] Strunz H, Nickel EH. Strunz Mineralogical Tables: Chemical–Structural Mineral Classification System. 9th ed., Stuttgart: Schweizerbart, 2001. ISBN 978-3-510-65188-7

[10] Valsami-Jones E. Phosphorus in Environmental Technologies: Principles and Applications: Principles and Applications (Integrated Environmental Technology). 1st ed., London: IWA Publishing, 2004. ISBN 1-84339-001-9

[11] Haldar SK, Tišljar J. Introduction to Mineralogy and Petrology. 1st ed., Amsterdam: Elsevier, 2013. ISBN 978-0-12-408133-8

[12] Manku GS. Theoretical Principles of Inorganic Chemistry. Tata McGraw-Hill Education, 1980. ISBN 0-07-096500-5

[13] von Zelewsky A. Stereochemistry of Coordination Compounds: A Textbook Series (Book 3). 1st ed., New York: John Wiley & Sons, 1996. ISBN 0-471-95599-X

[14] Agricola G. Fluores, in Bermannus, Sive De Re Metallica, ædibus Frobenianis (Basel) 1530; 125–127.

[15] Cockbain AG. The crystal chemistry of the apatites. Mineralogical Magazine 1968;36(281) 654–660.

[16] Wyckoff RWG. Crystal Structures: Vol. 3. Inorganic Compounds $R_x(MX_4)_y$, $R_x(M_nX_p)_y$, Hydrates and Ammoniates. 2nd ed., New York: John Wiley & Sons, 1965.

[17] Bishop AC, Woolley AR, Woolley WRH. Cambridge Guide to Minerals, Rocks and Fossils. 2nd ed., Cambridge: Cambridge University Press, 1999. ISBN 0-521-77881-6

[18] Dana JD. A System of Mineralogy: Including an Extended Treatise on Crystallography: With an Appendix, Containing the Application of Mathematics to Crystallographic Investigation, and a Mineralogical Bibliography. Durrie & Peck and Herrick & Noyes, 1837.

[19] Wenk Hans-R, Minerals Bulakh A.: Their Constitution and Origin. 1st ed., Cambridge: Cambridge University Press, 2004. ISBN 9780521529587

[20] Simmons R, Ahsian N, Raven H. The Book of Stones: Who They Are & What They Teach. Revised edition, Canada: North Atlantic Books, 2007. ISBN: 978-1-55643-668-0

[21] Miller WH. A Treatise on Crystallography. Cambridge: J. & J. J. Deighton, 1839. ASIN: B00896TYYA.

[22] Tareen JAK, Kutty TRN. A Basic Course in Crystallography. Universities Press, 2001. ISBN: 978-8173713606

[23] Tilley RJD. Understanding Solids: The Science of Materials. 2nd ed., New York: John Wiley & Sons, 2013. ISBN: 978-1-118-42328-8

[24] Hammond Ch. The Basics of Crystallography and Diffraction: International Lunion of Crystallography Texts on Crystallography (Book 12). 3rd ed., Oxford University Press, 2009. ISBN: 978-0199546459

[25] Klein C, Philpotts A. Earth Materials: Introduction to Mineralogy and Petrology. Cambridge University Press, 2012. ISBN: 9780521145213

[26] Bennett DW. Understanding Single-Crystal X-Ray Crystallography. New York: John Wiley & Sons, 2010. ISBN: 978-3-527-32677-8

[27] Nye JF. Physical Properties of Crystals: Their Representation by Tensors and Matrices. Oxford University Press, 1985. ISBN: 978-0198511656

[28] Julian MM. Foundations of Crystallography with Computer Applications. 2nd ed., CRC Press, 2014. ISBN: 978-1466552913

[29] Narasimhamurty TS. Photoelastic and Electro-Optic Properties of Crystals. Springer Science & Business Media, 2012. ISBN: 9781475700251

[30] Neuendorf KKE, Mehl JP, Jackson JA. Glossary of Geology. 5th revised ed., Springer Science & Business Media, 2011. ISBN: 978-0922152896

[31] De Graef M, McHenry ME. Structure of Materials: An Introduction to Crystallography, Diffraction and Symmetry. 2nd ed., Cambridge University Press, 2012. ISBN: 9781107005877

[32] Douglas B, Ho Shi-M. Structure and Chemistry of Crystalline Solids. Springer Science & Business Media, 2007. ISBN: 9780387366876

[33] Hughes JM, Cameron M, Crowley KD. Structural variations in natural F, OH, and Cl apatites. American Mineralogist 1989;74 870–876.

[34] Zhu Ch, Sverjensky DA. F-Cl-OH partitioning between biotite and apatite. Geochimica et Cosmochimica Acta 1992;56 3435–3467.

[35] Kucera M. Industrial Minerals and Rocks: Developments in Economic Geology. Elsevier, 2013. ISBN: 978-0444597502

[36] Douce AEP, Roden M. Apatite as a probe of halogen and water fugacities in the terrestrial planets. Geochimica et Cosmochimica Acta 2006;70(12) 3173–3196.

[37] Drouet Ch. A comprehensive guide to experimental and predicted thermodynamic properties of phosphate apatite minerals in view of applicative purposes. Journal of Chemical Thermodynamics 2015;81 143–159.

[38] Elliot JC. Studies in Inorganic Chemistry 18. Structure and Chemistry of the Apatites and Other Calcium Orthophosphates. Elsevier, 1994. ISBN: 978-0-444-81582-8

[39] Deer WA. Rock-Forming Minerals: Non-silicates, volume 5B, 2nd ed., Geological Society of London, 1998. ISBN: 978-1897799901

[40] McConnell D. Apatite: Its Crystal Chemistry, Mineralogy, Utilization, and Geologic and Biologic Occurrences: Applied Mineralogy (Volume 5). Springer Science & Business Media, 2012. ISBN: 978-3709183144

[41] Dorozhkin SV. A review on the dissolution models of calcium apatites. Progress in Crystal Growth and Characterization of Materials 2002;44(1) 45–61

[42] Ragot S, Zeyer J, Zehnder L, Reusser E, Brandl H, Lazzaro A. Bacterial community structures of an alpine apatite deposit. Geoderma 2013;202–203 30–37.

[43] Werner GA. Arragonischer Apatit. Bergmannische Journal 1788;1 95

[44] Terpstra RA, Bennema P, Hartman P, Woensdregt CF, Perdok WG, Senechal ML. F faces of apatite and its morphology: theory and observation. Journal of Crystal Growth 1986;78 468–478.

[45] Pasero M, Kampf AR, Ferraris C, Pekov IV, Rakovan JR, White TJ. Nomenclature of the apatite supergroup minerals. European Journal of Mineralogy 2010;22 163–179.

[46] Dorozhkin SV. Calcium Orthophosphates: Applications in Nature, Biology, and Medicine. 1st ed., CRC Press, 2012. ISBN: 978-9814316620

[47] Levinson AA. A system of nomenclature for rare-earth minerals. American Mineralogist 1966; 51 152–158.

[48] Bayliss P, Levinson AA. A system of nomenclature for rare-earth mineral species: Revision and extension. American Mineralogist 1988;73 422–423.

[49] Burke EAJ. Tidying up mineral names: an IMACNMNC scheme for suffixes, hyphens and diacritical marks. Mineralogical Record 2008;39: 131–135.

[50] Levinson AA. A system of nomenclature for rare-earth minerals. American Mineralogist 1966;51 152–158.

[51] Nickel EH. Standardization of polytype suffixes. Canadian Mineralogist 1993;31 767–768.

[52] Hausmann JFL. 1. Polychrom. Pyromorphit, in Handbuch der Mineralogie, Göttingen 1813; 1090–1096.

[53] Newberry NG, Essene EJ, Peacor DR. Alforsite, a new member of the apatite group: the barium analogue of chlorapatite. American Mineralogist 1981;66 1050–1053.

[54] Hata M, Marumo F, Iwai S, Aoki H. Structure of barium chlorapatite. Acta Crystallographica, Section B 1979;35 2382–2384.

[55] Beudant FS. Mimetèse, plomb arséniaté, in Traité Élémentaire de Minéralogie. 2nd ed., Paris 1832, p. 594–595.

[56] Winther C. Britholite, a new mineral. Meddelelser om Grønland 1901;24 190–196.

[57] Nickel EH, Mandarino JA. Procedures involving the IMA Commission on New Minerals and Mineral Names and guidelines on mineral nomenclature. American Mineralogist 1987; 72 1031–1042.

[58] Hata S. Abukumalite, a new yttrium mineral. Scientific Papers of the Institute of Physical and Chemical Research 1938;34 1018–1023.

[59] McConnell D. The substitution of SiO_4- and SO_4-groups for PO_4-groups in the apatite structure; ellestadite, the end-member. American Mineralogist 1937;22 977–986.

[60] Chesnokov BV, Bazhenova LF, Bushmakin AF. Fluorellestadite $Ca_{10}[(SO_4),(SiO_4)]_6F_2$—a new mineral, Zapiski Vsesoyuznogo. Mineralogicheskogo Obshchestva 1987;116(6) 743–746.

[61] Cavarretta G, Mottana A, Trece F. Cesanite, $Ca_2Na_3[(OH)(SO_4)_3]$, a sulphate isotypic to apatite, from the Cesano geothermal field (Latium, Italy). Mineralogical Magazine 1981;44 269–273.

[62] Frondel C. Two yttrium minerals: spencite and rowlandite. Canadian Mineralogist 1961; 6 576–581.

[63] Jaffe HW, Molinski VJ. Spencite, the yttrium analogue of tritomite from Sussex County, New Jersey. American Mineralogist 1962; 47 9–25.

[64] Knudsen AC, Gunter ME. Sedimentary phosphorites—an example: Phosphoria Formation, southeastern Idaho, U.S.A., in Reviews in Mineralogy and Geochemistry: Volume 48. Phosphates. (Kohn MJ, Rakovan J, Hughes JM, eds.), Mineralogical Society of America, Washington, DC, 2002, p. 363–389.

[65] Pan Y, Fleet ME. Composition of the apatite-group minerals: substitution mechanisms and controlling factors, in Reviews in Mineralogy and Geochemistry: Volume 48. Phosphates. (Kohn MJ, Rakovan J, Hughes JM, eds.), Mineralogical Society of America, Washington, DC, 2002, p.13–49.

[66] Fleet ME. Carbonated Hydroxyapatite: Materials, Synthesis, and Applications. CRC Press, 2014. ISBN: 978-9814463683

[67] Brown PW, Constanz B. Hydroxyapatite and Related Materials. CRC Press, 1994. ISBN: 978-0849347504

[68] Eastaugh N, Walsh V, Chaplin T. Pigment Compendium Set: Pigment Compendium: A Dictionary of Historical Pigments. 1st ed., Butterworth-Heinemann, 2005. ISBN: 978-0750657495.

[69] Brötgger WC, Bäckström H. Dahllite, a new mineral from Odegarden, Bamle, Norway. Abstracted in Zeitschrift für Kristallographie und Mineralogie 1890;17 426.

[70] Schaller WT. Mineralogical notes: Series 2. Bulletin 509, U.S. G.P.O., 1912.

[71] Stein CA. Jahrb. Vet. Naturk. Nassau, 1866, vol. 19–20, p. 41; Jahrb. Min., 1866, p. 716.

[72] Comptes rendus de I'Académie des sciences de I'URSS, Volume 31. L'Académie, 1963.

[73] Rogers AF. A new locality for Voelckerite and the validity of Voelckerite as a mineral species. Mineralogical Magazine 1914;17 155–162.

[74] Voelcker JA. Die chemische Zusammensetzung des Apatits nach eigenen vollständigen Analysen. Berichte der deutschen chemischen Gesellschaft 1883;16(2) 2460–2464.

[75] Chirvinsky PN. Jb. Min.: II. 1911;61 71.

[76] Chirvinsky PN. Materials for the knowledge of the natural productive forces of Russia. published in commission for the Russian Academy of Science, Petrograd: 1919;30 52 p.

[77] Eakle AS, RogersAF. Wilkeite, a new mineral of the apatite group and okenite, its alteration product. American Journal of Science 1914;37 262–267.

[78] Siewert MW. Zeitschrift für Naturwissenschaften, Halle 1874;10: 339 (as Manganapatit).

[79] Dana JD, Brush GJ. A System of Mineralogy: Descriptive Mineralogy, Comprising the Most Recent Discoveries. 5th ed., New York: John Wiley & Sons, 1868.

[80] Cleaveland P. An Elementary Treatise on Mineralogy and Geology. 2nd ed., Cummings and Hilliard, 1822.

[81] Jameson R. A System of Mineralogy. Neill & Company, 1816.

[82] Foit FFJr. Crystal chemistry of alkalideficient schorl and tourmaline structural relationships. American Mineralogist 1989;74 422–431.

[83] Bloodaxe ES, Hughes JM, Dyar MD, Grew ES, Guidotti ChV. Linking structure and chemistry in the Schorl–Dravite series. American Mineralogist 1999;84 922–928.

[84] Bnown GEJr., Mills BA. High-temperature structure and crystal chemistry of hydrous alkali-rich beryl. American Mineralogist 1986;71 547–556.

[85] Aurisicchio C, Grubessi GFO, Zanazzi PF. Reappraisal of the crystal chemistry of beryl. American Mineralogist 1988;73 826–837.

[86] Jameson R. A System of Mineralogy. Volume 2. 2nd. ed., A. Constable & Co., 1816.

[87] Bulletin of the Bureau of Mineral Resources, Geology and Geophysics. 69th ed., 1964, p. 172.

[88] Knight Ch. Natural History: Or, Second Division of "The English Encyclopedia". Bradbury, Evans & Company, 1866.

[89] Tschirwinsky W. Podolite, a new mineral: Centralbl. Min., Geol. u. Pal., 1907; abstracted in Zeitschr. Kryst. Min., 1909;46 296.

[90] Hunt WF. The American Mineralogist. Volume 27 Mineralogical Society of America, 1942.

[91] Winchell AN, Winchell NH. Elements of optical mineralogy: an introduction to microscopic petrography. Volume 2. 4th ed., New York: John Wiley & Sons, 1959.

[92] Dunn PJ. Dehrnite and lewistonite: discredited. Mineralogical Magazine 1978;42 282–284.

[93] Hey MH, Gottardi. On the use of names, prefixes and suffixes, and adjectival modifiers in the mineralogical nomenclature. Canadian Mineralogist 1980;18 261–262.

[94] Gmelin L. Rare Earth Elements: Main Volume. 8th ed., Verlag Chemie, 1984.

[95] Volkova MI, Melentiev BN. Compt. Rend. (Doklady) Acad. Sci. URSS 1939;25 122 (strontium apatite, saamite).

[96] Cámara F, Sokolova E, Abdu YA, Hawthorne FC. Saamite, Ba□TiNbNa$_3$Ti(Si$_2$O$_7$)$_2$O$_2$ (OH)$_2$(H$_2$O)$_2$, a group-III Ti-disilicate mineral from the Khibiny alkaline massif, Kola Peninsula, Russia: description and crystal structure. Canadian Mineralogist 2014;52 745–761.

[97] Stamatakis MG. Phosphate deposits of Neogene age in Greece. Mineralogy, geochemistry and genetic implications. Chemie der Erde–Geochemistry 2004;64(4) 329–357.

[98] Rogers AF. Mineralogy and petrography of fossil bone. Bulletin of the Geological Society of America 1924;35(3) 535–556. doi: 10.1130/GSAB-35-535

[99] Herman DZ. Fossilization type of *Elephas hysudrindicus* from blora on the basis of petrographic and scanning electron microscopic analyses. Jurnal Geologi Indonesia 2011;6(2) 75–84.

[100] Conybeare CEB. Lithostratigraphic Analysis of Sedimentary Basins. Elsevier Science, 2013. ISBN: 978-1483268606

[101] Jébrak M. Hydrothermal breccias in vein-type ore deposits: a review of mechanisms, morphology and size distribution. Ore Geology Reviews 1997;12(3) 111–134.

[102] Breithaupt A. Bestimmung neuer mineral-specien, hedyphan. Journal für Chemie und Physik 1830;60 308–316.

[103] von Kobell F. Vanadinit, in Grundzüge der Mineralogie (Nürnberg) 1838; 283–283.

[104] Fleet ME. Infrared spectra of carbonate apatites: ν2-Region bands. Biomaterials 2009;30(8) 1473–1481.

[105] Fleet ME, Liu X. Local structure of channel ions in carbonate apatite. Biomaterials 2005;26(36) 7548–7554.

[106] Derek EC. Chemistry, Biochemistry and Technology, 6th ed., CRC Press, 2013. ISBN: 978-1439840887

[107] Mehmel M. Uber die Struktur des Apatits. I. Zeitschrift für Kristallographie, 1930;75 323–331.

[108] Náray-Szabó S. The structure of apatite (CaF)Ca$_4$(PO$_4$)$_3$. Zeitschrift für Kristallographie 1930;75 387–398.

[109] Baikie T, Pramana SS, Ferraris C, Huang Y, Kendrick E, Knight K, Ahmad Z, White TJ. Polysomatic apatites. Acta Crystallographica Section B 2010;66(1) 1–16.

[110] Borodin LS, Kazakova ME. Belovite—a new mineral from an alkaline pegmatite, Doklady Akademii Nauk SSSR 1954;96 613–616.

[111] Pekov IV, Kulikova IM, Kabalov YK, Eletskaya OV, Chukanov NV, Menshikov YP, Khomyakov AP. Belovite–(La) Sr$_3$Na(La,Ce)[PO$_4$]$_3$(F,OH)—a new rare earth mineral in

the apatite group. Zapiski Vserossijskogo Mineralogicheskogo Obshchestva 1996;125(3) 101–109.

[112] Khomyakov AP, Kulikova IM, Rastsvetaeva RK. Fluorcaphite Ca(Sr,Na,Ca)(Ca,Sr, $Ce)_3(PO_4)_3F$—a new mineral with the apatite structural motif. Zapiski Vserossijskogo Mineralogicheskogo Obshchestva 1997;126(3) 87–97.

[113] Khomyakov AP, Kulikova IM, Rastsvetaeva RK. Fluorcaphite Ca(Sr,Na,Ca)(Ca,Sr, $Ce)_3(PO_4)_3F$—a new mineral with the apatite structural motif. Zapiski Vserossijskogo Mineralogicheskogo Obshchestva 1997;126(3) 87–97.

[114] Habelitz S, Pascual L, Durán A. Nitrogen-containing apatite. Journal of the European Ceramic Society 1999;19(15) 2685–2694.

[115] Luo Y, Hughes JM, Rakovan J, Pan YM. Site preference of U and Th in Cl, F, and Sr apatites. American Mineralogist 2009;94 345–351.

[116] Hughes JM, Cameron M, Crowley KD. Crystal structures of natural ternary apatites: Solid solution in the $Ca_5(PO_4)_3X$ (X = F, OH, Cl) system. American Mineralogist 1990;75 295–304.

[117] Boyer L, Carpena J, Lacout JL. Synthesis of phosphate-silicate apatites at atmospheric pressure. Solid State Ionics 1997;95(1–2) 121–129.

[118] Deer WA. Rock-forming minerals: non-silicates, volume 5B, 2nd ed., Geological Society of London, 1998. ISBN: 978-1897799901

[119] Baikie T, Ahmad Z, Srinivasan M, Maignan A, Pramana SS, White TJ. The crystallographic and magnetic characteristics of Sr_2CrO_4 (K_2NiF_4-type) and $Sr_{10}(CrO_4)_6F_2$ (apatite-type). Journal of Solid State Chemistry 2007;180(5) 1538–1546.

[120] Chakhmouradian AR, Medici L. Clinohydroxylapatite: a new apatite-group mineral from northwestern Ontario (Canada), and new data on the extent of Na-S substitution in natural apatites. European Journal of Mineralogy 2006;18 105–112.

[121] Nordenskiöld AE. Åtskilliga nya mineralfynd vid Nordmarken och Långban, Geologiska Föreningens i Stockholm Förhandlingar 1876;3 119–123.

[122] Dunn PJ, Peacor DR, Valley JW, Randall CA. Ganomalite from Franklin, New Jersey, and Jakobsberg, Sweden: new chemical and crystallographic data. Mineralogical Magazine 1985;49 579–582.

[123] Carlson S, Norrestam R, Holtstam D, Spengler R. The crystal structure of ganomalite, $Pb_9Ca_{5.44}Mn_{0.56}Si_9O_{33}$. Zeitschrift für Kristallographie 1997;212(3) 208–212.

[124] Penfield SL, Warren CH. Some new minerals from the zinc mines at Franklin, N.J., and note concerning the chemical composition of ganomalite. American Journal of Science 1899;8 339–353.

[125] Engel G, Krieg F, Reif G. Mischkristallbildung und Kationenordnung im System Bleihydroxylapatit-Calciumhydroxylapatit. Journal of Solid State Chemistry 1975;15(2) 117–126.

[126] Newnham RE, Wolfe RW, Darlington CNW. Prototype structure of Pb5Ge3O11. Journal of Solid State Chemistry 1973;6(3) 378–383.

[127] Thompson JBJr. Biopyriboles and polysomatic series. American Mineralogist 1978;55 239–249.

[128] Veblen DR. Polysomatism and polysomatic series: a review and applications American Mineralogist 1991;76 801–826.

[129] Croce F, Scrosati B, Alvani C, Casadio S. Mechanism of oxygen incorporation in the $YBa_2Cu_3O_{6.9}$ superconductor. Solid State Ionics 1989;36(1–2) 85–88.

[130] Kuzmann E, Homonnay Z, Nagy S, Vértes A, Halász I, Gál M. Structural investigation of the $EuBa_2Cu_3O_{7-\delta}$ high TC superconductor by ^{151}Eu, ^{119}Sn, ^{57}Fe and 57Co Mössbauer spectroscopy. Spectrochimica Acta Part A: Molecular Spectroscopy 1992;48(1) 51–64.

[131] Liang KC, Nowick AS. High-temperature protonic conduction in mixed perovskite ceramics. Solid State Ionics 1993;61(1–3) 77–81.

[132] Shen WQ, Dong ML, Li JH, Du SB, Li GQ, Zheng GG, Zheng JQ, Liu JX, Guan WY. Temperature dependence of quadrupole interaction in high Tc superconductor $YBa_2(Cu_{0.9}Ag_{0.1})_3O_{7-\delta}$. Solid State Communications 1990;73(10) 701–709.

[133] van Berkel FPF, Ijdo DJW. The orthorhombic gluorite related compounds Ln_3RuO_7, Ln, Nd, Sm and Eu. Materials Research Bulletin 1986;21(9) 1103–1106.

[134] Sktani ZDI, Azhar AZA, Ratnam MM, Ahmad ZA. The influence of in-situ formation of hibonite on the properties of zirconia toughened alumina (ZTA) composites. Ceramics International 2014;40(4) 6211–6217.

[135] Song Jun-H, Park Sang-Y. In situ co-textured microstructure of alumina/alumina: Ca-hexaluminate multilayer composites. Ceramics International 2001;27(4) 443–449.

[136] Domínguez C, Chevalier J, Torrecillas R, Gremillard L, Fantozzi G. Thermomechanical properties and fracture mechanisms of calcium hexaaluminate. Journal of the European Ceramic Society 2001;21(7) 907–917.

[137] Hoskyns-Abrahall JL. Inaug.-Diss., Munich, 1889. Abstract in Zeits. Kryst. Min., 16, p. 389.

[138] Rammelsberg CF, Neues Jahrbuch für Mineralogie 1897;2 37.

[139] Groth P. Tab. Tabellarische Übersicht der Mineralien nach ihren krystallographish-chemischen Beziehungen. 4th ed., Braunschweig, 1898, p.87.

[140] Carnot A. Sur les variations observers dans la compositions des apatites. Bull. Soc. Franc. Min., 1896;19 135.

[141] Teodorovich GI. Authigenic Minerals in Sedimentary Rocks. Springer Science & Business Media, 2012. ISBN: 978-1468406528

[142] Abouzeid Abdel-Z.M. Physical and thermal treatment of phosphate ores—an overview. International Journal of Mineral Processing 2008;85(4) 59–84.

[143] Dave SCh. Growth Kinetics of Fluorapatite Deposition on Synthetic Hydroxyapatite. University of Michigan, 1987.

[144] Center for Archaeological Sciences Norman Herz Professor of Geology and Director, Society of Archaelogical Sciences both at University of Georgia Ervan G. Garrison Associate Professor of Anthropology and Geology and President. Geological Methods for Archaeology. Oxford University Press, 1997. ISBN: 978-0198025115

[145] Somasundaran P. Encyclopedia of Surface and Colloid Science, 2004 Update Supplement, Volume 5. CRC Press, 2004. ISBN: 978-0824721541

[146] Njema H, Debbichi M, Boughzala K, Said M, Bouzouita K. Structural, electronic and thermodynamic properties of britholites $Ca_{10-x}La_x(PO_4)_{6-x}(SiO_4)_xF_2$ $(0 \leq x \leq 6)$: experiment and theory. Materials Research Bulletin 2014;51 210–216.

[147] Ivanova TI, Frank-Kamenetskaya OV, Kol'tsov AB, Ugolkov VL. Crystal structure of calcium-deficient carbonated hydroxyapatite. Thermal decomposition. Journal of Solid State Chemistry 2001;160(2) 340–349.

[148] Tressaud A, Haufe G. Fluorine and Health: Molecular Imaging, Biomedical Materials and Pharmaceuticals. Elsevier, 2008. ISBN: 978-0080558110

[149] Kreidler ER, Hummel FA. The crystal chemistry of apatite: Structure fields of fluor- and chlorapatite. Journal of Physics and Chemistry of Solids 2007;68 1863–1871.

[150] Posner AS, Perloff A, Diorio AF. Refinement of the hydroxyapatite structure. Acta Crystallographica 1958;11 308–309.

[151] Vallet-Regi M, Gonzalez-Calbet JM. Calcium phosphates as substitution of bone tissues. Progress in Solid State Chemistry 2004;32 1–31.

[152] Mohandes F, Salavati-Niasari M, Fathi M, Fereshteh Z. Hydroxyapatite nanocrystals: simple preparation, characterization and formation mechanism. Materials Science and Engineering: C 2014;45 29–36.

[153] Zhou H, Lee J. Nanoscale hydroxyapatite particles for bone tissue engineering. Acta Biomaterialia 2011;7(7) 2769–2781.

[154] Sopyan I, Mel M, Ramesh S, Khalid KA. Porous hydroxyapatite for artificial bone applications. Science and Technology of Advanced Materials 2007;8(1–2) 116–123.

[155] Landi E, Riccobelli S, Sangiorgi N, Sanson A, Doghieri F, Miccio F. Porous apatites as novel high temperature sorbents for carbon dioxide. Chemical Engineering Journal 2014;254 586–596.

[156] Olajire AA. CO_2 capture and separation technologies for end-of-pipe applications—a review. Energy 2010;35(6) 2610–2628.

[157] Ropp RC. Encyclopedia of the Alkaline Earth Compounds. Newnes, 2012. ISBN: 978-0444595539.

[158] Ries H, Watson TL. Engineering geology. Рипол Классик, 1915. ISBN: 978-5877730748

[159] Sjögren H. Svabit, ett mineral af apatitgruppen från Harstigsgrufvan, Geologiska Föreningens i Stockholm Förhandlingar 1891;13 789–796.

[160] Beudant FS. Smithsonite, zinc carbonaté, in Traité Élémentaire de Minéralogie, 2nd ed., Paris, 1832, 354–357.

[161] Emsley J. Nature's Building Blocks: An A–Z Guide to the Elements. Oxford University Press, 2001. ISBN: 978-0198503415

[162] Palache Ch. The Minerals of Franklin and Sterling Hill, Sussex County, New Jersey Geological Survey professional paper. U.S. Government Printing Office, 1937.

[163] Dunn PJ, Petersen EU, Peacor DR. Turneaureite, a new member of the apatite group from Franklin, New Jersey, Balmat, New York and Långban, Sweden. Canadian Mineralogist 1985;23, 251–254.

[164] Dana ES. Dana's New Mineralogy: The System of Mineralogy of James Dwight Dana and Edward Salisbury Dana. Wiley, 1819.

[165] Dunn PJ, Peacor DR, Newberry N. Johnbaumite, a new member of the apatite group from Franklin, New Jersey, American Mineralogist 1980;65 1143–1145.

[166] Lee YJ, Stephens PW, Tang Y, Li W, Phillips WB, Parise JB, Reeder R. Arsenate substitution in hydroxylapatite: structural characterization of the $Ca_5(P_xAs_{1-x}O_4)_3OH$ solid solution. American Mineralogist 2009;94 666–675.

[167] Smith GFH, Prior GT. On fermorite, a new arsenate and phosphate of lime and strontian, and Tilasite, from manganese-ore deposits of India. Mineralogical Magazine 1911;16 84–96

[168] Frenzel A. Mineralogisches Lexicon: für das Königreich Sachsen. BoD—Books on Demand, 2013. ISBN: 978-3845701745

[169] Hoffmann CAS. Handbuch der Mineralogie von C.A.S. Hoffmann, erster-vierten Band: 4. Sapienza—Università di Roma, Biblioteca di Scienze della Terra, 1817.

[170] Frost RL, Reddy BJ, Palmer SJ. The structure of mimetite, arsenian pyromorphite and hedyphane—a near-infrared spectroscopic study. Polyhedron 2008;27(6) 1747–1753.

[171] Reddy BJ, Frost RL, Palmer SJ. A near-infrared spectroscopic study of the phosphate mineral pyromorphite $Pb_5(PO_4)_3Cl$. Spectrochimica Acta Part A: Molecular and Biomolecular Spectroscopy 2008;71(2) 430–435.

[172] Cooper MP, Stanley CJ. Pyromorphite Group Minerals from the Caldbeck Fells, Cumbria, England: Mineralogical Record 1991;22(2) 105–121.

[173] Clarke FW. The Data of Geochemistry. Bulletin (Geological Survey (U.S.)), U.S. Government Printing Office, 1920.

[174] Nriagu JO. Lead orthophosphates-III. Stabilities of fluoropyromorphite and bromopyromorphite at 25°C. Geochimica et Cosmochimica Acta 1973;37(7) 1735–1743.

[175] Chevrier G, Giester G, Heger G, Jarosch D, Wildner M, Zemann J. Neutron single-crystal refinement of cerussite, $PbCO_3$, and comparison with other aragonite-type carbonates. Zeitschrift für Kristallographie 1992;199 67–74.

[176] McDonald WS, Cruickshank DWJ. Refinement of the structure of hemimorphite. Zeitschrift für Kristallographie 1967;124(3) 180–191.

[177] Kenngott A. Hemimorphit, in Das Mohs'sche Mineralsystem, Verlag und Druck (Wien), 1853; 67–68

[178] Hurlbut CS, Sharp WE. Dana's Minerals and How to Study Them (After Edward Salisbury Dana). 4th. ed., John Wiley & Sons, 1998. ISBN: 978-0471156772

[179] Pekov IV, Britvin SN, Zubkova NV, Pushcharovsky DY, Pasero M, Merlino S. Stronadelphite, $Sr_5(PO_4)_3F$, a new apatite-group mineral. European Journal of Mineralogy 2010;22(6) 869–874.

[180] Bell AMT, Henderson CMB, Wendlandt RF, Harrison WJ. Rietveld refinement of $Ba_5(AsO_4)_3Cl$ from high-resolution synchrotron data. Acta Crystallographica Section E. Structure Reports Online. ISSN 1600–5368

[181] Calos NJ, Kennard CHL, Davis RL. Crystal structure of mimetite, $Pb_5(AsO_4)_3Cl$. Zeitschrift für Kristallographie—Crystalline Materials 1990;191(1–4) 125–129. DOI: 10.1524/zkri.1990.191.14.125

[182] Yang Z, Ding K, de Fourestier J, He Li. The crystal structure of mimetite-2M, a new polymorph of mimetite from Xianghualing tin-polymetallic orefield, Hunan Province, P. R. China. Neues Jahrbuch für Mineralogie—Abhandlungen 2013;190(2) 229–235.

[183] Ramsdell LS, The crystal structure of some metralic sulfides. Journal Mineralogical Society of America 1925;10 281–304.

[184] Frost RL, Reddy BJ, Palmer SJ. The structure of mimetite, arsenian pyromorphite and hedyphane—a near-infrared spectroscopic study. Polyhedron 2008;27(6) 1747–1753.

[185] Dai Y, Hughes JM, Moore PB. The crystal structures of mimetite and clinomimetite, $Pb_5(AsO_4)_3Cl$. Canadian Mineralogis 1991;29 369–376.

[186] Kostov I. Crystal habits of minerals. 1st. ed., Prof. Marin Drinov Academic Pub. House, 1999. ISBN: 978-9544306274

[187] Trotter J, Barnes WH. The structure of vanadinite. Canadian Mineralogist 1958;6 161–173.

[188] Dai Y, Hughes JM. Crystal-structure refinements of vanadinite and pyromorphite. Canadian Mineralogist 1989;27 189–192.

[189] Prakash S. Advanced Inorganic Chemistry: Volume 2. S. Chand, 2000. ISBN: 978-8121917872

[190] Damour AA. Notice sur la descloizite, nouvelle espèce minérale. Annales de Chimie et de Physique 1854;41 72–78.

[191] Qurashi MM, Barnes WH. The structures of the minerals of the descloizite and adelite groups: V—Descloizite and Conichalcite (Part 3). The structure of descloizite. Canadian Mineralogist 1964;8 23–39.

[192] Roscoe HE. On two new vanadiium minerals. Proceedings of the Royal Society of London 1876;25 109–112.

[193] Cooper MA, Hawthorne FC. The crystal structure of mottramite, and the nature of Cu↔Zn solid solution in the mottramite-descloizite series. Canadian Mineralogist 1995;33 1119–1124.

[194] Cesbron F, Fritsche J. La mounanaïte, nouveau vanadate de fer et de plomb hydraté, Bulletin de la Société Française de Minéralogie et de Cristallographie 1969;92 196–202.

[195] Fleischer M. New mineral names. American Mineralogist 1969;54 1737–1742.

[196] Brugger J, Armbruster T, Criddle AJ, Berlepsch P, Graeser S, Reeves S. Description, crystal structure, and paragenesis of krettnichite, $PbMn^{3+}_2(VO_4)_2(OH)_2$, the Mn^{3+} analogue of mounanaite. European Journal of Mineralogy 2001;13(1) 145–158.

[197] Jambor JL, Roberts AC. New mineral names. American Mineralogist 2001;86 1534–1537.

[198] Laufek F, Skála R, Haloda J, Císařová I. Crystal structure of vanadinite: refinement of anisotropic displacement parameters. Journal of the Czech Geological Society 2006;51(3–4) 271–275.

[199] McKetta JJJr. Encyclopedia of Chemical Processing and Design: Volume 61—Vacuum System Design to Velocity: Terminal in Setting: Estimation. CRC Press, 1997. ISBN: 978-0824726126

[200] Palache C, Berman H, Frondel C. The System of Mineralogy of James Dwight Dana and Edward Salisbury Dana, Yale University 1837–1892. Volume II. 7th revised and enlarged ed., New York: John Wiley and Sons, Inc., 1951; 398–399.

[201] Lasky SG. Geology and ore deposits of the Bayard area, Central Mining District, New Mexico. Bulletin (Volume 870). U.S. Government Printing Office, 1936.

[202] Geological Survey (U.S.). Bulletin—United States Geological Survey, Volume 868–870, The Survey, 1937.

[203] Masaoka M, Kyono A. Single crystal growth of lead vanado-chlorapatite $Pb_5(VO_4)_3Cl$ using CsCl flux method. Materials Letters 2006;60(29–30) 3922–3926.

[204] Geological Survey (U.S.). Geological Survey Professional Paper. Volume 926. U.S. Government Printing Office, 1975.

[205] Moskalyk RR, Alfantazi AM. Processing of vanadium: a review. Minerals Engineering 2003;16(9) 793–805.

[206] Wills BA. Mineral Processing Technology: An Introduction to the Practical Aspects of Ore Treatment and Mineral Recovery (In SI/Metric Units). International Series on Materials Science and Technology. 4th ed., Elsevier, 2013. ISBN 978-483286679

[207] Mukherjee S. Applied Mineralogy: Applications in Industry and Environment. Springer Science & Business Media, 2012. ISBN: 978-9400711624

[208] Durand G. Synthesis of vanadinite. Comptes Rendus. 1957;244 2621–2622.

[209] Bayley WS. Descriptive Mineralogy. New York London, D. Appleton and company, 1917.

[210] Nordberg GF, Fowler BA, Nordberg M. Handbook on the Toxicology of Metals. 4th ed., Academic Press, 2014. ISBN 978-0123973399

[211] van Arkel EA, de Boer JH, Wilhelmus K. Process of precipitating metals on an incandescent body. Patent US1891124 A, 1932.

[212] Wiberg E, Wiberg N. Inorganic Chemistry. Academic Press, 2001. ISBN: 978-0123526519

[213] Waseda Y, Isshiki M. Purification Process and Characterization of Ultra High Purity Metals: Application of Basic Science to Metallurgical Processing. Springer Science & Business Media, 2001. ISBN: 978-3540413226

[214] Binnewies M, Glaum R, Schmidt M, Schmidt P. Chemical Vapor Transport Reactions. Walter de Gruyter, 2012. ISBN: 978-3110254655

[215] Berry KO, Smardzewski RR, McCarley RE. Vaporization Reactions of Vanadium Iodides and Evidence for Gaseous Vanadium(IV) Iodide. Inorganic Chemistry 1969;8(9) 1994–1997.

[216] van Arkel AE, de Boer JH. Darstellung von reinem Titanium-, Zirkonium-, Hafnium- und Thoriummetall. Zeitschrift für anorganische und allgemeine Chemie 1925;148(1) 345–350.

Chapter 2

Other Minerals from the Supergroup of Apatite

Petr Ptáček

Additional information is available at the end of the chapter

http://dx.doi.org/10.5772/62210

Abstract

The supergroup of apatite is divided into five groups of minerals. Therefore, minerals from the group of apatite were described in the first chapter, the second chapter of this book continues with description of minerals from the other four groups, i.e. minerals from the group of britholite, belovite, ellestadite and hedyphane. The structure, properties and known localities of these minerals were described. Although carbonate-apatite species are discredited from the IMA list of minerals, the chapter ends with description of structure and properties of carbonate-hydroxylapatite, carbonate-fluorapatite, and carbonate-rich varieties of apatite, i.e. francolite, dahlite, kurskite and collophane. The introduced three basic types of carbonate-apatites, i.e. type A, B and AB) are then discussed in Chapter 10 in depth.

Keywords: Apatite, Britholite, Belovite, Ellestadite, Hedyphane, Carbonate-apatite, Francollite, Dahlite

As was mentioned in **Section 1.1**, the supergroup of apatite is divided into five groups. The most important minerals form the group of britholite (**Section 2.3**), belovite (**Section 2.2**), ellestadite (**Section 2.4**) and hedyphane (**Section 2.1**), which are described in this chapter together with carbonate fluorapatite and hydroxylapatite (**Section 2.6**).

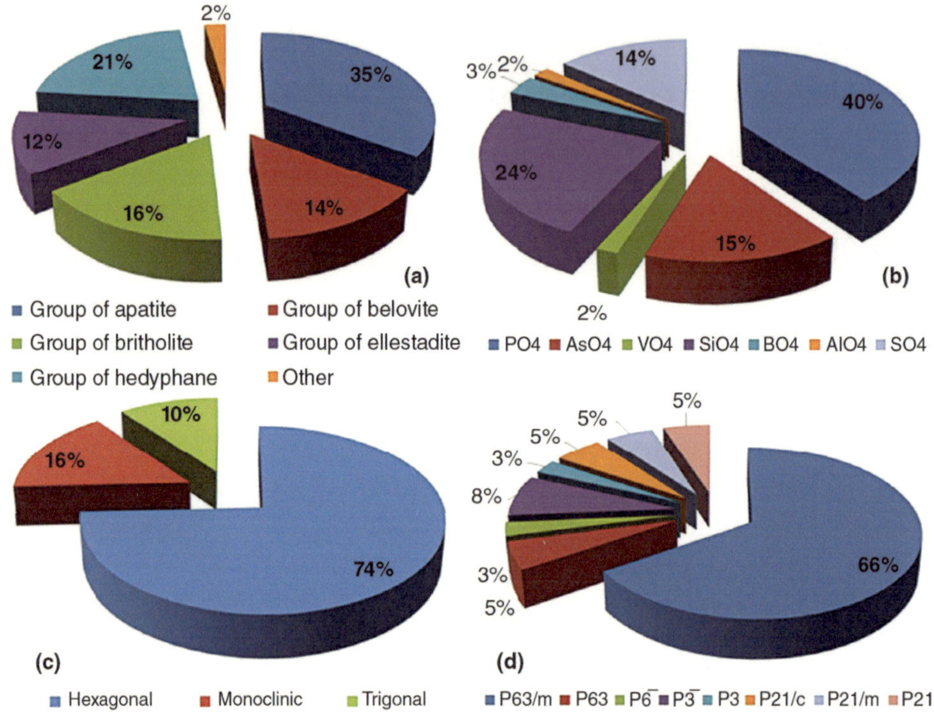

Fig. 1 Distribution of described minerals from the supergroup of apatite (discredited species are also included) among individual groups (a) and distribution of kind of XO_4 tetrahedra (b), crystal system (c) and space group (d) among these species.

Other minerals from the supergroup of apatite include 65%, i.e. 28 described mineral species **Fig. 1(a)**, which predominantly crystallize in hexagonal system (**c**) and in the space group $P6_3/M$ (**d**). The $[PO_4]^{3-}$ unit is the most frequent ortho-oxyanion for the supergroup of apatite in general (**b**), but its content in individual groups varies strongly (**Fig. 2**).

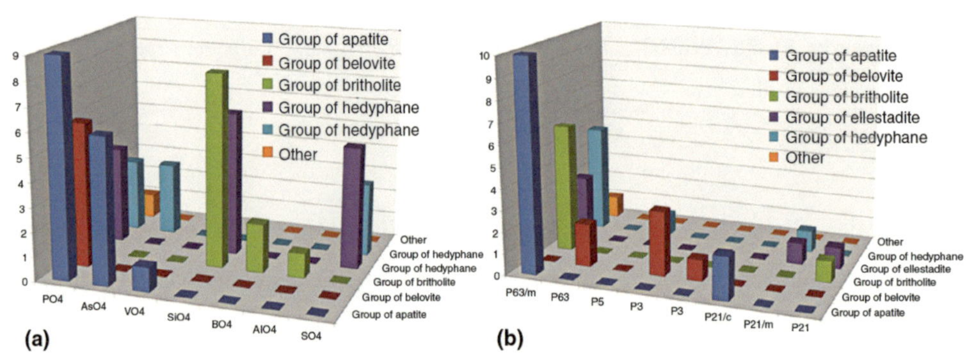

Fig. 2 Frequency of XO_4 ions and point groups for individual groups from the supergroup of apatite.

http://dx.doi.org/10.5772/62210

2.1. The group of hedyphane

2.1.1. Hedyphane

Hedyphane ($Ca_2Pb_3(AsO_4)_3Cl$), calcium-lead chloroarsenate [1],[2],[3],[4] is a mineral that was originally described from Langban, Sweden. It also occurs at the Harstig Mine, Pajsberg, Sweden and was moderately abundant at the Franklin Mine, Franklin, Sussex County, New Jersey. The mineral occurs in the localities introduced in **Fig. 3**. The mineral was named in 1830 by German mineralogist Johann Friedrich August Breithaupt [5] and its Greek name is usually translated as "pleasant appearance or beautifully bright."

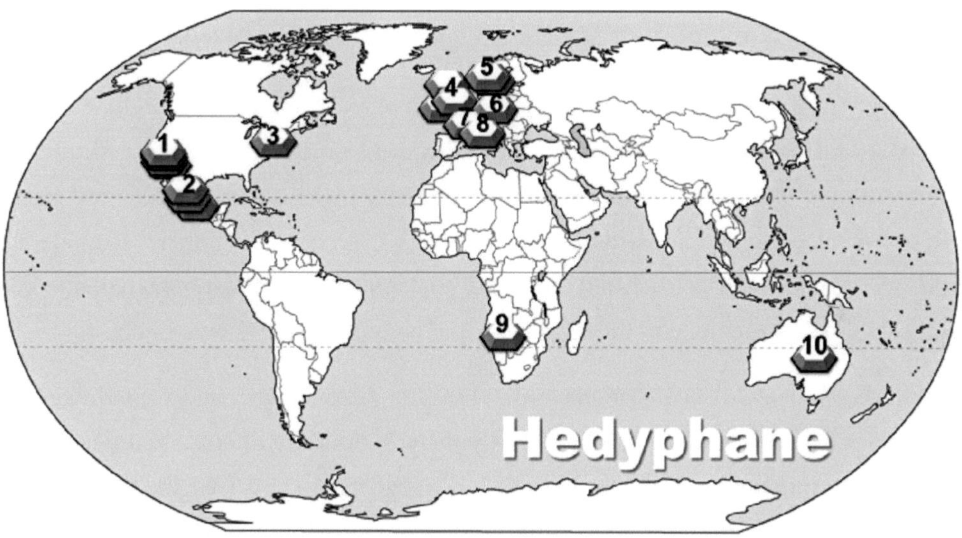

1) Reward Mine (Brown Monster Mine; Ruth; Hirsch and Telescope group; Brown Monster-Reward; Graham-Jones; Eclipse; Golden West claims; Hidden Treasure group; F. D. Roosevelt claims), Reward, Russ District, Inyo Mts (Inyo Range), Inyo Co., California, USA (a), Bagdad-Chase Mine (Barstow mill; Camp Rochester; Pacific Mine; Pacific Mines; Roosevelt Consolidated Mine; Buckeye-Stedman-Bagdad Chase Mine), Stedman, Stedman District, San Bernardino Co., California, USA (b), Rowley Mine (Rawley Mine; Reliance Mine; Reliance Copper Mine; Rainbow Mine; Theba Mine; San Carlos patented claim #4524), Theba, Painted Rock District, Painted Rock Mts, Maricopa Co., Arizona, USA (c) and Glove Mine (Sunrise Mine), Glove Mine group (Zombie & Zeco claims; Festiago-Franklin; Blacksmith adit), Devil's Cash Box ridge, Amado, Cottonwood Canyon, Tyndall District, Santa Rita Mts, Santa Cruz Co., Arizona, USA (d).
2) Potosí Mine (El Potosí Mine), Francisco Portillo, West Camp, Santa Eulalia District, Mun. de Aquiles Serdán, Chihuahua, Mexico (a), Ojuela Mine, Mapimí, Mun. de Mapimí, Durango, Mexico (b) and Bilbao Mine, La Blanca, Mun. de Ojo Caliente, Zacatecas, Mexico (c)
3) Franklin Mine, Franklin, Franklin Mining District, Sussex Co., New Jersey, USA
4) Tynagh Mine, Killimor, Co. Galway, Ireland (a), Brandy Gill Mine, Carrock Fell, Caldbeck Fells, North and Western Region (Cumberland), Cumbria, England, UK (b) and Torr Works Quarry (Merehead Quarry), Cranmore, Somerset, England, UK (c)
5) Jakobsberg Mine, Jakobsberg ore field, Nordmark district, Filipstad, Värmland, Sweden (a), Långban, Filipstad, Värmland, Sweden (b), Harstigen Mine, Pajsberg, Persberg district, Filipstad, Värmland, Sweden (c) and Sjögruvan, Grythyttan, Hällefors, Västmanland, Sweden (d)
6) Alexander Shaft (Alexander Mine), Vrančice, Příbram, Central Bohemia Region, Bohemia (Böhmen; Boehmen), Czech Republic
7) Poggio San Vittore asbestos mine, Balangero, Lanzo Valleys, Torino Province, Piedmont, Italy
8) Maffei Mine, Botro ai Marmi, Campiglia Marittima, Livorno Province, Tuscany, Italy (b)
9) Kombat Mine (Klein Otavi; Asis), Kombat, Grootfontein District, Otjozondjupa Region, Namibia
10) Aroona Mine, Beltana Mine (E.Z. Mine; Beltana deposit; Beltana-Aroona deposit; Puttapa mine), Puttapa, Leigh Creek, North Flinders Ranges, Flinders Ranges, South Australia, Australia (a) and Beltana Mine (E.Z. Mine; Beltana deposit; Beltana-Aroona deposit; Puttapa mine), Puttapa, Leigh Creek, North Flinders Ranges, Flinders Ranges, South Australia, Australia (b)

Fig. 3 Known localities for the mineral hedyphane.

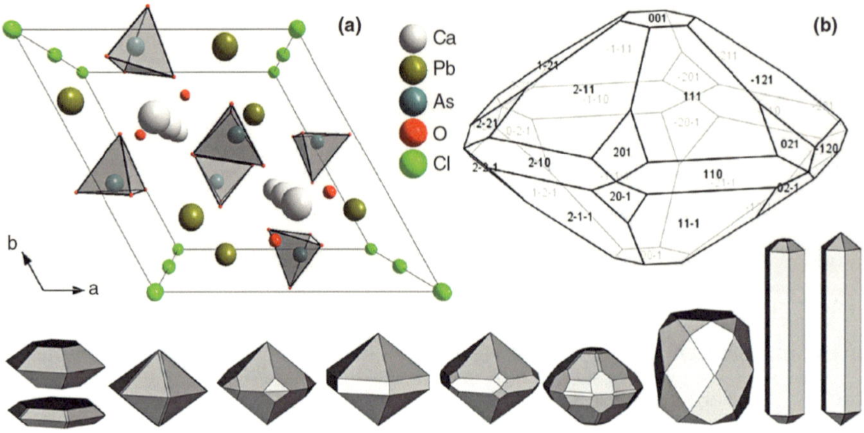

Fig. 4 The structure (perspective view according to the *c*-axis; a) and the crystal habit of hedyphane crystals (b).

The neotype (refer to **Footnote 48** in **Chapter 1**) hedyphane occurs as light yellow or white[1] euhedral crystal[2] approximately 1 mm in maximum dimension. Hedyphane is translucent with white streaks and has greasy to vitreous or (sub)resinous luster on crystal faces and fracture surfaces. The cleavage is indiscernible, the fracture is even, and the mineral is moderately brittle. Hedyphane is isostructural (arsenate analogue) with phosphohedyphane (**Section 2.1.3)** [3]. The structure and the crystal habit of hedyphane are shown in **Fig. 4**.

The unit cell parameters of hedyphane are $a = 10.14$, $c = 7.185$ Å, $Z = 2$ and $V = 639.78$ Å^3. The average density of mineral is 5.81 g·cm^{-1}. Hedyphane is optically uniaxial and positive. The hardness of the mineral on the Mohs scale is in the range from 4 to 5. Despite the fact that the formal charges of Ca^{2+} and Pb^{2+} are the same, the Pb^{2+} ion predominantly occupies M(2) sites in the structure of hedyphane. The exclusive presence of Pb in the M(2) site is probably due to the presence of stereoactive lone pair of electrons (please see **Fig. 7** and **Fig. 9**), which is characteristic for the Pb^{2+} ion in many compounds [6],[3],[7].

The average length of Ca-O(n), bond where $n = 1$, 2 and 3, in CaO_9 polyhedron is 2.60 Å. The structure of PbO_5Cl_2 polyhedron with interatomic distances is shown in **Fig. 5**. The average value of O(n)-Pb-O(3) angle, where $n = 2$ and 3, is 99.3°. The average interatomic distance and O-As-O angle in AsO_4 tetrahedron are 1.69 Å and 109°, respectively [3].

[1] Also described as "white variety" of "green lead ore" [4], i.e. the mineral mimetite (**Section 1.6.7**). It should be noted that "white lead ore" (grayish-white color, glassy luster and crystallized in small prisms) is related to the mineral cerussite ($PbCO_3$).

[2] Crystal habit shows well-formed and easily recognized faces. On the contrary, crystal faces that are not well formed are termed as anhedral. The intermediate texture between euhedral and anhedral is called subhedral.

http://dx.doi.org/10.5772/62210

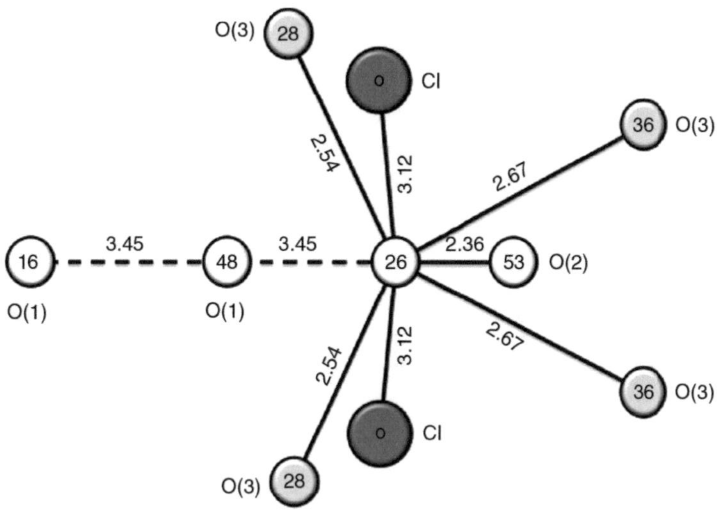

Fig. 5 The coordination polyhedron around the Pb site in hedyphane (projection on 001). The number within the circle denotes *x* coordinates of atoms [3].

2.1.2. Fluorphosphohedyphane

Fluorphosphohedyphane ($Ca_2Pb_3(PO_4)_3F$, [1],[8]): occurs in the oxidation zone of a small Pb-Cu-Zn-Ag deposit in the Blue Bell claims, about 11 km west from Baker, San Bernardino County, California (**Fig. 6**). Fluorphosphohedyphane is a fluor-analogue of phosphohedyphane, forms subparallel intergrowths and irregular clusters of transparent, colorless, highly lustrous, hexagonal prisms with pyramidal terminations.

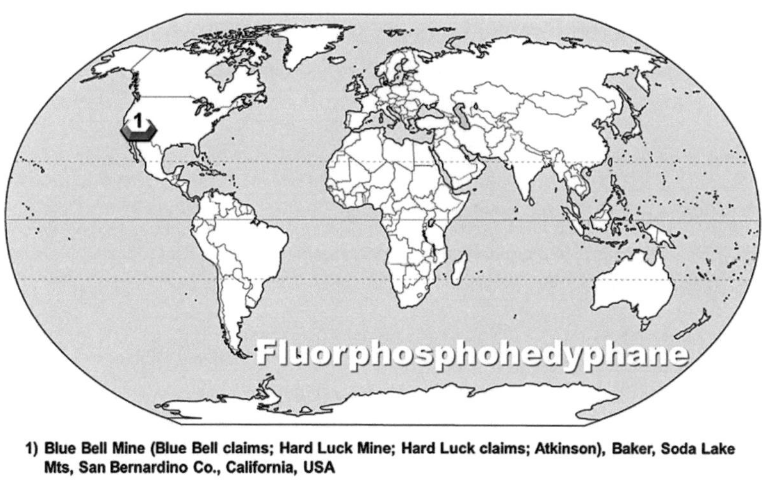

Fig. 6 The locality for the mineral phosphohedyphane.

Fluorphosphohedyphane is found in cracks and narrow veins in highly siliceous hornfels[3] [9] in association with cerussite1, chrysocolla[4] ($(Cu_{2-x}Al_x)H_{2-x}Si_2O_5(OH)_4{\cdot}nH_2O$ [10]), fluorite, fluorapatite, goethite, gypsum, mimetite, opal ($SiO_2{\cdot}nH_2O$ [11],[12]), phosphohedyphane, plumbogummite ($PbAl_3(PO_4)(PO_3OH)(OH)_6$ [13],[14],[15]), plumbophyllite ($Pb_2Si_4O_{10}{\cdot}H_2O$ [16]), plumbotsumite ($Pb_5Si_4O_8(OH)_{10}$ [17],[18],[19], pyromorphite (**Section 1.6.4**), quartz and wulfenite ($PbMoO_4$ [20]). The streak of the new mineral is white; the luster is subadamantine [8]. The structure and the crystal habit of fluorphosphohedyphane hedyphane are shown in **Fig. 7**.

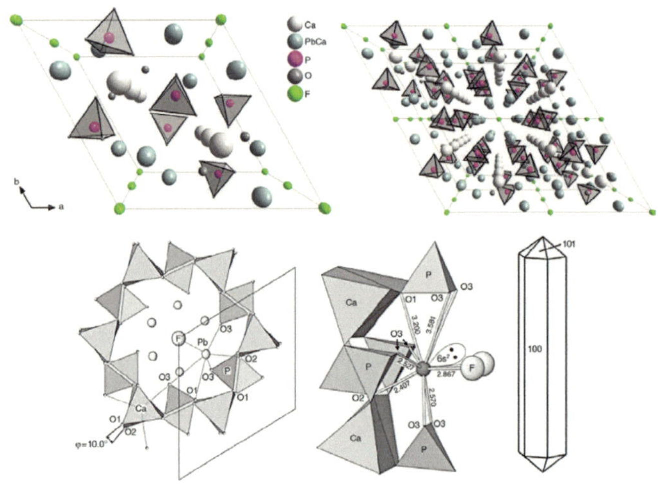

Fig. 7 The structure (perspective view along the *c*-axis), the coordination of Pb with likely approximate location of lone-electron pair and the crystal habit of fluorphosphohedyphane in clinographic projection[5] [8].

Fluorphosphohedyphane has the apatite structure with the ordering of Ca and Pb in two cation sites, as in hedyphane and phosphohedyphane. The Pb^{2+} cation exhibits a stereoactive $6s^2$ lone electron pair[6] [21] (**Fig. 7**). The Z anion site at (0, 0,½) is fully occupied by F forming six bonds of 2.867 Å to Pb atoms, in contrast to six Pb-Cl bonds of 3.068 Å in phosphohedyphane. For

---

[3] The name for this type of contact metamorphic rock was given by K. von Leonhardt. The name originates from the designation of the highest peaks in the Alps but it can be also derived from ancient mining term from Saxony (Germany) which was used to describe hard, compact metamorphic rock developed at the margin of an igneous body. These rocks possess outstanding toughness due to fine-grained nonaligned crystals of platy or prismatic habit. Hornfels are sometimes banded, but their texture can be also porphyroblastic, i.e. they occur as large crystals within fine ground groundmass of metamorphic rock [9].

[4] The name of the mineral comes from the Greek words *chyrosos* (gold) and *kolla* (glue). The mineral is also named as bisbeeite (blue mineral of the composition of $(Cu,Mg)SiO_3{\cdot}nH_2O$ named after Bisbee Cochise County, Arizona).

[5] In the clinographic projection the crystal is turned by angle Θ around a vertical axis in order to make the front- and the right-hand faces visible. Other forms are orthographic projection and perspective projection.

[6] The electronic configuration for Pb is [Xe] $4f^{14}\,5d^{10}\,6s^2\,6p^2$. Cations with formal $ns^2\,np^6$ electronic configuration usually display novel properties and it is widely believed that the so-called ns^2 lone pair is responsible for the stereochemical activity that causes the Jahn–Teller geometry distortion, specific optical properties, and ferroelectricity. Lone electron pair is also used for the explanation of anisotropies of thermal expansion coefficient, piezoelectric and elastic properties, and optoelectronic properties [21].

http://dx.doi.org/10.5772/62210

fluorphosphorhedyphane, phosphohedyphane and hedyphane in which Ca^{2+} occupies the M(1) site and Pb^{2+} occupies the M(2) site, the M(1) metaprism twist angles are notably smaller, 10.0°, 8.6° and 5.2°, respectively [8].

The mineral is brittle with subconchoidal fracture and no cleavage. Based on the empirical formula, the calculated density is 5.45 g·cm^{-3}. Fluorphosphohedyphane is hexagonal with the space group $P6_3/M$ and the cell parameters a = 9.6402, c = 7.0121 Å, a:c = 1:0.727, V = 564.4 Å^3 and Z = 2. The hardness of the mineral on the Mohs scale is 4.

2.1.3. Phosphohedyphane

Phosphohedyphane ($Ca_2Pb_3(PO_4)_3Cl$ [1],[22]): the mineral from the Capitana mine, Copiapó, Atacama Province, Chile, discovered in 2004. Known localities for the mineral phosphohedyphane are introduced in **Fig. 8**. Phosphohedyphane is brittle with subconchoidal fracture and no cleavage. Phosphohedyphane is hexagonal with the space group $P6_3/M$ and the cell parameters a = 9.857, c = 7.13 Å, V = 599.94 Å^3 and Z = 2. The hardness of the mineral on the Mohs scale is 4. The mineral is closely associated with quartz, duftite ($PbCuAsO_4(OH)$ [23]) and bayldonite ($Cu_3PbO(AsO_3OH)_2(OH)_2$ [24]).

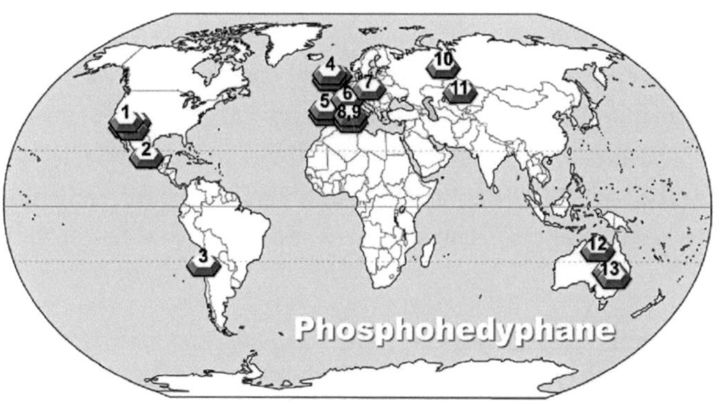

1) Silver Coin Mine, Valmy, Iron Point District, Humboldt Co., Nevada, USA (a), Reward Mine (Brown Monster Mine; Ruth; Hirsch and Telescope group; Brown Monster-Reward; Graham-Jones; Eclipse; Golden West claims; Hidden Treasure group; F. D. Roosevelt claims), Reward, Russ District, Inyo Mts (Inyo Range), Inyo Co., California, USA (b), Library Mine, Otto Mountain, Baker, San Bernardino Co., California, USA (c), Tonopah-Belmont Mine (East Vulture Mine; Belmont-McNeil Mine; Economy Mining Co. Mine), Belmont Mountain, Tonopah, Osborn District, Big Horn Mts, Maricopa Co., Arizona, USA (d), C and B Mine (C & B Mine; C. and B. Mine; C & B Group Mine; Vanadium Mine; Vanadium property; International property; Cutler and Bywater Mine), Christmas, Christmas area, Banner District, Dripping Spring Mts, Gila Co., Arizona, USA (e), Mammoth-Saint Anthony Mine (Mammoth-St Anthony Mine; Mammoth Mine; St. Anthony Mine), St. Anthony deposit, Tiger, Mammoth District, Pinal Co., Arizona, USA (f), Hardshell Mine (Manto Mine; Eagle-Picher properties), Hardshell Gulch, Harshaw, Harshaw District, Patagonia Mts, Santa Cruz Co., Arizona, USA (g), Reef Mine (Tungsten Reef Mine; Exposed Reef Mine; Sitric Mine; Reef group of claims), Carr Canyon, Hartford District (Huachuca Mountains District), Huachuca Mts, Cochise Co., Arizona, USA (h) and Manila Mine, Tombstone District, Tombstone Hills, Cochise Co., Arizona, USA (ch)
2) Bilbao Mine, La Blanca, Mun. de Ojo Caliente, Zacatecas, Mexico
3) Maria Catalina mine, Pampa Larga district, Tierra Amarilla, Copiapó Province, Atacama Region, Chile
4) Wilsons Vein, Whytes Cleuch, Wanlockhead, Dumfries & Galloway (Dumfriesshire), Scotland, UK (a), Higher Roughton Gill (Balliway Rigg), Roughton Gill, Caldbeck Fells, North and Western Region (Cumberland), Cumbria, England, UK (b), Whitwell Quarry, Whitwell, Derbyshire, England, UK (c), Wheal Mary Ann, Menheniot, Liskeard District, Cornwall, England, UK (d) and Wheal Mexico (Trebellan Mine; Incl. Treworthy Mine), Wheal Cubert (Cubert United Mine), Cubert, St Agnes District, Cornwall, England, UK (e)
5) La Paloma Mine, Zarza la Mayor, Cáceres, Extremadura, Spain (a) and Preguiça Mine, Sobral da Adiça, Moura, Beja District, Portugal (b)
6) Nuizière (Nuissière Mine), Chenelette, Rhône, Rhône-Alpes, France
7) Příbram, Central Bohemia Region, Bohemia (Böhmen; Boehmen), Czech Republic
8) Sa Duchessa Mine, Oridda, Domusnovas, Carbonia-Iglesias Province, Sardinia, Italy
9) Djebel Sekarna Mine (Sekarna deposit), Djebel Sekarna, Siliana, Siliana Governorate, Tunisia
10) Berezovskoe Au Deposit (Berezovsk Mines), Berezovskii (Berezovskii Zavod), Ekaterinburg (Sverdlovsk), Sverdlovskaya Oblast', Middle Urals, Urals Region, Russia
11) Kyzylespe (Kyzyl-Espe), Balkhash Region (Balqash; Karatas; Prebalkhashie), Karagandy Province (Qaragandy Oblysy; Karaganda Oblast'), Kazakhstan
12) Tennant Creek, Barkly Region, Northern Territory, Australia
13) Broken Hill, Yancowinna Co., New South Wales, Australia (a) and Cave Hill (David Mitchell Limestone Quarry), Lilydale, Victoria, Australia (b)

Fig. 8 Known localities for the mineral phosphohedyphane.

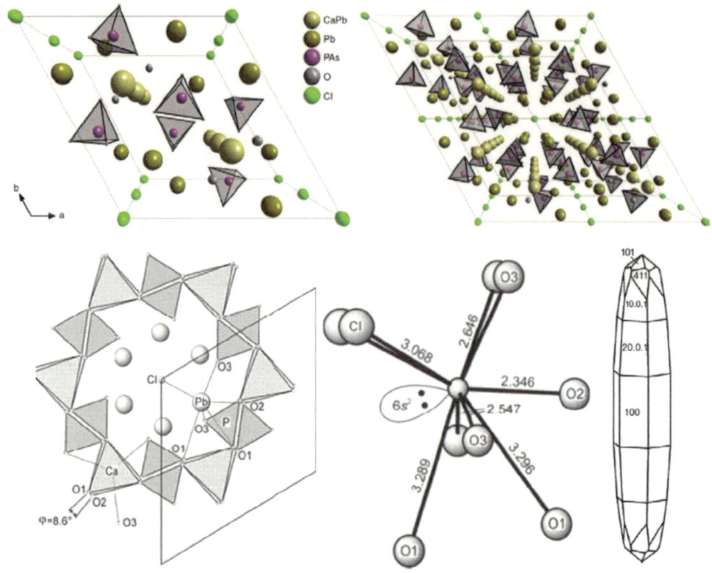

Fig. 9 The structure (perspective view along the *c*-axis), the coordination of Pb with approximate location of lone-electron pair and the crystal habit of phosphohedyphane [22].

The mineral is a phosphate analogue of hedyphane and possesses an apatite structure with the ordering of Ca and Pb in two nonequivalent large cation sites. The structure refinement indicates that the Ca(2) sites are completely occupied by Pb and the Ca(1) sites contain 92% Ca and 8% Pb. The tetrahedral site refines to 91% P and 9% As. The refinement indicates the 0,0,0 position to be fully occupied by Cl. The structure and the crystal habit of phoshohedyphane are shown in **Fig. 9**.

Other secondary minerals identified in the oxidized zone together with phosphohedyphane are: anglesite ($PbSO_4$ [25]), arsentsumebite ($Pb_2Cu(AsO_4)(SO_4)(OH)$ [26],[27]), azurite ($Cu_3(CO_3)_2(OH)_2$ [28]), beaverite[7] ($PbCu^{2+}Fe^{3+}{}_2(SO_4)_2(OH)_6$ [29],[30],[31]), calcite ($CaCO_3$, hexagonal with the space group $R\bar{3}c$[8] [32]), cerussite, mimetite (**Section 1.6.7**), malachite ($Cu_2CO_3(OH)_2$ [33]), mottramite and perroudite ($Ag_4Hg_5S_5(I,Br)_2Cl_2$ [34]) [22].

2.1.4. Morelandite

Morelandite ($Ca_2Ba_3(AsO_4)_3Cl$, (Ba, Ca, Pb)$_5$(AsO_4, PO_4)$_3$Cl [1],[35],[36]), is a mineral that was named in 1978 according to Moreland. It occurs as small irregular masses associated with hausmannite ($Mn^{2+}Mn^{3+}{}_2O_4$ [37]) and calcite in the Jakobsberg mine, near Nordmark, Sweden (**Fig. 10**). The structure of morelandite is shown in **Fig. 11**.

---

[7] The minerals beaverite-(Cu) and beaverite-(Zn), i.e. $PbZnFe^{3+}{}_2(SO_4)_2(OH)_6$ [29], were recognized. Beaverite is an old name for the mineral beaverite-(Cu).

[8] There is also an orthorhombic polymorph (PMCN) aragonite.

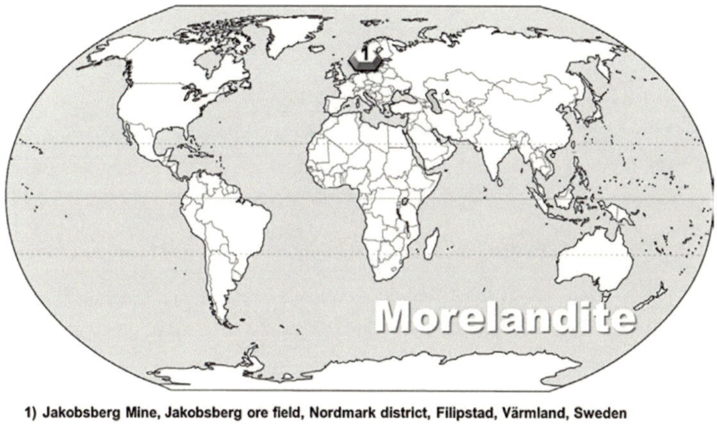

Fig. 10 Known locality of the mineral morelandite.

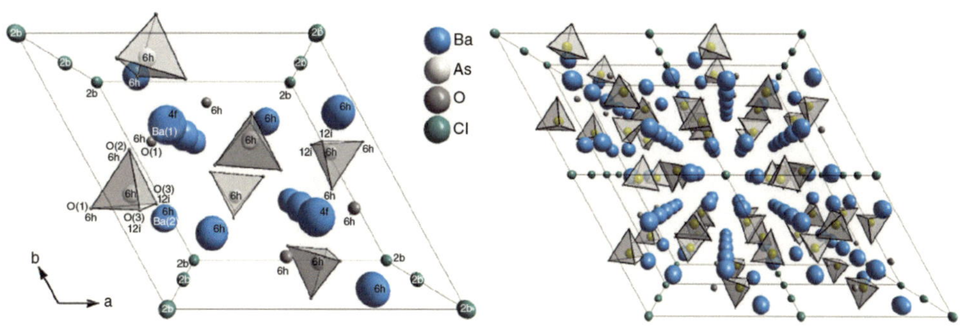

Fig. 11 The structure of morelandite (perspective view according to the *c*-axis).

The mineral is gray to light yellow with white streaks, greasy to vitreous luster, and shows poor cleavage on {001}. Morelandite is hexagonal with the space group $P6_3/\text{M}$ and the cell parameters $a = 10.169$, $c = 7.315$ Å, $V = 655.09$ Å^3 and $Z = 2$. The hardness of the mineral on the Mohs scale is 4½.

2.1.5. Aiolosite

Aiolosite ($Na_2(Na_2Bi)(SO_4)_3Cl$, ideally $Na_4Bi(SO_4)_3Cl$ [7]): hexagonal mineral with the space group $P6_3/\text{M}$ and the cell parameters $a = 9.626$, $c = 6.88$ Å, $V = 552.09$ Å^3 and $Z = 2$. The calculated density of the mineral is 3.59 g·cm^{-3}. Aiolosite is a sulfate mineral isotopic with apatite, which was found in an active medium-temperature intracrater fumarole at La Fossa crater, Vulcano Island, Aeolian archipelago, Sicily, Italy (**Fig. 12**). It occurs as acicular to slender prismatic crystals up to 0.5 mm long in an altered pyroclastic breccia (refer to **Footnote 27** in **Section 1.1**), together with alunite, anhydrite ($CaSO_4$ [38]), demicheleite-(Br) (BiSBr [39]), bismuthinite (Bi_2S_3 [40]) and panichiite ($(NH_4)_2SnCl_6$ [41]). Aiolosite is colorless to white, with white streaks and nonfluorescent. The luster is vitreous.

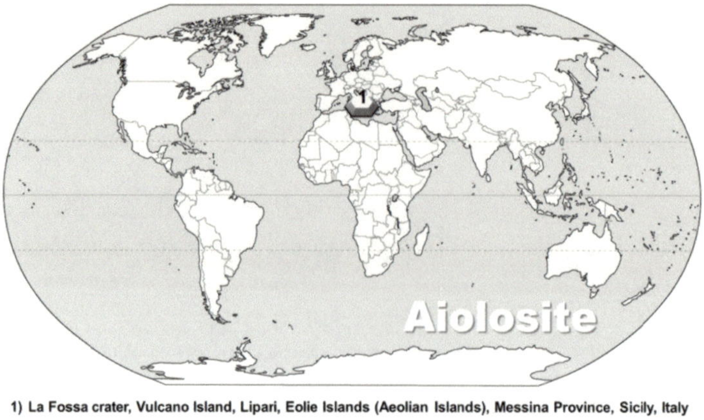

Fig. 12 The locality of the mineral aiolosite.

The structure of the mineral aiolosite is shown in **Fig. 13**.

The structure of aiolosite shows two independent cationic sites M(1) and M(2). Due to close similarity in ionic radii of Na^+ and Bi^{3+}, Bi exclusively prefers the M(2) site instead of M(1), which can be ascribed mainly to the Coulombic effect, in view of the higher charge of Bi^{3+} compared to Na^+, since the average M(2)-O distance (2.516 Å) is shorter than that of M(1)-O (2.617 Å). A similar effect also affects the distribution of Na^+ and Ca^{2+} sites in cesanite (**Section 2.1.7**) [7].

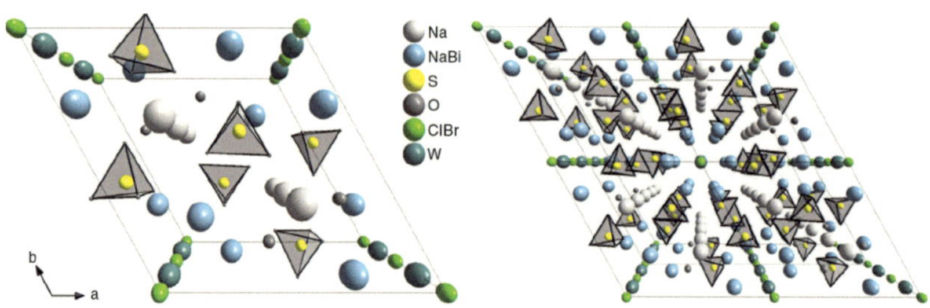

Fig. 13 The structure of aiolosite (perspective view according to the *c*-axis).

2.1.6. Caracolite

Caracolite ($Na_2(Pb_2Na)(SO_4)_3Cl$, sodium lead hydroxylchlorosulfate [1],[42],[43],[44]), is a vitreous colorless or grayish mineral from Beatriz mine, Caracoles, Chile, which was reported by WEBSKY in 1886. Known localities and the structure of the mineral caracolite are shown in **Fig. 14**. It occurs as crystalline incrustations with imperfect pseudohexagonal crystals up to 1 mm large. The crystals have the form of hexagonal pyramids with the base and the prism, but they are supposed to be pseudohexagonal. The mineral exhibits complex polysynthetic twinning with rather large extinction angles.

http://dx.doi.org/10.5772/62210

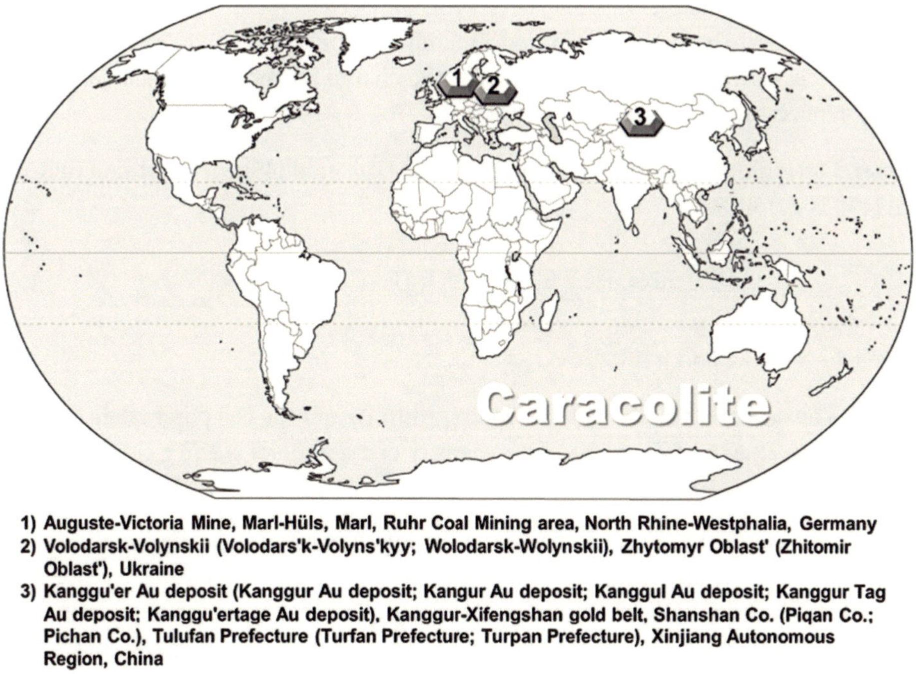

Fig. 14 The localities of the mineral aiolosite.

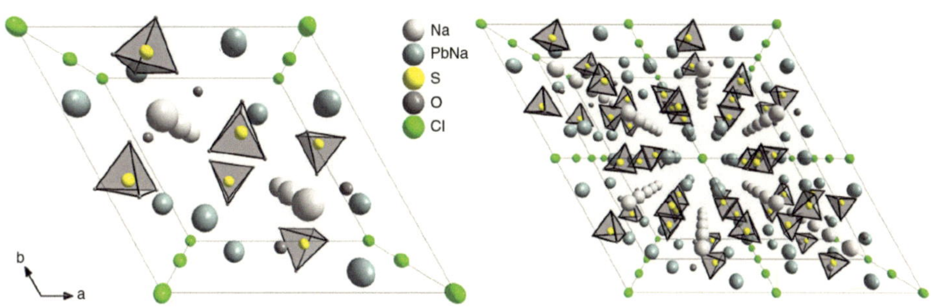

Fig. 15 The structure of caracolite (perspective view according to the *c*-axis).

Caracolite is monoclinic mineral with the space group P2$_1$/M and the cell parameters a = 19.62, b = 7.14, c = 9.81 Å and β = 120°, V = 1190.14 Å^3, Z = 4. Calculated density is 4.50 g·cm^{-3}. The hardness of the mineral on the Mohs scale is 4½. The structure of caracolite is shown in **Fig. 15**.

2.1.7. Cesanite

Cesanite ($Ca_2Na_3(SO_4)_3OH$ [45],[46],[47]) is a colorless, medium to coarse-grained, soft mineral which occurs both as a solid vein (1 cm thick) and as cavity-filling of an explosive breccia in core samples of the Cesano-I geothermal well (Cesano area, Latium, Italy). Cesanite was recognized as new mineral by CAVARRETA et al [47]. The crystal structure determination confirms that cesanite has to be considered a member of the apatite-wilkeite-ellestadite series, where $(PO_4)^{3-}$ is entirely substituted by $(SO_4)^{2-}$, the charge balance being made up by partial substitution of Na^+ for Ca^{2+} and H_2O for (OH^-, Cl^-, F^-).

The general formula of this series, proposed by HARADA et al [48] and modified by CAVARRETA et al [47], is as follows:

$$Ca_{5-w}Na_w(Si_y, S_z, P_{3-y-z})O_{12}(F, Cl, OH)_x nH_2O;$$

where $w = 1 - x - y + z$ and $n \leq 1 - x$.

Cesanite is a hexagonal mineral with the space group $P\bar{6}$ and the cell parameters $a = 9.463$, $c = 6.9088$ Å, $V = 535.79$ Å^3 and Z = 1. Calculated density of the mineral is 2.75 g·cm^{-3}. The hardness of the mineral on the Mohs scale ranges from 2 to 3.

The structure of cesanite is shown in **Fig. 16**. Synthetic and natural cesanite show typical elements of the apatite structure, but the reduction of symmetry from the centrosymmetric space group $P6_3/M$ to the noncentrosymmetric space group $P\bar{6}$ leads to a doubling of the number of crystallographically independent sites. Na and Ca cations are distributed over four independent sites. They are coordinated either by six O atoms and one hydroxyl ion or by water molecule (M(1), M(2)) or nine O atoms (M(3), M(4)) [46].

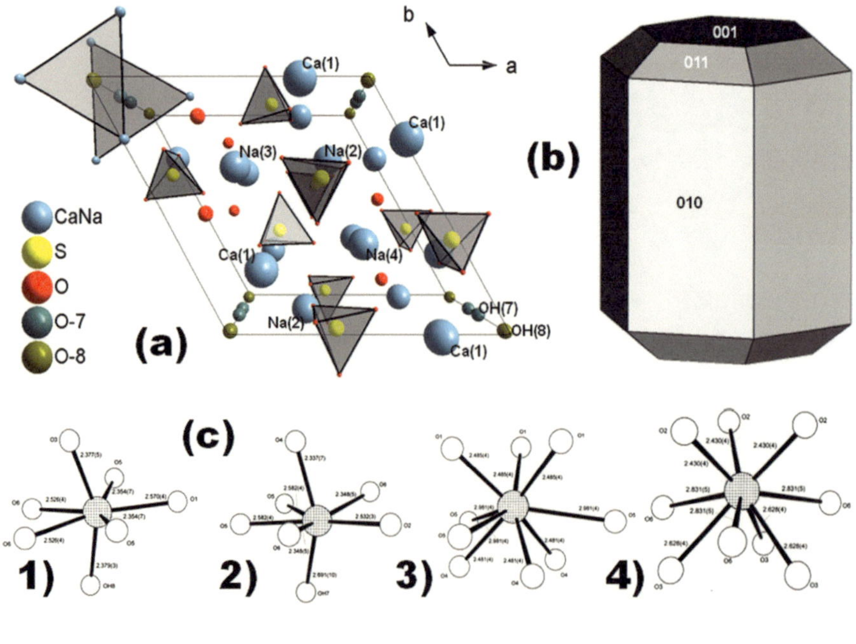

Fig. 16 The structure of cesanite (perspective view according to the *c*-axis; a), crystal habit (b) and the coordination polyhedra for M(1) (1), M(2) (2), M(3) (3) and M(4) (4) in synthetic analogue of the mineral cesanite (c) [47].

http://dx.doi.org/10.5772/62210

2.2. The group of belovite

The minerals from the group of belovite are cation ordered. Strontium substitutes for Ca in the M(2) site, and Na + REE substitute for Ca in the M(1) site. This results in lowering of symmetry from $P6_3/M$ (the space group of the apatite archetype structure) to $P6_3$ (fluorstrophite, fluorcaphite), $P\bar{3}$ (belovites), or P3 (deloneite) [1].

2.2.1. Belovite-(Ce)

Belovite-(Ce) ($NaCeSr_3(PO_4)_3F$ [49],[50],[51]), is a mineral from alkaline pegmatite in differentiated alkalic massifs which was named in 1954 by L.S. Borodin and M.E. Kazakova according to Russian mineralogist and crystallographer N.V. Belov. The mineral is found in Russia, on Mts. Punkaruaiv, Lepkhe-Nelm, Sengischorr, Karnasurt, Kedykvyrpakhk and Alluaiv, Lovozero massif; and on Mts. Kukisvumchorr and Koashva, Khibiny massif, Kola Peninsula. The localities of belovite-(Ce) are shown in **Fig. 17**.

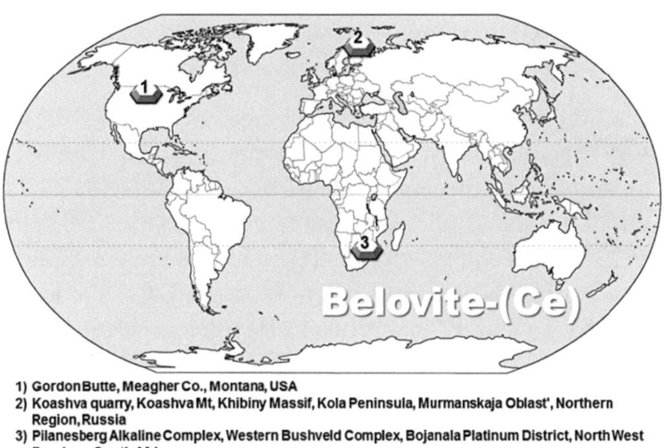

Fig. 17 The localities of the mineral belovite-(Ce).

The mineral belovite-(Ce) is usually associated with ussingite ($Na_2AlSi_3O_8(OH)$ [52]), natrolite ($(Na_2(Si_3Al_2)O_{10}{\cdot}2H_2O$ [53]), chkalovite ($Na_2BeSi_2O_6$ [54]), epistolite ($Na_4TiNb_2(Si_2O_7)_2O_2(OH)_2{\cdot}4H_2O$ [55]), tugtupite ($Na_4BeAlSi_4O_{12}Cl$ [56]), manganneptunite ($Na_2KLi(Mn^{2+},Fe^{2+})_2Ti_2[Si_8O_{24}]$ and manganoneptunite [57] (the mineral is isostructural with neptunite [58],[59]), murmanite ($Na_2Ti_2(Si_2O_7)O_2{\cdot}2H_2O$ [60]), gaidonnayite ($Na_2ZrSi_3O_9{\cdot}2H_2O$ [61]), nordite-(La) ($Na_3SrLaZnSi_6O_{17}$), lamprophyllite ($Na_3(Sr,Na)Ti_3(Si_2O_7)_2O_2(OH)_2$ [62]), fluorcaphite, lomonosovite, deloneite-(Ce), sitinakite ($KNa_2Ti_4Si_2O_{13}(OH){\cdot}4H_2O$ [63]), aegirine ($NaFe^{3+}Si_2O_6$ [64]), sodalite ($Na_4Si_3Al_3O_{12}Cl$ [65]), microcline[9] ($KAlSi_3O_8$ [66],[67]) and

[9] Originally, the mineral was named as mikroklin [66],[67]: triclinic mineral, space group $C\bar{1}$ with the unit cell parameters: a = 8.5784 Å, b = 12.9600 Å, c = 7.2112 Å, α = 90.30°, β =116.03°, and γ = 89.125°.

($NaAlSiO_4$ [68]). The morphology of belovite-(Ce) crystals and its structure are shown in **Fig. 18**.

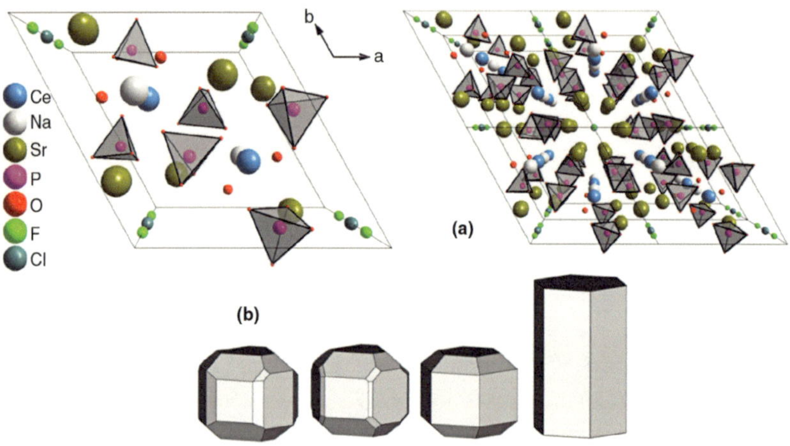

Fig. 18 The structure of belovite-(Ce) (perspective view according to the *c*-axis; a) and the shape of belovite-(Ce) crystals (b).

Belovite-(Ce) is the cerium analogue of belovite-(La) (**Section 2.2.2**) and the strontium analogue of kuannersuite-(Ce) (**Section 2.2.7**). The ideal formula of belovite-(Ce) is $Sr_6(Na_2REE_2)(PO_4)_6O_{24}(OH,F,Cl)_2$, and it is equivalent to apatite *sensu stricto*[10] with the following substitution of $Ca(2)_{-6}Sr_{+6}$ and $Ca(1)_{-4}Na_{+2}REE_{+2}$. Strontium overcomes the REE in the competition for Ca(2) sites of apatite. The sites equivalent to Ca(1) of apatite must respond to the occupation by essentially equal amounts of Na and REE. Unlike single Ca(1) site in apatite *sensu stricto*, low symmetry in the space group $P\bar{3}$ yields two Ca(1) subequivalents, one dominated by Na and the other one by REE [51].

Belovite-(Ce) is a brittle mineral with a honey-yellow or greenish color that crystallizes in trigonal system with the unit cell parameters $a = 9.692$ and $c = 7.201$Å, $a{:}c = 1 : 0.743$, $V = 585.80$ Å^3 and $Z = 2$. It has white streaks, (sub-)vitreous, resinous or greasy luster and a hardness on the Mohs scale of 5. Calculated and measured densities of the mineral are 4.23 and 4.19 g·cm^{-3}, respectively. It has imperfect prismatic and pinacoidal cleavage.[11]

2.2.2. Belovite-(La)

Belovite-(La) ($NaLaSr_3(PO_4)_3F$ [1],[69]) was named according to N.V. Belov (**Section 2.2.1**) with respect to higher content of La than Ce, i.e. the mineral is the lanthanum analogue of belovite-(Ce) described above and $NaSr_3La$ analogue of fluorapatite (**Section 1.5.1**). It occurs as prismatic crystals, up to 3 cm large, and it may also be granular. The structure and the crystal habit of the mineral belovite-(La) are shown in **Fig. 19**.

[10] Latin phrase (abbreviated as s.s.) used, which means "in exact sense."

[11] Cleavage that is parallel to the orientation {0001}, i.e. to the base of crystal.

http://dx.doi.org/10.5772/62210

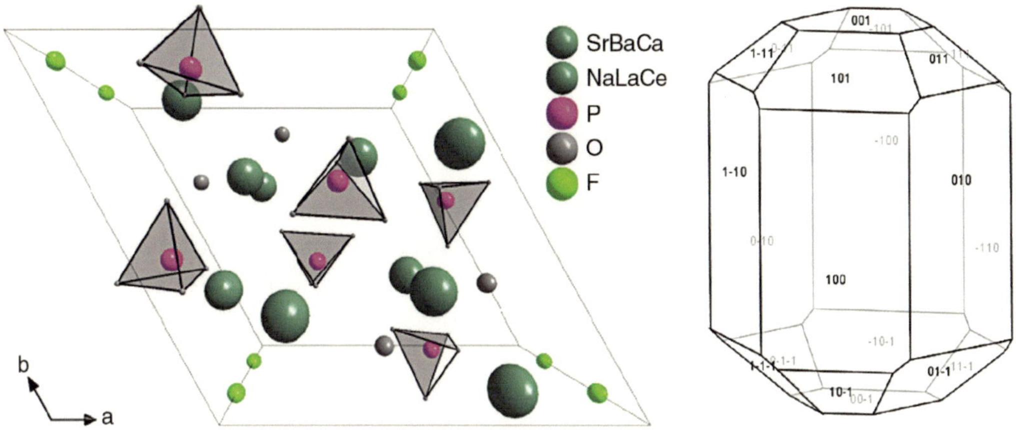

Fig. 19 The structure (shown along the *c*-axis) and the crystal habit of the mineral belovite-(La).

The mineral belovite-(La) crystallizes as trigonal in the space group P$\bar{3}$with the cell parameters a = 9.647 and c = 7.17 Å, a:c = 1:0.743, V = 577.88 Å^3 and Z = 2. Belovite-(La) is a very brittle mineral with yellow or greenish-yellow color and vitreous luster that does not show apparent cleavage. It has measured and calculated densities of 4.19 and 4.05 g·cm^{-3}, respectively. It has white streaks and the hardness of the mineral on the Mohs scale is equal to 5.

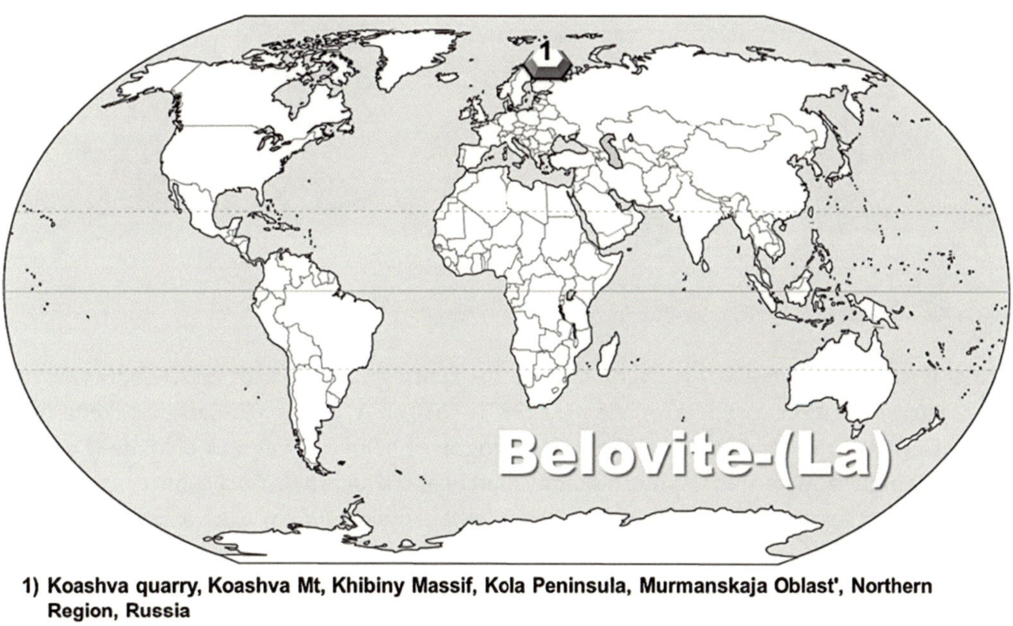

Fig. 20 The locality of belovite-(La).

Belovite-(La) can be found in natrolite veinlets[12] in pegmatites in a differentiated alkalic massif [70]), lamprophyllite, murmanite, aegirine, pectolite ($NaCa_2Si_3O_8(OH)$ [71]), microcline and natrolite.

2.2.3. Carlgieseckeite-(Nd)

Carlgieseckeite-(Nd)[13] ($NaNdCa_3(PO_4)_3F$ [31],[72],[73]) was named according to the mineralogist and polar explorer Giesecke with respect to the content of Nd that is higher than the content of other REE. The mineral was found in Kuannersuit (formerly Kvanefjeld) Plateau, northern section of the Ilímaussaq alkaline complex, South Greenland, Denmark. It is associated with albite ($NaAlSi_3O_8$ [74]), analcime ($NaAlSi_2O_6{\cdot}H_2O$ [75]) and fluorapatite in the cavities of albite vein cross-cutting augite syenite. Carlgieseckeite-(Nd) forms hexagonal tabular crystals up to 0.25 × 1 × 1.3 mm, and their parallel intergrowth up to 0.7 × 1.3 mm is found epitactically overgrown on prismatic crystals of fluorapatite. A phase with idealized formula $Na_{1.5}Nd_{1.5}Ca_2(PO_4)_3F$ epitactically overgrows some crystals of carlgieseckeite-(Nd).

The structure of the mineral (**Fig. 21**) is representative of the structure type of belovite *sensu stricto*. In this structure, large M cations occupy three sites with different coordination numbers: 9-fold polyhedra M(1) (average distance <M(1)–O> is 2.522 Å), 6-fold polyhedral M(1)′ (reduced 9-fold polyhedra, <M(1)′–O> with average interatomic distance 2.445 Å) and 7-fold polyhedra M(2) (<M(2)–O,F > 2.486 or 2.560 Å in the case of F1) [72].

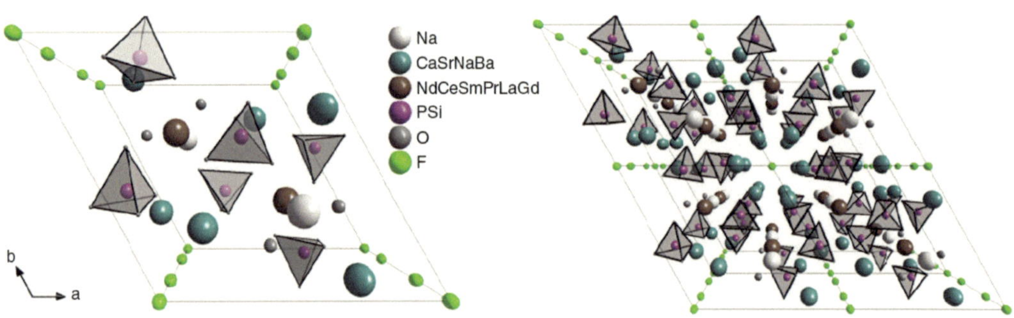

Fig. 21 The structure of carlgieseckeite -(Nd) (perspective view according to the *c*-axis).

Mineral carlgieseckeite-(Nd) is trigonal, from the space group $P\bar{3}$ with crystallographic cell parameters a = 9.4553, c = 6.9825 Å, a:c = 1:0.738, V = 540.62 Å^3 and Z = 2. Carlgieseckeite-(Nd) is the isostructural Ca- and Nd-dominant analogue of belovite-(Ce) and belovite-(La). The mineral carlgieseckeite-(Nd) is colorless, transparent and shows a distinct color-change effect, from almost colorless with a greenish hue in daylight to pink in yellow electric light. The luster is vitreous. The Mohs hardness is about 5. The mineral is brittle with no observed cleavage [72].

[12] Sheetlike body of minerals which crystallize within the rock.

[13] The holotype material is deposited in the Fersman Mineralogical Museum of Russian Academy of Sciences, Moscow [72].

http://dx.doi.org/10.5772/62210

2.2.4. Deloneite

Deloneite ($(Na_{0.5}REE_{0.25}Ca_{0.25})(Ca_{0.75}REE_{0.25})Sr_{1.5}(CaNa_{0.25}REE_{0.25})(PO_4)_3F_{0.5}(OH)_{0.5}$ [1],[76]): the name of the mineral was changed from deloneite-(Ce) to deloneite. The mineral was named by KHOMYAKOV, LISITIN, KULIKOVA and RASTSVETAEVA in 1996 according to Russian mathematical crystallographer BORIS NIKOLAEVICH DELONE. The mineral usually occurs as anhedral to subhedral[2] crystals in the matrix. The locality and the structure of the mineral are shown in **Fig. 22** and **Fig. 23**, respectively.

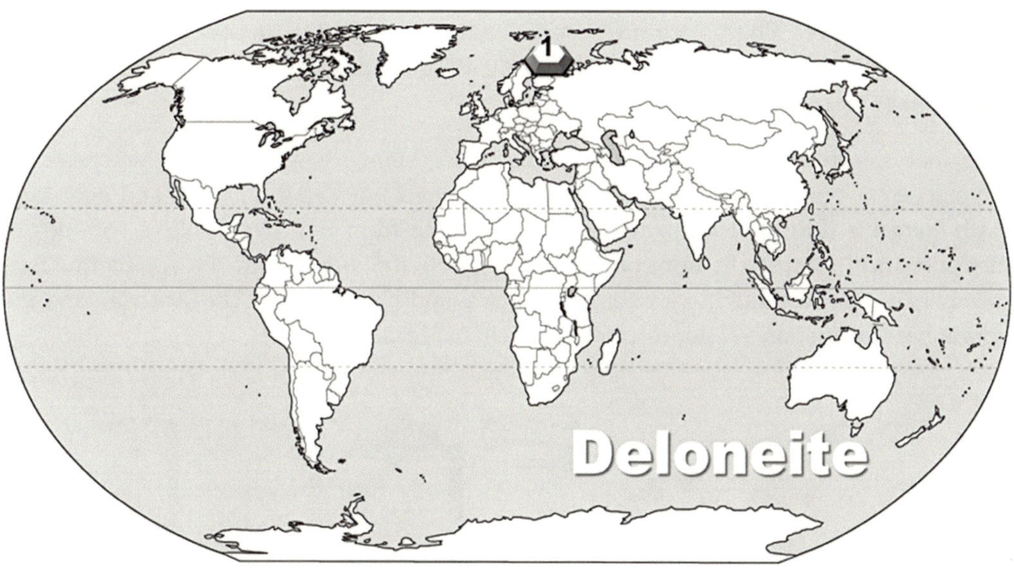

1) **Koashva quarry, Koashva Mt, Khibiny Massif, Kola Peninsula, Murmanskaja Oblast', Northern Region, Russia**

Fig. 22 The locality of deloneite.

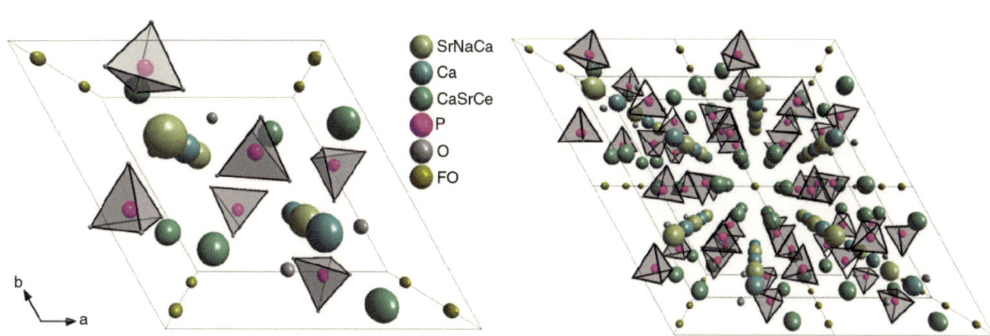

Fig. 23 The structure of the mineral deloneite (perspective view according to the *c*-axis).

Deloneite is a bright yellow mineral which crystallizes in trigonal systems with the unit cell crystallographic parameters $a = 9.51$, $c = 7.01$ Å, $a{:}c = 1{:}0737$, $V = 549.05$ Å^3 and Z = 2. The mineral is brittle, with a vitreous luster, white streak, an average density of 3.93 g·cm^{-3} and a hardness on the Mohs scale that is equal to 5.

2.2.5. Fluorcaphite

Fluorcaphite ($SrCaCa_3(PO_4)_3F$ [1],[77],[78]): the name of this mineral is an acronym for its elemental composition, i.e. fluorine, calcium and phosphorus. Fluorcaphite is a common accessory mineral in albitite,[14] which developed at the contact between quartzite and peralkaline nepheline syenites[15] of the Lovozero complex, in northwestern Russia. The rock consists predominantly of albite, aegirine, sodic amphibole (arfvedsonite ($NaNa_2(Fe^{2+}{}_4Fe^{3+})Si_8O_{22}(OH)_2$) [79] – magnesio-arfvedsonite ($NaNa_2(Mg_4Fe^{3+})Si_8O_{22}(OH)_2$) [80],[81] and narsarsukite ($Na_2(Ti,Fe,Zr)Si_4(O,F)_{11}$) [82].

Fluorcaphite forms euhedral prismatic crystals up to 0.3 mm in length. Most of the crystals are homogeneous, but a few contain resorbed core relatively depleted in Sr, Na and light rare-earth elements (LREE). This pattern of zoning arose from two overprinting episodes of metasomatism[16] [83],[84]. In terms of composition, both the core and the rim are intermediate members of a solid solution between fluorapatite and belovite-(Ce). The structure and the crystal habit of the mineral fluorcaphite is shown in **Fig. 24**.

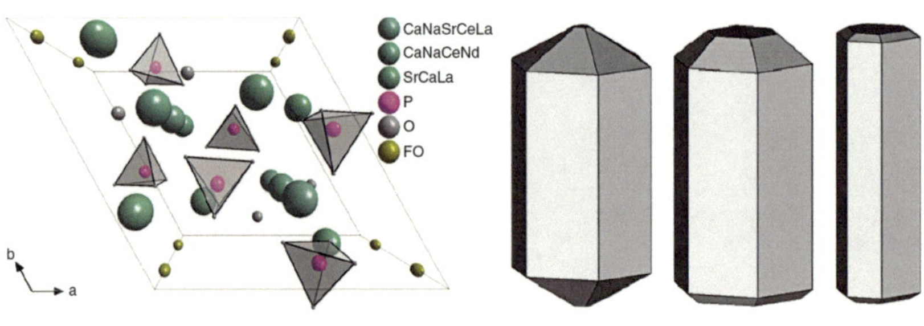

Fig. 24 The structure (view along *c*-axis) and the crystal habit of the mineral fluorcaphite.

Fluorcaphite is light or bright yellow hexagonal mineral which crystallizes in the space group $P6_3$ with the crystallographic parameters $a = 9.485$, $c = 7.000$ Å, $a{:}c = 1{:}0.738$, $V = 545.39$ Å^3 and Z = 2. It has white streaks, vitreous luster and the hardness on the Mohs scale is 5. Calculated and measured densities are 4.09 and 3.6 g·cm^{-3}, respectively. Fluorcaphite does not show any cleavage, the mineral is brittle with the formation of subconchoidal fractures.

[14] Granular rock essential consisting of the mineral albite.

[15] Coarse-grained intrusive rock crystallized slowly under conditions similar to granite, but is deficient of quartz.

[16] The term was introduced by Neumann [83]. Metasomatism is a metamorphic process by which the chemical composition of a rock or rock portion is altered in a pervasive manner and which involves the introduction and/or removal of chemical components as the results of the interaction of the rock with aqueous fluids (solutions). During the metasomatism, the rock remains in a solid state.

2.2.6. Fluorstrophite

Fluorstrophite ($SrCaSr_3(PO_4)_3F$ [1],[85],[86]) formerly "strontium-apatite" [87] and later changed to apatite-(SrOH) [1]. It possesses massive, coarse granular to compact morphology. The crystal forms include short to long hexagonal prisms, they can also be thick and tabular. Similar to fluorcaphite, the name of the mineral reflects its chemical composition (fluorine, strontium and phosphorus). The localities of the mineral fluorstrophite are shown in **Fig. 25**.

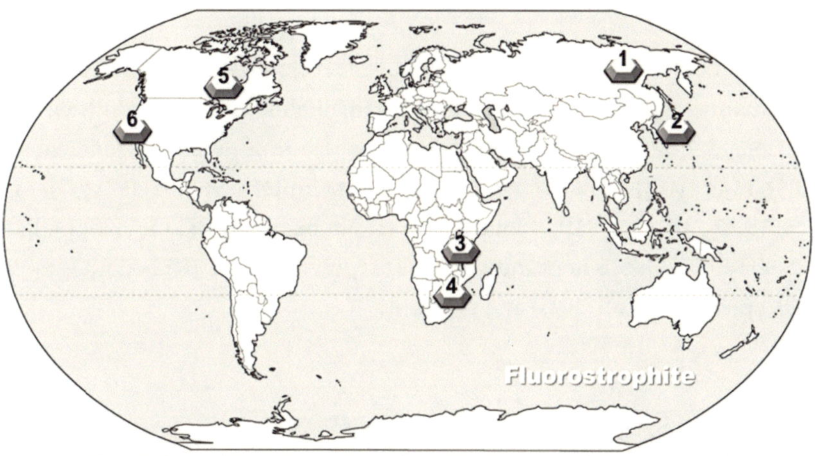

1) Murunskii Massif, Chara and Tokko Rivers Confluence, Aldan Shield, Sakha Republic (Saha Republic; Yakutia), Eastern-Siberian Region, Russia
2) Itoigawa City, Niigata Prefecture, Chubu Region, Honshu Island, Japan
3) Nkumbwa Hill Carbonatite (Nkombwa Hill Carbonatite), Isoka, Isoka District, Muchinga Mts., Northern Province, Zambia
4) Pilanesberg Alkaline Complex, Western Bushveld Complex, Bojanala Platinum District, North West Province, South Africa
5) Chipman Lake carbonatite, O'Meara Township, Thunder Bay District, Ontario, Canada
6) Victor claim (Victor Mine), Clear Creek area, Picacho Peak, New Idria District, Diablo Range, San Benito Co., California, USA

Fig. 25 Localities for the mineral fluorstrophite.

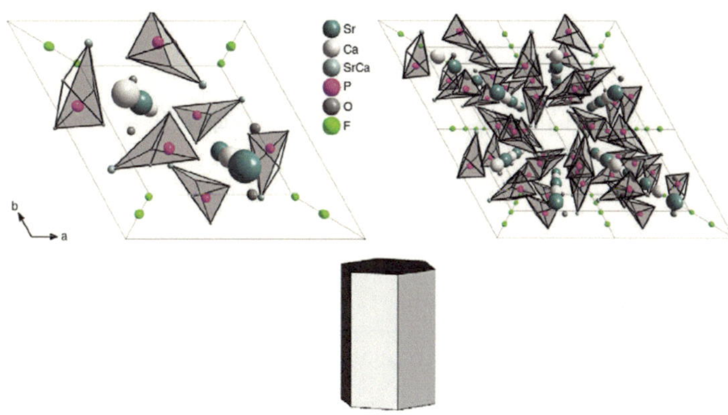

Fig. 26 The structure and the crystal habit of the mineral fluorstrophite.

The structure and the crystal habit of the mineral fluorstrophite are shown in **Fig. 26**. It is a green, yellow-green or colorless mineral with vitreous-greasy luster that crystallizes in hexagonal system with the space group $P6_3/M$ or $P6_3$. The crystallographic parameters of the unit cell are a = 9.565 and c = 7.115 Å, the ration $a:c$ = 1:0.744, V = 563.74 Å^3 and Z = 2. The hardness of the mineral on the Mohs scale is 5. Calculated and measured densities are 3.74 and 3.84 g·cm^{-3}, respectively. It has imperfect cleavage to {1010}.

2.2.7. Kuannersuite-(Ce)

The mineral kuannersuite-(Ce) ($Na_2Ce_2Ba_6(PO_4)_6FCl$ [88]) was found and named according to the locality (Kuannersuit plateau) in the Ilímaussaq alkaline complex, South Greenland (**Fig. 27**). It occurs associated with the minerals including aegirine, analcime, beryllite ($Be_3SiO_4(OH)_2{\cdot}H_2O$ [89]), chkalovite, galena, gmelinite[17] [90],[91],[92],[93], gonnardite ($(Na,Ca)_2(Si,Al)_5O_{10}{\cdot}3H_2O$ [94]), lovdarite ($K_2Na_6Be_4Si_{14}O_{36}{\cdot}9H_2O$ [95],[96]), nabesite ($Na_2BeSi_4O_{10}{\cdot}4H_2O$ [97],[98]), neptunite, pectolite, polylithionite ($KLi_2AlSi_4O_{10}F_2$ [99]), pyrochlore[18] [100], sphalerite (ZnS [101]) and tugtupite.

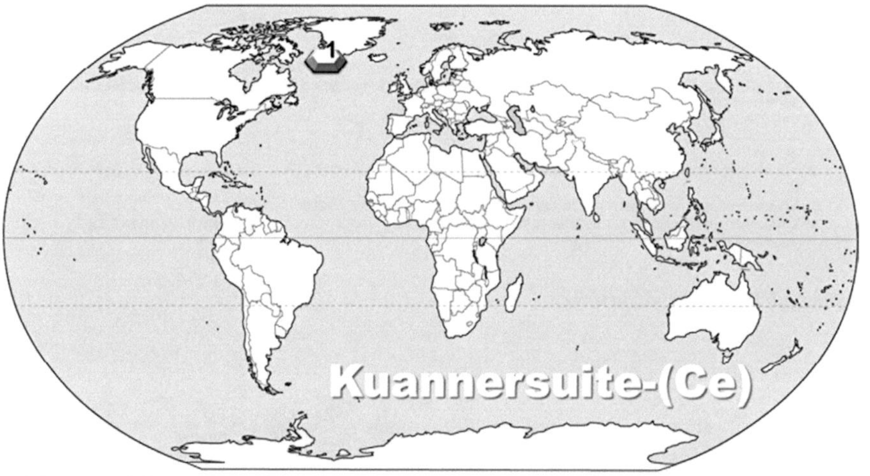

Fig. 27 The locality for the mineral kuannersuite-(Ce).

---

[17] There are three minerals: gmelinite-(Ca), gmelinite-(K), and gmelinite-(Na) with the composition of $Ca_2(Si_8Al_4)O_{24}{\cdot}11H_2O$ [90],[91], $K_4(Si_8Al_4)O_{24}{\cdot}11H_2O$ [93], and $Na_4(Si_8Al_4)O_{24}{\cdot}11H_2O$ [91],[93], respectively.

[18] The member of the pyrochlore group ($(Na,Ca)_2Nb_2O_6(OH,F)$). A new scheme of nomenclature for the pyrochlore supergroup, approved by the CNMNC–IMA, is based on the ions at the A, B, and Y sites. The subgroups should be changed to the groups: pyrochlore (1), microlite (2), roméite (3), betafite (4), and elsmoreite (5). The new names are composed of two prefixes and one root name (identical to the name of the group). The first prefix refers to the dominant anion (or cation) of the dominant valence [either H_2O or □] at the Y site. The second prefix refers to the dominant cation of the dominant valence [either H_2O or □] at the A site. The prefix "keno–" represents "vacancy." Where the first and the second prefix are equal, only one prefix is applied [100].

http://dx.doi.org/10.5772/62210

It occurs as light rose-colored hexagonal prismatic crystals, up to 1.5 mm long, with a white streak and a vitreous luster. It is a barium analogue of belovite-(Ce) (**Section 2.2.1**) and $NaCeBa_3$ analogue of fluorapatite (**Section 1.5.1**). The mineral is brittle and shows poor cleavage along {001} and {100}. The structure and the crystal habit of the mineral kuannersuite-(Ce) are shown in **Fig. 28**.

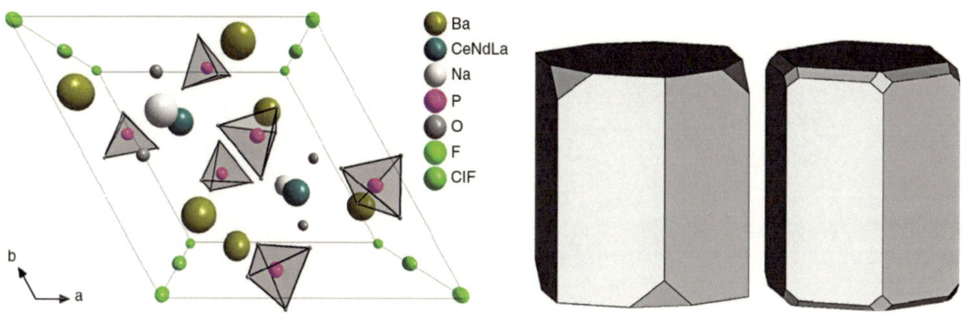

Fig. 28 The structure and the crystal habit of the mineral kuannersuite-(Ce).

Kuannersuite-(Ce) crystallizes in trigonal systems with the space group P$\bar{3}$. The parameters of unit cell are: a = 9.9097 and c = 7.4026 Å, V = 629.558 Å^3 and Z = 2. There is no fluorescence under ultraviolet light (long or short wave). The Mohs hardness of kuannersuite is between 4½ and 5½, and calculated density is 4.5 g·cm^{-3}.

2.3. The group of britholite

Britholites are typically phosphorus-bearing silicates with apatite structure and general formula: $(REE,Ca)_5[(Si,P)O_4]_3$**Z**, where REE is usually yttrium and **Z** = OH^-, F^- or Cl^-. The minerals from the group of britholite usually contain significant impurities of thorium and sometimes also uranium. These minerals are widespread in alkaline rocks such as pegmatites and metasomites[19] related to syenite[15] and nepheline–syenite complexes [102]. The name of this group is derived from the Greek word *brithos* for weight in order to refer to the high density of the mineral. The following minerals are described below.

The structure and the crystallographic data of some of the minerals from the group of britholite were introduced in **Fig. 29** and **Table 1**, respectively. The structural, thermodynamic and electronic properties of britholites were investigated by Njema et al [103].

[19] The series of metamorphic processes whereby chemical changes occur in minerals or rocks as the result of the introduction of material, often in hot aqueous solutions, from external sources.

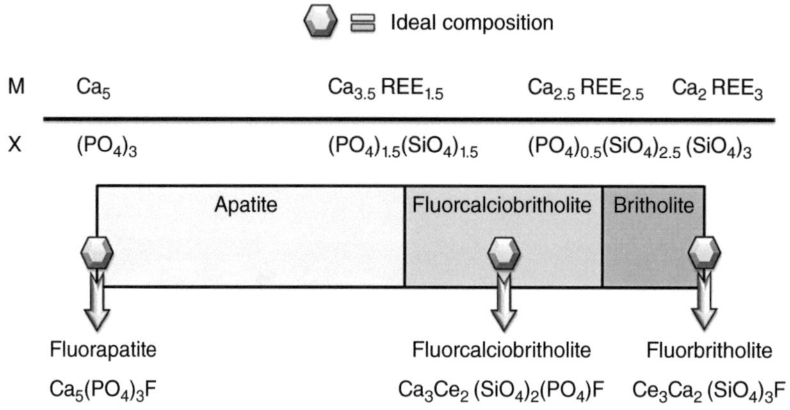

Fig. 29 The structure (the view according to axis *c*) of mineral: (a) britholite-(Ce), (b) britholite-(Y) and (c) fluorbritholite-(Ce). Coupled heterovalent substitution at M and T site in the series apatite–calciobritholite–britholite [1].

Mineral name	Crystallographic parameters							Hardness (Mohs)
	a	*c*	*a*:*c*	*Z*	*V*	SG	Density*	
	[Å]		–	–	[Å³]	–	[g·cm⁻³]	
Britholite-(Ce)	9.63	7.03	1:0.730	2	564.60	$P6_3/M$	4.45/4.49	5½
Britholite-(Y)	9.43	6.81	1:0.722		524.45		4.25/4.07	5.0
Fluorbritholite-(Ce)	9.52	6.98	1:0.734		547.74		4.66/4.67	
Fluorbritholite-(Y)	9.44	6.82	1:0.722		526.68		–/4.61	5½
Fluorcalciobritholite	9.58	6.99	1:0.729		555.17		4.20/4.25	5½
Tritomite-(Ce)	9.35	6.88	1:0.736		520.89		4.20/5.02	5.5
Tritomite-(Y)	9.32	6.84	1:0.734		514.54		3.22/4.48	3.5-6.5

Measured/calculated

Table 1 The crystallographic data of minerals from the group of britholite

2.3.1. Britholite-(Ce)

The britholite-(Ce) (Lessignite-(Ce), $(Ce,Ca)_5(SiO_4)_3OH$) [104],[105],[106]) mineral (**Fig. 30**) was first recognized as the new mineral by G. Flink (1897) in the pegmatite form of the nepheline–syenite at Naujakasik, Ilímaussaq complex, Greenland. Known localities for the mineral britholite are shown in **Fig. 31**.

The specimen was named and described by Chr. Winther [104] as opaque, brown crystals of the composition of $3[4SiO_2,2(Ce,La,Di,Fe)_2O_3,3(Ca,Mg)O,H_2O,NaF],2[P_2O_5,Ce_2O_3]$, which are apparently hexagonal prisms with pyramids, but it actually consists of biaxial orthorhombic individuals twined together as in aragonite. The Th-rich britholite-(Ce) was also known as fenghuangshite [107]. Britholite-(Ce) (first described as britholite) is the forefather of the

http://dx.doi.org/10.5772/62210

Fig. 30 The crystal (13 mm) of britholite-(Ce) from Ostkogen, Tvedalen, Norway.

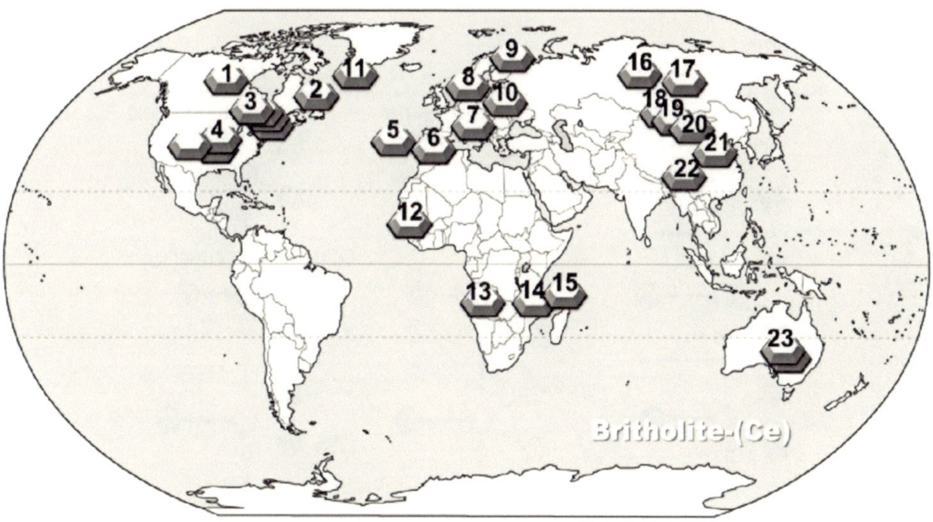

1) Thor Lake syenite complex, Blachford Lake alkaline complex, Mackenzie District, Northwest Territories, Canada
2) Red Wine Alkaline Complex, Labrador,Newfoundland and Labrador, Canada
3) Coldwell complex, Thunder Bay District, Ontario, Canada (a), Kipawa alkaline complex, Les Lacs-du-Témiscamingue, Témiscamingue RCM, Abitibi-Témiscamingue, Québec, Canada (b), Manny zone (Dufresne property), Oka complex, Oka, Deux-Montagnes RCM, Laurentides, Québec, Canada (c) and Fallon Brothers Quarry (Quincy Quarry; Ballou Quarry; Old Colony Crushed Stone Company Quarry), Quincy, Norfolk Co., Massachusetts, USA (d).
4) Pea Ridge Mine, Sullivan, Washington Co., Missouri, USA (a), Magnet Cove, Hot Spring Co., Arkansas, USA (b) and Iron Hill (Iron Hill carbonatite complex), White Earth District (Powderhorn District), Gunnison Co., Colorado, USA (c).
5) Água de Pau volcano (Fogo volcano), San Miguel Island, Azores District, Portugal
6) Jebel Bou-Agrao, Tamazeght Mountain Range, Khénifra Province, Meknès-Tafilalet Region, Morocco (a) and Sierra de las Cabras, Cancarix, Hellín, Albacete, Castile-La Mancha, Spain (b)
7) San Vito quarry, San Vito, Ercolano, Monte Somma, Somma-Vesuvius Complex, Naples Province, Campania, Italy
8) Norra Kärr, Gränna, Jönköping, Småland, Sweden (a) and Midtfjellet Quarry (Mellomfyllet Quarry), Malerød, Larvik, Vestfold, Norway (b).
9) Afrikanda Massif, Afrikanda, Kola Peninsula, Murmanskaja Oblast', Northern Region, Russia
10) Yastrebets Massif (Yastrebets'kiy; Yastrubetsky), Ukrainian Shield, Dnipropetrovs'k Oblast' (Dnepropetrovsk Oblast'), Ukraine
11) Tuperssuatsiat Bay, Tunulliarfik Fjord (Tunugdliarfik), Ilímaussaq complex, Narsaq, Kujalleq, Greenland
12) Los Archipelago (Los Islands), Guinea
13) Virulundo, Tombwa City Council, Namibe Province, Angola
14) Junguni Hill, Balaka District, Malawi
15) Ampasibitika, Ampasindava Peninsula, Ambanja District, Diana Region (Northern Region), Antsiranana Province, Madagascar
16) Kiisk alkaline massif (Kiyskoe), Enisei Range (Yenisei Ridge; Enisei Ridge), Krasnoyarsk Territory (Krasnoyarsk Kray; Krasnoyarskii Krai), Eastern-Siberian Region, Russia
17) Burpala alkaline Massif, Maigunda River, Mama River Basin, Buriatia (Buryatia), Prebaikalia (Pribaikal'e), Eastern-Siberian Region, Russia
18) Khaldzan Buragtag massif, Altai Mts, Hovd Aimag (Khovd Aimag), Mongolia
19) Mushgai Khudag carbonatite complex (Mushgai Hudag; Mushugai Khuduk; Mushugay Khuduk), Shara Hulasa Hiid, Ömnögovi Aimag, Mongolia
20) Bayan Obo deposit (Bayun-Obo deposit; Baiyunebo deposit), Bayan Obo, Bayan Obo Mining District, Baotou League (Baotou Prefecture), Inner Mongolia Autonomous Region, China
21) Weishan Mine (Chisan Mine; Xishan Mine; Deposit No. 101), Weishan Co., Jining Prefecture, Shandong Province, China
22) Maoniuping Mine, Mianning Co., Liangshan Autonomous Prefecture, Sichuan Province, China
23) Olympic Dam Mine, Roxby Downs, Stuart Shelf, South Australia, Australia (a) and Paratoo copper mine, Yunta, Olary Province, South Australia, Australia (b).

Fig. 31 The localities for the mineral britholite-(Ce).

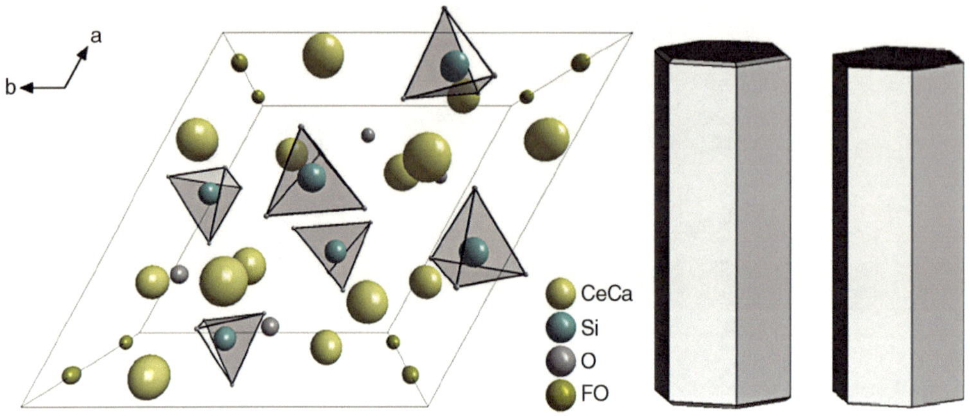

Fig. 32 The crystal structure (perspective view according to *c*-axis) of britholite-(Ce) and some common crystal shapes.

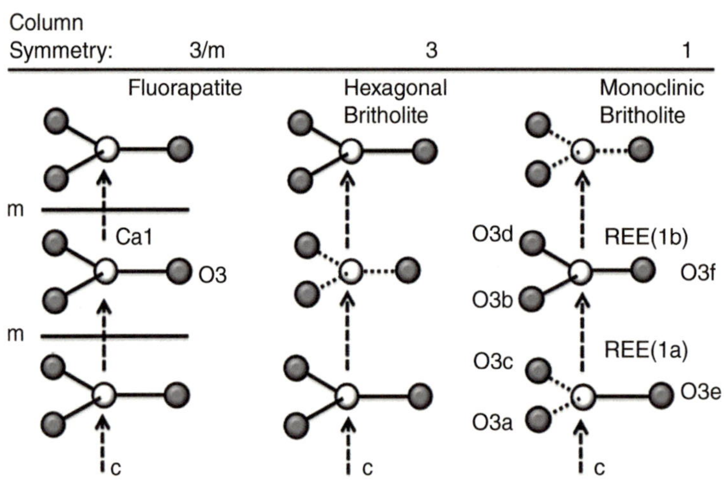

Fig. 33 Depiction of Ca(1)-O(3) triangles in apatite (a) and REE(1)-O(3) triangles in hexagonal (b) and monoclinic britholite (*c*). The dashed and solid lines in the structure of britholite indicate short and long REE-O bonds, respectively [106].

britholite group [108]. The structure of monoclinic britholite-(Ce) is shown in **Fig. 32** and the crystallographic data are listed in **Table 1**.

The crystal structure of monoclinic dimorphs **Fig. 33** of the mineral britholite-(Ce) (and also of britholite-(Y)described below) was solved in $\mathbf{P2_1}$ space group by Noe et al [106]. The monoclinic britholite dimorph differs from its hexagonal counterpart principally in the ligation of the REE equivalent of the apatite Ca(l) site. Whereas in $\mathbf{P6_3}$ britholite each Ca(l) equivalent has either three short or three long REE-O(3) bonds; in the $\mathbf{P2_1}$ dimorph, the Ca(1) equivalents have either one long and two short REE-O(3) bonds or one short and two long REE-O(3) bonds. Arrangement of short and long bonds leads to $\mathbf{P6_3}$ symmetry in hexagonal britholite due to removal of m from symmetry elements of apatite, and $\mathbf{P2_1}$ symmetry in monoclinic

http://dx.doi.org/10.5772/62210

britholite due to removal of symmetry elements **3** and **M**. The reduction in symmetry explains the common observation of biaxial optical characteristics of britholite samples [106].

2.3.2. Britholite-(Y)

The mineral britholite-(Y) ($(Y,Ca)_5(SiO_4)_3OH$, abukumalite, [105], [109]) occurs similarly to britholite-(Ce) in granite, alkaline rocks, skarns and hydrothermal veins [107]. The structure of monoclinic ($P2_1$, refer the discussion to **Fig. 33**) britholite-(Y) and known localities are shown in **Fig. 35** and **Fig. 34**, respectively.

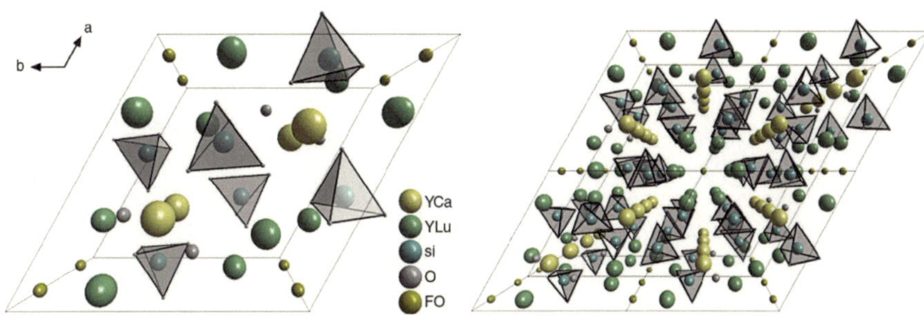

Fig. 34 The structure (perspective view along the *c*-axis) of britholite-(Y).

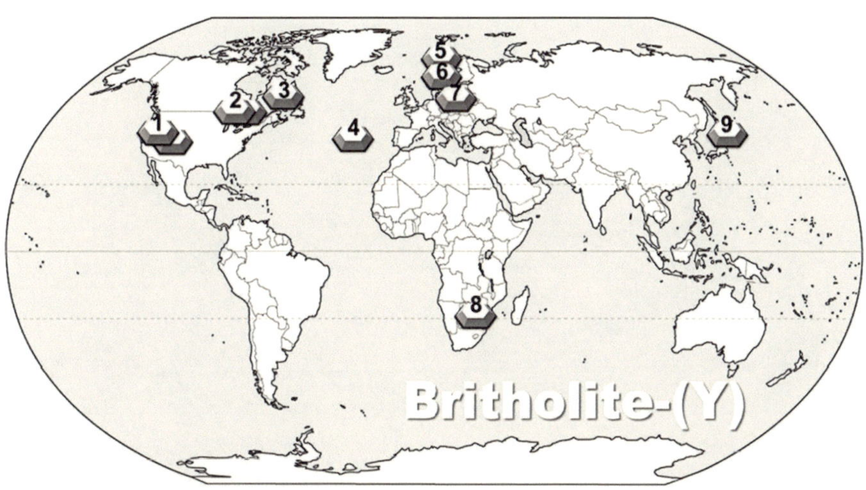

1) Buff Peak prospect, Bottle Creek District, Humboldt Co., Nevada, USA (a) and Iron Hill (Iron Hill carbonatite complex), White Earth District (Powderhorn District), Gunnison Co., Colorado, USA (b)
2) Black River pegmatite, Republic, Marquette Co., Michigan, USA (a) and Kipawa alkaline complex, Les Lacs-du-Témiscamingue, Témiscamingue RCM, Abitibi-Témiscamingue, Québec, Canada (b)
3) Kwyjibo deposit, Sept-Rivières RCM, Côte-Nord, Québec, Canada
4) Água de Pau volcano (Fogo volcano), San Miguel Island, Azores District, Portugal
5) Stetind pegmatite, Tysfjord, Nordland, Norway (a) and Hundholmen, Tysfjord, Nordland, Norway (b)
6) Norra Kärr, Gränna, Jönköping, Småland, Sweden
7) Stupné, Považská Bystrica Co., Trenčín Region, Slovakia
8) Pilanesberg Mtn, Pilanesberg Alkaline Complex, Western Bushveld Complex, Bojanala Platinum District, North West Province, South Africa
9) Fusamata, Kawamata, Fukushima Prefecture, Tohoku Region, Honshu Island, Japan

Fig. 35 The localities for the mineral britholite-(Y).

Britholite-(Y) is very brittle mineral with reddish brown or black color, pale brown streak and resinous luster that crystallizes as hexagonal in the space group $P6_3/M$ with the unit cell parameters $a = 9.43$ and $c = 6.81$ Å, $a{:}c = 1{:}0.722$, $V = 524.45$ Å^3 and $Z = 2$. Calculated and measured densities of the mineral are 4.07 and 4.25 g·cm^{-3}, respectively.

2.3.3. Fluorbritholite-(Ce)

The mineral fluorbritholite-(Ce) ($(Ce,Ca)_5(SiO_4)_3F$) [1],[110]) is the fluorine-rich analogue of britholite-(Ce). The structure and known localities of the mineral fluorbritholite-(Ce) are shown in **Fig. 36** and **Fig. 37**, respectively. The mineral has a yellow, reddish-brown color, or

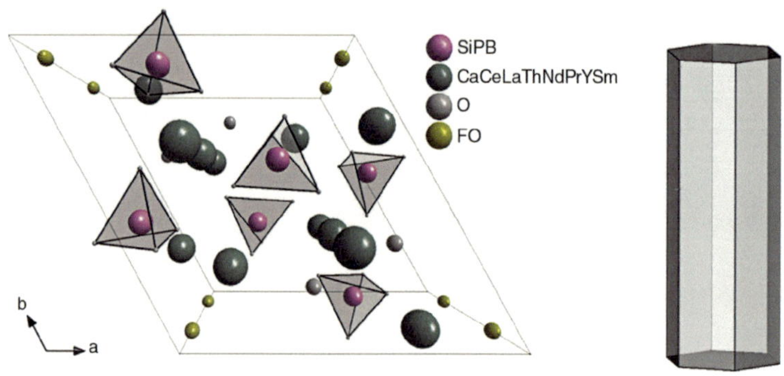

Fig. 36 The structure (perspective view along the *c*-axis) and the crystal habit of fluorbritholite-(Ce).

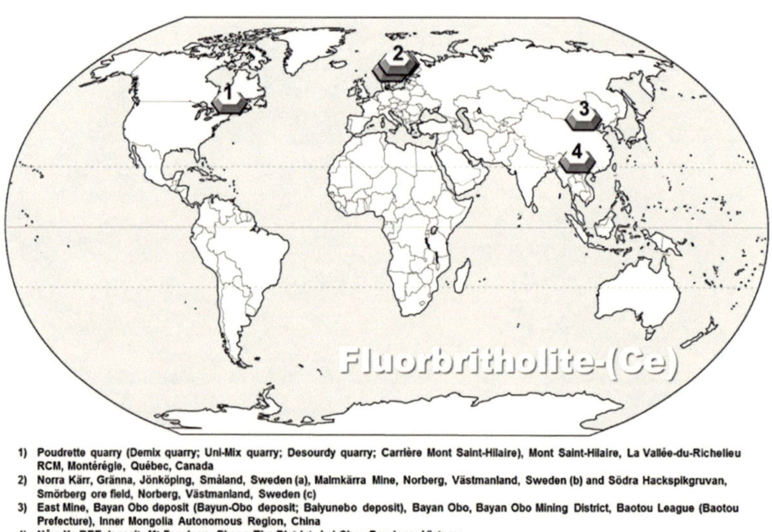

Fig. 37 The localities for the mineral fluorbritholite-(Ce).

http://dx.doi.org/10.5772/62210

it may be colorless. Its hardness on the Mohs scale is 5. Measured and calculated densities of fluorbritholite-(Ce) are 6.67 and 4.66 $g \cdot cm^{-3}$, respectively.

Fluorbritholite-(Ce) is a very brittle mineral that crystallizes as hexagonal in the space group $P6_3/m$. The unit cell shows following crystallographic parameters: $a = 9.517$ and $c = 6.983$ Å, a:c = 1:0.734, $V = 547.74$ Å^3 and $Z = 2$.

2.3.4. Fluorbritholite-(Y)

The mineral fluorbritholite-(Y) ($(Y,Ca)_5(SiO_4)_3F$) [108]) was named as the fluorine-dominant analogue of britholite-(Y), where the Levinson-type suffix modifier, -(Y), indicates the dominance of yttrium among rare-earth elements. It forms irregular grains, hexagonal to tabular crystals and short-prismatic to thick-tabular crystals. The known localities and structures of the mineral fluorbritholite-(Y) are shown in **Fig. 38** and **Fig. 39**, respectively.

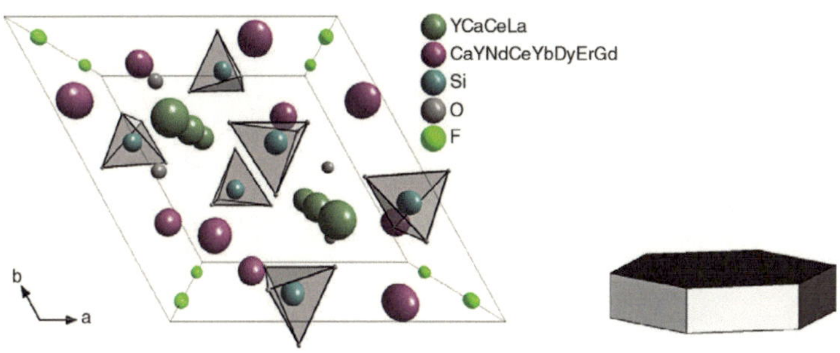

Fig. 38 The structure (perspective view along the *c*-axis) and the crystal habit of fluorbritholite-(Y).

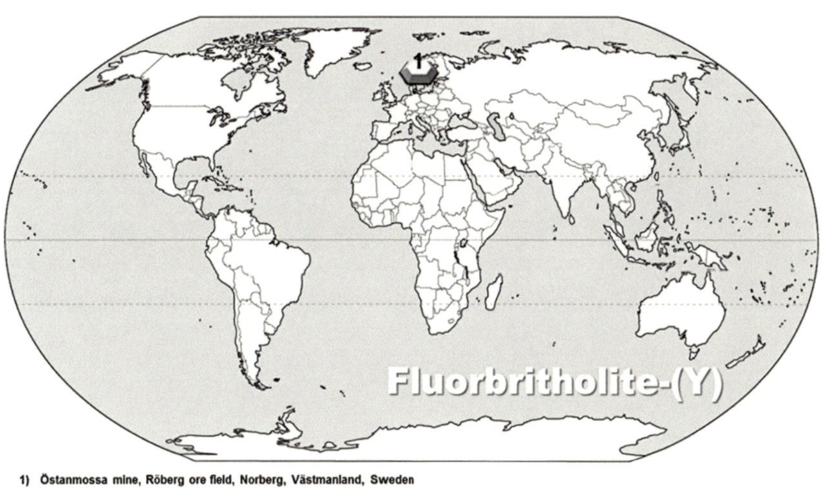

Fig. 39 The localities for the mineral fluorbritholite-(Y).

The mineral fluorbritholite crystallizes in hexagonal systems of the space group $P6_3/m$ with the crystallographic parameters of unit cell $a = 9.444$ and $c = 6.819$ Å, $a:c = 1:0.722$, $V = 526.68$ Å^3 and $Z = 2$. It is a brittle mineral of light-pink or brown color and calculated density of 4.61 g·cm $^{-3}$. It has a pale brownish or white streak and its hardness on the Mohs scale is 5.

2.3.5. Fluorcalciobritholite

The mineral fluorcalciobritholite ($(Ca,REE)_5(SiO_4,PO_4)_3F$; [1],[102]) was found at Mount Kukisvumchorr, Khibiny alkaline complex, Kola Peninsula, Russia and differs from fluorbritholite and fluorapatite in the content of calcium (Ca > Σ REE) and phosphorus (Si > P), respectively. The main crystal form is a hexagonal prism. The mineral is transparent, with a pale pinkish to brown color and a white streak. The structure and the locality of fluorcalciobritholite is shown in **Fig. 40** and **Fig. 41**, respectively.

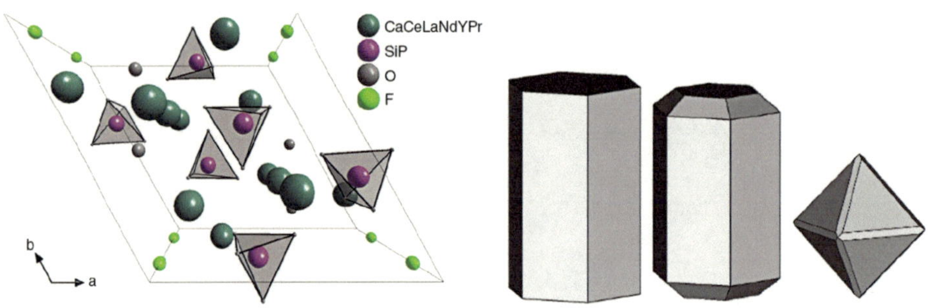

Fig. 40 The structure (perspective view along the *c*-axis) and the crystal habit of the mineral fluorcalciobritholite.

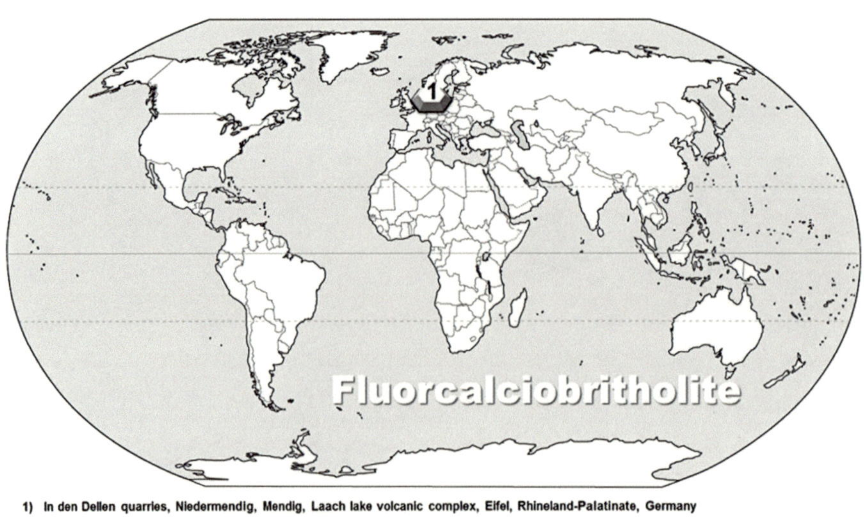

Fig. 41 The localities for the mineral fluorcalciobritholite.

The ideal chemical formula for fluorcalciobritholite may be written as $(Ca_3REE_2)[(SiO_4)_2(PO_4)]F$. In the view of coupled heterovalent substitutions occurring at the M and T sites in the series apatite–calciobritholite–britholite, it is more practical in this case for nomenclature purposes to consider the total abundance of M cations as a single, composite site [1].

Pale pinkish brown or brown mineral fluorcalciobritholite crystallizes as hexagonal in the space group $P6_3/M$ with the crystallographic parameters $a = 9.58$ and $c = 6.985$ Å, a:c = 1:0.729, $V = 555.17$ Å^3 and Z = 2. Calculated and measured densities of the mineral are 4.25 and 4.2 g·cm^{-3}, respectively. It has white streak and vitreous luster. The mineral is brittle and its hardness on the Mohs scale is equal to 5½.

2.3.6. Tritomite-(Ce)

Tritomite-(Ce) ($Ce_5(SiO_4,BO_4)_3(OH,O)$ [105],[111]) was first found by Weibye in 1849 at the island of Låven in Langesundsfjord as dark tetrahedral crystals in leucophanite ($(Na,Ca)_2BeSi_2(O,OH,F)_7$ [112] or analcime. Chemically and structurally, it is very similar to melanocerite (melanocerite-(Ce),[20] $Ce_5(SiO_4,BO_4)_3$ (OH,O) [1],[105],[111]) and caryocerite [111]. The pyramidal crystal of mineral tritomite-(Ce) is shown in **Fig. 42**.

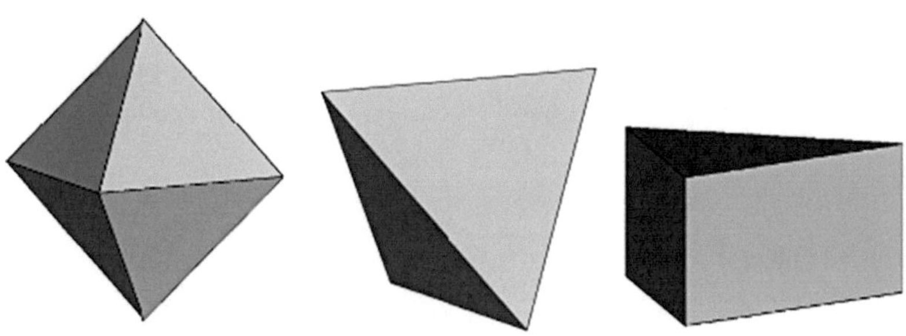

Fig. 42 The crystal habit of the mineral tritomite-(Ce) and tritomite-(Y).

The mineral was named from the Greek *τριτομος* meaning "cut in three parts" in allusion to the triangular and pseudo-tetrahedral crystal habit [113],[114]. Known localities for the mineral tritomite-(Ce) are shown in **Fig. 43**.

Tritomite-(Ce) crystallizes as hexagonal mineral in the space group $P6_3/M$ with crystallographic parameters $a = 9.35$ and $c = 6.88$ Å, a:c = 1: 0.736, $V = 520.89$ Å^3 and Z = 2. It is a very brittle mineral of dark brown color with a hardness (on the Mohs scale) of 5½.

[20] Since the mineral is equal to tritomite-(Ce), the name of melanocerite-(Ce) is discredited [1].

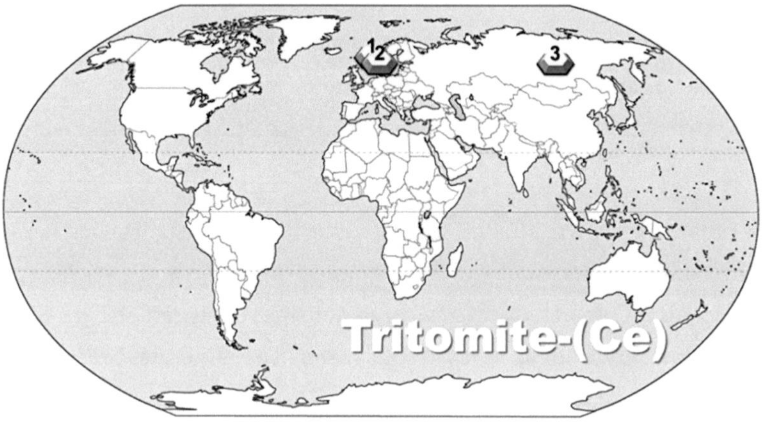

1) Østskogen Quarry, Tvedalen, Larvik, Vestfold, Norway (a), Vevja Quarry, Tvedalen, Larvik, Vestfold, Norway (b), Saga 1 Quarry, Sagåsen, Mørje, Porsgrunn, Telemark, Norway (c), Arøya, Langesundsfjorden, Larvik, Vestfold, Norway (d) and Låven (Skådön; Lamö; Lamanskjaer), Langesundsfjorden, Larvik, Vestfold, Norway (e)
2) Norra Kärr, Gränna, Jönköping, Småland, Sweden
3) Burpala alkaline Massif, Maigunda River, Mama River Basin, Buriatia (Buryatia), Prebaikalia (Pribaikal'e), Eastern-Siberian Region, Russia

Fig. 43 Known localities for the mineral tritomite-(Ce).

2.3.7. Tritomite-(Y)

The mineral tritomite-(Y) (($Y_5(SiO_4,BO_4)_3(OH,O,F)$, $[Y^{3+}(Cr, Pr, Th)^{4+}Ca](Si_2B)O_{12}O$ [111],[113],[115]) was first described by Frondel. It is also known as the hexagonal mineral spencite (named after Canadian geologist H.S. Spence) [116]. The mineral tritomite-(Y) is formed in the nepheline syenite pegmatites of the area, which carries rare earths predominantly from the yttrium group. Known localities of mineral tritomite-(Y) are introduced in **Fig. 44**.

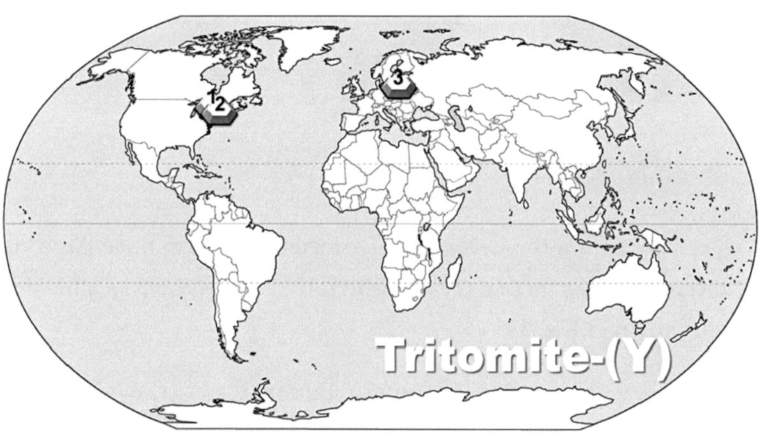

1) Cardiff Uranium Mine, Cardiff Township, Haliburton Co., Ontario, Canada
2) Madawaska Mine (Faraday Mine), Faraday Township, Hastings Co., Ontario, Canada
3) Pincina maar, Lučenec Co., Banská Bystrica Region, Slovakia

Fig. 44 Known localities for the mineral tritomite-(Y).

http://dx.doi.org/10.5772/62210

When heated in air to temperatures ranging from 600°C to 1000°C, tritomite-(Y) recrystallizes to the structure of apatite and amorphous phase, presumably to a calcium borosilicate glass [60]. The pyramidal crystals of the mineral tritomite-(Y) are similar to tritomite-(Ce) ones, which are shown in **Fig. 42**.

Tritomite-(Y) crystallizes as hexagonal in the space group $P6_3/M$ with unit cell parameters a = 9.32 and c = 6.84 Å, a:c = 1:0.734, V = 514.54 Å^3 and Z = 2. It is dark green-black, red-brown, nearly black mineral with vitreous or resinous luster and average density of 3.22 $g \cdot cm^{-3}$. It is a brittle mineral forming small fragments with conchoidal fracture. The hardness of tritomite-(Y) on the Mohs scale ranges from 3.5 to 6.5 [111],[113],[115].

2.4. The group of ellestadite

Ellestadites *sensu lato* are sulfato-silicates. For stoichiometric reasons, the incorporation of the sulfate anion $(SO_4)^{2-}$ in the structure of apatite in the place of $(PO_4)^{3-}$ or $(AsO_4)^{3-}$ must be coupled with a concurrent substitution by silicate anions $(SiO_4)^{4-}$. This holds in all cases in which the M sites are occupied by divalent cations. Pure sulfates with an apatite structure may occur only by reducing overall positive charge associated with the **M** cations, as is the case in cesanite and caracolite from the group of hedyphane (**Section 2.1**) [1].

The structural formula of ellestadite and (with slight modification) of wilkeite can be expressed as follows [117]:

$$[Ca_6(F, Cl, O, OH)_2][(S, Si, P, C)O_4]_6(Ca, C)_4$$

This formula indicates that two-fifths of the Ca^{2+} ions are located on threefold axes and can be replaced by carbon. Three-fifths of the Ca^{2+} ions are tied to F^-, Cl^- and O^- anions or OH^- groups and cannot be replaced by carbon. All Ca^{2+} ions are tied to O-ions, which are arranged in tetrahedral coordination with S-, Si-, P- or C-ions at the centers.

2.4.1. Fluorellestadite

The mineral fluorellestadite (formerly called ellestadite-(F) [85],[118],[119] $Ca_5(SiO_4)_{1.5}(SO_4)_{1.5}F$ [1],[120]) is a rare mineral found in nature in skarns or metamorphosed limestones[21] [121]. It was named according to American analytical chemist R.B. Ellestad and fluorine in the chemical composition. The mineral occurs as needles, as hexagonal prismatic, poorly terminated crystals up to 3 mm long, and as fine-grained aggregates. Thin needles are colorless, crystals are transparent and aggregates are translucent. Known localities of fluorellestadite are introduced in **Fig. 45**. The structure of mineral ellestadite is shown on **Fig. 46**.

[21] Limestone is a name used for sedimentary rock composed mainly of calcium carbonate, usually in the form of calcite (trigonal $CaCO_3$) or aragonite (orthorhombic $CaCO_3$), but there could also be considerable amounts of magnesium carbonate ($MgCO_3$, trigonal mineral magnesite) or dolomite (trigonal $CaMg(CO_3)_2$) [121].

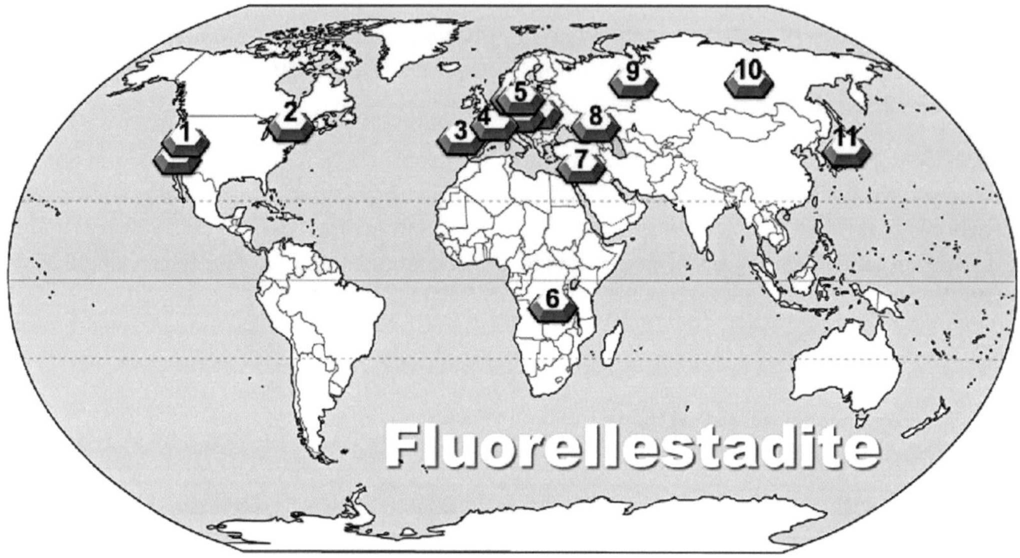

1) Crestmore quarries, Crestmore, Riverside Co., California, USA (a) and Champion Mine (White Mountain Mine; Jeffrey Mine; Diaspore; Vulcanus; Black Eagle; Champion Sillimanite Mine; Champion Andalusite Mine; Vulcanite Mine; Vulcanite & Vulcanus No. 156; Black Eagle & Vulcanus Nos. 8 and 9 No. 157), White Mountain Peak, White Mts, Mono Co., California, USA (b)
2) Oka complex, Oka, Deux-Montagnes RCM, Laurentides, Québec, Canada
3) Bendada Mines, Bendada, Sabugal, Guarda District, Portugal
4) Lapanouse-de-Sévérac slag locality, Aveyron, Midi-Pyrénées, France
5) Caspar quarry, Bellerberg volcano, Ettringen, Mayen, Eifel, Rhineland-Palatinate, Germany (a), Zeilberg Quarry, Maroldsweisach, Franconia, Bavaria, Germany (b), Basalt quarry, Klöch, Bad Radkersburg, Styria, Austria (c) and Zlatno, Vysoká, Banská Štiavnica Co., Banská Bystrica Region, Slovakia (d)
6) Kampijimpanga, Solwezi District, North-Western Province, Zambia
7) Hatrurim (Hatrurim Basin), Negev, Israel (a) and Har Ye'elim, Southern District (HaDarom District), Israel (b)
8) Upper Chegem volcanic caldera, Baksan Valley, Kabardino-Balkarian Republic, Northern Caucasus Region, Russia
9) Kopeisk, Chelyabinsk coal basin, Chelyabinsk Oblast', Southern Urals, Urals Region, Russia (a) and Chelyabinsk coal basin, Chelyabinsk Oblast', Southern Urals, Urals Region, Russia (b)
10) Yoko-Dovyrensky Massif, Dovyren Highlands, Buriatia Republic (Buryatia), Transbaikalia (Zabaykalye), Eastern-Siberian Region, Russia
11) Chichibu mining district (Chichibu mine), Nakatsugawa, Ohtaki-mura, Chichibu City, Saitama Prefecture, Kanto Region, Honshu Island, Japan

Fig. 45 Known localities for the mineral fluorellestadite.

Fluorellestadite is colorless, blue or pale bluish green hexagonal mineral belonging to the space group $P6_3/M$. The unit cell parameters are a = 9.485, c = 6.916 Å, Z = 2 and V = 538.84 Å^3. Calculated density is 3.10 g·cm^{-3}. The hardness of the mineral on the Mohs scale is 4½.

The mineral is also known from burned coal dumps, where its formation is possible in the presence of carbonaceous and carbonate rocks such as the rests of pyrometamorphism[22] [9] of sedimentary rocks. The generalized formula of this mineral can be expressed as $Ca_{10}(SiO_4)_{3-x}(SO_4)_{3-x}(PO_4)_{2x}$ $(OH,F,Cl)_2$, where the parameter x varies from 0 (ellestadite) to 3 (apatite).

[22] The term pyrometamorphism, which is derived from the Greek word *pyr/pyro* (fire), *meta* (change), and *morph* (shape or form) was first used by Brauns to describe high-temperature changes which take place at immediate contact of magma and country rock with or without interchanges of material. Tyrrell defined pyrometamorphism as pertaining to the "effect of the highest degree of heat possible without actual fusion." There are a number of rock terms commonly used in association with the phenomenon of pyrometamorphism including hornfels,[3] buchite, porcellanite, sanidinite, emery, paralava, clinker, fulgurite, or with other general terms such as fused or burnt rock [121].

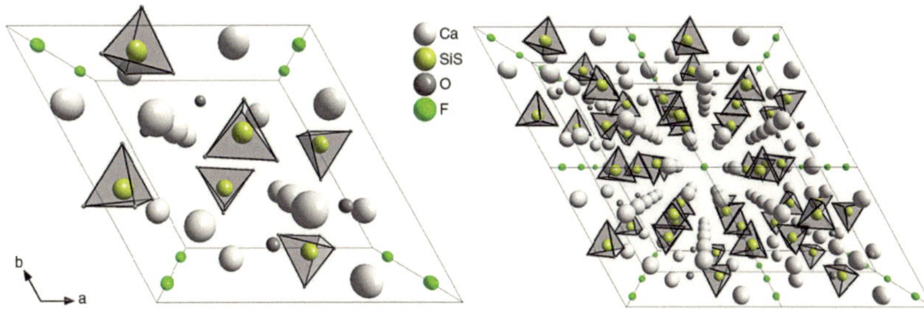

Fig. 46 The structure of the mineral fluorellestadite (perspective view along the *c*-axis).

2.4.2. Hydroxylellestadite

Hydroxylellestadite (formerly called ellestadite-(OH) [85], $Ca_5(SiO_4)_{1.5}(SO_4)_{1.5}(OH)$ [1],[117],[122],[123]) was first reported at cornet Hill by Pascal et al [124] and Marincea et al [125]. Natural hydroxylellestadite[23] occurrences were reported from pegmatite veins, skarn and pyrometamorphic deposits and from mine dumps, but this mineral has never been reported from a cave. The mineral forms aggregates of xenomorphic crystals which have a maximum length of 0.5 mm and a maximum width of about 0.1 mm.

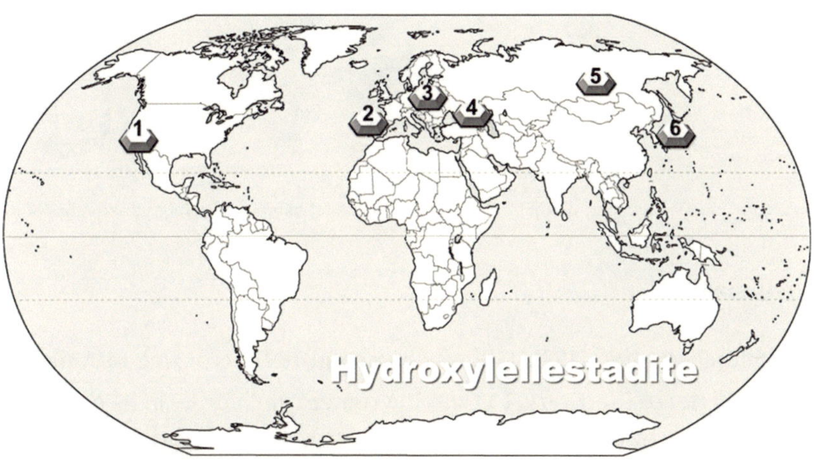

1) Crestmore quarries, Crestmore, Riverside Co., California, USA
2) Bendada Mines, Bendada, Sabugal, Guarda District, Portugal
3) Zlatno, Vysoká, Banská Štiavnica Co., Banská Bystrica Region, Slovakia
4) Upper Chegem volcanic caldera, Baksan Valley, Kabardino-Balkarian Republic, Northern Caucasus Region, Russia
5) Yoko-Dovyrensky Massif, Dovyren Highlands, Buriatia Republic (Buryatia), Transbaikalia (Zabaykalye), Eastern-Siberian Region, Russia
6) Chichibu mining district (Chichibu mine), Nakatsugawa, Ohtaki-mura, Chichibu City, Saitama Prefecture, Kanto Region, Honshu Island, Japan

Fig. 47 Known localities for the mineral hydroxylellestadite.

[23] Synthetic analogs are known as "technical products," such as burnt industrial waste and cement [122].

Hydroxylellestadite is associated with berlinite[24] ($AlPO_4$ [126]) , another high-temperature mineral. It is likely to have formed within highly phosphatized, silicate-rich, carbonate-mudstone sediments heavily compacted and thermally transformed due to *in situ* bat guano combustion. Known localities, where the mineral hydroxylellestadite can be found, and its structure are shown in **Fig. 47** and **Fig. 48**, respectively.

Hydroxylellestadite is a pink or purple-gray hexagonal mineral, which belongs to the space group $P6_3/M$. The unit cell parameters are a = 9.491, c = 6.921 Å, Z = 2 and V = 539.91 Å^3. Calculated density is 3.11 $g \cdot cm^{-3}$. The hardness of the mineral on the Mohs scale is in the range of 3½ to 4½. Hydroxylellestadite shows faded white-yellow fluorescence when irradiated with UV light, independently of the excitation frequency [122].

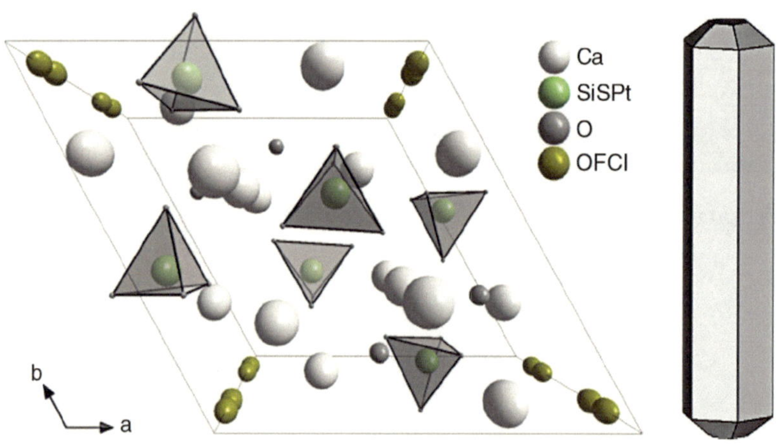

Fig. 48 The structure (perspective view along the *c*-axis) and the crystal habit of mineral hydroxylellestadite.

2.4.3. Chlorellestadite

The mineral chlorellestadite[25] [127],[128] was named in 1892 according to American analytical chemist R.B. Ellestad (**Section 2.4.1**) and the content of chlorine in its chemical composition in veinlets cutting contact with metamorphosed limestone. The structure of the mineral ellestadite is shown in **Fig. 49**.

The mineral occurs as a compact mass. The mineral chlorellestadite is associated with diopside, wollastonite, vesuvianite ($Ca_{10}Mg_2Al_4(SiO4)_5$ $(Si_2O_7)_2(OH)_4$ [129]), monticellite ($CaMgSiO_4$ [130]) and calcite.

---

[24] The mineral was named after Swedish pharmacologist N.J. Berlin. The mineral is Al-P analogue of quartz.

[25] The IMA status of the mineral was discredited in 2010.

http://dx.doi.org/10.5772/62210

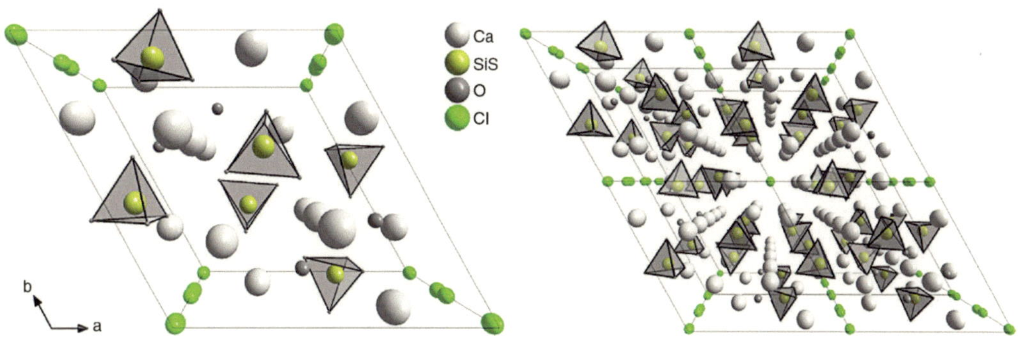

Fig. 49 The structure of the mineral chlorellestadite (perspective view along the *c*-axis).

Chlorellestadite is a hexagonal mineral that crystallizes in the space group $P6_3/M$ with crystallographic parameters $a = 9.53$ and $c = 6.91$ Å, $a{:}c = 1{:}0.725$, $V = 543{,}49$ Å^3 and $Z = 2$. It has white streaks and a vitreous luster. The color of the mineral is pink, yellowish green, pale rose, orange, but it can also be colorless. The hardness on the Mohs scale is 4½. Calculated and measured densities of the mineral are 3.068 and 3.113 $g \cdot cm^{-3}$, respectively.

2.4.4. Mattheddleite

Mattheddleite [131],[132],[133],[134] is a mineral with the composition $Pb_{10}(SiO_4)_3(SO_4)_3Cl_2$ (LIVINGSTONE et al [131]) or $Pb_5(Si_{1.5}S_{1.5})O_{12}(Cl_{0.57}OH_{0.43})$ (STELLE et al [134]) which is a lead member of apatite supergroup where phosphorus is totally replaced by sulfur and silicon: $Si^{4+} + S^{6+} \leftrightarrow 2P^{5+}$. Mattheddleite was first recognized in typical Pb mineral region at Leadhills, Scotland and named after Scottish mineralogist MATTHEW FORSTER HEDDLE (1828–1897). The **Z** = $OH^- + Cl^-$ anion position is zoned from an OH-rich interior to a Cl-rich exterior. Known localities, where the mineral hydroxylellestadite can be found, and its structures are shown in **Fig. 50** and **Fig. 51**, respectively.

Mattheddleite is a colorless or white hexagonal mineral belonging to the space group $P6_3/M$. The unit cell parameters are $a = 9.963$ (10.0056 [134]) and $c = 7.464$ (7.4960 [134]) Å, $Z = ½$ ($Z = 2$ [134]) and $V = 641.63$ (649.9 [134]) Å^3. Calculated density is 6.96 (6.822 [134]) $g \cdot cm^{-3}$. The hardness of the mineral on the Mohs scale is 4½.

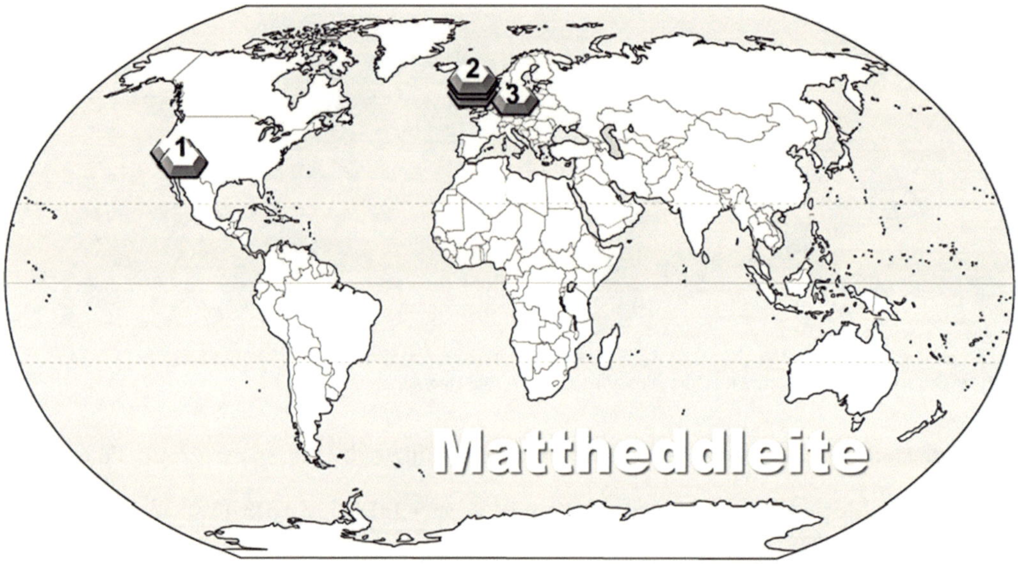

1) Bird Nest E3, Otto Mountain, Baker, San Bernardino Co., California, USA (a) and Evening Star Mine (Silver Queen Mine; Old Queen group), Tiger Wash, Belmont Mountain, Tonopah, Osborn District, Big Horn Mts, Maricopa Co., Arizona, USA (b)
2) Penberthy Croft Mine, St Hilary, Mount's Bay District, Cornwall, England, UK (a), The Gannel Smelter slag locality, Crantock, St Agnes District, Cornwall, England, UK (b), Torr Works Quarry (Merehead Quarry), Cranmore, Somerset, England, UK (c), Machen Quarry, Machen, Caerphilly (Mid Glamorgan; Monmouthshire), Wales, UK (d), Esgair Mwyn Mine, Ysbyty Ystwyth, Ceredigion (Dyfed; Cardiganshire), Wales, UK (e), Hendre Felen Mine, Ysbyty Ystwyth, Ceredigion (Dyfed; Cardiganshire), Wales, UK (f), Frongoch Mine (Bron-y-Goch Mine; Llawynwnwch Mine), Pontrhydygroes, Upper Llanfihangell-y-Creuddyn, Ceredigion (Dyfed; Cardiganshire), Wales, UK (g), Daren Mine (Old Daren Mine; Darren Consolidated Mine; Daren United Mine; Great Darren Mine; Darren Mine), Pen-bont Rhydybeddau, Trefeirig, Ceredigion (Dyfed; Cardiganshire), Wales, UK (h) and several other localities in UK
3) Glücksrad Mine, Oberschulenberg, Bockswies veins, Clausthal-Zellerfeld, Harz, Lower Saxony, Germany

Fig. 50 Known localities for the mineral mattheddleite.

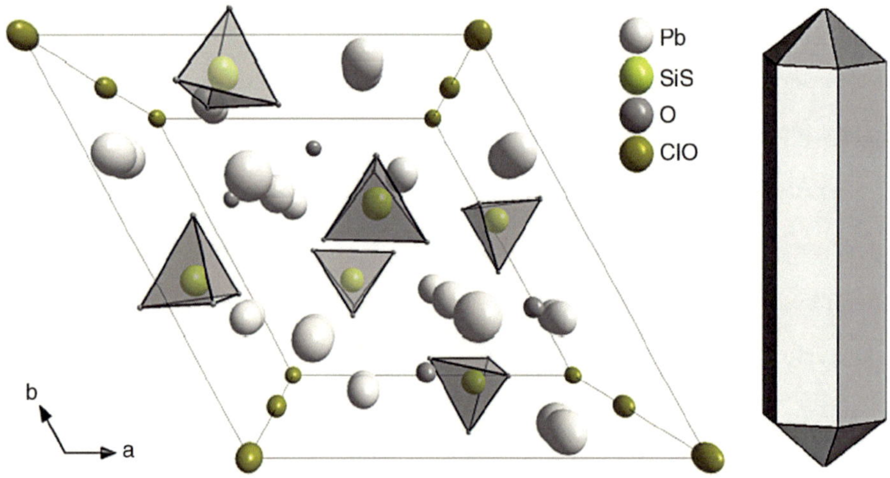

Fig. 51 The structure and the crystal habit of the mineral mattheddleite (perspective view along the *c*-axis).

http://dx.doi.org/10.5772/62210

2.5. Pieczkaite

The mineral pieczkaite ($Mn_5^{2+}(PO_4)_3Cl$ [135],[136]) was found in the Southeastern shoreline of a small, unnamed island in Cross Lake, Manitoba, Canada (54°41′N, 97°49′W; **Fig. 52**) and classified as the member of the supergroup of apatite. It is isostructural with calcium fluorapatite (**Section 1.5.1**). The approximate composition of hydrothermally grown manganese chlorapatite is $Mn_5(PO_4)_3Cl_{0.9}(OH)_{0.1}$ [136].

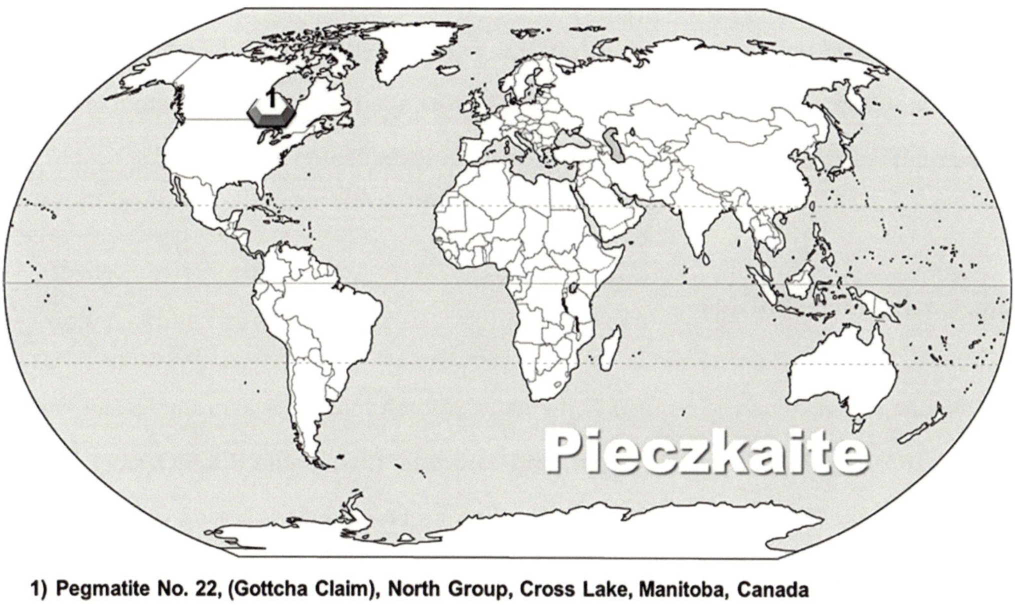

Fig. 52 Locality for the mineral pieczkaite.

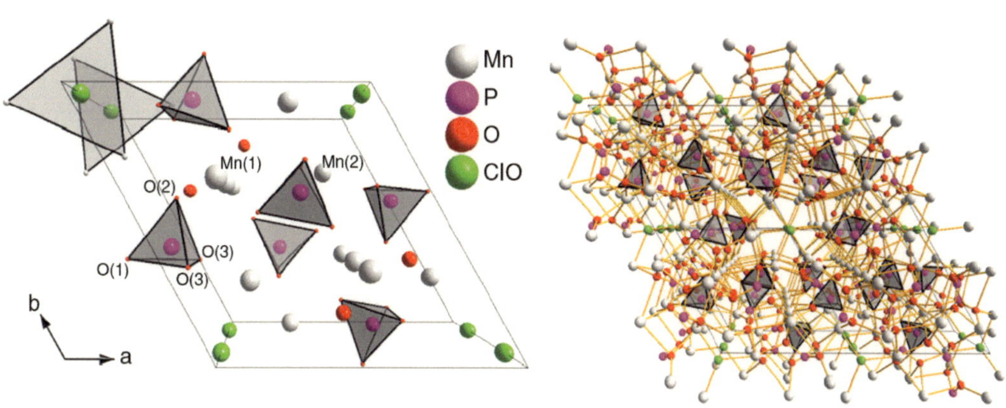

Fig. 53 The structure of the mineral pieczkaite (perspective view along the *c*-axis).

It is a hexagonal mineral that crystallizes in the space group $P6_3/M$ with the crystallographic parameters of unit cell $a = 9.532$ and $c = 6.199$ Å, $a{:}c = 1{:}0.6501$, $V = 587.78$ Å^3 and Z = 2. Calculated density of pieczkaite is 3.783 g·cm^{-3}. The hardness of the mineral on the Mohs scale varies in the range from 4 to 5. The structure of the mineral pieczkaite is shown in **Fig. 53**.

The coordination polyhedron around Mn(1) has the point-group symmetry 3 and is a trigonal prism in which the two triangles of oxygen atoms are slightly rotated relative to each other. The coordination polyhedron around Mn(2) is a severely distorted octahedron. The phosphate group is more distorted than in any of the other apatites. The chlorine atom is located in the center of an equilateral triangle formed by three Mn(2) atoms [136].

2.6. Carbonate-apatites

As mentioned previously (**Section 1.1**) the name of both most typical examples, i.e. carbonate-hydroxylapatite ($Ca_5(PO_4,CO_3)_3OH$) and carbonate-fluorapatite ($Ca_5(PO_4,CO_3)_3F$), was discredited from the IMA list of minerals [1]. The structure and the crystal shape of carbonate-apatite and carbonate-fluorapatite are shown in **Fig. 54**. The carbonate-apatites, the properties of which are listed in **Table 7** (**Chapter 1**), are intensively studied as the mineral constituents of bones and teeth as described in **Section 10.9.**

The carbonate-rich apatites are:

1. **Francolite** ($Ca_{10-x-y}Na_xMg_y(PO_4)_{6-z}(CO_3)_zF_{0.4z}F_2$ or $Ca_5(PO_4,CO_3)_3F$) [137] is the name used for massive, cryptocrystalline or amorphous varieties of carbonate-rich hydroxyl- and fluorapatite. Francolite and staffelite are the synonyms for carbonate-fluorapatite.

   This complex carbonate-substituted apatite is found only in marine environments, and, to a much smaller extent, in weathered deposits, for instance above carbonatites [138]. The mineral was named according to its occurrence at Wheal Franco, Whitchurch, Tavistock District, Devon, England.

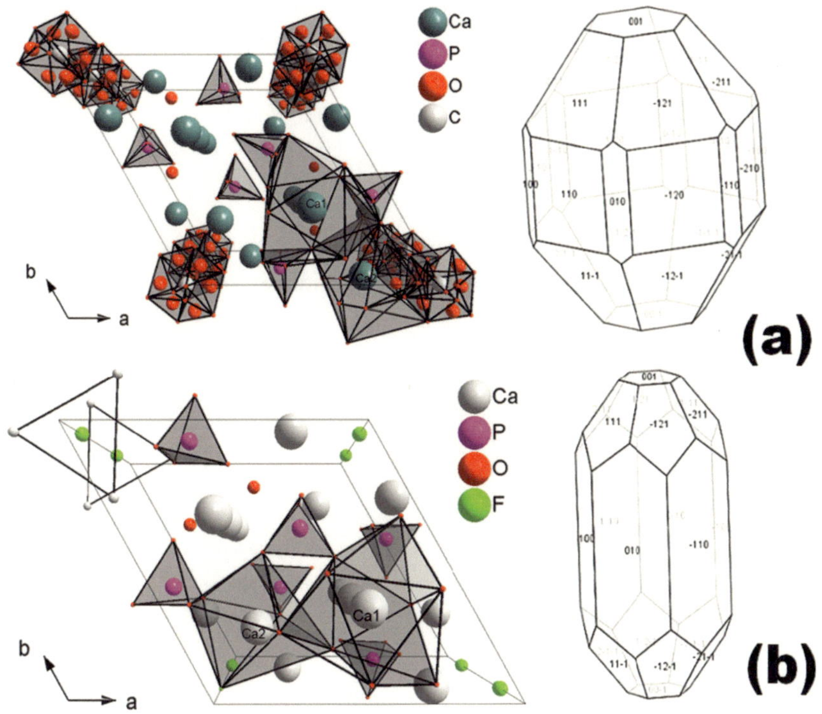

Fig. 54 The structure and the crystal shape of carbonate-hydroxylapatite (a) and fluorapatite (b).

2. **Dahlite** (carbonate-hydroxylapatite, podolite, $3Ca_3(PO_4)_2{\cdot}CaCO_3$ or $(Na,Ca)_5(PO_4, CO_3)_3OH$) [137],[139]. This phosphate structure is found in marine sediments [140],[138].

3. **Kurskite** ($Ca_{10}P_{4.8}C_{1.2}O_{22.8}F_2(OH)_{1.2}$) [137],[141],[142] forms nodular or platform-type phosphorites, widespread within Russia. It is a carbonate-rich mineral that can be found in two varieties:

   - Radiating (previously incorrectly termed as staffelite)
   - Optically amorphous

   The mineral is usually gray or brown due to the content of organic, humic or ferruginous impurities. Sometimes, it is white or black colored. Pure kurskite has a specific gravity of 3 $g{\cdot}cm^{-3}$.

4. **Collophane** ($3Ca_3(PO_4)_2{\cdot}nCa(CO_3,F_2,O){\cdot}xH_2O$ [137]) this type of phosphate minerals is typical for marine phosphate sediments [138]. Apatite is a principal constituent of fossil bones and other organic matter. The name cellophane is sometimes used for such phosphatic material [143].

According to the accommodation of carbonate ion in the apatite structure, three basic types of apatites (**Fig. 55**) can be recognized [144]:

i. **Type A**: carbonate ion of ideal geometry in upright open configuration (bisector of $[CO_3]^{2-}$ triangle parallel to *c*-axis); configuration with apical oxygen located at the position of OH^- (a)

ii. **Type B**: closed (bisector normal to *c*-axis) configuration of type A1 carbonate ion in the space group $P6_3/M$ (b)

iii. **Type AB**: open (and inverted) type A2 carbonate ion and the location of type B carbonate ion close to the sloping face of substituted $[PO_4]^{3-}$ tetrahedron (*c*).

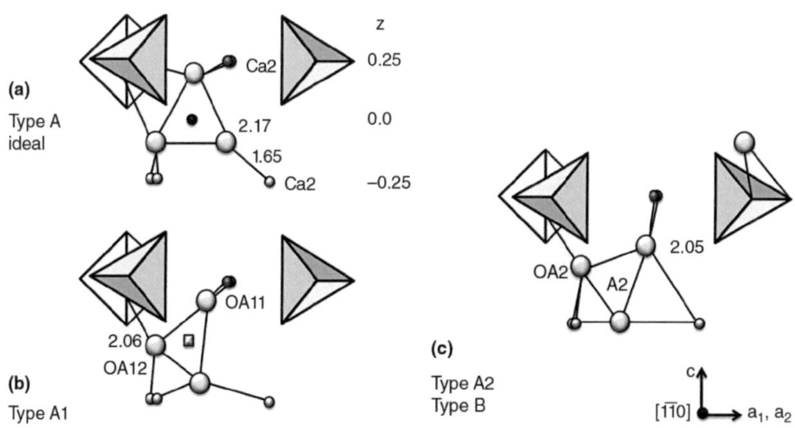

Fig. 55 Part of the *c*-axis channel showing the accommodation of carbonate ion in the structure of hydroxylapatite [144].

Individual types of carbonate apatite and their importance for bone and dental enamel are described in **Section 10.9.2.**

Carbonate apatites have distinctive X-ray patterns and rather small cell parameter *a*. An empirical relationship between the content of CO_2 in apatite and the separation (Δ [Å]) of the 211 and 112 X-ray diffraction lines has been given by O'Brien et al [145],[146]:

$$CO_2 \left[\text{wt.\%}\right] = 17.335 - \left(615.524 \cdot \Delta\right) \tag{1}$$

Author details

Petr Ptáček

Brno University of Technology, Czech Republic

References

[1] Pasero M, Kampf AR, Ferraris C, Pekov IV, Rakovan JR, White TJ. Nomenclature of the apatite supergroup minerals. European Journal of Mineralogy 2010;22, 163–179.

[2] Breithaupt A. Bestimmung neuer mineral-specien, hedyphan, Journal für Chemie und Physik 1830;60 308–316.

[3] Rouse RC. Hedyphane from Franklin, New Jersey and Långan, Sweden: cation ordering in an arsenate apatite. American Mineralogist 1984;69 920–927.

[4] Watts H. A Dictionary of Chemistry and the Allied Branches of Other Sciences. Volume 3. Longmans, Green, and Company, 1873.

[5] Breithaupt JFA. Vollständiges handbuch der mineralogy. Volume 2. Arnoldische Buchhandlung Dresden and Leipzig, 1841.

[6] Palache Ch. The Minerals of Franklin and Sterling Hill, Sussex County, New Jersey Geological Survey professional paper. U.S. Government Printing Office, 1937.

[7] Demartin F, Gramaccioli CM, Campostrini I, Pilati T. Aiolosite, $Na_2(Na_2Bi)(SO_4)_3Cl$, a new sulfate isotypic to apatite from La Fossa Crater, Vulcano, Aeolian Islands, Italy. American Mineralogist 2010;95 382–385.

[8] Kampf AR, Housley RM. Fluorphosphohedyphane, $Ca_2Pb_3(PO_4)_3F$, the first apatite supergroup mineral with essential Pb and F, American Mineralogist 2011;96 423–429.

[9] Grapes R. Pyrometamorphism. 2nd ed., Springer Science & Business Media, 2010. ISBN: 978-3642155888

[10] Theophrastus (315 BC) Chrysokolla, in Theophrastus On Stones translated by Caley ER and Richards JFC (1956), Ohio State University (Columbus, Ohio) 25–27.

[11] Guthrie GD, Bish DL, Reynolds RC. Modeling the X-ray diffraction pattern of opal-CT. American Mineralogist 1995;80 869–872.

[12] Nagase T, Akizuki M. Texture and structure of opal-CT and opal-C in volcanic rocks. The Canadian Mineralogist 1997;35 947–958.

[13] Berzelius JJ. Diaspore, in Nouveau Système de Minéralogie, Méquignon-Marvis (Paris) 1819;30 282–285.

[14] Shepard CU. Plumbo-gummite, in Treatise on Mineralogy: Second Part, Hezekiah Howe & Co. (New Haven) 1835; 113–113.

[15] Scott KM. Solid solution in, and classification of, gossan-derived members of the alunite-jarosite family, northwest Queensland, Australia. American Mineralogist 1987;72 178–187.

[16] Kampf AR, Rossman GR, Housley RM. Plumbophyllite, a new species from the Blue Bell claims near Baker, San Bernardino County, California. American Mineralogist 2009;94: 1198–1204.

[17] Keller P, Dunn PJ. Plumbotsumite, $Pb_5(OH)_{10}Si_4O_8$, a new lead silicate from Tsumeb, Namibia, Chemie der Erde 1982;41 1–6.

[18] Fleischer M, Cabri LJ, Chao GY, Mandarino JA, Pabst A. New mineral names, American Mineralogist 1982;67 1074–1082.

[19] Marty J, Kampf AR, Housley RM, Mills SJ, Weiβ S. Seltene neue Tellurmineralien aus Kalifornien, Utah, Arizona und New Mexico (USA). Lapis 2010;35() 42–51.(66)

[20] Haidinger W. Zweite Klasse: Geogenide. II. Ordnung. Baryte. VII. Bleibaryt. Wulfenit., in Handbuch der Bestimmenden Mineralogie, Bei Braumüller and Seidel (Wien) 1845; 499–506.

[21] Du Y, Ding Hang-Ch, Sheng L, Savrasov SY, Wan X, Duan Chun-G. Microscopic mechanism of stereochemically active lone pair studied from orbital selective external potential calculation. Journal of Physics Condensed Matter 2014;26(2) 025503. doi: 10.1088/0953-8984/26/2/025503

[22] Kampf AR, Steele IM, Jenkins. Phosphohedyphane, $Ca_2Pb_3(PO_{4)3}Cl$, the phosphate analog of hedyphane: description and crystal structure. American Mineralogist 2006;91 1909–1917.

[23] von Pufahl O. Mitteilungen über mineralien und erze von Südwestafrika, besonders solche von Tsumeb. Centralblatt fur Mineralogie, Geologie und Palaontologie. 1920; 289–296.

[24] Church AH. Chemical researches on some new and rare Cornish minerals. Journal of the Chemical Society 1865;18, 259–268.

[25] Beudant FS. Anglesite, plomb sulfaté, in Traité Élémentaire de Minéralogie, 2nd ed., (Paris) 1832; 459–460.

[26] Bideaux RA, Nichols MC, Williams SA. The arsenate analog of tsumebite, a new mineral. American Mineralogist 1966;51 258–259.

[27] Vésignié JPL. Présentation d'échantillons. Bulletin de la Société Française de Minéralogie 1935;58 4–5.

[28] Beudant FS. Azurite, in Traité Élémentaire de Minéralogie, 2nd ed., (Paris) 1832; 373–374.

[29] Sato E, Nakai I, Terada Y, Tsutsumi Y, Yokoyama K, Miyawaki R, Matsubara S. Beaverite-(Zn), $Pb(Fe_2Zn)(SO_4)_2(OH)$, a new member of the alunite group, from Mikawa Mine, Niigata Prefecture. Japan, Mineralogical Magazine 2011;75 375–377.

[30] Butler BS, Schaller WT. Beaverite, a new mineral. Journal of the Washington Academy of Sciences 1911;1 26–27.

[31] Bayliss P, Kolitsch U, Nickel EH, Pring A, Alunite supergroup: recommended nomenclature. Mineralogical Magazine 2010;74 919–927.

[32] Mineral known since antiquity under a number of names, but it appears to have first been called calcite in this publication: Freiesleben JC. Calcit, Magazin für die Oryktographie von Sachsen 1836;7 118–121.

[33] Süsse P. Verfeinerung der Kristallstruktur des Malachits, $Cu_2(OH)_2CO_3$. Acta Crystallographica 1967;22(1) 146–151. doi:10.1107/S0365110X67000222

[34] Sarp H, Birch WD, Hlava PF, Pring A, Sewell DKB, Nickel EH. Perroudite, a new sulfide-halide of Hg and Ag from Cap-Garonne, Var, France, and from Broken Hill, New South Wales, and Coppin Pool, Western Australia. American Mineralogist 1987;72 1251–1256.

[35] Dunn PJ, Rouse RC. Morelandite, a new barium arsenate chloride member of the apatite group. The Canadian Mineralogist 1978;16 601–604.

[36] Fleischer M, Pabst A, Mandarino JA. New mineral names, American Mineralogist 1980;65 205–210.

[37] Turner E. Chemical examination of the oxides of manganese. Part II. On the composition of the ores of manganese described by Mr. Haidinger. The Philosophical Magazine 1828;4 96–104.

[38] Ludwig CF. A. G. Werners Mineral-System, Erste Klasse Erdige Fossilien, VI. Kalk-Geschlecht, in Handbuch der Mineralogie nach A. G. Werner Volume 2, Siegfried Lebrecht Crusius (Leipzig), 1804, 209–212.

[39] Demartin F, Gramaccioli CM, Campostrini I. Demicheleite-(Cl), BiSCl, a new mineral from La Fossa crater, Vulcano, Aeolian Islands, Italy. American Mineralogist 2009;94 1045–1048.

[40] Beudant FS. Sulfures de bismuth. Bismuthine, in Traité Élémentaire de Minéralogie, 2nd ed., (Paris), 1832; 418–421.

[41] Demartin F, Campostrini I, Gramaccioli CM. Panichiite, natural ammonium hexachlorostannate(IV), $(NH_4)_2SnCl_6$, from La Fossa crater, Vulcano, Aeolian Islands, Italy. The Canadian Mineralogist 2009;47 367–372.

[42] Websky M. Über caracolite und percylit, Sitzungsberichte der Königlich Preussischen Akademie der Wissenschaften 1886;48 1045–1050.

[43] Nickel EH, Nichols MC. Mineral Reference Manual. Springer Science & Business Media, 1991. ISBN: 978-0442003449

[44] Mellor JW. A Comprehensive Treatise on Inorganic and Theoretical Chemistry. Volume 7. Longmans, Green and Company, 1957.

[45] Tazzoli V. The crystal structure of cesanite $Ca_{1+x}Na_{4-x}(SO_4)_3(OH)_x{\cdot}(1-x)H_2O$, a sulphate isotypic to apatite. Mineralogical Magazine 1983;47 59–63.

[46] Piotrowski A , Kahlenberg V , Fischer RX , Lee Y , Parise JB. The crystal structures of cesanite and its synthetic analogue—A comparison. American Mineralogist 2002;87 715–720.

[47] Cavarretta G, Mottana A, Tecce F. Cesanite, $Ca_2Na_3[(OH)(SO_4)_3]$, a sulphate isotypic to apatite, from the Cesano geothermal field (Latium, Italy). Mineralogical Magazine 1981;44 269–273.

[48] Harada K, Nagashima N, Nakao K, Kato A. Hydroxylellestadite, a new apatite from Chichibu mine, Saitama prefecture, Japan. American Mineralogist 1971;56 1507–1518.

[49] Borodin LS, Kazakova ME. Belovite—a new mineral from an alkaline pegmatite, Doklady Akademii Nauk SSSR 1954;96 613–616.

[50] Pekov IV, Chukanov NV, Beletskaya OV, Khomyakov AP, Menshikov YP. Belovite-(Ce): New data, refined formula and relationship to other minerals of the apatite group. Zapiski Vserossijskogo Mineralogicheskogo Obshchestva 1995;124 98–110.

[51] Rakovan JF, Hughes JM. Strontium in the apatite structure: Strontian fluorapatite and belovite-(Ce). The Canadian Mineralogist 2000;38 839–845.

[52] Böggild OB. Ussingit, ein neues Mineral von Kangerdluarsuk. Zeitschrift für Krystallographie und Mineralogie 1915; 54 120–126.

[53] Klaproth MH. Chemische untersuchung des natroliths. Naturforschender Freunde zu Berlin, Neue Schriften 1803);4 243–248.

[54] Gerasimovsky VI. Chkalovite, Comptes Rendus (Doklady) de l'Académie des Sciences de l'URSS 1939;22 259–263.

[55] Boeggild OB. Epistolite, a new mineral. Meddelelser om Grønland 1901; 24 183–190.

[56] Fleischer M. New mineral names. American Mineralogist 1963;48 1178–1184.

[57] Fersman AE. The Chibina Massiv of Kola Island, Transactions of the Northern Scientific and Economic Expedition 1923;16, 16–73.

[58] Flink G. Om några mineral från Grönland, Geologiska Föreningens i Stockholm Förhandlingar 1893;15 195–208.

[59] Cannillo E, Mazzi F, Rossi G. The crystal structure of neptunite. Acta Crystallographica 1966;21 200–208.

[60] Sokolova E. From structure topology to chemical composition. I. Structural hierarchy and stereochemistry in titanium disilicate minerals. The Canadian Mineralogist 2006;44 1273–1330.

[61] Chao GY, Watkinson DH. Gaidonnayite, $Na_2ZrSi_3O_9{\cdot}2H_2O$, a new mineral from Mont St. Hilaire, Quebec, The Canadian Mineralogist 1974;12 316–319.

[62] Hackman V. Petrographische beschreibung des nephelinsyenites vom Umptek und einiger ihn begleitenden gesteine, Fennia 1894;11 101–196.

[63] Men'shikov YP, Sokolova EV, Egorov-Tismenko YK, Khomyakov AP, Polezhaeva LI. Sitinakite $Na_2KTi_4Si_2O_{13}(OH)\cdot 4H_2O$—a new mineral. Zapiski Vserossijskogo Mineralogicheskogo Obshchestva 1992;121(1) 94–99.

[64] Berzelius J. (Untitled note on aegirine), Neues Jahrbuch für Mineralogie, Geognosie, Geologie und Petrefaktenkunde 1835; 184–185.

[65] Thomson T. A chemical analysis of sodalite, a new mineral from Greenland, Transactions of the Royal Society of Edinburgh 1812;6 387–395.

[66] Breithaupt A. Ueber die Felsite und einige deue Specien ihres Geschlechts. Journal für Chemie und Physik 1830;60 316–330

[67] Bailey SW. Refinement of an intermediate microcline structure. American Mineralogist 1969;54 1540–1545.

[68] Haüy RJ. Népheline. Traité de minéralogie 1801;3 186–190.

[69] Pekov IV, Kulikova IM, Kabalov YK, Eletskaya OV, Chukanov NV, Menshikov YP, Khomyakov AP. Belovite-(La) $Sr_3Na(La,Ce)[PO_4]_3(F,OH)$—a new rare earth mineral in the apatite group. Zapiski Vserossijskogo Mineralogicheskogo Obshchestva 1996;125(3) 101–109.

[70] Semenov EI. Oxides and hydroxides of titanium and niobium in the Lovozero alkalic massif, Institute of Mineralogy, Geochemistry, Crystal Chemistry and Trace Elements. Akademiya Nauk CCCP, Trudy 1957;1 41–59.

[71] von Kobell F. Ueber den Pektolith, Archiv für die Gesammte Naturlehre 1828;13 385–393.

[72] Pekov IV, Zubkova NV, Husdal TA, Kononkova NN, Agakhanov AA, Zadov AE, Pushcharovsky DY. Carlgieseckeite-(Nd), $NaNdCa_3(PO_4)_3F$, a new belovite-group mineral species from the Ilímaussaq alkaline complex, South Greenland. Canadian Mineralogist 2012;50 571–580.

[73] Pekov IV, Zubkova NV, Husdal TA, Agakhanov AA, Zadov AE, Pushcharovsky, DY. Carlgieseckeite-(Nd), IMA 2010-036. CNMNC Newsletter, 2010, p. 901; Mineralogical Magazine 74 899–902.

[74] Gahn JG, Berzelius J. Underfökning af nagra i grannskapet af Fahlun funna fossilier, Afhandlingar i Fysik, Kemi och Mineralogi 1815;4 148–216.

[75] Haüy RJ. Analcime. Journal des Mines 1797;5 278–279.

[76] Khomyakov AP, Lisitsin DV, Kulikova IM, Rastsvetaeva RK. Deloneite-(Ce) $NaCa_2SrCe(PO_4)_3F$ – a new mineral with a belovite-like structure. Zapiski Vserossijskogo Mineralogicheskogo Obshchestva 1996;125(5) 83–94.

[77] Khomyakov AP, Kulikova IM, Rastsvetaeva RK. Fluorcaphite Ca(Sr,Na,Ca)(Ca,Sr, $Ce)_3(PO4)_3F$—a new mineral with the apatite structural motif. Zapiski Vserossijskogo Mineralogicheskogo Obshchestva 1997;126(3) 87–97.

[78] Chakhmouradian AR, Hughes JM, Rakovan R. Fluorocaphite a second occurrence and detailed structural analysis: simultaneous accommodation of Ca, Sr, Na, and LREE in the apatite atomic arrangement. The Canadian Mineralogist 2005;43(2) 735–746.

[79] Brooke HJ. A description of the crystalline form of some new minerals. Arfwedsonite, The Annals of Philosophy 1823;5 381–384.

[80] Miyashiro A. The chemistry, optics, and genesis of the alkali-amphiboles. Journal of Faculty of Science, University of Tokyo, Section II 1957;11 57–83.

[81] Oberti R, Boiocchi M, Hawthorne FC, Ball NA, Harlow GE. Magnesio-arfvedsonite, IMA 2013-137. CNMNC Newsletter No. 20, June 2014, page 553; Mineralogical Magazine 2014;78 549–558.

[82] Flink G. On the minerals from Narsarsuk on the Firth of Tunugdliarfik in Southern Greenland, Meddelelser om Grønland 1901;24 9–180.

[83] Naumann KF. Lehrbuch der Mineralogie. Leipzig, Engelman, 1826.

[84] Zharikov VA, Pertsev NN, Rusinov VL, Callegari E, Fettes DJ. Metasomatism and metasomatic rocks. A systematic nomenclature for metamorphic rocks: 9. Metasomatic rocks. Recommendations by the IUGS Subcommission on the Systematics of Metamorphic Rocks. Recommendations, web version of 01.02.2007.

[85] Burke EAJ. Tidying up mineral names: an IMACNMNC scheme for suffixes, hyphens and diacritical marks. Mineralogical Record 2008;39: 131–135.

[86] Efimov AF, Kravchenko SM, Vasil'eva ZV. Strontium-apatite—a new mineral, Doklady Akademii Nauk SSSR 1962;142 439–442.

[87] Pekov IV, Britvin SN, Zubkova NV, Pushcharovsky DY, Pasero M, Merlino S. Stronadelphite, $Sr_5(PO_4)_3F$, a new apatite-group mineral. European Journal of Mineralogy 2010;22(6) 869–874.

[88] Friis H, Balić-Žunić T, Pekov IV, Petersen OV. Kuannersuite-(Ce), Ba6Na2$REE_2(PO_4)_6$FCl, a new member of the apatite group, from the Ilímaussaq alkaline complex, South Greenland: description and crystal chemistry. The Canadian Mineralogist 2004;42 95–106.

[89] Kuzmenko MV. Beryllite—a new mineral. Doklady Akademii Nauk SSSR 1954; 99 451–454.

[90] Passaglia E, Pongiluppi D, Vezzalini G. The crystal chemistry of gmelinites. Neues Jahrbuch für Mineralogie, Monatshefte 1978; 310–324.

[91] Coombs DS, Alberti A, Armbruster T, Artioli G, Colella C, Galli E, Grice JD, Liebau F, Mandarino JA, Minato H, Nickel EH, Passaglia E, Peacor DR, Quartieri S, Rinaldi R, Ross M, Sheppard RA, Tillmanns E, Vezzalini G. Recommended nomenclature for zeolite minerals: report of the Subcommittee on Zeolites of the International Mineralogical Association, Commission on New Minerals and Mineral Names, The Canadian Mineralogist 1997;35 1571–1606.

[92] Khomyakov AP, Polezhaeva LI, Malinovsky YA. Gmelinite-K $(K,Na,Ca)_6$ $[Al_7Si_{17}O_{48}]$ $\cdot 22H_2O$, a new zeolite mineral from Lovozero alkaline massif, Kola Peninsola, Russia. Zapiski Vserossijskogo Mineralogicheskogo Obshchestva 2001;130(3) 65–71.

[93] Brewster D. Description of gmelinite, a new mineral species. Edinburgh Journal of Science 1825;2 262–267.

[94] Lacroix A. Sur la gonnardite. Bulletin de la Société Française de Minéralogie 1896;19 426–429.

[95] Men'shikov YP, Denisov AP, Uspenskaya YI, Lipatova EA. Lovdarite, a new hydrous alkali-beryllium silcate. Doklady Akademii Nauk SSSR 1973;213 130–133.

[96] Fleischer M. New mineral names. American Mineralogist 1974;59 873–875.

[97] Petersen OV, Giester G, Brandstätter F, Niedermayr G. Nabesite, $Na_2BeSi_4O_{10}\cdot 4H_2O$, a new mineral species from the Ilímaussaq alkaline complex, South Greenland. The Canadian Mineralogist 2002;40 173–181.

[98] Jambor JL, Roberts AC. New mineral names, American Mineralogist 2003;88 251–255.

[99] Lorenzen J. Untersuchung einiger mineralien aus Kangerdluarsuk in Grönland, Zeitschrift für Krystallographie und Mineralogie 1884;9 243–254.

[100] Atencio D, Andrade MB, Christy AG, Gieré R, Kartashov P. The pyrochlore supergoroup of minerals: nomenclature. Canadian Mineralogist 2010;48 673–698.

[101] Glocker EF. Ordo IV. Cinnabaritae. IV. Cinnabaritae sphaleritoidei. 12. Sphalerites, in Generum et Specierum Mineralium, Secundum Ordines Naturales Digestorum Synopsis, Apud Eduardum Anton 1847; 13–18.

[102] Pekov IV, Pasero M, Yaskovskaya AN, Chukanov NV, Pushcharovsky DY, Merlino S, Zubkova NV, Kononkova NN, Men'shikov YP, Zadov AE. Fluorcalciobritholite, $(Ca,REE)_5[(Si,P)O_4]_3F$, a new mineral: description and crystal chemistry. European Journal of Mineralogy 2007;19 95–103.

[103] Njema H, Debbichi M, Boughzala K, Said M, Bouzouita K. Structural, electronic and thermodynamic properties of britholites $Ca_{10-x}La_x(PO_4)_{6-x}(SiO_4)_xF_2$ $(0 \leq x \leq 6)$: Experiment and theory. Materials Research Bulletin 2014;51 210–216.

[104] Winther C. Britholite, a new mineral. Meddelelser om Grønland 1901;24 190–196.

[105] Nickel EH, Mandarino JA. Procedures involving the IMA Commission on New Minerals and Mineral Names and guidelines on mineral nomenclature. American Mineralogist 1987; 72 1031–1042.

[106] Noe DC, Hughes JM, Mariano AN, Drexler JW, Kato A. The crystal structure of monoclinic britholite-(Ce) and britholite-(Y). Zeitschrift für Kristallographie 1993;206 233–246.

[107] Peishan Z, Xueming Y, Kejie T. Mineralogy And Geology of Rare Earths in China. Series of solid earth sciences research in China, VSP, 1996. ISBN: 978-9067642200

[108] Pekov IV, Zubkova NV, Chukanov NV, Husdal TA, Zadov AE, Pushcharovsky DY. Fluorbritholite-(Y), $(Y,Ca,Ln)_5[(Si,P)O_4]_3F$, a new mineral of the britholite group. Neues Jahrbuch für Mineralogie, Abhandlungen 2011;188 191–197.

[109] Hata S. Abukumalite, a new yttrium mineral. Scientific Papers of the Institute of Physical and Chemical Research 1938;34 1018–1023.

[110] Gu J, Chao GY, Tang S. A new mineral—fluorbritholite-(Ce). Journal of Wuhan University of Technology 1994;9(3) 9–14.

[111] Jones AP, Wall F, Williams CT. Rare Earth Minerals: Chemistry, Origin and Ore Deposits. Mineralogical Society of Great Britain and Ireland: The Mineralogical Society series, Volume 7. Springer Science & Business Media, 1996. ISBN: 978-0412610301

[112] Grice JD, Hawthorne FC. Refinement of the crystal structure of leucophanite. The Canadian Mineralogist 1989;27 193–197.

[113] Jaffe HW, Molinski VJ. Spencite, the yttrium analogue of tritomite from Sussex County, New Jersey. American Mineralogist 1962; 47 9–25.

[114] Henderson P. Rare Earth Element Geochemistry (Developments in Geochemistry). Elsevier, 2013. ISBN: 978-1483289779.

[115] Frondel C. Two yttrium minerals: spencite and rowlandite. The Canadian Mineralogist 1961; 6 576–581.

[116] Bayliss P, Levinson AA. A system of nomenclature for rare-earth mineral species: revision and extension. American Mineralogist 1988;73 422–423.

[117] McConnell D. The substitution of SiO_4- and SO_4-groups for PO_4-groups in the apatite structure; ellestadite, the end-member. American Mineralogist 1937;22 977–986.

[118] Ciesielszuk J. Ellestadite-(F) – a mineral formed in the overburned coal dump (Upper Silesian Coal Basin). Mineralogia – Special Papers 2008;32 54.

[119] Jambor JL. New mineral names. American Mineralogist 1989;74 500–505.

[120] Chesnokov BV, Bazhenova LF, Bushmakin AF. Fluorellestadite $Ca_{10}[(SO_4),(SiO_4)]_6F_2$—a new mineral, Zapiski Vsesoyuznogo. Mineralogicheskogo Obshchestva 1987;116(6) 743–746.

[121] Allen NK. Limestone and Other Sedimentary Rocks. The Rosen Publishing Group, 2009. ISBN: 978-1435827592

[122] Onac BP, Effenberger H, Ettinger K, Panzaru SC. Hydroxylellestadite from Cioclovina Cave (Romania): microanalytical, structural, and vibrational spectroscopy data. American Mineralogist 2006;91 1927–1931.

[123] Marincea S, Dumitraş Delia-G, Călin N, Anason AM, Fransolet AM, Hatert F. Spurrite, tilleyite and associated minerals in the exoskarn zone from cornet hill (Metaliferi Massif, Apuseni Mountains, Romania). The Canadian Mineralogist 2013;51 359–375.

[124] Pascal Marie-L, Fonteilles M, Verkaeren J, Piret R, Marincea Ş. The melilite-bearing high-temperature skarns of the Apuseni Mountains, Carpathians, Romania. Canadian Mineralogist 2001;39 1405–1434.

[125] Marincea, Ş, Bilal E, Verkaeren J, Pascal ML, Fonteilles M. Superposed parageneses in the spurrite-, tilleyite-, and gehlenite-bearing skarns from Cornet Hill, Apuseni Mountains, Romania. Canadian Mineralogist 2001;39 1435–1453.

[126] Muraoka Y, Kihara K. The temperature dependence of the crystal structure of berlinite, a quartz-type form of $AlPO_4$, Sample: T = 25°C. Physics and Chemistry of Minerals 1997;24 243–253.

[127] Rouse RC, Dunn PJ. A contribution to the crystal chemistry of ellestadite and the silicate sulfate apatites. American Mineralogist1982;67 90–96.

[128] Saint-Jean SJ, Hansen S. Nonstoichiometry in chlorellestadite, locality: synthetic, sample: nonstoichiometric. Solid State Sciences 2005;7 97–102.

[129] Fitzgerald S, Rheingold AL, Leavens PB. Crystal structure of Cu-bearing vesuvianite. American Mineralogist1986;71 1011–1014.

[130] Merlino S. Okenite, $Ca_{10}Si_{18}O_{46}{\cdot}18H_2O$: the first example of a chain and sheet silicate. American Mineralogist 1983;68 614–622.

[131] Livingstone A, Ryback G, Fejer EE, Stanley CJ. Mattheddleite, a new mineral of the apatite group from Leadhills, Strathclyde region. Scottish Journal of Geology 1987;23 1–8.

[132] Jambor JL, Bladh KW, Ercit TS, Grice JD, Grew ES. New mineral names, American Mineralogist 1988;73 927–935.

[133] Essene EJ, Henderson CE, Livingstone A. The missing sulphur in mattheddleite, sulphur analysis of sulphates, and paragenetic relations at Leadhills, Scotland. The Mineralogical Society 2006;70(3) 265–280.

[134] Steele M, Pluth JJ, Livingstone A. Crystal structure of mattheddleite: a Pb, S, Si phase with the apatite structure. Mineralogical Magazine 2000;64(5) 915–921.

[135] Tait K, Hawthorne FC, Ball N, Abdu Y. (2014) Pieczkaite, IMA 2014-005. CNMNC Newsletter No. 20, June 2014, page 555; Mineralogical Magazine 2014;78 549–558.

[136] Engel G, Pretzsch J, Gramlich V, Baur WH. The crystal structure of hydrothermally grown manganese chlorapatite, $Mn_5(PO_4)_3Cl_{0.9}(OH)_{0.1}$. Acta Crystallographica 1975;B31 1854–1860.

[137] Teodorovich GI. Authigenic Minerals in Sedimentary Rocks. Springer Science & Business Media, 2012. ISBN: 978-1468406528

[138] Abouzeid Abdel-Z.M. Physical and thermal treatment of phosphate ores—An overview. International Journal of Mineral Processing 2008;85(4) 59–84.

[139] Greenwood NN, Earnshaw A. Chemistry of the Elements. 2nd ed., Elsevier, 2012. ISBN: 978-0080501093

[140] Valsami-Jones E. Phosphorus in Environmental Technologies: Principles and Applications: Principles and Applications (Integrated Environmental Technology). 1st ed., London: IWA Publishing, 2004. ISBN 1-84339-001-9.

[141] Chirvinsky PN. Materials for the knowledge of the natural productive forces of Russia. Published in commission for the Russian Acad. Sci., Petrograd: 1919;30 52 p.

[142] Notholt AJG, Sheldon RP, Davidson DF. Phosphate Deposits of the World: Volume 2, Phosphate Rock Resources. Cambridge Earth Science Series. Revised edition, Cambridge University Press, 2005. ISBN: 978-0521673334

[143] Bishop AC, Woolley AR, Woolley WRH. Cambridge Guide to Minerals, Rocks and Fossils. 2nd ed., Cambridge: Cambridge University Press, 1999. ISBN 0-521-77881-6

[144] Fleet ME, Liu X, Penelope LK. Accommodation of the carbonate ion in apatite: An FTIR and X-ray structure study of crystals synthesized at 2–4 GPa. American Mineralogist 2004;89 1422–1432.

[145] O'Brien GW, Milnes AR, Veeh HH, Heggie DT, Riggs SR, Cullen DJ, Marshall JF, Cook PJ. Sedimentation dynamics and redox iron-cycling: controlling factors for the apatite-glauconite association on the East Australian continental margin. Book Chapter in: In: Notholt, A J G and Jarvis, I (eds.). Phosphorite Research and Development. Geological Society of Australia. Special publication. Geological Society of Australia, 1990.

[146] Deer WA. Rock-Forming Minerals: Non-silicates, volume 5B, second edition. Geological Society of London, 1998. ISBN: 978-1897799901

Chapter 3

Identification, Characterization and Properties of Apatites

Petr Ptáček

Additional information is available at the end of the chapter

http://dx.doi.org/10.5772/62211

Abstract

The parameters of unit cell, structure, refractive index, solubility data, PO_4/CO_3 ratio, surface area, etc., are important parameters for characterization of phosphate rocks. Third chapter of this book introduces methods for identification, characterization and properties of apatites in four main sections. The first part describes techniques used for identification and investigation of properties of phosphate minerals, including X-ray diffraction analysis, powder neutron diffraction, X-ray fluorescence as well as spectroscopic and microscopic methods. Some of these techniques are then demonstrated on the fluorapatite specimen in the second part. The third part of this chapter deals with thermodynamic properties of apatite-type compounds and introduces some of thermodynamic predictive methods. The fourth part is dedicated to dissolution of apatite, where the reaction between solids and liquids according to different dissolution models is described. Chapter ends with methods for the evaluation of reactivity of phosphate rocks.

Keywords: Apatite, Carbonate to Phosphate Ratio, Dissolution of Apatite, Reactivity of Apatite, Citrate Solubility

3.1 Techniques used for identification of phosphate minerals

There are several parameters which are used in characterization of phosphate rocks, including the following [1],[2]:

a. Unit-cell parameters

b. Refractive index

c. Solubility data based on chemical extraction methods, e.g. neutral ammonium citrate solubility (NAC, **Section 3.4.3**)

d. PO_4/CO_3 ratio as a measure of carbonate substitution in phosphate minerals

e. Surface area and pore size distribution indicating the potential reactivity

The most important techniques used for the identification and characterization of phosphate minerals include methods for identification of phase composition, chemical composition, structure, surface properties, etc. A few often applied methods are introduced in this chapter [1],[2].

3.1.1. X-ray diffraction analysis

The diffraction of a beam of X-rays by a crystalline material is the process of beam scattering[1] by electrons associated with atoms in the crystal and of the interference of these scattered X-rays because of the periodic arrangement of atoms in the crystal and its symmetry.[2] X-ray diffraction analysis (XRD) is used for the determination of mineralogical composition and quantitative X-ray diffraction analysis (Rietveld method) for the refinements of structure of apatite from measured data using specialized software [3],[4],[5],[6].

Following the discovery of X-rays by Röentgen[3] [8],[10],[11] in 1895 and the proof that X-rays have the wave properties and diffract from a periodic atomic array by von Laue [12] and his students in 1912–1913, the analytical application of X-ray diffraction has developed slowly over the next 20 years. Most of the earliness efforts were aimed at the solution of crystal structures of common phases. Debye and Sheerer (1916) and Hull (1917) suggested that powder diffraction patterns could be used for the identification of quantification of crystalline compounds. However, because most of the early developments were directed toward solving single-crystal structures, it was really the middle 1930s when the powder diffraction method began to attract the follower with the publication of the procedure of Hanawalt and Rinn and the database of patterns by Hanawalt, Rinn and Frevel (1938). With the conversion of data sets into the first set of the Powder Diffraction File in 1941, the phase identification applications expanded, and the modern counter diffractometer was developed by Parrish, Hamacher and Lowitzsch [5].

The phase identification was one of the first applications to grow to useful level. Other major applications of diffraction analysis include following phase changes under nonambient conditions and atmospheres. The first diffraction experiments were actually done on single crystals. The method is primarily directed toward determining the crystal periodicity and symmetry and solving the arrangement of atoms in the material because this information is

---

[1] Scattering is the process where the beam of radiation or particles is deviated from its initial trajectory by the inhomogeneity in the medium which it transverses [7].

[2] Other kinds of radiation commonly used for diffraction analysis are neutrons (**Section 3.1.2**) and electrons (**Chapter 3.1.10**).

[3] Wilhelm Conrad Röentgen (1845–1923) was the rector of the University of Würzburg [8]. The first X-ray photography was published in 1896 [9].

http://dx.doi.org/10.5772/62211

difficult to obtain from powder experiment. Powder diffraction is one of the most important material characterization techniques in the material research and industry [5].

Single-crystal diffraction studies are not limited to crystal structure analysis. The diffraction topography is a large field that has provided much information on the perfection of crystals used in industry as integral parts of devices. The examples include crystals used in sensing and control devices, substrates for electronic components, tools and dies, turbine blades and many other applications [5].

3.1.2. Powder neutron diffraction

Neutron powder diffraction (PND) or elastic neutron scattering enables to determine nuclear and magnetic structure of solids. Most of the information on the nature of ordered magnetic phases or magnetic structures comes from neutron diffraction experiments. Neutrons have no electric charge and interact with the nuclei rather than with the charge distribution of atoms in matter. They have the wavelength in the range of interatomic distances. They have magnetic moment and interact with the magnetic moment of atoms in matter. The mass of neutrons is similar to that of atomic nuclei; hence, they have energy and momentum similar to those of atoms in solid and fluid materials [13],[14]. The first neutron diffraction experiments were performed in 1945 by EO Wollan in the graphite reactor at Oak Ridge National Laboratory, USA [15].

Neutron scattering (NS) results from the interactions with atomic nuclei, i.e. overscattering lengths (distances) of the order of 10^{-15} m (1 fm).[4] Although scattering amplitude decreases greatly with the scattering vector (it is inverse to the scattering length), there are insignificant variations of scattering amplitude in the same range of scattering vector for neutrons. Consequently, powder diffraction with neutrons can resolve very fine structural and textural details of complex atomic structures. Moreover, the weak interaction of neutrons with matter results in very low attenuation offering a unique advantage for nondestructive, in situ work and bulk analysis (for polycrystalline materials, no crushing is required to obtain the patterns) [14].

Neutron diffraction was used to determine atomic arrangement in material. Inelastic neutron scattering measures the vibrations of atoms and small-angle neutron scattering[5] (SANS) is used to study larger structures such as polymers and colloids. The technique of surface reflection (reflectometry) was used to study layered materials. The technique of SANS provides the information about the size, shape and domain orientations; conformational changes and/or flexibility; and molecular associations is solution. For the structural studies, the elastic scattering effects, where there is no energy exchange between the radiation and atoms, are exploited [7],[14],[16].

Neutron powder diffraction is a method often used for the structure refinement of apatite or apatite type compounds from measured data using specialized software [17],[18],[19],[20],

[4] The interaction with electrons during X-ray analysis takes place over the distances of 10^{-10} m (1 Å) [14].

[5] Small angle scattering (SAS) of X-rays is abbreviated as SAXS [7].

[21],[22],[23],[24] and the effect of substitutions in the apatite structure [25]. Neutron diffraction data enable to explain the oxygen over-stoichiometry in the structure of $La_{9.67}(SiO_4)_6O_{2.5}$ apatite [17]. Since neutrons make possible the accurate determination of the thermal factors and provide the visualization of the diffusion paths in ionic conductors, powder neutron diffraction is also used for the characterization of solid oxide fuel cell materials [26]. This method is also used to investigate apatite in hard tissues where it provides the evidence about the deficiency of hydroxyl ion in bone apatites [27] and reconfirms that the inorganic portion is basically a hydroxylapatite-like material [27],[28].

3.1.3. X-ray fluorescence analysis and total-reflection X-ray fluorescence analysis

X-ray fluorescent analysis (XRF) is a method for the determination of sample composition [29],[30]. The origin of characteristic X-ray spectra can be described as follows. When sufficient energy is introduced into the atom, the electrons may be knocked out of one the inner shells. The atom is then in an excited (ionized) state and returns to the ground state within 10^{-8} s. The place of the missing electron is filled by an electron from a neighboring other shell, the place of which, in turn, is filled by an electron from more outer shell. The atom then returns to the ground state in steps. In every step, i.e. in every electron jump, the electron from a higher energy level goes into a lower energy level emitting excess energy in the form of an X-ray quantum. The energy of emitted radiation is characteristic for the atomic number of emitting element as well as for particular electron transition taking place within the electron shell of the atom. By measuring the energy or the wavelength of emitted radiation, the particle element can be identified unambiguously [31].

The energy that is necessary for the atom to get to excited state can be introduced either by the collision with a high-energy electron (sample is bombarded by electrons which are accelerated by high-voltage) or by the absorption of an energy-rich photon, i.e. the X-ray quantum (sample is irradiated by X-ray or gamma rays). In modern X-ray fluorescence analysis, the sample is irradiated by polychromatic radiation from an X-ray tube. In analogy to the optical case, this technique is referred to as fluorescence, which is responsible for the name X-ray fluorescence analysis as the technique of spectrochemical analysis with X-rays [31].

There are two types of instruments (**Fig. 1**) used for X-ray fluorescence spectrometry [32],[33]:

1. **Wavelength-dispersive XRF (WDXRF) or total reflection XRF (TRXRF):** the method is also often abbreviated as XRF. X-rays impinge on the sample (**Fig. 1**(**a**)) and generate fluorescent X-rays. These are then diffracted on a crystal. A goniometer selects the geometry between the crystal and detector that controls the detection of X-ray from the element of interest. Different crystals have different sensitivities. Many of commercial WDXRF instruments have two detectors and up to six crystals to optimize the conditions for each element.

2. **Energy-dispersive XRF (EDXRF):** the method is also abbreviated as EDX. The EDXRF instruments use much less energetic X-ray tube. Emitted X-ray radiation from the sample impinges directly on a detector, typically Si(Li), which generates pulses on an incident beam. These pulses are sorted and counted by a multichannel analyzer (**Fig. 1**(**b**)).

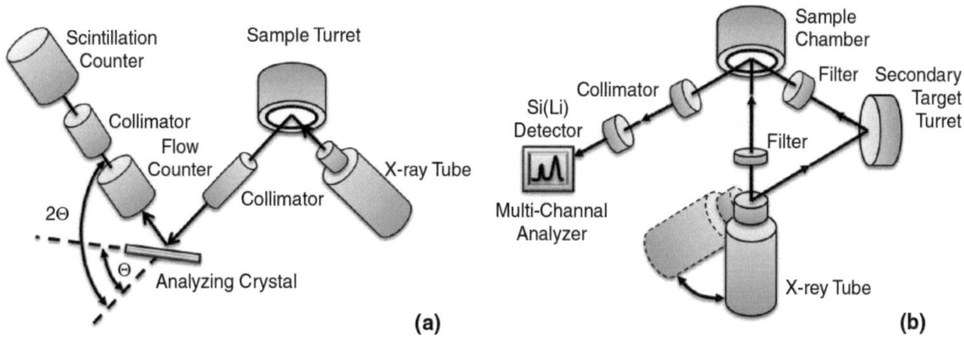

Fig. 1. Schematic representation of X-ray fluorescence analyzer: (a) wavelength-dispersive (XRF) and (b) energy dispersive (XRF).

Simultaneous determination for all elements, the atomic number of which is greater than Mg is possible.

The resolution and sensitivity of EDXRF is typically an order of magnitude worse than that for WDXRF.

Synchrotron radiation X-ray fluorescence (SRXRF) microprobe, a promising technique, is a nondestructive and qualitative to semiquantitative analysis of minerals and single fluid inclusions [34]. Synchrotron radiation (SR) is a powerful advanced light source (synchrotron radiation source, SRS) compared to conventional X-ray tube radiation and has many unique properties, such as high intensity, natural collimations, well-defined polarization, wide spectral range and energy tenability [35]. SRXRF is a widely applied technique for microscopic analysis of chemical elements. High-resolution requirements can be achieved using microbeam synchrotron radiation X-ray fluorescence (μ-SRXRF). Synchrotron radiation X-ray fluorescence can also provide the information about the oxidation state and coordination environment of metals using techniques known as X-ray absorption of near-edge structure (XANEX) or by micro-XANEX spectroscopy [37]. The unique tool for studying, the local structure around selected elements is X-ray absorption fine structure (XAFS) [38].

X-ray fluorescence is usually used to investigate the composition of apatite rocks and minerals for the purpose of their characterization [36],[39],[40], estimation of naturally occurring radionuclides in fertilizers [41] and analysis of phosphate ore at various stage of processing [42], e.g. flotation [43],[44].

3.1.4. Inductively coupled plasma spectrometry

Prior to inductively coupled plasma[6] (ICP), the flame, direct current-arc and controlled-waveform spark were used for the atomization (i.e. decomposition of sample to individual

[6] Plasma is defined as an electrically neutral gas which consists of positive ions and free electrons. Plasma have sufficiently high energy to atomize, ionize, and excite virtually all elements in the periodic table, which are intentionally introduced into it for the purpose of elemental chemical analysis [45].

atoms) and excitation of sample in elemental analysis. ICP denotes the technique that uses atmospheric pressure argon inductively coupled plasma[7] (ICP) for the atomization and excitation of sample. This plasma is a highly energetic media consisting of inert ionized gas with equivalent temperatures from 7000 to 10,000 K. Inductively coupled plasma are formed by coupling energy produced by RF generator (typically 700–1500 W) to the plasma support gas with an electromagnetic fields [45]. The cross-section of typical ICP torch is shown in **Fig. 2**.

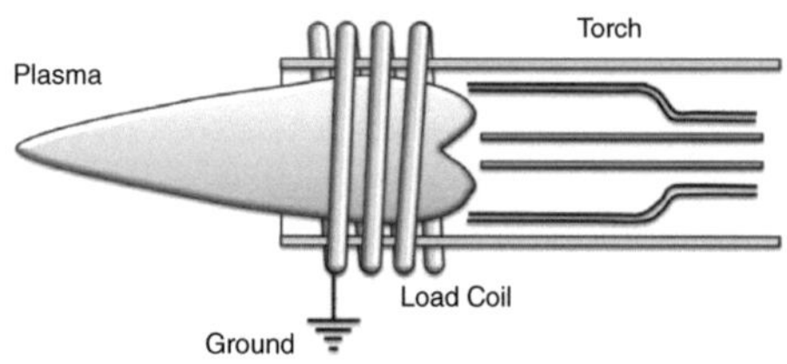

Fig. 2. Plasma with torch assembly and load coil [45].

The treatment of sample before the quantification includes vaporization, atomization, excitation and ionization.[8] The introduction of analyzed sample into inductively coupled plasma was applied in analytical techniques including [29],[45],[46],[47],[48]:

i. **Inductively coupled plasma atomic emission spectroscopy** (ICP-AES): the method connects atomic emission spectrometry with ICP. AES is based on spontaneous emission of free atoms or ions when the excitation is performed by thermal or electric energy. The method can identify and determine the concentration of up to 40 elements simultaneously with the detection limit of parts per billion (ppb). Serious limitations of this technique are the spectral interferences. Despite rapid growth of ICP-MS, ICP-AES still plays a dominant role in elemental analysis of geological, environmental, biological and other materials.

ii. **Inductively coupled plasma atomic fluorescence spectroscopy** (ICP-AFS): ASF is an analytical technique for the determination of elements in small quantities. It is based on the emission of free atoms when the excitation is performed by radiation energy.

iii. **Inductively coupled plasma-mass spectroscopy** (ICP-MS): is analytical technique for the determination of elemental composition of virtually any material. A sample, usually in the form of an aqueous solution, is converted to an aerosol by a neutralization process and transported to the plasma by an argon gas stream. In the plasma, the elements of analyte are atomized, followed immediately by ionization. The composition of ion population in the plasma is proportional to the concentration of

[7] Although there are many types of plasma, such as direct current, microwave induced, etc., the ICP is considered the most useful technique for analytical spectroscopy [45].

[8] Each element has characteristics first and second ionization potential, which depends on specific electronic structure of given element. Higher ionization potential means that more externally applied energy is required for ionization (thermal radiation, collision with other ion or electron, or exposure to high-energy photons) [45].

analyte species in original sample solution. Ions produced by ICP are representatively sampled and extracted from the plasma; next they are separated and measured by a quadrupole or time-of-flight mass spectrometer

The method known as laser ablation—inductively coupled plasma—mass spectrometry (LA-ICP-MS) is a coupling technique of laser ablation with ICP-MS technique [49],[50],[51],[52],[53]. Multicollector inductively coupled plasma mass spectrometry (MC-ICP-MS) was applied as the benchmark method for isotopic analysis [54] and for the determination of heavy rare earth elements in apatites [55],[56],[57].

3.1.5. Thermal ionization mass spectroscopy

Thermal ionization mass spectroscopy (TIMS) is highly specialized technique of mass spectroscopy used for very precise determination of isotope ratios and, as such, is widely used for the determination of stable isotope ratios in isotope geology and for the analysis of nuclear materials [58]. In this method, the solid sample is thermally ionized in solid-source mass spectrometer. Ions are accelerated into the mass analyzer and then transported to the detector [59]. Multicollector thermal ionization mass spectrometry (MC-TIMS) uses laminated magnetic sector filet for high speed peak jumping and low hysteresis for the mass of ion beams [60],[61].

The basic variants of the method of thermal ionization mass spectroscopy are as follows:

1. Isotope dilution–thermal ionization mass spectrometry method (ID-TIMS) [62],[63] is used for accurate determination of element concentration and is generally considered to be the definitive method to other techniques [58]. The method was applied for direct measurement of uranium, thorium, lead, etc., concentrations in the determination of single grain fission-track ages (**Section 7.3.3**).
2. Chemical abrasion–thermal ionization mass spectrometry method (CA-TIMS)

The ability for precise determination of isotope ratio in apatite predetermines this method for geochronological investigations, i.e. studying of chronologic records in accessory minerals of igneous rocks [61],[63],[64],[65].

3.1.6. Secondary ion mass spectrometry

Secondary ion mass spectroscopy (SIMS) is a method[9] for the characterization of solid surface elemental composition and isotope distribution. The technique can be applied to all elements and allows quantitative analysis of solid surfaces, including monolayers. Energetic ion bombardment of a solid surface (primary ions, e.g. Ar^+, Cs^+, O^{2+}, …) causes that atoms of the sample are shifted from their original states as positive and negative ions which are termed as secondary ions. These ions are then analyzed by mass spectrometer (e.g. quadrupole mass spectrometer) to determine the composition of the surface of sample [66],[67]. The applica-

[9] The method was originally developed in the 1950s and 1960s by Herzog et al and Honig et al to analyze metals and oxides [69], but the basis of SIMS can be traced back to the beginning of the twentieth century with the first experimental evidence of secondary ions given by JJ Thomson in 1910 [68].

tions of SIMS can be broadly subdivided into static and dynamic SIMS. Static SIMS (SSIMS) is used to investigate the composition of the outermost monolayer on any solid. Dynamic SIMS (DSIMS) examines the concentration profile as the function of depth [68],[69],[70].

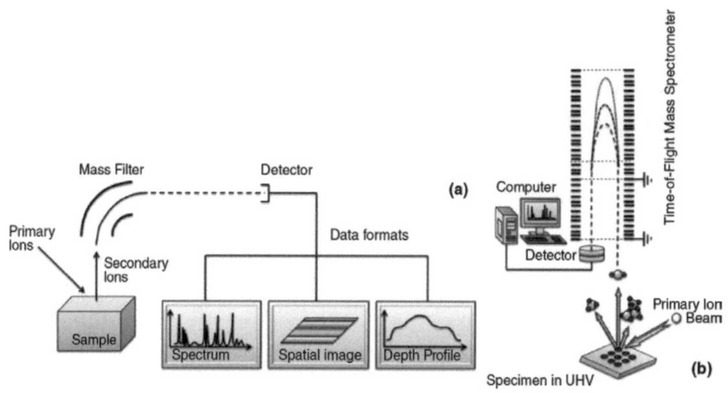

Fig. 3. Main components of SIMS [68] and schematic representation of ion microprobe imaging time-of-flight secondary ion mass spectrometry with reflectron-based mass analyser [69].

The mass filter of mass spectrometer of SIMS instruments defines the type of instruments [68], [70]:

1. Magnetic sector SIMS (M-SIMS) instruments (**Fig. 3**(**b**), similar to those used in original mass spectrometer)
2. Quadrupole SIMS (Q-SIMS) instruments (**Fig. 4**(**a**), first appeared in the 1970)
3. Time-of-flight SIMS (ToF-SIMS) instruments (**Fig. 4**(**b**), first appeared in the 1980)

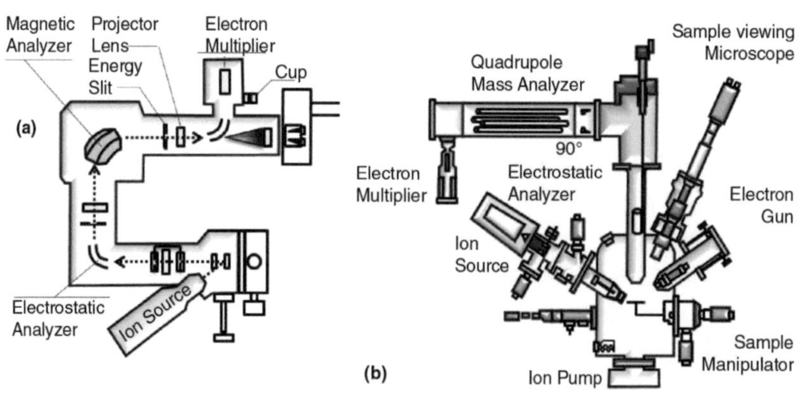

Fig. 4. Schematic representation of magnetic sector and quadrupole SIMS instruments [70].

Secondary ion emission begins when the primary ions energy exceeds some threshold level. This energy ranges from 30 to 80 eV and is much higher than the energies for sputtering of neutral particles. The most important features of the method are very low sensitivity limit for majority of elements, the possibility to determine the concentration profile, the possibility of isotope analysis and the identification of all the elements and isotopes starting from hydrogen [66].

The SIMS method is often used for the measurement of deuterium/hydrogen (D/H) ratio and hydroxyl (OH) content in anhydrous minerals and melt inclusions in Martian meteorites [71] and lunar materials (apatite grains, glass beds, melt inclusions and agglutinates in soils) [72]. The method is widely used to investigate biological apatites and collagen apatite composites [73] and the analysis of micrometer-sized samples like, e.g. interplanetary dust, presolar grains and small inclusions in meteorites, has become more and more important in cosmochemistry [74]. The method is also utilized for U-Th-Pb dating of apatites as common accessory in igneous rocks (**Chapter 8**), based on the radioactive decay of U and Th [75],[76],[77].

3.1.7. Laser secondary neutral mass spectrometry

Laser secondary neutral mass spectroscopy (laser-SNMS) can be further divided to nonresonant laser-SNMS (NR-laser-SNMS) and resonant laser-SNMS (R-laser-SNMS). In NR-laser-SNMS (**Fig. 5(b)**), an intense laser beam is used to nonselective ionization of all atoms and molecules within the volume intersected by the laser beam. Sufficient laser power density, which is necessary to saturate the ionization process, is typically achieved in a small volume. It limits the sensitivity of the method and leads to the problems with quantification due to the differences between effective ionization volumes of different elements. Laser-SNMS method has significantly improved ionization efficiency over SIMS (**a**) [78],[79].

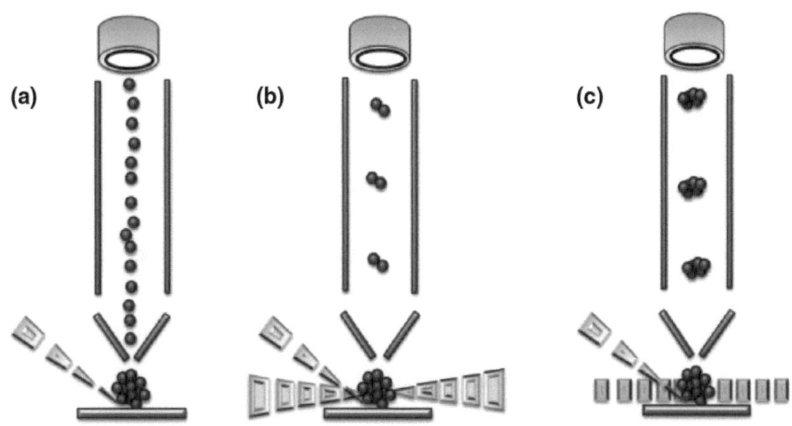

Fig. 5. Comparison of ToF-SIMS and laser-SNMS: (a) direct analysis of secondary ions (ToF-SIMS), (b) nonresonant laser-post-ionization of secondary neutrals (NR-laser-SNMS) and (c) resonant post-ionization of secondary neutrals (R-laser-SNMS) [78].

Resonant laser-SNMS uses a resonance laser ionization process, which selectively and efficiently ionizes atoms and molecules over a relatively large volume (**Fig. 5(c)**). The method has unit ionization efficiency for over 80% of elements in periodic table, i.e. the overall efficiency is greater than that for NR-laser-SNMS, the quantification is also simpler and extremely high selectivity prevents almost all isobaric and molecular interferences [78],[79].

The method of NR-laser-SNMS was used by Dambach et al [80] to investigate different states of biomineralization in vitro. The results indicate that in the vicinity of single osteoblasts, extracellular enrichment of potassium typically occurs during initial stages of mineralization. Potassium may interact with matrix macromolecules and prevent an uncontrolled apatite deposition. However, apatite biomineral formation is correlated with a potassium release. In conclusion, potassium seems to be involved in the process of extracellular matrix biomineralization.

3.1.8. Electron paramagnetic resonance

The concept of electron paramagnetic resonance (EPR) spectroscopy is very similar to more familiar nuclear magnetic resonance (NMR). Both methods deal with the interaction between electromagnetic radiation and magnetic moments. In the case of EPR, the magnetic moments arise from electron rather than nuclei. The term EPR was introduced as a designation taking into account contributions from electron orbital as well as spin angular momentum. The term electron spin resonance (ESR) was also widely used because in most cases the absorption is linked primarily to the electron-spin angular momentum [82],[83]. EPR spectrum is a diagram in which the absorption of microwave frequency radiation is plotted against the magnetic field intensity [83].

The technique of electron paramagnetic resonance spectroscopy may be regarded as the consequence of the Stern–Gerlach experiment. They showed (in 1920) that an electron magnetic moment in an atom can take on only discrete orientation in a magnetic field, despite the sphericity of the atom. Subsequently, Uhlenbeck and Goudsmit liked the electron magnetic moment with the concept of electron spin angular momentum. In hydrogen atom, there is additional angular momentum arising from the proton nucleus. Breit and Rabi described the resultant energy levels of hydrogen atom in a magnetic field. Rabi et al [81] studied the transition between levels induced by an oscillating magnetic field, and this experiment was the first observation of magnetic resonance. The first observation of electron paramagnetic resonance peak was made in 1945 by Zavoisky, who detected the radiofrequency absorption line from $CuCl_2 \cdot 2H_2O$ sample using the radiofrequency (RF) source at 133 MHz [82].

The major components of EPR spectrometer are shown in **Fig. 6**. The microwave bridge supplies the microwaves at controlled frequency and power, which are transmitted to the sample cavity via the waveguide. The sample cavity is placed perpendicular to applied magnetic field, which can be varied in controlled way. In addition to this main magnetic field, a controlled but smaller oscillating magnetic field is superimposed on the cavity via the Zeeman modulation frequency. The ideal way to perform the experiment would be to apply a fixed magnetic field and vary the microwave frequency. However, microwave generators are only tunable over very limited ranges. Thus, the microwave frequency is fixed and applied magnetic field is varied. The magnetic field is applied until it reaches the value at which the sample will absorb some of the microwave energy, i.e. and EPR transition occurs [84],[85].

http://dx.doi.org/10.5772/62211

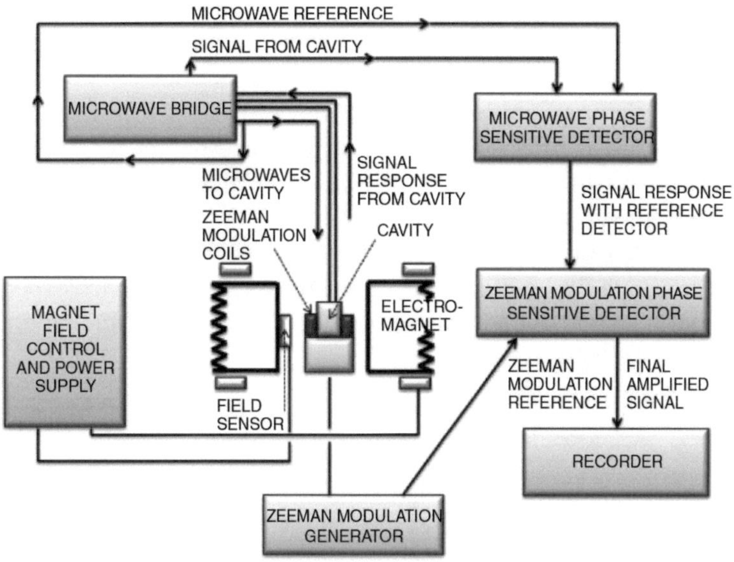

Fig. 6. The block diagram for typical continuous wave EPR spectrometer [84].

Electron paramagnetic resonance (EPR) spectrum of X-irradiated sodium and carbonate containing synthetic apatites has been studied by Moens et al [86]. Observed spectra were decomposed in terms of five theoretical curves representing O^- radical, two CO^{3-} radicals (surface and bulk) and two CO^{2-} radicals (surface and bulk). These species were also described in A-type and B-type carbonate-apatites [87],[88], tooth enamel [89],[90],[91],[92], [93] and bone [94],[95], apatites, renal stones [96], etc.

3.1.9. Nuclear magnetic resonance

Solid-state nuclear magnetic resonance (NMR) is a technique for accurate measurement of nuclear magnetic moments where the resonance frequency depends on its chemical environment [97],[98],[99]. The method can provide useful information on the number of molecules in the asymmetric unit and on the site symmetry of the molecule in the lattice to assist in the refinement of powder X-ray diffraction (**Section 3.1.1**) data. The method can distinguish between different polymorphs. Alternatively, solid-state NMR can be used for direct and accurate measurement of internuclear distances. For amorphous and disordered solids, such as inorganic glasses and organic polymers, solid-state NMR provides structural information that cannot be obtained by any other technique [100],[101]. NMR is also the diagnostic method used in veterinary science and medicine particularly in clinical research of human brain by magnetic resonance imaging (MRI) [102].

The solution-state NMR method was developed for the investigation of structure of soluble proteins [103]. Solution and solid-state NMR are both excellent methods for the determination of chemical composition [100].

The structural information of apatites is usually investigated from ^{1}H, ^{19}F and ^{31}P NMR spectra of apatites [104]. The ^{31}P solid-state NMR spectroscopy is a useful tool to investigate structural information about apatites on bone organic and inorganic mineral components, as well as to investigate the crystallinity and compositional changes in carbonated apatites [105]. Intact bone is a demanding tissue for structural studies. Serious experimental problems arise from the morphological diversity of bone and from the co-existence, interrelationship and great complexity of its organic and inorganic components. Furthermore, one has to perform noninvasive analysis because bone samples are very sensitive to physical effects and chemical treatment. Solid-state ^{31}P NMR gives us a unique opportunity to look specifically at the minerals of whole bone without any chemical pretreatment, thus avoiding the intervention into the bone structure [106].

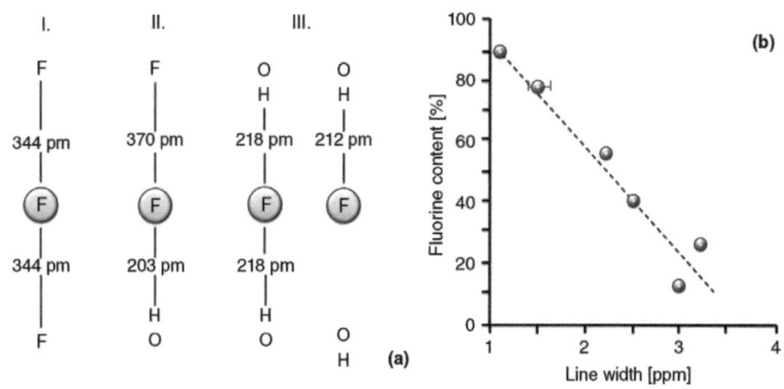

Fig. 7. Nuclear separation along parallel chains (the crystallographic c-axis) in various apatites (a): FFF group (I), FFH group (II) and HFH group (III). Correlation between observed 19F line width and fluorine content of fluorinated hydroxylapatite (b) [104].

The ^{19}F NMR spectrum of fluorinated calcium hydroxylapatite ($Ca_{10}(PO_4)_6F_{2x}(OH)_{2-2x}$, where x is the fraction of OH^- replaced by F^-) indicates the correlation between ^{19}F chemical shift tensor parameters and the content of fluorine in apatite. The presence of OH^- groups induces perturbations of fluorine environments, involving the displacements of both fluorine and hydroxyl groups from their normal positions. This leads to a distortion of the electronic environment with regard to the investigated fluorine nucleus and gives reasons for observed change in the ^{19}F chemical shift tensor of fluoridated hydroxylapatite with different fluorine content. Furthermore, the presence of OH-group destroys the fluoride long-range structure and that results in an isotropic chemical shift distribution. This leads to observed increase in the ^{19}F line width in the case of low fluorine content [104],[107].

3.1.10. Scanning electron microscopy, structure and elemental analysis

The scanning microscope (SEM) permits the observation and characterization of heterogeneous organic and inorganic materials on a nanometer (nm) to micrometer (μm) scale. In SEM, the area to be examined or the volume to be analyzed is irradiated with finely focused electron beam, which may be swept in a raster across the surface of the specimen to form an imager or

http://dx.doi.org/10.5772/62211

may be static to obtain the analysis at the position. The type of signals produced from the interaction of the electron beam (primary electron, PE) with the sample (**Fig. 8(a)**) includes secondary electrons (SE, with energy ≤50 eV), backscattered electrons (BSE, $E > 50$ eV), Auger electrons (AE), X-ray characteristics (X) and other photons of various energies such as continuum X-rays and heat. Low-loss electrons (LLE) show the energy losses of a few hundreds of eV. These signals are obtained from specific emission volumes within the sample and can be used to examine many characteristics of the sample such as surface topography, crystallography, composition, etc. [108],[109],[110].

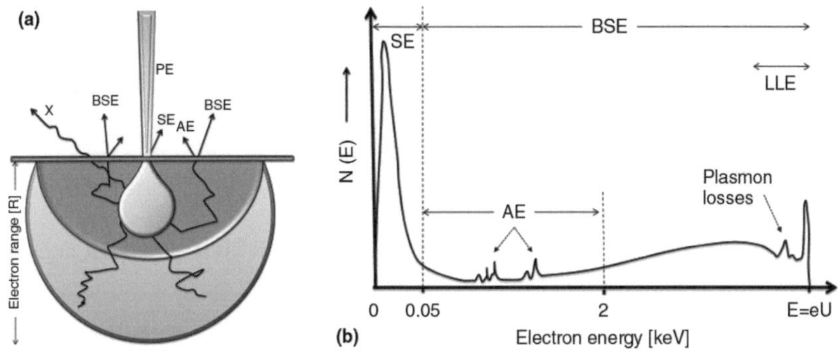

Fig. 8. Electron–specimen interaction (a) and schematic energy spectrum (a) [109].

Secondary and Auger electrons are highly susceptible to elastic and inelastic scattering and can leave the specimen only from a very thin surface layer of the thickness of a few nanometers. The most probable energy of BSE falls into the broad part of the spectrum in **Fig. 8(b)**, but they also show more or less pronounced elastic peak followed by plasnom losses, which depend on the primary energy, the take-off angle and the tilt of the specimen. Continuously slowing-down approximation assumes that the mean electron energy decreases smoothly with decreasing path length of the electron trajectories inside the specimen. The maximum information depth of BSE is of the order of half the electron range. Characteristic X-rays will only be excited in the volume in which the electron energy exceeds the ionization energy of the inner shell involved. Inelastic scattering in semiconductors results in the generation of electron-hole pairs. The recombination can take place without radiation but may result in the emission of light quanta (cathodoluminescence, CL) [111].

The method known as electron backscattering diffraction (EBDS) enables to determine the crystal structure and grain orientation of crystals on the surface of specimen. To collect maximum intensity in the diffraction pattern, the surface of specimen is stipple tilted at an angle of typically 70° from the horizontal (**Fig. 22(a)**). The intensity of backscatter **Kikuchi patterns** (please see the pattern of fluorapatite in **Fig. 23**) is rather low, as is the contrast of the signal, so extremely sensitive cameras and contrast enhancement facilities are required. This pattern allows to identify the phases and shows the misorientation across the grain boundaries [108].

Scanning electron microscope can be also used to determine compositional information using characteristic X-ray. The development of instruments for obtaining localized chemical analysis of solid samples, i.e. electron probe microanalyzer (EMPA), occurred at the same time as the development of SEM.

Scanning electron microscopy (SEM) is used for grain interactions and spot analysis [98],[112], electron microprobe microanalysis (EPMA) for the distribution of elements in the matrix, investigation of the effects of impurities on the properties of apatites and investigation of reaction interface [113],[114],[115],[116].

3.1.11. Fourier transform infrared and Raman spectroscopy

Infrared (IR) spectroscopy is one of the most important analytical techniques that can be used for the investigation of any sample in any state. Liquids, solutions, pastes, powders, films, fibers, gases and surfaces can be examined with judicious choice of sampling technique. Infrared spectrometers have been commercially available since the 1940s [117].

Fourier transform infrared (FT-IR or FTIR) spectroscopy is divided into three regions according to the increasing wavelength [118]:

1. Near-IR (NIR) spectroscopy, abbreviated as FT-NIR
2. Mid-IR (MIR) spectroscopy, abbreviated as FT-MIR
3. Far-IR (FAR) spectroscopy, abbreviated as FT-FAR

The spectral ranges of near-, mid- and far-infrared spectroscopy are shown in **Fig. 9(a)**.

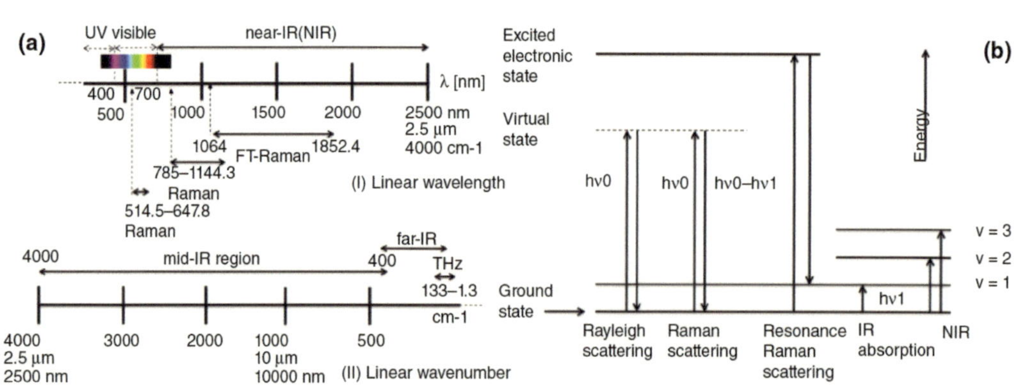

Fig. 9. Schematic illustration of relationships between the ranges of (a) vibrational spectroscopy and electromagnetic spectrum [118] and (b) spectroscopic transitions underlying several types of vibrational spectroscopy. v_0 indicates the laser frequency, while v is the vibrational quantum number. The virtual state is a short-lived distortion of the electron distribution by the electric field of the incident light [119].

The background for Raman spectroscopy was given by the discovery of Raman scattering by Krishna and Raman in 1928. Until approximately 1986 when Fourier transform (FT)–Raman was introduced, physical and structural investigations dominated in literature over relatively few reports of Raman spectroscopy applied in chemical analysis [119],[120].

When monochromatic light with the energy $h\nu_0$ encounters the matter (gas, solid or liquid), there is a small probability that it will be scattered at the same frequency (**Fig. 9(b)**). If the object, e.g. molecule is much smaller than the wavelength of the light, the scattering is Rayleigh scattering. The "*virtual state*" is not necessarily a true quantum state of the molecule but can be considered a very short-lived distortion of the electron cloud caused by oscillating electric field of the light. Since blue light is more efficiently scattered than red one, Rayleigh scattering is responsible for the blue color of sky. The electron cloud of the molecule is also perturbed by molecular vibrations, and it is possible for the optical and vibration oscillations to interact, leading to Raman scattering. Raman scattering is shown in (**Fig. 9(b)**) in which the scattered photon is lower in energy by an amount equal to the vibration transition. Raman spectrum consist of scattered intensity plotted versus energy and each peak corresponds to given Raman shift from the incident light energy $h\nu_0$ [119].

Just like Rayleigh scattering, Raman scattering depends on the polarizability of scattering molecules. IR band, on the other hand, arises from a change in the dipole moment. In many cases, the transitions that are allowed in Raman are forbidden in IR, so these techniques are often complementary (please compare **Fig. 18(a)** and **(b)**). In polarizable molecules, incident light can excite the vibrational modes, leading to scattered light diminished in energy by the amount of vibrational transition energies (same as in fluorescence). Scattered light under these conditions reveals the satellite lines below the Rayleigh scattering peak at the incident frequency–Stokes lines (Stokes part of spectrum). If there is enough energy, it is also possible to see anti-Stokes lines. Since anti-Stokes lines are usually weaker than Stokes lines, only the Stokes part of spectrum is usually measured [121].

The method combining Raman spectrometer with microscopic tools, typically an optical microscope, is known as micro-Raman spectroscopy (μRS) or also Raman microscopy. The μRS is nondestructive and noncontact method for the characterization of organic and inorganic materials [122].

Infrared [97],[98],[123],[112],[124],[125],[126], Raman [97],[98] and micro-Raman spectroscopies [125] were often used to identify and investigate the structure and extent of substitution and to optimize the synthesis conditions of minerals from the supergroup of apatite. Since carbonate ions exhibit clear vibrational signature in infrared spectrum, infrared spectroscopy is widely used to investigate the structure and to evaluate the carbonate/phosphate ratio ($r_{c/p}$) and the amount of carbonate ions in carbonate-apatites [127]:

$$CO_3^{2-}\ [\text{wt.\%}] = 28.62\ r_{c/p} + 0.0843 \tag{1}$$

The example of infrared and Raman spectrum of fluorapatite is described in **Section 3.2.3**.

Infrared spectra of phosphate minerals in the pyromorphite series are described by Adler [128]. In the pyromorphite series, the equilibrium internuclear X-O distance in XO_4^{3-} ion (PO_4^{3-},

AsO_4^{3-} and VO_4^{3-}) is primarily a function of the ionic radius of X atom. Since Pb, in this case, is always the dominant externally coordinated cation, for various members, there is no significant change in the interaction between the molecular vibration and the external environment. Bradger's equation [128],[129],

$$k_0 = 1.86 \cdot 10^5 / \left(R - d_{ij}\right)^3 \tag{2}$$

although specifically applicable to internuclear distances in diatomic molecules, reflects generally the inverse relationship between the force constant $\boldsymbol{k_0}$ and the internuclear distance $\boldsymbol{R}$. Symbol $\boldsymbol{d_{ij}}$ denotes the constant the values of which depend on the nature of bonded atoms. The molecular vibration frequency $\boldsymbol{v}$ is dependent on the restoring forces, measured in terms of $\boldsymbol{k_0}$, between participating atoms as well as on the masses of these atoms. The relationship may be expressed approximately by the equation:

$$v = \frac{1}{2}\pi c \sqrt{\frac{k}{u}} \tag{3}$$

where the vibration frequency $\boldsymbol{v}$ is a function of the force constant $\boldsymbol{k}$ and the reduced mass $\boldsymbol{u}$ of vibrating atoms, all other terms being invariant.

The spectral frequency differences between pyromorphite, mimetite and vanadinite are explicable and to a considerable degree predictable in terms of these parameters. On complete substitution of As or V for P the effect of reduced force constants is reinforced by increases in mass, thereby shifting $\boldsymbol{v_3}$ and $\boldsymbol{v_1}$ to lower frequencies. Because of opposing mass and force-constant effects and perhaps also because of dissimilarities in orbital configurations, the relative positions of absorption bands are less predictable for mimetite and vanadinite than for pyromorphite and mimetite. The theoretical frequency trends are depicted in **Fig. 10** [128].

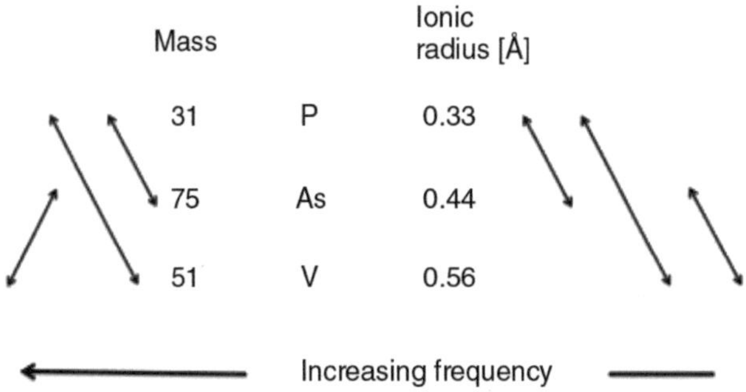

Fig. 10. Theoretical effect of change in mass and ionic radius on infrared vibration frequency of tetrahedral XO_4^{3-} ions, where $X = P^{5+}$, As^{5+} of V^{5+} [128].

http://dx.doi.org/10.5772/62211

3.1.12. Thermal analysis

Thermal analysis (TA) refers to a group of techniques[10] in which the property of a sample is monitored against time or temperature while the temperature of the sample, in a specified atmosphere[11] is programmed. These methods study the relationship between sample property and its temperature as the sample is heated or cooled in a controlled manner. The individual techniques are divided according to the measured property[12] [131],[132], [130] as is introduced in **Table** 1.

Property	Technique and abbreviation		Notes
Heat	Scanning calorimetry	–	–
Temperature	Thermometry	–	Also can be described as heating or cooling curve
Temperature difference	Differential thermal DMA analysis	DTA	A technique where the temperature difference between the sample and reference material is measured
Heat flow rate difference	Differential scanning calorimetry	DSC	A technique where the difference between heat flow rates into the sample and reference material is measured
Mass	Thermogravimetry or thermogravimetric analysis	TG or TGA	The abbreviation TG has been used, but should be avoided, so that it was not confused with T_g (temperature of glass transition)
Dimensional and mechanical properties	Dynamic mechanical analysis	DMA	Moduli (storage/loss) are determined
	Thermomechanical analysis	TMA	Deformations are measured
	Thermodilatometry	TD	Dimensions are measured
Electrical properties	Dielectric thermal analysis	DEA	Dielectric constant/dielectric loss are measured
	Thermally stimulated current	TSC	Current is measured
Magnetic properties	Thermomagnetometry	TM	Often combined with TGA
Gas flow	Evolved gas analysis	EGA	The composition and/or the amount of gas/vapor is determined
	Emanation thermal analysis	ETA	Trapped radioactive gas within the sample is released and measured
Pressure	Thermomanometry	–	Evolution of gas is detected by pressure change

[10] The definition of terms in thermal analysis was developed by ICTAC (Confederation for Thermal Analysis and Calorimetry).

[11] Gaseous environment of the sample, which may be controlled by the instrumentation or generated by the sample [130].

[12] Resulting dependence, i.e. any graph of any combination of property vs. time or temperature derived from a thermal analysis technique, should be termed as thermal curve, which is a simplified form of more correct term thermoanalytical curve. The first mathematical derivation of any curve with respect to temperature or time leads to the derivative thermoanalytical curve [130]. Since the name thermogram has medical usage, the thermal analysis curve should not be termed as thermogram [131].

Property	Technique and abbreviation		Notes
	Thermobarometry	–	Pressure exerted by dense sample on the walls of a constant volume cell is studied
Optical properties	Thermoptometry or thermos optical analysis	TOA	A family of techniques in which optical characteristics or property of the sample is studied
	Thermoluminescence	TL	Light emitted by the sample is measured
Acoustic properties	Thermosonimetry or thermoacoustimetry	TS	Emitted (sonimetry) or absorbed (acoustimetry) sound is measured.
Structure	Thermodiffractometry	–	Techniques where the compositional or chemical nature of the sample is studied.
	Thermospectrometry		

Table 1 Methods of thermal analysis according to measured property or physical quantity.

The measurement should be performed as follows:

1. **Combined**: the application of two or more techniques to different samples at the same time. This can include thermal and nonthermal analytical techniques.

2. **Simultaneous**: indicates the measurement of two or more properties of a single sample at the same time.

The sample-controlled method where the feedback used to control the heating is the rate of transformation is termed as controlled-rate thermal analysis (CRTA) [130].

Simultaneous thermogravimetry (thermogravimetric analysis) and differential thermal analysis (TG-DTA) are mostly used to investigate the course of synthesis and the characterization of prepared apatites or to investigate the process of thermal decomposition of apatites, i.e. the processes such as dehydroxylation (e.g. **Section 1.5.2**), defluorination (**Section 3.2.4** and **8.6**), decarbonation (thermal decomposition of carbonate-apatites, **Section 4.6.1**), etc.

3.1.13. Optical properties

The analysis of optical properties is essential for each mineral examined, and through the use of microscopy, the optical properties of individual minerals may be interpreted in great detail. Optical mineralogy investigates the interaction of light (usually is limited to visible light) with minerals and rocks. Optical mineralogy concerns mainly the use of polarizing (petrographic) microscope which has two Nicol prisms, polarizer and analyzer (polarizer below the stage and the analyzer above the objective) [133]. Human eye is the most sensitive for viewing a solid in the wavelength symmetrically spread in intensity around 550 nm (**Fig. 11(a)**). Optical microscopy in visible light (from 700 nm (red) to 420 nm (violet), **Fig. 11(b,c)**) helps study the objects of smaller sizes up to lower limit of ~ 1 μm (c). **Fig. 11(d)** shows the scope of spectroscopic techniques associated with electromagnetic spectrum [134].

http://dx.doi.org/10.5772/62211

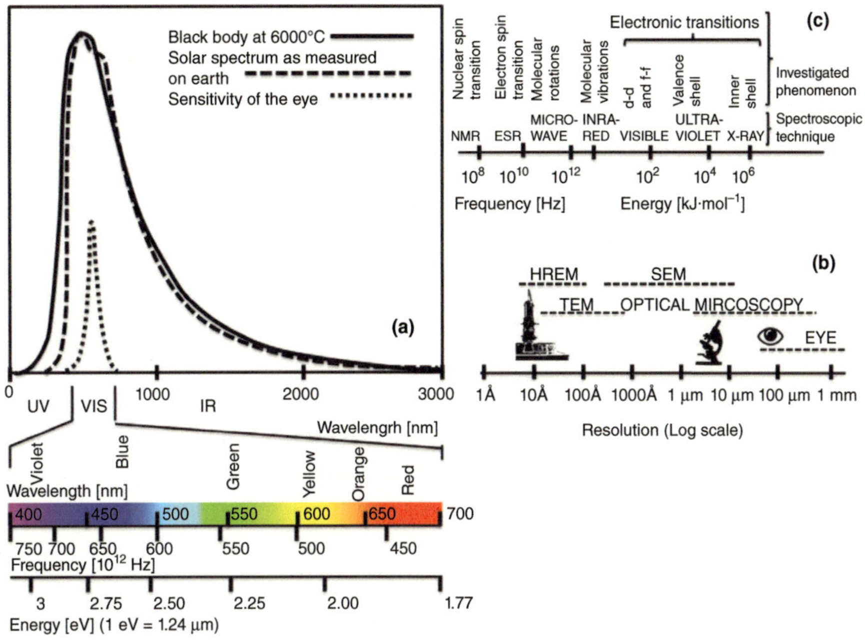

Fig. 11. Electromagnetic spectrum of solar radiation and black body (a), the resolution of various techniques (b) and the regions associated with spectroscopic techniques (c) [134].

Refractive index (n) is related to the angle of incidence (i) and the angle of refraction (r) according to the Snell's law:

$$n = \sin i / \sin r \tag{4}$$

The refractive index increases as the wavelength of light decreases [134]. The absorption coefficient is related to the imaginary part of the refractive index. It was found that since the refractive index of a medium depends on the density of electrons in that medium, the index increases with the density of matter [135].

The luster of mineral (R) depends of the way in which the light is reflected from the surface of a mineral. The reflection is again dependent on the refractive index. Normally, the greater is the index of refraction, the brighter is the luster. The luster is classified into two broad classes [134]:

a. Nonmetallic luster: results from the interaction of light with dielectric semiconducting and poorly conducting substances

b. Metallic luster: occurs in minerals having metallic bonding (e.g. native metals) or high degree of covalent bonding (sulfides, sulfosalts, etc.)

Optical properties of apatites can be determined using the complex dielectric function [136], [137],[138]

$$\varepsilon(\omega) = \varepsilon_1(\omega) + i\varepsilon_2(\omega) \tag{5}$$

in the range of linear response. By calculating the wave function matrix and using Kramers–Krönig relations, the imaginary and real part of the dielectric function $\varepsilon_1(\omega)$ and $\varepsilon_2(\omega)$ can be derived respectively as follows:

$$\begin{aligned}\varepsilon_2(\hbar\omega) = \frac{e^2}{\pi m\omega^2} \int_{BZ} dk^3 \sum_{n,l} \left| \langle \Psi_n(\bar{k},\bar{r}) | \rangle - i\hbar | \langle \Psi_1(\bar{k},\bar{r}) \rangle \right|^2 f_1(\bar{k}) \\ \left[1 - f_n(\bar{k})\right] \delta\left[E_n(\bar{k}) - E_1(\bar{k}) - \hbar\omega\right]\end{aligned} \tag{6}$$

where $f(\vec{k})$ is the Fermi distribution function, ***l*** and ***n*** mark occupied state and unoccupied state, respectively. $\Psi_n(\vec{k},\vec{r})$ is the Bloch wave function for the *n*th band with the energy $E_n(\vec{k})$ at the Brillouin zone point ***k***. The matrix element momentum transition corresponds to the term $|\langle\Psi_n(\bar{k},\bar{r})\rangle| - i\hbar|\langle\Psi_1(\bar{k},\bar{r})\rangle|^2$.

For each apatite, the real part ε_1 and imaginary part ε_2 of dielectric function have similar features with some subtle differences. The real part ε_1 has two main peaks:

1. First peak lies near the energy of 7 eV. Obviously, the first peak is caused by the transitions from O–*2p* and P–*3p* levels.

2. The second peak lying at ~25 eV results from the transitions from Ca–*3p* levels.

The ε_2 curve in the energy range from 5 to 15 eV characterizes three main peaks:

1. The first peak lies in 7.10, 7.16 and 7.43 eV for FAP, ClAP and BrAP, respectively.

2. The second peak is located at 8.51, 8.62 and 8.76 eV for FAP, ClAP and BrAP, respectively.

3. The third peak is situated at 10.99, 11.08 and 11.45 eV for FAP, ClAP and BrAP, respectively.

The refractive index *n* can be obtained by $n = \sqrt{\varepsilon_1(0)}$. The $\varepsilon_1(0)$ for FAP, ClAP and BrAP are 1.38, 1.41 and 1.46, respectively. Therefore, the refractive indexes for FAP, ClAP and BrAP are 1.17, 1.19 and 1.21. Moreover, analogous dielectric function curves and similar refractive index values show that the optical property of each apatite has some independence from the *c*-axis ion [136].

3.1.14. Measuring of surface area, porosity and pore size distribution

There is a conventional mathematical idealization that asserts that a cube of edge length ***a*** possesses a surface area of $6a^2$ and that a sphere of radius ***r*** exhibits $4\pi r^2$. In reality, however, mathematical perfect or ideal geometric forms are unattainable (**Fig. 12**) since all real surfaces exhibit flaws under microscopic examination. Real surface irregularities (voids, pores, steps, etc.) make the real surface area greater than corresponding theoretical area. When the

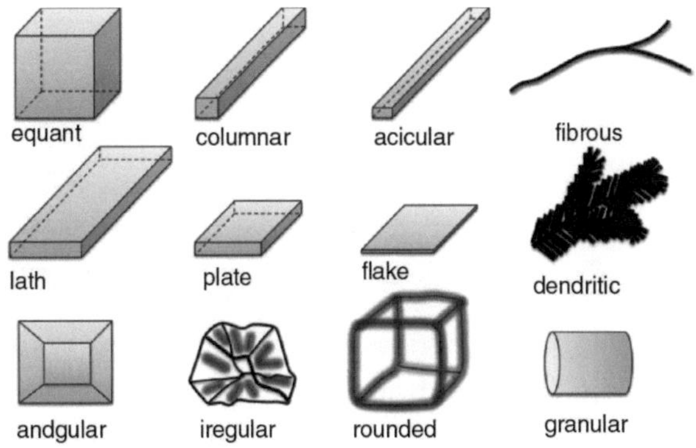

Fig. 12. Description of particle shapes [140].

cube, real or imaginary, of one meter edge length is subdivided into smaller cubes each one micrometer (10^{-6} m) in length, there will be 10^{18} particles formed, each exposing an area of 6×10^{-12} m^2. Thus, the total area of all particles is 6×10^6 m^2. This operation increases by million-fold the exposed area of fine powder compared to undivided material. Whenever the matter is divided into smaller particles, new surfaces must be produced with corresponding increase in surface area [139],[140].

The particle size distribution (PSD) was usually determined by sieve analysis, sedimentation methods (gravitational or centrifugal), microscopic techniques, light scattering, multiangle laser light scattering (MALLS), etc. [141].

The range of specific surface area, i.e. area per gram of matter, can vary widely depending on the particles' size, shape and porosity. The influence of pores can often overwhelm the size and external shape factors. The powder consisting of spherical particles exhibits total surface (S_t) and volume ($V = M/\rho$) [139]:

$$S_t = 4\pi \left(r_1^2 N_1 + r_2^2 N_2 + \ldots + r_n^2 N_n\right) = 4\pi \sum_{i=1} r_i^2 N_i \tag{7}$$

$$V = \frac{4}{3}\pi \left(r_1^3 N_1 + r_2^3 N_2 + \ldots + r_n^3 N_n\right) = \frac{4}{3}\pi \sum_{i=1} r_i^3 N_i \tag{8}$$

$$S = \frac{S_t}{M} = \frac{3\sum_{i=1} N_i r_i^2}{\rho \sum_{i=1} N_i r_i^3} \tag{9}$$

where r_i and N_i are the average radii and numbers of particles in the fraction i. For spheres of uniform size, **Eq. 9** becomes the law:

$$S = \frac{3}{\rho r} \tag{10}$$

Besides the calculation of specific surface from its geometry and PSD curve, the adsorption isotherm (predominantly Langmuir and BET[13]), air permeability methods,[14] and monolayer sorption methods were used to determine the specific surface area [142],[143].

The porosity is defined as the ration of pore volume to total volume. Porous material is defined as solids containing pores **Fig. 13(a)**, which are classified into two major types: open and closed pores. Penetrating open pores (interconnected pores) are permeable for fluid and therefore are important in applications such as filters. Pores accessible from only one end are referred to as dead-end pores. Noninterconnected (closed) pores are not accessible at all. The classification of pores according to their size is shown in **Fig. 13(b)**. Pores can be also classified as the pores among agglomerates and pores among primary particles (**Fig. 13(c)**) [144],[145].

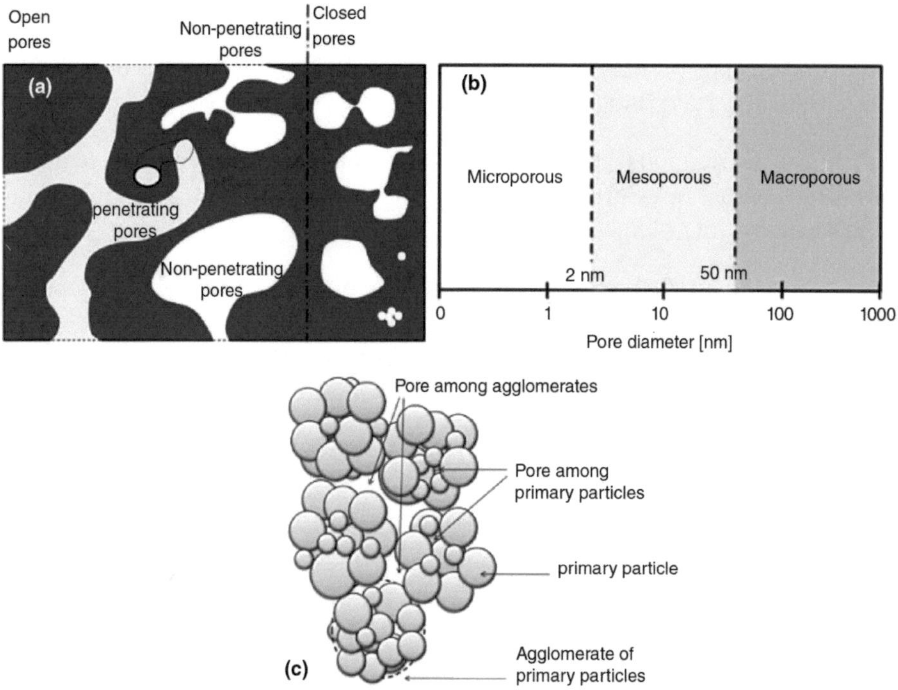

Fig. 13. Schematic illustration of different morphology of pores (a) and classification of porous materials based on pore size (b) and schematic illustration of pores among agglomerates and primary particles (c) [144].

---

[13] Brunauer–Emmett–Teller (BET) theory of multilayered physical adsorption of gas molecules on a solid surface [143].

[14] For example, the Blaine method, where fixed volume of air passes through the bed at steadily decreasing rate, which is controlled and measured by the movement of oil in a manometer, the time required being measured. The method is widely used for the determination of specific surface area of cements [143].

http://dx.doi.org/10.5772/62211

Pore size distribution and permeability[15] are a very desirable quantities for the characterization of structure of porous solids, which can be determined by the following [145],[146],[147]:

i. Stationary fluid (static) method

ii. Capillary pressure (HASSLER[16]) method

iii. Quasi-steady-state methods, including the gas drive method and the solution-gas drive method.

The surface area and the porosity of apatites have effect on the floatability (**Section 8.7**) of apatites of different type and origin [148]. The surface area, the porosity and the pore size distribution are properties of great importance for the preparation of biological apatites in tissue engineering (**Section 10.9**) and tailoring their mechanical properties, solubility and bioactivity [149],[150],[151],[152]. Ionic surfactants, such as decyltrimethylammonium bromide ($CH_3(CH_2)_9N(CH_3)_3(Br)$, C(10)TAB), hexadecyltrimethylammonium bromide ($(CH_3(CH_2)_{15}N$ $(Br)(CH_3)_3$, C(16)TAB), as well as nonionic surfactant, can be used to control the pore (pore size and volume) and surface characteristics of mesoporous apatite materials under maintained pH [153]. The porosity also affects electrical properties of oxyapatites (**Chapter 5**) [154],[155],[156].

3.2. Investigation and characterization of apatite specimen

3.2.1. Identification of the specimen

Some techniques mentioned above will be demonstrated on the specimen of apatite sample (**Fig. 14(a)**) from Sljudjanka, Bajkal. The translucent specimen with glassy luster is greenish blue colored and brittle as can be seen from large amount of smoothly curving conchoidal fractures on the surface (please see also **Fig.** 20). When scratched by a single crystal of corundum (**Fig. 14(b)**), the sample shows white colored scratch. Since the sample surface can be also scratched by feldspar (**Fig. 14(c)**) but not by fluorite, the hardness in the Mohs scale[17] is equal to 5, i.e. corresponds to apatite.

[15] The flow of fluids through porous materials is of great importance in the fields of industrial chemistry, oil technology, and agriculture. In general, it may be stated that the principal interest is in the transport through reactive materials [147].

[16] Experimental techniques for the measurement of relative permeability can be divided to steady- (1) and unsteady-state (displacement) methods (2). The steady-state method was developed by HASSLER (1944). Semipermeable membranes are provided at each end which keeps the fluids separated, except inside the core where the fluids flow simultaneously. The pressure is measured in each phase through semipermeable barriers, and the pressure differences between the phases are maintained constant throughout the medium so as to eliminate the capillary end effect as well as to ensure a uniform saturation along the core. The saturation can be altered by applying capillary pressures across the nonwetting phase ports and wetting phase semipermeable membranes [146].

[17] The Mohs scale of mineral hardness is graduated as follows: talc (1), gypsum (2), calcite (3), fluorite (4), apatite (5), orthoclase (6), quartz (7), topaz (8), corundum (9), and diamond (10). Apatite should be also scratched by steel knife (up to 5.5) and glass (up to 6).

Fig. 14. Specimen of apatite from Sljudjanka, Bajkal (a) shows white color of scratch (b). The sample can be also scratched by minerals with hardness ≥6 (feldspar and higher) on the Mohs scale (c).

The microhardness of apatite sample was then determined by Vickers microhardness test.[18] **Fig. 15** shows the replica of diamond pyramid base[19] on the surface after the indentation of sample. The average hardness of sample was determined to be 552 (±30) HV 0.05/10. The formation of radial cracks on the corners indicates brittle material [157]. According to the mineral hardness conversion charts, the measured value is in good agreement with the tabular value of apatite (535 HV [158]).

Moreover, the sample does not show any luminescence when elucidated by long UV light (**Fig. 16(a)**). The specific gravity of the specimen was assessed by hydrostatic weighting (b) and pycnometric technique (c) to be 3.18 and 3.16 ± 0.20 g·cm^{-3}, respectively. These values are in a good agreement with average density of apatite (3.19 g·cm^{-3}, **Chapter 1**). All properties of investigated specimen mentioned above identify it as apatite, but the exact kind of apatite mineral and its chemical composition must still be determined yet.

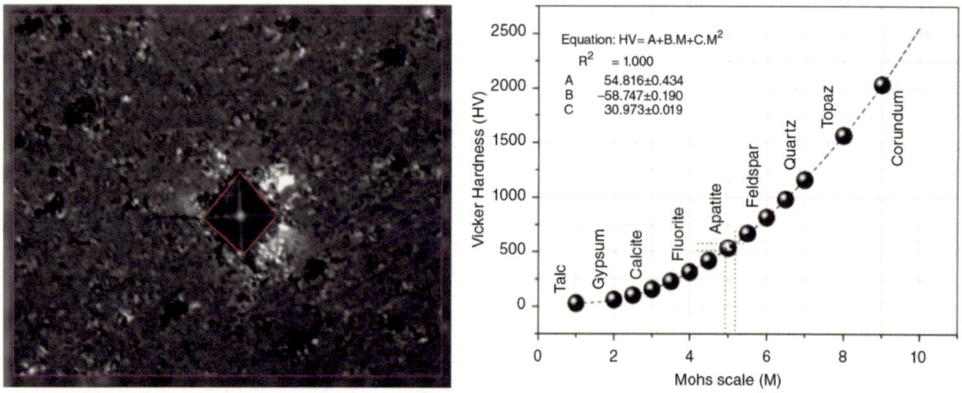

Fig. 15. The Vickers microhardness test with the load of 0.05 kgf for the time of 10 s and the correlation of results with the Mohs scale.

[18] Hardness tester LECO AMH 43. The method is also known as the Vickers pyramid number (HV) or the diamond pyramid hardness (DPH).

[19] Diamond pyramid with apical angle of 136°.

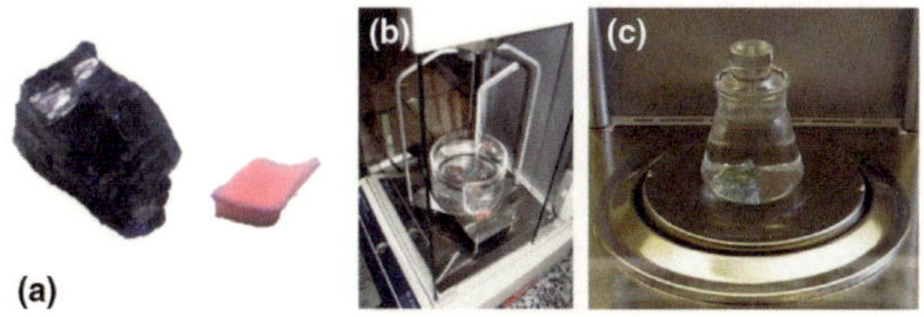

Fig. 16. The specimen of apatite under UV light (compared with the fragment of red luminescence of alumina).

The sample was next treated to fine powder via milling in stain-less steal laboratory vibration mill. The apatite mineral was then determined by X-ray diffraction analysis (**Fig. 17**), mid-infrared spectroscopy (**Fig. 18**) and EBDS (**Fig. 23**) as fluorapatite ($Ca_5(PO_4)_3F$, ref. [159]) with small amount (1%) of accessory mineral calcite[20] ($CaCO_3$). Since the crystallographic parameters of identified hexagonal apatite are $a = 9.3917$, $c = 6.8826$ Å and **Z** = 2, it is possible to calculate the axial ratio (**Eq. 11**), the volume of cell (**Eq. 12**) and the density (**Eq. 13**) as follows:

$$a{:}c = 1{:}0.7328 \tag{11}$$

$$V = a^2\ c\ \sin(60) = 9.3917^2 6.8826\ \sin(60) = 525.74\ \overset{\circ}{A}^3 \tag{12}$$

$$\rho_{\text{calculated}} = \frac{M_{\text{apatite}} Z}{V \dfrac{N_A}{1\times10^{24}}} = \frac{504.31\times 2}{525.74 \dfrac{6.025\times10^{23}}{1\cdot10^{24}}} = 3.19\ \text{g}\cdot\text{cm}^{3} \tag{13}$$

where M is the molecular weight of fluorapatite (**Table 7** in **Chapter 1**) and N_A is the Avogadro constant. The reconstruction of the cell of investigated apatite specimen is shown in **Fig. 24**.

3.2.2. X-ray diffraction analysis

Powder X-ray diffraction analysis of the apatite specimen (**Fig. 17**) identified it as fluorapatite. According to quantitative Rietveld analysis, the sample contains 99% of fluorapatite. There is also small amount (1%) of calcite[21] that occurs on the surface of apatite specimen. Since the fluorapatite specimen (**Fig. 14**) is single crystal (**Section 3.2.6**), the crystal faces cannot be recognized, i.e. the crystal habit is anhedral (refer to **Footnote 2** in **Chapter 2**). Nevertheless, XRD pattern shows that the most intensive diffraction possesses the Miller index (211), which corresponds to the Miller-Bravais indices (21–31), i.e. dihexagonal dipyramid (**Chapter 1**).

[20] Since it is present as "free carbonate," the sample cannot be considered as carbonate-fluorapatite (**Section 2.6** and **4.6.1**).

[21] As the results of thermal analysis revealed (**Section 3.2.4**), the Rietveld analysis slightly overestimates the content of free carbonate in the sample.

There is also basal pinacoid {1000}, first-order {10–10} and second-order {11–20} hexagonal prism and dihexagonal prism {21–30}.

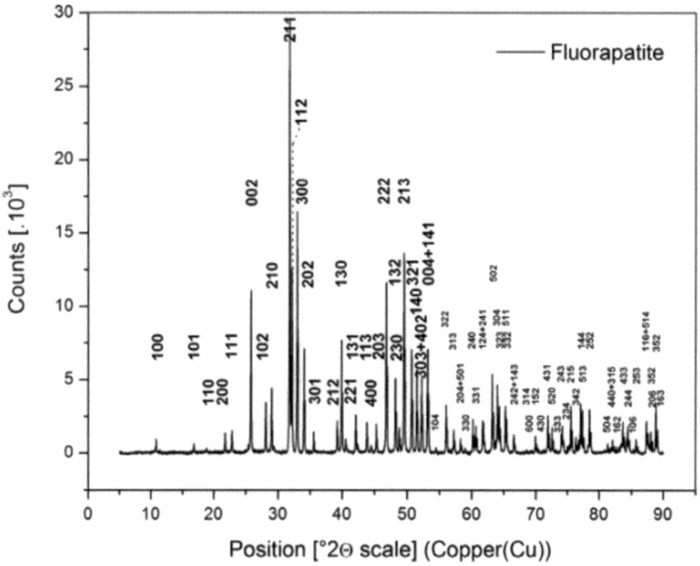

Fig. 17. X-ray diffraction analysis of investigated specimen of apatite.

It is obvious that fluorapatite belongs to the hexagonal-dipyramidal crystal system, but the estimation of crystal habit of corresponding euhedral crystals from these results is highly speculative due to possible combination of pinacoid (*c*), first-order (m) and second-order hexagonal prisms (a) and first-order (*p*) and second-order dipyramids (*s*) with dihexagonal dipyramid (*v*) faces in the single crystal.

3.2.3. Infrared and Raman spectroscopy

Infrared (mid-FT-IR[22]) and Raman spectrum of fluorapatite is shown in **Fig. 18**(**a**) and (**b**), respectively. The most expressive infrared bands are attributed to fundamental frequencies of tetrahedral phosphate ion $[PO_4]^{3-}$. The structure of apatite leads to the reduction of ion symmetry from T_d (four fundamental frequencies with IR inactive ν_1 mode) to C_s, where ν_1 mode becomes IR active [128],[160],[161],[162],[163]:

1. The $\nu_1(PO_4)$ mode is very weak (*vw*[23]) band that is related to symmetric stretching of phosphate ion.
2. Bending: ν_2 mode (*vw*).

[22] Baseline corrected spectrum measured by KBr technique.

[23] Abbreviation used for the expression of intensity and width of peak in the spectrum: very weak (*vw*), weak (*w*), middle (*m*), strong (*s*) and very strong (*vs*), shoulder (*sh*), broad (*b*), very broad (*vb*), and sharp (*sp*). Spectral bands related to impurities are abbreviated as *imp* [162].

3. Anti-symmetric stretching: ν_3 mode is the strongest (*vs, vb*) band in the infrared spectrum appearing in the spectral region from 1000 to 1150 cm^{-1}.

4. Bending: ν_4 mode (*m, vb*) is observed between 540 and 620 cm^{-1}.

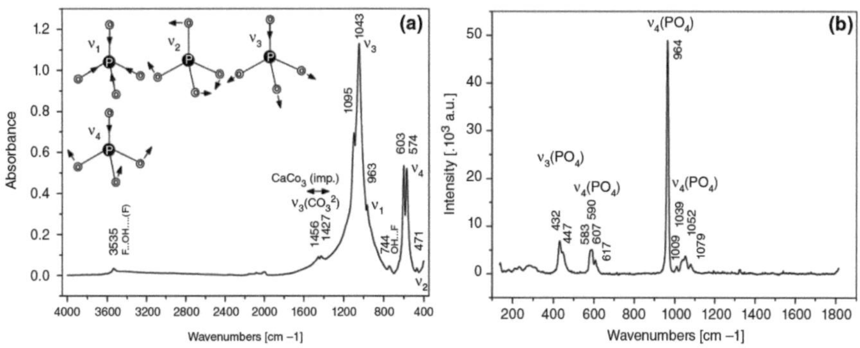

Fig. 18. Infrared (a) and Raman (b) spectrum of investigated specimen of fluorapatite.

The crystallinity of natural and synthetic apatite samples is often determined from the broadening of $\nu_4(PO_4)$ infrared absorption bands [160]. The assignment of bands in infrared and Raman spectrum of fluorapatite is listed in **Table** 2.

Vibration mode	Assignment C_{6h} factor group symmetry	Raman shift [cm^{-1}]	IR
ν_1	A_g, E_{2g}, E_{1u}	956	965
ν_2	E_{1g}	432	–
	E_{2g}	–	–
	A_g	449	–
	E_{1u}	–	460
	A_u	–	470
ν_3	A_u, E_{2g}	1034	1032
	E_{1g}, E_{1u}	1042	1040
	A_g	1053	–
	E_{2g}	1061	–
	A_g	1081	–
	E_{1u}	–	1090
ν_4	A_u	–	560
	E_{1u}	–	575
	E_{2g}	581	–

Vibration mode	Assignment C_{6h} factor group symmetry	Raman shift [cm^{-1}]	IR
	A_g, E_{1g}	592	–
	E_{1u}	–	601
	A_g	608	–
	E_{2g}	617	–

Table 2. The interpretation of infrared and Raman bands in the spectrum of stoichiometric fluorapatite [164].

Factor group analysis of the hexagonal $P6_3/M$ space group fluorapatite structure (Z = 2) yields an irreducible representation for optically active vibration of [165]:

$$\Gamma = 12\ A_g(R) + 7A_u(\mathrm{IR}) + 8E_{1g}(R) + 11E_{1u}(\mathrm{IR}) + 13E_{2g}(R) \tag{14}$$

where IR and R denote infrared and Raman activity, respectively. The influence of pressure on the infrared and Raman spectra of fluorapatite was investigated by Willams and Knittle [165]. Fluorapatite remains stable under pressures of at least 25 GPa at 300 K. Local environment of phosphate groups in fluorapatite becomes progressively less distorted from tetrahedral symmetry under the compression, as manifested by progressively smaller site-group.

The Davydov (factor group) splitting also decreases under the compression. This decrease is consistent with nondipole effects playing a primary role in the Davydov splitting of apatite; indeed, the magnitude of the Davydov splitting appears to be modulated by increases of the site symmetry of phosphate group under the compression [165].

The spectrum **Fig. 18(a)** shows weak peak located in the domain of OH stretching modes (from 3500 to 3600 cm^{-1}) at the wave number of 3535 cm^{-1}. This band belongs to the OH stretching mode in the hydrogen bond F…OH…(F) [166]. According to Freund and Knobel [167], the band at ~744 cm^{-1} belongs to the vibration of OH…F bond. According to Knubovets [168], the bands in the range from 745 to 720 cm^{-1} in apatite spectra could also be attributed to symmetric valence oscillations of the P-O-P bridge bonds, formed by the condensation of the PO_4^{3-} tetrahedron. The presence of calcite causes that antisymmetric stretching mode (ν_3) of planar CO_3^{2-} ion appears in the infrared spectrum of investigated sample [169].

3.2.4. Thermal analysis

The results of simultaneous TG-DTA of investigated fluorapatite specimen are shown in **Fig. 19**. The sample is heated with the rate of 10°C·min^{-1} up to the temperature of 1425°C. The mass of sample is reduced by 1.25% during TG-DTA when the final temperature is reached. The most important features are the thermal decomposition of $CaCO_3$ and the thermal decomposition of fluorapatite.

http://dx.doi.org/10.5772/62211

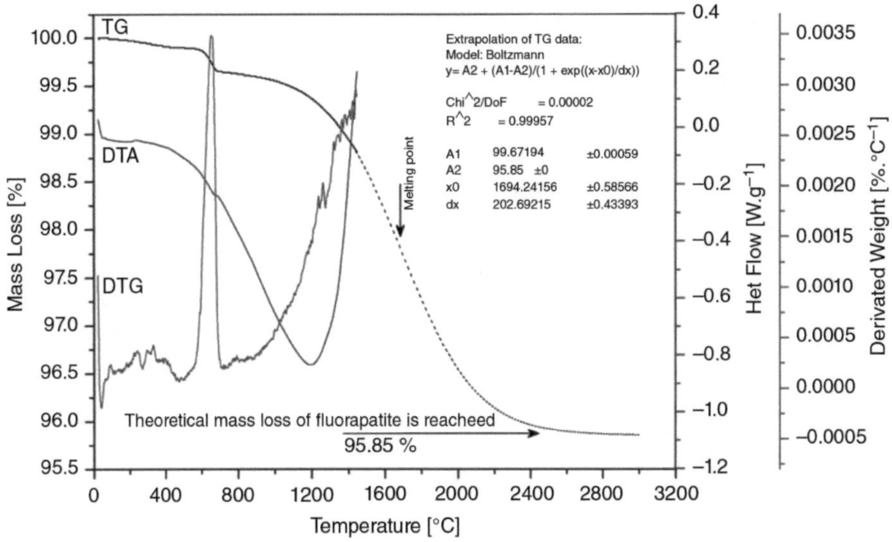

Fig. 19. Thermal analysis of investigated specimen of fluorapatite.

The content of calcite was verified by thermal analysis. The weight of sample was reduced by 0.26% during the thermal decomposition of calcite. Since the theoretical mass loss of calcite is (100 × 44.09) / 100.086 = 43.97%,[24] the content of calcite in the investigated sample of fluorapatite is (100 × 0.26) / 43.97 = 0.59%. Although this value is lower than the content of calcite determined by Rietveld analysis, there is still good agreement of both methods. The DTG curve shows that the process starts at the temperature of 565°C and wide of peak is of 135°C. The maximum rate decarbonation is reached at the temperature of 656°C.

At temperatures higher than 900°C, the weight of sample is reduced by the thermal decomposition of fluorapatite. The extrapolated beginning of the defluorination process (**Section 8.6**) was determined to be 1199°C. At the temperature of 1425°C, the defluorination process is still not complete. The extrapolation of experimental data[25] shows that the thermal decomposition is most probably not complete before the temperature of melting point is reached (**Table 7** in **Chapter 1**).

3.2.5. Scanning electron microscopy and WDX analysis

The microphotographs from SEM analysis of break plane of investigated fluorapatite specimen are shown in **Fig. 20**. The series of conchoidal fractures on the surface show brittle fracture as further characteristic properties of apatite.

---

[24] It was calculated as 100 × molar mass (molar weight) of CO_2/molar mass of $CaCO_3$.

[25] Theoretical mass loss (3.77%) of fluorapatite was used to set the fixed value of parameter A_2.

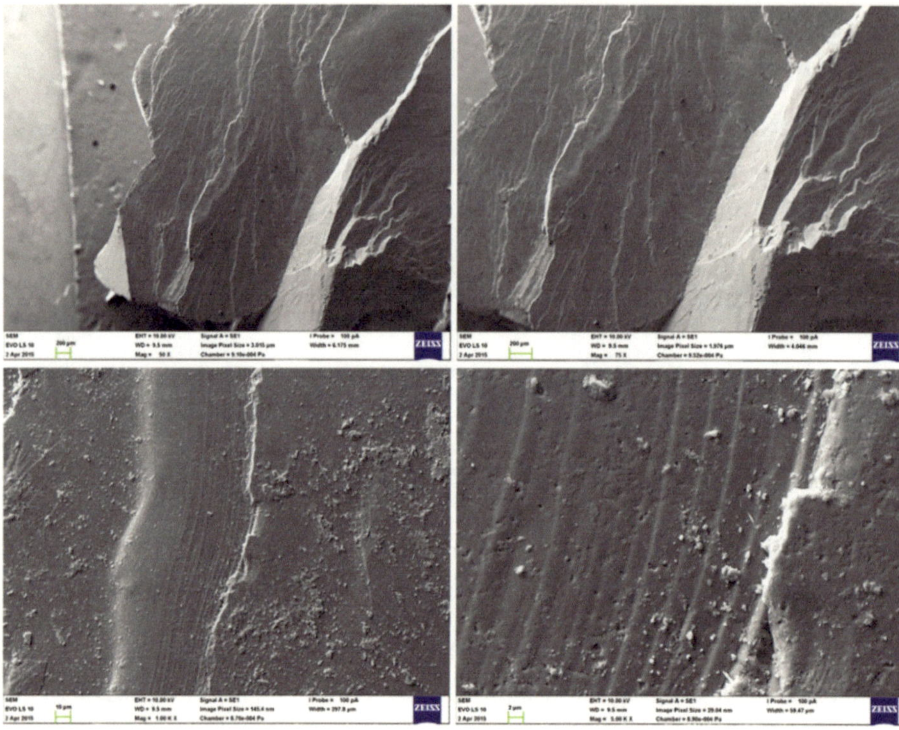

Fig. 20. SEM image of the break plane of investigated fluorapatite specimen: 50× (a), 75× (b), 1 000× (c) and 5 000× (d).

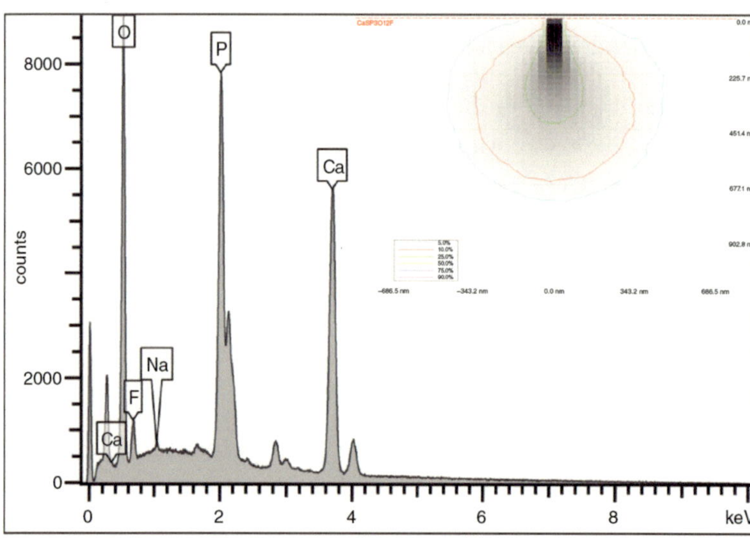

Fig. 21. Typical results of WDX analysis of the fluorapatite specimen and simulation of electron–specimen interaction.

The chemical composition of sample was determined by WDX analysis (**Fig. 21**). The average results are introduced in **Table** 3.

Element	Ca	P	O	F	Na
Weight [%]	38.64 ± 0.96	16.86 ± 0.47	40.55 ± 1.37	3.75 ± 0.08	0.23 ± 0.03
Ideal composition*	39.74	18.43	38.07	3.77	–
Atomic [%]	22.67 ± 0.85	12.81 ± 0.52	59.63 ± 1.31	4.65 ± 0.05	0.24 ± 0.03
Atomic ratio	4.91 ± 0.31	2.78 ± 0.18	12.77 ± 0.19	1	0.05 ± 0.01
Ideal composition[26]	5	3	12	1	–

Table 3. Average chemical composition of investigated specimen of fluorapatite.

The composition of fluorapatite ($Ca_5(PO_4)_3F \rightarrow Ca_5P_3O_{12}F$) corresponds to the element ratio: Ca:P:O:F = 5:3:12:1. The results of WDX analysis of investigated specimen (**Table** 3) are in good agreement with the composition of ideal fluorapatite and fulfill all the criteria mentioned in **Section 1.4**.

3.2.6. EBDS analysis

The fragments of investigated fluorapatite specimens (**Fig. 22**) were further investigated by EBDS analysis.

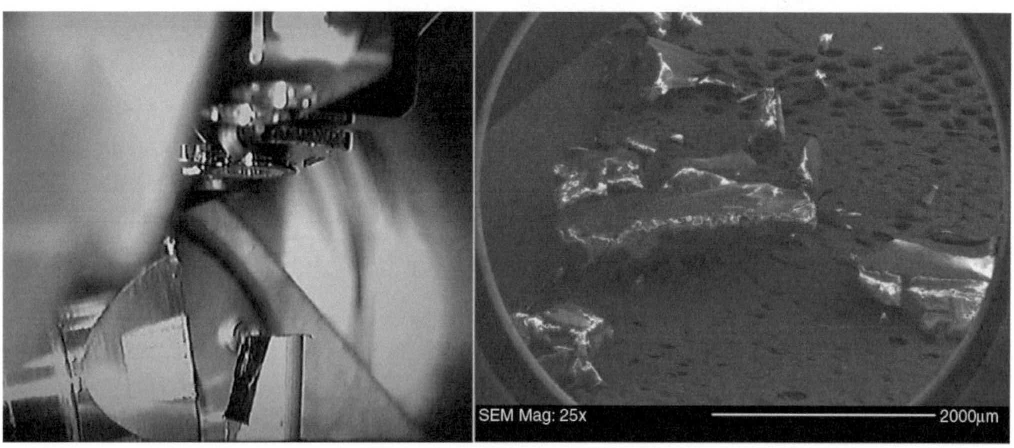

Fig. 22. Sample stage (a) and microphotograph of fluorapatite fragments used for EBDS (b).

The Kikuchi patterns confirm the sample as single crystal of fluorapatite **Fig. 23**.

[26]. Resulting from the apatite formula (**Section 1.4**).

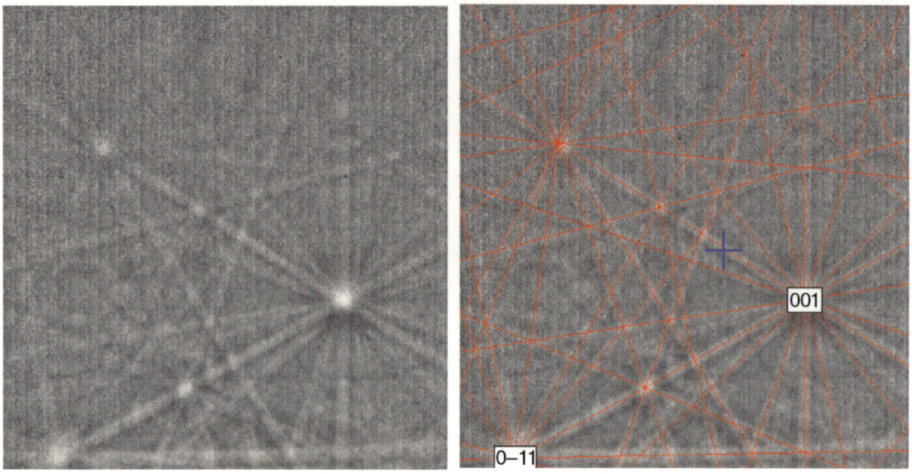

Fig. 23. EBDS analysis of investigated fluorapatite specimen.

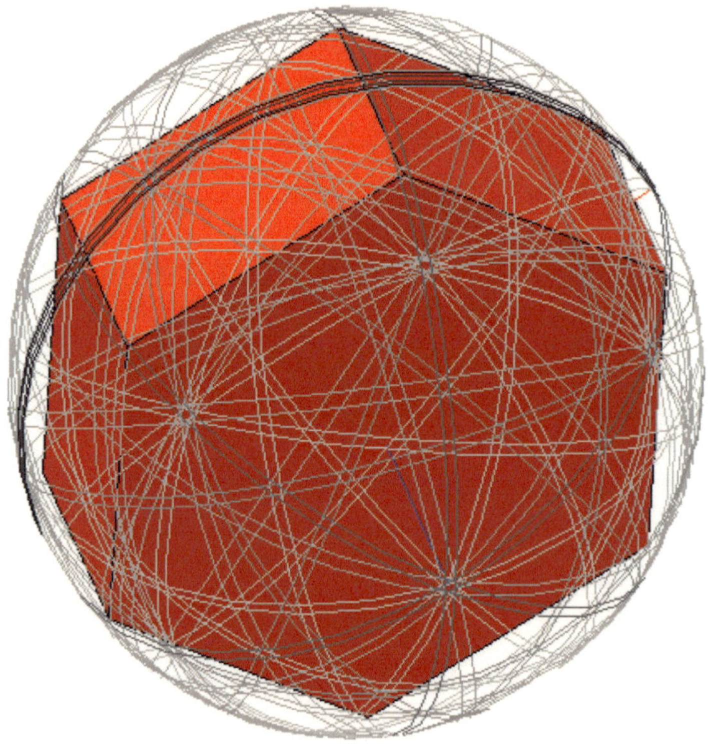

Fig. 24. The Kikuchi sphere showing the orientation of apatite crystal break plane.

The Kikuchi sphere in **Fig. 24** shows the orientation of the sample that is in agreement with indistinct cleavage of fluorapatite to the direction given by Miller-Bravais indices of {0001}.

3.3. Thermodynamic properties of apatite

Drouet [170] gives the comprehensive review on experimental and predicted thermodynamic properties of phosphate apatites and oxyapatites, where O^{2-} ion replaces $2X^-$ in general $M_{10}(PO_4)_6X_2$ formula of apatite phase and publishes the summary of available thermodynamic data including standard formation Gibbs energy ($\Delta G_f°$), $\Delta H_f°$ and $S°$ at the temperature of 298 K (25°C) and the pressure of 1 bar (10^5 Pa), which are listed in the periodic table of phosphate apatites in **Fig. 25**.

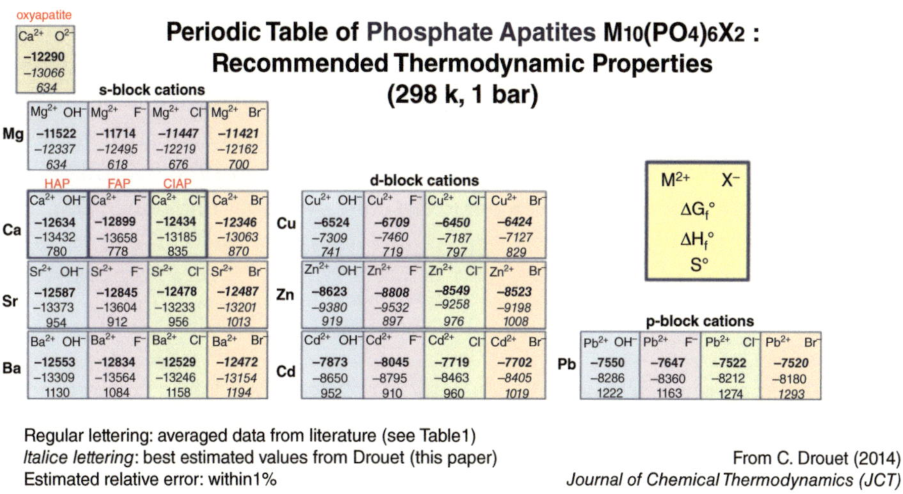

Fig. 25. Comprehensive periodic table of phosphate apatites provided by C. Drouet [170].

The discrepancies between published thermodynamic data probably arise from variable crystallinity states, polymorphs (either hexagonal or monoclinic, those not being systematically identified in literature reports), nonstoichiometry, hydration state and/or the presence of undetected impurities. When experimental-based data are not available (or are questionable), the so-called prediction of thermodynamic properties of solids becomes relevant. For example, it may allow an understanding of some unsuccessful experiment aiming at obtaining a desired hypothetical composition, or it may fill the gap between reported and needed thermodynamic values for the evaluation of equilibria constants or for the establishment of phase diagrams [170]. There are many methods developed for this purpose the summary of which can be found in works [171],[172].

For double oxides, $A_xB_yO_z$ in the system AO-BO was established the dependence [171]:

$$\Delta H^{\circ}_{f,cc} = \left(A_xB_yO_z\right) = f\left(\Delta \bar{H}^{\circ}_{f}\right) \tag{15}$$

where $\Delta H_{fcc}°(A_xB_yO_z)$ is the standard enthalpy of the formation of double oxide AxByOz from the component oxides AO and BO and $\Delta \bar{H}_f$ represents the sum of molar fraction enthalpies of component oxides AO and BO according to the following relationship:

$$\Delta \bar{H}_f^\circ = x_{AO}\Delta H_f^\circ(AO) + x_{BO}\Delta H_f^\circ(BO) \tag{16}$$

$\Delta H°(AO)$ and $\Delta H°(BO)$ are the standard enthalpies of the formation of component oxides from the elements, and X_{AO} and X_{BO} are the molar fractions of component oxides in the double oxide $A_xB_yO_z$ with a given composition [171].

The entropy of a solid compound is a function of masses of constituent atoms and the forces acting between these atoms: the greater the mass and the lower the force, the larger the entropy. The entropy of ionic solid will also depend upon the magnitude of the ionic charges. For compounds, the specific heat of which has reached the Dulong and Petit [173] value of 6 cal. per gram-atom [174],[175],[27] the mass is the principal factor, and in 1921, the authors gave an equation for the contribution of each element to the entropy of the compound [176].

$$S^\circ(298\,\mathrm{K}) = {}^3\!/_2 R\ \ln \mathrm{at.wt.} - 0.94 \tag{17}$$

For simple salts, such as alkali halides, the entropy may be estimated with fair accuracy as the sum of the entropies of constituent elements as given by this equation. However, the forces in solid salts are largely the ionic attractions, and the effect of the ionic radii upon the force constants and the vibrational frequencies is appreciable; in general, the entropy of a large ion is increased and the entropy of a small ion is decreased compared to the values given by **Eq. 17**. [176].

3.3.1. Volume-based thermodynamic predictive method

The volume-based thermodynamic approach (VTB), the so-called first-order method, has especially received much attention because the method is rather easy to use and has been shown in some cases to lead to output data well related to experimental results [170].

[27] One calorie is 4.184 J (joules). Gram-atom [gm] (and gram-molecule) was used to specify the amount of chemical elements or compound. These units had a direct relation with "atomic weights" and "molecular weights," which are in fact relative masses. "Atomic weights" were originally referred to the atomic weight of oxygen, by general agreement taken as 16. Although physicists separated the isotopes in a mass spectrometer and attributed the value of 16 to one of the isotopes of oxygen, chemists attributed the same value to the (slightly variable) mixture of isotopes 16, 17, and 18, which was for them naturally occurring element oxygen. Finally, an agreement between the International Union of Pure and Applied Physics (IUPAP) and the International Union of Pure and Applied Chemistry (IUPAC) brought this duality to an end in 1959/1960. Physicists and chemists have ever since agreed to assign the value 12, exactly, to the so-called atomic weight of the isotope of carbon with the mass number 12 (carbon 12, ^{12}C), correctly called the relative atomic mass $Ar(^{12}C)$. The unified scale thus obtained gives the relative atomic and molecular masses, also known as the atomic and molecular weights, respectively [174]. The law is also known as Dulong and Petit principle, which can be expressed in modern unit as: atomic weigh × specific heat $\approx \partial(3\boldsymbol{kTN}_A)/\partial T\, 3kN_A \approx \boldsymbol{c}_V \approx 25\,\mathrm{J{\cdot}K^{-1}mol^{-1}}$, i.e. the atomic weight of solid element multiplied by its molar specific heat is a constant [175].

3.3.2. Additive estimation methods

Additive estimation or contributive methods are probably the simplest approach based on the following [170],[177]:

1. **Atomic and ionic contribution**: the technique based on the method proposed by KELLOGG [178]:

$$C^{\circ}_{\mathrm{pm}}\left(A_xB_y,298.15\ \mathrm{K}\right)=x\ C^{\circ}_{\mathrm{pm}}\left(A\right)+y\ C^{\circ}_{pm}\left(B\right)\quad \text{e.g.:}$$
$$C^{\circ}_{\mathrm{pm}}\left(\mathrm{BaCl_2},298.15\ \mathrm{K}\right)=C^{\circ}_{\mathrm{pm}}\left(\mathrm{Ba}\right)+2\ C^{\circ}_{\mathrm{pm}}\left(\mathrm{Cl}\right) \tag{18}$$

The approach was later revised by KUBASCHEWSKI [179],[180]. These authors also proposed the method for the estimation of parameters A, B and C in the temperature dependence of $C_{\mathrm{pm}}°(\mathrm{T})$[28]:

$$C^{\circ}_{\mathrm{pm}}\left(T\right)=A+BT+\frac{C}{T^2} \tag{19}$$

$$A=\frac{10^{-3}T_m\left[C^{\circ}_{\mathrm{pm}}\left(298.15\ \mathrm{K}\right)+4.7n\right]-1.25n\cdot10^5\left(T_m\right)^{-2}-9.05n}{10^{-3}T_m-0.298} \tag{20}$$

$$B=\frac{25.6n+4.2n\cdot10^5\left(T_m\right)^{-2}-C^{\circ}_{\mathrm{pm}}\left(298.15\ \mathrm{K}\right)}{10^{-3}T_m-0.298} \tag{21}$$

$$C=-4.2n \tag{22}$$

where n is the number of ions (contributions) in the formula unit. The described approach is worthy for the substances with melting point temperatures (T_{m}) bellow 2300 K. The data on cationic and anionic contributions to heat capacity at 298 K are published in works [177], [179],[181],[182],[183].

For ionic compounds, the entropy can be calculated[29] from additive data given in **Table** 4, empirically found for cation and anion constituents of the compound (increments method of LATIMER [184]) [172].

	Element	C_{pm}°	S_{m}°	Element	C_{pm}°	S_{m}°	Element	C_{pm}°	S_{m}°
Contribution of cations	Ag	25.76	57.6	Hf	25.52	53,0	Pr	24.27	61.1

[28] The full equation for the temperature dependence is $C_{\mathrm{pm}}° = \mathrm{A + BT + C/T^2 + DT^2 + F/T^{1/2}}\ [\mathrm{J{\cdot}K^{-1}{\cdot}mol^{-1}}]$.

	Element	$C_{pm}{}^{o}$	$S_{m}{}^{o}$	Element	$C_{pm}{}^{o}$	$S_{m}{}^{o}$	Element	$C_{pm}{}^{o}$	$S_{m}{}^{o}$
[J·K^{-1}·mol^{-1}]									
	Al	19.66	23.4	Hg	25.10	59.4	Rb	26.36	59.2
	As	25.10	45.2	Ho	23.01	56.0	Sb	23.85	58.9
	Au	–	58.5	In	24.27	55.0	Se	21.34	60.5
	B	–	23.5	Ir	(23.85)	50.0	Si	–	35.2
	Ba	26.36	62.7	K	25.94	46.4	Sm	25.10	60.2
	Be	(9.62)	12.6	La	(25.52)	62.3	Sn	23.43	58.2
	Bi	26.78	65.0	Li	19.66	14.6	Sr	25.52	48.7
	Ca	24.69	39.1	Lu	–	51.5	Ta	23.01	53.8
	Cd	23.01	50.7	Mg	19.66	23.4	Te	–	69.0
	Ce	23.43	61.9	Mn	23.43	43.8	Th	25.52	59.9
	Co	28.03	34.1	Mo	–	35.9	Ti	21.76	39.3
	Cr	23.01	32.9	Na	25.94	37.2	Tl	27.61	72.1

[29]**Table 4** [172] refers to the values of entropy contribution from work [176] where some data for PO_4^{3-} ion are given in brackets. In order to verify this value (for the charge of cation, i.e. Ca^{2+}, the contribution of PO_4^{3-} anion is 17 [calories] × 4.184 = 71.13 J·K^{-1}·mol^{-1}, it is possible to calculate it from recommended thermodynamic data from **Fig. 25**. If (for example) three apatite end members were used, it is possible to calculate the contribution to PO_4^{3-} anion in hydroxylapatite, fluorapatite, chlorapatite, and bromapatite as follows:

$$\text{HAP}:\ (10\cdot 39.1)+(6x)+(2\cdot 18.83)=780\ \Rightarrow x=58.56\ \text{J}\cdot\text{K}^{-1}\cdot\text{mol}^{-1}$$

$$\text{FAP}:\ (10\cdot 39.1)+(6x)+(2\cdot 17.00)=728\ \Rightarrow x=50.50\ \text{J}\cdot\text{K}^{-1}\cdot\text{mol}^{-1}$$

$$\text{ClAP}:\ (10\cdot 39.1)+(6x)+(2\cdot 31.80)=835\ \Rightarrow x=63.40\ \text{J}\cdot\text{K}^{-1}\cdot\text{mol}^{-1}$$

$$\text{BrAP}:\ (10\cdot 39.1)+(6x)+(2\cdot 45.70)=870\ \Rightarrow x=64.60\ \text{J}\cdot\text{K}^{-1}\cdot\text{mol}^{-1}$$

It is also possible to calculate it from the contribution data for $Ca_3(PO_4)_2$ or $Mg_3(PO_4)_2$, where S°(298.15K) = 235.998 and 189.2 J·K^{-1}mol^{-1} (HSC software v.7.1), respectively:

$$(3\cdot 39.1)+(2x)=235.998\ \Rightarrow x=59.35\ \text{J}\cdot\text{K}^{-1}\cdot\text{mol}^{-1}$$

$$(3\cdot 23.40)+(2x)=189.2\ \Rightarrow x=59.5\ \text{J}\cdot\text{K}^{-1}\cdot\text{mol}^{-1}$$

Therefore, it is suggested to use average value from these calculations, i.e. PO_4^{3-} (for M^{2+} cation) ≈ 59 J·K^{-1}mol^{-1}. It is then possible that the application of contribution techniques to apatite leads to positive error in estimated thermodynamic data.

http://dx.doi.org/10.5772/62211

	Element	$C_{pm}°$	$S_m°$	Element	$C_{pm}°$	$S_m°$	Element	$C_{pm}°$	$S_m°$
	Cs	26.36	67.9	Nb	23.01	48.1	U	26.78	64.0
	Cu	25.10	44.0	Nd	24.27	60.7	V	22.18	36.8
	Fe	24.94	35.0	Ni	(27.61)	35.1	Y	(25.10)	50.4
	Ga	(20.92)	40.0	Os	–	50.0	Zn	21.76	42.8
	Gd	23.43	56.0	P	14.23	39.5	Zr	23.85	37.2
	Ge	20.08	49.8	Pb	26.78	72.2	–	–	–

	Anion	$C_{pm}°$	$S_m°$ for oxidation charge of cation			
			I	II	III	IV
Contribution of anions J·K^{-1}·mol^{-1}]	H^-	8.79	–	–	–	–
	F^-	22.80	20.8	17.0	18.3	20.4
	Cl^-	24.69	36.3	31.8	30.3	34.4
	Br^-	25.94	50.3	45.7	44.7	50.8
	I^-	26.36	58.3	53.5	54.8	53.9
	O^{2-}	18.41	4.5	2.9	2.4	3.2
	S^{2-}	24.48	20.6	18.4	20.1	17.0
	Se^{2-}	26.78	35.5	32.8	34.1	30.9
	Te^{2-}	27.20	38.1	41.9	44.1	40.1
	OH^-	30.96	(20.9)	(18.8)	(12.6)	–
	SO_4^{2-}	76.57	80.0	69.5	64.2	(41.8)
	SO_3^{2-}	–	42.9	–	–	–
	NO_3^-	64.43	86.0	74.0	–	–
	NO_2^-	–	70.6	–	–	–
	CO_3^{2-}	58.58	62.4	46.6	–	–
	CrO_4^{2-}	90.76	–	–	–	–
	MoO_4^{2-}	90.37	–	–	–	–
	WO_4^{2-}	97.49	–	–	–	–
	SiO_3^{2-}	59.3	60.7	43.9	29.3	–
	SiO_4^{4-}	73.5	–	–	–	–
	PO_4^{3-}	73.90	(100.4)	(71.1)	(50.2)	–
	UO_4^{2-}	107.11	–	–	–	–

Table 4. The contribution of cations and anions to $C_{pm}°$ and $S_m°$ (298.15 K) [172],[180].

2. **Structural and simple oxide contribution**: the value of $C_{pm}°(298.15\ K)$ or the parameters of the temperature dependence in **Eq. 19** can be calculated from the contribution of constituent oxides (Neumann–Kopp rule, NKR):

$$x\ A(s) + yB(s) = A_xB_y(s), \quad \text{e.g.,} \quad \mathrm{SrO}(s) + \mathrm{CuO}(s) = \mathrm{SrCuO_2}(s) \tag{23}$$

$$\begin{aligned} &C_{pm}^{\circ}(A_xB_y, 298.15\,\mathrm{K}) = x\ C_{pm}^{\circ}(A, 298.15\,\mathrm{K}) + y\ C_{pm}^{\circ}(B, 298.15\,\mathrm{K}) \quad \text{e.g.,} \\ &C_{pm}^{\circ}(\mathrm{SrCuO_2}, 298.15\,\mathrm{K}) = C_{pm}^{\circ}(\mathrm{SrO}, 298.15\,\mathrm{K}) + C_{pm}^{\circ}(\mathrm{CuO}, 298.15\,\mathrm{K}) \end{aligned} \tag{24}$$

or from the structural contribution [179].

3. **Prediction method for homological series and groups of chemically related substances (oxides)**: based on the approach of ALDABERGENOV et al [185] and GOSPODINOV and MIHOV [186]. The molar heat capacity in homological series as $A_m(B_xO_y)_n$, is a linear function of n, i.e. the coefficient which specifies the number of complex anions $(B_xO_y)^{z-}$ in the formula unit. For example, for the series of alkaline aluminate, it can be written as

$$\mathrm{AlO_2} \xrightarrow{+\mathrm{AlO_2}} \mathrm{Al_2O_4}^{2-} \xrightarrow{+\mathrm{AlO_2}} \mathrm{Al_3O_6}^{3} \ \ldots \tag{25}$$

Since each higher anion is formed by the addition of primary ion $(AlO_2)^-$ unit, higher anion is considered to form n-multiples of primary ion, the value of which is determined from the available experimental data for $KAlO_2$, $LiAlO_2$ and $NaAlO_2$ and from ions contribution for cation K^+, Li^+ and Na^+ obtained from their standard entropies in an infinitely diluted solution [179].

Apatite phase may be treated in the first approximation as the sum of contributions arising from the constitution of binary oxides/compounds. For example, in the case of fluorapatite ($Ca_{10}(PO_4)_6F_2$), a decomposition into contribution of $9CaO + 3P_2O_5 + CaF_2$ could be considered. It can be generalized to any end-member in the form $9CaO + 3P_2O_5 + XF_2$ [170].

3.4. Dissolution of apatites

At the fundamental level, the reactions between solids and liquids involve a coupled sequence of mass transport, adsorption/desorption phenomena, heterogeneous reactions, chemical transformations of intermediates, etc., the identification, separation and kinetic quantification of which are all necessary if the mechanism of the process is to be fully understood and described [111],[187]. It was generally accepted that the process during the dissolution of lattice ions includes the following [187],[188]:

a. Detachment of species (ion) from a kink site

http://dx.doi.org/10.5772/62211

b. Surface diffusion from the crystal steps

c. Desorption from the surface

d. Diffusion into the bulk solution

The dissolution of apatites under steady-state conditions, in pure water or in aqueous acidic media, includes the following simultaneous steps [187],[189]:

i. Diffusion of chemical reagents (H^+) from bulk solution to the solid/liquid interface. In the case of acidic dissolution, the diffusion of acid anions (A^{n-}) must be taken into account as well.

ii. Adsorption of H^+ onto the surface of apatite.

iii. Chemical transformation of the surface.

iv. Desorption of ions of fluoride, calcium and phosphate from the crystal surface.

v. Adsorption of chemicals from solution back onto the surface of apatite.

These steps are likely to be more complicated, e.g. the processes I and V include chemical transformation of ionic species during diffusion because the pH of solution is known to depend on the distance from the solid/liquid interface. In other words, the value of pH is higher near to the surface of apatite and decreases with increasing distance from the surface.

When apatite gets in contact with undersaturated solution, the dissolution states from 1 to 5 mentioned above take place. In order to provide detailed description of the process, the following assumption must be introduced [187]:

a. Stoichiometric apatite is dissolved, and neither nonstoichiometric layer nor other ions except for calcium, phosphate, hydroxide and fluoride are initially present in the crystal lattice, whereas the volume and surface defects (dislocations and dislocations outlets, respectively) might be present and, when present, they are distributed randomly.

b. Except for the presence of dislocation outlets, the initial surface of apatite is perfect (molecularly sooth). Otherwise, each imperfection might be the dissolution nucleus.

c. Despite the limitations and drawbacks discussed above, all models are correct and complementary to each other.

d. In some cases, anions of acid might have an influence on apatite dissolution due to the specific affinity (e.g. citrate) by means of formation of insoluble compounds (e.g. sulfate). That is the reason why anions are not specified either here or below. It is just assumed that the dissolution of apatite happens in acid H_nA, where A^{n-} is an anion.

e. The hydration effect on all ions and molecules involved as well as that on crystal surface of apatite is omitted for simplicity.

f. All crystal faces of solid apatite are equal and have similar ionic arrangement. There is not any considered specific influence of different crystal faces on the dissolution mechanism.

Since the dissolution models have limitations and drawbacks, none of them was able to describe the dissolution of apatite in general. Furthermore, the most of models were elaborated for the apatite dissolution in slightly acidic or nearly neutral solution ($4 < pH < 8$), for relatively small values of solution undersaturation and for the temperatures in the range from 25°C to 37°C, nothing is known about their validity for the dissolution of apatite in strong inorganic acid such as HCl, HNO_3 and H_2SO_4 and at temperatures above 70°C [187].

The classification of congruent (stoichiometric)/incongruent (nonstoichiometric) dissolution is based on direct measurements of either ionic concentrations in the solution or the surface composition of apatite during the dissolution [187]:

1. **Congruent dissolution**: ions in solids are dissolved simultaneously with the dissolution rates proportional to their molar concentrations, e.g. for $Ca_5(PO_4)_3Z$ it should be written [190],[191]:

$$\frac{[\mathrm{Ca}]_t}{[\mathrm{P}]_t - [\mathrm{P}]_0} - R = 0 \tag{26}$$

   where $[Ca]_t$, $[P]_t$ and $[P]_0$ denotes actual (at time $\boldsymbol{t}$) concentration of calcium and phosphorus in the solution and initial concentration of phosphorus (at time $t = 0$), respectively. The value of R is given by ideal stoichiometric ration of Ca:P = 5/3 in the formula of apatite (**Table 7** in **Chapter 1**).

2. **Incongruent dissolution**: the dissolution rate is different for each ion. That leads to the formation of surface layer with chemical composition different from the bulk of solid apatite phase.

The behavior of surface of apatite during the dissolution according to Dorozhkin [112] is shown in **Fig. 26**. Fluorine from fluorapatite or hydroxyl from hydroxyapatite dissolves most probably as the first. This can be explained by their position in the channels of crystal lattice. The dissolution starts with replacements of fluorine for water. Proton(s), chemisorbed on the nearest phosphate group(s), most probably catalyze this process. Local positive charge on apatite is formed as the result (*b*). Obtained local positive charge is removed by the detachment of one of the nearest calcium cations: Ca(2) is more likely to be detached first (*c*) since Ca(1) is located rather far from the channel. Acidic anions present in the solution most probably participate in this process. Later, proton(s) from the bulk solution replace other calcium cation(s) around the nearest phosphate group. Very thin surface layer of acidic calcium phosphates is formed as the result [112].

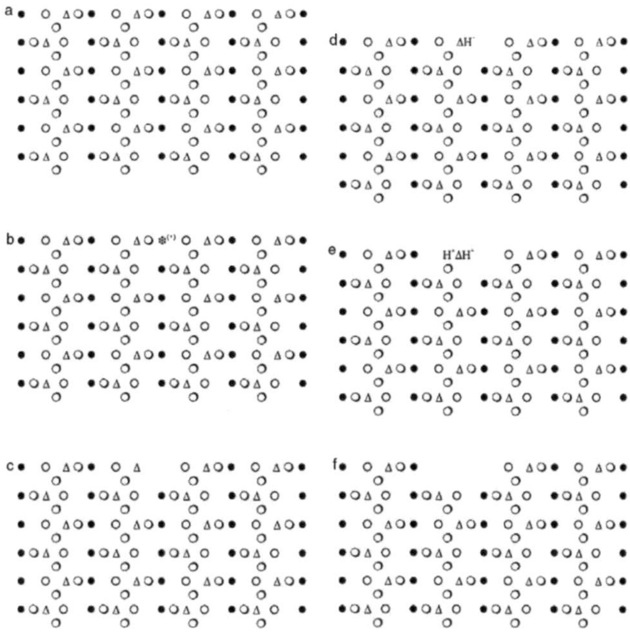

Fig. 26. Schematic illustration of the surface dissolution mechanism of apatite at the nanolevel: (a) part of the initial surface of apatite; (b) replacement of one fluorine (or hydroxyl) anion with water molecule resulting in local positive charge formation; (c) removal of one of the nearest calcium cations; (d) sorption of next proton; (e) removal of another calcium cation with simultaneous formation of acidic calcium phosphate; (f) detachment of one phosphate anion together (or simultaneously) with third calcium cation. A jump-wise shift of the dissolution step occurs simultaneously at stage f. (●) Fluorine for fluorapatite or hydroxyl for hydroxylapatite; (○) Ca(II) on the first plane, (○) Ca(II) on the back plane and Ca(I) on the back plane; ($^{*(+)}$) molecule of water and local positive charge; (Δ) PO_4^{3-} tetrahedra; $H^+\Delta H^+$ and ΔH^+ represent the surface tetrahedral anions of $H_2PO_4^-$ and HPO_4^{2-}, respectively. Chemisorbed protons, water molecules and acidic anions are omitted for simplicity. Note that crystal structure of apatite is shown very schematically: it should be hexagonal, while here it looks more or less like cubic [112].

When all (or almost all) the nearest calcium cations have been replaced with protons according to the reactions [112]:

$$CaHPO_4 + H^+ \rightarrow Ca^{2+} + H_2PO_4^- \tag{27}$$

$$CaHPO_4 + 2H^+ \rightarrow Ca^{2+} + H_3PO_4 \tag{28}$$

phosphate anions ($H_2PO_4^-$, $CaH_2PO_4^+$, or H_3PO_4) also detach (*f*). As the result, the dissolution step moves forward jump-wise over a distance equal to the dimension of phosphate anion, of approximately 3 Å. The detachment of phosphate anions and calcium cations results in the formation of hole. The dimension of this hole should be close to the lattice parameters of apatite. Most probably, it is a dissolution nucleus on which the polynuclear dissolution mechanism is based [112].

3.4.1. Dissolution of fluorapatite

The adsorption of H^+ onto the surface of apatite (**Fig. 26(b)–(d)**) resulted in the aqueous pH increasing from 5.60 to 8.45 within the first hour of dissolution. Ions of H^+ were adsorbed onto oxygen ions of phosphate groups as well as onto ions of fluoride [187],[191].

The reaction of stoichiometric (congruent) dissolution of pure stoichiometric apatite can be expressed by the reaction[30] [112],[187],[192]:

$$Ca_5(PO_4)_3Z \rightarrow 5\ Ca^{2+} + 3\ PO_4^{3-} + Z^- \tag{29}$$

where $\boldsymbol{Z}$ = OH^- and F^-. Assuming the unit activity of the solid phase ($a_{Ca_5(PO_4)_3Z}$ = 1), the equilibrium constant of dissolution (K) can be expressed via the solubility product or ion activity product (IAP) of apatite[31] [187],[193],[194],[195],[196]:

$$K = \text{IAP} = a_{Ca^{2+}}^{5}\, a_{PO_4^{3-}}^{3}\, a_{Z} \qquad [\text{eq.}] \tag{30}$$

where $\boldsymbol{a}_i$ denotes the thermodynamic activity of aqueous species. The standard Gibbs (free) energy of the reaction related to the standard temperature (298.15 K) and pressure (0.101 MPa) is given by the formula[32]:

---

[30] Reaction **Eq. 7** and **Eq. 12** are widely used for the description of dissolution process of apatite [112] using stoichiometry pertinent single or double apatite formula and Z = F or OH.

[31] Out of equilibrium state, ***IAP*** is not equal to ***K*** (see the discussion to **Eq. 20**). Double formula of apatite is assumed then with respect to the apatite stoichiometry; the law for ionic activity product has the following form:

$\boldsymbol{K}_{sp} = \boldsymbol{a}_{(Ca^{2+})}{}^{10}\, \boldsymbol{a}_{(PO_4^{3-})}{}^{6}\, \boldsymbol{a}_{(Z-)}{}^{2}$.

For example, in hydroxylapatite, where Z^- = OH^-, the activity of OH^- anion can be expressed by using ionic product of water (25°C): $\boldsymbol{K}_w = \boldsymbol{a}_{(H^+)}\, \boldsymbol{a}_{(OH^-)} = 1{\cdot}10^{-14}\ (mol{\cdot}dm^{-3})^2$ and then $\boldsymbol{a}_{(OH^-)} = \boldsymbol{K}_w / \boldsymbol{a}_{(H^+)}$. Since $pH = -log\ [H^+]$ and $pOH = -log\ [OH^-]$ the $[H^+] = 10^{-pH}$ and $[OH^-] = 10^{-pOH}$ and $pK_w = pH + pOH = 14$:

$\boldsymbol{K}_{sp} = \boldsymbol{a}_{(Ca^{2+})}{}^{10}\, \boldsymbol{a}_{(PO_4^{3-})}{}^{6}\, 10^{-2pOH} = \boldsymbol{a}_{(Ca^{2+})}{}^{10}\, \boldsymbol{a}_{(PO_4^{3-})}{}^{6}\, 10^{-2(14-pH)} = \boldsymbol{a}_{(Ca^{2+})}{}^{10}\, \boldsymbol{a}_{(PO_4^{3-})}{}^{6}\, \boldsymbol{K}_w{}^{2} 10^{2pH}$.

The activity of ionic species is the product of ion molar concentration ([196]) and ion activity coefficient, e.g. ($\boldsymbol{a}_{(Ca^{2+})}$ = $([Ca^{2+}] / [Ca^{2+}]^{\circ})\, \gamma_{(Ca^{2+})}$, where the standard state $[Ca^{2+}]^{\circ} = 1\ mol{\cdot}dm^{-3}$ can be chosen), which can be calculated, e.g. the example via **Debye–Hückel**, **extended Debye–Hückel**, or **modified Davies equation** (in dependence on ionic strength):

$\log \gamma_i = -\boldsymbol{A}\boldsymbol{Z}_i^2 \sqrt{\boldsymbol{I}}\ (\boldsymbol{I} < 10^{-3})$,

$\log \gamma_i = [-\boldsymbol{A}\boldsymbol{Z}_i^2 \sqrt{\boldsymbol{I}} / (1+\boldsymbol{B}{\cdot}\boldsymbol{\alpha}_i \sqrt{\boldsymbol{I}})]\ (10^{-3} < \boldsymbol{I} < 0.1)$,

or $\log \gamma_i = [-\boldsymbol{A}\boldsymbol{Z}_i^2 \sqrt{\boldsymbol{I}} / (1+\boldsymbol{B}{\cdot}\boldsymbol{\alpha}_i \sqrt{\boldsymbol{I}})] + 0.3\boldsymbol{I}\ (\boldsymbol{I} > 0.1)$, respectively.

A and ***B*** are the temperature-dependent constants, $\boldsymbol{Z}_i$ is the charge number of ***i***th ion, $\boldsymbol{\alpha}_i$ is the radius of hydrated ***i***th ion, and ***I*** is the number of ions: $\boldsymbol{I} = \frac{1}{2} \sum ([C_i]Z_i^2)$.

The solubility product defined as the product of concentration of compound constituents ions, which are released during the dissociation is often used: $\boldsymbol{K}_{sp} = [Ca^{2+}]^{10}\, [PO_4^{3-}]^{6}\, [Z^-]^{2}$.

[32] $\Delta_r G^{\circ}$ = -RT ln K (in the equilibrium state) and ln K = ln 10 log K. From that, it should be derived that $\Delta_r G^{\circ}$ = -8.314·298.15 ln 10 log K = -5708 log K [J·mol^{-1}], where ln 10 ≈ 2.303. Since $\log_{10} K = \log K = \lg K$, it can also be written as $\Delta_r G^{\circ} = -5708 \lg K$.

http://dx.doi.org/10.5772/62211

$$\Delta_r G^\circ = -5.707 \ \log \ \mathrm{K} \qquad \left[\mathrm{kJ \cdot mol^{-1}}\right] \tag{31}$$

For reaction 29, the following equation[33] can be derived [187]:

$$\begin{aligned}\Delta_r G^\circ &= 5 \ \Delta G_f^\circ\left(\mathrm{Ca^{2+}}\right) + 3 \ \Delta G_f^\circ\left(\mathrm{PO_4^{3-}}\right) + \Delta G_f^\circ\left(\mathrm{Z^-}\right) \\ &-\Delta G_f^\circ\left(\mathrm{Ca_5\left(PO_4\right)_3 Z}\right)\end{aligned} \tag{32}$$

Eq. 32 can be further treated as follows:

$$\begin{aligned}\Delta G_f^\circ\left(\mathrm{Ca_5\left(PO_4\right)_3 Z}\right) &= 5 \ \Delta G_f^\circ\left(\mathrm{Ca^{2+}}\right) + 3 \ \Delta G_f^\circ\left(\mathrm{PO_4^{3-}}\right) \\ &+\Delta G_f^\circ\left(\mathrm{Z^-}\right) - \Delta_r G^\circ\end{aligned} \tag{33}$$

Harouiya et al [197] assumes that the dissolution of apatite in the temperature range from 5°C to 50°C, and the pH from 1 to 6 can be expressed by the following formula:

$$\mathrm{Ca_5\left(PO_4\right)_3} F + 3 \ \mathrm{H^+} \rightarrow 5 \ \mathrm{Ca^{2+}} + 3 \ \mathrm{HPO_4^{2-}} + F^- \tag{34}$$

With regard to assumed standard state, the equilibrium constant of reaction (**Eq. 34**) can be written as[34]

$$K' = a_{\mathrm{Ca^{2+}}}^5 \ a_{\mathrm{HPO_4^2}}^3 \ a_{\mathrm{F}} \ a_{\mathrm{H^+}}^{-3} \tag{35}$$

The chemical affinity[35] (A) of **Eq. 34** is given by the law:

$$A = -RT \ \ln\left(\frac{K' a_{\mathrm{H^+}}^3}{a_{\mathrm{Ca^{2+}}}^5 \ a_{\mathrm{HPO_4^2}}^3 \ a_{\mathrm{F}}}\right) \tag{36}$$

In the closed-system experiment, the dissolution rates are generally obtained from the slope of concentration of reactive solution versus the time:

---

33 $\Delta_r G^\circ = \sum \nu_i \ \Delta G^\circ_{f,i}$, where ν_i denotes the stoichiometric coefficient of given species and $\Delta G^\circ_{f,i}$ its standard enthalpy of formation.

34 Since the saturation of solution with respect to $Ca_5(PO_4)_3Z$ means that $K' = 0$ (equilibrium state), it can be derived that $a_{(Ca^{2+})}^5 \ a_{(PO_4^{3-})}^3 \ a_{(Z-)} - K' a_{(H+)}^3 = 0$ and then $\Delta_r G^\circ = -RT \ln (a_{(Ca^{2+})}^5 \ a_{(PO_4^{3-})}^3 \ a_{(Z-)} / K' a_{(H^+)}^3)$. Since $A = -\Delta_r G^\circ$ (please see note 35), $A = -RT \ln (K' a_{(H^+)}^3 / a_{(Ca^{2+})}^5 \ a_{(PO_4^{3-})}^3 \ a_{(Z-)})$.

35 The relationship between the reaction Gibbs energy and chemical affinity: $A = -\Delta_r G^\circ$ was introduced by T. de Donder.

$$r = \frac{\partial c_i}{\partial t} \frac{M_r}{v_i s} \tag{37}$$

where ***r*** refers to the dissolution rate of apatite, c_i denotes the concentration of *i*th element, *t* is the time, M_r designates the mass of fluid in the reactor, v_I is the stoichiometric coefficient and *s* designates the total surface area of sample in the reactor. The slope of the plot may not be constant and may increase or decrease with time from the following reasons [195],[197]:

1. Changes in the reactive fluid volume, which may occur due to the evaporation of solvent or regular sampling of reactive fluid
2. Nonzero order reaction kinetics
3. Approach to equilibrium, where the dissolution rate decreases and reaches zero at equilibrium. This approach is described by the transition state theory as follows:

$$r = r_+ \left(1 - \exp\left[\frac{-A}{\sigma RT}\right]\right) \tag{38}$$

The symbol r_+ symbolizes the far from equilibrium dissolution rate, which may depend on the composition of solution, *A* is the affinity of reaction of dissolution, *σ* stands for the Temkin's average stoichiometric number equal to the ration of rate of destruction of the activated or precursor complex relative to overall rate, *R* designates universal gas constant and *T* is the temperature on the absolute scale. Overall rate (*r*) is equal to forward rate (r_+) when $A >> \sigma RT$. As one of the approaches of equilibrium, overall rates gradually decrease and reach zero at equilibrium where $A = 0$. The value of *r* is within 10% of r_+ when $A/\sigma RT > 2.3$ which is equivalent to $A > 1.36\ \sigma$ kcal·mol^{-1}. It indicates that the parameter *σ* plays a crucial role in the variation of dissolution rates at near to equilibrium conditions [195],[197].

The value of r_+ depends on the pH according to the following equation [197],[198]:

$$r_+ = k\ \mathrm{pH}^{-n} \tag{39}$$

where *k* refers to the tare constant and *n* stands for the reaction order determined as the slope of linear dependence of $ln\ r_+$ on pH. The dependence of *k* on the temperature is given by the Arrhenius law:

$$k = A_\mathrm{A}\ \exp\left(\frac{-E_\mathrm{A}}{\mathrm{RT}}\right) \tag{40}$$

where A_A is the preexponential (frequency) factor and E_A is the activation energy of the process. The combination of **Eqs. 38, 39** and **40** leads to the equation for the dissolution rate as follows [197]:

$$r = A_A a_{H^+}^n \exp\left(E_A / RT\right)\left(1 - \exp\left[\frac{-A}{\sigma RT}\right]\right) \tag{41}$$

For acidic dissolution of calcium fluorapatite, ions of *F* were found to dissolve faster (or prior to) when compared to calcium and phosphate. A similar phenomenon of prior (or faster) dissolution of calcium when compared to that of phosphate was also found [187],[199]. The release of calcium and phosphate ions from the surface of apatite seems to be affected by the presence of salts, such as Na_2SO_4, CH_3COONa, or NaCl, in the solution. The concentration of phosphate in the solution increases in the following order [187]:

$$Na_2SO_4 > CH_3COON_a > NaCl.$$

On the contrary, the concentration of calcium ions decreases in the same order.

The undersaturation (US) and relative undersaturation (US_r) of apatite solvent dissolved upon is defined as follows[36] [193]:

$$US_r = 1 - US = 1 - \left(\frac{IAP}{K}\right)^{1/18} \tag{42}$$

where *K* is the equilibrium constant of reaction 29 and IAP is the ion activity product. The law is written with regard to the stoichiometry of double formula of apatite, where $\sum \nu_i$ in **Eq. 30** is 2 × (5 + 3 + 1) = 18. The value of *IAP/K* ratio is as follows:

a. IAP/*K* > 1, the reaction proceeds to the left (precipitation, supersaturated solution)

b. IAP/*K* = 1, the reaction is in the equilibrium state (saturated solution)

c. IAP/*K* < 1, the reaction proceeds to the right (dissolution, undersaturated solution)

This ration is also used to calculate the saturation index (SI) for the reaction of dissolution [200]:

$$SI = \log\left(\frac{IAP}{K}\right) \tag{43}$$

Depending on the saturation index, the following states of solutions are recognized:

[36] Since the system is not in the equilibrium state IAP ≠ K.

i. SI < 0 undersaturation

ii. SI = 0 saturation, i.e. mineral[37] or salt is in equilibrium with solution

iii. SI > 0 supersaturation

3.4.2. Classification of the dissolution models

The models, which are usually applied for the description of dissolution of apatites, include the following [187]:

1. **Diffusion and kinetically controlled models**: the dissolution of apatite was found to be the diffusion (transport) controlled in some cases [201],[202], kinetically (surface) controlled in other ones [203],[204], and even intermediate [205], i.e. both diffusion and kinetically controlled. Both models usually operate with the so-called driving force which means either the concentration gradient within the Nernst diffusion layer (the diffusion controlled model) or the gradient of ionic chemical potentials between the apatite crystal surface and bulk solution (the kinetically controlled model). Moreover, the results obtained on these models are valid only within the experimental conditions studied; no extrapolation can be made beyond the tested ranges. For example, after, let's say, a slight agitation decrease or temperature increase, an initially kinetically controlled dissolution might be controlled by the diffusion. Thus, high sensitivity to applied experimental conditions appears to be the main drawback of these models [187].

2. **Polynuclear model**: is based on the study of dissolution and kinetics of growth of apatite under constant composition conditions [191],[193],[206],[207],[208]. Polydispersed samples of apatite were put into a stirred undersaturated (for dissolution experiments) or supersaturated (for those on crystal growth) solutions, and the pH of solution and the amount of added chemicals (an acid for the dissolution experiments and a base for those on crystal growth) were permanently recorded as the functions of time. The results obtained were plotted versus either undersaturation or supersaturation values: straight lines were obtained in the specific logarithmic coordinates typical for this model. According to the model, the dissolution nuclei, i.e., the collections of vacant sites for Ca^{2+}, PO_4^{3-}, and OH^- ions, are formed on the crystal surface of apatite and spread over the surface with a definite lateral rate.

   The nucleation rate is assumed as a function of mean ion activity. The lateral growth rate of nuclei is assumed proportional to the difference between total concentration of calcium ions in the saturated solution and in a solution, while the rate constant is related to the frequency for calcium ions to make a diffusion jump into a kink and, simultaneously, partly dehydrate. Recent investigation reveals that the rate-determining step was not the diffusion but two-dimensional surface nucleation [187].

3. **Self-inhibition model**: assumes the formation of self-inhibition calcium-rich layer on the surface of apatite during the dissolution. According to this model, apatite is dissolved by

[37] The ranges of Si near zero are generally considered to be within the equilibrium zone for the mineral. The ranges of SI = 0 ± 0.5 and 0 ± (5%) (lg $K_{mineral}$) were used in various studies [200].

ionic detachment of calcium and phosphate ions from the surface to a solution. When an initial portion of apatite has been dissolved, some amount of calcium cations (probably, in connection with anionic counter ions) is returned from the solution and adsorbed back onto the surface of apatite. The latter process results in the formation of a semipermeable ionic membrane, which is formed from positively charged layer containing strongly adsorbed calcium ions, i.e. calcium-rich layer [187], [201],[209],[210],[211].

4. **Congruent and incongruent dissolution model**: was already described above.

5. **Chemical model**: This model was developed from the self-evident supposition that it would be highly unlikely if apatite were dissolved by the detachment of "single molecules" equal to the unit cells and consisting of 18 ions. Moreover, in the crystal lattice, practically all ions are shared with neighboring unit cells and often cannot be attributed to given "single molecule." Based on the experimental results obtained on one hand, and on analysis of the data found in references on the other hand, a sequence of four successive chemical reactions was proposed to describe the process of apatite dissolution [112], [187],[212],[213]:

$$Ca_5(PO_4)_3(F,OH) + H_2O + H^+ \rightarrow Ca_5(PO_4)_3(H_2O^+) + HF,\ H_2O \quad (44)$$

$$2\ Ca_5(PO_4)_3(H_2O^+) \rightarrow 3\ Ca_3(PO_4)_2 + Ca^{2+} + 2\ H_2O \quad (45)$$

$$Ca_3(PO_4)_2 + 2H^+ \rightarrow Ca^{2+} + 2\ CaHPO_4 \quad (46)$$

$$CaHPO_4 + H^+ \rightarrow Ca^{2+} + H_2PO_4 \quad (47)$$

Eqs. 44–47 can be used instead well-known net reactions [187]:

$$Ca_5(PO_4)_3(F,OH) + H_2O + 7H^+ \rightarrow 5\ Ca^{2+} + 3\ HPO_4 \\ +HF,\ H_2O \quad (48)$$

$$Ca_5(PO_4)_3(F,OH) \rightarrow 5\ Ca^{2+} + 3\ PO_4^{3-} + F,\ OH^- \quad (49)$$

In principle, the dissolution process could also happen according to reaction 49 followed by chemical interaction in the solution among ions of apatite and acid near the crystal surface [214]:

$$5\ Ca^{2+} + 3\ PO_4^{3-} + F^-, OH^- + 7\ H^+ \rightarrow 5\ Ca^{2+} + 3\ H_2PO_4^- + HF,\ H_2O \quad (50)$$

1. **Etch pit formation**: the process of etch pit formation describes the dissolution of apatite crystals containing structural defects (dislocations and inclusions). The presence of dislocations accelerates the dissolution because they give rise to continuous steps on the surface and the strain energy, which they cause in crystals, favors the etch pit formation. The pits appear at the dislocation outlets; they are 0.1–10 pm in size (the dimensions depend on the dissolution kinetics and dissolution time: they increase when the dissolution progresses) and usually have a hexagonal shape according to the crystal symmetry **$P6_3/M$** of pure apatite [187],[215],[216],[217],[218],[219].
2. **Ion exchange model**: based on a supposition about the adsorption of protons and anions of acid (e.g. citrate anions) from a solution onto the surface of apatite and removing of ions of calcium and phosphate into the solution instead [187].
3. **Hydrogen catalytic model**: based on a reasonable suggestion about the adsorption of protons onto negatively charged oxygen ions of phosphate groups of apatite. The sorption of protons results in the transformation of surface PO_4^{3-} groups into HPO_4^{2-} and catalyzes the dissolution process [187].

3.4.3 Methods for the evaluation of reactivity of phosphate rocks

Chemical methods are used for the evaluation of reactivity of different phosphate rocks from which the fertilizers are manufactured for their possible direct application as fertilizers via empirical solubility test. Citric acid, formic acid, neutral ammonium citrate,[38] and alkaline ammonium citrate are used as solvents for the extraction of P_2O_5. The latter is used mainly for the evaluation of calcined aluminum phosphates. Most of these reagents were not originally intended to evaluate the reactivity of phosphate rocks. For instance, neutral and alkaline ammonium citrate solutions were originally intended to separate chemical reaction products in superphosphate and other fertilizers from unreacted rock on the assumption that unreacted rock was insoluble in these reagents. The citric acid extraction was developed to evaluate basic slag, a popular fertilizer material in European countries. The formic acid extraction was developed specifically for phosphate rocks [220].

Nearly all extraction methods use the ratio of sample weight to extraction volume 1 g:100 ml. [39] The extraction time usually ranges from 30 min to 1 hour. The temperature and the agitation during extraction test may be specified. For example, the AOAC method uses neutral ammonium citrate[40] of specified concentration (1 g of sample and 100 ml of solution) with the extraction time of 30 min at 65°C. The Wagner method uses 2% solution of citric acid, the

---

[38] Neutral ammonium citrate is prepared by dissolving required amount of citric acid and neutralizing it with ammonium hydroxide. The pH of the reagent is adjusted to neutral [221].

[39] The amount of used solution is also expressed in the name of the method, e.g. 100 ml method or 150 ml method [222].

[40] The neutral ammonium citrate test was used as the official method in the United States, and the test by acidic acid was developed for the comparison [222].

extraction time of 30 min and the temperature of 17.5°C [220],[221],[222],[223]. Unavailable phosphoric acid is usually expressed as the portion of fertilizer, which is *insoluble* in neutral ammonium citrate [769]. Phosphate removed during the neutral ammonium test is termed as *citrate-soluble*. The sum of *water-soluble* and *citrate-soluble* phosphate is termed as *available* [224].

Fused magnesium phosphate (FMP) is highly soluble in 2% citric acid[41] but is less soluble in neutral citrate, while calcined defluorinated phosphate (CDP) and Thomas slag (Thomas phosphate) are fairly soluble in both citric acid and citrate. Actually, FMP dissolves fairly rapidly in neutral citrate at the beginning, but the dissolution is hindered by gelatinous silica, which forms on the surface of the FMP particles. This layer can be broken by vigorous stirring [222].

One disadvantage of all these methods is that the percentage of leached P_2O_5 depends on the grade of the rock, especially when the rock contains inert gangue minerals such as silica. In order to eliminate the adventitious effect of grade, the concept of absolute citrate solubility index (ASC) was developed [225],[226]:

$$\mathrm{ASC} = \frac{\text{AOAC citrate solubility } P_2O_5[\%]}{\text{Theoretical } P_2O_5 \text{ in apatite } [\%]} \tag{51}$$

The percentage of dissolved P_2O_5 is expressed as the gangue-free apatite [220]. If rocks contain free calcium of magnesium carbonate, these carbonates should be removed by the extraction with a suitable reagent before carrying out the test in order to obtain correct indication of reactivity [227].

It was also found that the length of the *a*-axis in the apatite unit cell ($\boldsymbol{a}_0$) is statistically related to the ASC according to the relationship [225]:

$$\mathrm{ASC} = 421.4\,(9.369 - a_0) \tag{52}$$

Author details

Petr Ptáček

Brno University of Technology, Czech Republic

---

[41] It is believed that weak citric acid solution imitates the condition near the plant roots [222].

References

[1] Straaten PV. Rocks for Crops, Agro Minerals of Sub-Sahara Africa. (CD), ICRAF, Nairobi, Kenya, 2002. ISBN: 0-88955-512-5.

[2] Abouzeid Abdel-Z.M. Physical and thermal treatment of phosphate ores—an overview. International Journal of Mineral Processing 2008;85(4) 59–84.

[3] Comodi P, Liu Y, Zanazzi PF, Montagnoli M. Structural and vibrational behaviour of fluorapatite with pressure. Part 1: in situ single-crystal X-ray diffraction investigation. Physics and Chemistry of Minerals 2001;28(4) 219–224.

[4] Wilson RM, Elliott JC, Dowker SEP, Smith RI. Rietveld structure refinement of precipitated carbonate apatite using neutron diffraction data. Biomaterials 2004;25(11) 2205–2213.

[5] Smith F. Industrial Applications of X-Ray Diffraction. CRC Press, 1999. ISBN: 978-0824719920

[6] Elliott JC. Structure and Chemistry of the Apatites and Other Calcium Orthophosphates, Elsevier Press, Amsterdam, 1994. ISBN: 978-0-444-81582-8

[7] Svergun DI, Koch MHJ, Timmins PA, May RP. Small Angle X-Ray and Neutron Scattering from Solutions of Biological Macromolecules. IUCr (International Union of Crystallography) texts on crystallography—Volume 19. Oxford University Press, 2013. ISBN: 978-0199639533

[8] Nitske WR. The Life of Wilhelm Conrad Röntgen, Discoverer of the X ray. University of Arizona Press, 1971. ISBN: 978-0816502592

[9] Daubrée A. Experiences sur la production artficielle de l'apatite, de la topaze, at de quelques autres métax fruorifés. Comptes rendus de l'Académie des Sciences 1851;32 625.

[10] Röntgen WC. On a new kind of rays. Nature 1896;53 274–276. doi: 10.1038/053274b0

[11] Weber AS. Nineteenth-Century Science: An Anthology. Broadview Press, 2000. ISBN: 978-1551111650

[12] Rosenthal-Schneider I. Begegnungen mit Einstein, von Laue und Planck: Realität und wissenschaftliche Wahrheit, Facetten der Physik—Volume 12. Springer-Verlag, 2013. ISBN: 978-3663018841

[13] Chatterji T. Neutron Scattering from Magnetic Materials. Gulf Professional Publishing, 2005. ISBN: 978-0080457055

[14] Liang L, Rinaldi R, Schober H. Neutron Applications in Earth, Energy and Environmental Sciences. Neutron Scattering Applications and Techniques. Springer Science & Business Media, 2008. ISBN: 978-0387094168

[15] Kolb U, Shankland K, Meshi L, Avilov A, David WIF. Uniting Electron Crystallography and Powder Diffraction, NATO Science for Peace and Security Series B: Physics and Biophysics. Springer, 2012. ISBN: 978-9400755802

[16] Fitter J, Gutberlet T, Katsaras J. Neutron Scattering in Biology: Techniques and Applications. Biological and Medical Physics, Biomedical Engineering. Springer Science & Business Media, 2006. ISBN: 978-3540291114

[17] Guillot S, Beaudet-Savignat S, Lambert S, Vannier Rose-N, Roussel P, Porcher F. Evidence of local defects in the oxygen excess apatite $La_{9.67}(SiO_4)_6O_{2.5}$ from high resolution neutron powder diffraction. Journal of Solid State Chemistry 2009;182(12) 3358–3364.

[18] Lambert S, Vincent A, Bruneton E, Beaudet-Savignat S, Guillet F, Minot B, Bouree F. Structural investigation of $La_{9.33}Si_6O_{26}$- and $La_9AESi_6O_{26+\delta}$-doped apatites-type lanthanum silicate (AE = Ba, Sr and Ca) by neutron powder diffraction. Journal of Solid State Chemistry 2006;179(8) 2602–2608.

[19] Sansom JHJ, Richings D, Slater PR. A powder neutron diffraction study of the oxide-ion-conducting apatite-type phases, $La_{9.33}Si_6O_{26}$ and $La_8Sr_2Si_6O_{26}$. Solid State Ionics 2001;139(3–4) 205–210

[20] Matsushita Y, Izumi F, Kobayashi K, Igawa N, Kitazawa H, Oyama Y, Miyoshi S, Yamaguchi S. Powder neutron diffraction of La-apatite under low temperature. Nuclear Instruments and Methods in Physics Research Section A: Accelerators, Spectrometers, Detectors and Associated Equipment 2009;600(1) 319–321.

[21] Tolchard JR, Slater PR. A high temperature powder neutron diffraction structural study of the apatite-type oxide ion conductor, $La_{9.67}Si_6O_{26.5}$. Journal of Physics and Chemistry of Solids 2008;69(10) 2433–2439

[22] Mazza D, Tribaudino M, Delmastro A, Lebech B. Synthesis and Neutron Diffraction Study of $La_5Si_2BO_{13}$, an Analog of the Apatite Mineral. Journal of Solid State Chemistry 2000;155(2) 389–393.

[23] O'Donnell MD, Hill RG, Fong SK. Neutron diffraction of chlorine substituted fluorapatite. Materials Letters 2009;63(15) 1347–1349.

[24] Masubuchi Y, Higuchi M, Takeda T, Kikkawa S. Oxide ion conduction mechanism in $RE_{9.33}(SiO_4)_6O_2$ and $Sr_2RE_8(SiO_4)_6O_2$ (RE = La, Nd) from neutron powder diffraction. Solid State Ionics 2006;177(3–4) 263–268.

[25] Leventouri Th, Bunaciu CE, Perdikatsis V. Neutron powder diffraction studies of silicon-substituted hydroxyapatite. Biomaterials 2003;24(23) 4205–4211.

[26] Alonso JA, Martínez-Lope MJ, Aguadero A, Daza L. Neutron powder diffraction as a characterization tool of solid oxide fuel cell materials. Progress in Solid State Chemistry 2008;36(1–2) 134–150.

[27] Loong C-K, Rey C, Kuhn LT, Combes C, Wu Y, Chen S-H, Glimcher MJ. Evidence of hydroxyl-ion deficiency in bone apatites: an inelastic neutron-scattering study. Bone 2000;26(6) 599–602.

[28] Young RA, Spooner S. Neutron diffraction studies of human tooth enamel. Archives of Oral Biology 1970;15(1) 47–63.

[29] Ibrahim DM, Mostafa AA, Korowash SI. Chemical characterization of some substituted hydroxyapatites. Chemistry Central Journal 2011;5(1) 74.

[30] Lachance GR, Claisse F. Quantitative X-ray fluorescence analysis: theory and application. Wiley, 1995. ISBN: 978-0471951674

[31] Muller R. Spectrochemical Analysis by X-Ray Fluorescence. Springer Science & Business Media, 2012. ISBN: 978-1468417975

[32] Crompton TR. Polymer Reference Book. Smithers Rapra Publishing, 2006. ISBN: 978-1859574928

[33] Nadkarni RA. Modern Instrumental Methods of Elemental Analysis of Petroleum Products and Lubricants, ASTM special technical publication—Volume 1109. ASTM (American Society for Testing and Materials) International, 1991. ISBN: 978-0803114166

[34] Tang YH, Han CM, Bao ZK, Huang YY, He W, Hua W. Analysis of apatite crystals and their fluid inclusions by synchrotron radiation X-ray flourescence microprobe. Spectrochimica Acta Part B: Atomic Spectroscopy 2005;60(4) 439–446.

[35] Uda M, Demortier G, Nakai I. X-rays for Archaeology. NATO Science Series: Mathematics, Physics, and Chemistry—Volume 224. Springer Science & Business Media, 2005. ISBN: 978-1402035814

[36] Nutman AP, Friend CRL. Petrography and geochemistry of apatites in banded iron formation, Akilia, W. Greenland: consequences for oldest life evidence. Precambrian Research 2006;147(1–2) 100–106.

[37] Chen Ch, Chai Z, Gao Y. Nuclear Analytical Techniques for Metallomics and Metalloproteomics. Royal Society of Chemistry, 2010. ISBN: 978-1847559012

[38] Bunker G. Introduction to XAFS: A Practical Guide to X-ray Absorption Fine Structure Spectroscopy. Cambridge University Press, 2010. ISBN: 978-0521767750

[39] Parat F, Dungan MA, Streck MJ. Anhydrite, pyrrhotite, and sulfur-rich apatite: tracing the sulfur evolution of an Oligocene andesite (Eagle Mountain, CO, USA). Lithos 2002;64(3–4) 63–75.

[40] Klemm W. Anomale Wertigkeiten. Angewandte Chemie 1951;63(17–18) 396–402.

[41] Chauhan P, Chauhan RP, Gupta M. Estimation of naturally occurring radionuclides in fertilizers using gamma spectrometry and elemental analysis by XRF and XRD techniques. Microchemical Journal 2013;106; 73–78.

[42] Hasikova J, Sokolov A, Titov V, Dirba A. On-line XRF analysis of phosphate materials at various stages of processing. Procedia Engineering 2014;83 455–461.

[43] Zhou F, Wang L, Xu Z, Liu Q, Chi R. Reactive oily bubble technology for flotation of apatite, dolomite and quartz. International Journal of Mineral Processing 2015;134 74–81.

[44] Oliveira MS, Santana RC, Ataíde CH, Barrozo MAS. Recovery of apatite from flotation tailings. Separation and Purification Technology 2011;79(1) 79–84.

[45] Taylor HE. Inductively Coupled Plasma–Mass Spectrometry: Practices and Techniques. Academic Press, 2001. ISBN: 978-0126838657

[46] Montaser A. Inductively Coupled Plasma Mass Spectrometry. John Wiley & Sons, 1998. ISBN: 978-0471186205

[47] Levinson R. More Modern Chemical Techniques. Royal Society of Chemistry, 2001. ISBN: 978-0854049295

[48] Lajunen LHJ, Perämäki P. Spectrochemical Analysis by Atomic Absorption and Emission. Royal Society of Chemistry, 2004. ISBN: 978-0854046249

[49] Hasebe N, Barbarand J, Jarvis K, Carter A, Hurford AJ. Apatite fission-track chronometry using laser ablation ICP-MS. Chemical Geology 2004;207(3–4) 135–145.

[50] Johnstone S, Hourigan J, Gallagher Ch. LA-ICP-MS depth profile analysis of apatite: Protocol and implications for (U–Th)/He thermochronometry. Geochimica et Cosmochimica Acta 2013;109 143–161.

[51] Guzmics T, Zajacz Z, Kodolányi J, Halter W, Szabó C. LA-ICP-MS study of apatite- and K feldspar-hosted primary carbonatite melt inclusions in clinopyroxenite xenoliths from lamprophyres, Hungary: Implications for significance of carbonatite melts in the Earth's mantle. Geochimica et Cosmochimica Acta 2008;72(7) 1864–1886.

[52] Chew DM, Sylvester PJ, Tubrett MN. U–Pb and Th–Pb dating of apatite by LA-ICPMS. Chemical Geology 2011;280(1–2) 200–216.

[53] Donelick RA, O'Sullivan PB, Ketcham RA, Hendriks BWH, Redfield TF. Relative U and Th concentrations from LA-ICP-MS for apatite fission-track grain-age dating. Geochimica et Cosmochimica Acta 2006;70(18) A143.

[54] Fisher ChM, McFarlane ChRM, Hanchar JM, Schmitz MD, Sylvester PJ, Lam R, Longerich HP. Sm-Nd isotope systematics by laser ablation-multicollector-inductively coupled plasma mass spectrometry: methods and potential natural and synthetic reference materials. Chemical Geology 2001;284(1–2) 1–20.

[55] Yang Yue-H Wu Fu-Y, Xie Lie-W, Chu Zhu-Y, Yang Jin-H. Re-evaluation of interferences of doubly charged ions of heavy rare earth elements on Sr isotopic analysis using multi-collector inductively coupled plasma mass spectrometry. Spectrochimica Acta Part B: Atomic Spectroscopy 2014;97 118–123.

[56] Yang Yue-H Wu Fu-Y, Chu Zhu-Y, Xie Lie-W, Yang Jin-H. High-precision simultaneous determination of 147Sm/144Nd and 143Nd/144Nd ratios in Sm-Nd mixtures using multi-collector inductively coupled plasma mass spectrometry and its comparison to isotope dilution analysis. Spectrochimica Acta Part B: Atomic Spectroscopy 2013;79–80 82–87.

[57] Pourmand A, Dauphas N, Ireland TJ. A novel extraction chromatography and MC-ICP-MS technique for rapid analysis of REE, Sc and Y: revising CI-chondrite and Post-Archean Australian Shale (PAAS) abundances. Chemical Geology 2012;291 38–54.

[58] Boutton T. Mass Spectrometry of Soils (Books in Soils, Plants, and the Environment). CRC Press, 1996. ISBN: 978-0824796990

[59] Casas JS, Sordo J. Lead: Chemistry, analytical aspects, environmental impact and health effects. Elsevier, 2011. ISBN: 978-0080463889

[60] Becker S. Inorganic Mass Spectrometry: Principles and Applications. John Wiley & Sons, 2008. ISBN: 978-0470517208

[61] Hellebrand SJE. Textural, geochronological and chemical constraints from polygenetic titanite and monogenetic apatite from a mid-crustal shear zone: an integrated EPMA, SIMS, and TIMS study. Chemical Geology 2007;241(1–2) 88–107.

[62] Schoene B, Latkoczy C, Schaltegger U, Günther D. A new method integrating high-precision U–Pb geochronology with zircon trace element analysis (U–Pb TIMS-TEA). Geochimica et Cosmochimica Acta 2010;74(24) 7144–7159.

[63] Teixeira W, D'Agrella-Filho MS, Hamilton MA, Ernst RE, Girardi VAV, Mazzucchelli M, Bettencourt JS. U–Pb (ID-TIMS) baddeleyite ages and paleomagnetism of 1.79 and 1.59 Ga tholeiitic dyke swarms, and position of the Rio de la Plata Craton within the Columbia supercontinent. Lithos 2013;174 157–174.

[64] Metcalfe I, Crowley JL, Nicoll RS, Schmitz M. High-precision U-Pb CA-TIMS calibration of middle Permian to lower Triassic sequences, mass extinction and extreme climate-change in eastern Australian Gondwana. Gondwana Research, In Press, Corrected Proof, Available online 16 September 2014. doi:10.1016/j.gr.2014.09.002

[65] Yang Yue-H Wu Fu-Y, Yang Jin-H, Chew DM, Xie Lie-W, Chu Zhu-Y, Zhang Yan-B, Huang Ch. Sr and Nd isotopic compositions of apatite reference materials used in U–Th–Pb geochronology. Chemical Geology 2014;385 35–55.

[66] Cherepin VT. Secondary Ion Mass Spectroscopy of Solid Surfaces. VSP, 1987. ISBN: 978-9067640787

[67] O'Connor J, Sexton BA, Smart RSC. Surface Analysis Methods in Materials Science Physics and Astronomy. Springer Series in Surface Sciences—Volume 23. Springer Science & Business Media, 2003. ISBN: 978-3540413301

[68] van der Heide P. Secondary Ion Mass Spectrometry: An Introduction to Principles and Practices. John Wiley & Sons, 2014. ISBN: 978-1118916773

[69] Kuo J. Electron Microscopy: Methods and Protocols—Volume 369. Springer Science & Business Media, 2007. ISBN: 978-1597452946

[70] Microelectronics Failure Analysis: Desk Reference. ASM International, 2004. ISBN: 978-0871708045

[71] Watson LL, Hutcheon ID, Epstein S, Stolper EM, Water on Mars: clues from deuterium/hydrogen andwater contents of hydrous phases in SNCmeteorites. Science 1994;265 86–90.

[72] Barnes JJ, Franchi IA, Anand M, Tartèse R, Starkey NA, Koike M, Sano Y, Russell SS. Accurate and precise measurements of the D/H ratio and hydroxyl content in lunar apatites using NanoSIMS. Chemical Geology 2013;337–338 48–55.

[73] Gotliv Bat-A, Robach JS, Veis A. The composition and structure of bovine peritubular dentin: mapping by time of flight secondary ion mass spectroscopy. Journal of Structural Biology 2006;156(2) 320–333.

[74] Stephan T. TOF-SIMS in cosmochemistry. Planetary and Space Science 200149(1) 859–906.

[75] Li Qiu-L, Li Xian-H, Wu Fu-Y, Yin Qing-Z, Ye Hai-M, Liu Y, Tang Guo-Q, Zhang Chuan-L. In-situ SIMS U-Pb dating of phanerozoic apatite with low U and high common Pb. Gondwana Research 2012;21(4) 745–756.

[76] Schaltegger U, Schmitt AK, Horstwood MSA. U-Th-Pb zircon geochronology by ID-TIMS, SIMS, and laser ablation ICP-MS: recipes, interpretations, and opportunities. Chemical Geology 2015;402 89–110.

[77] Ling Xiao-X, Schmädicke E, Li Qiu-L, Gose J, Wu Rui-H, Wang Shi-Q, Liu Y, Tang Guo-Q, Li Xian-H. Age determination of nephrite by in-situ SIMS U–Pb dating syngenetic titanite: a case study of the nephrite deposit from Luanchuan, Henan, China. Lithos 2015;220–223 289–299.

[78] Friedbacher G, Bubert H. Surface and Thin Film Analysis: A Compendium of Principles, Instrumentation, and Applications. John Wiley & Sons, 2011. ISBN: 978-3527636945

[79] Lee MS. Mass Spectrometry Handbook. Wiley Series on Pharmaceutical Science and Biotechnology: Practices, Applications and Methods—Volume 7. John Wiley & Sons, 2012. ISBN: 978-1118180723

[80] Dambach S, Fartmann M, Kriegeskotte C, Brüning C, Wiesmann HP, Lipinsky D, Arlinghaus HF. Laser-SNMS analysis of apatite formation in vitro. Applied Surface Science 2004;231–232 506–509.

[81] Rabi II, Zacharias S, Millman P. Kusch. A new method of measuring nuclear magnetic moment. Physical Review 1938;53(4) 318.

[82] Weil JA, Bolton JR. Electron Paramagnetic Resonance: Elementary Theory and Practical Applications. 2nd ed., John Wiley & Sons, 2007. ISBN: 978-0470084984

[83] Brustolon M. Electron Paramagnetic Resonance: A Practitioner's Toolkit. John Wiley & Sons, 2009. ISBN: 978-0470432228

[84] Mabbs FE, Collison D. Electron Paramagnetic Resonance of d Transition Metal Compounds Studies in Inorganic Chemistry. Elsevier, 2013. ISBN: 978-1483291499

[85] Misra SK. Handbook of Multifrequency Electron Paramagnetic Resonance: Data and Techniques. John Wiley & Sons, 2014. ISBN: 978-3527672455

[86] Moens PDW, Callens FJ, Verbeeck RMH, Naessens DE. An EPR spectrum decomposition study of precipitated carbonated apatites (NCAP) dried at 25°C: adsorption of molecules from the atmosphere on the apatite powders. Applied Radiation and Isotopes 1993;44(1–2) 279–285.

[87] de Oliveira LM, Rossi AM, Lopes RT. Dose response of A-type carbonated apatites prepared under different conditions. Radiation Physics and Chemistry 2001;61(3–6) 485–487.

[88] Schramm DU, Rossi AM. Electron spin resonance (ESR) studies of CO_2^- radicals in irradiated A and B-type carbonate-containing apatites. Applied Radiation and Isotopes 2000;52(5) 1085–1091.

[89] Vugman NV, Rossi AM, Rigby SEJ. EPR dating CO_2^- sites in tooth enamel apatites by ENDOR and triple resonance. Applied Radiation and Isotopes 1995;46(5) 311–315.

[90] Fattibene P, Callens F. EPR dosimetry with tooth enamel: a review. Applied Radiation and Isotopes 2010;68(11) 2033–2116.

[91] Rudko VV, Ishchenko SS, Vorona IP, Baran NP. Isotropic radical image in biological apatites. Radiation Measurements 2007;42(9) 1580–1582.

[92] Ishchenko SS, Vorona IP, Okulov SM, Baran NP. ^{13}C hyperfine interactions of CO_2^- in irradiated tooth enamel as studied by EPR. Applied Radiation and Isotopes 2002;56(6) 815–819.

[93] Nosenko VV, Vorona IP, Baran NP, Ishchenko SS, Vysotskyi BV, Krakhmalnaya TV, Semenov YA. Comparative EPR study CO_2^- radicals in modern and fossil tooth enamel. Radiation Measurements 2015;78 53–57.

[94] Rudko VV, Vorona IP, Baran NP, Ishchenko SS. Thermally stimulated transformation of the EPR spectra in γ-irradiated bone tissue. Radiation Measurements 2009;44(3) 239–242.

[95] Ishchenko SS, Vorona IP, Okulov SM, Baran NP, Rudko VV. ENDOR study of CO_2^- radicals in hydroxyapatite of γ-irradiated bone. Radiation Measurements 2011;46(1) 37–39

[96] Köseogˇlu R, Köseogˇlu E, Köksal F, Başaran E, Demirci D. EPR of some irradiated renal stones. Radiation Measurements 2005;40(1) 65–68.

[97] Fleet ME. Infrared spectra of carbonate apatites: ν2-Region bands. Biomaterials 2009;30(8) 1473–1481.

[98] Yu H, Zhang H, Wang X, Gu Z, Li X, Deng F. Local structure of hydroxy–peroxy apatite: a combined XRD, FT-IR, Raman, SEM, and solid-state NMR study. Journal of Physics and Chemistry of Solids 2007;68(10) 1863–1871.

[99] Klinowski J. New Techniques in Solid-State NMR, Volume 246. Springer Science & Business Media, 2004. ISBN: 978-3540221685

[100] Duer MJ. Solid State NMR Spectroscopy: Principles and Applications. John Wiley & Sons, 2008. ISBN: 978-0470999387

[101] Ohannesian L, Streeter A. Handbook of Pharmaceutical Analysis Drugs and the Pharmaceutical Sciences. CRC Press, 2001. ISBN: 978-0824741945

[102] Bakhmutov VI. Solid-State NMR in Materials Science: Principles and Applications. CRC Press, 2011. ISBN: 978-1439869635

[103] Burnell E, de Lange CA. NMR of Ordered Liquids. Springer Science & Business Media, 2003. ISBN: 978-1402013430

[104] Braun M, Jana C. 19F NMR spectroscopy of fluoridated apatites. Chemical Physics Letters 1995;245(1) 19–22.

[105] McElderry John-DP, Zhu P, Mroue KH, Xu J, Pavan B, Fang M, Zhao G, McNerny E, Kohn DH, Franceschi RT, Holl MMB, Tecklenburg MMJ, Ramamoorthy A, Morris MD. Crystallinity and compositional changes in carbonated apatites: Evidence from 31P solid-state NMR, Raman, and AFM analysis. Journal of Solid State Chemistry 2013;206 192–198.

[106] Kaflak A, Kolodziejski W. Phosphorus-31 spin-lattice NMR relaxation in bone apatite and its mineral standards. Solid State Nuclear Magnetic Resonance 2007;31(4) 174–183.

[107] Yesinowski JP, Eckert H. Hydrogen environments in calcium phosphates: proton MAS NMR at high spinning speeds. Journal of the American Chemical Society 1987;109(21) 6274–6282.

[108] Goldstein J, Newbury DE, Joy DC, Lyman ChE, Echlin P, Lifshin E, Sawyer L, Michael JR. Scanning Electron Microscopy and X-ray Microanalysis: Third Edition. Springer Science & Business Media, 2012. ISBN: 978-1461502159

[109] Reimer L. Scanning Electron Microscopy: Physics of Image Formation and Microanalysis. Springer Science & Business Media, 1998. ISBN: 978-3540639763

[110] Krinsley DH, Pye K, Boggs S, Tovey NK Jr. Backscattered Scanning Electron Microscopy and Image Analysis of Sediments and Sedimentary Rocks. Cambridge University Press, 2005. ISBN: 978-0521019743

[111] Compton RG, Harding MS, Pluck MR, Atherton JH, Brennan CM. Journal of Physical Chemistry 1993;97(40) 10416–10420.

[112] Dorozhkin SV. Surface Reactions of Apatite Dissolution. Journal of Colloid and Interface Science 1997;191(2) 489–497.

[113] Xu Wen-G, Fan Hong-R, Hu Fang-F, Santosh M, Yang Kui-F, Lan Ting-G. In situ chemical and Sr–Nd–O isotopic compositions of apatite from the Tongshi intrusive complex in the southern part of the North China Craton: implications for petrogenesis and metallogeny. Journal of Asian Earth Sciences 2015;105 208–222.

[114] Xing Chang-M, Wang ChY, Li C. Trace element compositions of apatite from the middle zone of the Panzhihua layered intrusion, SW China: insights into the differentiation of a P- and Si-rich melt. Lithos 2014;204 188–202.

[115] Roman-Lopez J, Correcher V, Garcia-Guinea J, Prado-Herrero P, Rivera T, Lozano IB. Effect of the chemical impurities on the luminescence emission of natural apatites. Spectrochimica Acta Part A: Molecular and Biomolecular Spectroscopy 2014;126 142–147.

[116] Kamiishi E, Utsunomiya S. Nano-scale reaction processes at the interface between apatite and aqueous lead. Chemical Geology 2013;340 121–130.

[117] Stuart BH. Infrared Spectroscopy: Fundamentals and Applications. Vydavatel John Wiley & Sons, 2004. ISBN: 978-0470011133

[118] Chalmers JM, Edwards HGM, Hargreaves MD. Infrared and Raman Spectroscopy in Forensic Science. John Wiley & Sons, 2012. ISBN: 978-0470749067

[119] McCreery RL. Raman Spectroscopy for Chemical Analysis. Chemical Analysis: A Series of Monographs on Analytical Chemistry and Its Applications—Volume 225. John Wiley & Sons, 2005. ISBN: 978-0471231875

[120] Ferraro JR, Nakamoto K. Introductory Raman Spectroscopy. Academic Press, 2012. ISBN: 978-0323137904

[121] Madou MJ. Solid-State Physics, Fluidics, and Analytical Techniques in Micro- and Nanotechnology. Fundamentals of Microfabrication and Nanotechnology—Volume 1. CRC Press, 2011. ISBN: 978-1420055115

[122] Kumar ChSSR. Raman Spectroscopy for Nanomaterials Characterization. Springer Science & Business Media, 2012. ISBN: 978-3642206207

[123] Fleet ME, Liu X. Local structure of channel ions in carbonate apatite. Biomaterials 2005;26(36) 7548–7554.

[124] Errassifi F, Sarda S, Barroug A, Legrouri A, Sfihi H, Rey C. Infrared, Raman and NMR investigations of risedronate adsorption on nanocrystalline apatites. Journal of Colloid and Interface Science 2014;420 101–111.

[125] Antonakos A, Liarokapis E, Leventouri T. Micro-Raman and FTIR studies of synthetic and natural apatites. Biomaterials 2007;28(19) 3043–3054.

[126] Gómez-Morales J, Iafisco Mi, Delgado-López JM, Sarda S, Drouet C. Progress on the preparation of nanocrystalline apatites and surface characterization: overview of

fundamental and applied aspects. Progress in Crystal Growth and Characterization of Materials 2013;59(1) 1–46.

[127] Grunenwald A, Keyser C, Sautereau AM, Crubézy E, Ludes B, Drouet C. Revisiting carbonate quantification in apatite (bio)minerals: a validated FTIR methodology. Journal of Archaeological Science 2014;49 134–141.

[128] Adler HH. Infrared spectra of phosphate minerals: symmetry and substitutional effects in the pyromorphite series. American Mineralogist 1964;49 1002–1015.

[129] Bencnn RM. A relation between internuclear distances and bond force constants. Journal of Chemical Physiscs 1934;2 128–131.

[130] Lever T, Haines P, Rouquerol J, Charsley EL, Van Eckeren P, Burlett DJ. ICTAC nomenclature of thermal analysis, 2014 (IUPAC Recommendations 2014). Pure and Applied Chemistry 2014;86(4) 545–553. DOI 10.1515/pac-2012-0609

[131] Brown ME. Introduction to Thermal Analysis: Techniques and Applications. Hot Topics in Thermal Analysis and Calorimetry—Volume 1. Springer Science & Business Media, 2001. ISBN: 978-1402002113

[132] Brown ME, Gallagher PK. Handbook of Thermal Analysis and Calorimetry: Recent Advances, Techniques and Applications. Handbook of Thermal Analysis and Calorimetry—Volume 5. Elsevier, 2011. ISBN: 978-0080556314

[133] Kerr PF. Optical Mineralogy. 4nd ed., McGraw-Hill, 1977. ISBN: 978-0070342187

[134] Mitra S. Fundamentals Of Optical, Spectroscopic And X-Ray Mineralogy. New Age International, 1996. ISBN: 978-8122409826

[135] Jonasz M, Fournier GR. Chapter 6—Refractive indices and morphologies of aquatic particles. Light Scattering by Particles in Water, 2007; 447–558.

[136] Li Cheng-X, Duan Yong-H, Hu Wen-Ch. Electronic structure, elastic anisotropy, thermal conductivity and optical properties of calcium apatite $Ca_5(PO_4)_3X$ (X=F, Cl or Br). Journal of Alloys and Compounds 2015;619 66–77.

[137] Duan YH, Sun Y. First-principles calculations of optical properties of Mg_2Pb. Science China Physics, Mechanics and Astronomy 2014;57(2) 233–238.

[138] Rulis P, Ouyang L, Ching WY. Electronic structure and bonding in calcium apatite crystals: Hydroxyapatite, fluorapatite, chlorapatite, and cromapatite. Physical Review B 70, 155104 (2004). DOI: 10.1103/PhysRevB.70.155104.

[139] Lowell S, Shields JE, Thomas MA, Thommes M. Characterization of Porous Solids and Powders: Surface Area, Pore Size and Density. Particle Technology Series—Volume 16. Springer Science & Business Media, 2012. ISBN: 978-1402023033

[140] Merkus HG. Particle Size Measurements: Fundamentals, Practice, Quality Particle Technology Series—Volume 17. Springer Science & Business Media, 2009. ISBN: 978-1402090165

[141] Allen T. Powder Sampling and Particle Size Determination. Elsevier, 2003. ISBN: 978-0080539324

[142] LeRoy Jackson M, Barak P. Soil Chemical Analysis: Advanced Course. UW-Madison Libraries Parallel Press, 2005. ISBN: 978-1893311473

[143] Taylor HFW. Cement Chemistry. Revised edition, Thomas Telford, 1997. ISBN: 978-0727725929

[144] Ishizaki K, Komarneni S, Nanko M. Porous Materials: Process Technology and Applications. Materials Technology Series—Volume 4. ISBN: 978-1461558118

[145] Borch J, Lyne MB, Mark RE, Habeger Ch. Handbook of Physical Testing of Paper, Volume 2. 2nd ed., CRC Press, 2001. ISBN: 978-0203910498

[146] Donaldson EC, Chilingarian GV, Yen TF. Enhanced Oil Recovery, I: Fundamentals and Analyses. Developments in Petroleum Science. Elsevier, 1985. ISBN: 978-0080868721

[147] Millington RJ, Quirk JP. Permeability of porous solids. Transactions of the Faraday Society 1961;57 1200–1207.

[148] Zhong K, Vasudevan TV, Somasundaran P. Floatability of apatites of different type and origin: role of surface area and porosity. International Journal of Mineral Processing 1993;38(3–4) 177–188.

[149] Cardoso L, Fritton SP, Gailani G, Benalla M, Cowin SC. Advances in assessment of bone porosity, permeability and interstitial fluid flow. Journal of Biomechanics 2013;46(2) 253–265.

[150] Tang H, Guo Y, Jia D, Zhou Y. High bone-like apatite-forming ability of mesoporous titania films. Microporous and Mesoporous Materials 2010;131(1–3) 366–372.

[151] Habibovic P, Juhl MV, Clyens S, Martinetti R, Dolcini L, Theilgaard N, van Blitterswijk CA. Comparison of two carbonated apatite ceramics in vivo. Acta Biomaterialia 2010;6(6) 2219–2226.

[152] Liu Dean-M. Influence of porosity and pore size on the compressive strength of porous hydroxyapatite ceramic. Ceramics International 1997;23(2) 135–139.

[153] Mohammad NF, Othman R, Yeoh FY. Controlling the pore characteristics of mesoporous apatite materials: hydroxyapatite and carbonate apatite. Ceramics International, In Press, Accepted Manuscript, Available online 9 May 2015. doi:10.1016/j.ceramint.2015.04.162

[154] Panteix PJ, Julien I, Abélard P, Bernache-Assollant D. Influence of porosity on the electrical properties of $La_{9.33}(SiO_4)_6O_2$ oxyapatite. Ceramics International 2008;34(7) 1579–1586.

[155] Pons A, Jouin J, Béchade E, Julien I, Masson O, Geffroy PM, Mayet R, Thomas P, Fukuda K, Kagomiya I. Study of the formation of the apatite-type phases $La_{9.33+x}(SiO_4)_6O_{2+3x/2}$

synthesized from a lanthanum oxycarbonate $La_2O_2CO_3$. Solid State Sciences 2014;38 150–155.

[156] Li B, Liu W, Pan W. Synthesis and electrical properties of apatite-type $La_{10}Si_6O_{27}$. Journal of Power Sources 2010;195(8) 2196–2201.

[157] Medeiros EE, Dias AMS. Experimental and numerical analysis of Vickers hardness testing. IJRRAS 2013;17(1) 9–18.

[158] CiDRA Precision Services, LLC. Mineral Hardness Conversion Chart. http://www.cidraprecisionservices.com/mohs-conversion.html

[159] Saenger AT, Kuhs WF. Structural disorder in hydroxyapatite. Zeitschrift für Kristallographie 1992;199 123–148.

[160] Bhatnagar VM. IR-spectra of fluorapatite and fluorchlorapatite. Experientia 1967;24(1) 10–12

[161] Balan E, Delattre S, Roche D, Segalen L, Morin G, Guillaumet M, Blanchard M, Lazzeri M, Brouder Ch, Salje EKH. Line-broadening effects in the powder infrared spectrum of apatite. Physics and Chemistry of Minerals 2011;38(2) 111–112.

[162] Foil AM, Wilkins ChH. infrared spectra and characteristic frequencies of inorganic ions. Their use in qualitative analysis. Analytical Chemistry 1952;24(8) 1253–1294.

[163] Bonner CE Jr, Chess ChC, Meegoda Ch, Stefanos S, Loutts GB. Volume effects on the Raman frequencies of phosphate ions in fluorapatite crystals. Optical Materials 2004;26 17–22.

[164] Tressaud A, Haufe G. Fluorine and Health: Molecular Imaging, Biomedical Materials and Pharmaceuticals. 1st ed., Elsevier, 2008. ISBN: 978–0444530868

[165] Williams Q, Knittle E. Infrared and raman spectra of $Ca_5(PO_4)_3F_2$-fluorapatite at high pressures: compression-induced changes in phosphate site and Davydov splittings. Journal of Physics and Chemistry of Solids 199657(4) 417–422.

[166] Veiderma M, Knubovets R, Tõnsuaadu K. Structural properties of apatites from Finland studied by FTIR spectroscopy. Bulletin of the Geological Society of Finland 1998;1–2 69–75.

[167] Freund F, Knobel RM. Distribution of fluorine in hydroxyapatite studied by infrared spectroscopy. Journal of the Chemical Society, Dalton Transactions 1977;11 1136–1140.

[168] Knubovets R. Structural mineralogy and properties of natural phosphates. Reviews in Chemical Engineering 1993;9(3–4) 161–216.

[169] Ptáček P, Bartoníčková E, Švec J, Opravil T, Šoukal F, Frajkorová F. The kinetics and mechanism of thermal decomposition of $SrCO_3$ polymorphs. Ceramics International 2015;41(1) 115–126.

[170] Drouet Ch. A comprehensive guide to experimental and predicted thermodynamic properties of phosphate apatite minerals in view of applicative purposes. Journal of Chemical Thermodynamics 2015;81 143–159.

[171] Moiseev G, Leitner J, Sestak J, Zhukovsky V. Empirical dependences of the standard enthalpy of formation for related inorganic compounds enhancing glass formers. Thermochimica Acta 1996;280–281 511–521.

[172] Moiseev GK, Sestak J. Some calculations methods for estimation of thermodynamical and thermochemical properties of inorganic compounds. Progress in Crystal Growth and Characterization of Materials 1995;30(1) 23–81.

[173] Petit A.-T, Dulong P-L. Recherches sur quelques points importants de la Théorie de la Chaleur. Annales de Chimie et de Physique (in French) 1819;10 395–413.

[174] Bureau International des Poids et Mesures the International System of Units (SI). 8th ed., Organisation Intergouvernementale de la Convention du Mètre, STEDI Media, Paris, 2006. ISBN: 92-822-2213-6

[175] Rogers D. Einstein's Other Theory: The Planck-Bose-Einstein Theory of Heat Capacity. Princeton University Press, 2005. ISBN: 978-0691118260

[176] Latimer WM. Methods of estimating the entropies of solid compounds. Journal of the American Chemical Society 1951;73(4) 1480–1482.

[177] Leitner J, Chuchvalec P, Sedmidubský D, Strejc A, Abrman P. Estimation of heat capacities of solid mixed oxides. Thermochemica Acta 2003;395 47–46.

[178] Kellogg HH, in Fitterer GR (Ed.). Problems in the non-ferrous industries and a note on estimation of heat capacity, in: Applications of Fundamental Thermodynamics to Metallurgical Processes. Gorden and Breach, London, 1967, p.357

[179] Kubaschewski O, Ünal H. An empirical estimation of the heat capacities of inorganic compounds. High Temperatures—High Pressures 1977;9(3) 361–365.

[180] Kubaschewski O, Alock CB, Spencer PJ. Materials thermochemistry. Materials Science Monographs. 6th ed., Pergamon Press, 1993. ISBN: 9780080418896

[181] Spencer PJ. Estimation of thermodynamic data for metallurgical applications. Thermochimica Acta 1988;314(1–2) 1–21.

[182] Moiseev GK, Šesták J. Some calculations methods for estimation of thermodynamical and thermochemical properties of inorganic compounds. Progress in Crystal Growth and Characterization of Materials 1995;30 23–81.

[183] Kumok VN. Problem of Correlation Methods for Evaluation of Thermodynamic and Characteristics, Direct and Reverse Task of Chemical Thermodynamics. Nauka, Novosibirsk, 1987, p.108.

[184] Latimer WM. Methods of estimating the entropies of solid compounds. Journal of the American Chemical Society 1951;73 1480–1482.

[185] Aldabergenov MK, Balakaeva GT, Kokibasova GT. Calculation of thermodynamic functions of aluminates of alkaline and alkaline earth elements. Journal of Physical chemistry 1998;72(5) 808–811.

[186] Gospodinov GG, Mihov DI. Molar heat capacities of the tellurites $MnTeO_3$, $MnTe_2O_5$, and $Mn_2Te_3O_8$. Journal of Chemical Thermodynamics 1993;25(10) 1249–1252.

[187] Dorozhkin SV. A review on the dissolution models of calcium apatites. Progress in Crystal Growth and Characterization of Materials 2001;44(1) 45–61.

[188] Zhang J, Nancolias GH. Unexpected pH dependence of dissolution kinetics of dicalcium phosphate dihydrate. Journal of Physical Chemistry 1994;98(6) 1689–1694.

[189] Zhu Y, Zhang X, Chen Y, Xie Q, Lan J, Qian M, He N. A comparative study on the dissolution and solubility of hydroxylapatite and fluorapatite at 25°C and 45°C. Chemical Geology 2009;268(1–2) 89–96.

[190] Patel PR, Gregory TM, Brown WE. Solubility of $CaHPO_4{\cdot}2H_2O$ in the quaternary system $Ca(OH)_2$-H_3PO_4-NaCl-H_2O at 25°C. Journal of Research of the National Bureau of Standards—A. Physics and Chemistry 1974;78(6) 675–681.

[191] Christoffersen J, Christoffersen MR. Kinetics of dissolution of calcium hydroxyapatite V. The acidity constant for the hydrogen phosphate surface complex. Journal of Crystal Growth 1982;57 21–26.

[192] McDowell H, Gregory TM, Brown WE. Solubility of $Ca_5(PO_4)_3OH$ in the System $Ca(OH)_2$-H_3PO_4-H_2O at 5, 15, 25 and 37°C. Journal of Research of the National Bureau of Standards —A. Physics and Chemistry 1977;81(2–3) 273–281.

[193] Tang R, Henneman ZJ, Nancollas GH. Constant composition kinetics study of carbonated apatite dissolution. Journal of Crystal Growth 2003;249(3–4) 614–624.

[194] Shellis RP, Lee AR, Wilson RM. Observations on the apparent solubility of carbonate-apatites. Journal of Colloid and Interface Science 1999;218(2) 351–358.

[195] Chaïrat C, Schott J, Oelkers EH, Lartigue Jean-E, Harouiya N. Kinetics and mechanism of natural fluorapatite dissolution at 25°C and pH from 3 to 12. Geochimica et Cosmochimica Acta 2007;71(24) 5901–5912.

[196] Wang LK, Hung Yung-T, Shammas NK. Physicochemical Treatment Processes. Handbook of Environmental Engineering—Volume 3. Springer Science & Business Media, 2007. ISBN: 978-1592598205

[197] Harouiya N, Chaïrat C, Köhler SJ, Gout R, Oelkers EH. The dissolution kinetics and apparent solubility of natural apatite in closed reactors at temperatures from 5 to 50°C and pH from 1 to 6. Chemical Geology 2007;244(3–4) 554–568.

[198] Chaïrat C, Oelkers EH, Schott J. Lartigue J-E Harouiya N. Fluorapatite surface composition in aqueous solution deduced from potentiometric, electrokinetic, and solubility measurements, and spectroscopic observations. Geochimica Cosmochimica Acta 2007;71(24) 5888–5900.

[199] Smith AN, Posner AM, Quirk JP. Incongruent dissolution and surface complexes of hydroxyapatite. Journal of Colloid and Interface Science 1974;48(3) 442–449.

[200] Deutsch WJ, Siegel R. Groundwater Geochemistry: Fundamentals and Applications to Contamination. CRC Press, 1997. ISBN: 978-0873713085

[201] Thomann JM, Voegel JC, Gramain P. Kinetics of dissolution of calcium hydroxyapatite powder. III: pH and sample conditioning effects. Calcified Tissue International 1990;46(2) 121–129.

[202] Chin KOA, Nancollas GH. Dissolution of fluorapatite. A constant-composition kinetics study. Langmuir 1991;7 2175–2179.

[203] Margolis HC, Moreno EC. Kinetics of hydroxyapatite dissolution in acetic, lactic, and phosphoric acid solutions. Calcified Tissue International 1992;50(2) 137–143.

[204] Paschalis EP, Wikiel K, Nancollas GH. Dual constant composition kinetics characterization of apatitic surfaces. Journal of biomedical materials research 1994;28(12) 1411–1418.

[205] Nelson DGA, Featherstone JDB, Duncan JF, Cutress TW. Effect of carbonate and fluoride on the dissolution behaviour of synthetic apatites. Caries Research 1983;17(3) 200–211.

[206] Christoffersen MR, Christoffersen J. Possible mechanisms for the growth of the biomaterial, calcium hydroxyapatite microcrystals. Journal of Crystal Growth 1992;121(4) 617–630.

[207] Christoffersen J, Christoffersen MR, Johansen T. Kinetics of growth and dissolution of fluorapatite. Journal of Crystal Growth 1996;163(3) 295–303.

[208] Christoffersen MR, Dohrup J, Christoffersen J. Kinetics of growth and dissolution of calcium hydroxyapatite in suspensions with variable calcium to phosphate ratio. Journal of Crystal Growth 1998;186 283–290.

[209] Mafe S, Manzanares JA, Reis H, Thomann JM, Gramain P. Model for the dissolution of calcium hydroxyapatite powder. Journal of Physical Chemistry 1992;96(2) 861–866.

[210] Gasser P, Voegel JC, Gramain P. Surface reactions on hydroxyapatite in the presence of fluoride ions 1. Saturated and congruent conditions. Colloids and Surfaces A: Physicochemical and Engineering Aspects 1993;74(2–3) 275–286.

[211] Schaad P, Poumier F, Voegel JC, Gramain P. Analysis of calcium hydroxyapatite dissolution in non-stoichiometric solutions. Colloids and Surfaces A: Physicochemical and Engineering Aspects 1997;121(2–3) 217–228.

[212] Dorozhkin SV. Acidic dissolution mechanism of natural fluorapatite. II. Nanolevel of investigations. Journal of Crystal Growth 1997;182(1–2) 133–140.

[213] Dorozhkin SV. Inorganic chemistry of the dissolution phenomenon: the dissolution mechanism of calcium apatites at the atomic (ionic) level. Comments on Inorganic Chemistry 1999;20(4–6) 285–299.

[214] Pearce EIF. On the dissolution of hydroxyapatite in acid solutions. Journal of Dental Research 1988;67(7) 1056–1058.

[215] Lowell LC. Dislocation etch pits in apatite. Acta Metallurgica 1958;6(12) 775–778.

[216] Jongebloed WL, Molenaar I, Arends J. Orientation-dependent etchpit penetration and dissolution of fluoroapatite. Caries Research 1973;7(2) 154–165.

[217] Jongebloed WL, van den Berg PJ, Arends J. The dissolution of single crystals of hydroxyapatite in citric and lactic acids. Calcified Tissue Research1974;15 1–9.

[218] Arends J, Jongebloed WL. Dislocations and dissolution in apatites: theoretical considerations. Caries Research 1977;11(3) 186–188.

[219] Arends J, Jongebloed WL. Ultrastructural studies of synthetic apatite crystals Journal of Dental Research 1979;58 (Spec. Issue B) 837–843.

[220] Hignett TP. Developments in Plant and Soil Sciences—Volume 15. Springer Science & Business Media, 2013. ISBN: 978-9401715386

[221] Somani L.L. Dictionary of agricultural and allied sciences—Volume 4. Mittal Publications, 1989. ISBN: 978-8170991410

[222] Nielsson FT. Manual of Fertilizer Processing. Fertilizer Science and Technology—Volume 5. CRC Press, 1986. ISBN: 978-0824775223

[223] Manual Fertilizer, UN Industrial Development Organization, International Fertilizer Development Center. Springer Science & Business Media, 1998. ISBN: 978-0792350323

[224] Naidu S. International Encyclopaedia of Agricultural Science and Technology: Soils and composts. International Encyclopaedia of Agricultural Science and Technology—Volume 4. Mittal Publications, 2008. ISBN: 978-8183242233

[225] Espinosa J. Influence of Rock Phosphate on Available Phosphorus as Measured by Plant Uptake and Soil Extractants. Thesis, INIAP Archivo Historico, 1979

[226] Espinosa J. Abstract Influence of Rock Phosphate on Available Phosphorus as Measured by Plant Uptake and Soil Extractants. Thesis, INIAP Archivo Historico.

[227] Shemilt LW. Chemistry and World Food Supplies: The New Frontiers, Chemrawn II: Invited papers presented at the International Conference on Chemistry and World Food Supplies, Manila, Philippines, 1982. ISBN: 0-08-029243-7.

Chapter 4

Synthetic Phase with the Structure of Apatite

Petr Ptáček

Additional information is available at the end of the chapter

http://dx.doi.org/10.5772/62212

Abstract

The previous chapters were dedicated to description of structure and properties of minerals from supergroup of apatite and introduction of method for identification and investigation of properties of phosphate minerals. The first synthesis of apatite was performed by Daubrée by passing the PCl_3 vapor over red-hot lime. The fourth chapter of this book introduces techniques for the preparation of synthetic analogs of apatite minerals including solid-state synthesis, wet chemical methods, hydrothermal synthesis as well as methods for preparation of single crystals. Chapter continues with description of structure and properties of synthetic compounds of apatite type and ends with incorporation of 3d-metal ions into the hexagonal channel of apatite.

Keywords: Apatite, Solid-State Synthesis, Wet Chemical Methods, Hydrothermal Synthesis, Single crystal, Lacunar Apatite

The first synthesis of apatite was performed by DAUBRÉE [1], who obtained it in crystals by passing the vapor of phosphorus trichloride (PCl_3) over red-hot lime. The synthetic mineral analogues of chlorapatite, fluorapatite, or the mixtures of these phases were prepared by MANROSS [2] via fusing of sodium phosphate either with calcium chloride, calcium fluoride, or both together. The similar process was also successfully used by BRIEGLEB [3]. FORCHHAMMER [4] prepared chlorapatite by fusing calcium phosphate with sodium chloride. When bone ash or marl was used instead of artificial calcium phosphate, mixed apatite was formed. Similar results were reported by DEVILLE and CARON [5], who fused bone ash with ammonium chloride and either calcium chloride or fluoride, and also by DITTE [6], who repeated Forchhammer's experiment [7]. ZAMBONINI and FERRUCIO [8] found that the fusion of $Ca_3(PO_4)_2$ with $CaCl_2$ produced apatite with very weak birefringence. Fusing $Ca_3(PO_4)_2$ with an excess of NaCl gave crystals with the birefringence of 0.0050–0.0058.

By heating calcium phosphate with calcium chloride and water, under the pressure at 250°C, DEBRAY prepared chlorapatite [9]. WEINSHENK [10] also prepared chlorapatite by heating calcium chloride, ammonium phosphate and ammonium chloride at the temperature of 150 to 180°C in a sealed tube.[1] CAMERON AND MCCAUGHEY [11] prepared fluorapatite by dissolving calcium fluoride in fused disodium phosphate and lixiviating[2] the cooled melt. Artificial fluorapatite sometimes exhibits the peculiarity of re-entrant pyramidal ends or phantom crystals. Spodiosite ($Ca_2(PO_4)F$) is orthorhombic, chlorspodiosite ($Ca_2(PO_4)Cl$ [12],[13]) being much less developed along the *c*-axis than fluorspodiosite [14]. Chlorapatite was formed when dicalcium phosphate was added in excess to molten calcium chloride. When precipitated calcium phosphate was used, chlorspodiosite[3] ($Ca_3(PO_4)_2{\cdot}CaCl_2$, $Ca_2(PO_4)Cl$ [14]) was obtained. Apatite was reported by HUTCHINS [15] and VOGHT [16] as present in lead-furnace slag [7].

The history of synthesis of various apatite compounds and substitution in the apatite structure including the preparation of didymium bearing chlorapatite and chlorspodiosite is described in the work of ZAMBONINI and FERRUCIO [17]. The paper describes three of those apatite syntheses using the mixture of $Ca_3(PO_4)_2$, $DiPO_4$ [4] and $CaCl_2$ in the weight ratio of 1:0.07:2.19 (1), 1:0.15:8.76 (2) and 1:0.67:2 (3). These mixtures were heated to the temperature of 1180, 1000 and 1100°C, respectively. Transparent, colorless apatite with the content of 3% $DiPO_4$ results from the first experiment. The second experiment leads to the pale-violet crystal of chlorspodiosite3. The crystals of $DiPO_4$ and only little amount of chlorapatite and chlorspodiosite with 9% of $DiPO_4$ were recognized in the third mixture.

HENDRICKS et al [19] reported the preparation of hydroxylapatite by the hydrolysis of tricalcium phosphate[5] with neutral ammonium citrate. Oxoapatite was then prepared by the ignition of hydroxylapatite to constant weight at 50°C.[6]

---

[1] The method is described in **Section 4.1.1**.

[2] The separation techniques based on leaching of minerals from a solid by dissolving them in a liquid.

[3] Chlorine analog of spodiosite [14]. Calcium chlorspodiosite is colorless crystalline compound structurally related to the mineral wagnerite ($Mg_2(PO_4)F$). It was reported by NACKEN in his study of the phase relationships which was obtained in the system $CaCl_2$-CaO-P_2O_5 [12]. The synthesis of the compound by dissolving $Ca_3(PO_4)_2$ in fused $CaCl_2$ was reported by KLEMENT and GEMBRUCH [13]. Since the mineral was recognized as the mixture of fluorapatite, calcite and serpentine, it was discredited (IMA action 2003-03-B).

[4] Didymium (Di) was recognized as the mixture of element of neodymium and praseodymium [18].

[5] Tricalcium phosphate was prepared by slow addition of Na_3PO_4 solution to the solution with the excess of $Ca(NO_3)_2$. The precipitate was washed with saturated solution of $Ca_3(PO_4)_2$ until the filtrate was free of nitrates. The salt was then dried at 50°C [19],[18].

[6] Hydroxylapatite is very difficult to dehydrate, even at high temperature (**Section 1.5.2**). As a consequence of this, many workers have mistaken it for oxyapatite [18].

http://dx.doi.org/10.5772/62212

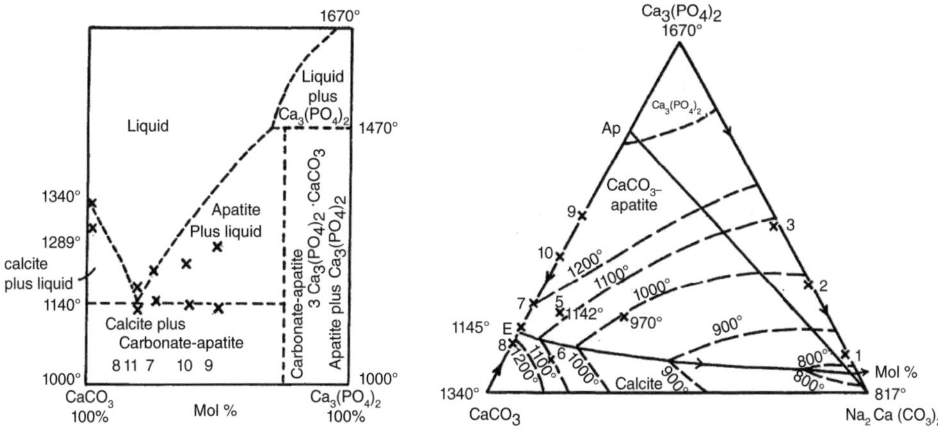

Fig. 1. The phase equilibrium in carbonate-apatite–$CaCO_3$ (a) and carbonate-apatite–$CaCO_3$–Na_2CO_3 systems (b) [20].

Eitel [20] investigated the binary systems $Ca_3(PO_4)_2$–$CaCO_3$ (**Fig. 1(a)**) and $Ca_3(PO_4)_2$–$CaCO_3$–Na_2CO_3[7] (**b**) and claimed to have crystallized carbonate-apatite from a melt, but he determined CO_2 only qualitatively and on a sample that may have been impure. The structure of prepared carbonate-apatite is not known [18].

The equilibrium given by the following scheme was predicted- for carbonate-apatite [20]:

$$CaCO_3 + \text{liquid} \leftrightarrow \text{carbonate} - \text{apatite} + Na_2Ca(CO_3)_2 \quad (1)$$

$$\text{liquid} \leftrightarrow \text{carbonate} - \text{apatite} + Na_2Ca(CO_3)_2 + Na_2(CO_3)_2 \ (\text{mix of crystals}) \quad (2)$$

The prepared crystals are typically long or short lengths with hexagonal prism $\{10\bar{1}0\}$, dipyramid $\{10\bar{1}1\}$ and pinacoid {0001}. The skeletal crystals had inclusions along the central canals parallel to the *c*-axis. The refractive index is uniaxial (-), $\varepsilon = 1.635$, $\omega = 1.626$, $\delta = 0.009$.

4.1. Common synthetic techniques for the preparation of apatites

The literature on the preparation of synthetic analogues of minerals from the supergroup of apatite[8] (**Section 1.1**) is very extensive, but can be divided into three main categories [18],[21],[22],[23],[24]:

[7] Introduced as double salt of $Na_2Ca(CO_3)_2$. Prepared mixtures were heated under the pressure of CO_2 in the range from 55 to 100 and from 23 to 54 kg·cm^{-2} (kg·cm^{-2} = 98066.5 Pa) for binary and ternary system, respectively [20].

[8] In older literature termed as apatite-like substances.

1. **Solid-State Synthesis**[9] (**Dry Method**) usually requires rather high temperature ≥ 1200°C and the product is characterized by a gradient of composition in the grain of material. The method and special techniques, conditions and devices used to control the product properties and to improve the reaction rate and homogeneity of the products are described in **Section 4.1.1**.

2. **Wet chemical method**[10] (precipitation from the solution): requires a long time period (10 hour or more) and often results in amorphous and non-stoichiometric products.

3. **Hydrothermal synthesis**[11]: involves heating of reactants with water in closed vessel, an autoclave

Other methods (**Section 4.1.1**) such as microwave synthesis, combustion synthesis and high pressure method or deposition techniques are used much rarely.

The pressure-temperature ranges of these methods are shown in **Fig. 2**. Some of the most applied techniques are described in this chapter. The methods for the preparation of single crystals are described separately in the next **Section 4.2**.

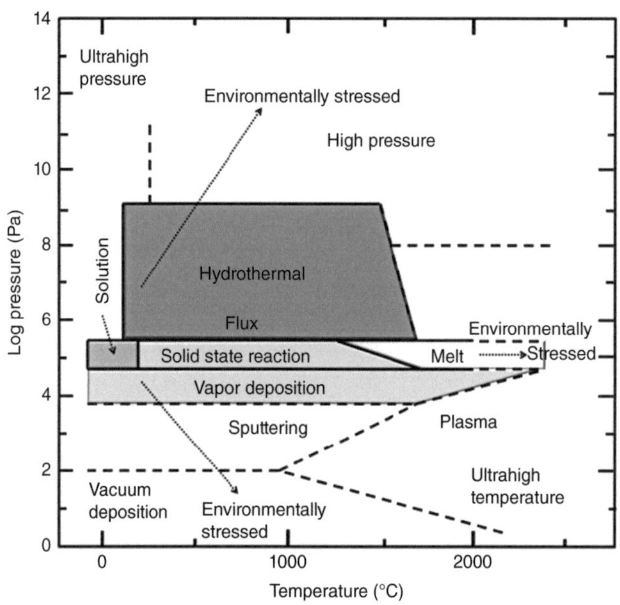

Fig. 2. Pressure-temperature range for the material preparation [21].

---

[9] Various alternative names, such as Shake 'n Bake Methods or Beat 'n Heat, are used for the solid state synthesis (reaction) in literature [23].

[10] JAFFE [18] recognized the precipitation by metathesis and the precipitation by hydrolysis.

[11] Sometimes are included among high-pressure methods [24].

4.1.1. Solid-State Synthesis

Solid-phase reactions (syntheses) are usually activated by high-temperature treatment [25]. The preparation of materials in solid-state is rather different form the synthesis of discrete molecules. The process involves the treatment of the whole lattice. Often, the post-synthesis purification of materials is not possible due to low solubility of formed phases. Hence, all effort must be made to avoid the excess of reagents. These methods are usually slow due to entire reaction which occurs in the solid state and requires the diffusion across the points of contact in a mixture [26],[27]. Special techniques can also be used based on the reduction of particle size or on the preparation of precursor in order to reduce the particle size, improve the product homogeneity and lower the temperature of thermal treatment [24].

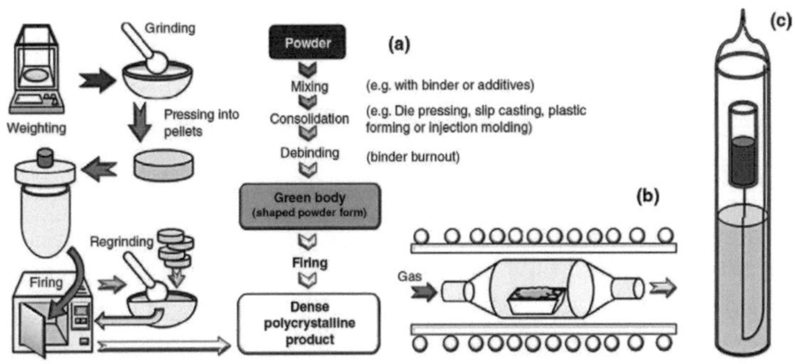

Fig. 3. Reaction scheme for solid-state synthesis: (a) ceramic method [26], (b) reaction of solids under special gas atmosphere [24] and (c) sealed glass tube [21],[33].

The most applied techniques[12] in solid-state synthesis are [23],[26],[24],[29],[28],[30],[31],[32]:

a. **Ceramic method**[13]: is the most common way of preparation of solids (metal oxides, aluminosilicates, sulfides, nitrides, etc.) that is based on thermal treatment of compacted powder[14] of two or more nonvolatile solids, which react to reach required composition and desired microstructure of the product **Fig. 3(a)**. Since the reaction can occur only at the interface of solids, once the surface layer has reacted, the reaction continues as the reactants diffuse from the bulk to the interface. The rate of reaction is then often limited by the diffusion hence it is important to prepare raw meal from fine and well mixed particles in order to maximize the contact area and minimize the diffusion path. It also decreases the temperature required for the thermal treatment. The repeating of regrind-

[12] Solid-state synthesis is classified among physical methods (together with vapor phase synthesis, laser ablation, etc.) some of other techniques listed below (sol-gel process, precipitation method, etc.) are considered as chemical methods [28].

[13] Since ceramic can be fabricated by a variety of methods, some of which have their origin in early civilizations, ceramic methods must be distinguished from the ceramic fabrication processes.

[14] The consolidation of ceramic powder to produce a **green body** is commonly referred to as forming. The main methods include dry of semidry pressing (1), plastic forming with water or organic polymers (2) and casting from a concentrated suspension or slurry (3) [30].

ing of the sample and the repeating of its thermal treatment is usually required to improve the homogeneity of the product.

The nucleation of a new phase, the epitactic[15] and topotactic[16] phenomenon's (oriented nucleation), crystal growth, phase transformation[17] and the sintration[18] are common during the thermal treatment.[19]

b. **Sealed tube (pipe) method (reaction)**: is applied in the cases when direct reaction under ambient conditions (in air at one atmosphere pressure) cannot be performed due to high volatility of reactants, air sensitivity of starting materials and/or products, or the desire to form a compound with a metal in an unusually low oxidation state. Typically for this type of reactions, the components are loaded into a glass (method was first applied by De Sénarmont [33], **Fig. 3(c)**) or quartz ampoule (tube) in a glass box, evacuated and sealed off by melting the glass/quartz using a blow torch. The whole tube is heated to required temperature and time. Cooled tube is then broken up to get the product. The reaction of material inside with the tube may cause that the side-wall of tube is weaken. In the combination with pressure in the tube, these reactions can be hazardous. The synthesis can be also performed with metal capsules (solvothermal reactions, **Section 4.1.3**) sealed by welding.

c. **Synthesis under controlled (special) atmosphere**: the preparation of some compounds must be carried out under a special atmosphere, but not necessarily at high pressures. The oxidation (O_2), inert (Ar, N_2) or reduction atmosphere (CO/CO_2 or H_2/H_2O) is used to prepare the compounds in required oxidation state.

d. **Solid-state synthesis under high-pressure**: high pressure can be applied [24],[30],[34],[35]:

- **Directly applied pressure** (external pressure, pressure sintering, or pressure-assisted sintering): includes the techniques known as hot pressing (pressure is applied uniax-

---

[15] There is a structural similarity between the substrate and the nucleus that is limited to 2D interface and referred to as epitaxy [23].

[16] There is a structural similarity between the substrate and the nucleus (like for epitaxy[15])) that extends to 3D for topotaxy [23].

[17] Phase transformation has usually significant effects on the reaction rate. The reaction rate is strongly increased at the temperatures near the phase transformation because the mobility of atoms is also increased. This phenomenon is termed as **Hedvall effect** [29].

[18] Sintering is defined as the bonding of adjacent surfaces in a mass of powder or a compact by heating [29]. In general there are three types of sintering process including Solid-Phase (Dry) Sintering (1), Liquid-Phase Sintering (2) and Reactive (Reaction) Sintering (3) [29]. The process can also be divided according to applied conditions and densification practice to Conventional Sintering (1), Microwave Sintering (2) and Pressure Sintering (3). The stages of the sintering process include: (1) initial stage (formation and growth of necks), (2) intermediate stage (pores reached their equilibrium shapes, continuous porosity), (3) final stage (pores reached their equilibrium shapes, isolated (enclosed) porosity) [30]. The process can also be divided according to the mass transport mechanism to viscous sintering and diffusion sintering (further divided according to dominant type of diffusion to surface diffusion, volume diffusion, intergranular diffusion, grain-boundary diffusion, but gas transport (diffusion) of matter can also occur).

[19] Generally, the term **firing** is used when the processes occurring during thermal treatment of green body are fairly complex, as in many traditional ceramics produced from clay-based materials. In less complex cases the term sintering[18] is used [30].

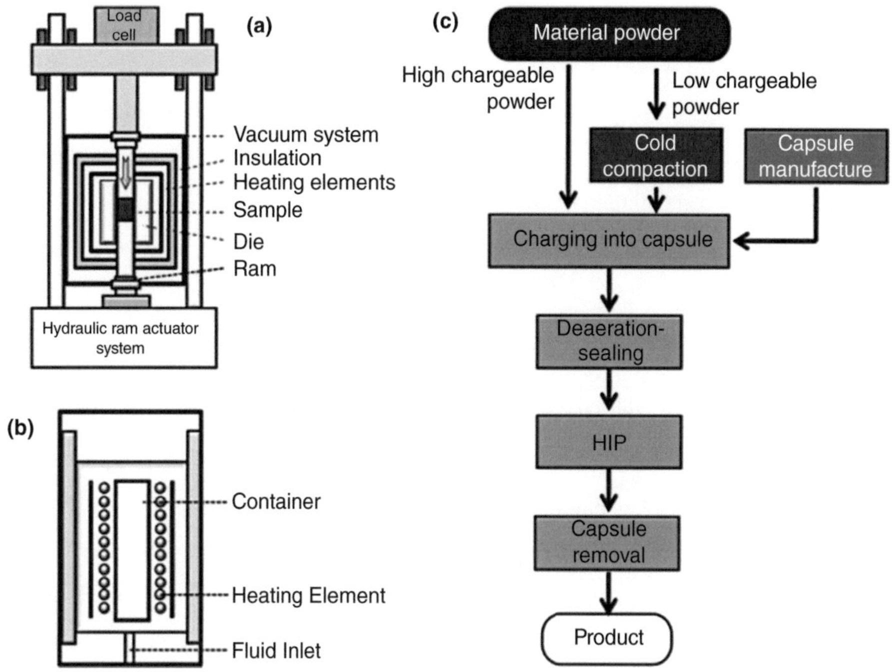

Fig. 4. Schematic diagram of hot pressing process [30], hot isostatic pressing [36] and the example of metallic capsule method [37].

ially or biaxially to the powder in a die, **Fig. 4(a)**) and sinter forging (similar to hot-pressing but without confining the sample in a die).

- **Indirectly applied pressure** can be applied through inert (hot isostatic pressing, HIP, **Fig. 4(b)**) or reactive gas, through ultrasound, by milling equipment (mechanochemical (powder) synthesis and activation) and by detonation.

The method is particularly important for the preparation of dense sample of ceramics with high degree of covalent bonding such as SiC and Si_3N_4.

Normal process of compaction of powder material involves uniaxial pressing in a die followed by sintering for densification. However better densification can be achieved by exerting a uniform pressure from all directions through a fluid medium onto the powder material retained in container (die), i.e. the pressure is generated by heating of medium. The process is termed as isostatic compaction, and as **hot isostatic pressing** (HIP) if performed at high temperature. The diffusion of medium from the container to the sample is avoided by its encapsulation by metal or glass (metallic or glass capsule method, **Fig. 4(c)** [36],[37].

In the case of **ultrasound (cavitation) methods** the phenomenon known as cavitation[20] takes place. Traveling the ultrasound wave leads to high pressure volume (compression) of the liquid which is followed by low pressure. Sudden expansion (rarefaction) leads to the formation of tiny bubbles. The bubbles expand to an unsustainable size and then collapse. The expansion and the collapse of bubbles create very localized hot spots, which reach instantaneous pressures of more than 100 MPa and temperatures of up to 5000°C (**Section 9.2.2**) [24], [28].

By **detonation methods** usually the nano-sized particles are prepared. The detonation is a superfast (with the velocity exceeding that of sound) exothermic reaction through the substance. The detonation wave consisting of the shock front, chemical reaction zone and the region where the products are scattered, spreads at constant rate due to continuous supply of energy from the chemical transformation of new portions of the explosive to the shock wave. The temperature (in the range from 2000 to 5000 K) and the pressure (shock wave) are reached by detonation in a suitably strong vessel. For example, the change of pressure at the end of the reaction zone ranges from 9.5 to 30 GPa for hexogen. This method was used for the preparation of nano-sized synthetic diamond,[21] graphite, boron nitride, etc. [24],[38].

Mechanochemical synthesis[22] is a solid-state synthesis method that takes advantage of the perturbation of surface-bonded species by pressure to enhance the thermodynamic and kinetic reactions between solids. The pressure can be applied at ambient temperature by friction and impact via milling equipment (**Fig. 5**) ranging from low energy ball mills to high energy stirred mills.[23] The main advantage of this method is the simplicity and low cost. The method was successfully used for the synthesis of oxides, phosphates, carbides, complexes, intermetalides [44],[45],[46],[47],[48],[24] alloys, etc.

---

[20] Generally, the cavitation can be divided into four types on the basis of the mode of generation of cavitation conditions: (1) acoustic cavitation (sound waves of high frequency 16 kHz–1 MHz), (2) Hydrodynamic cavitation (pressure variation is obtained by changing the geometry of the system), (3) optic cavitation (passing of photons of high intensity) and (4) particle cavitation (produced by the bombardments of other types of elementary particles, e.g. protons) [28].

[21] The method for the detonation transformation of graphite into diamond was earlier developed at the Institute of Chemical Physics of the Academy of Sciences of the USSR [38].

[22] Mechanochemical synthesis (reaction) of solids in the presence of water can be considered as hydrothermal one [25].

[23] Mechanochemistry is a branch of chemistry which is concerned with chemical and physico-chemical changes of substances of all stages of aggregation due to the influence of mechanical energy (Ostwald [39]). Colloid mills can be classified into three main groups with regard to the mechanism utilized for production of dispersion: beater-type mills (1), the smooth-surface type (2) and the rough-surface type. Beater-type mills include the original Plauson machine and some modified mills [48].

[24] Also intermetallic compounds, i.e. substances composed of two or more metallic elements with given stoichiometry and structure. Different atomic species occupy different lattice sites [44].

http://dx.doi.org/10.5772/62212

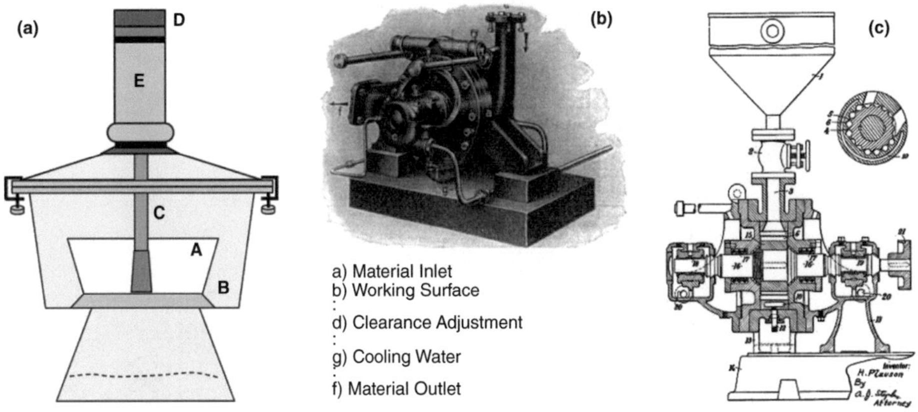

Fig. 5. The first mechanochemical reactor[25] (a): mortar (A), iron collar (B), pestle (C), handle (D) and rubber tube (E) [32],[39], Plauson-Oderberg (beater-type [48][23]) colloid mill for wet milling (b) and longitudinal section through Plauson colloidal mill [47] (c).

Mechanochemical activation involves the dispersion of solids and their deformation. These processes cause the generation of defects in solids, and also accelerate the migration of defects in the bulk, increase the number of contacts between particles and renew the contacts [25],[32],[34],[39],[40],[41],[42],[43],[49].

The advanced preparation techniques used for the reduction of particle size, activation of starting material or preparation of nano-scale[26] precursor are a broad group of methods with complicated nomenclature. The most known methods are:

i. **Sol-gel process** [26],[30],[50],[51],[52]: in this method, a solution of metal compounds (usually alkoxides of $M(OR)_4$ type, such as TMOS ($Si(OCH_3)_4$), TEOS ($Si(OC_2H_5)_4$), TEOT ($Ti(OC_2H_5)_4$), etc.,[27] or $(RO)_xMR'_{4-x}$ type, such as $(H_3CO)_3SiCH_3$, $(H_3CO)_3Si(CH{=}CH_2)$, etc.[28]) or suspension of very fine particles in a liquid (sol) is converted into rigid gel by removing the solvent or by adding a component which causes the gel to solidify. Two different sol-gel processes can be distinguished,[29] depending on whether the sol or the solution of alkoxides (alkoxide methods) is used **Fig. 6(a,b)**.

---

[25] Constructed by PARKER [39] for the study of solid–solid reaction of the type: $Na_2CO_3 + BaSO_4 \rightarrow Na_2SO_4 + BaCO_3$ [41]. Earlier works were concerned with the decomposition of solids by high pressure and by grinding in a mortar with the pestle [42],[43].

[26] Nano it the Greek word for dwarf. In the International System of Units (SI) it is the decimal multiple 10^{-9} used as prefix. Nanoscience refers to the range from one to several hundred manometers and the nanotechnologies are the technologies in which atoms are manipulated in quantities of one to several thousand atoms. Nanoscience probably first gained the attention in 1959 in the lecture of the American Nobel laureate of 1965 in physics R. FEYNMAN, who stated: *...that the day was no far off, when substances could be assembled at an atomic level* [27].

[27] Tetrametoxysilane, tetraetoxysilane, tetraethyl orthotitanate, etc.

[28] Methyltrimethoxysilane, vinyltrimethodxysilane, etc. **R**′ can also be reactive (polymerizable) groups such as (3-aminopropyl)trietoxysilane (APTES) or (3-glycidopropyl)trimetoxysilane (GLYMO) which can be used to prepare interconnected inorganic-organic networks. These materials are known as ORMOCERS (organically modified ceramics).

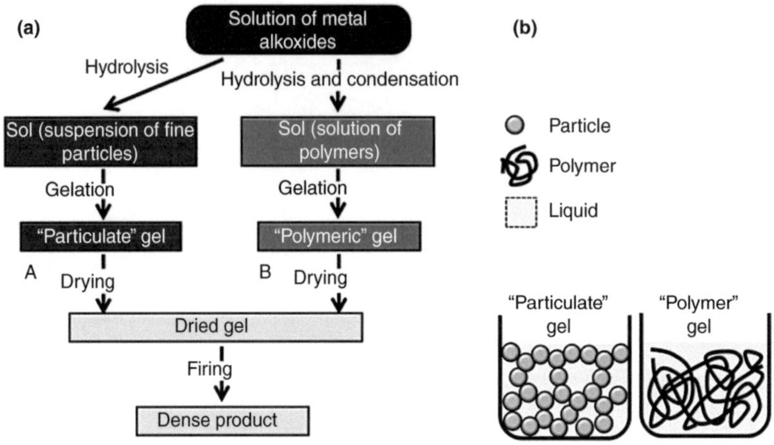

Fig. 6. Basic flow charts for sol-gel processes using a sol (a) and a solution of alkoxides (b). Schematic diagram of the structure of particulate gel formed from a sol (c) and polymeric gel from a solution (d) [30].

Starting with a sol, gelled material consists of identifiable colloidal particles which were joined together by the surface forces to form a network **Fig. 6(c)**. When the solution of metal-alkoxides is used (**d**), the gel consist of a network of polymer chains formed via the reaction of hydrolysis (**Eq. 3**[30]) and condensation (**Eq. 4**[31]):

$$M(OR)_4 + H_2O' \leftrightarrow M(OR)_3 O'H + ROH \tag{3}$$

[29] This classification [30] does not set aside often used group of semi-alkoxide methods (using the mixture of soluble salt and metal alkoxide), Pechini type polymeric gel methods (liquid mix techniques) as well as modified Pechini methods. The Pechini process usually uses soluble nitrates, acetates, chlorides, carbonates, isopropoxides or other metal compounds which are dissolved in the solution of citric acid (in general in polycarboxylic acids) and ethylene glycol (in general glycol). The polycondensation reaction leads to the polymeric gel accommodating the stable chelates of metal cations [50]:

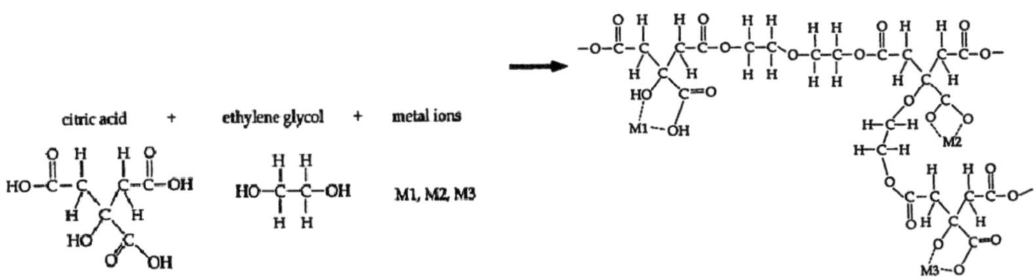

[30] The first step is the hydrolysis, the equilibrium constant for higher degree of hydrolysis decreases depending on the nature of –OR (increasing length of hydrocarbon chain and its branching). The process is also affected by the temperature, the solvent composition and applied water to alkoxide ratio, the type of catalysis, the application of ultrasound energy, etc. [52].

[31] Due to the formation of ROH molecule is termed as "alcohol producing condensation", i.e. the reaction between alkoxy (nonhydrolyzed) and hydrolyzed groups. The reaction with two hydrolyzed groups leads to the formation of –M–O–M– bridge and water, i.e. is termed as "water producing condensation".

http://dx.doi.org/10.5772/62212

$$M(OR)_3OH + M(OR)_4 \leftrightarrow (RO)_3 - M - O - M - (RO)_3 + ROH \quad (4)$$

Drying of gel leads to xerogel. The process is usually followed by shrinkage and formation of cracks. The thermal treatment of xerogel often involves the pyrolysis and calcination. If monolith is needed (aerogel [51]) the supercritical drying is usually applied.

ii. **Precipitation (co-precipitation) methods**[32] [26],[53]: the main reason why the precipitation is used to make ceramic powder is that it gives pure solid product, rejecting to the supernatant most of the impurities. In addition the particle morphology and the particle size distribution can be controlled to some degree. Mixed ion solution can be precipitated to produce a solid precursor containing required ions, although care must be taken to ensure correct ration of ions in the precipitate. The precipitation of powder involves the nucleation and the growth from supersaturated solution.

iii. **Solvation of metal salts** [26]: this method is based on dissolving the metal salt (often nitrates, hydroxides or oxalates) in suitable solvent followed by evaporating the mixture to dryness. Dried residue is then processed as in the ceramic method. In the case that the components have similar solubility, the method ensures better mixing than usual ceramic method.

iv. **Polymer pyrolysis** [30],[54]: reefers to the pyrolytic decomposition of metal-organic polymeric compounds (so-called preceramic polymers) to produce ceramics. The properties of the products depend on the nature of polymer and applied pyrolysis conditions. The polymer pyrolysis is an extension of well-known method for the production of carbon fibers by the pyrolysis of carbon-based polymers.

v. **Combustion methods** [55],[56],[57]: are characterized by high-temperatures, fast heating rates and short reaction times. The techniques involve an exothermic decomposition of fuel-oxidant precursor which results in either finely dispersed powder of precursor or direct product. Nitrates of required metals and urea, glycin or glucose as the most applied fuel are used as starting materials. The advantages of the sol-gel process and the combustion method are combined in the **gel combustion**.

vi. **Self-propagating high temperature synthesis (SHS)** [57],[58],[59]: was discovered in 1967 by MERZHANOV [60],[61] and is based on the exothermic reaction[33] between two or more reagents that after the initiation to ignition temperature (T_{ig}) does not require any external source of heat and rapidly propagates thorough the reaction mixture.[34]

[32] The precipitation method is used for the preparation of precursor that is further treated by solid-state synthesis. The techniques of direct precipitation of apatite are described in **Section 4.1.2**.

[33] Therefore often associated with combustion techniques [56].

[34] The process is analogous to the process of frontal polymerization in which localized polymerization reaction zone propagates thorough the mixture of solution of a monomer and initiator due to the heat diffusion and the occurrence of exothermic reaction [59].

vii. **Spray pyrolysis** [62],[63]: small droplets of a solution containing desired precursor are introduced into the hot zone to the furnace to obtain required product. According to applied conditions in different stages of the thermal cycle (**Fig.** 7) the aerosol will form non-coherent powder and solid particles.

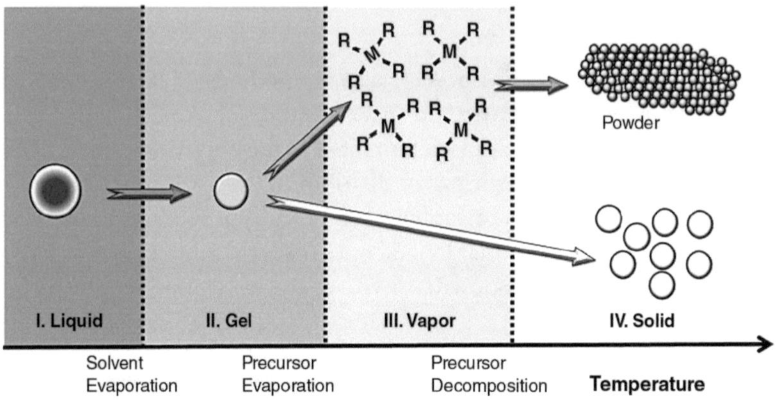

Fig. 7. Thermal cycle of spray pyrolysis [63].

viii. **Freeze drying** [35],[64]: the method discovered in 1965 that consists of fast freezing of the precursor solution (ensures preservation of maximum chemical homogeneity achieved in the starting solution), sublimation of solvent and final calcination. The main characteristics of the freeze drying method are that the drying is not accompanied by the coagulation of particles and the shrinkage of particles usually does not occur. The method produces fine and reactive powder of high purity

The reaction schemes for the solid-state synthesis of apatite structured compounds were published by KNYAZEV et al [65]:

$$4.5\ M^{II}(NO_3)_2 \cdot nH_2O + 0.5\ M^{II}L_2 \cdot mH_2O + 3(NH_4)_2\,HPO_4 \rightarrow \\ M_5^{II}(PO_4)_3L + 9\ NO_2 + 2.25\ O_2 + 6\ NH_3 + (4.5 + 4.5n + 0.5m)\ H_2O \tag{5}$$

where M^{II} = Ca, Sr, Ba, Cd, Pb and L = OH, F, Cl, Br and I.

$$3\ M^{II}(NO_3)_2 \cdot nH_2O + 2\ NH_4VO_3 \rightarrow M_3^{II}(VO_4)_2 + 6\ NO_2 + 1.5\ O_2 \\ +2\ NH_3 + (3n+1)\ H_2O \tag{6}$$

$$1.5\ M_3^{II}(VO_4)_2 + 0.5\ M^{II}L_2 \cdot mH_2O \rightarrow M_5^{II}(VO_4)_3L + 0.5m\ H_2O \tag{7}$$

where M^{II} = Ca, Sr, Ba, Cd, Pb and L = F, Cl and Br.

http://dx.doi.org/10.5772/62212

$$4.5\ M^{II}CO_3 + 3\ CrO_3 + 0.5\ M^{II}L_2 \cdot mH_2O \rightarrow M_5^{II}(CrO_4)_3 L + 4.5\ CO_2 + 0.75\ O_2 + 0.5m\ H_2O \quad (8)$$

where M^{II} = Ca, Sr and L = F and Cl.

$$4.5\ Ba^{II}CO_3 + 3\ MnO_2 + 0.5\ Ba^{II}L_2 \cdot mH_2O + 0.75\ O_2 \rightarrow Ba_5^{II}(MnO_4)_3 L + 4.5\ CO_2 + 0.5m\ H_2O \quad (9)$$

The temperature effect (**Table 1**) observed during the synthesis includes [65]:

1. synthesis;
2. polymorphic transition;
3. thermal decomposition; and
4. melting.

Compound	T_s	T_r	T_d	T_m
	[K]			
$Ca_5(PO_4)_3OH$	310	992	–	–
	1173	–	1523	–
$Ca_5(PO_4)_3Cl$	1073	953	–	>1723
$Ca_5(VO_4)_3Cl$	1073	794	–	>1723
$Ca_5(CrO_4)_3Cl$	1123	9007	–	1616
$Sr_5(PO_4)_3Br$	1473	–	1661	–
$Sr_5(VO_4)_3Cl$	1023	–	–	>1723
$Sr_5(CrO_4)_3F$	1373	–	–	1705
$Ba_5(VO_4)_3Cl$	1073	–	–	>1723
$Ba_5(MnO_4)_3F$	1123	–	1163	–
$Pb_5(PO_4)_3I$	973	–	–	256
$Pb_5(VO_4)_3F$	923	–	1044	–

Table 1. Temperature of synthesis (T_s), polymorphic transition (T_{tr}), decomposition (T_d) and melting (T_m) of some apatite-structured compounds [65].

The temperatures of these effects for some apatites are listed in **Table 1**.

The phase transformation and the thermal expansion coefficient of apatite-structured compound with the composition given by the formula $\mathbf{M_5(XO_4)_3Z_q}$ (M = Ca, Sr, Cd, Ba, Pb) were investigated by Chernorukov et al [66]. Pb-containing apatites are shown to undergo the phase transition involving the reduction in unit-cell symmetry from hexagonal to monoclinic. The thermal expansion anisotropy in the hexagonal phases increases in the order:

$$Ca<Sr<Ba<Pb<Cd,$$

and the monoclinic phases are less anisotropic but have larger thermal expansion coefficients in comparison with the hexagonal phases.

4.1.2. Precipitation method

The precipitation method is based on the combination reaction(s) when cations and anions in the solution combine to form insoluble ionic solid, so-called precipitate. The method can be divided to [35]:

i. **Direct precipitation method** is based on the reaction of neutralization and precipitation. The precipitate is then separated from the solution via filtration.

ii. **Homogeneous precipitation method** does not need precipitants because decomposed chemical acts as the precipitant, e.g. urea is decomposed in ammonium hydroxide and formed NH_4OH acts as the precipitant:

$$(NH_2)_2CO + 3\ H_2O \xrightarrow{70°C} 2\ NH_4OH + CO_2 \quad (10)$$

iii. **Coprecipitation method** is initiated by the addition of precipitant to mixed-salt solution.

iv. **Compound precipitation method** is the precipitation of stoichiometric compounds from the solution.

Wet techniques of apatite preparation are based on the precipitation from solution at ambient temperature [67]. The preparation techniques based on aqueous precipitation at moderate temperatures often lead to non-stoichiometric apatites [68]. Hydroxylapatite close to the ideal formula, can be precipitated by the addition of $Ca(OH)_2$ to diluted phosphoric acid and complete neutralization at the boiling point [69]

$$6\ H_3PO_4 + 10\ Ca(OH)_2 \rightarrow Ca_{10}(PO_4)_6(OH)_2 + 18\ H_2O \quad (11)$$

Precipitated hydroxylapatite shows extremely small crystal sizes (hexagonal plates ~200Å sides) and large surface area from 50 to 200 $m^2 \cdot g^{-1}$.

The Eh-pH diagrams for the Ca-P-H_2O system at 25 and 300°C for 1.67 mol activity of Ca and 1 mol activity of P (a_{Ca} = 1.67a_P) under the pressure of 1 bar are shown in **Fig. 8(a)** and **(b)**, respectively. The Pa_{Ca}-pH diagrams, where Pa_{Ca} = $-log\ a_{Ca}$ for this system show that the pH of minimum solubility of HAP clearly decreases with increasing temperature. At each tempera-

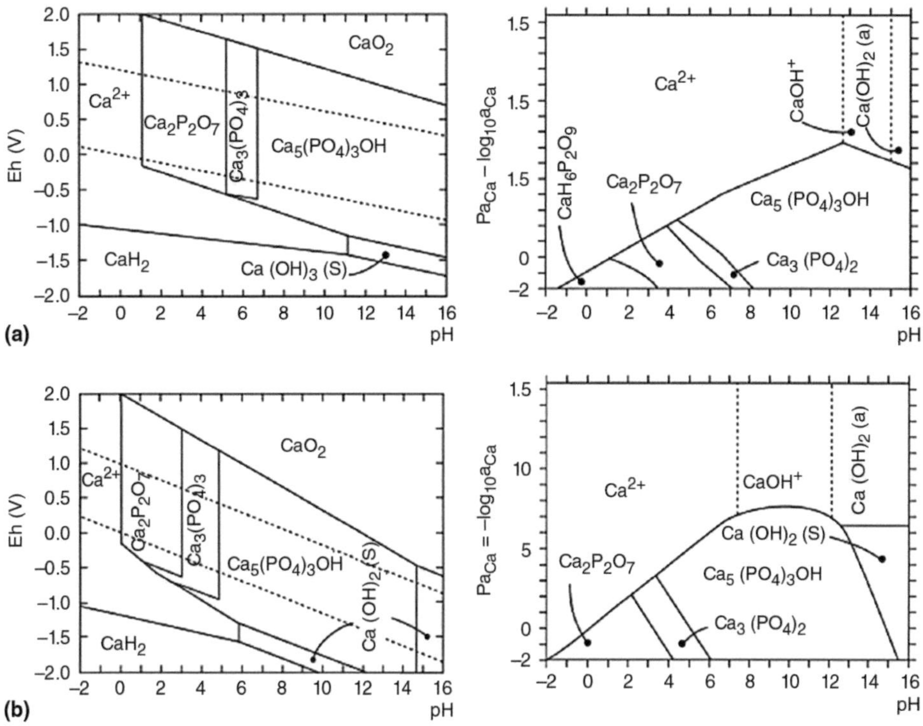

Fig. 8. Eh-pH and Pa_{Ca}-pH diagram of Ca-P-H_2O system at 25°C (a) and 300°C (b).

ture, HAP predominates in higher pH range, while $Ca_3(PO_4)_2$, $Ca_2P_2O_7$ and $CaH_6P_2O_9$ have predominates at lower pH [21].

The stability of calcium phosphates at higher temperatures is shown in **Fig. 9**. The equation numbers refer to the following reactions [21]:

$$3\ CaHPO_4 \leftrightarrow Ca_3\left(PO_4\right)_2 + H_3PO_4 \tag{12}$$

$$3\ Ca\left(H_2PO_4\right)_2 \leftrightarrow Ca_3\left(PO_4\right)_2 + 4\ H_3PO_4 \tag{13}$$

$$Ca_{10}H_2P_6O_{26} \leftrightarrow 3\ Ca_3\left(PO_4\right)_2 + CaO + H_2O \tag{14}$$

$$CaH_5PO_6 \leftrightarrow CaHPO_4 + 2\ H_2O \tag{15}$$

$$CaH_6P_2O_9 \leftrightarrow Ca\left(H_2PO_4\right)_2 + H_2O \tag{16}$$

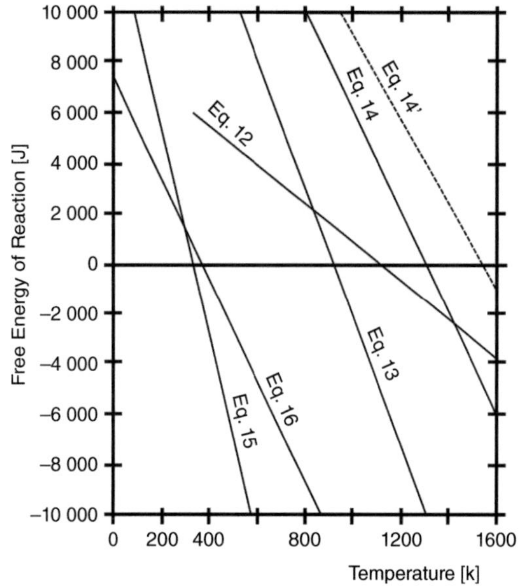

Fig. 9. Temperature dependence of free energy of reaction for some calcium phosphates according to **Eqs. 12–23** for water vapor fugacity equal to 0.03 atm., except for the dashed line **Eq. 14′**, which corresponds to water vapor fugacity equal to 1 atm. [21].

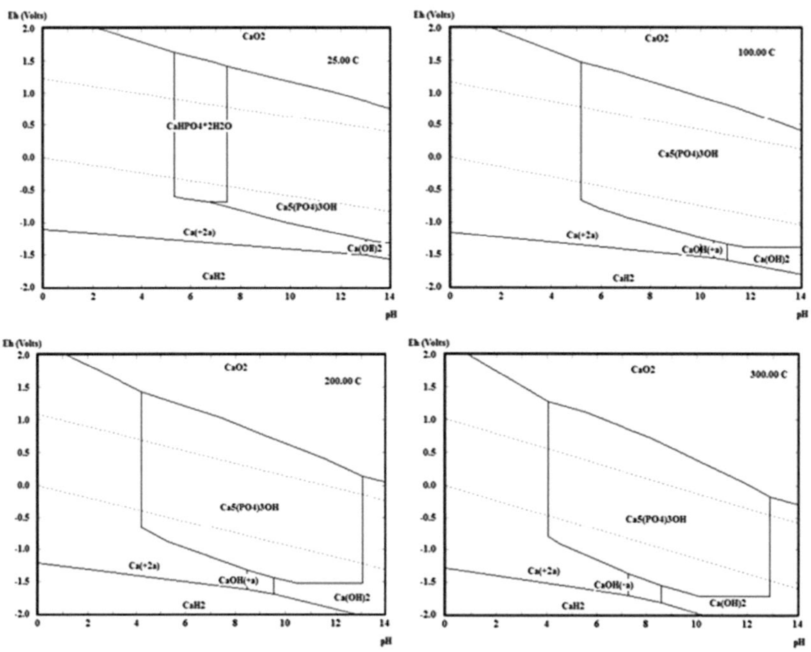

Fig. 10. Eh-pH diagram of Ca-P-H_2O system at 25°C (a), 100°C (b), 200°C (c) and 300°C (d).

For the purpose of this book the calculation of Eh-pH diagram for the solution where the concentration of ions (Ca^{2+}, PO_4^{3-} and OH^-) is equivalent to the system containing $5{\cdot}10^{-3}$ mol·dm $^{-3}$ of apatite was performed. The ionic strength (refer to **Footnote 31** in **Section 3.4.1**) of that solution enables the calculation of activity of Ca and P using the activity coefficient estimated from modified Davies equation (refer to **Footnote 31** in **Section 3.4.1**) and the concentration of calcium as follows: $\boldsymbol{a}_{Ca} = 1.062{\cdot}10^{-3}$ and $\boldsymbol{a}_{P} = 1.67{\cdot}\boldsymbol{a}_{Ca} = 6.36{\cdot}10^{-4}$. If the activity of ions is used instead of its concentration, the Eh-pH in **Fig. 10** can be calculated.

In this system,[35] $CaHPO_4{\cdot}2H_2O$ is stable under ambient temperature and nearly neutral pH. Hydroxylapatite becomes stable at the pH higher than 7.5. With increasing temperature, the formation of HAP instead of $CaHPO_4{\cdot}2H_2O$ is more probable. Minimal solubility of hydroxylapatite is then shifted to significantly lower pH than for **Fig. 8**.

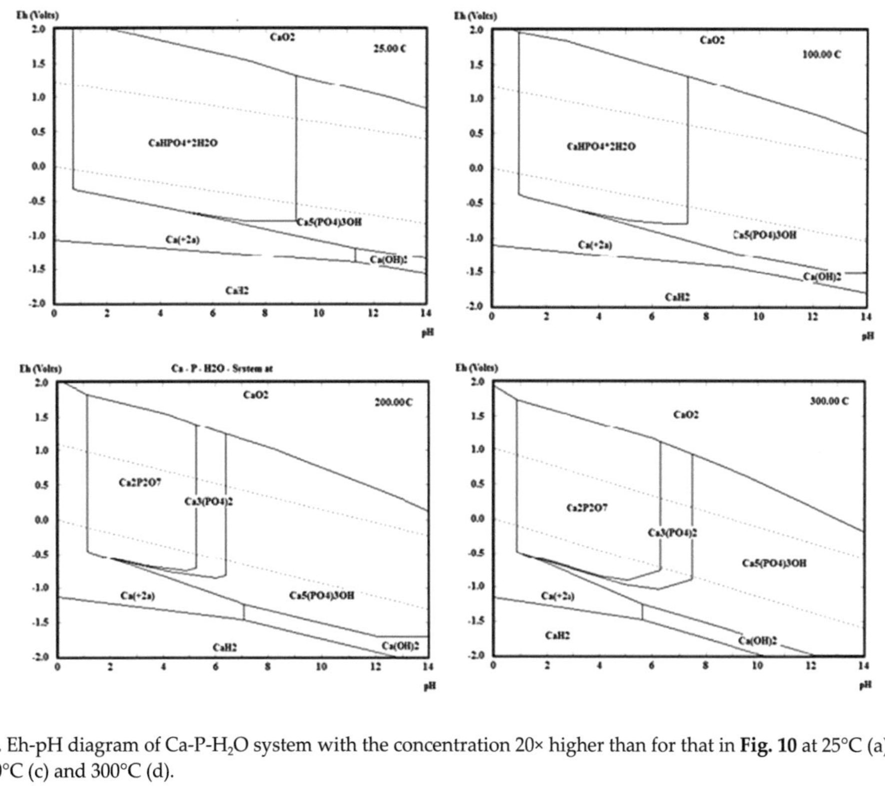

Fig. 11. Eh-pH diagram of Ca-P-H_2O system with the concentration 20× higher than for that in **Fig. 10** at 25°C (a), 100°C (b), 200°C (c) and 300°C (d).

Other difference is a fact, that the field of stability of $Ca(OH)_2$ starts at the pH = 13 for the system with elevated temperature. The formation of $CaHPO_4{\cdot}2H_2O$, $Ca_2P_2O_7$ and $Ca_3(PO_4)_2$ was not predicted.

The calculation for 20-times higher concentration **Fig. 11** than for the system mentioned above shows broadening field of $CaHPO_4{\cdot}2H_2O$. $Ca_2P_2O_7$ was formed by the thermal condensation

[35] The main difference against to the systems on **Fig. 8** and **Fig. 11** is significantly lower ionic strength.

of $CaHPO_4$ at temperatures higher than 164°C in acidic environment and $Ca_3(PO_4)_2$ precipitated from the solution at nearly neutral conditions. Hydroxylapatite again predominates at higher pH and $Ca(OH)_2$ does not appear at higher temperatures and the pH below 14 (the same as for **Fig. 8**).

The phase equilibrium in the system CaO-P_2O_5-H_2O was extensively studied by the solid-state reaction method under the atmospheric pressure of water vapor by VAN WAZRER [70] and in aqueous systems at temperatures lower than 100°C by BROWN et al [71],[72]. BIGGAR [73] studied the CaO-P_2O_5-H_2O system in the temperature range from 700 to 950°C and the pressure of 1 kbar. FENG and ROCKETT [74] studied the system CaO-P_2O_5-H_2O at 1000 bar with 50%wt. and 200°C (**Fig. 12**).

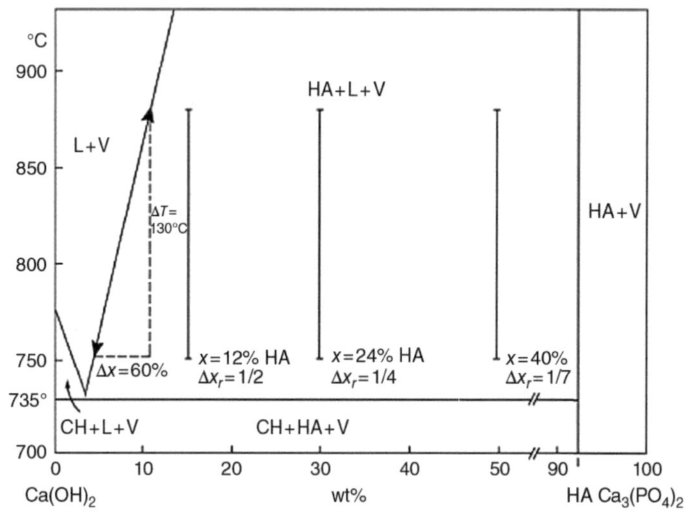

Fig. 12. Phase diagram of $Ca(OH)_2$-$Ca_3(PO_4)_2$-H_2O system [21].

The **molten (fused) salts**[36] **precipitation method** uses the precursor mixed with low melting point salt such as NaCl, KCl, or their eutectics. Upon the melting of the mixture, reactant oxides dissolve in the salt and desired compound precipitates due to its low solubility in molten salt.

[36] Fused salts are widely used in many industrial processes requiring to free the limitations arising from the use of aqueous solutions. Their thermal stability and generally low vapor pressure enable fast reaction rates and ability to dissolve many inorganic compounds making them useful solvents in electrometallurgy, metal coating, treatment of by-products, and energy conversion. It is recalled that one of the most important chemicals produced worldwide, sulfuric acid, is made by the molten salt catalysis. The electrolysis of molten salt is a technique used by H. MOISSAN for the isolation of element fluorine from the melt of KF·2HF (Moissan's method is used for industrial production of fluorine). It was also used by H. DAVY to discover several new elements (sodium, potassium, alkali metals) and to prove the chlorine as a new element (originally discovered by C.W. SHEELLE who considered it as "*dephlogisticated marine acid*"). Today the industrial production of Li and Na is based on the electrolysis of eutectic melt of LiCl–KCl (or $CaCl_2$) and NaCl–KCl (or $CaCl_2$), respectively. The production of K, Rb and Cs is based on the reduction of molten KCl, RbCl and CsCl by Na at the temperature of 600°C. Molten salt method also plays significant role in the development of energy resources, including the reprocessing of nuclear wastes, molten carbonate/solid oxide fuel cells (**Section 10.4**), and high temperature molten salt batteries. Fused alkali nitrates/nitrites are valuable materials for the heat transport and storage in solar plants. Molten salt bathes remain of large use in industry for the treatment of steel and variety of other metals as well as nonmetals, such as glass, plastics and rubber [75].

http://dx.doi.org/10.5772/62212

The melt is cooled down, and the salt is dissolved to yield the powder product of synthesis [75], [76].

The structure of a molten salt is characterized by an alteration of positively and negatively charged ionic solvation shells around a given ion. This arises from the predominance of Coulombic effects, which results in a strong attraction between oppositely charged species and a strong repulsion otherwise [75].

The utilization of molten salt precipitation method for the synthesis of apatites at "moderate temperatures" in the range from 500 to 700°C was also reported. Based on its principle, the method combines the advantages of thermal hydrolysis ("dry method") and the precipitation from the solution ("wet method"). As a reaction media, the chloride melt of the equimolar NaCl-KCl (665°C) composition as well as eutectic melt (390°C) in the system Li_2CO_3 (27)–Na_2CO_3 (28)–K_2CO_3 (45% mol.) can be used. The most probable reactions are estimated from the thermodynamic consideration as follows [22]:

$$6\ KPO_3 + 6\ CaO + 3\ CaCO_3 + Ca(OH)_2 \rightarrow HAP + 3\ K_2CO_3 \tag{17}$$

$$6\ NaPO_3 + 6\ CaO + 3\ CaCO_3 + Ca(OH)_2 \rightarrow HAP + 3\ Na_2CO_3 \tag{18}$$

$$6\ KPO_3 + 6\ CaO + 3\ CaCO_3 + CaF_2 \rightarrow FAP + 3\ K_2CO_3 \tag{19}$$

$$6\ NaPO_3 + 6\ CaO + 3\ CaCO_3 + CaF_2 \rightarrow FAP + 3\ Na_2CO_3 \tag{20}$$

4.1.3 Hydrothermal synthesis

The original hydrothermal[37] method involves heating of the reactants in a closed vessel, an autoclave, with water (heterogeneous reaction). Autoclave is usually constructed from thick stainless steel to withstand the high pressures and is fitted with safety valves; it may be lined with nonreactive materials, such as noble metals, quartz or Teflon. When the autoclave is heated, the pressure increases and the water remains liquid above its normal boiling temperature of 100°C, so-called superheated water. These conditions, in which the pressure is raised above atmospheric pressure and the temperature is raised above the boiling temperature of water are known as hydrothermal conditions (high-pressure-high-temperature, HPHT). HPHT conditions enable to dissolve and recrystallize (recover) the materials which are relatively insoluble under ordinary conditions. The methods enable [21],[24],[67]:

1. Synthesis of new phases or stabilization of new complexes.

[37] The term hydrothermal is of purely geological origin. It was first used by British geologist, Sir Roderick Murchison, to describe the action of water at elevated temperature and pressure in bringing about changes in the Earth's crust, and leading to the formation of various rocks and minerals. Materials scientists popularized the technique, particularly during 1940s. The first hydrothermal synthesis was performed by Schafhautl in Papin's digester, who obtained quartz crystals upon hydrothermal treatment of freshly precipitated silic acid [21].

2. Crystal growth of several inorganic compounds.

3. Preparation of finely divided materials and microcrystallites with well-defined size and morphology for specific applications.

4. In situ fabrication of materials with desired size, shape and also dispersity in case of nanomaterials.

5. Leaching of ores in metal extraction.

6. Decomposition, alteration, corrosion and technique.

Several definitions of hydrothermal synthesis use aqueous solvent under HPHT conditions [21]:

a. In hydrothermal synthesis the material is subjected to the action of water, at temperatures generally near, though often considerably above the critical temperature[38] of water (~370°C) in closed bombs, and therefore, under the corresponding high pressures developed by such solution [77].

b. Hydrothermal synthesis is a heterogeneous reaction in aqueous media above 100°C and the pressure higher than 1 bar [78].

c. Hydrothermal synthesis involves water as a catalyst and occasionally as a component of solid phases in the synthesis at elevated temperature (>100°C) and pressure greater than a few atmospheres [79].

Depending on the type of solvent used in the heterogeneous reaction the glycothermal, alcothermal, ammonothermal, lyothermal, carbothermal, etc., methods are recognized. According to applied solvent and condition, the hydrothermal methods can be further divided as follows [21].

i. **Conventional hydrothermal techniques**, which use aqueous solvent.

ii. **Solvothermal techniques or methods**, which use nonaqueous solvent.

iii. **Supercritical hydrothermal** methods use aqueous and nonaqueous solvent under critical to supercritical conditions.

iv. **Multienergy hydrothermal methods** combine hydrothermal method with additional microwave, electrochemical, sonar, mechanochemical, etc. energy.

Hydrothermal conditions exist in nature, and numerous minerals including naturally occurring zeolites and gemstones, are formed by this process. The term has been extended to other systems with moderately raised conditions and temperatures lower than those typically used in ceramics and sol-gel syntheses. Lower temperatures used are one of the advantages of the method. Other methods include the preparation of compounds in unusual oxidation states or phases, which are stabilized by raised temperature and pressure [24].

[38] The temperature and the pressure at critical point of water are 373.946°C and 22.064 MPa, respectively.

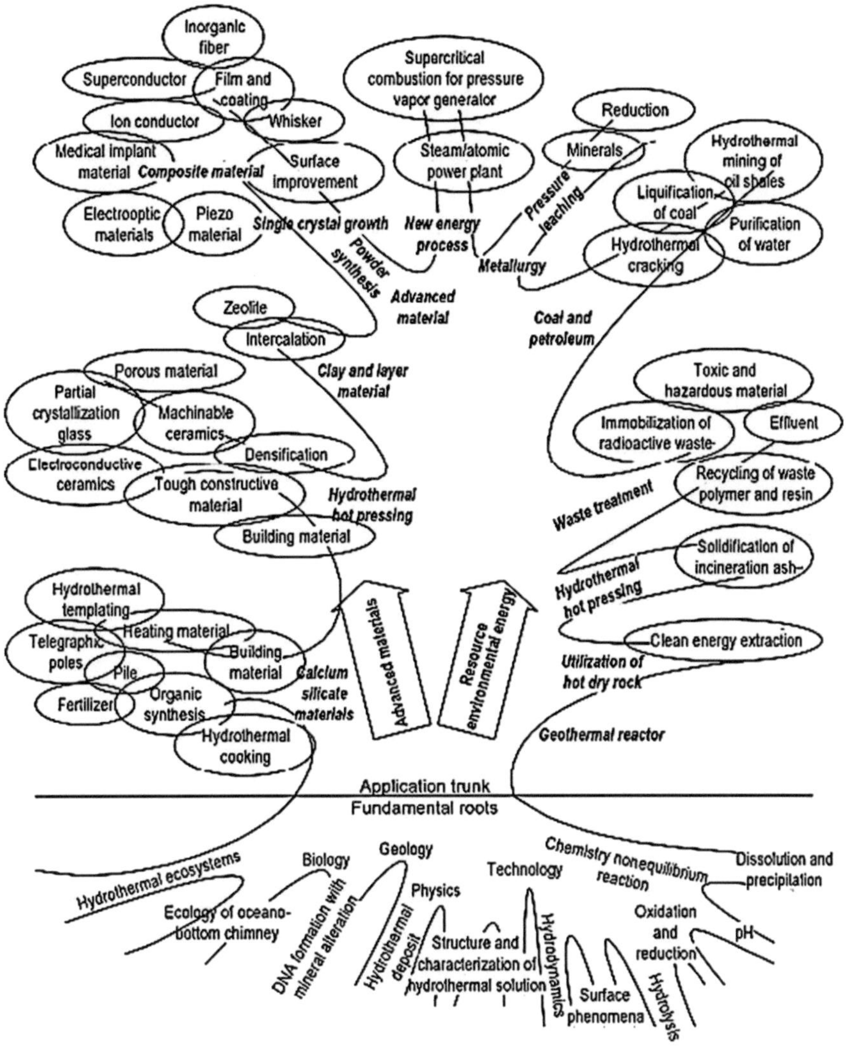

Fig. 13. Tree showing the interdisciplinary nature of hydrothermal technology [21].

Hydrothermal synthesis was used industrially[39] to prepare large crystals of quartz and synthetic gemstones. It is useful in metal oxide systems, where oxides are not soluble in water at atmospheric pressure but dissolve in superheated water under hydrothermal conditions. Where even these temperatures and pressures are insufficient to dissolve the starting materials, alkali or metal salts as mineralizers can be added, the anions of which form complexes with the solid and render it soluble [24].

[39] The first successful commercial application of hydrothermal technology was in mineral extraction or in ore beneficiation. The method was used to leach bauxite by sodium hydroxide by KARL JOSEF BAYER in 1892. The product of so-called Bayer ´s process, aluminum hydroxide, is then converted to Al_2O_3 and used to produce aluminum metal or in ceramics [21].

Throughout the course of evolution of hydrothermal synthesis from the geoscientific applications to modern technologies, the hydrothermal technique has captured the attention of scientists and technologists from different branches of science. The hydrothermal technique is popularly used by geologists, biologists, physicists, chemists, ceramists, hydro-metallurgists, materials scientists, engineers, etc. **Fig. 13** shows different branches of science either emerging out from the hydrothermal technique or closely linked up with the hydrothermal technique. One could firmly say that this family tree will keep expanding its branches and roots in the years to come [21].

The hydrothermal techniques for the preparation of compounds with the structure of apatite should be divided to:

1. low-temperature hydrothermal synthesis (LHS);
2. high-temperature hydrothermal synthesis.

The hydrothermal synthesis of all three normal apatite end-members was reported by Baumer and Argiolas [80]. They prepared crystallites of sizes from 50 to 500 μm. The synthesis of chlorapatite at 400°C and the pressure <3 kbar proceeds via the reaction:

$$10\ CaCl_2 + 6\ H_3PO_4 \Leftrightarrow Ca_{10}\left(PO_4\right)_6 Cl_2 + 18\ HCl \tag{21}$$

The synthesis and the stability of carbonate-fluorapatite were examined by Jahnke [81]. The carbonate-apatite phase is stable in solutions relatively rich in carbonate such as sea-water. When exposed to low-carbonate solutions, the carbonate-apatite should lose the CO_3^{2-} ion [82].

During the hydrothermal synthesis of HAP whisker, the acetamide was used by Zhang and Darvell [83] as an agent to drive homogeneous precipitation at temperatures below 100°C. Acetamide shows low hydrolysis rate in both acidic and basic conditions, releasing acetate and ammonia:

$$CH_3CONH_2 + H_3O^+ \rightarrow CH_3CO_2H + NH_4^+ \tag{22}$$

which do no substitute in HAP lattice. The precipitation of hydroxylapatite from the solution of $Ca(NO_3)_2{\cdot}4H_2O$ and $(NH_4)_2HPO_4$ in 0.05 mol·dm^{-3} (Ca:P = 1.67) treated to the temperature of 180°C for 10–15 hours yielded to large rod-like and well-crystallized particles of hydroxylapatite.

The stoichiometric single crystals of hydroxylapatite nanorods with mono-dispersion and narrow-size distribution in diameter were successfully synthesized by Lin et al [84] via the hydrothermal microemulsion method [85].[40] The microemulsion was prepared using CTAB as the surfactant and *n*-pentanol as the cosurfactant. First, 0.5 M $Ca(NO_3)_2$ solutions and 0.3 M $(NH_4)_2HPO_4$ solutions were obtained by dissolving $Ca(NO_3)_2{\cdot}4H_2O$ and $(NH_4)_2HPO_4$ in

[40] The emulsification consists in dispersing of one fluid in another, non-miscible one, via the creation of interface [85].

distilled water, respectively, and the pH of both solutions was adjusted to 11.0 by adding ammonia solution. These aqueous solutions were used as the water phase and n-hexane was used as oil phase. The mixture of surfactants [86][41] and $Ca(NO_3)_2$ solution was stirred, ultrasonicated and optically transparent microemulsion was obtained. The solution of $(NH_4)_2HPO_4$ was drop wisely added into the $Ca(NO_3)_2$ microemulsion solution to obtain a suspension, and the pH of the suspension was maintained at 11.0 using ammonia solution. Then the suspension of the microemulsion was transferred into stainless steel autoclaves and maintained at 180°C for 18 h. Washed hydroxylapatite powder was then calcined at 600°C for 2 h [84].

4.2 Preparation of single crystals

The first technique that was used for the production of crystals **Fig. 14(a)** was described by Verneuil [90],[91],[92],[93] at the turn of the 20th century,[42] but there is an evidence that the so-called Geneva ruby had been grown by a similar technique almost 20 years earlier. The second technique for single crystal growth was introduced by Czochralsky [94] few years later, who needed materials with small dimensions[43] in order to study the growth kinetics of metal (**Fig. 14(b)**). This technique is based on pulling thin wires from the melt at various speeds and obtaining single crystals. Beginning in 1950s, it eventually developed into the complex technology required in order to obtain large-diameter prefect crystals which are a raw material for the electronics industry, but controlling the dimension of the crystal was very difficult. The idea of using a shaping device floating on the melt surface to stabilize the crystal growth was introduced by Gomperz [95], who used a drilled mica plate. Since that time, numerous types of shaping devices have been used to get crystals of various shapes [26],[87],[88].

Kyropoulos developed the melt growth techniques (**Fig. 15**) for growing large crystal from the melt using a cooled seed in 1926 [96],[97],[98]. The method was demonstrated via the production of large single crystals of alkali halides [99].

---

41 The name surfactant is a contraction of the term: **surf**ace-**act**ive-**agent**. It can be defined as the substance that, even if present at low concentration, has ability to be adsorbed onto the surface or interface of the system and significantly alter (usually decrease) its surface or interface free energy. While the term **surface** usually means the interface between condensed phase and gas, the interface is considered as a boundary between two immiscible phases. The molecular structure of surfactant contains lyophobic group (little attraction for solvent) and lyophilic group (strong attraction for solvent), i.e. amphipathic structure [86].

42 Verneuil in fact wished to study the properties of ruby and other alumina-based crystals and was aware of very high melting temperatures of these materials, which prevented the use of any crucible material known in that time. This problem was solved- by melting alumina powder in a hydrogen-oxygen flame and solidifying the droplets on a colder seed. Nowadays this technique is used for the production of single crystals of sapphire (single crystal of Al_2O_3 in **Chapter 3** (**Fig. 14**) was prepared by this method) and spinel with only little changes [87]. The crystal grows from the melt film, which thickness is defined by the crystal diameter and the thermal conditions at the crystallization front [93].

The scheme of Verneuil's growth unit [87],[90],[93]: electromagnet (A, or camshaft) operating the hammer (M), supply chamber of fine Al_2O_3 powder (P), feeder (C, D), oxygen (O) and hydrogen (H) inlet, growing crystal (R), crystal holder (S) and device for the crystal adjustment (V).

43 Small dimension is necessary to dissipate the latent heat of solidification efficiently and rapidly [87].

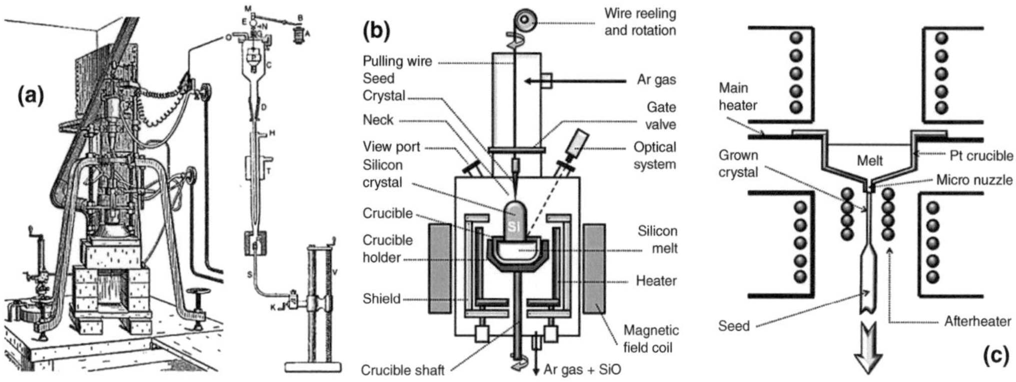

Fig. 14. Scheme of Verneuil's method[42] (a) [87],[90], Czochralsky growth apparatus (b) [88] and μ-PD apparatus (c) [89].

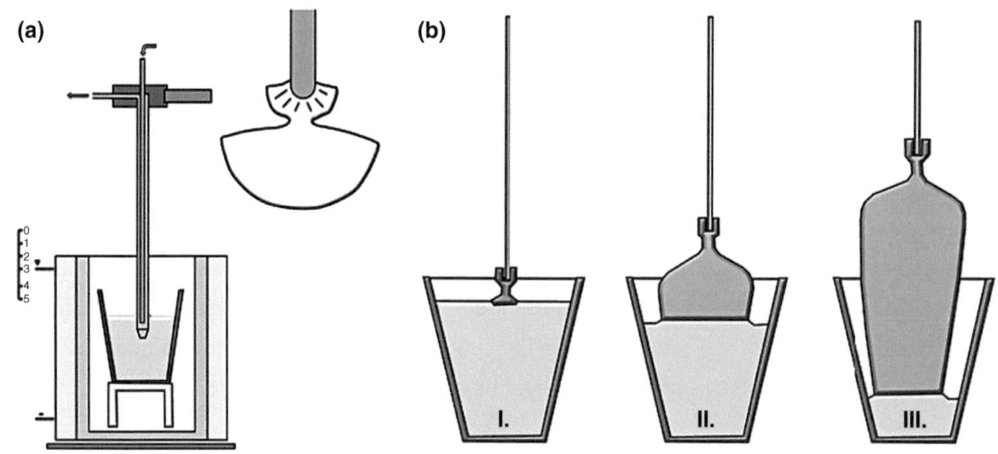

Fig. 15. Schematic illustration of Kyropoulos method [96] and three stages of Kyropoulos method [98].

After important growth processes based on capillarity, historically the next development was the BRIDGMAN method [100], aiming at increasing the crystal size and consisting in growing the crystal in crucible. The next method to be invented in 1952 by PHANN [101] was the floating zone (FZ) technique.[44] This method is capillary-based technique that was originally developed for the material purification [87]. The schematic representation of convection in the molten zone according to HIGUCHI et al [102] is shown in **Fig. 16**. The Marangoni convection[45] in molten zone leads to the formation of tiny bubbles, which are not arranged randomly, but form a ring inside the crystal.

[44] The floating zone is generated by means of water-cooled induction coil fed by radio frequency power in the megahertz range [103].

[45] Marangoni convection, which is caused by the differences in the surface tension over the melt surface, flows along the interface from the surface to a central region of the melt. On the other hand, forced convection, which is caused by the crystal rotation, flows towards the periphery from the center [102].

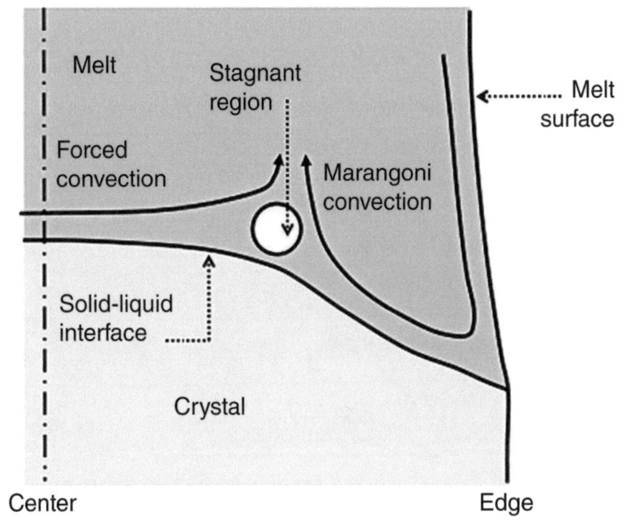

Fig. 16. Schematic representation of convection in the molten zone [102].

Since then various modifications of these basic methods have been proposed, such as the pedestal growth, edge-defined film-fed growth (EFG) process, inverted EFG process, micro-pulling down (μ-PD, **Fig. 14(c)**), etc., all based on the use of capillary force in order to maintain and shape the liquid. **Fig. 17** show the classification of these methods based on the presence or absence of the crucible or shaping die in contact with molten material and on the direction of pulling [87],[89],[103],[104].

Whiskers can be described as long filamentary defect-free single crystals of great mechanical strength, which is attributed to their high structural perfection. The explanation of whisker growth is based on the screw dislocation theory. The dislocation appears only along the whisker axis, while in another two dimensions the faces will stay perfect. Consequently, no growth will occur at an appreciable rate on the side faces of whisker. Due to the presence of axial screw dislocation the whisker grows only along its axis [105]. Apatite whiskers are usually prepared by hydrothermal synthesis [83],[106],[107],[108], molten salt method [109] and also via the precipitation method [110].

Dendrites[46] are normally single crystals, and the branches follow definite crystallographic orientation. The branches are regularly oriented and the opposite sides of the primary stem show marked symmetry. The growth of dendritic crystal is controlled by the diffusion of latent heat from growing crystal-melt interface [110],[111]. The dendritic growth of apatite crystal is described in glass ceramics [111],[112] and the formation of comb-shaped acicular and dendritic apatite was also observed as a product of quenching of trapped phosphate melt inclusions [113].

---

[46] The name was derived from the Greek word "tree like".

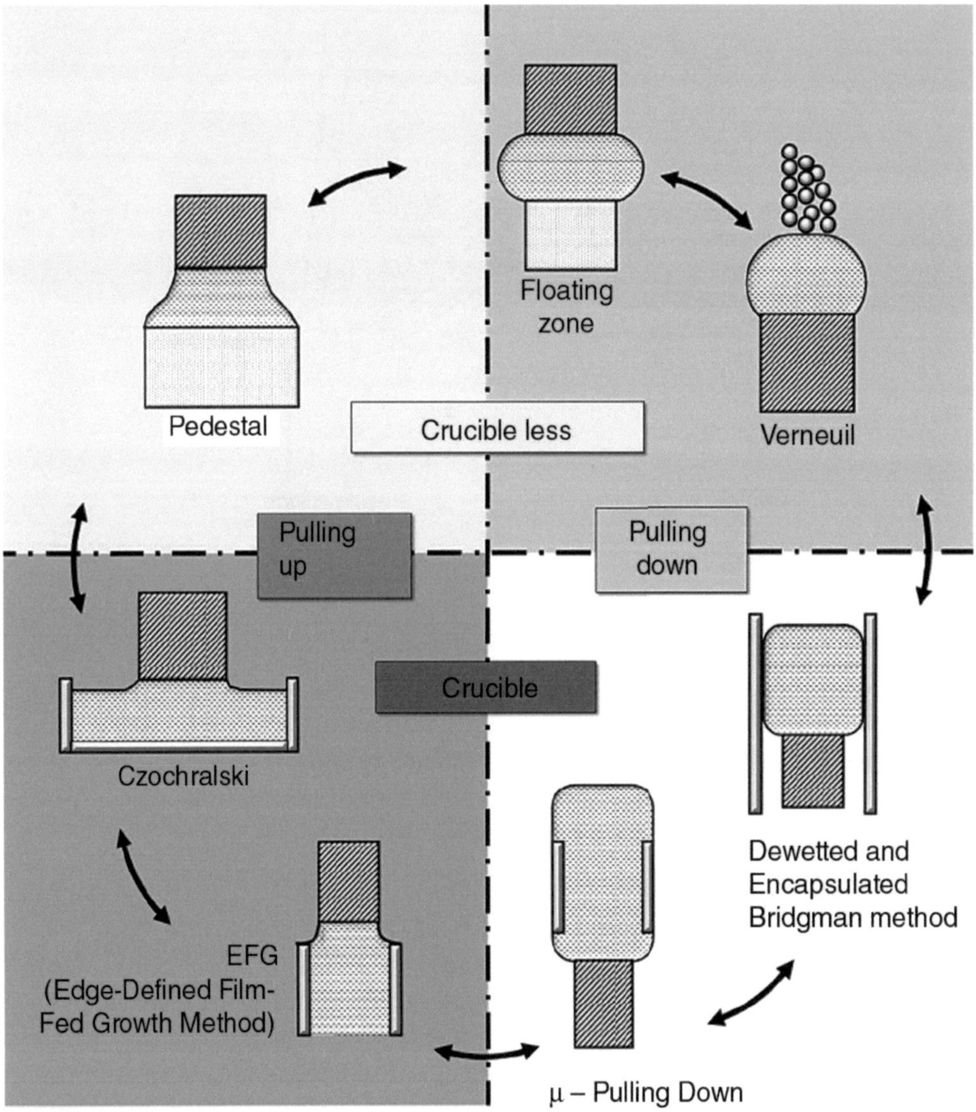

Fig. 17. Classification of various crystal growth processes using capillary forces for maintaining or shaping the molten material [87].

4.2.1 Fluorapatite

Single crystals of fluorapatite up to 5 cm long and of 1 cm maximum diameter were first prepared via the Kyropoulos method (pulling the crystal from the melt) by Johnson [114]. The Czochralsky method was used by Mazelski et al [114],[115] to grow the crystals up to 30 cm long. The ratio of CaF_2 to $Ca_3(PO_4)_2$ as determined by chemical analysis of crystals depends upon the value of the same ration in the melt. Even if the melt had correct stoichiometric

http://dx.doi.org/10.5772/62212

composition, grown fluorapatite would appear to have the deficiency of CaF_2 of about 5%. Fluorapatite as well as chlorapatite crystal with the length from 5 to 6 mm were grown by PRENER [116] from the solutions of apatite in molten calcium fluoride and chloride, respectively. The analyses of these flux-grown crystals agreed with theoretical values within 0.1% [114], [117].

4.2.2 Other compounds of apatite type

Single crystals of apatite-type $Nd_{9.33}(SiO_4)_6O_2$, $Pr_{9.33}(SiO_4)_6O_2$ and $Sm_{9.33}(SiO_4)_6O_2$ were prepared by HIGUCHI et al [102],[118],[119] from the stoichiometric mixture of Nd_2O_3, Pr_6O_{11} and Sm_2O_3 with SiO_2 (9.33 : 6) via the floating zone method. The crystal growth using the optical floating zone technique (a) was extensively used to grow a variety of bulk crystals, particularly single crystals of metal oxides [120],[121].

The pseudobinary phase diagram for the Nd_2O_3–SiO_2 system around the apatite phase is shown in **Fig. 19** [118],[122]. With the except of the end-member Nd_2O_3 and SiO_2, the apatite phase (Nd_2O_3 : SiO_2 = 7:9), Nd_2SiO_5 and $Nd_2Si_2O_7$ are observed. Both, Nd_2SiO_5 and $Nd_2Si_2O_7$ melt incongruently, while the apatite phase melts congruently.

YOSHIKAWA et al [123] prepared <0001> oriented $Ca_8La_2(PO_4)_6O_2$ (CLPA) single crystals with the apatite structure, which were grown by the Czochralsky method. This material can be used as substrate for the growth of <0001> GaN epitaxial layers.

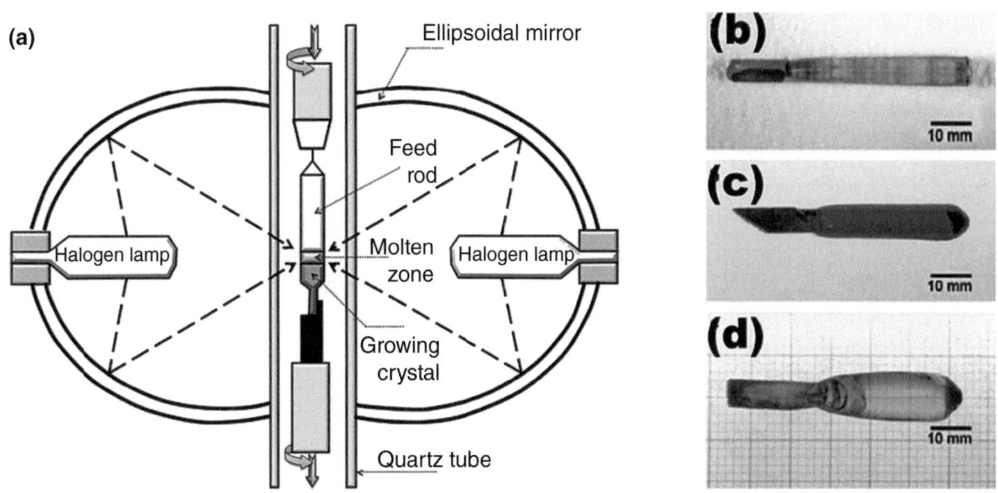

Fig. 18. Schematic diagram of the furnace with double ellipsoidal mirrors (a) and single grown crystals of $Pr_{9.33}(SiO_4)_6O_2$ (b), $Nd_{9.33}(SiO_4)_6O_2$ (c) and $Sm_{9.33}(SiO_4)_6O_2$ (d) [118].

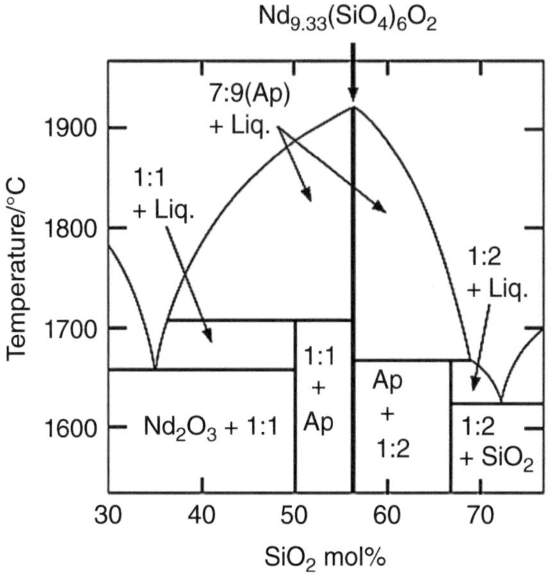

Fig. 19. Reconstructed pseudobinary phase diagram around the apatite phase $Nd_{9.33}(SiO_4)_6O_2$ in the Nd_2O_3-SiO system [118].

The growth of single crystal of synthetic analogue of vanadinite (lead vanado-chlorapatite, $Pb_5(VO_4)_3Cl$) using the CsCl flux method was performed by Masaoka and Kyono [124]. No impurity phases were formed from this crystal growth method. Crystals obtained via this method exhibit well-developed hexagonal prismatic form of the size of several millimeters along the [0001] direction. The largest crystals were approximately 6×1×1 mm.

The first hydrothermal growth of single crystals of chlorapatite was reported by Roufosse et al [125]. Crystals grown from the system chlorapatite-HCl-H_2O at 50 000 psi and pH = 1 with the growth zone at 465°C and dissolution zone at 360°C were found to be of high stoichiometry.

4.3 Synthetic analogues of the mineral hydroxylapatite

The synthetic analogue of the mineral hydroxylapatite can by prepared by the reaction [126]:

$$\begin{aligned}&10\ Ca(NO_3)_2 + 6\ (NH_4)_2HPO_4 + 8\ NH_4OH \rightarrow Ca_{10}(PO_4)_6(OH)_2\\&+20\ NH_4NO_3 + 6\ H_2O\end{aligned} \quad (23)$$

Aqueous solutions of 0.167 mol·cm^{-3} of $Ca(NO_3)_2$ and 0.100 mol·cm^{-3} of $(NH_4)_2HPO_4$ were prepared, and their pH values were adjusted to above 8 by the addition of ammonium hydroxide. $(NH_4)_2HPO_4$ solution was heated to about 85°C and then slowly dropped into equal volume of vigorously stirred solution of $Ca(NO_3)_2$. The temperature of the reaction mixture was kept at 85°C and stirring was maintained for further 3 days. In order to remove CO_2, the flow of N_2 was introduced to the suspension in reaction vessel. The suspension was then filtered and washed.

The survey of known chemical reactions successfully used for the synthesis of hydroxylapatite was provided by Shojai et al [32]. Depending on applied method (**Table 2**) and conditions (**Fig. 20**), different shapes of apatite particles can be prepared.

Shape*		Approximated size range	Method(s) of synthesis**
	Irregular, formless, sphere	5 nm–200 μm	ss, mch, cc, hl, sg, hth, em, sch, ht, bs, cp
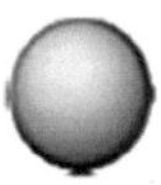	Sphere, microsphere, nanosphere, ball	10 nm–1000 μm	mch, cc, sg, hth, em, sch, ht, bs, cp
	Rod, needle, tube, filament, fibber, wire, whisker, prism, worm, hexagonal prism, platelet, lath, strip	length: 10 nm–150 μm diameter: 3 nm–50 μm aspect ratio: 2–1200	ss, mch, cc, hl, sg, hth, em, sch, ht, bs, cp
	Plate, flake, sheet	length: 40 nm–50 μm width: 20 nm–35 μm thickness: 3 nm–5 μm	cc, hth, bs, cp
	Self-assembled nanorods, bundles of nanorods, oriented bundle, oriented raft, enamel prism-like structures, clusters of nanotubes, oriented array of bundled needles, packed nanorods	length: 200 nm–80 μm width: 100 nm–50 μm (oriented nanorods of 10 nm–13 μm diameter and 600 nm–5 μm length).	cc, hl, tht, bs, cp
	Dandelion, chrysanthemum, flower, feathery structure, bundle of fibers, self-assembled nanorods, rosette	1–8 μm (oriented nanorods of 80–500 nm diameter and 600 nm–5 μm length)	hth, em, bs, cp
	Leaf, flake, sheet, plate	800 nm–10 μm (organized nanoplates of 20–100 nm thickness)	cc, hl, cp
	Flower	700 nm–60 μm (organized petals of 20 nm–10 μm width and 180 nm–50 μm length)	cc, hth, bs

Shape*		Approximated size range	Method(s) of synthesis**
	Porous microsphere, mesoporous sphere	0.5–7 μm (pores 20–150 nm)	hth, cp
	Bowknot, self-assembled nanorods	1.5–2.5 μm (organized nanorods of 100–150 nm diameter and 1–2 μm length)	cp
	Dumbbell	2–3 μm (organized nanoparticles of ~50 nm size)	cc

* Consult with **Section 3.1.14**.
** Solid-state synthesis (ss), mechanochemical method (mch), conventional chemical precipitation (cc), hydrolysis method (hl), sol-gel method (sg), hydrothermal method (hth), emulsion method (em), sonochemical method (sch), high-temperature processes (ht), synthesis from biogenic sources (bs), combination procedures (cp).

Table 2. Shape of hydroxylapatite particles prepared by given synthesis methods [32].

The influence of conditions on the morphology of hydroxylapatite particles is shown in **Fig. 20**. During hydrothermal synthesis, the particle size of HAP decreases with increasing pH value [32],[127],[128].

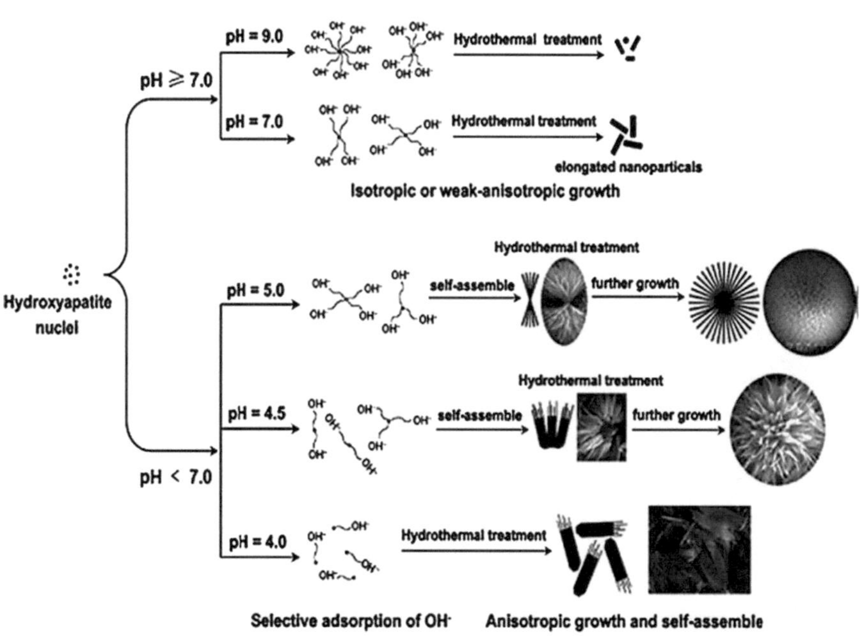

Fig. 20. The formation and the morphology evolution mechanism of $Ca_5(PO_4)_3OH$ samples with various morphologies based upon different pH values [127],[128].

http://dx.doi.org/10.5772/62212

Complete replacement of halogen occurs when either fluorapatite or chlorapatite is heated in the steam of H_2 or H_2O at high temperatures [69]:

$$Ca_{10}(PO_4)_6F_2 \xrightarrow{H_2,\ 1630°C} Ca_{10}(PO_4)_6(OH)_2 \xleftarrow{H_2O,\ 800°C} Ca_{10}(PO_4)_6Cl_2 \tag{24}$$

4.4 Fluorapatites

4.4.1 Synthetic analogues of the mineral fluorapatite

In literature various routes for the preparation of synthetic analogues of fluorapatite are described which include solid-state reactions of the type [129]:

$$3\ Ca_3(PO_4)_2 + CaF_2 \rightarrow Ca_{10}(PO_4)_6F_2 \tag{25}$$

At the temperature of 900°C hydroxylapatite reacts with calcium fluoride to give fluorapatite [69]:

$$Ca_{10}(PO_4)_6(OH)_2 + CaF_2 \rightarrow Ca_{10}(PO_4)_6F_2 + CaO + H_2O(g) \tag{26}$$

Fluorapatite can be also prepared directly by firing a mix of $3Ca_3(PO_4)_2$ with CaF_2 at 1600°C, or from calcium pyrophosphate and calcium fluoride:

$$18\ Ca_2P_2O_7 + 14\ CaF_2 \rightarrow 5\ Ca_{10}(PO_4)_6F_2 + 6\ POF_3 \tag{27}$$

Chlorapatite can be prepared by similar method using calcium chloride. It can also be produced in the reversible reaction according to **Eq. 21**.

Original phase diagram **Fig. 21**(a) for the section $Ca_3(PO_4)_2$–CaF_2 of the ternary system CaO–P_2O_5–CaF_2 was published by NACKEN [130]. The range of compositions was extended by BERAK [131] (b), and further refined (**Fig. 22**) by BERAK and T.-HUDINA [132]. Important features are congruent melting of $Ca_{10}(PO_4)_3F_2$ at 1650°C, eutectics with $Ca_3(PO_4)_2$ at 1620°C and second one with CaF_2 at 1203°C. Sufficiently precise phase diagram enables to determine necessary information on the flux growth of fluorapatite, so the crystal with only slight deficiency in fluorine compared to the theoretical one can be prepared. The problem concerning possible stable existence of spodiosite ($Ca_2(PO_4)F$) analogous to naturally occurring mineral remains unsettled, but it appears unlikely to be stable at liquidus temperatures [133].

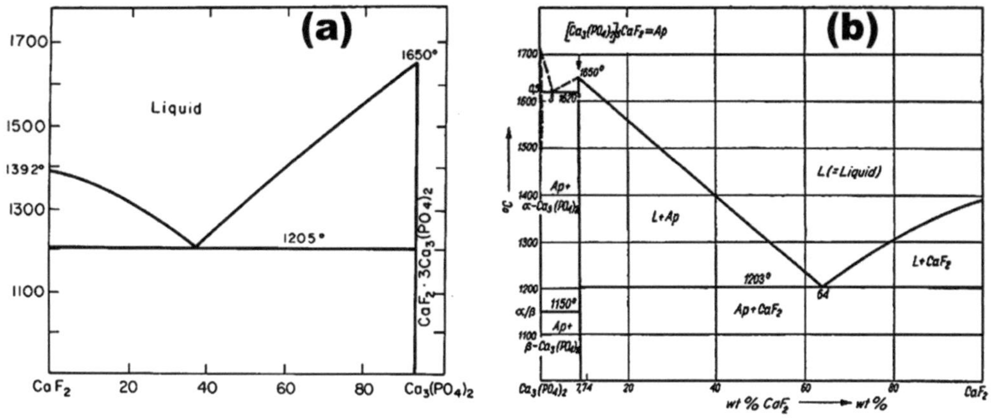

Fig. 21. Phase diagram of $Ca_3(PO_4)_2$–CaF_2 section by Nacken [130] (a) and Berak [131] (b).

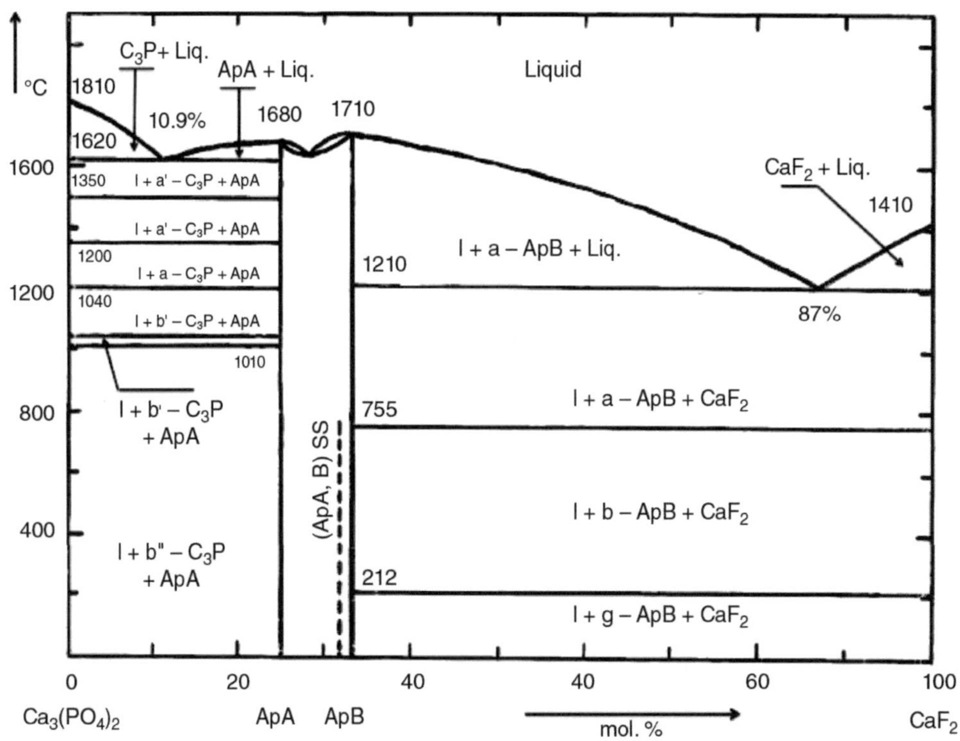

Fig. 22. Phase equilibrium in the system $Ca_3(PO_4)_2$–CaF_2: $Ca_{10}(PO_4)_6F_2$ (ApA) and $Ca_7(PO_4)_4F_2$ (ApB) [132].

The implication of the crystal growth of apatite and calcite in the systems $Ca_3(PO_4)_2$–$CaCO_3$–$Ca(OH)_2$–CaF_2 (**Fig. 23(a)**) and $Ca_3(PO_4)_2$–$Ca(OH)_2$–CaF_2–H_2O (**b**) enables the quaternary phase diagram provided by Willie [134].

http://dx.doi.org/10.5772/62212

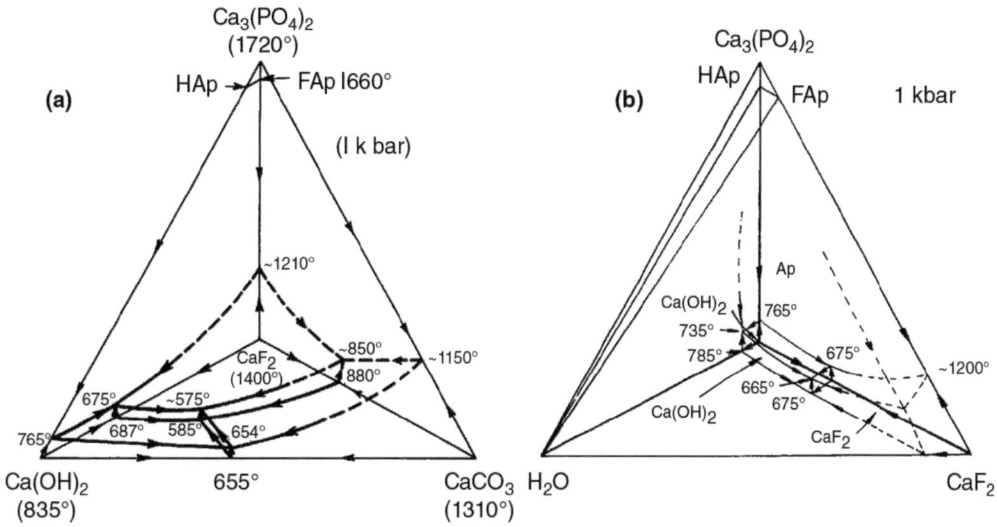

Fig. 23. System $Ca_3(PO_4)_2$–$CaCO_3$–$Ca(OH)_2$–CaF_2 (a) and $Ca_3(PO_4)_2$–$Ca(OH)_2$–CaF_2–H_2O (b) at the pressure of 1 kbar [134].

Long and uniform HAP whiskers with high crystallinity, controlled morphology and high aspect ratio were synthesized by Zhang and Darwell [83] via the hydrothermal method using acetamide. Compared to urea as an additive, which is commonly used to raise the pH in order to drive the nucleation and growth of HA crystals [106], acetamide has low hydrolysis rate under required hydrothermal conditions. This allows better and easier control, giving rise to rapid growth of whiskers at low supersaturation. The whisker length and width were in turn given by the solution conditions, including the concentration of Ca and PO_4 [83].

4.4.2 Barium fluorapatite

Barium apatite can be prepared by solid-state reaction [135]:

$$6\ BaHPO_4 + 3BaCO_3 + BaF_2 \xrightarrow{heat} Ba_{10}\left(PO_4\right)_6 F_2 + 3CO_2 + 3H_2O \tag{28}$$

It possesses typical hexagonal structure with the space group $P6_3/M$ and $a = 10.153$ Å, $c = 7.733$ Å, c:a = 1:0.722, $\boldsymbol{V} = 10.153$ Å^3 and $\mathbf{Z} = 2$ [135].

Ba(1) atoms are located in columns on three threefold axes and are coordinated by nine oxygen atoms. The Ba(2) sites form triangles around the F site and are coordinated by six oxygen atoms and one fluoride ion. Fluoride ions are statistically displaced by ~0.25 Å from the Ba(2) triangles. This displacement of F ions is analogous to the displacement of OH ion in $Ca_{10}(PO_4)_6(OH)_2$ [138].

4.5 Chlorapatites

4.5.1 Synthetic analogues of the mineral chlorapatite

The stoichiometric Ca:P ratio in the composition of chlorapatite, the mole ratio of calcium to phosphorous was equal to 1.67 [139]. The reaction of $CaCl_2$ with H_3PO_4 under hydrothermal conditions including the temperature of 400°C and the pressure < 3 kbar leads to chlorapatite (**Eq. 21**) [82],[140].

The mechanochemical synthesis of chlorapatite in a high energy planetary mill should be described by the reaction [139]:

$$9CaO + CaCl_2 + 3\ P_2O_5 \rightarrow Ca_{10}(PO_4)_6Cl_2 \tag{29}$$

NACKEN [141] determined the phase diagram for the section $Ca_3(PO_4)_2$–$CaCl_2$ of the ternary system CaO–P_2O_5–$CaCl_2$ (**Fig. 24**). Chlorapatite crystallized from melts of its own composition is highly deficient in Cl, while lower temperatures near 1040°C lead to the crystallization of stoichiometric chlorapatite [133].

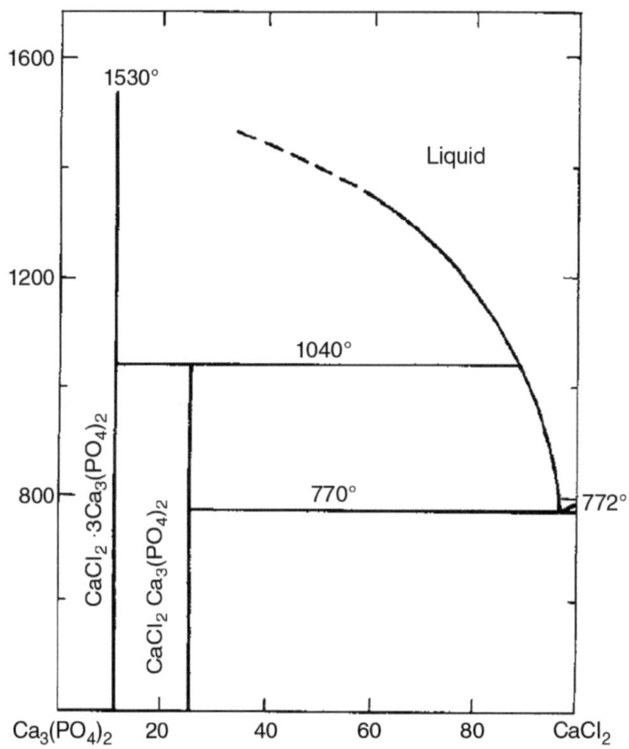

Fig. 24. Phase equilibrium in the system $Ca_3(PO_4)_2$–$CaCl_2$ by NACKEN [133].

The mechanosynthesis and the characterization of chlorapatite nanopowders were performed by Fahami et al [139]. The formation of chlorapatite takes place according to the reaction 29. At the beginning of milling, the main products were stoichiometrically deficient chlorapatite and calcium oxide. Eventually, high crystalline CAP nanopowder was obtained after 300 min of milling. By increasing the milling time to 300 min, the lattice strain significantly increased.

4.5.2 Cadmium chlorapatite

Large crystals of $Cd_5(PO_4)_3Cl$ (space group $P6_3/m$, a = 9.633 Å, c = 6.484 Å and Z = 2) grow hydrothermally at 500°C and 800–1400 atm. from Na- and NH_4-containing solutions [142], [143]. The phase transition in cadmium chlorapatite from $P6_3/m$ to $P6_c/mcm$ was confirmed through the temperature dependent Raman measurements. The phase transition temperature from lower temperature phase ($P6_3/m$) to high temperature phase ($P6_3/mcm$) is approximately 700°C and was detected through the disappearance of low-temperature phase A_g Raman bands as the temperature approached the transformation temperature [144].

4.5.3 Other chlorapatites

The structure of apatite phase of the composition $Ba_5(OsO_5)_3Cl$ ($P6_3cm$, $\boldsymbol{a}$ = 10.928 Å, $\boldsymbol{c}$ = 7.824 Å, $\boldsymbol{V}$=809.2 Å^3, Z = 2 and ρ = 6.29 g·cm^{-3}) where PO_4 tetrahedra are replaced by pyramidal OsO_5 groups was reported by Plaisier et al [145] as isomorphous with $Ba_5(ReO_5)_3Cl$ (Besse et al [146]) and $Ba_5(ReO_5)_3I$ (Baud et al [147]). The structure (**Fig. 25**) consists of columns of Ba(1) atoms parallel to the $\boldsymbol{c}$-axis and chains of $ClBa_6$ octahedra with common faces along the $\boldsymbol{c}$-axis. Among these are isolated pyramidal OsO_5 groups. Ba(1) atoms lie on the threefold-axis and are surrounded by nine oxygen atoms. Ba(1)-O distances vary between 2.74 and 2.76 Å. Atom of Ba(2) is surrounded by seven oxygen atoms and two atoms of chlorine.

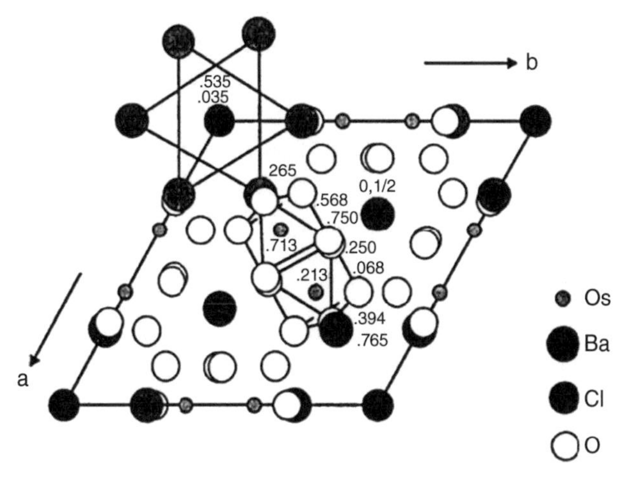

Fig. 25. Projection of the structure of $Ba_5(OsO_5)_3Cl$ along the c-axis [145].

Suzuky and Kibe [148] used the NaCl flux method to prepare barium ($Ba_5(PO_4)_3Cl$) and strontium chlorapatite ($Sr_5(PO_4)_3Cl$) crystals and modified.[47] Wilhelmy method [149],[150],[151][48] for the determination of surface free energy (~26 mN·m^{-1} for both apatite crystals[49]). The determination of specific surface free energies (surface tension) for single crystal of $Sr_5(PO_4)_3Cl$ [152] (aspect ratio is 3.2) via the measurement of contact angles of water and formamide (CH_3NO) shows that ideal flat surface without a step should have uniform specific surface free energy, estimated to ≤ 26 and ≤ 50 mN·m^{-1} for (101◎0) and (101◎1) faces,[50] respectively. Experimentally obtained specific surface free energies roughly satisfy the Wulff's relationship [153],[154]:

$$\frac{\gamma_i}{h_i} = \text{const}, \tag{30}$$

where γ_i is the specific surface free energy of the i-th face of the crystal and h_i is the distance of face from the Wulff's (central) point of crystal.

4.6 Carbonated (biological) apatites

It seems now to be generally accepted that CO_3^{2-} dominantly replaces PO_4^{3-} in biological apatite (BAP, BAp) [155]. Carbonate-hydroxyl-apatite ($Ca_{10}(PO_4,CO_3)_6(OH)_2$), can be found mainly on islands and in caves, as a part of bird and bat excrements, guano [156].

4.6.1 Carbonated hydroxyl$^-$ and fluorapatite

Carbonated hydroxylapatite is the most important mineral in human dental enamel and bone [157],[158],[159],[160],[161],[162],[163],[164]. The presence of highly carbonated apatite was also proposed as a marker of the presence of bacteria (infectious microorganism) in kidney stones[51] [165],[166]. According to the position of planar bivalent carbonate ion (CO_3^{2-}) with anionic radius of 0.176 nm in the structure of apatite, three kinds of carbonate apatite are recognized in literature [157],[158],[167],[168],[169].

---

[47] The weight of liquid was measured instead of the weight of crystal.

[48] Since measured parameter is force (**F**, measured by tensiometer or microbalance), the Wilhelmy plate (or rod) method can be easily applied for small contact angles. The surface tension (γ) is calculated from the equation: $\gamma = F/(l \cos \Theta)$, where l = 2×length + 2×width of the plate, and wetting angle Θ is usually not determined but its value is taken for zero (complete wetting is assumed) or the value from either literature is used [149],[150],[151].

[49] The value should be affected by the estimation of value of aspect ratio.

[50] First-order hexagonal prism and first-order dipyramid, respectively.

[51] From the medical point of view, pathological calcifications refer to a concretion, e.g. a kidney stone, often associated with the tissue alteration. Additionally, normal physiological calcifications such as bone may become pathological through the influence of diseases such as arthrosis or osteoporosis Different chemical phases constitute the pathological calcifications, but calcium phosphate apatites are present in most of them [166]. Biological apatites (BAP) are described in **Section 7.1.3**.

1. **Type A**: CO_3^{2-} substitutes for OH^- (or Z^- anion in general) in the apatite channel at z ≈ 0.5 (**Fig. 27(a)**) by two triad clusters of Ca^{2+} atoms z ≈ 0.25 (1/4) and 0.75 (3/4). The composition of CCAP is given by the formula: $Ca_{10}(PO_4)_6(CO_3)_x(OH)_{2-2x}$. The TYPE-A substitution can be described as follows [170]:

$$\begin{array}{l} Ca_{10}\left(PO_4\right)_6\left(OH\right)_2 + CO_3^{2-} \rightarrow Ca_{10}\left(PO_4\right)_6 CO_3 + 2\ OH^- \\ \text{or}\ \left[{}^{A}CO_3^{2-} + {}^{A}[V] \Leftrightarrow 2\left(OH^-, F^-, Cl^-\right)\right] \end{array} \tag{31}$$

The crystal structure of type-A carbonate apatite is controversial [172]. There are three different structures: with the space group$P\bar{6}$in hexagonal symmetry (**Fig. 27**) [171], with the space group$P\bar{3}$[159],[173] and with the space group Pb in monoclinic symmetry [174], [175].

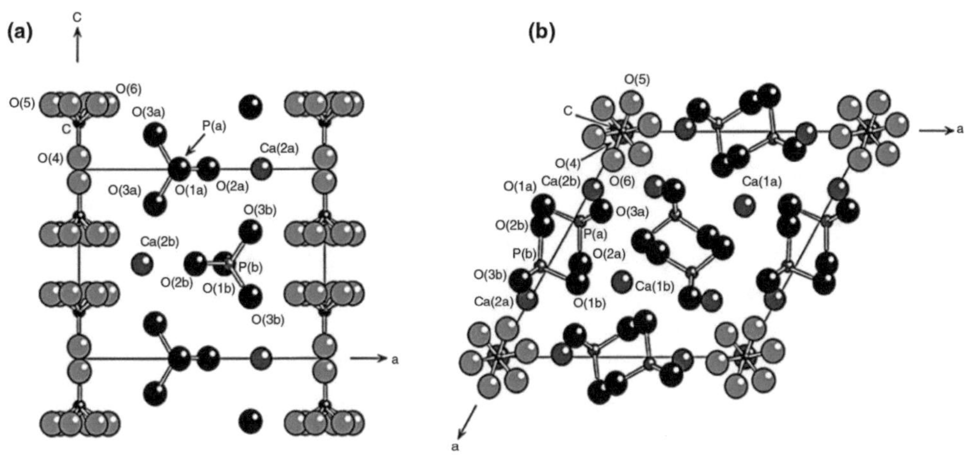

Fig. 26. The atomic configuration of type-A carbonated apatite on the (010) plane (a) and (001) plane (b) according to SUETSUGU et al [171].

2. **Type B**: CO_3^{2-} substitutes for phosphate ion (PO_4^{3-}). Different chemical formula is used to describe B-type carbonate apatite, the simplest and often used is [176]: $Ca_{10-x}(PO_4)_{6-x}(CO_3)_x(OH)_{2-x}$. The interpretation of the location of type-B carbonate ion is also problematical [158]. The carbonate ion is located in the vicinity of substituted phosphate group and occupies as many phosphate oxygen sites as possible. Some examples of the B-TYPE substitutions are described by equations [170],[177]:

$$\begin{array}{l} Ca_{10}\left(PO_4\right)_6\left(OH\right)_2 + Na^+ + CO_3^{2-} \rightarrow Ca_9Na\left(PO_4\right)_5 CO_3\left(OH\right)_2 \\ + Ca^{2+} + PO_4^{3-} \end{array} \tag{32}$$

$$\begin{aligned}&Ca_{10}(PO_4)_6(OH)_2 + CO_3^{2-} \rightarrow Ca_9[V_{Ca}](PO_4)_5CO_3(OH)[V_{OH}]\\&+PO_4^{3-} + OH^- + Ca^{2+}\end{aligned} \tag{33}$$

$$\begin{aligned}&Ca_{10}(PO_4)_6(OH)_2 + HPO_4^{2-} + CO_3^{2-} \rightarrow\\&Ca_9[V_{Ca}]HPO_4(PO_4)_4CO_3(OH)[V_{OH}] + 2\ PO_4^{3-} + Ca^{2+}\end{aligned} \tag{34}$$

where ***V*** denotes the vacancy. There is no clear energetic preference of CO_3^{2-} to substitute for any particular PO_4^{3-} group [178]. Sodium (**Eq. 33**) or other alkali metal cation (AM = Li, Na, K, Rb and Cs) is also known to increase the maximum ratio of carbonate substitution in B-site because its incorporation in calcium sites induces favorable electrical charge balance [176]:

$$\left[{}^{B}CO_3^{2-} + AM^+ \Leftrightarrow Ca^{2+} + PO_4^{3-}\right] \tag{35}$$

3. **Type AB**: mixed A-B type of apatite, where the composition can be described as: $Ca_{10}(PO_4)_{6-y}(CO_3)_{x+(3/2)y}(OH)_{2-2x}$. If both PO_4^{3-} (*z* ≈ 0.25 or 0.75) and OH^- anions were replaced by two CO_3^{2-}, the process can be described as follows:

$$\begin{aligned}&Ca_{10}(PO_4)_6(OH)_2 + 2\ CO_3^{2-} \rightarrow Ca_{10}(PO_4)_5(CO_3)_2(OH)\\&+OH^- + PO_4^{3-}\end{aligned} \tag{36}$$

Two different structural roles of CO_3^{2-} anion result in characteristic infrared (IR) signatures: type A carbonate having a doublet band at about 1545 and 1450 cm^{-1} (asymmetric stretching vibration, ν_3) and a singlet band at 880 cm^{-1} (out-of-plane bending vibration, ν_2), and type B having these bands at about 1455, 1410 and 875 cm^{-1}, respectively [158].

Published structural studies [158] of carbonated apatites were performed with the synthetic phases. Ren et al [171] investigated the structure of carbonated apatite using the ab initio simulation (**Fig. 27**) with the conclusion that the most energetically stable substitutions is type-AB in which two carbonate ions replace one phosphate group and one hydroxyl group respectively. The crystal structure of A-type of carbonated apatite is energetically more favorable than B-type of substitution. The most stable configuration of type-A is carbonate triangular plane almost parallel to the *c*-axis at *z* = 0.46. Type-A substitution tends to increase the lattice parameter ***a*** but decreases ***c*** whereas type-B substitution shows the opposite effect. The lowest energy configuration of type-B has calcium ion replaced by a sodium ion to balance the charge (**Eq. 33**) and the carbonate lying almost flat on the ***b***/***c*** plane.

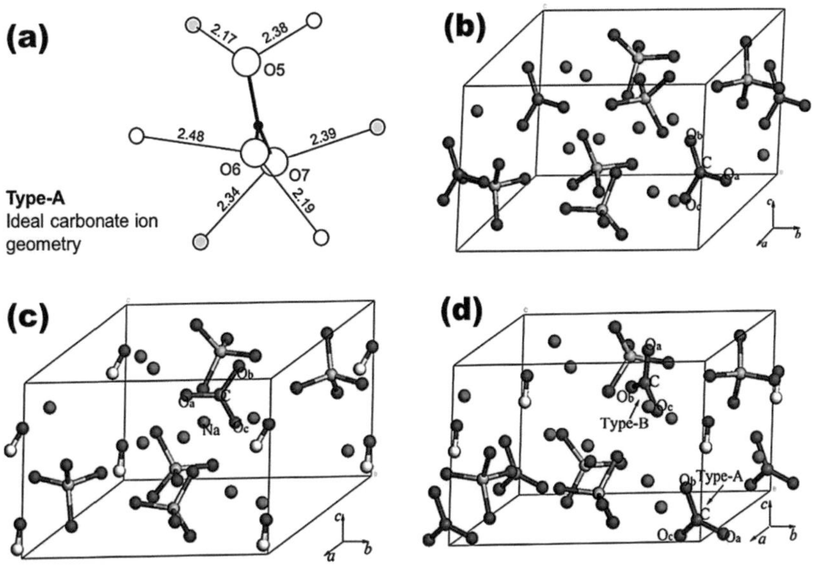

Fig. 27. The model of channel structure for ideal carbonate ion geometry of TYPE-A (a) [101] and the energetically favored configuration of TYPE-A carbonated apatite (b), the most stable structure of TYPE-B carbonated apatite (c) and the configuration of TYPE-AB after the geometry optimization (d) [170].

The type-A of carbonated apatite in which carbonate ion was completely substituted for the hydroxyl site, was synthesized by heating low crystalline and stoichiometric synthetic analogue of hydroxylapatite powder in the flow of dry carbon dioxide gas at 1000°C for 24 h by TONEGAWA et al [172]. The chemical composition of this phase can be described by the formula: $Ca_{10}(PO_4)_6(CO_3)_{0.93\pm0.06}$. The crystal structure was determined to be of monoclinic symmetry with the space group Pb in the temperature range from 25 to 500°C.

The synthesis of type-A carbonate apatite can be performed by heating of pure HAP at temperatures from 800 to 1000°C for several hours in dry CO_2 atmosphere according to the reaction [176],[179]:

$$Ca_{10}(PO_4)_6(OH)_2 + y\ CO_2 \leftrightarrow Ca_{10}(PO_4)_6(OH)_{2-y}(CO_3)_y + y\ H_2O \quad (37)$$

Type-B carbonated apatite powders are generally synthesized from the precipitation reaction in aqueous media [176]. The reaction **38** (or **Eq. 4** as was described in **Section 1.5.2**) can be used for the capture of carbon dioxide at high temperature over the operation limit of CaO-based sorbents.

The mechanochemical synthesis of B-type carbonated fluorapatite under argon atmosphere using high-energy planetary ball mill was described by N.-TABRIZI and FAHAMI [180]. The process can be described by the reaction:

$$(9-x/2)\ CaCO_3+(3-x/2)\ P_2O_5+CaF_2\rightarrow Ca_{10-x/2}(PO_4)_{6-x}(CO_3)_x F_2 +(9-3x/2)CO_2 \qquad (38)$$

4.6.2 Carbonate-chlorapatites

Carbonated chlorapatite nanopowders can be synthesized by the mechanochemical process under argon atmosphere using the mixture of calcite ($CaCO_3$), phosphorus pentoxide (P_2O_5) and calcium chloride ($CaCl_2$) as raw materials [181],[182]:

$$(9-x/2)\ CaCO_3+(3-x/2)\ P_2O_5+CaCl_2\rightarrow Ca_{10-x/2}(PO_4)_{6-x}(CO_3)_x Cl_2+(9-3x/2)\ CO_2 \qquad (39)$$

The substitution degree of PO_4^{3-} was given by the x value in the general formula of TYPE-B of $Ca_{10-x/2}(PO_4)_{6-x}(CO_3)_xCl_2$.

The high-pressure (1 GPa) synthesis of sodium-bearing carbonate chlorapatite of TYPE A-B (CCLAP, $Ca_{10-(y+z)}Na_y[V]_z[(PO_4)_{6-(y+2z)}(CO_3)_{y+2z}][Cl_{2-2x}(CO_3)_x]$, where $x \approx y \approx 4z \approx 0.4$) from carbonate rich melt in the temperature range from 1000 to 1350°C, was described by FLEET et al [169].

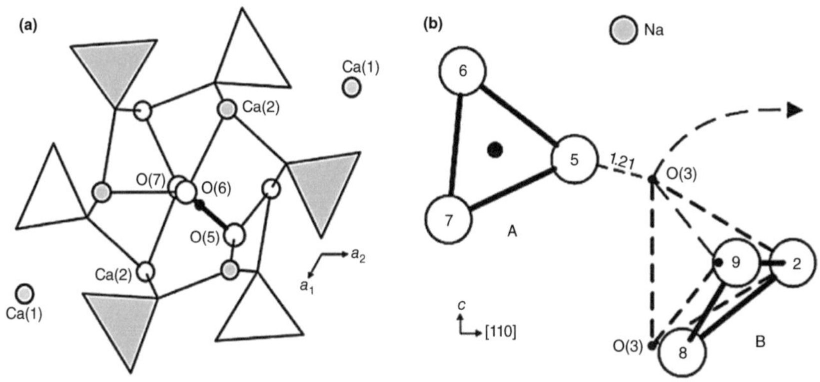

Fig. 28. The structure of carbonate chlorapatite showing one of 12 possible orientations of the type-A carbonate ion in apatite channel: the unit-cell origin is in the center of figure, shaded phosphate polyhedra and Ca(2) atoms are centered at z = 3/4 (a) and the fragment of CCLAP structure showing the location of B carbonate ion close to the sloping faces of substituted phosphate tetrahedron (b) according to FLEET AND LIU [169].

The structure of Na-bearing CCLAP crystals (**Fig. 28(a)**) with the contents of Na and A and B-type of carbonate ranges between those of Na-bearing carbonated fluorapatite (CFAP) and carbonated hydroxylapatite (CHAP). The stoichiometric amount of Na and A-type of carbonate is consistent with the near linear (1:1) correlation reported for CHAP and CFAP and provides the evidence of active role of Na in the substitution of carbonate into the apatite channel, even if Na does not appear in usual charge-balanced substitution scheme [169]:

$$\left[{}^{A}CO_3^{2-} \Leftrightarrow 2\ Cl^-\right] \tag{40}$$

On the other hand, the B : Na ratio is higher than one (approximately 1 : 1.5) and is located between the values determined for CHAP (B : Na = 1) and CFAP (B : Na = 2). The substitutions of B carbonate ion into CCLAP seem to be more complex than those into CHAP, which is expressed by **Fig. 28** or by the following charge-balanced substitutions scheme:

$$\left[Na^+ + {}^{B}CO_3^{2-} \Leftrightarrow Ca^{2+} + PO_4^{3-}\right] \tag{41}$$

There should be additional vacancies including the charge-balancing mechanism:

$$\left[1/2\ [V] + {}^{B}CO_3^{2-} \Leftrightarrow 1/2\ Ca^{2+} + PO_4^{3-}\right] \tag{42}$$

This leads to the formula of sodium-bearing carbonate chlorapatite mentioned above.

Similar profiles of ν_3 bands in FT-IR spectra for all carbonate apatite composition series and carbonated contents, together with common X-ray structure suggest that Na cation and A and B carbonate ion substituents are present as randomly distributed defect clusters within host apatite structure. The defect cluster depicted in **Fig. 28** (**b**) facilitates local charge compensation by Na-for-Ca substitution, explains the linear 1:1 correlation between Na and A carbonate, and minimizes the effects of spatial accommodation [169].

The synthesis of hydroxyl-chlorapatite solid solution via the precipitation method can be presented as follows [183]:

$$\begin{aligned} &10\ Ca(NO_3)_2 + 6\ (NH_4)_2HPO_4 + 2\ NH_4Cl + 6\ NH_4OH \rightarrow \\ &Ca_{10}(PO_4)_6\left[(OH)_{2-x}Cl_x\right] + 20\ NH_4NO_3 + H_2O \end{aligned} \tag{43}$$

Also fluorine and chlorine co-substituted hydroxylapatites can be prepared by aqueous precipitation method [184]:

$$\begin{aligned} &10\ Ca(NO_3)_2 + 6\ (NH_4)_2HPO_4 + NH_4F + NH_4Cl + 6\ NH_4OH \rightarrow \\ &Ca_{10}(PO_4)_6(F,Cl) + 20\ NH_4NO_3 + 6\ H_2O \end{aligned} \tag{44}$$

$$\begin{aligned} &10\ Ca(NO_3)_2 + 6\ (NH_4)_2HPO_4 + x\ NH_4F + \\ &y\ NH_4Cl + \left[8 - (x+y)\right]\ NH_4OH \rightarrow Ca_{10}(PO_4)_6\left(OH_{2-(x+y)}F_xCl_y\right) \\ &+20\ NH_4NO_3 + 6\ H_2O \end{aligned} \tag{45}$$

Carbonate can be introduced into the structure of carbonated barium-chlorapatite by stirring apatite in an $(NH_4)_2CO_3$ solution for 1 week [185]:

$$Ba_{10}(PO_4)_6Cl_2 + CO_3^{2-} \rightarrow Ba_9(PO_4)_5Cl + PO_4^{3-} + Cl^- + Ba^{2+} \tag{46}$$

The attempts to prepare carbonated barium-chlorapatite in a one-step synthesis results in a mixture of $BaCO_3$ and $Ba_3(PO_4)_3$. The variations in the manner in which carbonate was added to the reaction mixture, such as co-titrating a carbonate solution along with $BaCl_2$ and $NH_4H_2PO_4$, pre-mixing it with $NH_4H_2PO_4$, or adding it first or last did not eliminate the precipitation of simple salts. The inability at 60°C and at the pH of 10 to produce carbonated barium-chlorapatite at any ratio of carbonate to phosphate in the aqueous solution is probably due to close molar solubility of simple salts [185].

4.7 Bromoapatites

4.7.1 Calcium bromapatite

Calcium bromapatite has typical hexagonal apatite structure with the space group $P6_3/M$, $\boldsymbol{a}$ = 9.761 Å, $\boldsymbol{c}$ = 6.739 Å, c:a = 1: 0.6904, $\boldsymbol{V}$ = 556.06 Å^3 and $\boldsymbol{Z}$ = 2 (**Fig. 29**) [186]. The synthesis of **calcium bromapatite** ($Ca_{10}(PO_4)_6Br_2$) in the tubular quartz reactor (sealed-tube method) can be described by the reaction [135],[187],[188],[189]:

$$Ca_{10}(PO_4)_6OH_2 + 2\ HBr \xrightarrow{1193\ K,\ 15\ h} Ca_{10}(PO_4)_6Br_2 + 2\ H_2O \tag{47}$$

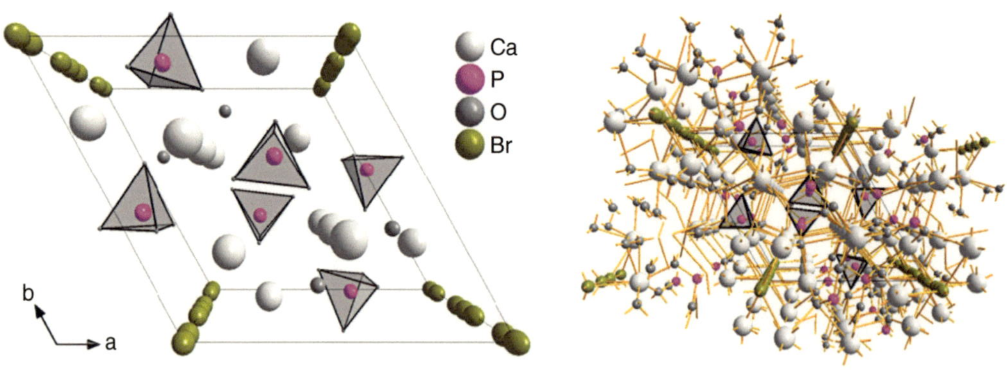

Fig. 29. The structure of $Ca_5(PO_4)_3Br$ (perspective view along the c-axis).

The phase can also be prepared via solid-state synthesis reaction [135]:

http://dx.doi.org/10.5772/62212

$$6\ CaHPO_4 + 3\ CaCO_3 + CaBr_2 \xrightarrow{heat} Ca_{10}(PO_4)_6 Br_2 + 3\ CO_2 + 3\ H_2O \quad (48)$$

4.7.2 Lead bromapatite

The synthesis of **lead bromapatite** ($Pb_{10}(PO_4)_6Br_2$) by solid-state synthesis via sintering of equal amount of $Pb_9(PO_4)_6$ and $PbBr_2$ at the temperature of 250°C in a platinum tube was described by Bhatnagar [191]. Br^- (1.95 Å) anions at the **Z**-site in general formula of apatite ($Pb_{10}(PO_4)_6Z_2$) can be readily substituted by other usual monovalent anions (Cl^-, F^- and OH^-).

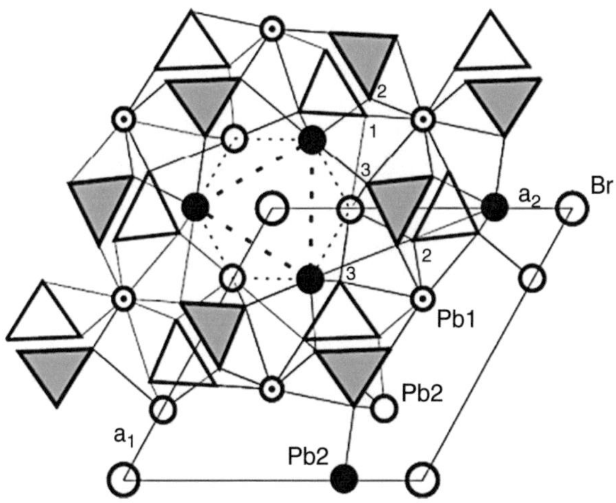

Fig. 30. The structure of lead bromapatite identifying the channel polyhedron (broken lines) formed by Pb(2) cations in apatite channel wall: open Pb(2) circles are at the height z = 1/4 and closed circles are at z = 3/4; triangles are PO_4 tetrahedra centered at z = 1/4 (open) and z = ¾ (shaded); numbers (1, 2, 3) identify oxygen atoms forming the corners of tetrahedra [190]

Liu et al [190] prepared synthetic lead bromapatite via solid-state reaction at pressure up to 1 GPa. In the structure of this phase (**Fig. 30**), isolated PO_4 tetrahedra centered at z = 1/4, 3/4 are linked by Pb(1) in nine-fold (6 + 3) coordination and Pb(2) in an irregular sevenfold (6 + 1) coordination. A prominent feature is large c-axis channel which is defined by tri-clusters of M(2) cations at z = 1/4, ¾ and accommodates a variety of anion components.

4.7.3 Strontium bromapatite

Strontium bromapatite (strontium bromoapatite) can be prepared via solid-state reaction (**Eq. 50**) and wet (solution) method (**Eq. 51**) according to the following reactions [135]:

$$6\ SrHPO_4 + 3\ SrCO_3 + SrBr_2 \xrightarrow{heat} Sr_{10}(PO_4)_6 Br_2 + 3\ CO_2 + 3\ H_2O \quad (49)$$

$$10\ SrBr_2(aq) + 6\ Na_3PO_4(aq) \rightarrow Sr_{10}(PO_4)_6Br_2(s) + 18\ NaBr(aq) \quad (50)$$

Since the precipitate contains Na^+ ions, it must be washed thoroughly to obtain pure product. A small amount of hydroxylapatite may also be present.

Strontium bromapatite forms softer crystal than fluorapatite or strontium chlorapatite. Since it is not stable under the mercury-vapor discharge in fluorescent lamp (**Section 10.6**), strontium bromapatite cannot be used for the production of lighting phosphor [135].

4.7.4 Other bromapatites

Other bromapatites are $Cd_5(PO_4)_3Br$ (a = 9.733 Å, c = 6.468 Å and Z = 2), $Cd_5(AsO_4)_3Br$ (a = 10.100 Å, c = 6.519 Å and Z = 2) and $Cd_5(VO_4)_3Br$ (a = 10.173 Å, c = 6.532 Å and Z = 2). They can be grown from melt in platinum crucible filled with $Cd_3(MO_4)_2$ (M = As, V and P) and the excess of $CdBr_2$. All phases belong to the space group of $P6_3/M$ [142],[143],[186].

4.8 Iodoapatites

Since, the apatite structure is capable of accommodating monovalent anions, strontium iodoapatites were investigated as a potential waste form to immobilize radioactive iodine [135],[192].

4.8.1 Calcium iodoapatite

Calcium iodoapatite ($Ca_5(PO_4)_3I$) does not exist as a separate phase but as oxoapatite. Iodo-oxyapatite (pentadecacalcium iodide oxide nanophosphate, $Ca_{15}(PO_4)_9(I,O)$) was synthesized by the flux method (**Section 4.2**). The crystal structure was refined in the space group $P6_3/M$ with lattice parameters $\boldsymbol{a}$ = 9.567 Å, $\boldsymbol{b}$ = 20.754 Å and $\boldsymbol{Z}$ = 2. Iodo-oxyapatite has typical hexagonal structure but the unit cell is tripled along the hexad (refer to **Footnote 16** in **Chapter 1**) and oxide ions along this direction [135].

4.8.2 Strontium Iodoapatite

Strontium iodoapatite (strontium iodoapatite, strontium iodine-apatite) is of academic interest due to large size of I^- ions compared to other halide ions. However, the thermodynamic functions determined for the alkaline earth apatite series preclude the formation of stable iodoapatite because the cationic size of Sr^{2+} or Ba^{2+} is too small relating to that of iodide ion which must fit upon the c-axis of the structure next to the triads of Sr^{2+} or Ba^{2+} at (000), (010), and (001) positions in the lattice. Such crystals relevant to radioactive waste management include fluorapatite, and, in the end, iodoapatite which should be able to immobilize the radioactive species [135].

http://dx.doi.org/10.5772/62212

4.8.3 Lead vanado-iodoapatite

The preparation of lead vanado-iodoapatite ($Pb_{10}(VO_4)_6I_2$) by hot pressing (HP), isostatic hot pressing (HIP) and sealed-tube method (**Section 4.1.1**) were described in literature. $Pb_{10}(VO_4)_6I_2$ is thermally stable up to about 800 K. The thermal conductivity of hot-pressed sample, with the theoretical density of 82%, increases gradually with increasing temperature from 0.65 W·m $^{-1}$·K^{-1} at room temperature to 0.78 W·m^{-1}·K^{-1} at 523 K. The leaching rate of iodine for apatite was two orders of magnitude higher than that of AgI glass waste form. Despite the high leaching rate (compared to AgI embedded in glass), high chemical stability up to 800 K and acceptable mechanical properties of this apatite suggest that it be a good waste form when embedded in a suitable matrix material [135],[193].

Facile low temperature solid-state synthesis of iodoapatite by high-energy ball milling of PbI_2, PbO and V_2O_5 was described by Lu et al [195]. As-milled iodoapatite is in the form of amorphous matrix embedded with nanocrystals and can be readily crystallized by subsequent thermal annealing at low temperature of 200°C with minimal iodine loss.

4.8.4 Cadmium vanado-iodoapatite

Synthetic cadmium apatites containing iodine such as $Cd_5(VO_4)_3I$ (space group $P6_3/M$, $a = 10.307$ Å, $c = 6.496$ Å and Z = 2 [142],[143]) were prepared, but other apatites containing iodine are unknown [142],[194],[186]. $Cd_5(VO_4)_3I$ can be grown form the flux of $Cd_3(VO_4)_2$ and the excess of CdI_2 in the platinum crucible [142].

4.8.5 Radium iodoapatite

Radium iodoapatite, if could be formed, would have the formula $Ra_5(PO_4)_3I$ but it has not been prepared yet. This salt would be probably best prepared by solid-stare reaction [135]:

$$6\ \mathrm{RaHPO_4} + 3\quad \mathrm{RaCO_3} + \mathrm{RaI_2} \rightarrow \mathrm{Ra_{10}}\left(\mathrm{PO_4}\right)_6 \mathrm{I_2} + 3\ \mathrm{CO_2} + 3\ \mathrm{H_2O} \tag{51}$$

Whether this compound can be formed remains speculative.[52]

4.9 Chalcogenide phosphate apatites

The preparation and the structure of chalcogenide phosphate apatites of the composition $Ca_{10}(PO_4)_6S$ (calcium sulfoapatite), $Sr_{10}(PO_4)_6S$ (strontium sulfoapatite), $Ba_{10}(PO_4)_6S$ (barium sulfoapatite) and $Ca_{10}(PO_4)_6Se$ (calcium selenoapatite) was reported by Henning et al [196]. These apatite phases are isostructural and crystallize in the trigonal space group$P\bar{3}$.with the

[52] In human body radium behaves in a similar way as calcium. When ingested, it is readily adsorbed in bone where it may directly irradiate the bone and other tissues. This exposition may result in fatal disease as the tragic story of "*Radium girls*" working with luminous paint trade named: Undark (the mix of radium and zinc sulfide produced by U.S. Radium Corporation between 1917 and 1938) shows [136],[137].

chalcogenide ion positioned at (0 0 ½). The sulfoapatites show no ability to absorb H_2S in the way oxyapatite absorbs H_2O at elevated temperatures. This can be attributed to the position of sulfide ion and the way it influences the crystal structure around the vacant chalcogenide position at (000). Calcium sulfoapatites, $Ca_{10}(PO_4)_6S$, can be successfully synthesized only using oxide starting materials with sulfur vapor under H_2 atmosphere instead of toxic H_2S gas [197].

4.9.1 Anion deficient lacunar lead apatites

The apatitic structure can accommodate a great variety of other substituent's and vacancies in anionic sites (**Chapter 6**). Previous studies on apatites showed that the only system, where the compounds with the apatite structure could be prepared without **Z** anion, is the lead system. These apatites have the vacancies in Z anion sites (**Fig. 31**) with general formula $APb_4(XO_4)_3$, where A is monovalent cation Na, Ag, K, etc. Pb^{2+}on plays a crucial role allowing to preserve the ideal apatitic network. This role is related to the presence of lone $6s^2$pairs (**Section 2.1.2**) which can compensate for the Coulomb imbalance due to the existence of anion gaps in the tunnels of apatites [198],[199].

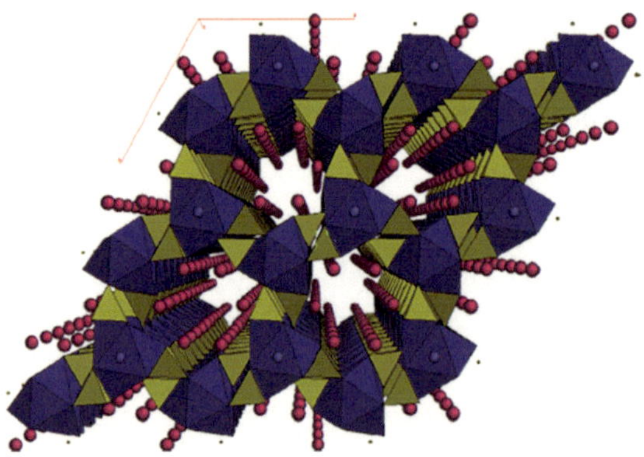

Fig. 31. Polyhedral view in the ab-plane of the crystal structure of $NaCaPb_3(PO_4)_3$ showing the tunnels [199].

Lead in apatite is of great interest from two points of view. First, lead is known as a "bone seeker" as it accumulates in bones and teeth, second, it may contribute to the deviation from the general formula of apatites. Moreover a new voltammetric sensor for the quantification of mercury based on $NaCaPb_3(PO_4)_3$ modified carbon paste electrode can be prepared. Because of the importance of these types of lacunar apatites and the problems which they may cause in biomaterial applications, particular attention has been paid during past few years to synthesize new lacunar anionic apatites [198],[199].

Silver lead apatite ($Ag_2Pb_8(PO_4)_6$, $P6_3/M$, a = 9.765 and c = 7.198 Å) and sodium lead apatite ($Na_2Pb_8(PO_4)_6$, $P6_3/M$, a = 9.731, c = 7.200 Å and Z = 2) were prepared by solid-state synthesis by TERNANE et al [200] from the stoichiometric mixture of $Pb_3(PO_4)_2$ with Ag_3PO_4 (at 215°C and 100 atm.) and PbO, Na_2CO_3 and $(NH_4)_2HPO_4$, respectively. The synthesis of sodium lead apatite can be described by chemical equation:

$$8\ PbO + Na_2CO_3 + 6\left(NH_4\right)_2 HPO_4 \xrightarrow[700°C/2\ days]{500°C/3\ days} Pb_8Na_2\left(PO_4\right)_6 + 12\ NH_3 + 9H_2O + CO_2 \quad (52)$$

The unit cell contains eight divalent Pb^{2+}cations, two monovalent cations (Na^+ or Ag^+) and six $[PO_4]^{3-}$ ions. The triangle sites, *6h*, are occupied by Pb^{2+}ions only while the column positions, *4f*, are occupied by nearly equal amounts of Pb^{2+} and monovalent ions (Na^+ or Ag^+).

The structure of this phase was also investigated by Koumiri et al [201]: ***a*** = 9.7249 Å, ***c*** = 7.190 Å and Z = 2. From the interatomic distances it appears that lead cations in the lacunar apatite $NaPb_4(PO_4)_3$ behave in two different ways:

i. Pb2(II) cations with stereochemically inactive lone $6s^2$ pairs are engaged in almost totally ionic bond Pb(2)/Na–O at the mixed site.

ii. Pb1(II) is engaged in a Pb(1)–O bond with more covalent character, where its lone $6s^2$ pair is stereochemically active and constitutes the seventh ligand of lead cation.

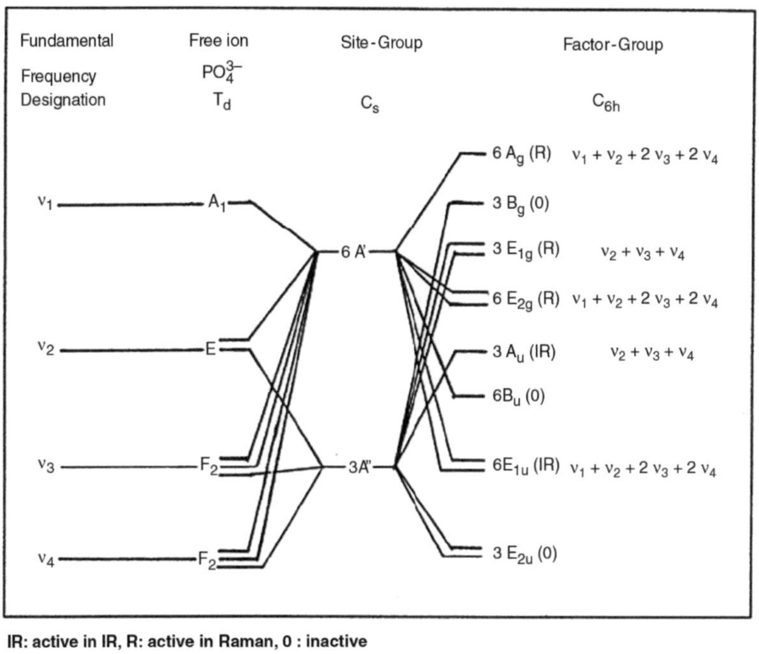

Fig. 32. The correlation chart for PO_4^{3-} fundamental modes under free-ion, site-group and factor group analyses in $Pb_8M_2(PO_4)_6$ where M = Ag and Na [200].

All $[PO_4]^{3-}$ groups are crystallographically equivalent in the cell and have C_s as the site group. P, O(1) and O(2) atoms are in *6h* positions; remaining O(3) oxygen atoms occupy the (121) positions. On this basis, the optical modes at k = 0 are distributed as follows [200]:

$$\Gamma_{opt} = 12\ A_g + 8\ E_{1g} + 13\ E_{2g} + 8\ A_u + 12\ E_{1u} \tag{53}$$

where A_g, E_{1g} and E_{2g} are Raman-active normal modes, A_u and E_{1u} normal modes are infra-red active. The internal modes from tetrahedral phosphate ions, six per unit cell, can be deduced by the group factor analysis given in **Fig. 32**.

Naddari et al [202] performed the solid-state synthesis of colorless calcium-lithium lead apatite ($Pb_6Li_2Ca_2(PO_4)_6$, LCPbAp, $P6_3/M$, $a = 9.679$ and $c = 7.113$ Å, $V = 577.09$ Å^3, $Z = 1$ and $\rho_{calc.} = 5.48$ g·cm^{-3}) via the thermal treatment of mixture of Li_2CO_3, $(NH_4)_2HPO_4$, $CaCO_3$ and PbO at 800°C in air for 12 h and subsequently at 900°C for 12 hours. The structure of $Pb_6Li_2Ca_2(PO_4)_6$ is shown in the perspective view in **Fig. 33(a)**. Site (I) is occupied by 0.88 Ca^{2+}, 1.96 Li^+ and 1.148 Pb^{2+}. These cations are coordinated to nine oxide anions forming a tricapped trigonal prism. In the tunnel set around the c axis, site (II) is occupied by 4.98 Pb^{2+} and 1.02 Ca^{2+}. These cations constitute the walls of the tunnel and are arranged in equilateral triangles (**Fig. 33(b)**).

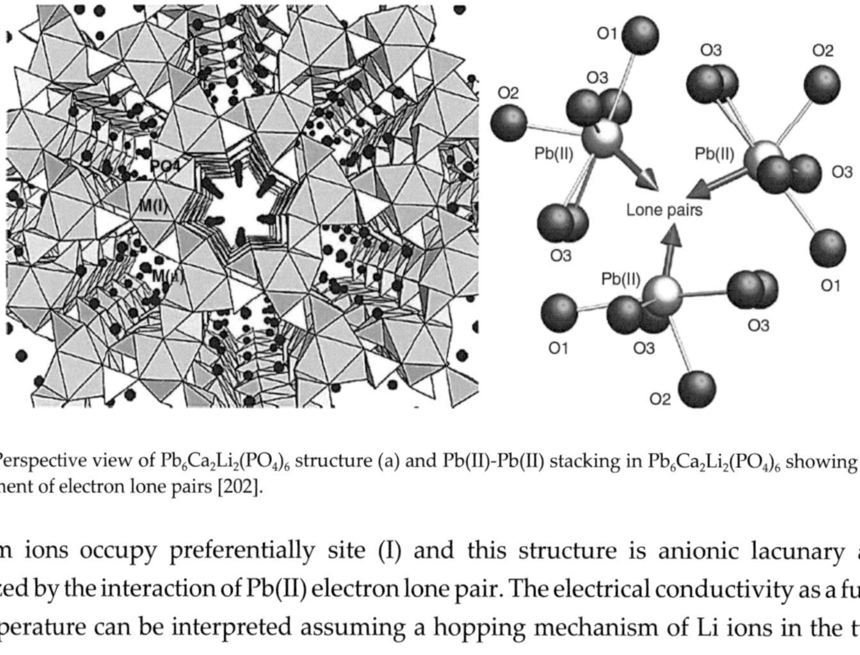

Fig. 33. Perspective view of $Pb_6Ca_2Li_2(PO_4)_6$ structure (a) and Pb(II)-Pb(II) stacking in $Pb_6Ca_2Li_2(PO_4)_6$ showing possible arrangement of electron lone pairs [202].

Lithium ions occupy preferentially site (I) and this structure is anionic lacunary apatite stabilized by the interaction of Pb(II) electron lone pair. The electrical conductivity as a function of temperature can be interpreted assuming a hopping mechanism of Li ions in the tunnels [202].

Tricationic lacunar apatites $Na_{1-x}K_xPb_4(AsO_4)_3$ ($0 \le x \le 1$) were synthesized as single phases by solid-state method at 700°C (48 h) by Manoun et al [198]:

$$\begin{gathered} 4\ PbO + 0.5x\ K_2CO_3 + 0.5(1-x)\ Na_2CO_3 + 3\ (NH_4)_2\ HAsO_4 \rightarrow \\ Na_{1-x}K_xPb_4(AsO_4)_3 + 0.5\ CO_2 + 1.5\ H_2O + 3\ NH_3 \end{gathered} \tag{54}$$

It was found that Pb(II) ions in the solid solutions preferentially occupied the M(1) and M(2) sites in the lacunar anionic apatite structure. The structure contains the channels running along the *c*-axis and centered at (00z). The channels are most probably occupied by lone electron pairs of Pb^{2+} cations.

The factor group analysis [198] of the hexagonal structure ($P6_3/M$) shows that the normal modes of vibration can be classified among the irreducible representations of C_{6h} as follows:

$$\Gamma = 12A_g + 8E_{1g} + 13E_{2g} + 9A_u + 12B_u + 9B_g + 13E_{1u} + 8E_{2u} \tag{55}$$

where the internal mode contribution of (AsO_4) groups to the IR- and Raman-active vibrations is:

$$\begin{aligned} \Gamma_{AsO_4} &= 6A_g\left(\nu_1 + \nu_2 + 2\nu_3 + 2\nu_4\right) + 3E_{1g}\left(\nu_2 + \nu_3 + \nu_4\right) + \\ &6E_{2g}\left(\nu_1 + \nu_2 + 2\nu_3 + 2\nu_4\right) + 3A_u\left(\nu_2 + \nu_3 + \nu_4\right) + \\ &6E_{1u}\left(\nu_1 + \nu_2 + 2\nu_3 + 2\nu_4\right) \end{aligned} \tag{56}$$

where *g* and *u* modes are Raman-and IR-active, respectively [198],[203],[204].

The syntheses of apatites, $Na_{1-x}K_xCaPb_3(PO_4)_3$ ($0 \le x \le 1$), with anion vacancy were carried out using the solid-state reactions at 700°C for 48 h [199]:

$$\begin{aligned} &\frac{1-x}{2}\ Na_2CO_3 + \frac{x}{2}K_2CO_3 + 3\ PbO + Ca\left(NO_3\right)_2 \cdot 4H_2O + \\ &3\left(NH_4\right)H_2PO_4 \rightarrow Na_{1-x}K_xCaPb_3\left(PO_4\right)_3 + 0.5\ CO_2 + \\ &3\ NH_3 + NO_2 + NO_3 + 8.5\ H_2O \end{aligned} \tag{57}$$

The lattice constants of the solid solutions varied linearly with x. It was found that Pb ions in the solid solutions occupied M(1) and M(2) sites in the lacunar apatite structure. The structure was described as built up from $[PO_4]^{3-}$ tetrahedra and Pb^{2+} of six-fold coordination cavities (*6h* positions), which delimit void hexagonal tunnels running along [001]. The tunnels are connected by cations of mixed sites (*4f*) half occupied by Pb^{2+} and half by Na^+/K^+ mixed alkali cations.

The factor group analysis of the hexagonal structure ($P6_3/M$) shows that the normal modes of vibration can be classified among the irreducible representations of C_{6h} by **Eq. 56** where the internal mode contribution of (PO_4) groups to the IR and Raman-active vibrations is [199]:

$$\Gamma_{PO_4} = 6A_g(\nu_1+\nu_2+2\nu_3+2\nu_4)+3E_{1g}(\nu_2+\nu_3+\nu_4)$$
$$+6E_{2g}(\nu_1+\nu_2+2\nu_3+2\nu_4)+3A_u(\nu_2+\nu_3+\nu_4) \quad (58)$$
$$+6E_{1u}(\nu_1+\nu_2+2\nu_3+2\nu_4)$$

4.10 Synthetic analogues of other minerals from the supergroup of apatite

These minerals were usually prepared in order to elucidate the structure of naturally occurring minerals or due to its potential applications in immobilization of nuclear and toxic waste (**Chapter 6**) and electrical properties (ionic conductivity). Synthetic analogues of ellestadite and britholite weren't included because they are described in **Chapter 5** and **6**, respectively. Rare earth apatites are described separately in next **Chapter 5**.

4.10.1 Cesanite

Synthetic cesanite as an analogue of mineral with the composition $Na_3Ca_2(SO_4)_3(OH)$ (**Section 2.1.7**) shows typical features of the apatite structure, as shown in **Fig. 34**. The symmetry reduction from the centrosymmetric space group $P6_3/M$ to the non-centrosymmetric space group $P\bar{6}$leads to a doubling of the number of crystallographically independent sites. The origin of the unit cell is shifted by Z + ¼ relating to the origin in the space group $P6_3/M$. Alternating pairs of isolated tetrahedral anions (the sulfate-groups) form ribbons running parallel to the *c*-axis. As the sulfur atoms are located in special Wyckoff positions 3*j* and 3*k*, the tetrahedra have the point group symmetry *m* [205].

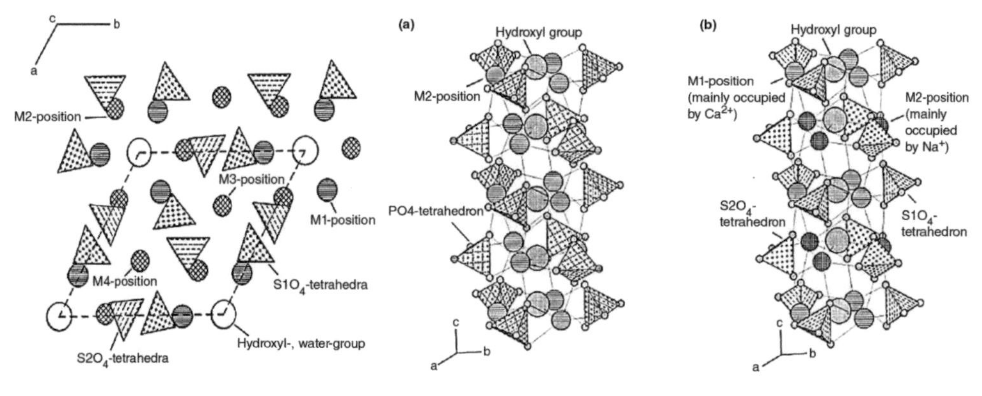

Fig. 34. The projection of the crystal structure of synthetic cesanite parallel to (001) (1) and the arrangement of cations and sulfate tetrahedra around the 6_3 and the 6◎ axes, respectively (2): phosphate apatite (a) and synthetic cesanite (b) [205].

Small spread in the S-O distances and O-S-O angles indicates only minor deviations from ideal tetrahedral symmetry. The sub-structure of the array of sulfate tetrahedra shows a distinct pseudo-symmetry, closely mimicking $P6_3/M$. Maximal deviations from this symmetry occur at O(4) atom, which is shifted by 0.16 Å (synthetic) and 0.02 (natural) from its position in $P6_3/M$. Na and Ca cations are distributed either by six O atoms and one hydroxyl ion or water molecule (M(1) and M(2)) or by nine O atoms (M(3) and M(4) [205].

Synthetic analogues of minerals cesanite Halide sulfates have general formula [206]:

$$M_3^{+}M_2^{2+}(SO_4)_3Z \tag{59}$$

where Z = OH, F and Cl. Klement [207] synthesized sodium-calcium sulfatapatite, $Na_6Ca_4(SO_4)_6F_2$, by full substitution of S^{6+} for P^{5+} through the substitution scheme [208]:

$$S^{6+} + M^{+}(\text{e.g.,Na}) \Leftrightarrow P^{5+} + Ca^{2+} \tag{60}$$

where the hydroxyl equivalent is the equivalent to mineral cesanite, $Na_6Ca_4(SO_4)_6(OH)_2$. Kreidler and Hummel [209] also synthesized $Na_6Ca_4(SO_4)_6F_2$ and $Na_6Pb_4(SO_4)_6F_2$ apatite-like phases. Knyazev et al [206] prepared the compounds of the composition of $Na_3Ca_2(SO_4)_3F$, $Na_3Cd_2(SO_4)_3Cl$, and $Na_3Pb_2(SO_4)_3Cl$ with the structure of apatite via the solid-state reactions:

$$Na_2SO_4 + 2\ CaSO_4 + NaF \rightarrow Na_3Ca_2(SO_4)_3F \tag{61}$$

$$Na_2SO_4 + 2\ CdSO_4 + NaCl \rightarrow Na_3Cd_2(SO_4)_3Cl \tag{62}$$

$$Na_2SO_4 + 2\ PbSO_4 + NaCl \rightarrow Na_3Pb_2(SO_4)_3Cl \tag{63}$$

from the stoichiometric reaction mixture in a porcelain crucible. The mixtures of components were calcined in several steps at the temperatures of 570, 770 and 1020 K for 10 h, with intermediate grindings in agate mortar every 2 h [206].

The $Na_3Ca_2(SO_4)_3F{:}Ce^{3+}$ phosphor was prepared by Nikhare et al [210] via the solid-state method according to the following reaction:

$$\begin{aligned} &2\ NaNO_3 + CaNO_3 + NaF + 3\ (NH_4)_2SO_4 + Ce(NO_3)_3 \rightarrow \\ &Na_3Ca_2(SO_4)_3F:Ce + 4\ NO + 8\ H_2O \end{aligned} \tag{64}$$

The pigment shows a single high intensity emission peak at 307 nm when excited by UV light of the wavelength of 278 nm

The compound having the formula: $K_3Ca_2(SO_4)_3F$, was identified in coatings of heat recovery cyclones of Portland clinker kiln. The structure of this phase (noncentrosymmetric, space group $Pn2_1a$, $a = 13.415$, $b = 10.943$ and $c = 9.127$ Å, **V** = 1284.75 Å^3, **Z** = 4 and $\rho = 2.61$ g·cm^{-3}) was reported as very distorted analogue of apatite where fluoride atoms are oriented along the pseudo-screw *a*-axis [211],[212].

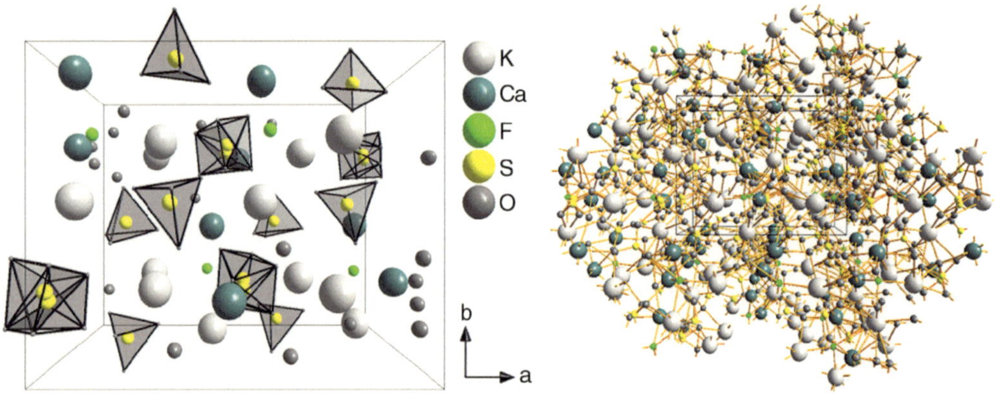

Fig. 35. The structure of $K_3Ca_2(SO_4)_3F$ according to FAYOS et al [212] phase in the perspective view along the c-axis.

The activation by Eu or Ce leads to the phosphor: $K_3Ca_2(SO_4)_3F$:Ce, Eu, which was prepared by PODDAR et al [213] via the precipitation method. The $K_3Ca_2(SO_4)_3F$:Ce luminescent pigment shows the emission at 334 nm when excited at 278 nm due to the $5d \rightarrow 4f$ transition of Ce^{3+} ions. The phases $K_3Ca_2(SO_4)_3F:Eu^{2+}$ and $K_3Ca_2(SO_4)_3F:Eu^{3+}$ show the emissions at 440 nm, and 596 and 615 nm via the transitions of $^5D_0 \rightarrow {}^7F_1$ and $^5D_0 \rightarrow {}^7F_2$ of Eu^{3+} ion, which are in blue and red region of the visible spectrum, respectively.

4.10.2 Bismuth calcium oxyapatites

The synthesis, the characterization and ionic conductivity of $Ca_{8-x}Sr_xBi_2(PO_4)_6O_2$ phase where $x = 3$, 4 and 5 was reported by TRABELSI et al [214]. Sr^{2+} ions were noted to occupy two sites (4*f*) and (6*h*), with a strong preference for (6*h*) sites. Heavy Bi^{3+} atoms preferentially occupied the (6*h*) site.

New bismuth calcium silicon oxide $Ca_4Bi_{4.3}(SiO_4)(HSiO_4)_5O_{0.95}$, with the apatite structure was synthesized by UVAROV et al [215]. The structure was refined from the powder X-ray diffraction data. The refinement revealed that the phase had the $P6_3/m$ space group with the unit cell parameters $a = 9.6090$ Å, $c = 7.0521$ Å, $V = 563.9$ Å^3 and $c{:}a = 0.734$.

Also the structure of bismuth calcium vanadium oxide ($BiCa_4V_3O_{13}$, $BiCa_4(VO_4)_3O$) was reported by HUANG and SLEIGHT [216] as apatite without an inversion center. The phase crystallizes in hexagonal system with the space group $P6_3$, $a = 9.819$ Å, $b = 7.033$ Å, **V** = 587.2 Å^3 and **Z** = 2. Calculated density is 4.129 g·cm^{-3}. Lower symmetry of the structure may be related to the site selective distribution of Bi atoms at the Ca sites.

http://dx.doi.org/10.5772/62212

4.10.3 Incorporation of 3d-metal ions to the hexagonal channel of apatite

The hexagonal channel in the structure of apatite can accommodate infinite linear chains of [-Me-O-]$_n^{n-}$, where Me = Cu, Zn, Ni, Co… (**Fig. 36**). The incorporation of *3d*-metal ions in the hexagonal channels of strontium phosphate apatite ($Sr_5(PO_4)_3OH$) was investigated by KAZIN et al [217].

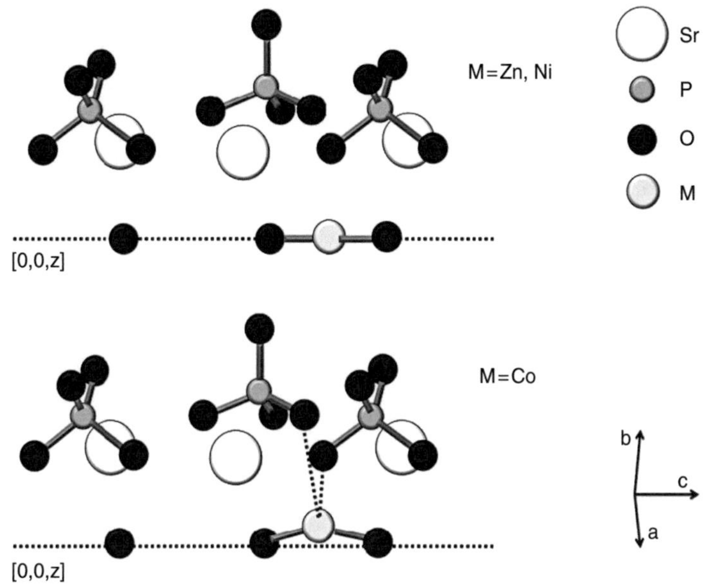

Fig. 36. Crystal structure fragments of doped apatite showing the atomic arrangement at the hexagonal channel where 3-d metal atoms are located [217].

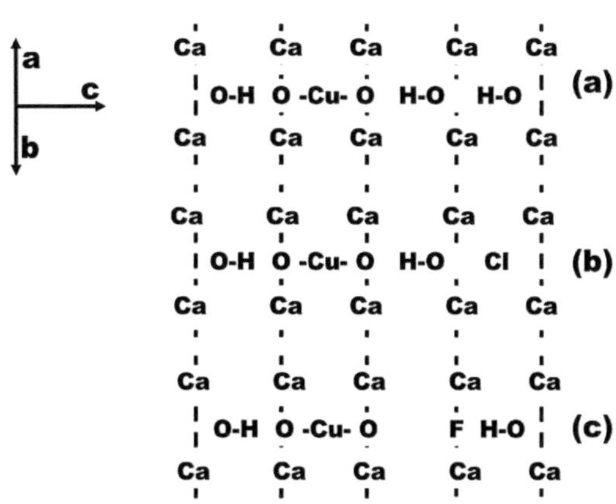

Fig. 37. Depiction of the structure in hexagonal channels (along the c-axis) in the lattice of copper-containing apatite: hydroxylapatite (a), hydroxylapatite with OH partially replaced by Cl and hydroxylapatite with OH partially replaced by F (c). The planes passing through Ca atoms depict the channel walls.

Doping of $Sr_5(PO_4)_3OH$ with ZnO, NiO and CoO at 1400°C in air resulted in the incorporation of 3d-ions entering the hexagonal channels of the apatite structure, formally substituting for protons in the OH groups. The structure of apatite channels in the phases with the composition of $Sr_5(PO_4)_3Zn_{0.15}O_{0.3}(OH)_{0.7}$ (white and shade of green), $Sr_5(PO_4)_3Ni_{0.2}O_{0.4}(OH)_{0.6}$ (green) and $Sr_5(PO_4)_3Co_{0.2}O_{0.5}(OH)_{0.4}$ (dark-violet) contains O-Me-O fragments separated by OH groups. Co atoms were localized in the position shifted by 0.5 Å from the center of channel. Their coordination can be described as distorted from linear O-Co-O probably by additional coordination to phosphate oxygen atoms [217].

In other work of Kazim et al [218], the synthesis and the properties of three compounds possessing the apatite structure with the composition of $Ca_5(PO_4)_3Cu_yO_{y+\delta}(OH)_{0.5-y-d}X_{0.5}$, where the parameter $y = 0.01 - 0.3$ for X = OH, $y = 0.01 - 0.1$ for X = F and $y = 0.1$ for X = Cl. Similar copper containing vanadates $Sr_5(VO_4)_3CuO$ are formed as co-products in the synthesis of vanadium doped superconductors [218],[219].

Author details

Petr Ptáček

Brno University of Technology, Czech Republic

References

[1] Daubrée A. Experiences sur la production artficielle de l'apatite, de la topaze, at de quelques autres métax fruorifés. Comptes rendus de l'Académie des Sciences 1851;32 625.

[2] Manross NS. Justus Liebig's Annalen der Chemie 1852;82 358.

[3] Briegleb H. Ueber die Einwirkung des phosphorsauren Natrons auf Flussspath in der Glühhitze. Justus Liebig's Annalen der Chemie 1856;97(1) 95–119.

[4] Forchhammer JG. Ann. d. Chem. u. Pharm. 1854;90 77.

[5] Deville HS-Claire, Caron H. Comptes Rendus 1858;17 985.

[6] Ditte A. Comptes Rendus 1882;94 1592.

[7] Clarke FW. The Data of Geochemistry. Bulletin (Geological Survey (U.S.)), U.S. Government Printing Office, 1920.

[8] Zambonini F. Quelques observations sur la composition des apatites. Acad. Sci. Paris Comptes Rendus 1916;162 919–921.

[9] Debray H. Comptes rendus 1861;52 44.

[10] Weinshenk E. Zeitschrift für Krystallographie und Mineralogie 1890;17 489.

[11] Cameron FK, McCaughey WJ. Jour. Phys. Chem. 1911;15 464–470.

[12] Hickok RL. Synthesis of calcium chlorspodiosite. United States Patent Office 3,378,354 Patented Apr. 16, 1968.

[13] Klement R, Gembruch F. Versuche über Isomorphie in der Wagneritgruppe. Naturwissenschaften 1941;29: 301–302.

[14] Mineralogical chemistry. Journal of the Chemical Society, Abstracts 1911;100 ii733-ii737. DOI: 10.1039/CA9110005733

[15] Hutchins WM. Occurrence of Apatite in Slag. Nature 1887;36 460.

[16] Voght JHL. Beiträge zur Kenntniss der Gesetze der Mineralbildung in Schmelzmassen und in den Neovulkanischen Ergussgesteinen. Kristiania, Alb. Cammermeyer´s verlag, Oslo 1852.

[17] Zambonini F. Uber die Mischristalle, welche die Verbindungen des Calciums, Strontiums, Bariums, und Bleis mit jenen der seltenen Erden bilden. Zeitschr. Kristallographie Band 1923;28 226–292.

[18] Jaffe EB. Abstract of the literature on synthesis of apatites and some related phosphates. Geological survey circular 135, 1951.

[19] Hendricks S, Jefferson ME, Mosely VM. The crystal structure of some natural and synthetic apatite-like substances. Zeitschr. Kristallographie 1932;81 352–369.

[20] Eitel W. Uber Karbonatphosphate der Apatitgruppe: Sehr. Königsberger Gelehrter Gesell. Naturwiss. 1924;4 159–177.

[21] Byrappa K, Yoshimura M. Handbook of Hydrothermal Technology. Materials Science and Process Technology. 2nd ed., William Andrew, 2012. ISBN: 978-1437778366

[22] Katashev A, Dekhtyar Y, Spigulis J. 14th Nordic-Baltic Conference on Biomedical Engineering and Medical Physics: NBC 2008. 16–20 June 2008. Riga, Latvia. IFMBE Proceedings-Volume 20. Springer Science & Business Media, 2008. ISBN: 978-3540693673

[23] West AR. Solid State Chemistry and its Applications. 2nd ed., John Wiley & Sons, 2014. ISBN: 978-1118676257

[24] Smart LE, Moore EA. Solid State Chemistry: An Introduction, Fourth Edition. CRC Press, 2012. ISBN: 978-1439847923

[25] Avvakumov GV, Senna M, Kosova NV. Soft Mechanochemical Synthesis: A Basis for New Chemical Technologies. Springer Science & Business Media, 2007. ISBN: 978-0306476464

[26] Dann SE. Reactions and Characterization of Solids. Tutorial chemistry texts - Volume 2. Royal Society of Chemistry, 2000. ISBN: 978-0854046126

[27] Mehrer H. Diffusion in Solids: Fundamentals, Methods, Materials, Diffusion-Controlled Processes. Springer Series in Solid-State Sciences - Volume 155. Springer Science & Business Media, 2007. ISBN: 978-3540714880

[28] Manickam S, Ashokkumar M. Cavitation: A Novel Energy-Efficient Technique for the Generation of Nanomaterials. CRC Press, 2014. ISBN: 978-9814411554

[29] van der Put PJ. The Inorganic Chemistry of Materials: How to Make Things Out of Elements. Plenum Series in Social/Clinical. Springer Science & Business Media., 1998. ISBN: 978-0306457319

[30] Rahaman MN. Ceramic Processing and Sintering. Materials Engineering. 2nd ed., CRC Press, 2003. ISBN: 978-0824709884

[31] Stojanovic BD, Skorokhod VV, Nikolic M. Advanced Science and Technology of Sintering. Springer Science & Business Media, 2011. ISBN: 978-1441986665

[32] Sadat-Shojai M, Mohammad-T Khorasani, Dinpanah-Khoshdargi E, Jamshidi A. Synthesis methods for nanosized hydroxyapatite with diverse structures. Acta Biomaterialia 2013;9() 7591–7621.(8)

[33] De Sénarmont HH. Sur la formation des mineraux par voie humide dans le gits metallifered concrectionnes. Annales de chimie et de physique 1851;32(3) 129–175.

[34] Acton QA. Magnesium Compounds - Advances in Research and Application: 2013 Edition. Scholarly Editions, 2013. ISBN: 978-1481677417

[35] Somiya S, Aldinger F, Spriggs RM, Uchino K, Koumoto K, Kaneno M. Handbook of Advanced Ceramics: Materials, Applications, Processing and Properties, Volume 2. Academic Press, 2003. ISBN: 978-0080532943

[36] Chattopadhyay R. Advanced Thermally Assisted Surface Engineering Processes. Springer Science & Business Media, 2004. ISBN: 978-1402076961

[37] Koizumi M, Nishihara M Isostatic Pressing: Technology and Applications. Elsevier applied science. Springer Science & Business Media. , 1991. ISBN: 978-1851665969

[38] Batsanov SS. Effects of Explosions on Materials: Modification and Synthesis Under High-Pressure Shock Compression High pressure shock compression of condensed matter. Shock Wave and High Pressure Phenomena. Springer Science & Business Media., 1994. ISBN: 978-0387941233

[39] Ostwald W. Handbuch der allgemeinen Chemie, Band I, Academische Verlagsgesellschaft, Leipzig 1919.

[40] Block B. Verbesserungen an der Kolloidmühle, Chemische Apparatur 14, Heft 13 und 14, 1927.

[41] Balaz P. Mechanochemistry in Nanoscience and Minerals Engineering. Springer Science & Business Media, 2008. ISBN: 978-3540748557

[42] Bowden FP, Tabor D. The Friction and Lubrication of Solids, Volume 1. Oxford classic texts in the physical sciences. The international series of monographs on physics - Volume 9. Clarendon Press, 2001. ISBN: 978-0198507772

[43] Lea MC. Disruption of silver halides by mechanical force. The Journal of American Science 1892;43 527–531.

[44] Westbrook JH, Fleischer RL. Intermetallic Compounds, Crystal Structures of Intermetallic Compounds - Volume 1. Wiley, 2000a. ISBN: 978-0471608806

[45] Westbrook JH, Fleischer RL. Intermetallic Compounds, Basic Mechanical Properties and Lattice Defects of Intermetallic Compounds - Volume 2. Wiley, 2000b. ISBN: 978-0471611752

[46] Stoloff NS, Sikka VK. Physical Metallurgy and processing of Intermetallic Compounds. Springer Science & Business Media , 2012. ISBN: 978-1461312154

[47] Plauson H. Colloidal mill. Patent US2225797 A, 1940.

[48] Kelly WJ. Plauson Colloid Mill. Industrial and Engineering Chemistry 1923;15(9) 926–928.

[49] Parker LH. Reactions by trituration. Journal of the Chemical Society, Transactions 1914;105 1504–1516

[50] Sakka S. Handbook of Sol-gel Science and Technology. 1. Sol-gel processing. Springer Science & Business Media, 2005. ISBN: 978-1402079696

[51] Aegerter MA, Leventis N, Koebel MM. Aerogels Handbook Advances in Sol-Gel Derived Materials and Technologies. Springer Science & Business Media. , 2011. ISBN: 978-1441975898

[52] Brinker CJ, Scherer GW. Sol-Gel Science: The Physics and Chemistry of Sol-Gel Processing. Academic Press, 2013. ISBN: 978-0080571034

[53] Ring TA. Fundamentals of Ceramic Powder Processing and Synthesis. Academic Press, 1996. ISBN: 978-0080532196

[54] Youssef HA, El-Hofy HA. Ahmed MH. Manufacturing Technology: Materials, Processes, and Equipment. CRC Press, 2011. ISBN: 978-1439810859

[55] Xu R, Pang W, Huo Q. Modern Inorganic Synthetic Chemistry. Elsevier, 2011. ISBN: 978-0444535993

[56] Association Management, Resources Information. Nanotechnology: Concepts, Methodologies, Tools, and Applications: Concepts, Methodologies, Tools, and Applications. IGI Global, 2014. ISBN: 978-1466651265

[57] Schubert U, Hüsing N. Synthesis of Inorganic Materials. John Wiley & Sons, 2012. ISBN: 978-3527327140

[58] Tavadze GF, Shteinberg A. Production of Advanced Materials by Methods of Self-Propagating High-Temperature Synthesis. Springer Science & Business Media, 2013. ISBN: 978-3642352058

[59] Borisov AA, De Luca LT, Merzhanov AG. Self-Propagating High-Temperature Synthesis of Materials. Combustion science and technology book series - Volume 5. CRC Press, 2002. ISBN: 978-1560329787

[60] U.S.S.R. Inventor's Certificate 225 221 (1967).

[61] Merzhanov AG, Borovinskaya IP. elf-propagated high-temperature synthesis of refractory inorganic compounds. Doklady Akademii Nauk SSSR 1972;204(2) 366–369.

[62] Rao CNR, Thomas PJ, Kulkarni GU. Nanocrystals: Synthesis, Properties and Applications. Springer Series in Materials Science - Volume 95. Springer Science & Business Media, 2007. ISBN: 978-3540687528

[63] Schneller T, Waser R, Kosec M, Payne D. Chemical Solution Deposition of Functional Oxide Thin Films. Springer Science & Business Media, 2014. ISBN: 978-3211993118

[64] Lee B, Komarneni S. Chemical Processing of Ceramics, 2nd ed., CRC Press, 2005. ISBN: 978-1420027334

[65] Knyazev AV, Chernorukov NG, Bulanov EN. Apatite-structured compounds: Synthesis and high-temperature investigation. Materials Chemistry and Physics 2012 132(2–3) 773–781.

[66] Chernorukov NG, Knyazev AV, Bulanov EN. Phase transitions and thermal expansion of apatite-structured compounds. Inorganic Materials 2011;47(2) 172–177.

[67] Frank-Kamenetskaya O, Kol'tsov A, Kuz'mina M, Zorina M, Poritskaya L. Ion substitutions and non-stoichiometry of carbonated apatite-(CaOH) synthesised by precipitation and hydrothermal methods. Journal of Molecular Structure 2011;992(1–3) 9–18.

[68] Drouet Ch. A comprehensive guide to experimental and predicted thermodynamic properties of phosphate apatite minerals in view of applicative purposes. The Journal of Chemical Thermodynamics 2015;81 143–159.

[69] Derek EC. Chemistry, Biochemistry and Technology, Sixth Edition. CRC Press, 2013. ISBN: 978-1439840887

[70] Van Wazer JR. Phosphorus and its Compounds, Volume 1. Wiley-Interscience, New York, 1985.

[71] Brown PW, Hocker N, Hoyle S. Variations in Solution Chemistry During the Low-Temperature Formation of Hydroxyapatite. Journal of the American Ceramic Society 1991;74(8) 1848–1854.

[72] Brown PW. Phase Relationships in the Ternary System $CaO\text{-}P_2O_5\text{-}H_2O$ at 25°C. Journal of the American Ceramic Society 1992;75(1) 17–22.

[73] Biggar GM. Experimental Studies of Apatite Crystallization in Parts of the System CaO-P_2O_5-H_2O at 1000 Bars Mineralogical Magazine 1966;35 1110–1115.

[74] Feng SS, Rockett TJ. The System CaO-P_2O_5-H_2O at 200°C. Journal of the American Ceramic Society 1976;62 619–620.

[75] Lantelme F, Groult H. Molten Salts Chemistry: From Lab to Applications. Newnes, 2013. ISBN: 978-0124017221

[76] Bengisu M. Engineering Ceramics. Engineering Materials Springer Science & Business Media, 2013. ISBN: 978-3662043509

[77] Morey GW, Niggli P. The Hydrothermal Formation of Silicates, A Review. Journal of the American Ceramic Society 1913;35 1086–1130.

[78] Rabenau A. The Role of Hydrothermal Synthesis in Preparative Chemistry. Angewandte Chemie International Edition in English 1985;24 1026–1040.

[79] Roy R. Acceleration the Kinetics of Low-Temperature Inorganic Syntheses. Journal of Solid State Chemistry 1994;111(1) 11–17.

[80] Baumer A, Argiolas R. Hydrothermal synthesis and characterization by X-ray of chlor, fluor or hydroxyapatite crystals. Neues Jahrbuch Fur Mineralogie-Monatshefte 1981a; 8 344–348.

[81] Jahnke RA. The synthesis and solubility of carbonate fluorapatite. American Journal of Science 1984;284 58–78.

[82] Deer WA. Rock-forming minerals: Non-silicates, volume 5B, second edition. Geological Society of London, 1998. ISBN: 978-1897799901

[83] Zhang H, Darvell BW. Synthesis and characterization of hydroxyapatite whiskers by hydrothermal homogeneous precipitation using acetamide. Acta Biomaterialia 2010;6(8) 3216–3222.

[84] Lin K, Chang J, Cheng R, Ruan M. Hydrothermal microemulsion synthesis of stoichiometric single crystal hydroxyapatite nanorods with mono-dispersion and narrow-size distribution. Materials Letters 2007;61(8–9) 1683–1687.

[85] Leal-Calderon F, Schmitt V, Bibette J. Emulsion Science: Basic Principles. Springer Science & Business Media, 2007. ISBN: 978-0387396835

[86] Rosen MJ, Kunjappu JT. Surfactants and Interfacial Phenomena. 4nd ed., John Wiley & Sons, 2012. ISBN: 978-1118229026

[87] Duffar T. Crystal Growth Processes Based on Capillarity: Czochralski, Floating Zone, Shaping and Crucible Techniques, John Wiley & Sons, 2010. ISBN: 978-1444320213

[88] Capper P. Bulk Crystal Growth of Electronic, Optical and Optoelectronic Materials. Wiley Series in Materials for Electronic & Optoelectronic Applications - Volume 14. John Wiley & Sons, 2005. ISBN: 978-0470012079

[89] Fukuda T, Rudolph P, Uda S. Fiber Crystal Growth from the Melt. Advances in Materials Research - Volume 6. Springer Science & Business Media, 2013. ISBN: 978-3662072141

[90] Verneuil A. Production artificielle du rubis par fusion. Comptes Rendus (Paris) 1902;135 791–794.

[91] Verneuil A. Mémoire sur la reproduction artificielle du rubis par fusion. Annales de chimie et de physique 1904;8(3) 20–48.

[92] Verneuil A. Comptes Rendus (Paris) 1910;151 131–132.

[93] Dobrovinskaya ER, Lytvynov LA, Pishchik V. Sapphire: Material, Manufacturing, Applications. Springer Science & Business Media, 2009. ISBN: 978-0387856957

[94] Czochralski J. Ein neues Verfahren zur Messung der Kristallisationsgeschwindigkeit der Metalle. Zeitschrift für Physikalische Chemie 1918;92 219–221.

[95] von Gomperz EV. Untersuchungen an Einkristalldrähten. Zeitschrift für Physikalische Chemie 1922;8 184–190.

[96] Kyropoulos S. Ein Verfahren zur Herstellung grosser Kristalle. Zeitschrift fur Anorganische und Allgemeine Chemie 1926;154 308–311.

[97] Kyropoulos S. Dielektrizitätskonstanten regulärer Kristalle. Zeitschrift fur Physik 1930;63 849–854.

[98] Morgenstern H.. Wachstumsbeobachtungen an Alkalihalogenidkristallen. Zeitschrift für Kristallographie 1938;100 221–227.

[99] Feigelson R. 50 Years Progress in Crystal Growth: A Reprint Collection. Elsevier, 2004. ISBN: 978-0080489933

[100] Bridgman PW. Proceedings of the American Academy of Arts and Sciences 1925;60 305–385.

[101] Pfann WG. Principles of Zone-Melting. Transactions of the Metallurgical Society of AIME 1952;154 747–753.

[102] Higuchi M, Katase H, Kodaira K, Nakayama S. Float zone growth and characterization of $Pr_{9.33}(SiO_4)_6O_2$ and $Sm_{9.33}(SiO_4)_6O_2$ single crystals with an apatite structure. Journal of Crystal Growth 2000;218(2–4) 282–286.

[103] Dietze W, Doering E, Glasow P, Langheinrich W, Schulz M, Ludsteck A, Mader H, Mühlbauer A, Weiss H, v. Münch W, Runge H, Schleicher L, Schnöller M, Sirtl E, Uden E, Zulehner W. Technology of Si, Ge, and SiC / Technologie Von Si, Ge und SiC. Landolt-Bornstein: New series. Numerical Data and Functional Relationshi - Volume 17, Part 3. Springer Science & Business Media, 1983. ISBN: 978-3540114741

[104] Rogers D. Einstein's Other Theory: The Planck-Bose-Einstein Theory of Heat Capacity. Princeton University Press, 2005. ISBN: 978-0691118260

[105] Bhat HL. Introduction to Crystal Growth: Principles and Practice. CRC Press, 2014. ISBN: 978-1439883334

[106] Jokić B, Mitrić M, Radmilović V, Drmanić S, Petrović R, Janaćković D. Synthesis and characterization of monetite and hydroxyapatite whiskers obtained by a hydrothermal method. Ceramics International 2011;37(1) 167–173.

[107] Ioku K, Yamauchi S, Fujimori H, Goto S, Yoshimura M. Hydrothermal preparation of fibrous apatite and apatite sheet. Solid State Ionics 2002;151(1–4) 147–150.

[108] Suchanek W, Yashima M, Kakihana M, Yoshimura M. Processing and mechanical properties of hydroxyapatite reinforced with hydroxyapatite whiskers. Biomaterials 1996;17(17) 1715–1723.

[109] Tas AC. Molten Salt Synthesis of Calcium Hydroxyapatite Whiskers. Journal of the American Ceramic Society 2001;84(2) 295–300

[110] Aizawa M, Ueno H, Itatani K, Okada I. Syntheses of calcium-deficient apatite fibres by a homogeneous precipitation method and their characterizations. Journal of the European Ceramic Society 2006;26(4–5) 501–507.

[111] Shyu JJ, Wu JM. Growth kinetics of spherulitic apatite in some MgO-CaO-SiO_2-P_2O_5 glasses. Journal of Materials Science 1994;29(12) 3167–3171.

[112] Stanton KT, O'Flynn KP, Kiernan S, Menuge J, Hill R. Spherulitic crystallization of apatite–mullite glass-ceramics: Mechanisms of formation and implications for fracture properties. Journal of Non-Crystalline Solids 2010;356(35–36) 1802–1813.

[113] Sommerauer J, Katz-Lehnert K. Trapped phosphate melt inclusions in silicate-carbonate-hydroxyapatite from comb-layer alvikites from the Kaiserstuhl carbonatite complex (SW-Germany). Contributions to Mineralogy and Petrology 1985;91(4) 354–359.

[114] Elliott JC. Recent progress in the chemistry, crystal chemistry and structure of the apatites. Calcified Tissue Research 1969;3(1) 293–307.

[115] Robert M, Robert OC. Fluorapatite laser material doped with neodymium and manganese. US patent US3504300 A, 1970.

[116] Prener JS. The growth and crystallographic properties of calcium fluor- and chlorapatite. Journal of the Electrochemical Society 1967;114(1) 77–83.

[117] Zeitschrift Für Kristallographie, Volume 207–208. Akademische Verlagsgessellschaft, 1993.

[118] Higuchi M, Masubuchi Y, Nakayama S, Kikkawa S, Kodaira K. Single crystal growth and oxide ion conductivity of apatite-type rare-earth silicates. Solid State Ionics 2004;174(1–4) 73–80.

[119] Higuchi M, Kodaira K, Nakayama S. Gowth of apatite-type neodymium silicate single crystals by the floating-zone method. Journal of Crystal Growth 1999;207(4) 298–302.

[120] Nakayama S, Sakamoto M, Higuchi M, Kodaira K, Sato M, Kakita S, Suzuki T, Itoh K. Oxide ionic conductivity of apatite type $Nd_{9.33}(SiO_4)_6O_2$ single crystal. Journal of the European Ceramic Society 1999;19(4) 507–510

[121] Koohpayeh SM, Fort D, Abell JS. The optical floating zone technique: A review of experimental procedures with special reference to oxides. Progress in Crystal Growth and Characterization of Materials 2008;54(3–4) 121–137.

[122] Masubuchi Y, Higuchi M, Kodaira K. Reinvestigation of phase relations around the oxyapatite phase in the Nd_2O_3-SiO_2 system 2003;247(1–2) 207–212.

[123] Yoshikawa A, Kochurikhin VV, Futagawa N, Shimamura K, Fukuda T. Growth of $Ca_8La_2(PO_4)_6O_2$ single crystals as substrates for GaN epitaxial growth. Journal of Crystal Growth 1999;204(3) 302–306.

[124] Masaoka M, Kyono A. Single crystal growth of lead vanado-chlorapatite $Pb_5(VO_4)_3Cl$ using CsCl flux method. Materials Letters 2006;60(29–30) 3922–3926.

[125] Roufosse A, Harvill ML, Gilliam OR, Kostiner E. The hydrothermal crystal growth of chlorapatite. Journal of Crystal Growth 1973;19(3) 211–212.

[126] Yu H, Zhang H, Wang X, Gu Z, Li X, Deng F. Local structure of hydroxyl-peroxy apatite: A combined XRD FT-IR, Raman, SEM and solid-state NMR study. Journal of Physics and Chemistry of Solids 2007;68(10) 1863–1871.

[127] Zhou H, Lee J. Nanoscale hydroxyapatite particles for bone tissue engineering. Acta Biomaterialia 2011;7(7) 2769–2781.

[128] Zhang C, Yang J, Quan Z, Yang P, Li C, Hou Z, Lin J. Hydroxyapatite nano-and microcrystals with multiform morphologies: controllable synthesis and luminescence properties. Cryst Growth Design 2009;9(6) 2725–2733.

[129] Kreidler ER, Hummel FA. The crystal chemistry of apatite: Structure fields of fluor- and chlorapatite. Journal of Physics and Chemistry of Solids 2007;68 1863–1871.

[130] Nacken R. The formation of apatite I. Zentralbl. Mineral., Geol. Paleontol. 1912; 545–559.

[131] Berak J. Rocz. Chem. 1961;35 23–30.

[132] Berak J, Tomczak-Hudina I. Phase equilibrium in the system of tricalcium phosphate-calcium fluoride. Rocz. Chem., 1972;46(12) 2157–2164.

[133] Alper A. Phase Diagrams 6-V: Materials Science and Technology, Volume - 5. Refractory materials - volume 6. Elsevier, 2012. ISBN: 978-0323154895.

[134] Wyllie PJ. Phase equilibria in system CaO-CO_2-H_2O and related systems, with implications form crystal growth of calcite and apatite. Journal of the American Ceramic Society 1967;50(1) 43–46.

[135] Ropp RC. Encyclopedia of the Alkaline Earth Compounds. Newnes, 2012. ISBN: 978-0444595539

[136] Sibtain A, Morgan A, MacDougall N. Physics for Clinical Oncology. OUP Oxford, 2012. ISBN: 978-0199573356

[137] Coleman ND. Expert Report Writing in Toxicology: Forensic, Scientific and Legal Aspects. John Wiley & Sons, 2014. ISBN: 978-1118432334

[138] Mathew M, Mayer I, Dickens B, Schroeder LW. Substitution in barium-fluoride apatite: The crystal structures of $Ba_{10}(PO_4)_6F_2$, $Ba_6La_2Na_2(PO_4)_6F_2$ and $Ba_4Nd_3Na_3(PO_4)_6F_2$. Journal of Solid State Chemistry 1979;28(1) 79–95.

[139] Fahami A, Nasiri-Tabrizi B, Ebrahimi-Kahrizsangi R. Mechanosynthesis and characterization of chlorapatite nanopowders. Materials Letters 2013;110 117–121.

[140] Baumer A, Argiolas R. Synthèses hydrothermals et determinations RX d´apatites chlorée, fluorée ou hydroxylée. Neues Jb Miner Mh 1981b;8 344–348.

[141] Nacken R. The formation of apatite I. Zentralbl.Mineral., Geol. Paleontol., 1912: 545–559.

[142] Sudarsanan K, Young RA, Wilson AJC. The structures of some cadmium `apatites' $Cd_5(MO_4)_3X$. I. Determination of the structures of $Cd_5(VO_4)_3I$, $Cd_5(PO_4)_3Br$, $Cd_3(AsO_4)_3Br$ and $Cd_5(VO_4)_3Br$. Acta Crystallographica Section B 1977;33(10) 3136–3142.

[143] Engel G. Z. anorg, allgem. Chem. 1968;362 273–280.

[144] Park HL, Jang MS. Phase transition in cadmium chlorapatite [$Cd_5(PO_4)_3Cl$]. Solid State Communications 1983;48(2) 109–110.

[145] Plaisier JR, de Graaff RAG, IJdo DJW. Structure determination of a new apatite: $Ba_5(OsO_5)_3Cl$. Materials Research Bulletin 1995;30(10) 1249–1252.

[146] Besse J.-P, Baud G, Levasseur G, Chevalier R. Structure cristalline de $Ba_5(ReO_5)_3C_1$: une nouvelle apatite contenant l'ion $(ReO_5)^{3-}$. Acta Crystallographica Section B 1979;35(8) 1756–1759.

[147] Baud G., Besse J.-P, Sueur G, Chevalier R. Structure de nouvelles apatites au rhenium contenant des anions volumineux: $Ba_{10}(ReO_5)_6X_2$ (X = Br, I). Materials Research Bulletin 1979;14(5) 675–682.

[148] Suzuki T, Kibe T. Specific surface free energy and morphology of chlorapatite crystals studied by modified Wilhelmy method. Journal of Crystal Growth 2010;312(20) 3025–3028.

[149] Ghosh P. Colloid and Interface Science. PHI Learning Pvt. Ltd., 2009. ISBN: 978-8120338579

[150] Han KN. Fundamentals of Aqueous Metallurgy. SME, 2002. ISBN: 978-0873352154

[151] Li NN, Fane AG, Ho WSW, Matsuura T. Advanced Membrane Technology and Applications, John Wiley & Sons, 2011. ISBN: 978-1118211540

[152] Suzuki T, Kumeda I, Teshima K, Oishi S. Specific surface free energies of strontium chlorapatite single crystal determined by contact angles of liquid droplets. Chemical Physics Letters 2006;421(4–6) 343–347.

[153] Wulff G.. Zur Frage dee Geschwindigkeit des Wachstums und der Auflösung de Krystallflächen. Zeitschrift für Kristallographie und Mineralogie 1901;34 449–530.

[154] Mutaftschiev B. The Atomistic Nature of Crystal Growth. Springer Series in Materials Science - Volume 43. Springer Science & Business Media, 2013. ISBN: 978-3662045916.

[155] Wopenka B, Pasteris JD. A mineralogical perspective on the apatite in bone. Materials Science and Engineering: C 2005;25(2) 131–143.

[156] Abouzeid Abdel-Z.M. Physical and thermal treatment of phosphate ores - An overview. International Journal of Mineral Processing 2008;85(4) 59–84.

[157] Pasero M, Kampf AR, Ferraris C, Pekov IV, Rakovan JR, White TJ. Nomenclature of the apatite supergroup minerals. European Journal of Mineralogy 2010;22 163–179.

[158] Fleet ME. Infrared spectra of carbonate apatites: ν2-Region bands. Biomaterials 2009;30(8) 1473–1481.

[159] Fleet ME, Liu X. Local structure of channel ions in carbonate apatite. Biomaterials. 2005;26(36) 7548–7554.

[160] Barralet J, Best S, Bonfield W. Carbonate substitution in precipitated hydroxyapatite: an investigation into the effects of reaction temperature and bicarbonate ion concentration. Journal of Biomedical Materials Research Part A 1998;41(1) 79–86.

[161] Penel G, Leroy G, Rey C, Bres E. MicroRaman spectral study of the PO_4 and CO_3 vibrational modes in synthetic and biological apatites. Calcified Tissue International 1998;63(6) 475–481.

[162] Dowker SEP, Anderson P, Elliott JC, Gao XJ. Crystal chemistry and dissolution of calcium phosphate in dental enamel. Mineralogical Magazine 1999;63(6) 791–800.

[163] Suchanek WL, Shuk P, Byrappa K, Riman RE, TenHuisen KS, Janas VF. Mechanochemical-hydrothermal synthesis of carbonated apatite powders at room temperature. Biomaterials 2002;23(3) 699–710.

[164] McClellan GH, Lehr JR. Crystal chemical investigation of natural apatites. The American Mineralogis 1969;54 1374–1391.

[165] Englert KM, McAteer JA, Lingeman JE. Williams Jr JC. High carbonate level of apatite in kidney stones implies infection, but is it predictive? Urolithiasis 2013;41(5) 389–394.

[166] Bazin D, Carpentier X, Brocheriou I, Dorfmuller P, Aubert S, Chappard C, Thiaudière D, Reguer S, Waychunas G, Jungers P, Daudon M. Revisiting the localisation of Zn^{2+} cations sorbed on pathological apatite calcifications made through X-ray absorption spectroscopy. Biochimie 2009;91(10) 1294–1300.

[167] Fleet ME, Liu X, Penelope LK. Accommodation of the carbonate ion in apatite: An FTIR and X-ray structure study of crystals synthesized at 2–4 GPa. American Mineralogist 2004;89 1422–1432.

[168] Grunenwald A, Keyser C, Sautereau AM, Crubézy E, Ludes B, Drouet C. Revisiting carbonate quantification in apatite (bio)minerals: a validated FTIR methodology. Journal of Archaeological Science 2014;49 134–141.

[169] Fleet ME, Liu X. Location of type B carbonate ion in type A–B carbonate apatite synthesized at high pressure. Journal of Solid State Chemistry 2004;177(9) 3174–3182.

[170] Ren F, Lu X, Leng Y. Ab initio simulation on the crystal structure and elastic properties of carbonated apatite. Journal of the Mechanical Behavior of Biomedical Materials 2013;26 59–67.

[171] Suetsugu Y, Takahashi Y, Okamura FP, Tanaka J. Structure Analysis of A-Type Carbonate Apatite by a Single-Crystal X-Ray Diffraction Method. Journal of Solid State Chemistry 2000;155(2) 292–297.

[172] Tonegawa T, Ikoma T, Suetsugu Y, Igawa N, Matsushita Y, Yoshioka T, Hanagata N, Tanaka J. Thermal expansion of type A carbonate apatite. Materials Science and Engineering: B 2010;173(1–3) 171–175

[173] Fleet ME, Liu X. Carbonate apatite type A synthesized at high pressure: new space group (P3¯) and orientation of channel carbonate ion. Journal of Solid State Chemistry 2003;174(2) 412–417.

[174] Elliott JC, Bonel G, Trombe JC. Space group and lattice constants of $Ca_{10}(PO_4)_6CO_3$. Journal of Applied Crystallography 1980;13(6) 618–621.

[175] Elliott JC. Structure and Chemistry of the Apatites and Other Calcium Orthophosphates, Elsevier Press, Amsterdam, 1994. ISBN: 978-0-444-81582-8

[176] Lafon JP, Champion E, Bernache-Assollant D. Processing of AB-type carbonated hydroxyapatite $Ca_{10-x}(PO_4)_{6-x}(CO_3)_x(OH)_{2-x-2y}(CO_3)_y$ ceramics with controlled composition. Journal of the European Ceramic Society 2008;28(1) 139–147.

[177] Yao F, LeGeros JP, LeGeros RZ. Simultaneous incorporation of carbonate and fluoride in synthetic apatites: Effect on crystallographic and physico-chemical properties. Acta Biomaterialia 2009;5(6) 2169–2177.

[178] Peroos S, Du Z, de Leeuw NH. A computer modelling study of the uptake, structure and distribution of carbonate defects in hydroxy-apatite. Biomaterials 2006;27(9) 2150–2161.

[179] Elliott JC. Some observations on the crystal chemistry of carbonate-containing apatites. Archives of Oral Biology 1962;7 277–282.

[180] Nasiri-Tabrizi B, Fahami A. Mechanosynthesis of nanosized B-type carbonated fluorapatite. Materials Letters 2014;134 42–46.

[181] Fahami A, Beall GW, Nasiri-Tabrizi B, Pingguan-Murphy B. Effect of high-energy ball milling on the formation and microstructural features of carbonated chlorapatite nanopowders. Ceramics International 2015a;41(3) 4750–4758.

[182] Fahami A, Nasiri-Tabrizi B, Beall GW, Pingguan-Murphy B. Effect of ion concentration on mechanosynthesis of carbonated chlorapatite nanopowders. Materials Letters 2015b;146 16–19.

[183] Kannan S, Rocha JHG, Ferreira JMF. Synthesis of hydroxy-chlorapatites solid solutions. Materials Letters 2006a;60(7) 864–868.

[184] Kannan S, Rebelo A, Ferreira JMF. Novel synthesis and structural characterization of fluorine and chlorine co-substituted hydroxyapatites. Journal of Inorganic Biochemistry 2006b;100(10) 1692–1697.

[185] Yoder CH, Pasteris JD, Krol KA, Weidner VL, Schaeffer RW. Synthesis, structure, and solubility of carbonated barium chlor- and hydroxylapatites. Polyhedron 2012;44(1) 143–149.

[186] Elliott JC, Dykes E, Mackie PE. Structure of bromapatite and the radius of the bromide ion. Acta Crystallographica Section B 1981;37(2) 435–438.

[187] Cruz FJAL, da Piedade MEM, Calado JCG. Standard molar enthalpies of formation of hydroxy-, chlor-, and bromapatite. The Journal of Chemical Thermodynamics 2005;37(10) 1061–1070.

[188] Cruz FJAL, da Piedade MEM, Calado JCG. Erratum to "Standard molar enthalpies of formation of hydroxy-, chlor-, and bromapatite" J. Chem. Thermodyn. 37 (2005) 1061–1070]. The Journal of Chemical Thermodynamics 2006;38(7) 938.

[189] Dykes E. Preparation and characterisation of calcium bromapatite. Materials Research Bulletin 1974;9(9) 1227–1236.

[190] Liu X, Fleet ME, Shieh SR, He Q. Synthetic lead bromapatite: X-ray structure at ambient pressure and compressibility up to about 20 GPa. Physics and Chemistry of Minerals 2011;38(5) 397–406.

[191] Bhatnagar VM. Lead bromapatite, $Pb_{10}(PO_4)_6Br_2$. Inorganic and Nuclear Chemistry Letters 1971;7(3) 231–232.

[192] Ojovan MI. Handbook of Advanced Radioactive Waste Conditioning Technologies. Woodhead Publishing Series in Energy. Elsevier, 2011. ISBN: 978-0857090959

[193] Uno M, Shinohara M, Yamanaka S, Kurosaki K. Some properties of a lead vanado-iodoapatite $Pb_{10}(VO_4)_6I_2$. Journal of Nuclear Materials 2001;294(1–2) 119–122.

[194] Cockbain AG. The crystal chemistry of the apatites. Mineralogical Magazine 1968;36(281) 654–660.

[195] Lu F, Yao T, Xu J, Wang J, Scott S, Dong Z, Ewingc RC, Lian J. Facile low temperature solid state synthesis of iodoapatite by high-energy ball milling. RSC Advances 2014;4 38718–38725.

[196] Henning PA, Erik Adolfsson E, Grins J. The chalcogenide phosphate apatites $Ca_{10}(PO_4)_6S$, $Sr_{10}(PO_4)_6S$, $Ba_{10}(PO_4)_6S$ and $Ca_{10}(PO_4)_6Se$. Zeitschrift für Kristallographie International journal for structural, physical, and chemical aspects of crystalline materials 2008;215(4) 226–230.

[197] Junga SH, Hana SH, Leeb JS, Kimb YJ. Synthesis and Phase Transformation of $Ca_{10}(PO_4)_6S$ Sulfoapatite by Solid-state Reaction with Sulfur Vapor. Synthesis and Reactivity in Inorganic Metal-Organic and Nano-Metal Chemistry 2008;38(3) 303–306.

[198] Manoun B, Azdouz M, Azrour M, Essehli R, Benmokhtar S, El Ammari L, Ezzahi A, Ider A, Lazor P. Synthesis, Rietveld refinements and Raman spectroscopic studies of tricationic lacunar apatites $Na_{1-x}K_xPb_4(AsO_4)_3$ $(0 \leq x \leq 1)$. Journal of Molecular Structure 2011;986(1–3) 1–9.

[199] Lahrich S, Elmhammedi MA, Manoun B, Tamraoui Y, Mirinioui F, Azrour M, Lazor P. Elaboration, Rietveld refinements and vibrational spectroscopic study of $Na_{1-x}K_xCaPb_3(PO_4)_3$ lacunar apatites $(0 \leq x \leq 1)$. Spectrochimica Acta Part A: Molecular and Biomolecular Spectroscopy 2015;145 493–499.

[200] Ternane R, Ferid M, Trabelsi-Ayedi M, Piriou B. Vibrational spectra of lead alkali apatites $Pb_8M_2(PO_4)_6$ with M = Ag and Na. Spectrochimica Acta Part A: Molecular and Biomolecular Spectroscopy 1999;55(9) 1793–1797.

[201] El Koumiri M, Oishi S, Sato S, El Ammari L, Elouadi B. The crystal structure of the lacunar apatite $NaPb_4(PO_4)_3$. Materials Research Bulletin 2000;35(4) 503–513.

[202] Naddari T, Savariault Jean-M., El Feki H, Salles P, Salah AB. Conductivity and Structural Investigations in Lacunary $Pb_6Ca_2Li_2(PO_4)_6$ Apatite. Journal of Solid State Chemistry 2002;166(1) 237–244.

[203] Azdouz M, Manoun B, Azrour M, Bih L, El Ammari L, Benmokhtar S, Lazor P. Synthesis, Rietveld refinements and Raman spectroscopy studies of the solid solution $Na_{1-x}K_xPb_4(VO_4)_3$ $(0 \leq x \leq 1)$. Journal of Molecular Structure 2010;963(2–3) 258–266.

[204] Toumi M, Smiri L, Bulou A. Crystal Structure and Polarized Raman Spectra of $Ca_6Sm_2Na_2(PO_4)_6F_2$. Journal of Solid State Chemistry 2000;149(2) 308–313.

[205] Piotrowski A , Kahlenberg V , Fischer RX , Lee Y , Parise JB. The crystal structures of cesanite and its synthetic analogue - A comparison. American Mineralogist 2002;87 715–720.

[206] Knyazev AV, Bulanov EN, Korokin VZ. Synthesis and Thermal Expansion of $(SO_4)_3L$ (L = Halogen) compounds with the Apatite Structure. Inorganic materials 3013;49(11) 1133–1137.

[207] Klement R. Sodium calcium sulfate apatite, $Na_6Ca_4(SO_4)_6F_2$. Naturwissenschaften 1939;27(33) 568.

[208] Sha Lian-K, Chappell BW. Apatite chemical composition, determined by electron microprobe and laser-ablation inductively coupled plasma mass spectrometry, as a probe into granite petrogenesis. Geochimica et Cosmochimica Acta 1999;63(22) 3861–3881.

[209] Kreidler ER, Hummel FR. The crystal chemistry of apatite: Structure fields of fluor- and chlorapatite. The American Mineralogist 1970;55 170–184.

[210] Nikhare GN, Gedam SC, Dhoble SJ. Luminescence characteristics of $Na_3Ca_2(SO_4)_3F$: Ce^{3+} fluoride material. Cement and Concrete Research 1985;15(4) 581–584.

[211] Vázquez FT. A new compound: $K_3Ca_2(SO_4)_3F$, identified in coatings of heat recover cyclones. Cement and Concrete Research 1985;115(4) 581–584.

[212] Fayos J, Watkin DJ, Perez-Mendez M. Crystal structure of the apatite-like compound $K_3Ca_2(SO_4)_3F$. American Mineralogist 1987;72 209–212.

[213] Poddar A, Gedamb SC, Dhoblec SJ. Luminescence of Eu and Ce in $K_3Ca_2(SO_4)_3F$ fluoride material. Luminescence 2014, DOI: 10.1002/bio.2839

[214] Trabelsi IT, Oueslati A, Mhiri T, Toumi M. Synthesis, characterization and ionic conductivity of $Ca_{8-x}Sr_xBi_2(PO_4)_6O_2$ for x = (3, 4, 5). Journal of Alloys and Compounds 2015;641 14–21.

[215] Uvarov V, Shenawi-Khalil S, Popov I. New bismuth calcium oxysilicate with apatite structure: Synthesis and structural characterization. Journal of Solid State Chemistry 2010;183(7) 1484–1489.

[216] Huang J, Sleight AW. The Apatite Structure without an Inversion Center in a New Bismuth Calcium Vanadium Oxide: $BiCa_4V_3O_{13}$. Journal of Solid State Chemistry 1993;104(1) 52–58.

[217] Kazin PE, Zykin MA, Tretyakov YD, Jansen M. Synthesis and properties of colored copper-containing apatites of composition $Ca_5(PO_4)_3Cu_yO_{y+\delta}(OH)_{0.5-y-\delta}X_{0.5}$ (X=OH, F, Cl). Synthesis and properties of inorganic compounds 2008;53(3) 409–414.

[218] Kazin PE, Gazizova OR, Karpov AS, Jansen M, Tretyakov YD. Incorporation of 3d-metal ions in the hexagonal channels of the $Sr_5(PO_4)_3OH$ apatite. Solid State Sciences 2007;9 82–87.

[219] Kazin PE, Uskova MA, Tretyakov YD, Jansen M, Scheurell S, Kemnitz E. Formation of Bi(Pb)-2223 with chemically compatible V-rich phase. Physica C 1998;301 185–191.

Chapter 5

Rare-earth Element-bearing Apatites and Oxyapatites

Petr Ptáček

Additional information is available at the end of the chapter

http://dx.doi.org/10.5772/62209

Abstract

A number of prepared alkaline-earth-rare-earth silicates and germanates also have the structure of apatite type. The fifth chapter of this book then continues with description of synthetic compounds of apatite structure. Attention will be directed to description of rare-earth element bearing apatites and oxyapatites. The structure, properties and preparation of apatite-type silicates, germanates and borates were described. This chapter gives also description of oxygen-rich apatites, which are promising material for electrolytes in solid oxide fuel cells and sensors and explain the basic concepts between structure and conductivity of these compounds. The additional information about application of apatites is given in the last chapter of this book. Furthermore, N-apatite, REE vanadocalcic apatite and apatite type yttrium phosphates were described.

Keywords: Apatite, Oxyapatite, Conductivity, Silicates, Germanates, Borates, N-apatite, Apatite-type yttrium phosphates

In this chapter, the preparation and the properties of synthetic phases of apatite are given; the geological role is described in **Section 7.3**. The ideal general formula of an apatite-type oxide may be written as $M_{10}(XO_4)_6O_2$ (M = alkaline-earth and/or rare-earth element, X = Si, Ge, P, V, ...). The structure (**Fig. 1**) can be described in terms of a "microporous"[1] [1] framework $(A(1)_4(XO_4)_6)$ composed of face sharing $M(1)O_6$ trigonal meta-prismatic columns, which are corner connected to MO_4 tetrahedra. This framework allows some flexibility to accommodate remaining $M(2)_6O_2$ units [2].

[1] Microporous material is defined as containing pores with the diameters >2 nm. The materials with the pore diameter in the range from 2 to 50 nm and higher than 50 nm are termed as mesoporous and macroporous, respectively. In combination with nanotechnology, the term nanoporous material is often used. Despite the fact that there is not clear definition, usually the pores with the size from 0.1 to 100 nm are considered. In other words, nanoporous covers the range from microporous to macroporous [1].

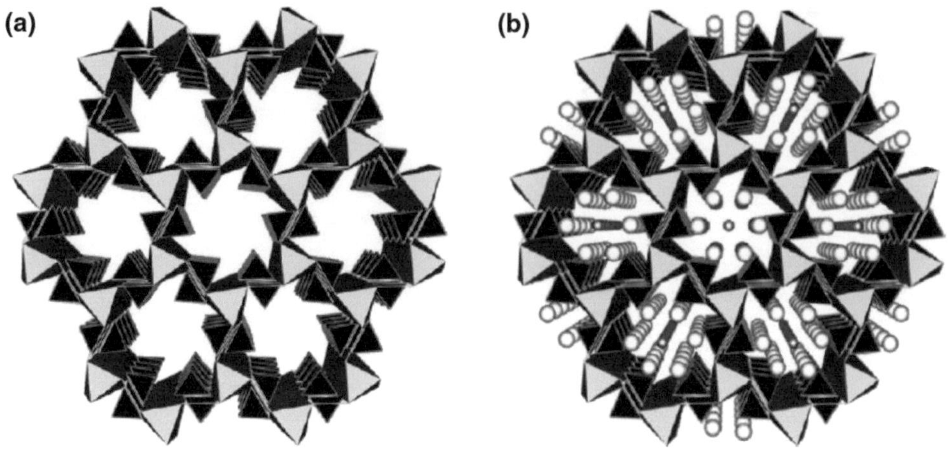

Fig. 1. Illustration of "microporous" $M(1)(XO_4)_6$ framework of the apatite ($M_{10}(XO_4)_6O_2$) structure (a): tetrahedra MO_4, M(1) cation at the center of trigonal meta-prism. Remaining $M(2)_6O_2$ units occupy the cavities within this framework (b): large spheres are M(2) cations and small spheres are O anions [2].

5.1 Apatite-type lanthanium silicates

During recent decades, oxyapatite[2]-type structure with the general formula: $REE_{9.33+x}Si_6O_{26+3x/2}$ (where REE is rare-earth element) [3],[4], $REE_{9.33}\square_{0.67}(SiO_4)_6O_2$ [5] or $REE_{10-x}(SiO_4)_6O_{2+y}$ [6] has attracted considerable attention as oxide ion conductors. Apatite-type oxides have attracted much attention as the material for electrolytes in solid oxide fuel cell and sensors (**Chapter 10**).

In low atomic number rare-earth silicate systems, an apatite phase occurs with a range of stability extending from $Ln_{4.67}(SiO_4)_3O$ to $Ln_4(SiO_4)_3$. The stability decreases as the rare-earth atomic number increases, with a mixture of Ln_2SiO_5 and $Ln_2Si_2O_7$ replacing apatite as the preferred phase assemblage [7],[8],[9],[10],[11].

Apatite-type rare-earth element (REE) silicates of the composition of $REE_{10-x}(SiO_4)_6O_{2+y}$, where REE = La, Nd, Gd and Dy, were prepared by Martínez-González et al [6] via the mechanochemical synthesis (stabilized zirconia planetary ball mill: ball-to-powder ratio ~10:1, 350 rpm for maximum time of 9 h) starting from the stoichiometric mixtures of constituent oxides, REE_2O_3 and SiO_2 (molar ratio = 4:5), followed by post-milling thermal treatment (1500°C for 3 h). The ionic conductivity increases with the increasing size of REE cations.[3] The mechano-

[2] Since the prefix "oxy-" can be explained as containing oxygen or additional oxygen, and the prefix "oxo-" is used for the functional group or substituent oxygen atom connected to another atom by a double bond (=O), the names oxyapatite, oxy-apatite, oxoapatite and oxo-apatite can be considered as synonyms. In the published literature, the name oxyapatite is the most frequently used (~90%), and the term oxy-apatite is the second (~8%). The names oxoapatite and oxo-apatite are used much rarely (only about ~2%).

Apatite-type silicates described in this chapter can be also named as oxy-britholites (oxybritholites) [84].

[3] This conclusion is in discrepancy with the findings of Higuchi et al [12] described below.

http://dx.doi.org/10.5772/62209

chemical synthesis of apatite-type lanthanum silicates from the mixture of La_2O_3 and amorphous silica without post-milling thermal treatment was described by FUENTES et al [4].

Rare-earth element-doped apatite-type lanthanum silicates of the composition of $La_9MSi_6O_{27}$, where M = Nd, Sm, Gd and Yb, were synthesized by the high-temperature solid-state reaction process by XIANG et al [3]. All rare-earth oxide powders (La_2O_3, Nd_2O_3, Sm_2O_3, Gd_2O_3 and Yb_2O_3) were firstly pre-calcined at 900°C for 2 h in order to achieve complete decarbonation and dehydroxylation before weighing. The stoichiometric mixtures were mechanically mixed in absolute ethanol for 24 h using zirconia milling media at the speed of 400 rpm and dried at 100°C in air. The powder mixture was calcined at 1350°C for 10 h and then ground by hand with an agate mortar and pestle to reduce the particle size. After that, the powders were uniaxially pressed at 20 MPa and then statically cold pressed at 200 MPa for 5 min. The compacts were pressureless sintered at 1650 K for 10 h in air.

Apatite-type phase	Lattice parameters [Å]		*M*	*V*	Density	*E*	σ_0
	a	*c*	[g·mol^{-1}]	[Å^3]	[g·cm^{-3}]	[eV]	[S·K·cm^{-1}]
$La_4SiO_3O_{12}$	9.376	6.761	831.87	514.73	–	–	–
$La_{10}Si_6O_{27}$	9.709	7.176	1989.552	585.82	5.168	0.76	$2.51·10^4$
$La_9NdSi_6O_{27}$	9.729	7.191	1994.886	589.46	5.344	0.79	$4.78·10^4$
$La_9SmSi_6O_{27}$	9.692	7.161	2001.046	582.55	5.405	0.82	$1.85·10^4$
$La_9GdSi_6O_{27}$	9.723	7.187	2007.896	588.41	5.485	0.87	$3.24·10^4$
$La_9YbSi_6O_{27}$	9.701	7.143	2023.686	582.16	5.459	0.98	$4.34·10^5$

Table 1. The properties of apatite-type lanthanium silicates [3],[8].

The lattice parameters and the properties of prepared apatite-type lanthanium silicates are listed in **Table 1**. All prepared compounds possess hexagonal apatite structure with the space group $P6_3/M$. The temperature dependence of total electrical conductivity for different compositions is determined using the Arrhenius equation[4] [13],[14],[15] in the following form [3],[5],[16],[17]:

$$\sigma\ T = \sigma_0\ \exp\left(-\frac{E}{k_B T}\right) = \sigma_0\ \exp\left(-\frac{\Delta H_m + \Delta H_a}{k_B T}\right) \tag{1}$$

where σ is the total electrical conductivity, σ_0 is the pre-exponential factor related to the effective number of mobile oxide ions, E is the activation energy for the electrical conduction process, k_B is the Boltzmann constant and T is absolute temperature. ΔH_m and ΔH_a denote the

[4] The equation of SVANTE AUGUST ARRHENIUS [13],[14], which predicts that the rate constant k depends on the temperature: $k = A\ exp\ (-E_a/RT)$, where A is the frequency (pre-exponential factor), E_a is the activation energy, R is universal gas constant (8.314 J·K^{-1}·mol^{-1}) and T is the thermodynamic temperature [13].

migration enthalpy of oxygen ion and the association enthalpy of defects, respectively. The determined activation energy and pre-exponential factor are listed in **Table 1**. It can be seen that the activation energy gradually increases from $La_{10}Si_6O_{27}$ to $La_9GdSi_6O_{27}$. Total electrical conductivity can be calculated from the following equation:

$$\sigma = \frac{h}{RS} \tag{2}$$

where *h* is the thickness of the specimen, *S* is the electrode area of the specimen surface and *R* is the total resistance including grain and grain boundary resistance. Lanthanum silicates doped with Nd or Yb cations exhibit higher total electrical conductivity than undoped lanthanum silicates. The highest total conductivity value obtained at 500°C is $4.31 \cdot 10^{-4}$ $S \cdot cm^{-1}$ for $La_9NdSi_6O_{27}$. The total electrical conductivity is also a function of partial pressure of oxygen [3].

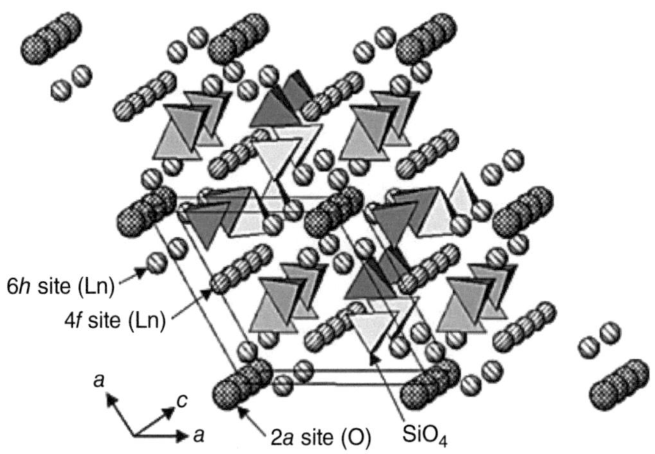

Fig. 2. Crystal structure of apatite-type rare-earth element silicate viewed along the c-axis [12].

The measurements using single crystals revealed definite anisotropy of the electrical conductivity of $Ln_{9.33}(SiO_4)_6O_2$, that is, the conductivity parallel to the c-axis is larger by one order of magnitude than that perpendicular to the c-axis. This fact clearly indicates that the channel oxide ions not bonded to silicon are the principal charge carriers in apatite-type lanthanum silicates. The structure of apatite-type rare-earth silicate is shown in **Fig. 2**. SiO_4 tetrahedra are isolated mutually, and Ln ions (REE ions, in general) at *6h* sites (sevenfold coordinated site (x,y,¼)) [18][5] form channels, in which oxide ions at *2a* sites are located (possess the threefold coordination with rare-earth ions at the *6h* sites in the same plane), along the c-axis.

http://dx.doi.org/10.5772/62209

These mobile ions at these sites have much larger anisotropic displacement parameters in the direction of the c-axis than those in the direction of the a-axis, even at room temperature, which reflects high oxide ion conduction along the c-axis. The ninefold coordinate position (4*f* site (1/3,2/3,*z*)) is the second site for the accommodation of REE cations in the structure of apatite-type REE silicate [6],[12],[19],[20].

Since the interstitial space provided by these rare-earth ions is the smallest throughout the channel along the c-axis of the apatite structure, the migration of oxide ions through the channel will not be affected significantly even if the sizes of rare-earth ions are varied. It is therefore reasonable that the electrical conductivities of apatite-type rare-earth silicates are independent on the kind of rare-earth elements [12].

Conventional oxide ion conductors are designed on the basis of the oxygen vacancy model by the introduction of aliovalent[6] [21] cations. In $Ln_{9.33}(SiO_4)_6O_2$, however, cation vacancies are present rather than oxygen vacancies. Therefore, the introduction of cation vacancies into the structure of an oxide material may induce high oxide ion conductivity if the structure has a channel or a plane that can be a path for the migration of oxide ions [12].

Apparent exchange of O(1), O(2) and O(3) oxide ions bonded to Si was observed by ^{17}O NMR measurement on $La_{9.33}Si_6O_{26}$ by Kiyino et al [22], while it was not observed for oxide ion on the isolated site O(4). The results indicate that oxide ions bonded to Si at the position O(1), O(2) and O(3) are the main diffusion species in the oxide ion conductivity.

Trivalent and divalent dopants[7] [23] have been introduced into the $La_{9.33}(SiO_4)_6O_2$ structure according to the following nominal mechanisms [24]:

$$Si^{4+} \rightarrow M^{3+} + \frac{1}{3}La^{3+} \tag{3}$$

$$\frac{2}{3}La^{3+} \rightarrow AEE^{2+} \tag{4}$$

$$\frac{1}{3}La^{3+} \rightarrow AEE^{2+} + \frac{1}{2}O^{2-} \tag{5}$$

---

[5] According to the Wyckoff notation: the specification of actual coordinates of atoms within the unit or primitive cell, which can be generated by the point-group operations or may be found by reference in the International Tables for Crystallography [18].

[6] Cation with different valence. Apatite structure shows large flexibility upon the substitution of other aliovalent cations at the 'Ca' sites, pentavalent and tetravalent ions such as V^{5+}, As^{5+} and Si^{4+} at the 'P' site and halide, oxide ions at the 'OH' site [21], as was described.

[7] Dopants are also termed as doping agents. It can be defined as an impurity element added to the material structure in low concentration (usually <1 wt.% [23]) in order to alter its properties.

where M = Al, Ga, B, Co, Fe, Mn, … and AEE denotes the alkaline-earth elements (Ca, Sr and Ba). Doping with Al, Ga and B according to the formula: $La_{9.33+x/3}(SiO_4)_{6-x}(MO_4)_xO_2$, via the mechanism in **Eq. 3**, causes that bulk conductivity increases in up to two orders of magnitude in the case of Al for $x = 1 - 1.5$. If, however, the sample is stoichiometric on both cation and anion sites, as for $La_8Sr_2(SiO_4)_6O_2$, the AEE doping reduces the conductivity and increases the activation energy for the conduction compared to $La_{9.33}(SiO_4)_6O_2$.

The effect of Fe doping on the electrical properties of lanthanum silicates of the composition of $La_{10}Si_{6-x}Fe_xO_{27-x/2}$ (where $x = 0.2, 0.4, 0.6, 0.8$ and 1.0) was performed by SHI and ZHANG [16] via the sol-gel process. Tetraethyl orthosilicate (TEOS), $La(NO_3)_3{\cdot}6H_2O$ and $Fe(NO_3)_3{\cdot}9H_2O$ were used as starting materials. Stoichiometric amounts of $Fe(NO_3)_3{\cdot}9H_2O$ and $La(NO_3)_3{\cdot}6H_2O$ were dissolved in the mixture of ethanol, acetic acid and distilled water. The appropriate amount of TEOS was added to the solution while continuous stirring. The solution became gradually a purple clear sol. After refluxing at 80°C for 1 – 2 h, the sol transferred to a clear gel. Then, the wet gel was dried at 100°C for 20 h. The gel was heated at 600°C for 4 h to remove water and organic components and to decompose nitrates. In order to get the desired phase, obtained precursor was then calcined at 1000°C for 4 h.

Apatite-type lanthanium silicate	**Lattice parameters [Å]**		***V***	***E* (600 – 800°C)**	***E* (400 – 550°C)**
	a	*c*	**[Å³]**	**[eV]**	
$La_{10}Si_{5.8}Fe_{0.2}O_{26.9}$	9.725	7.192	589.1	0.78	0.96
$La_{10}Si_{5.6}Fe_{0.4}O_{26.8}$	9.729	7.208	590.8	0.74	0.95
$La_{10}Si_{5.4}Fe_{0.6}O_{26.7}$	9.732	7.220	592.2	0.72	0.89
$La_{10}Si_{5.2}Fe_{0.8}O_{26.6}$	9.735	2.217	592.3	0.74	1.01
$La_{10}Si_5FeO_{26.5}$	9.743	7.229	593.5	0.75	1.02

Table 2. Lattice parameters of Fe-doped apatite-type lanthanium silicates [16].

All synthesized samples have hexagonal lattice structure with the space group of $P6_3/M$. The lattice parameters of prepared Fe-doped apatite-type lanthanium silicates and the activation energy of conductivities (**Eq. 1**) for different Fe contents are listed in **Table 2**. When $x = 0.6$, $La_{10}Si_{5.4}Fe_{0.6}O_{26.7}$ exhibits the lowest activation energy. The lattice parameters of $La_{10}Si_6O_{27}$ (**Table 1**) and doped specimen (**Table 2**) show that the values of ***a***, ***c*** and ***V*** increase with the content of iron. The conductivity of $La_{10}Si_{6-x}Fe_xO_{27-x/2}$ is independent of oxygen partial pressure in the range from 0 to 100 kPa, which indicates that the conductivity of all samples is mainly ionic [16].

The oxygen ionic and electronic transport in apatite ceramics with the composition of $La_{10}Si_{6-x}Fe_xO_{27-x/2}$ ($x = 1 - 2$) [25] and $La_{10-x}Si_{6-y}Al_yO_{27-3x/2-y/2}$ ($x = 0 - 33; y = 0.5 - 1.5$) [26],[27] was investigated by SHAULA et al In both cases, the essential role of oxygen content on the ionic conductivity of apatite phase was recognized. The ion transference number[8] [28] increases with decreasing partial pressure of oxygen. Such behavior indicates that the conduction under oxidizing condition is predominantly of p-type[9] (with respect to n-type of conductivity). Similar to the foundation of SHI and ZHANG [16], the conductivity of these phases is predominantly ionic and almost independent on partial pressure of oxygen. The ion transference numbers are higher than 0.99, while the p-type electronic contribution to total conductivity is about 3% (700 – 950°C, $La_{10}Si_4Fe_2O_{26}$). The oxygen ionic conductivity should increase with decreasing iron content due to higher concentration of oxygen interstitials.

Another important factor influencing the oxygen diffusion is M-site deficiency, which affects the unit cell volume and may cause the O(5) ion displacement into interstitial sites, thus creating the vacancies in the O(5) sites at fixed total oxygen content. In particular, an enhanced ionic conduction was found in the system $La_{9.33+x/3}Si_{6-x}Al_xO_{26}$, where Al doping is compensated by the A-site vacancy concentration without oxygen content variations [29],[30].

The incorporation of praseodymium in the apatite-type lattice of $La_{9.83-x}Pr_xSi_{4.5}Fe_{1.5}O_{26+\delta}$ (x = 0 – 6) decreases the unit cell volume, suppresses the Fe^{4+} formation according to Mössbauer spectroscopy[10] [31],[32],[33],[34] and increases p- and n-type electronic contributions to total conductivity under oxidizing conditions, while the level of oxygen ionic transport at temperatures above 1000 K remains unaffected [35].

Since the size of the conduction channel increases with the Mg doping, the enhancement of the ionic conductivity of lanthanum silicate-based apatites can be reached by optimizing the La content and the Mg doping level at the same time. The ionic conductivities of $La_{10}Si_{5.8}Mg_{0.2}O_{26.8}$ and $La_{9.8}Si_{5.7}Mg_{0.3}O_{26.4}$ at 800°C are 88 and 74 mS·cm^{-1} with the activation energy of 0.43 and 0.42 eV, respectively [36].

The ionic conduction in cation-deficient apatite $La_{9.33-2x/3}M_xSi_6O_{26}$, where M = Mg, Ca and Sr was investigated by YUAN et al [37]. The nature of dopant and the extent of substitution have a significant effect on the conductivity. The greatest decrease in conductivity is observed for Mg doping followed by Ca- and Sr-doped apatites. The effect is ultimately attributed to the amount of oxygen interstitials, which is affected by the crystal lattice distortion arising from the cation vacancies.

The incorporation of additional La_2O_3 into $La_{9.33}(SiO_4)_6O_2$ to form $La_{10}(SiO_4)_6O_3$ or intermediate compositions can most obviously be achieved by filling empty interstitial sites with oxygen. The only alternative scenario would involve the creation of cation vacancies on the Si sublattice, which is unlikely as Si is present as a complex anion. The incorporation of excess of La_2O_3 into $La_{9.33}(SiO_4)_6O_2$ can therefore be expressed as [10]:

---

[8] The fraction of total current that is transferred by a given ion is affected by its mobility. The sum of transport numbers for all ions in electrolyte is equal to one [28].

[9] The p-type carriers possess typically higher mobility [25].

[10] The technique is based on the Mössbauer effect of recoil-free nuclear resonance fluorescence [31], i.e. the phenomenon of emission or absorption of X-ray photon without the loss of energy. The Mössbauer effect has been detected in a total of 88 X-ray transitions in 72 isotopes of 42 different elements [32]. The ^{57}Fe Mössbauer isotope is the most frequently used [33]. The Mössbauer spectroscopy can be used to determine the oxidation states of iron in minerals and to identify the presence of some mineral species in samples of unknown composition [31].

$$La_{9.33}[\ ]_{0.67}(SiO_4)_6O_2 + 1/3\ La_2O_3 \rightarrow La_{10}(SiO_4)_6O_2O_i\left(La_{10}(SiO_4)O_3\right) \quad (6)$$

Thus, in the ideal pure $La_{10}(SiO_4)_6O_3$, the *4f* and *6h* sites are fully occupied by La^{3+} ions, while an extra oxygen interstitial is introduced into the lattice to maintain the electroneutrality. The oxygen interstitial may benefit the oxide ion transportation if it is located nearby the [001] direction c-axis of the conventional unit cell. From the space-filling consideration, the most appropriate sites for the oxygen occupation are in this position; however, some distortion of the O *2a* sites would be required to accommodate extra oxygen atoms. This could be achieved by decreasing the symmetry from $P6_3/M$ to $P6_3$ allowing oxygen to move from 0,0,1/4 to 0,0,x. Recent studies suggest that a range of partially occupied (0,0,x) sites may accommodate this extra interstitial oxygen. From this point of view, $La_{10}(SiO_4)_6O_3$ should exhibit higher conductivity than $La_{9.33}(SiO_4)_6O_2$ [10],[38].

Introducing Sr^{2+} cations to the La^{3+} atomic positions, as in the $La_{10}(SiO_4)_6O_3$ phase, leads to complete elimination of vacancies according to the substitution [39]:

$$La_{10}(SiO_4)_6O_3 \xrightarrow{+Sr^{2+}} La_9Sr_1(SiO_4)_6O_{2.5} \xrightarrow{+Sr^{2+}} La_8Sr_2(SiO_4)_6O_2 \quad (7)$$

The substitution of La_2O_3 by SrO, taking into account the charge balance and the oxygen content, can be represented as follows (Kröger-Vink notation[11] [40],[41]):

$$SrO + 1/2\ O_o^x + La_{La}^x \rightarrow Sr_{La}^{/} + 1/2\ V_o^{\bullet\bullet} + 1/2\ La_2O_3 \quad (8)$$

Lanthanum oxyapatite phases are substantially stable with respect to their binary oxides. The general trend in the formation enthalpies as a function of (La + Sr)/(La + Sr + Si) shows that the apatite phase becomes more energetically stable as the cation vacancy and oxygen excess concentrations decrease. The stoichiometric sample achieved by Sr^{2+} doping, with no cation vacancies or interstitial oxygen atoms, is the most stable composition. The energetics of lanthanum silicate apatite materials ($La_{9.33+x}(SiO_4)_6O_{2+3x/2}$ and $La_{10-x}Sr_x(SiO_4)_6O_{3-0.5x}$) depends on lanthanum deficiency and oxygen interstitial[12] [42],[43] concentrations, and the cation vacancy concentrations appear to be the dominant factor in energetics [39].

---

[11] The Kröger-Vink notation indicates the lattice position for the point defect species in the crystal and its effective electric charge relative to the perfect lattice: $\mathbf{M}_Y^{\ Z}$ is the atomic species **M** (or vacancy **V**) that occupies the lattice site **Y** and possesses the effective charge **Z**, where the symbols •, ' and × are used for the effective charge +1, -1 and neutral particle, respectively) [40]. For example, $\mathbf{Al}_i^{\bullet\bullet\bullet}$ is Al^{3+} ion at interstitial site (*i*), $\mathbf{V}_{Al}'''$ is Al^{3+} vacancy, $V_O^{\bullet\bullet}$ is O^{2-} vacancy, Sr_{La}' (**Eq. 8**) means Sr^{2+} ion replacing La^{3+} at lattice site, $Ti_{Al}^{\bullet}$ means Ti^{4+} replacing Al^{3+} at lattice site, *e*' is electron and $\boldsymbol{h}^{\bullet}$ is the hole. The equation must fulfill the following three rules: mass balance (1), electroneutrality or charge balance (2) and site ratio conservation balance (3) [41].

[12] Interstitial sites are sites between normal (equilibrium) atomic positions of ideal lattice atoms [42]. Interstitial atoms and vacancies (lattice site where atom is absent) are the simplest types of point defects in a crystal. A vacancy and interstitial atoms positioned close together are referred to as the Frenkel pair. Apart from the point defects, the line crystal defects (dislocation and disclination) are recognized [43].

http://dx.doi.org/10.5772/62209

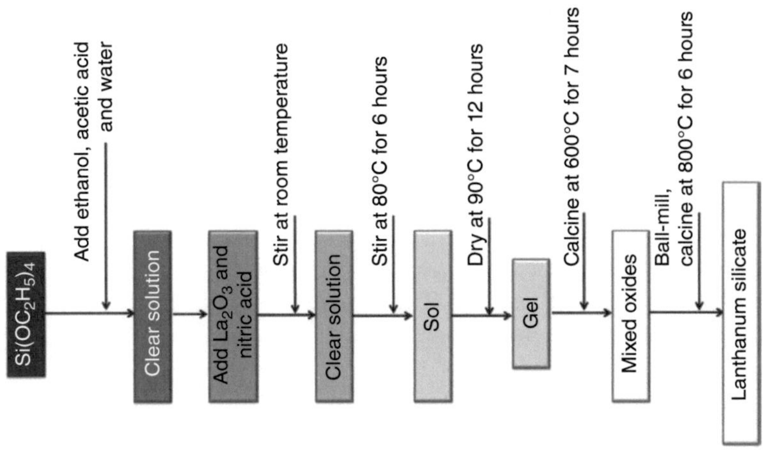

Fig. 3. Schematic diagram of preparation of lanthanum silicate by a sol-gel process [44].

The schematic diagram of the sol-gel process used by TAO and IRVINE [44] for the preparation of apatite-type lanthanium silicates is shown in **Fig. 3**. The room-temperature structure is hexagonal, the space group is $P6_3$ or $P6_3/M$, with $\boldsymbol{a} = 9.722$ and $\boldsymbol{c} = 7.182$ Å for $La_{10}(SiO_4)_6O_3$ and $\boldsymbol{a} = 9.717$ and $\boldsymbol{c} = 7.177$ Å for $La_{9.33}(SiO_4)_6O_2$, i.e. the cell volume of $La_{10}(SiO_4)_6O_3$ is a little greater than that of $La_{9.33}(SiO_4)_6O_2$. Both compositions exhibit high ionic conductivity, although the grain boundary resistance is the dominant feature in the impedance spectrum of both. In general, the conductivity of $La_{10}(SiO_4)_6O_3$ is higher than that of $La_{9.33}(SiO_4)_6O_2$ and this indicates that oxygen interstitials may be introduced into the apatite lattice of $La_{10}(SiO_4)_6O_3$, which may benefit the oxygen ion transportation [44].

The $La_{10}Si_6O_{27}$ nanopowders with apatite structure were synthesized by the LI et al [45] co-precipitation method. After the calcination at 900°C and then removing of La_2O_3 by acid washing, the pure stoichiometric $La_{10}Si_6O_{27}$ nanopowders are obtained. The oxyapatite ceramics with the density higher than 95% can be obtained at rather low sintering temperature of 1300°C, and it has comparable total conductivity with the samples sintered at 1650°C from the powders prepared by solid-state reaction.

La_2O_3 and TEOS in stoichiometric amount were used by MASUBUCHI et al [46] as the starting materials for the preparation of both powder and film of apatite-type $La_{9.33}(SiO_4)_6O_2$ via the alkoxide hydrolysis. Lanthanum oxide was dissolved in HNO_3 (6 mol·dm^{-3}) and mixed with ethanol. Then, stoichiometric amount of TEOS in ethanol was added (La:Si = 9.33:6) to this solution. The precursor solution was obtained by refluxing for one night. This solution was heated to gelating on the hot plate followed by calcination and annealing in powder preparation. Either quartz glass or Pt foil substrate was dipped to the gelatinous solution and dried for the film preparation. It was calcined at 500°C for 1 h to remove the organic contents and then fired at 1000°C for 10 h. This preparation steps were repeated to increase the film thickness. The film showed preferred orientation of the apatite crystal in thinner film. The conductivity of sintered body was lower in about one order of magnitude than the value of single crystal perpendicular to c-axis [46].

The synthesis and the conductivities of Ti-doped apatite-type phases of the composition of $(La/Ba)_{10-x}(Si/Ge)_6O_{26+z}$, where Ti substituted at the Si/Ge site, were reported by SANSOM et al [47]. The conductivities were shown to be the highest for the samples containing either cation vacancies or oxygen excess, which is consistent with previous studies of apatite-type oxide ion conductors. However, the Ti doping was shown to generally decrease the conductivity in comparison with equivalent samples containing only Si/Ge at the tetrahedral sites, with the greatest decrease for Si-containing samples.

Vanadium-doped oxyapatite phases of the composition of $La_{10-x}V_x(SiO_4)_6O_{3+x}$ were prepared by YUAN et al [48] via the sol-gel method. The apatite phase begins to form at 800°C, which is much lower than in the case of conventional solid-state synthesis method. The best conductivity of $La_9V(SiO_4)_6O_4$ is $1.67 \cdot 10^{-2}$ S·cm^{-1}, which is significantly higher than that for lanthanum silicate oxides ($1.19 \cdot 10^{-2}$ S·cm^{-1}). The valence ion V^{5+} doped for La^{3+} does lead to the formation of hexagonal apatites even with high oxygen contents.

The phase $La_5Si_2BO_{13}$ [49],[50],[51] crystallizes with apatite-related structure (**Fig. 4**) with the space group $P6_3/M$ and the cell parameters $\boldsymbol{a}$ = 9.5587 Å, $\boldsymbol{c}$ =7.2173 Å and $\boldsymbol{Z}$ = 2. The composition of these apatite-like compounds can be also expressed via the general formula: $La_{9.33+x}Si_{6-2y}B_{3y}O_6$, where $0 \le x \le 0.67$. At limiting compositions $x = 0.67$, La(2) site is fully occupied, and the formula referred to the unit cell is $La_{10}Si_4B_2O_{26}$ or, more simply, $La_5Si_2BO_{13}$.

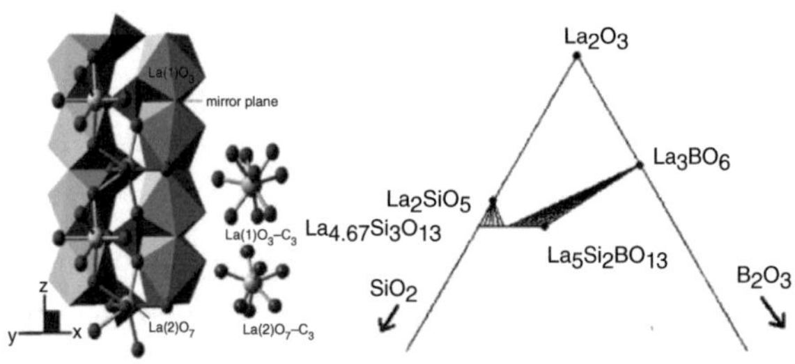

Fig. 4. The crystal structure of $La_5Si_2BO_{13}$ (a) [51] and the upper section of the ternary phase diagram La_2O_3-SiO_2-B_2O_3 at 1100°C (b) [49].

The comparison with other apatite-like structures shows lower distortion in the M(1) polyhedron and unusually short bond length from La in the M(2) site and O(4) oxygen in the column site (2.303 Å). These results can be explained in view of the presence of trivalent La and divalent O, respectively, in the M(1) and M(2) sites and in the column anion site, whereas, in apatites, these sites are occupied by divalent and monovalent ions, respectively [49].

The preparation of La-Si-O apatite-type thin films was described by VIEIRA et al [52] with Si/(La + Si) atomic ratios ranging from 0.36 to 0.43 being produced via the magnetron sputtering in reactive Ar/O discharge gas. The apatite-type lanthanum silicate phase was formed in all as-deposited films upon the annealing at 900°C for 1 h. The lanthanum silicate films obtained by annealing the as-deposited films with lower Si/(La + Si) atomic ratios have a preferential orientation with the c-axis perpendicular to the substrate, while low-intensity diffraction peaks ascribed to $La_2Si_2O_7$ phase were detected in the films deposited with higher

Si content. Preferentially oriented films have higher activation energy and lower ionic conductivity, as the ionic conductivity measurements were performed in the direction perpendicular to the c-axis. The highest ionic conductivity was obtained for the film deposited with a Si/(La + Si) atomic ratio of 0.42, with a value of 1.2×10^{-2} S·cm^{-1} at 750°C. By the incorporation of oxygen in the as-deposited films, the silicon segregation upon annealing was avoided.

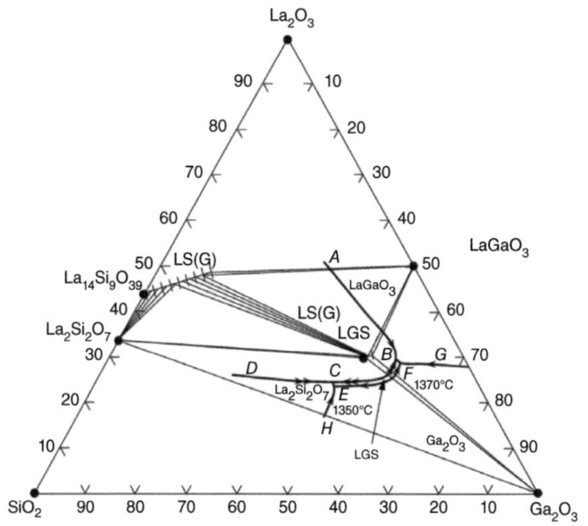

Fig. 5. Primary phase diagram of ternary system La_2O_3-Ga_2O_3-SiO_2 around LGS [53].

The formation of ternary compound with apatite structure in the system La_2O_3-Ga_2O_3-SiO_2 (**Fig. 5**) was first reported by Wang and Uda [53]. The apatite phase, which precipitates from the melt of the composition around that of stoichiometric $La_3Ga_5SiO_{14}$ (LGS), can be described by the formula: $La_{14}Ga_xSi_{9-x}O_{39-x/2}$, where $0 \leq x \leq 3.5$. Since there is a large field for the formation of solid solutions with the range extending from $La_{14}Si_9O_{39}$ to Ga_2O_3, some Si^{4+} sites are probably substituted by Ga^{3+}.

The liquidus surface of LS(G) was determined to be the field on the Ga_2O_3-poor side of boundary curve ABCD. The liquidus surface of LS(G) covers the stoichiometric composition of LGS. In this field, the crystallization of LS(G) aciculae was observed in all samples that were heated to temperatures above 1500°C. The liquidus volume of LGS is denoted by the field BCEF. It seems to be a narrow field in the composition between the liquidus surfaces of LS(G) and Ga_2O_3. E and F are eutectic points, where LGS + $LaGaO_3$ + Ga_2O_3 + liquid and LGS + Ga_2O_3 + $La_2Si_2O_7$ + liquid were found, respectively [53].

The CaO-La_2O_3-SiO_2-P_2O_5 phase diagram was investigated by El Ouenzerfi et al [54] in order to determine a domain inside which all points correspond to pure apatitic oxyphosphosilicates with the general formula: $Ca_xLa_y(SiO_4)_{6-u}(PO_4)_uO_t$. The defined domain (**Fig. 6**) is only a part of the whole existence domain of the apatitic structure, but it allows to prepare pure apatitic samples with well-controlled composition.

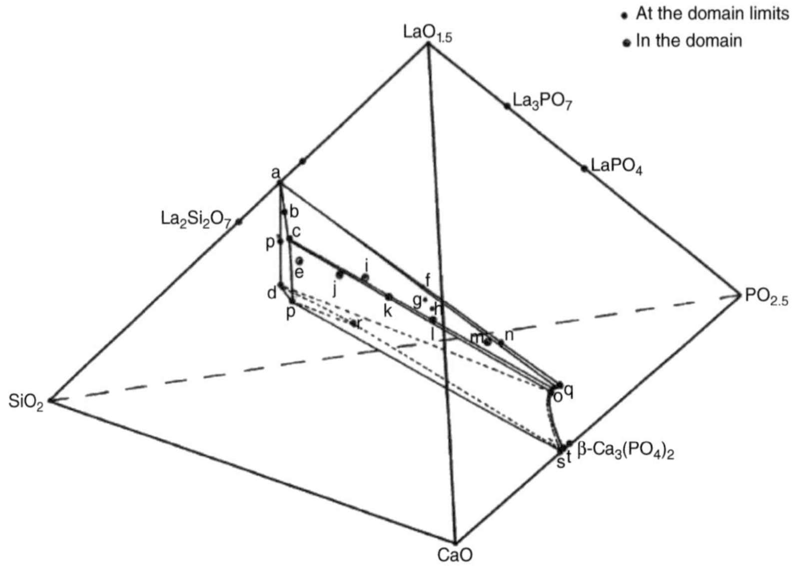

Fig. 6. Britholite stability domain in the CaO-SiO_2-La_2O_3-P_2O_5 quaternary system [54].

For these samples, the continuous change of the stoichiometry of each element proves that it exists as a solid solution including the oxygen content. This observation completes the literature data where britholites are presented as limited to three series corresponding to the stoichiometry [54],[55]:

1. $Sr_{2+x}La_{8-x}(SiO_4)_{6-x}(PO_4)_xO_2$ with $0 \leq x \leq 6$ [13];
2. $Sr_{3+x}La_{6-x}(SiO_4)_{6-x}(PO_4)_x$ with $0 \leq x \leq 1.5$;
3. $Sr_{4+x}La_{6-x}(SiO_4)_{6-x}(PO_4)_xO$ with $0 \leq x \leq 6$.

Inside this domain, the solid solution continuously varies between pure phosphate apatites $Ca_xLa_y(PO_4)_6O_t$ and pure silicate apatites $Ca_xLa_y(SiO_4)_6O_t$ and also between oxyapatites $Ca_xLa_y(SiO_4)_{6-u}(PO_4)_uO_t$ and nonoxyapatites $Ca_xLa_y(SiO_4)_{6-u}(PO_4)_u$.

During the investigation of the kinetics of solid-state sintering[14] of strontium-doped apatite-type lanthanum silicates ($Sr_xLa_{10-x}Si_6O_{27-x/2}$) under isothermal conditions (1250 – 1550°C), BONHOMME et al [56] recognized that the densification mechanism of the apatite ceramics was

[13]WANMAKER et al [55] reported the synthesis of apatite-type compounds of the composition:

(a) $M^{2+}(II)M(III)_{8-x}(SiO_4)_{6-x}(PO_4)_xO_2$, where $0 \leq x \leq 6$;

(b) $M(II)_{3+x}M(III)_{6-x}(SiO_4)_{6-x}(PO_4)_x$, where $0 \leq x \leq 1.4$;

(c) $M(II)_{4+x}M(III)_{6-x}(SiO_4)_{6-x}(PO_4)_xO$, where $0 \leq x \leq 6$.

where M(II) = Ca, Sr, Ba, Mg, Zn or Cd and M(III) Y or La. The paper also contains structural data for several other newly prepared oxy-britholites, including $Zn_2La_8(SiO_4)_6O_2$, $BaMgY_8(SiO_4)_6O_2$, $Zn_2Y_8(SiO_4)_6O_2$, $Cd_2Y_8(SiO_4)_6O_2$, $Ca_4La_5(SiO_4)_5(PO_4)$ and $Ba_4La_5(SiO_4)_5(PO_4)$.

controlled by the diffusion of rare-earth element (La) at the grain boundaries. This process showed the activation energy of 470 kJ·mol^{-1}.

The ternary phase diagram Al_2O_3-SiO_2-La_2O_3 at 1300°C (**Fig. 7**) was investigated by Mazza and Ronchetti [57]. $La_{14}Si_9O_{39}$ was described by Kuz'min and Belov [58] as an apatite-like structure of hexagonal symmetry (space group $P6_3/m$). Isomorphous compounds were also reported for Nd [59], Ce [60] and Sm [58]. The $La_{14}Si_9O_{39}$ compound extends its stability range in the interior of the phase diagram, forming the solid solution of the type $La_{14+1x/3}Si_{9-x}Al_xO_{39}$, which is stable from $x = 0$ to $x = 1.5$. This substitution stoichiometry (Al + 1/3La ↔ Si) can be described as a tetrahedral Al for Si substitution on the *6h* position and contemporary occupation of vacant La sites [57].

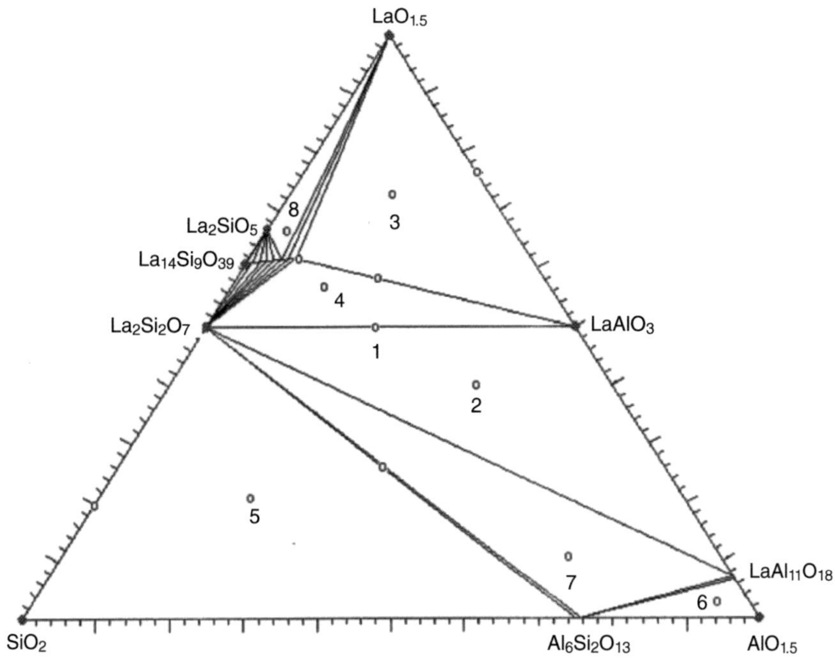

Fig. 7. Ternary phase diagram at 1300°C in air [57].

[14] The densification rate is considered as the function of temperature (*T*) and mean grain size (D_m). Constant ***A*** depends on the surface energy (γ_{sg}) of grains, on the apatite molar volume (Ω) and on average coefficient of diffusion of limiting species ***D***. This relationship can be written as follows [56]:

$$\text{(a)}\ \frac{d\rho}{dt} = \frac{AD}{T D_m^n}$$

The coefficient of diffusion ***D*** is thermally activated:

$$\text{(b)}\ D = D_0 \exp\left(-\frac{E_a}{RT}\right)$$

where D_0 is the pre-exponential coefficient of diffusion, ***R*** is the universal gas constant and E_a is the apparent activation energy of diffusion of the rate limiting process. Exponent ***n*** in **Eq. (a)** depends on the mechanism of transport of the limiting species governing the kinetics of densification.

5.2 Apatite-type lanthanium germanates

Lanthanum germanate and silicate apatite-based materials, both undoped and with partial substitution of, for example, Al, B instead of Si and Sr in place of La, are promising oxide ion conductors with potential applications as high-temperature solid electrolytes. Considerable uncertainties remain over the stoichiometry, the defect structure and the conductivity variations within various apatite systems, partly caused by the fact that the La:(Ge, Si) ratio is variable, giving rise to the solid solutions in the undoped systems as well as to the solid solutions formed by partial replacement of La and/or (Si, Ge) together with, depending on the solid solution mechanism, variations in oxygen content [24].

The preparation of single crystal of apatite-type lanthanium germanate of the composition of $La_{9.33}Ge_6O_{26}$ was reported by NAKAYAMA and SAKAMOTO [61]. The mixtures of La_2O_3 and GeO_2 were well mixed in ethanol under an atomic ratio of La:Ge = 9.33:6 using a ball mill. The mixture was dried and then calcined in air at 1000°C for 2 h. The resulting $La_{9.33}Ge_6O_{26}$ powders were further ball milled into finer powders. After the pre-sintering, the closely packed $La_{9.33}Ge_6O_{26}$ powders were heated at 1300°C for 2 h in air, and the surface of the specimen was mirror polished. The polished surface of polycrystalline $La_{9.33}Ge_6O_{26}$ ceramics was then bonded to a <001> face of $La_{9.33}Ge_6O_{26}$ seed crystal prepared by the CZOCHRALSKY (**Section 4.2**) method. On heating of the bonded sample at 1525 – 1550°C, continuous grain growth of polycrystalline $La_{9.33}Ge_6O_{26}$ occurred and the single crystal was gradually grown from the seed crystal into the polycrystalline region [61].

Apatite-type lanthanium germanate possesses hexagonal structure with the space group $P6_3/m$ and the lattice parameters: a = 9.9256 and c = 7.2900 Å, V = 621.97 Å^3 and **Z** = 2. The calculated density of the phase is 2.148 g·cm^{-3}. Similar to apatite-type lanthanium silicate ($La_{9.33}Si_6O_{26}$) described above, the structure of $La_{9.33}Ge_6O_{26}$ (**Fig. 8**) contains two different sites for atoms of La. The La(1) and La(2) sites are located at $4f$ and $6h$, respectively. While the $La_{9.33}Ge_6O_{26}$ single crystal showed little anisotropy in conductivity, the conductivity of $La_{9.33}Si_6O_{26}$ single crystal gave 100 times higher value parallel to the c-axis than that perpendicular to the c-axis at each temperature (**Section 5.1**) [61].

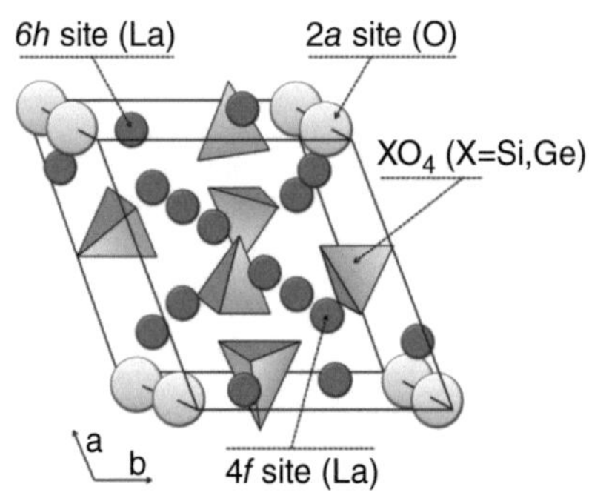

Fig. 8. Hexagonal structure proposed for the apatite-type phase of $La_{9.33}X_6O_{26}$, where X = Si and Ge [61].

The selective doping of $La_{9.33+x}(GeO_4)_6O_{2+3x/2}$ with Y leads to the stabilization of hexagonal lattice, even at high oxygen contents. Furthermore, this has the effect of enhancing the low-temperature conductivities [62]. Depending on the composition, the cell can be either hexagonal or triclinic, with the evidence of reduced low-temperature conductivities for the latter, attributed to increased defect trapping in this lower symmetry cell. In summary, it was shown that the series $La_8Y_2(GeO_4)_{6-x}(GaO_4)_xO_{3-x/2}$ can be prepared for $0 \le x \le 2$ with all samples showing the hexagonal symmetry, compared to the series without Y co-doping, $La_{10}(GeO_4)_{6-x}(GaO_4)_xO_{3-x/2}$, for which all compositions display the triclinic symmetry [24],[62], [63].

The effect of Ga doping of the oxygen stoichiometric series containing the cation vacancies, $La_{7.33+y/3}Y_2(GeO_4)_{6-y}(GaO_4)_yO_2$ $(0 \le y \le 2)$, single-phase samples was obtained for $y \ge 1.0$, with small impurities observed at lower Ga contents. The conductivities were shown to increase with increasing cation vacancy content, reaching the values of ≈0.02 S·cm^{-1} at 800°C, which are similar to the oxygen excess series. These results are in agreement with previous reports on the apatite systems, which showed that the oxide ion conductivity was maximized in samples containing the oxygen excess and/or the cation vacancies [24], [62],[63].

The series of apatite-type silicates/germanates of the composition of $La_{8+x}Sr_{2-x}Si_6O_{26+x/2}$ $(0 \le x \le 1)$ and $La_{8+x}Sr_{2-x}Ge_6O_{26+x/2}$ $(0 \le x \le 2)$ were prepared from high-purity La_2O_3, $SrCO_3$, SiO_2 and GeO_2 by ORERA et al [64] via the thermal treatment of these components mixed in the stoichiometric ratio.

The extent of, and the structural changes within, the apatite domain in the $LaO_{1.5}$-GeO_2-SrO ternary system at 1100°C was studied and the single-phase samples were obtained for $La_{9.33+x-2y/3}Sr_y(GeO_4)_6O_{2+1.5x}$ with x = 0.17 and 0.34. The hexagonal to triclinic transition is clearly associated with increasing oxygen content rather than with filling the La sites by the addition/substitution of Sr into the structure. The limits of undoped solid solution are ~$0.17 \le x \le 0.5$ at 1100°C [24].

The hydrothermal synthesis of apatite-type compound $NaRE_9(GeO_4)_6O_2$ (RE = Nd, Pr) with the hexagonal structure of the space group of $P6_3/M$ was described by EMIRDAG-EANES et al [65]. The structure is composed of REO_7 and REO_9 polyhedra as well as GeO_4 tetrahedra (**Fig. 10**). The unit cell dimensions are: a = 9.782(1) Å, c = 7.083(1)Å (T = 293 K) and V = 587.0(2)Å3 for REE = Nd and a = 9.802(1) Å, c = 7.116(1)Å (T = 293 K) and V = 592.1(2)Å3 for REE = Pr.

The high-temperature flux method for the preparation of single crystal of hexagonal $NaLa_9Ge_6O_{26}$ apatite-type germanate (space group $P6_3/M$, $\boldsymbol{a}$ = 9.883, $\boldsymbol{c}$ = 7.267 Å and $\mathbf{Z}$ = 1) was used by TAKAHASHI et al [66]. The crystal structure (**Fig. 11**) was found to be similar to that of silicate oxyapatite $NaY_9Si_6O_{26}$. The $4f$ cation sites are occupied disorderedly by La and Na. On the other hand, the $6h$ cation sites are occupied by La only. This compound constitutes a new member of the oxyapatite-type structure family with the composition given by general formula: $A_x Ln_{10-x}B_6O_{24}O_{3-x}$.

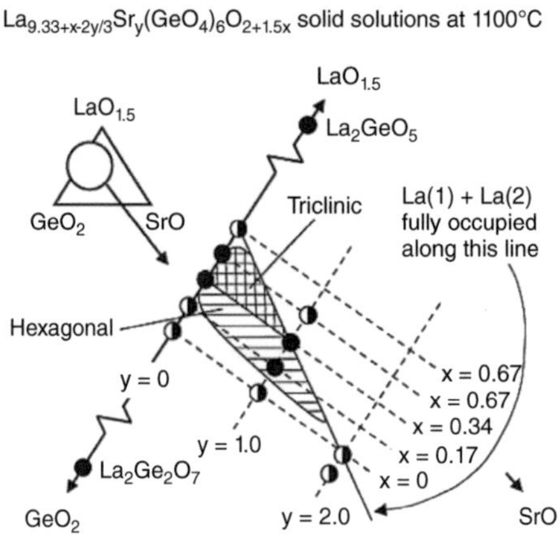

Fig. 9. Apatite solid solutions in the $LaO_{1.5}$-GeO_2-SrO ternary system [24]: pure phases are indicated by filled circles and the presence of secondary phases is shown by half-filled circles.

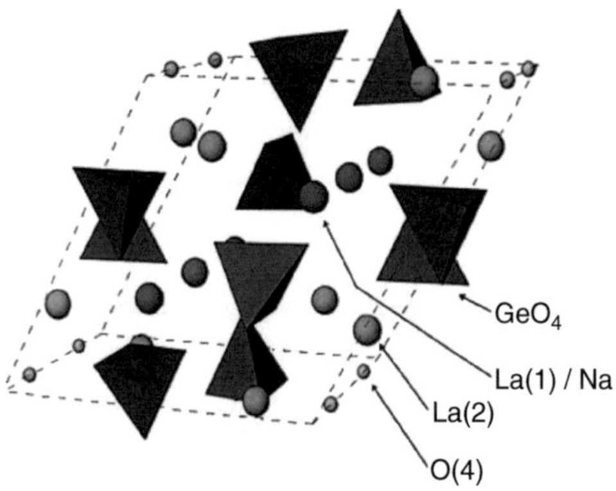

Fig. 11. The structure of sodium lanthanum germanate $NaLa_9Ge_6O_{26}$ [66].

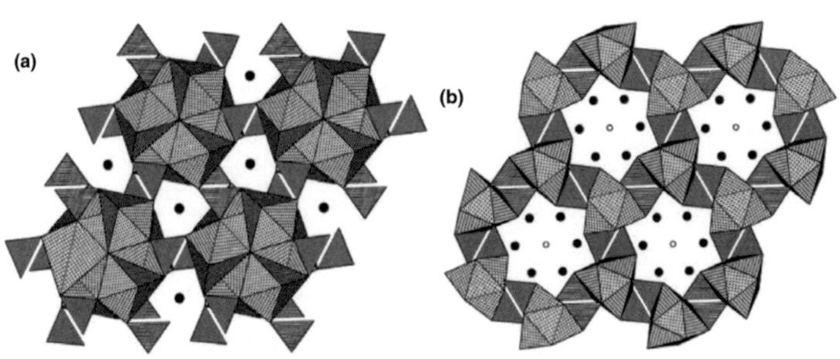

Fig. 10. The unit cell view with GeO_4 and NdO_7 (a) and GeO_4 and NdO_9 down the c-axis. NdO_7 and NdO_9 are dotted polyhedra, and GeO_4 are lined polyhedra [65].

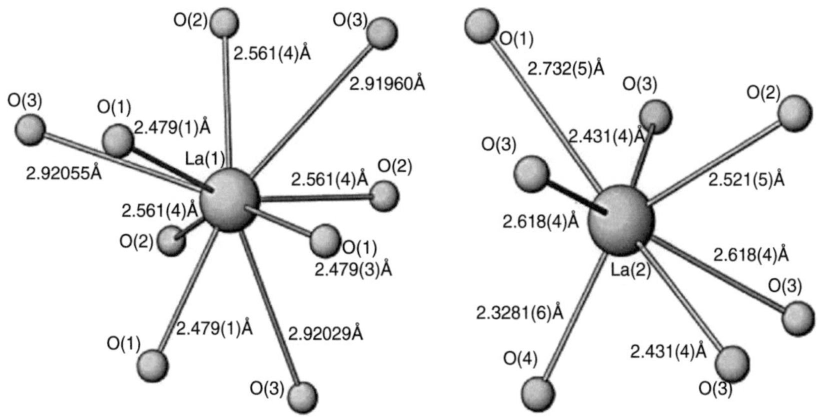

Fig. 12. Coordination environment of La atoms [66].

The coordination environments of La atoms by O atoms are shown in **Fig. 12**. La(1) atom on the 4*f* site is coordinated by nine O atoms. It is linked to three O(1) atoms in a distance of about 0.2479 nm, to three O(2) atoms in a distance of about 0.2561 nm and to three O(3) atoms in a distance of about 0.2924 nm. Because the distance between La(1) and O(3) is relatively large, La(1) atom can be also regarded to be in sixfold coordination, the environment of which is fairly distorted from an ideal octahedron. On the other hand, La(2) atom, which occupies the 6h position, is coordinated by seven O atoms, that is, O(1), O(2), O(4) and O(3). The distances of those two types of bonds between La(2) and O(3) atoms are 0.2618 nm×2 and 0.2431 nm×2, respectively. Those between La(2) and O(1), O(2) and O(4) are 0.2732, 0.2521 and 0.23281 nm, respectively [66].

5.3 Apatite-type borates

Two high terbium content apatites $Tb_5Si_2BO_{13}$ (a = 9.2569 Å, c = 6.8297 Å, **V** = 506.83 Å^3 and **Z** = 2) and $Tb_{4.66}Si_3O_{13}$ (a = 9.493 Å, c = 6.852 Å, **V** = 534.70 Å^3 and **Z** = 2) were prepared by Chen and Li [67] via spontaneous crystallization and synthesized with high purities and excellent crystallinities by the sol-gel process. Both compounds are isostructural with $P6_3/M$ space group and exhibit paramagnetic behavior down to 2 K. Owing to high Tb^{3+} ion concentrations and good transmittance in the range from 500 to 1500 nm, $Tb_5Si_2BO_{13}$ and $Tb_{4.66}Si_3O_{13}$ may be promising magneto-optical materials in the visible-near-IR range.

Both $Tb_5Si_2BO_{13}$ and $Tb_{4.66}Si_3O_{13}$ contain two distinct sites for Tb^{3+} cations, which are depicted in **Fig. 13**. Tb(2) is at the 4*f* site, which is on a threefold axis and coordinated by nine oxygen ions. However, for $Tb_{4.66}Si_3O_{13}$, the Tb(2) site is not fully occupied but leaves one of six Tb(2) positions randomly vacant while fully occupied in $Tb_5Si_2BO_{13}$. In contrast, Tb(1) at the 6*h* site is fully occupied and sevenfold coordinated in both $Tb_5Si_2BO_{13}$ and $Tb_{4.66}Si_3O_{13}$. Moreover, in $Tb_5Si_2BO_{13}$, one third of Si is disorderly occupied by B, which gives rise to extra 1/3 Tb^{3+} ion for the charge balance. The $Tb(2)O_9$ polyhedron consists of Tb(2) and nine oxygens along the c-

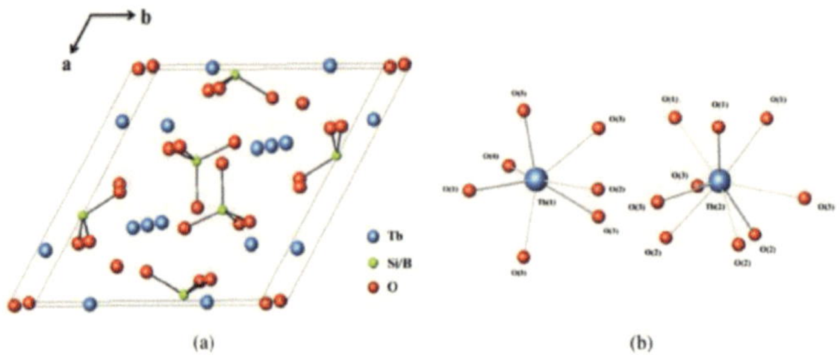

Fig. 13. Unit cell of $Tb_5Si_2BO_{13}$ (a) and coordination environment of kinds of Tb cations (b) [67].

axis. Six Tb(1) comprise a sixfold channel parallel to the c-axis. It is worth to note that the channel is considered to play an extremely important role in oxide ion conductivity [67].

The structure and optical properties of noncentrosymmetric borate $RbSr_4(BO_3)_3$ (RSBO) was described by Xia and Li [68]. RSBO can be viewed as a derivative of the apatite-like structure. Based on the anionic group approximation, the optical properties of the compound are compared to those of the structure-related apatite-like compounds with the formula "$A_5(TO_n)_3X$". When the structures of all apatite-like crystals are presented in orthorhombic unit cell, the arrangements of planar anionic BO_3 groups are all similar to one-third of the BO_3 groups aligned perfectly parallel at corner- and face-centered locations, whereas the other two-thirds of BO_3 groups are distributed differently.

Europium borate fluoride, $Eu_5(BO_3)_3F$ with an apatite-type structure, was synthesized by Kazmierczak and Höppe [69] as single-phase crystalline powder starting from europium oxide, europium fluoride and boron oxide at 1370 K. $Eu_5(BO_3)_3F$ crystallizes in the space group Pnma.

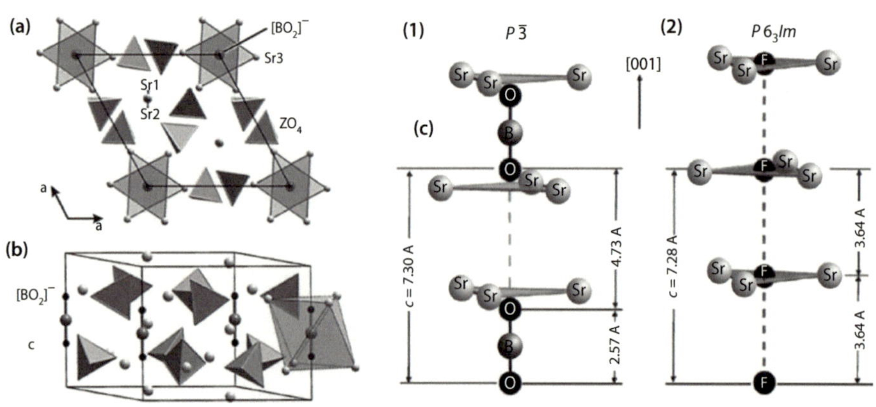

Fig. 14. The crystal structure of $Sr_{10}[(PO_4)_{5.5}(BO_4)_{0.5}](BO_2)$ (1): (a) the projection along [001] showing the channels formed by Sr3 (gray triangles) and the positions of the XO_4 tetrahedra (gray; Z = 11/12 P + 1/12B) and (b) the side view with emphasized $[BO_2]^-$ groups and the coordinating trigonal antiprism formed by Sr3. The comparison of the arrangement of $[BO_2]^-$ and F- ions within the Sr channels of $Sr_{10}(PO_4)_{5.5}(BO_4)_{0.5}(BO_2)$ (c(1)) and $Sr_{10}(PO_4)_6F_2$ (c(2)) [70].

http://dx.doi.org/10.5772/62209

The structure of single crystal of strontium phosphate orthoborate metaborate ($Sr_{10}[(PO_4)_{5.5}(BO_4)_{0.5}](BO_2)$) that was grown from the melt by CHEN et al [70] is shown in **Fig. 14**(a). The phase crystallizes in the space group P3 with the cell parameters $\boldsymbol{a}$ = 9.7973 Å, $\boldsymbol{c}$ = 7.3056 Å, $\boldsymbol{V}$ = 607.19 Å^3 and $\boldsymbol{Z}$ = 1. The crystal structure is closely related to apatite and contains linear metaborate groups, $[BO_2]^-$ (point group $D_{\infty h}$, B-O = 1.284 Å), taking positions within the channels running along the threefold inversion axis. Strontium sites are found to be fully occupied, while $[PO_4]^{3-}$ tetrahedra are partially replaced by$[BO_4]^{5-}$ groups.

The comparison of the nearest neighbors around $[BO_2]^-$ and F^- located within the channels is shown in **Fig. 14(b,c)**. F^- ions (0,0,1/4) are situated on the mirror plane in the center of Sr triangle. As a result, constant F…F distances of 3.64 Å ($\boldsymbol{a}$/2) are observed along [001] (**b**). In $Sr_{10}[(PO_4)_{5.5}(BO_4)_{0.5}](BO_2)$, the incorporation of boron atoms between two O atoms draws these atoms closer (d(O-B-O) = 2.57 Å) and at the same time increases the gaps between two neighboring $[BO_2]^-$ units (d(O…O) = 4.73 Å), which results in alternating O…O distances along the c-axis (**a**) [70].

By replacing Mn in $YCa_3(MnO)_3(BO_3)_4$ with trivalent Al and Ga, two new borates with the compositions of $YCa_3(MO)_3(BO_3)_4$ (M = Al, Ga) were prepared via the solid-state reaction by YU et al [71]. The phases are isostructural to gaudefroyite with the hexagonal space group $P6_3/M$. The cell parameters of $\boldsymbol{a}$ = 10.38775 Å, $\boldsymbol{c}$ = 5.69198Å for the Al-containing compound and $\boldsymbol{a}$ = 10:5167 Å, $\boldsymbol{c}$ = 5:8146 Å for the Ga analogue were obtained from the refinements. The structure is constituted of AlO_6 or GaO_6 octahedral chains interconnected by BO_3 groups in the $\boldsymbol{ab}$ plane to form a Kagomé-type lattice[15] [72],[73],[74], leaving trigonal and apatite-like tunnels. It was found that most rare-earth and Cr, Mn ions can be substituted into the Y^{3+} and M^{3+} sites, respectively, and the preference of rare-earth ions to be located in the trigonal tunnel is correlated to the sizes of the M^{3+} ions.

5.4 Other apatite-type REE silicates

Hexagonal apatite-type phase of the composition of $Pr_9K(SiO_4)_6O_2$ (space group $P6_3/M$, a = 9.6466 Å and c = 1136 Å, V = 573.28 Å^3, ρ_{calc} = 5.48 g·cm^{-3} and Z = 1) was synthesized by WERNER AND KUBEL [75] in a potassium fluoride flux. Potassium fills one (4f) of two metal positions present in the structure (**Fig. 15**) with the occupancy factor of 25%. The remaining positions of this site (Pr2/K2) are occupied by praseodymium.

[15] The Kagomé lattice (**d**) is one of the most interesting lattices in 2D, especially in materials in which the Kagome lattice is built from magnetic ions. Each of its vertices touches a triangle, hexagon, triangle and hexagon (the planes of corner-sharing equilateral triangles). The vertices correspond to the edges of the hexagonal (honeycomb) lattice (**c**), which in turn is the dual of triangular lattice (can be derived from triangular lattice by periodical removal of ¼ sites) (**b**). Since it has the same coordination number (z = 4), the Kagomé lattice is also related to the square lattice (**a**) [72], [73]. Numerous Kagomé compounds built from stacked Kagomé layers were found in Alunite (Jarosite, $KFe^{3+}_3(SO_4)_2(OH)_6$, (**e**)) family of minerals [74].

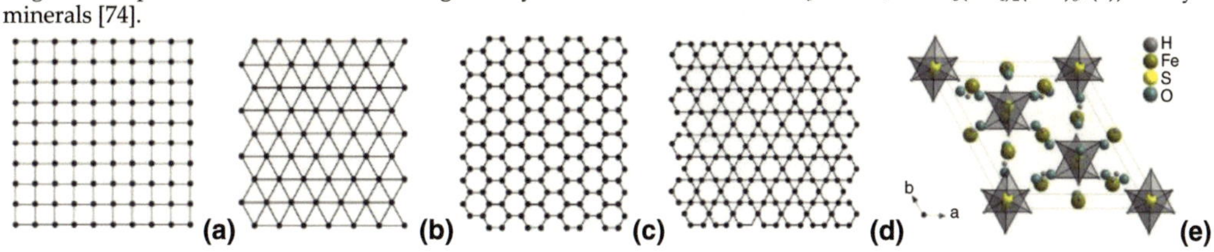

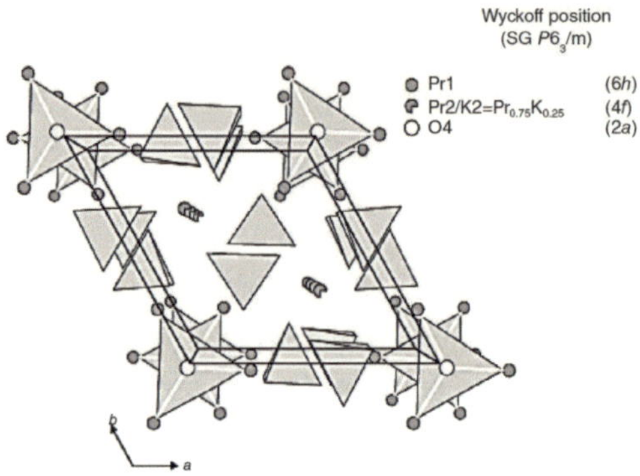

Fig. 15. The perspective view of $Pr_9K(SiO_4)_6O_2$ along the c-axis: regular planar coordination of the channel oxygen O(4) and the SiO_4 tetrahedra are marked in gray, and the unit cell is outlined [75].

Oxygen from the silicate groups forms a coordination polyhedron (ninefold) in the shape of a distorted threefold capped trigonal prism. These face sharing [(Pr2/K2)O_9]-polyhedra build up chains, which are interconnected via the SiO_4 groups. The resulting channel framework accommodates sevenfold oxygen-coordinated praseodymium (Pr1), attached to the inside of the tubes that are aligned parallel to the c-axis. Oxide ions O4, located on the longitudinal axis of the channels, exhibit anomalously high atomic displacement parameters along the c-direction [75].

Single crystals of apatite-type $Nd_{9.33}(SiO_4)_6O_2$, $Pr_{9.33}(SiO_4)_6O_2$ and $Sm_{9.33}(SiO_4)_6O_2$ were described in **Section 4.2.2**. The structure of samarium orthosilicate oxyapatite ($Sm_5(SiO_4)_3O$, **Fig. 16**) was resolved by Morgan et al [76]. The phase crystallizes in hexagonal system with the space group $P6_3/M$ and the cell parameters: $\boldsymbol{a}$ = 9.4959 Å, $\boldsymbol{c}$ = 7.0361 Å, $\boldsymbol{c}:\boldsymbol{a}$ = 0.7410 and $\boldsymbol{V}$ = 549.46 Å^3. The preparation and the structure of single crystal of strontium tetrapraseodymium tris(silicate) oxide ($SrPr_4(SiO_4)_3O$), which was grown by the self-flux method using $SrCl_2$, was described by Sakakura et al [77].

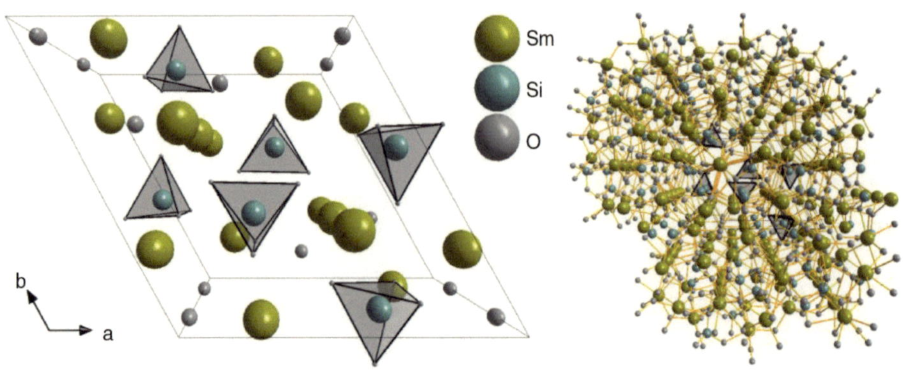

Fig. 16. The structure of $Sm_5(SiO_4)_3O$ viewed along the c-axis [76].

http://dx.doi.org/10.5772/62209

The M(2) sites are almost exclusively occupied by praseodymium. The complete series of apatite-like compounds $REE_{9.33}\square_{0.67}[SiO_4]_6O_2$, $LiREE_9[SiO_4]_6O_2$ and $NaREE_9[SiO_4]_6O_2$ were synthesized by Felsche [78] with REE: La → Lu. Apatite-type neodymium silicates doped with various cations at the Si site, $Nd_{10}Si_5BO_{27-\delta}$ (B = Mg, Al, Fe, Si), were synthesized by Xiang et al [79] via the solid-state reaction.

The crystal growth and the structure of three new neodymium-containing silicates, $Na_{0.50}Nd_{4.50}(SiO_4)_3O$, $Na_{0.63}Nd_{4.37}(SiO_4)_3O_{0.74}F_{0.26}$ and $Na_{4.74}Nd_{4.26}(O_{0.52}F_{0.48})[SiO_4]_4$, prepared using the eutectic mixture of KF/NaF were investigated by Latshaw et al [80]. $Na_{0.50}Nd_{4.50}(SiO_4)_3O$ and $Na_{0.63}Nd_{4.37}(SiO_4)_3O_{0.74}F_{0.26}$ adopt the apatite structure and crystallize in hexagonal space group $P6_3/m$, while $Na_{4.74}Nd_{4.26}(O_{0.52}F_{0.48})[SiO_4]_4$ crystallizes in tetragonal space group I-4 and exhibits rare-earth mixing on the sodium site. The unit cell parameters of the crystals are:

1. $Na_{0.50}Nd_{4.50}(SiO_4)_3O$: a = 9.5400 Å and c = 7.033 Å;

2. $Na_{0.63}Nd_{4.37}(SiO_4)_3O_{0.74}F_{0.26}$: a = 9.5533 Å and c = 7.0510 Å;

3. $Na_{4.74}Nd_{4.26}(O_{0.52}F_{0.48})[SiO_4]_4$: a = 12.1255 Å and c = 5.4656 Å.

Double REE silicate $Gd_{4.33}Ho_{4.33}(SiO_4)_6(OH)_2$ with the hydroxylapatite structure was synthesized by Wang et al [81] using the piston-cylinder high-pressure apparatus at the pressure of 2.0 GPa and the temperature of 1450°C. Since they have nearly identical chemical character, two REE cations (Ho and Gd) are distributed randomly among the M(1) and M(2) sites, and the charge balance is maintained by the cation vacancies in M(1). The presence of two different REE cations in the same compound might promote better understanding of the cooperative effects of ions under the solid-state conditions (**Fig. 17**).

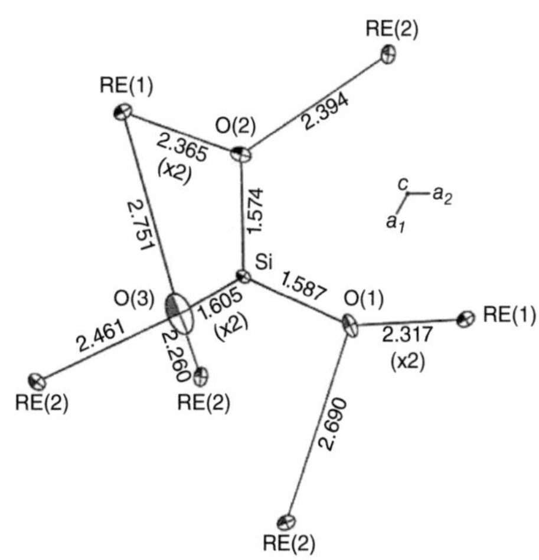

Fig. 17. Bond distances (Å) and anisotropic displacement in $Gd_{4.33}Ho_{4.33}(SiO_4)_6(OH)_2$: note that the exaggerated anisotropic displacement of O(3) is attributable to the high proportion of vacancies in REE(1) and near-equatorial distribution of strong bonds to Si and REE(2) [81].

The structure analysis reveals that the hexagonal compound crystallizes in usual apatite space group $P6_3/M$ with lattice parameters $\boldsymbol{a}$ = 9.3142 Å and $\boldsymbol{c}$ = 6.7010 Å. REE(1) atoms are connected to nine oxygen atoms with the REE(1)-O bond distances ranging from 2.317 to 2.751 Å (mean 2.478 Å) and REE(2) atoms are connected to seven oxygen atoms with the RE(2)-O bond distances ranging from 2.223 to 2.690 Å (mean 2.393 Å). Oxygen anion O(4) in the apatite channel is located on the 6_3 axis and coordinated with three REE(2) cations arranged in a tricluster perpendicular to the c-axis. An isotropic displacement parameter was used for O(4), and H atom was assumed to ride on it. OH^- anions are stacked in regular column in the apatite channel, and in locally ordered structure, their polar direction is flipped in neighboring channel [81].

The formation of apatite-type phases of the composition of $KNd_6(SiO_4)_6O_2$ from the glass precursor ($4K_2O$-Nd_2O_3-$17SiO_2$) during the hydrothermal experiments (500°C and 825 bar) carried out at KOH molarities of 6 or greater was reported by Haile et al [82]. High temperatures, high pressures and long times tended to favor the synthesis of this apatite-type phase over the $K_8Nd_3Si_{12}O_{32}OH$ phase. In comparison with the potassium system, the concentration of NaOH required for the synthesis of $NaNd_6(SiO_4)_6O_2$ phase (system $4Na_2O$-M_2O_3-$17SiO_2$, where M = Nd and Y) is very low. The formation of apatite-type phases in the $4Na_2O$-Y_2O_3-$17SiO_2$ system was not observed [83].

Two series of strontium-lanthanum apatites, $Sr_{10-x}La_x(PO_4)_{6-x}(SiO_4)_xF_2$ and $Sr_{10-x}La_x(PO_4)_{6-x}(SiO_4)_xO$ with $0 \leq x \leq 6$, were synthesized by Boughzala et al [84] via the solid-state reaction in the temperature range of 1200 – 1400°C:

$$3\ SrCO_3 + (6-x)/2\ Sr_2P_2O_7 + SrF_2 + x\ SiO_2 + x\ La(OH)_3 \rightarrow \\ Sr_{10-x}La_x(PO_4)_{6-x}(SiO_4)_x F_2 + 3\ CO_2 + 3x/2\ H_2O \tag{9}$$

and

$$4\ SrCO_3 + (6-x)/2\ Sr_2P_2O_7 + x\ SiO_2 + x\ La(OH)_3 \rightarrow \\ Sr_{10-x}La_x(PO_4)_{6-x}(SiO_4)_x O + 4\ CO_2 + 3x/2\ H_2O \tag{10}$$

where x = 0, 1, 2, 4 and 6. $Sr_2P_2O_7$ was synthesized by the following reaction at 900°C:

$$2\ SrCO_3 + 2\ (NH_4)_2HPO_4 \rightarrow Sr_2P_2O_7 + 2\ CO_2 + 3\ H_2O + 4\ NH_3 \tag{11}$$

The raw meal was prepared via mixing $SrCO_3$, La_2O_3, SiO_2, SrF_2 and $(NH_4)_2HPO_4$ in required stoichiometric amounts ($0 \leq x \leq 6$). The mixture was ground in an agate mortar, pressed to pellets and calcined at the temperature of 900°C for 12 h under the flow of argon ($Sr_{10-x}La_x(PO_4)_{6-x}(SiO_4)_xF_2$) and oxygen ($Sr_{10-x}La_x(PO_4)_{6-x}(SiO_4)_xO$). The product was ground and pressed again in order to improve its homogeneity. Next, thermal treatment was performed

http://dx.doi.org/10.5772/62209

at the temperature of 1200 and 1400°C (depending on the content of SiO_2) for 12 h. The samples were heated and cooled with the rate of 10°C·min^{-1}. The incorporation of La^{3+} and $SiO_4{}^{4-}$ ions into the apatite structures, i.e. the substitution of the pair La^{3+} and $SiO_4{}^{4-}$ for Sr^{2+} and $PO_4{}^{3-}$, induced an increase of parameter ***a*** and decrease of parameter ***c*** (**Fig. 18**) [84].

The formation of nanocrystalline Ce-Yb mixed silicate-type oxyapatite of the composition of $Yb_yCe_{9.33-y}(SiO_4)_6O_2$ via the solid-state synthesis was described by MAŁECKA and KĘPIŃSKI [85]. The phase was identified as an intermediate formed during the synthesis of Ce-Yb silicates.

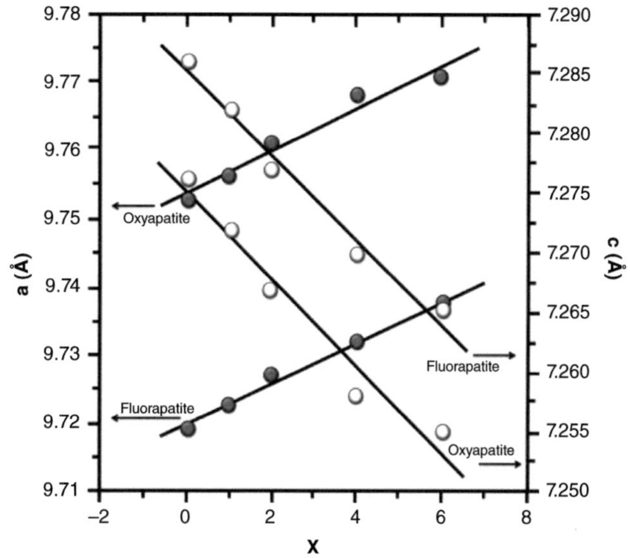

Fig. 18. The variation of *a* and *c* parameters of $Sr_{10-x}La_x(PO_4)_{6-x}(SiO_4)_xF_2$ and $Sr_{10-x}La_x(PO_4)_{6-x}(SiO_4)_xO$ phases with the value of *x* [84].

Different compositions of apatite-type $La_{10}Si_{6-x}W_xO_{27+\delta}$ ceramics were prepared successfully by XIANG et al [86] via the high-temperature solid-state reaction route. Doping with W^{6+} is beneficial to the removal of La_2SiO_5 impurity phase. When the doping content of W^{6+} is more than 0.1, the rod-like grains of $La_{10}Si_{6-x}W_xO_{27+\delta}$ ceramics are replaced gradually by equiaxed apatite-type grains, and randomly shaped convex $La_6W_2O_{15}$ particles appear at the grain boundaries. While doping with Nb^{5+} leads to the hexagonal-phase $La_{9.5}Ge_{5.5}Nb_{0.5}O_{26.5}$, the addition of MO^{6+} leads to the compound $La_{9.5}Ge_{5.5}Mo_{0.5}O_{26.75}$ with triclinic symmetry [87].

5.5 Apatite-type yttrium silicates

5.5.1 Yttrium silicates

The formation of the phase with the composition ($Y_4Si_3O_{12}$, $Y_4(SiO_4)_3$ or $2Y_2O_3{\cdot}3SiO_2$ [88]), which is stable between 1650 and 1950°C, was reported by TOROPOV and BONDAR [89] in the binary system of Y_2O_3-SiO_2 and by TOPOROV and FEDOROV [90] in the ternary system of CaO-Y_2O_3-SiO_2

(**Fig. 19(a)**). This was the first reported occurrence of the phase with the composition between yttrium orthosilicate (Y_2SiO_5, oxyorthodsilicate, YSO, Y_2O_3:SiO_2 = 1:1 [91]) and yttrium disilicate $Y_2Si_2O_7$ (yttrium pyrosilicate, YPS, 1:3). The structure of this phase was described as the garnet type [88].

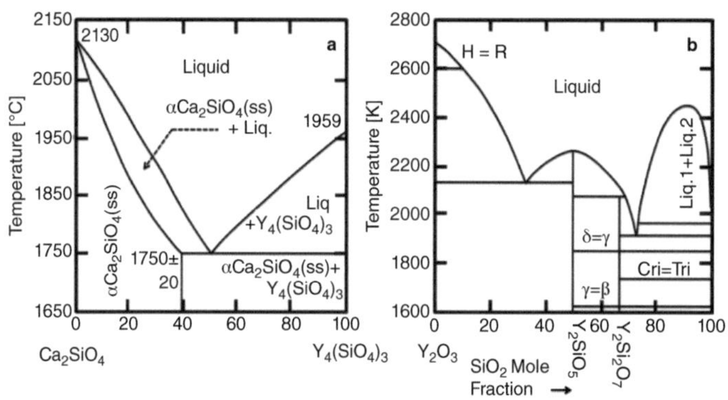

Fig. 19. Phase equilibrium in the system Ca_2SiO_4-$Y_4(SiO_4)_3$ according to Toporov and Fedorov [90] and Y_2O_3-SiO_2 phase diagram [92].

Since then, authors have disagreed about the existence of such a phase because the attempts to make it starting with yttria and silica powders resulted in the formation of only Y_2SiO_5 and $Y_2Si_2O_7$ [93]. This phase was not reported either by other studies of Y_2O_3-SiO_2 system [92],[94],[95], which contains two compounds. Y_2SiO_5 and$Y_2Si_2O_7$ were found, with two (A and B) and five (y, α, β, γ and δ, also called y, B, C, D and E [96]) polymorphs, respectively. The first has a congruent melting, whereas the second has an incongruent one (**Fig. 19(b)**).

Nevertheless, the formation of oxyapatite phase of the composition of $Y_{4.67}\square_{0.33}(SiO_4)_3O$ (7:9) prepared via the oxidation of nitrogen apatite $Y_5(SiO_4)_3N$ was reported by other authors [7],[97]. The apatite-like phase $Y_{4.67}(SiO_4)_3O$ possesses hexagonal structure with the space group P6$_3$/m, a = 9.368 and c = 6.735 Å [78],[98]. The specific gravity of the phase is 4.39 g·cm^{-3} and the hardness on the Mohs scale is 5 – 7 [99].

Since the structure of YSO containing two different types of anions includes the $[SiO_4]^{4-}$ complex ion and an additional non-silicon-bonded oxygen ion (NBO), it could be written as $Y_2(SiO_4)O$. This compound also displays two structure types of monoclinic symmetry with different linking of O-Y_4 tetrahedra. Low-temperature X_1 phase and high-temperature X_2 phase belong to the space groups of P2$_1$/c (Z = 2) and C2/c (Z = 8), respectively [91].

The samples of the composition of $Y_4(SiO_4)_3$, and similar ones containing small amount of iron oxide, corresponding to an overall composition of $Fe_{0.2}Y_4(SiO_4)_3O_{0.2}$, were produced by the mixed powder method and by the sol-gel route using yttrium nitrate ($Y(NO_3)_3 \cdot 5H_2O$), TEOS (tetraethylorthosilicate) and iron nitrate ($Fe(NO_3)_3 \cdot 9H_2O$) by Parmentier et al [7]. Nitrate was dissolved in ethanol/water mixture (volume ratio 7:3), the amount of the latter being controlled to give final Si concentration. Iron nitrate was added at this stage in calculated amounts corresponding to the final iron-doped apatite composition. The solution was stirred for a few hours and TEOS was added to give the appropriate silicon content and then the solution was placed in an oven at 60°C until the gelation occurred. The gel was dried at 80°C and calcined at 600°C for 1 h.

Powders prepared by the two routes were uniaxially pressed into pellets and treated to temperatures up to 1650°C in air in a Pt crucible, or for the heat treatments at 1700°C, carbon element furnace was used, and the samples were heated in a BN-lined crucible in nitrogen atmosphere. Iron appears to have two roles depending on the temperature; it stabilizes the apatite phase at high temperatures when produced by the sol-gel route and catalyzes the decomposition of sol-gel-derived apatite at low temperatures [7].

A new phase of yttrium magnesium silicate having the apatite structure was prepared by Suwa et al [100] at 1500°C in air. Its chemical composition can vary from $(Y_4Mg)Si_3O_{13}$ to $(Y_{4.33}Mg_{1.13})Si_{3.34}O_{13}$. The hexagonal unit cell dimensions $\boldsymbol{a}_0$ and c_0 of $(Y_4Mg)Si_3O_{13}$ are 9.298 ± 0.002A and 6.635 ± 0.001A, respectively, and its axial ratio c/a is 0.714. It is optically uniaxial negative with $\varepsilon = 1.810 \pm 0.005$, $\omega = 1.820 \pm 0.005$ and $\omega - \varepsilon = 0.010$. The cleavages parallel and perpendicular to the c-axis were recognized. The formation of the apatite-type phase of the composition of $NaY_9Si_6O_{26}$ in the ternary system Na_2O-Y_2O_3-SiO_2 was also reported by Lee et al [101].

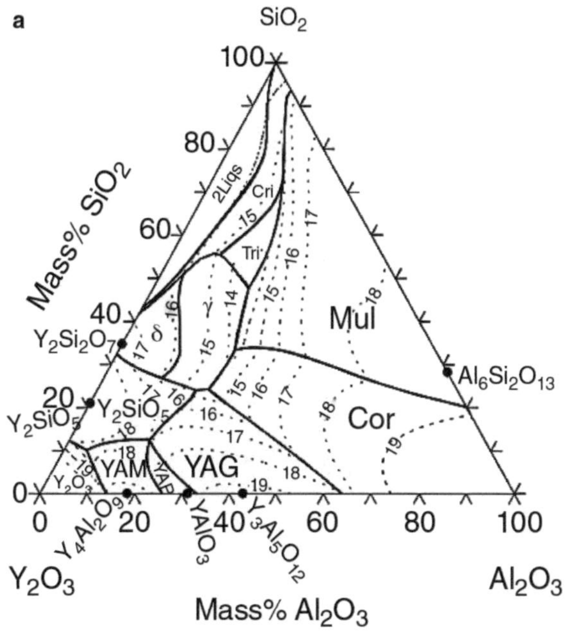

Fig. 20. Calculated liquidus surface of the Y_2O_3-Al_2O_3-SiO_2 system: three-phase equilibria with liquid phase (thick lines), liquidus surfaces for various solids (labeled area) and isothermal section (dotted lines, temperature in hundreds °C) [92].

The phase diagram of Al_2O_3-SiO_2-La_2O_3 system (**Fig. 20**) can be compared with the Y_2O_3-Al_2O_3-SiO_2 ternary diagram examined by Bondar and Galakhov in 1964 [102]. The latter represents the only other example of REE_2O_3-Al_2O_3-SiO_2 phase diagram found so far in the literature. The authors identified the liquidus surface of the whole ternary field, but they failed to elucidate the subsolidus phase relationships among different binary compounds. Due to the much smaller ionic size of Y^{3+} ion with respect to La^{3+} ion (1.18 and 1.015 Å for the eightfold coordination, respectively [103]) and lower bond-valence parameter (2.019 and 2.172 Å [104]), the stability of the binary compositions is substantially altered [57].

The β-alumina-like phase $LaAl_{11}O_{18}$ is no longer stable, while the garnet-like phase $Y_3Al_5O_{18}$ and $Y_4Al_2O_9$ monoclinic compound exist. The lacunar apatite-like phase $Y_{14}Si_9O_{39}$ reported by Wills et al [105] does not appear in the Y_2O_3-Al_2O_3-SiO_2 ternary diagram; however, a compound with similar Y/Si atomic ratio, namely $Y_4Si_3O_{12}$, also reported by Wills et al [105], does. Since the formation of $Y_4Si_3O_{12}$ phase was not confirmed, it may be stabilized by impurities [57],[92].

5.5.2 AM and AEE-yttrium orthosilicate oxyapatites

Alkaline metals (AM) and alkaline-earth element oxyapatites (oxybritholites) are described in this chapter. Phosphate minerals of the apatite supergroup possess strong affinity for strontium [106]. The apatite-type phase of the composition of $Sr_2Y_8Si_6O_{26}$ ($Sr_2Y_8(SiO_4)_6O_2$) was prepared by Zuev et al [107] via the two-stage calcination of mixture of $SrCO_3$, Y_2O_3, Eu_2O_3 and SiO_2 in air in order to investigate the spectral characteristics of $Sr_2Y_8(SiO_4)_6O_2$:Eu polycrystals. The structure of apatite phase is shown in **Fig. 21**. The sol-gel synthesis and the characterization of $Sr_2Y_{8(1-x)}Eu_{8x}Si_6O_{26}$ solid solution doped by Eu, where $x = 0.01 - 0.4$, was described by Karpov and Zuev [108]. The formation of calcium analogue ($Ca_2Y_8(SiO_4)_6O_2$, calcium-yttrium-silicate oxyapatite) was observed during the crystallization of SiO_2-Al_2O_3-CaO-Na_2O-K_2O-F-Y_2O_3 glass ceramics [109] and during the degradation of advanced environmental barrier coatings [110]. The phase $Ca_4Y_6(SiO_4)_6O_2$ was often prepared in order to investigate its luminescent properties [111],[112].

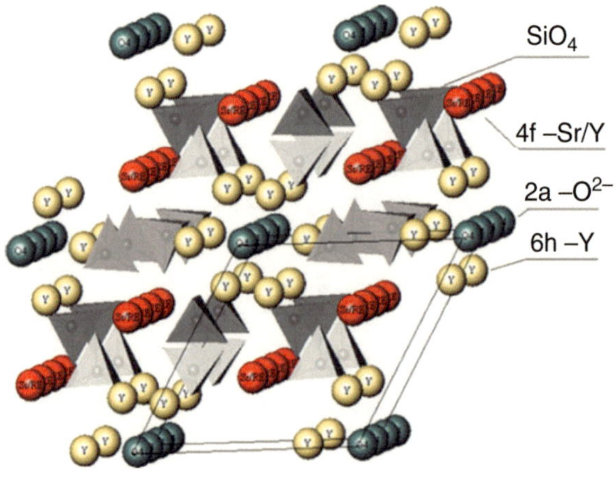

Fig. 21. The structure of $Sr_2Y_8(SiO_4)_6O_2$ [107].

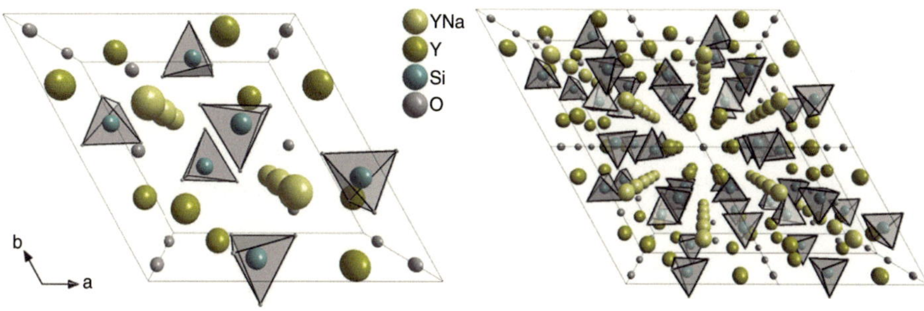

Fig. 22. The structure of $NaY_9(SiO_4)_6O_2$ oxyapatite (the perspective view along the c-axis) [114].

The precipitation of $NaY_9(SiO_4)_6O_2$ apatite-type compound (sodium nonayttrium hexakis(silicate) dioxide) in the SiO_2-B_2O_3-Al_2O_3-Y_2O_3-CaO-Na_2O-K_2O-F glass-ceramics system (**Section 10.3.8**) was described by van't Hoen et al [113]. The hexagonal structure of $NaY_9(SiO_4)_6O_2$ oxyapatite (**Fig. 22**) was resolved by Gunawardane et al [114]. The phase crystallizes in $P6_3/M$ space group with the cell parameters $\boldsymbol{a} = 9.334$, $\boldsymbol{c} = 6.759$ Å, $\boldsymbol{c}:\boldsymbol{a} = 0.7241$, $V = 509.97$ Å and $\mathbf{Z} = 1$).

Lithium yttrium orthosilicate ($LiY_9(SiO_4)_6O_2$, lithium nonayttrium hexakis(silicate) dioxide) crystallizes in centrosymmetric space group $P6_3/M$. The structure closely resembles those of fluorine apatite. There are two different crystallographic sites for Y^{3+} ion, which are coordinated by seven and nine O atoms. One-fourth of the nine-coordinated site is occupied by Li atoms, thus maintaining the charge balance. Si atom occupies the tetrahedral site [115].

The preparation, the properties and the effect of sintering additives of hexagonal ($P6_3/M$) strontium-yttrate-silicate oxyapatite (oxybritholite2) with the composition of $SrY_4(SiO_4)_3O$ as the main product of sinter-crystallization process, in which the non-equilibrium melt was formed in the temperature interval from 1300 to 1550°C in the SrO-Y_2O_3-SiO_2 system, was described by Ptáček et al [116]. The formation of non-equilibrium melt is facilitated by borate fluxes, alkaline fluxes and talc. The apparent activation energy and the frequency factor of the sinter-crystallization process were determined to be 1525 kJ mol^{-1} and $1.04 \cdot 10^{45}$ s^{-1}, respectively. The material shows low value of linear thermal expansion coefficient of $(1.1 \pm 0.1) \cdot 10^{-6}$°$C^{-1}$ in the temperature range from 25 to 850°C.

The course of synthesis can be expressed by the following reaction formula [116]:

$$AEECO_3 + 2\ Y_2O_3 + 3\ SiO_2 \rightarrow AEE\ Y_4[SiO_4]_3O + CO_2(g) \tag{12}$$

This **reaction 12** is too general to describe formed intermediates ($SrSiO_3$, Sr_2SiO_4, SrY_2O_4, ...[16]) and the process of sinter-crystallization of apatite:

[16] Detailed description of formed intermediates can be found in work [116].

$$AEECO_3 + 2\ Y_2O_3 + 3\ SiO_2 \rightarrow \text{intermediates} \rightarrow \\ \text{non-equilibrium melt} \rightarrow AEE\ Y_4[SiO_4]_3O + CO_2(g) \quad (13)$$

Since the formation of $SrY_4(SiO_4)_3O$ proceeds thorough non-equilibrium melt phase, the effect of sintering additives such as borate fluxes, fluorides and carbonates of alkaline metals as well as talc was investigated. Sintering additives facilitate the formation of melt phase and increase the length of sinter-crystallization interval. The expansion after the thermal decomposition of strontium carbonate is reduced as well. Calcinate, treated to the temperature lower than the temperatures of sinter-crystallization interval, has hydraulic activity. Therefore, it can be applied in special composite cements as an activator for latent hydraulic and pozzolanic materials.

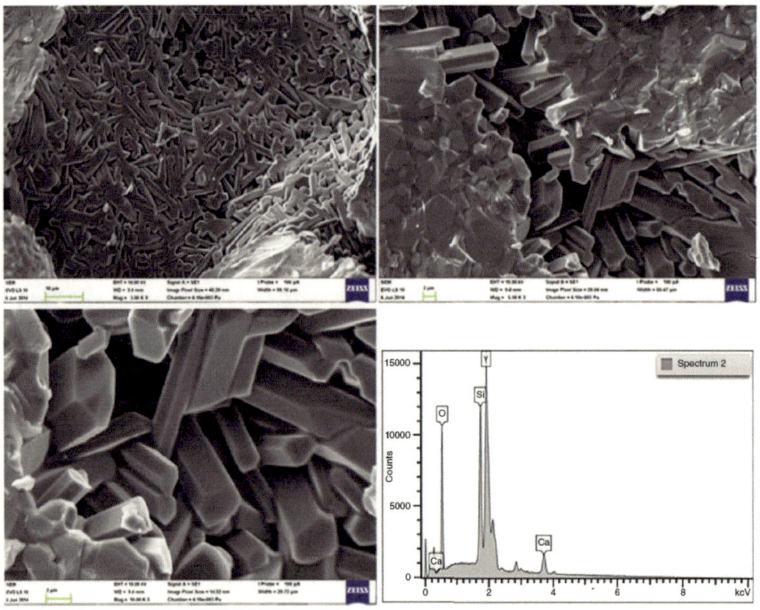

Fig. 23. SEM picture and WDX analysis of hexagonal crystal of $CaY_4(SiO_4)_3O$ apatite phase.

Effect on sinter-crystallization process	Decreasing intensity or temperature of effect →
Expansion before sinter-crystallization	Pure >> NaF >> Talc ≈ Li_2CO_3 ≈ $Li_2B_4O_7$ >> Na_2CO_3 > $LiBO_2$ > K_2CO_3
Firing shrinkage (sample treated to 1600°C)	$LiBO_2$ > $Li_2B_4O_7$ ≈ Pure > NaF > K_2CO_3 ≈ Na_2CO_3 > Talc > Li_2CO_3
Initial temperature of sinter-crystallization	Pure ≈ Na_2CO_3 ≈ Talc > $Li_2B_4O_7$ > K_2CO_3 ≈ NaF > $LiBO_2$ > Li_2CO_3
Maximum rate of sinter-crystallization	Li_2CO_3 > Pure > Li_2B4O_7 ≈ Talc ≈ NaF ≈ K_2CO_3 ≈ Na_2CO_3 > $LiBO_2$
Length of interval of sinter-crystallization	$LiBO_2$ >> $Li_2B_4O_7$ ≈ NaF > Li_2CO_3 > pure > Talc > Na_2CO_3 > K_2CO_3

Table 3. The influence of sintering additive on the behavior during thermal treatment [116].

After the process of sinter-crystallization, the reactivity of glassy phase with water drops. A significant benefit of talc is the fact that the glassy phase surrounding the crystals of apatite phase becomes resistant against the influence of water with this sintering additive. Furthermore, magnesium is not being incorporated into the structure of apatite phase during the crystallization of $SrY_4(SiO_4)_3O$ from non-equilibrium melt. The influence of sintering additives on the behavior during the thermal treatment is summarized in **Table 3** [116].

The important feature of this compound is the formation of colored center after the exposition to X-ray radiation (**Fig. 24**); hence, the prepared material is an important candidate for optical applications, sensors and dosimeters.

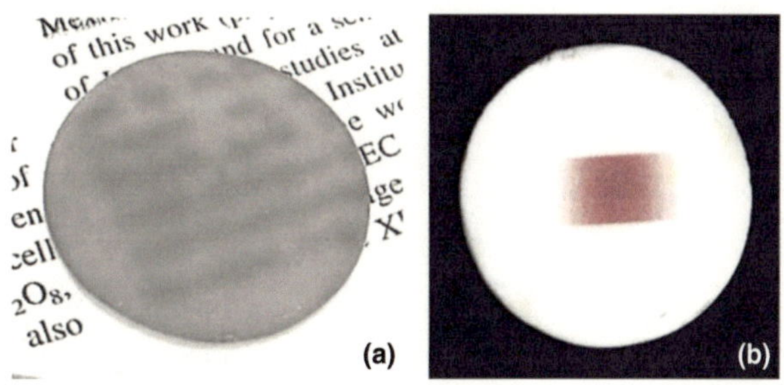

Fig. 24. Semitransparent disc of $SrY_4(SiO_3)_4O$ sintered specimen (a) and formation of colored center on X-ray irradiated area (Cu(Ka), 40 kV and 30 mA) [116].

On the other hand, this reaction also indicates that the synthesis of individual apatite analogues ($AEEY_4(SiO_4)_3O$, where AEE = Ca, Sr and Ba) and their solid solutions proceeds via similar ical intermediates formed in the temperature range, which is affected by the thermal stability of AEE carbonates that increases in the order: $CaCO_3$, $SrCO_3$ and $BaCO_3$.

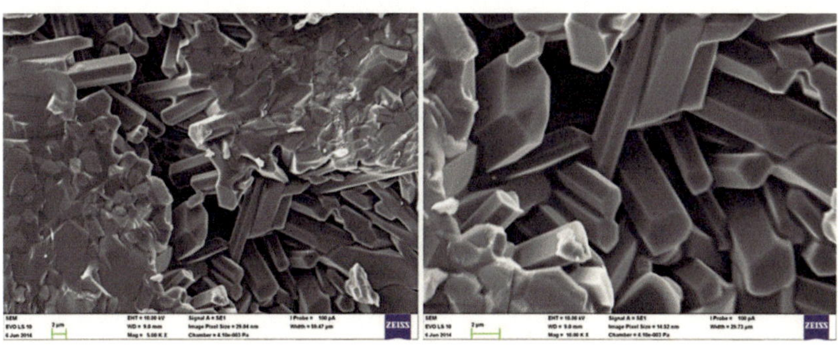

Fig. 25. SEM picture of hexagonal crystal of $CaY_4[SiO_4]_3O$ apatite phase.

While the synthesis of $CaY_4(SiO_4)_3O$ leads to well-developed hexagonal crystals (**Fig. 25**), the attempts for the preparation of $BaY_4(SiO_4)_3O$ phase were not successful. This synthesis leads to well-developed crystals of yttrium orthosilicate (Y_2SiO_5) surrounded by BaO-Y_2O_3-SiO_2 glassy phase (**Fig. 26**).

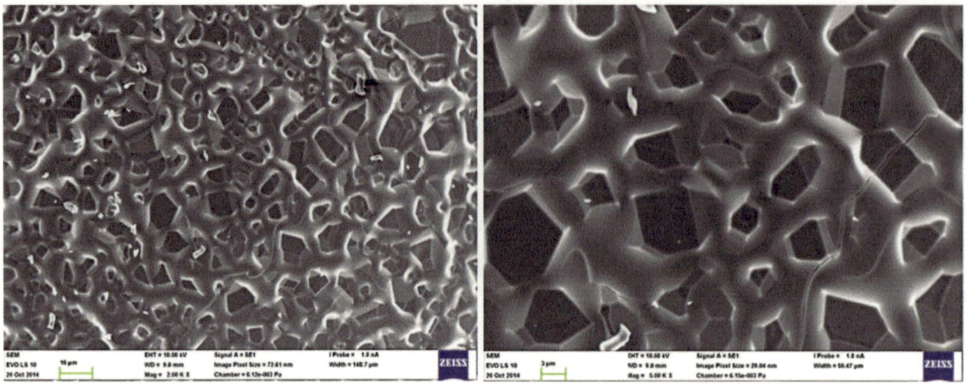

Fig. 26. SEM of Y_2SiO_5 crystal surrounded by barium-containing glassy phase.

The investigation of this system leads to the conclusion that non-limited $Ca^{2+} \leftrightarrow Sr^{2+}$ substitution can be performed in the binary system of (Ca-Sr)$Y_4[SiO_4]_3O$. On the contrary, the $BaY_4[SiO_4]_3O$ analogue of $CaY_4[SiO_4]_3O$ and $SrY_4[SiO_4]_3O$ apatite cannot be prepared; therefore, the extent of $Ca^{2+} \leftrightarrow Ba^{2+}$ and $Sr^{2+} \leftrightarrow Ba^{2+}$ substitutions is limited to 28 ± 4 and 38 ± 4%, respectively. The field of ternary solid solutions in the AEE$Y_4[SiO_4]_3O$ system, where AEE = Ca, Sr and Ba, is shown in **Fig. 27**.

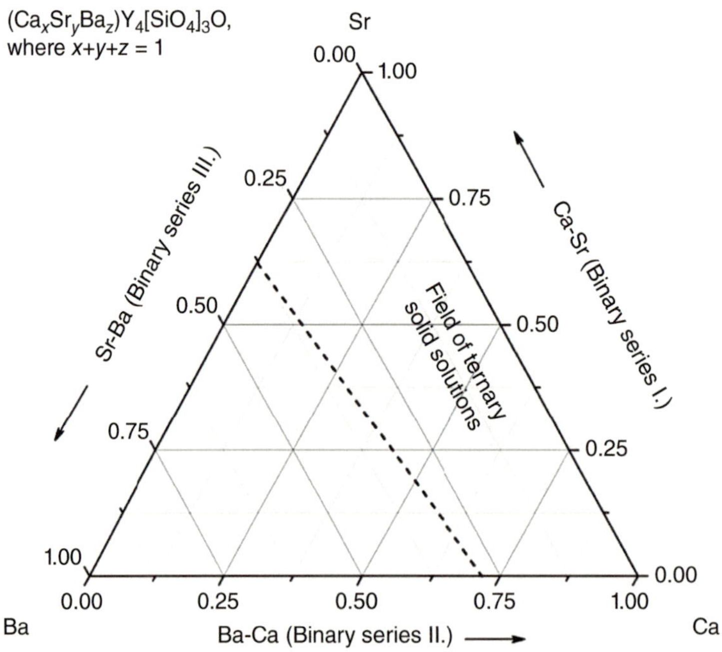

Fig. 27. The miscibility in the AEE $Y_4(SiO_4)_3O$ system.

http://dx.doi.org/10.5772/62209

5.5.3. N-apatite

The main secondary phases in Jänecke prism[17] [117] for Si_3N_4-Al_2O_3-SiO_2-Y_2O_3-YN-AlN system[18] are shown in **Fig. 28**. The formal exchange of oxygen by nitrogen leads to the compounds of N-apatite ($Y_{10}(SiO_4)_6N_2$, H-phase[19]), N-melilite[20] [118] ($Y_2Si_3O_3N_4$, M-phase), N-wollastonite ($YSiON_2$, K-phase) and N-woehlerite ($Y_4Si_2O_7N_2$, J-phase). The latter one forms a complete solid solution with $Y_4Al_2O_9$ (YAM) of the composition of $Y_4Si_{2-x}Al_xO_{7+x}N_{2-x}$ (Jss-phase) [119],[120],[121],[122],[123],[124].

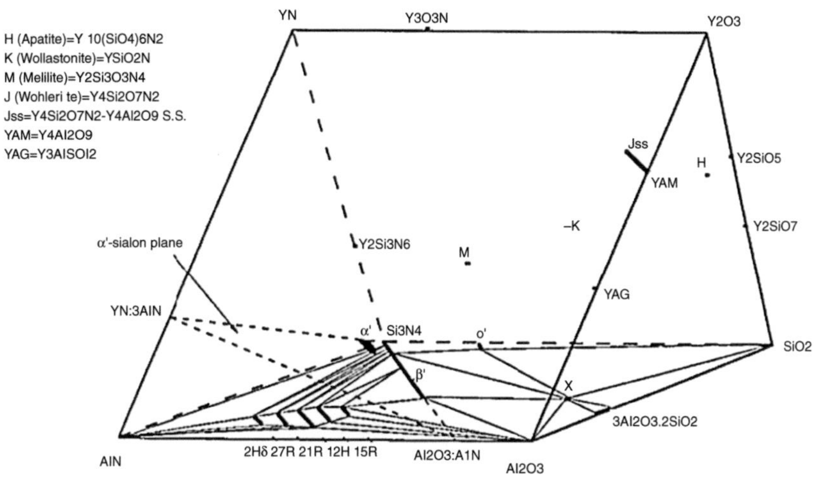

Fig. 28. Representation of the Si_3N_4-Al_2O_3-SiO_2-Y_2O_3-YN-AlN phase equilibrium.

$Y_{10}(SiO_4)_6N_2$ (N-apatite, H-phase, $(Y,Si,\square)_{10}[Si(O,N)_4]_6(O,N,\square)_2$ [125], silicon-yttrium oxynitride) was first identified by Rae et al [126] as a compound with the compositional mixture of $10Y_2O_3{\cdot}9SiO_2{\cdot}Si_3N_4$ that was stable up to 1750°C. There were other suggested compositions, such as $Y_{10}Si_7O_{23}N_4$ [127]. Later work by Gauckler et al [128] established N-apatite as a stoichiometric compound with the formula unit of $Y_{10}[SiO_4]_6N_2$ and the apatite structure (space group $P6_3/M$). The lattice constants of the hexagonal cell were reported to be ***a*** = 9.638

[17] Jänecke prism is used to visualize the phase relationships among α-sialon, β-sialon and other phases in the M-Si-Al-O-N system. α- and β-sialons are isostructural with α- and β-Si_3N_4, respectively. The substitution of Al-O for Si-N in β-Si_3N_4 yields β-sialon with general formula: $Si_{6-x}Al_zO_zN_{8-z}$ ($0 < z < 4.2$). The structure is built up by Si and Al tetrahedra coordinated with oxygen and nitrogen. The unit cell contains two Si_3N_4 units. α-sialons are solid solutions based on the α-Si_3N_4 structure, with the general formula: $M^{p+}{}_xSi_{12-(m+n)}Al_{(m+n)}O_nN_{16-n}$, where **M** is metal ion such as Li, Ca, Ba, Y and RE with a valence of **p+** and index **m** = **px** [117].

[18] Yttria is often used additive to improve the sintering behavior of Si_3N_4 [124].

[19] In dependence on the system composition, the general composition of H-phase (N-apatite) can be written as $(M,REE)_{10}(SiO_4)_6N_2$. The specification of cations then leads to the names such as Mg-Nd-N-apatite [123].

[20] The melilite-type structure (tetragonal mineral melilite $((Ca,Na)_2(Al,Mg,Fe^{2+})(Si,Al)_2O_7)$) is sorosilicate from the group of melilite, first described (Capo di Bove, near Rome, Italy) in 1976 and named from the Greek words *meli* "honey" and *lithos* "stone"). $Y_2Si_3O_3N_4$ was described by Fang et al [118]. N atoms fully occupy the bridging site (2*c*) and O atoms fully occupy the terminal site (4*e*) with 2 O and 6 N atoms at the bridging 8*f* site. The preferential distribution of O and N atoms at the 8*f* site results in two different local coordinations of Y and three different types of Si atoms.

Å and c = 6.355 Å. The electronic structure and bonding of the complex ceramic crystal $Y_{10}(SiO_4)_6N_2$ was studied by CHING et al [121]. This crystal is an insulator with direct band gap of 1.3 eV. It has some unique properties related to one-dimensional chain structure in the c-direction and planar N-Y bonding in the x-y plane.

The ternary phase diagrams of the Si_3N_4-Y_2O_3-SiO_2 [123] and Si_3N_4-La_2O_3-SiO_2 systems [129] are shown in **Fig. 29(a)** and **(b)**. The apatite phase is able to form various solid solutions that may influence the development of strength in silicon nitride densified by yttria [130].

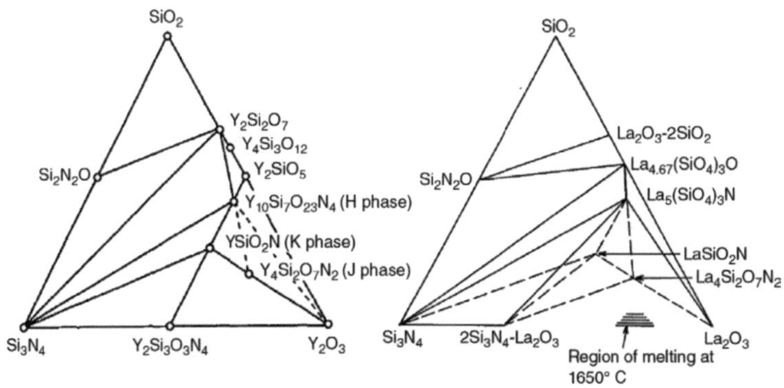

Fig. 29. Phase relationships in the systems Si_3N_4-SiO_2-Y_2O_3 [123] (a) and Si_3N_4-SiO_2-La_2O_3 at 1700°C and 1550°C (dashed lines) [129] (b).

The hexagonal lanthanum N-apatite phase of the composition of $La_5(SiO_4)_3N$ (isostructural with apatite) can be prepared from the mixture of La_2O_3 and Si powder sintered at temperatures in the range from 900 to 1200°C under the flow of nitrogen. The melting temperature of this phase was determined to be ~1600°C. It was observed that continuous heating and addition of Pd into the reaction mixture favored the formation of $La_5(SiO_4)_3N$. Prolonged heating of this compound yields $La_{4.67}(SiO_4)_3O$ [129],[131],[132],[133],[134],[135]. The absorption bands observed in infrared spectrum of lanthanum oxynitrides are introduced in **Table 4**.

Wavenumber [cm^{-1}]	Mode	Frequency [cm^{-1}]	[cm^{-1}]
225	δ (Si-O) (A_1)	730	ν (Si-N)
270	Si-O_2	800	SiO_2
327	δ (Si-O) (B_2)	840	Si-N
337	Si-O-Si (A_1)	872	SiO_4 (ν_1)
380	SiO_4 (ν_4)	909	ν (Si-N) vs
376 – 385	δ (Si-N) (sh)	930	Si-O-N_x
396	SiO_4 (ν_4)	940	ν Si-N vs
432	δ (Si-N) (sh)	960	
448	Si-N s or O-Si-O bend.	905	

Wavenumber [cm⁻¹]	Mode	Frequency [cm⁻¹]	[cm⁻¹]
462	SiO_4 (ν_4)	934	SiO_4 (s) (ν_3)
475		978	
490	ν Si-O (A_1)	980	ν Si-N (s)
542	ν Si-N (s)	1060	ν Si-O (ν_3)
600	ν Si-O	1090	ν Si-N (sh)
648	ν Si-O (w)	1130	ν Si-O (B_1)
679	ν Si-O m (B_2)	—	—

Table 4. The absorption bands observed in infrared spectrum of lanthanum oxynitrides [131].

Mitomo et al [129] used the mixture of Si_3N_4 and La_2O_3 powder heated in a 10 mm diameter carbon die using a hot-pressing apparatus. In some compositions and at temperatures above 1600°C, liquid phases were squeezed out of the die by applied pressure resulting in a change in the overall composition. The compacted powder was therefore heated at the pressure of 20 MPa up to 1400°C and the temperature was then raised to 1700°C without the pressure. The specimen was kept at 1700°C for 1 h and then allowed to cool.

The sintering temperature of Si_3N_4 with La_2O_3 additions is 1700 to 1800°C. Heating of powder mixture of various Si_3N_4/L_2O_3 ratios at 1700°C results in the formation of $2Si_3N_4{\cdot}La_2O_3$, $La_5(SiO_4)_3N$[21] and β-Si_3N_4 or a glass. The reactions occurring during heating were determined as follows [129]:

$$\begin{aligned} Si_3N_4 + La_2O_3 &\xrightarrow{1200\text{–}1250°C} Si_3N_4 + \left(La_4Si_2O_7N_2 + LaSiO_2N\right) \\ &\xrightarrow{1400\text{–}1500°C} LaSiO_2N + Si_3N_4 \\ &\xrightarrow{1650\text{–}1750°C} 2\,Si_3N_4 \cdot La_2O_3 + \text{liquid} \\ &\xrightarrow{20°C} 2\,Si_3N_4 \cdot La_2O_3 + La_5(SiO_4)_3N + \text{glass} \end{aligned} \tag{14}$$

The results of Sakai et al [136] indicate that N-apatite and N-diopside containing grain boundary phase may improve the oxidation resistance of silicon nitride. Since the oxidation of Si_3N_4 leads to the formation of protective SiO_2 layer on the surface:

$$Si_3N_4 + 3\ O_2 \rightarrow 3\ SiO_2 + 2\ N_2 \tag{15}$$

[21] The presence of $La_5(SiO_4)_3N$ is inevitable in the production of high-density materials by liquid-phase sintering; therefore, the amount of $La_5(SiO_4)_3N$ and glassy phase must be minimized to obtain materials with good high-temperature strengths [129].

the silicon nitride shows excellent oxidation resistance. Formed SiO_2 then reacts with the grain boundary constituents to form silicates:

$$SiO_2 + MeO_x \rightarrow MeSiO_{2+x} \quad (16)$$

but in the case of MgO, the formed layer did not act as protection [136]. $Y_{4.67}(SiO_4)_3O$ apatite (britholite phase[2]) is formed as the oxidation product of silicon yttrium oxynitride (H-phase) in the temperature range from 700 to 1400°C [137],[138].

5.6. REE vanadocalcic apatite

The synthesis and physicochemical study of rare-earth-containing vanadocalcic oxyapatites where the pair Ca^{2+} and □ was substituted by Ln^{3+} and $1/2O^{2-}$ was described by Benmoussa et al [139]. This substitution leads to lanthanum and praseodymium dioxyapatites $Ca_8Ln_2(VO_4)_6O_2$, where Ln = La and Pr. Regarding rare earths such as neodymium, samarium, europium, gadolinium and terbium, the Ln^{3+} content limit varies from one REE to another. It decreases when the REE ionic radius declines.

5.7. Apatite-type yttrium phosphates

The following compositions having the apatite structure were prepared by Wanmaker et al [140]:

a. $Me(II)_{2+x}Me(III)_{8-x}(SiO_4)_{6-x}(PO_4)_xO_2$, where $0 \le x \le 6$;

b. $Me(II)_{3+x}MeIII_{6-x}(SiO_4)_{6-x}(PO_4)_x$, where $0 \le x \le 1.5$;

c. $Me(II)_{4+x}Me(III)_{6-x}(SiO_4)_{6-x}(PO_4)_xO$, where $0 \le x \le 6$.

with Me(II) = Ca, Sr, Ba, Mg, Zn or Cd and Me(III) = Y or La. Among these, there are several new compounds, e.g. $Zn_2La_8(SiO_4)_6O_2$, $BaMgY_8(SiO_4)_6O_2$, $Zn_2Y_8(SiO_4)_6O_2$, $Cd_2Y_8(SiO_4)_6O_2$, $Ca_4La_5(SiO_4)_5(PO_4)$ and $Ba_4La_5(SiO_4)_5(PO_4)$. The crystallographic parameters were determined and their luminescence was studied. The most efficient activator proved to be trivalent antimony, especially in the compositions of type I. At 77°K, an emission band at about 400 nm was observed in many of these apatites.

The humidity-sensitivity of yttrium-substituted calcium oxyhydroxyapatites was studied by Owada et al [141]. The logarithm of the electrical resistance of present sensors decreased linearly with increasing relative humidity (RH) from 30 to 65%. The resistance of $[Ca_{9.0}Y_{1.0}](PO_4)_6[O_{1.5}\square_{0.5}]$ with the largest OH vacancy content was about one order of magnitude lower than that of calcium hydroxyapatite. It was found that the larger the ratio of surface hydroxyl groups per unit surface area in the sample, the lower the resistance and the higher the amount of OH vacancies.

A ceramic proton conductor was obtained in the solid solutions of yttrium-substituted oxyhydroxyapatite $(Ca_{10-x}Y_x)(PO_4)_6((OH)_{2-x-2y}O_{x+y}\square_y)$ [142]. Using the hydrogen concentration cells, it was confirmed that the specimens with the composition of $x \leq 0.65$ have the protonic transference number (t_i) is equal to one, while the values of t_i of specimens with $0.65 < x < 1$ were smaller than one. The conduction properties were also dependent on the composition of apatites. At $x = 0.65$, the conductivity (σ) showed the maximum value ($5\cdot10^{-4}$S·cm^{-1} at 800°C) in the relationship between σ and x, while the activation energy was the lowest (about 1.0 eV) at corresponding x. The applicability of proton conductive apatite for a fuel cell was discussed in **Section 10.4**.

Author details

Petr Ptáček

Brno University of Technology, Czech Republic

References

[1] Richards R. Surface and Nanomolecular Catalysis. CRC Press, 2006. ISBN: 978-1420015751

[2] Kendrick E, Slater PR. Synthesis of Ga-doped Ge-based apatites: Effect of dopant and cell symmetry on oxide ion conductivity. Materials Research Bulletin 2008;43(12) 3627–3632.

[3] Xiang J, Liu Zhan-G, Ouyang Jia-H, Yan Fu-Y. Synthesis, structure and electrical properties of rare-earth doped apatite-type lanthanum silicates. Electrochimica Acta 2012;65 251–256.

[4] Fuentes AF, Rodríguez-Reyna E, Martínez-González LG, Maczka M, Hanuza J, Amador U. Room-temperature synthesis of apatite-type lanthanum silicates by mechanically milling constituent oxides. Solid State Ionics 2006;177(19-25) 1869–1873.

[5] Tao Sh, Irvine JTS. Preparation and characterisation of apatite-type lanthanum silicates by a sol-gel proces. Materials Research Bulletin 2001;36(7-8) 1245–1258.

[6] Martínez-González LG, Rodríguez-Reyna E, Moreno KJ, Escalante-García JI, Fuentes AF. Ionic conductivity of apatite-type rare-earth silicates prepared by mechanical milling. Journal of Alloys and Compounds 2009; 476(1-2) 710–714.

[7] Parmentier J, Liddell K, Thompson DP, Lemercier H, Schneider N, Hampshire S, Bodart PR, Harris RK. Influence of iron on the synthesis and stability of yttrium silicate apatite. Solid State Sciences 2001;3(4) 495–502.

[8] Hamon C, Marchand R, Maunaye M, Gaudé J, Guyader J. Revue de chimie minérale 1975;12 259.

[9] Kuzmin EA, Belov NV. Doklady Akademii Nauk SSSR 1965;165 88.

[10] Felsche J. Naturwissenschaften 1967;29 325.

[11] McCarthy GJ, White WB, Roy RJ. Preparation of $Sm_4(SiO_4)_3$. Journal of Inorganic and Nuclear Chemistry 1967;29 253–254.

[12] Higuchi M, Masubuchi Y, Nakayama S, Kikkawa S, Kodaira K. Single crystal growth and oxide ion conductivity of apatite-type rare-earth silicates. Solid State Ionics 2004;174(1-4) 73–80.

[13] Arrhenius SA. Über die Dissoziationswärme und den Einfluss der Temperatur auf den Dissociationsgrad der Elektrolyte. Zeitschrift für Physikalische Chemie 1889;4 96–116.

[14] Arrhenius , SA. Über die Reaktionsgeschwindigkeit bei der Inversion von Rohrzucker durch Säuren. Zeitschrift für Physikalische Chemie 1889;4: 226–248.

[15] Clugston M, Flemming R. Advanced Chemistry. A New Mainstream Text for the New Specifications, Advanced Science. OUP Oxford, 2000. ISBN: 978-0199146338

[16] Shi Q, Zhang H. Electrical properties of iron doped apatite-type lanthanum silicates. Journal of Rare Earths 2012;30(12) 1235–1239.

[17] Takeda N, Itagaki Y, Aono H, Sadaoka Y. Preparation and characterization of $Ln_{9.33+x/3}Si_{6-x}Al_xO_{26}$ (Ln = La, Nd and Sm) with apatite-type structure and its application to a potentiometric O_2 gas sensor. Sensors and Actuators B: Chemical 2006;115(1) 455–459.

[18] El-Batanouny M, Wooten F. Symmetry and Condensed Matter Physics: A Computational Approach. Cambridge University Press, 2008. ISBN: 978-1139469517

[19] Lambert S, Vincent A, Bruneton E, Beaudet-Savignat S, Guillet F, Minot B, Bouree F. Structural investigation of $La_{9.33}Si_6O_{26}$- and $La_9AESi_6O_{26+\delta}$-doped apatites-type lanthanum silicate (AE = Ba, Sr and Ca) by neutron powder diffraction. Journal of Solid State Chemistry 2006;179(8) 2602–2608.

[20] Fukuda K, Asaka T, Okino M, Berghout A, Béchade E, Masson O, Julien I, Thomas P. Anisotropy of oxide-ion conduction in apatite-type lanthanum silicate. Solid State Ionics 2012;217 40–45.

[21] Sumathi S, Buvaneswari G. Synthesis of apatite structure based $BiNaCa_3(PO_4)_3OH$ and its application for condensation reaction. Ceramics International 2012;38(5) 3547–3552.

[22] Kiyono H, Matsuda Y, Shimada T, Ando M, Oikawa I, Maekawa H, Nakayama S, Ohki S, Tansho M, Shimizu T, Florian P, Massiot D. Oxygen-17 nuclear magnetic resonance measurements on apatite-type lanthanum silicate ($La_{9.33}(SiO_4)_6O_2$). Solid State Ionics 2012;228 64–69.

[23] Harman G. Wire Bonding in Microelectronics 3 Edition by George Harman: Wire Bonding in Microelectronics. McGraw-Hill Professional, 2010. ISBN: 978-0-07-164265-1.

[24] Abram EJ, Kirk CA, Sinclair DC, West AR. Synthesis and characterisation of lanthanum germanate-based apatite phases. Solid State Ionics 2005;176(23-24) 1941–1947.

[25] Shaula AL, Kharton VV, Waerenborgh JC, Rojas DP, Marques FMB. Oxygen ionic and electronic transport in apatite ceramics. Journal of the European Ceramic Society 2005;25(12) 2583–2586.

[26] Shaula AL, Kharton VV, Marques FMB. Oxygen ionic and electronic transport in apatite-type $La_{10-x}(Si,Al)_6O_{26\pm\delta}$. Journal of Solid State Chemistry 2005;178(6) 2050–2061.

[27] Shaula AL, Kharton VV, Marques FMB. Ionic and electronic conductivities, stability and thermal expansion of $La_{10-x}(Si,Al)_6O_{26\pm\delta}$ solid electrolytes. Solid State Ionics 2006;177(19-25) 1725–1728.

[28] Helfferich FG. Ion Exchange. Dover Science Books. McGraw-Hill Series in Advanced Chemistry. Courier Corporation, 1962. ISBN: 978-0486687841

[29] Abram EJ, Sinclair DC, West AR. A novel enhancement of ionic conductivity in the cation-deficient apatite $La_{9.33}(SiO_4)_6O_2$. Journal of Material Chemistry 2001;11 1978–1979.

[30] Kharton VV, Marques FMB, Atkinson A. Transport properties of solid oxide electrolyte ceramics: a brief review. Solid State Ionics 2004;174(1-4) 135–149.

[31] Mössbauer RL. Kernresonanzfluoreszenz von Gammastrahlung in Ir191. Zeitschrift für Physik A 1958;151(2) 124–143.

[32] Greenwood NN. Mössbauer Spectroscopy. Springer Science & Business Media, 2012. ISBN: 978-9400956971

[33] Dickson PE, Berry FJ. Mössbauer Spectroscopy. Cambridge University Press, 2005. ISBN: 978-0521018104

[34] Miglierini M, Petridis D. Mössbauer Spectroscopy in Materials Science. Nato Science Partnership Subseries: 3 — Volume 66. Springer Science & Business Media, 2013. ISBN: 978-9401145480

[35] Yaremchenko AA, Shaula AL, Kharton VV, Waerenborgh JC, Rojas DP, Patrakeev MV, Marques FMB. Ionic and electronic conductivity of $La_{9.83-x}Pr_xSi_{4.5}Fe_{1.5}O_{26\pm\delta}$ apatites. Solid State Ionics 2004;171(1-2) 51–59.

[36] Yoshioka H, Nojiri Y, Tanase Sh. Ionic conductivity and fuel cell properties of apatite-type lanthanum silicates doped with Mg and containing excess oxide ions. Solid State Ionics 2008;179(30) 2165–2169.

[37] Yuan W, Gu Y, Li L. Synthesis and ionic conduction of cation-deficient apatite $La_{9.332x/3}M_xSi_6O_{26}$ doped with Mg, Ca, Sr*. Chinese Journal of Chemical Engineering 2008;16(3) 488–491.

[38] Irvine JTS, Corcoran DJD, Cull PA. Structure and ionic conduction in solids. Nordic Workshop on Materials for Electrochemical Conversion, Geilo Norway 2000; 37-41.

[39] Hosseini SM, Shvareva T, Navrotsky A. Energetics of lanthanum silicate apatite: Influence of interstitial oxygen and cation vacancy concentrations in $La_{9.33+x}(SiO_4)_6O_{2+3x/2}$ and $La_{10-x}Sr_x(SiO_4)_6O_{3-0.5x}$. Solid State Ionics 2013;233 62–66.

[40] Karato S. Deformation of Earth Materials: An Introduction to the Rheology of Solid Earth. Cambridge University Press, 2008. ISBN: 978-1139469562

[41] Kong LB, Huang YZ, Que WX, Zhang TS, Li S, Zhang J, Dong ZL, Tang DY. Transparent Ceramics. Topics in Mining, Metallurgy and Materials Engineering. Springer, 2015. ISBN: 978-3319189567

[42] Chung Yip-W. Introduction to Materials Science and Engineering. CRC Press, 2006. ISBN: 978-0849392634

[43] Kosevich AM. The Crystal Lattice: Phonons, Solitons, Dislocations, Superlattices. John Wiley & Sons, 2006. ISBN: 978-3527606931

[44] Tao Sh, Irvine JTS. Preparation and characterization of apatite-type lanthanum silicates by a sol-gel proces. Materials Research Bulletin 2001;36(7-8) 1245–1258.

[45] Li B, Liu W, Pan W. Synthesis and electrical properties of apatite-type $La_{10}Si_6O_{27}$. Journal of Power Sources 2010;195(8) 2196–2201.

[46] Masubuchi Y, Higuchi M, Takeda T, Kikkawa Sh. Preparation of apatite-type $La_{9.33}(SiO_4)_6O_2$ oxide ion conductor by alcoxide-hydrolysis. Journal of Alloys and Compounds 2006;408-412 641-644.

[47] Sansom JEH, Sermon PA, Slater PR. Synthesis and conductivities of the Ti doped apatite-type phases $(La/Ba)_{10-x}(Si/Ge)_{6-y}Ti_yO_{26+z}$. Solid State Ionics 2005;176(19-22) 1765–1768.

[48] Yuan W, Gu Y, Li L. Synthesis and conductivity of oxyapatite ionic conductor $La_{10-x}V_x(SiO_4)_6O_{3+x}$. Chinese Journal of Chemical Engineering 2010;18(2) 328–332.

[49] Mazza D, Tribaudino M, Delmastro A, Lebech B. Synthesis and neutron diffraction study of $La_5Si_2BO_{13}$, an analog of the apatite mineral. Journal of Solid State Chemistry 2000;155(2) 389–393.

[50] Arbib EH, Eloudi E, Chaminade JP, Darriet J. The crystal structure of the phosphate eulytite $Ba_3Bi(PO_4)_3$. Materials Research Bulletin 2000;35(5) 761–773.

[51] Naidu SA, Varadaraju UV, Raveau B. Eu^{3+} luminescence in $La_5Si_2BO_{13}$ with apatite related structure and magnetic studies in $Ln_5Si_2BO_{13}$ (Ln=Gd, Dy). Journal of Solid State Chemistry 2010;183(8) 1847–1852.

[52] Vieira MM, Oliveira JC, Shaula AL, Trindade B, Cavaleiro A. Structure and ionic conductivity of reactively sputtered apatite-type lanthanum silicate thin films. Surface and Coatings Technology 2014;247 14–19.

[53] Wang Shou-Q, Uda S. Phase relations around langasite ($La_3Ga_5SiO_{14}$) in the system La_2O_3-Ga_2O_3-SiO_2 in air. Journal of Crystal Growth 2003;250(3-4) 463–470.

[54] El Ouenzerfi R, Goutaudier C, Panczer G, Moine B, Cohen-Adad MT, Trabelsi-Ayedi M, Kbir-Ariguib N. Investigation of the CaO-La_2O_3-SiO_2-P_2O_5 quaternary diagram. Synthesis, existence domain, and characterization of apatitic phosphosilicates. Solid State Ionics 2003;156(1-2) 209–222.

[55] Wanmaker WL, ter Vrugt JW, Verlijsdonk JG. Luminescence of alkaline earth yttrium and lanthanum phosphate-silicates with apatite structure. Journal of Solid State Chemistry 1971;3(3) 452–457.

[56] Bonhomme C, Beaudet-Savignat S, Chartier Th, Maître A, Sauvet Anne-L, Soulestin B. Sintering kinetics and oxide ion conduction in Sr-doped apatite-type lanthanum silicates, $La_9Sr_1Si_6O_{26.5}$. Solid State Ionics 2009;180(36-39) 1593–1598,

[57] Mazza D, Ronchetti S. Study on the Al_2O_3-SiO_2-La_2O_3 ternary system at 1300°C. Materials Research Bulletin 1999;34(9) 1375–1382.

[58] Kuz'min VN, Belov KP. Doklady Akademii Nauk SSSR 1965;165, 88.

[59] McCarthy GJ, White WB, Roy R. Preparation of $Sm_4(SiO_4)_3$. Journal of Inorganic and Nuclear Chemistry 1967;29(1) 253–254.

[60] Felsche J. Naturwissenschaften 1969;56 325.

[61] Nakayama S, Sakamoto M. Preparation of apatite-type $La_{9.33}Ge_6O_{26}$ single-crystal from sintered ceramics by a seeding method and its oxide ionic conduction. Solid State Ionics 2013;253 47–52.

[62] Kendrick E, Slater PR. Synthesis of hexagonal lanthanum germanate apatites through site selective isovalent doping with yttrium. Materials Research Bulletin 2008;43(8-9) 2509–2513.

[63] Najib A, Sansom JE, Tolchard JR, Slater PR, Islam MS. Doping strategies to optimise the oxide ion conductivity in apatite-type ionic conductors. Dalton Transactions 2004;7(19) 3106–3109.

[64] Orera A, Headspith D, Apperley DC, Francesconi MG, Slater PR. Formation of apatite oxynitrides by the reaction between apatite-type oxide ion conductors, $La_{8+x}Sr_{2-x}$(Si/Ge)$_6O_{26+x/2}$, and ammonia. Journal of Solid State Chemistry 2009;182(12) 3294–3298.

[65] Emirdag-Eanes M, Pennington WT, Kolis JW. Synthesis, structural characterization, and magnetic properties of $NaRE_9(GeO_4)_6O_2$ (RE = Nd, Pr). Journal of Alloys and Compounds 2004;366(1-2) 76–80.

[66] Takahashi M, Uematsu K, Ye Zuo-G, Sato M. Single-crystal growth and structure determination of a new oxide apatite, $NaLa_9(GeO_4)_6O_2$. Journal of Solid State Chemistry 1998;139(2) 304–309.

[67] Chen P, Li RK. Two high terbium content apatites: $Tb_5Si_2BO_{13}$ and $Tb_{4.66}Si_3O_{13}$. Journal of Alloys and Compounds 2015;622 859–864.

[68] Xia MJ. Li RK. Structure and optical properties of a noncentrosymmetric borate $RbSr_4(BO_3)_3$. Journal of Solid State Chemistry 2013;197 366–369.

[69] Kazmierczak K, Höppe HA. Synthesis, crystal structure and optical spectra of europium borate fluoride $Eu_5(BO_3)_3F$. European Journal of Inorganic Chemistry 2010;18 2678–2681.

[70] Chen S, Hoffmann S, Carrillo-Cabrera W, Akselrud LG, Prots Y, Schwarz U, Zhao Jing-T, Kniep R. $Sr_{10}[(PO_4)_{5.5}(BO_4)_{0.5}](BO_2)$: Growth and crystal structure of a strontium phosphate orthoborate metaborate closely related to the apatite-type crystal structure. Journal of Solid State Chemistry 2010;183(3) 658–661.

[71] Yu Y, Wu QS, Li RK. Structure of two new borates $YCa_3(AlO)_3(BO_3)_4$ and $YCa_3(GaO)_3(BO_3)_4$. Journal of Solid State Chemistry 2006;179(2) 429–432.

[72] Kim HK, Ao Sio-I, Rieger BB. IAENG Transactions on Engineering Technologies: Special Edition of the World Congress on Engineering and Computer Science 2011. Springer Science & Business Media, 2012. ISBN: 978-9400747852

[73] Khomskii D. Transition Metal Compounds. Cambridge University Press, 2014. ISBN: 978-1107020177

[74] Lacroix C, Mendels P, Mila F. Introduction to Frustrated Magnetism: Materials, Experiments, Theory. Springer Series in Solid-State Sciences — Volume 164. Springer Science & Business Media, 2011. ISBN: 978-3642105890

[75] Werner F, Kubel F. Apatite-type $Pr_9K(SiO_4)_6O_2$ — A potential oxide ion conductor. Materials Letters 2005;59(28) 3660–3665.

[76] Morgan MG, Wang M, Mar A. Samarium orthosilicate oxyapatite, $Sm_5(SiO_4)_3O$. Acta Crystallographica Section E, 2010. DOI: 10.1107/S1600536802011868

[77] Sakakura T, Kamoshita M, Iguchi H, Wang J, Ishizawa N. Apatite-type $SrPr_4(SiO_4)_3O$. Crystallographica Section E, 2010. DOI: 10.1107/S1600536810033349

[78] Felsche J. Rare earth silicates with the apatite structure. Journal of Solid State Chemistry 1972;5(2) 266–275.

[79] Xiang J, Liu Zhan-G, Ouyang Jia-H, Zhou Y, Yan Fu-Y. Influence of doping with various cations on electrical conductivity of apatite-type neodymium silicates. Ceramics International 2013; 39(5) 4847–4851.

[80] Latshaw AM, Smith MD, zur Loye Hans-C. Crystal growth and structure of three new neodymium containing silicates: $Na_{0.50}Nd_{4.50}(SiO_4)_3O$, $Na_{0.63}Nd_{4.37}(SiO_4)_3O_{0.74}F_{0.26}$ and $Na_{4.74}Nd_{4.26}(O_{0.52}F_{0.48})[SiO_4]_4$. Solid State Sciences 2014;35 28–32.

[81] Wang Ch, Liu X, Fleet ME, Feng Sh , Xu R. High-pressure synthesis and single-crystal structure refinement of gadolinium holmium silicate hydroxyapatite $Gd_{4.33}Ho_{4.33}(SiO_4)_6(OH)_2$. Journal of Solid State Chemistry 2006;179(7) 2245–2250.

[82] Haile SM, Wuensch BJ, Siegrist T, Laudise RA. Hydrothermal synthesis of new alkali silicates I. Potassium neodymium phases. Journal of Crystal Growth 1993;131(3-4) 352–372.

[83] Haile SM, Wuensch BJ, Laudise RA. Hydrothermal synthesis of new alkali silicates II. Sodium neodymium and sodium yttrium phases. Journal of Crystal Growth 1993;131(3-4) 373–386.

[84] Boughzala K, Salem BE, Chrifa AB, Gaudin E, Bouzouita K. Synthesis and characterization of strontium-lanthanum apatites. Materials Research Bulletin 2007;42(7) 1221–1229.

[85] Małecka MA, Kępiński L. Synthesis and structure of nanocrystalline mixed Ce-Yb silicates. Materials Research Bulletin 2013;48(7) 2571–2577.

[86] Xiang J, Ouyang Jia-H, Liu Zhan-G. Microstructure and electrical conductivity of apatite-type $La_{10}Si_{6-x}WxO_{27+\delta}$ electrolytes. Journal of Power Sources 2015;284 49–55.

[87] Wang Sea-F, Hsu Yung-F, Lin Wan-J. Effects of Nb^{5+}, Mo^{6+}, and W^{6+} dopants on the germanate-based apatites as electrolyte for use in solid oxide fuel cells. International Journal of Hydrogen Energy 2013;38(27) 12015–12023.

[88] Institut kristallografii im. Shubnikova AV, Sheftal' NN. Growth of Crystals, Volume 6, Part 1. Consultants Bureau, 1965.

[89] Toropov NA, Bondar IA. In: Levin EM, Robbins CR, McMurdie HF (Eds.), Phase Diagrams for Ceramists, 1969.

[90] Toporov NA, Fedorov NF. Composition diagram of the Ca_2SiO_4-$Y_4(SiO_4)_3$ system. Zhurnal Neorganicheskoi Khimii 1965;10 666–668.

[91] Laczai N, Péter Á, Kovács L, Bencs L, Lőrincz E. Synthesis and characterization of $Y_2(SiO_4)O$:Ce with LiF and NaF additives. Solid State Sciences 2015;45 23–29.

[92] Mao H, Selleby M, Fabrichnaya O. Thermodynamic reassessment of the Y_2O_3-Al_2O_3-SiO_2 and its subsystems. Calphad 32 399-412. DOI: 10.1016/j.calphad.2008.03.003.

[93] Liddell K, Thompson DP. X-ray diffraction data for yttrium silicates. British Ceramic Transactions 1986;85 17–22.

[94] Warshaw I, Roy R. In: Progress in Science and Technology of Rare Earths, Vol. 1. New York: Pergamum Press, 1964, p. 203.

[95] Ito J, Johnson H. Synthesis and study of yttrialite. American Mineralogist 1968;53 1940–1952.

[96] Escudero A, Alba MD, Becerro AI. Polymorphism in the $Sc_2Si_2O_7$-$Y_2Si_2O_7$ system. Journal of Solid State Chemistry 2007;180 1436–1445.

[97] Wills RR, Cunningham JA, Wimmer JM, Stewart RW. Stability of the silicon yttrium oxynitrides. Journal of the American Ceramic Society 1976;59(5-6) 269–270.

[98] Cannas C, Musinu A, Piccaluga G, Deidda C, Serra F, Bazzoni M, Enzo S. Advances in the structure and microstructure determination of yttrium silicates using the Rietveld method. Journal of Solid State Chemistry 2005;178(5) 1526–1532.

[99] McColm I. Dictionary of Ceramic Science and Engineering. 2nd ed., Springer Science & Business Media, 2013. ISBN: 978-1475723212

[100] Suwa Y, Naka S, Noda T. Preparation and properties of yttrium magnesium silicate with apatite structure. Materials Research Bulletin 1968;3(2) 139–147.

[101] Lee FC, Marr J, Glasser FP. Compounds in the $Na_2O{\cdot}Y_2O_3{\cdot}SiO_2$ system. Ceramics International 1981;7(2) 43–47.

[102] Bondar IA, Galakhov FY. Izvestiya Akademii Nauk: Seriya Khimicheskaya 1964;7 1325–1326.

[103] Shannon RD. Revised effective ionic radii and systematic studies of interatomic distances in halides and chalcogenides. Acta Crystallographica Section A 1976;32(5) 751–767.

[104] Brown ID, Altermatt D. Bond-valence parameters obtained from a systematic analysis of the inorganic crystal structure database. Acta Crystallographica Section B 1985;41 244–247.

[105] Wills RR, Cunningham JA, Wimmer JM, Stewart RW. Stability of the silicon yttrium oxynitrides. Journal of the American Ceramic Society 1976;59(5-6) 269–270.

[106] Rakovan JF, Hughes JM. Strontium in the apatite structure: Strontian fluorapatite and belovite-(Ce). The Canadian Mineralogist 2000;38 839–845.

[107] Zuev MG, Karpov AM, Shkvarin AS. Synthesis and spectral characteristics of $Sr_2Y_8(SiO_4)_6O_2$: Eu polycrystals. Journal of Solid State Chemistry 2011;184(1) 52–58.

[108] Karpov AM, Zuev MG. Sol-gel synthesis and spectral characteristics of crystal phosphors $Sr_2Y_{8(1-x)}Eu_{8x}Si_6O_{26}$. Glass Physics and Chemistry 2012;38(4) 431–436.

[109] Holand W, Beall GH. Glass Ceramic Technology. 2nd ed., John Wiley & Sons, 2012. ISBN: 978-1118265925

[110] Ahlborg NL, Zhu D. Calcium-Magnesium-Aluminosilicate (CMAS) Reactions and Degradation Mechanisms of Advanced Environmental Barrier Coatings. National Aeronautics and Space Administration. NASA/TM-2013-218091

[111] Li G, Zhang Y, Geng D, Shang M, Peng C, Cheng Z, Lin J. Single-composition trichromatic white-emitting $Ca_4Y_6(SiO_4)_6O$: $Ce^{3+}/Mn^{2+}/Tb^{3+}$ phosphor: Luminescence and energy transfer. ACS Applied Materials & Interfaces 2012;4(1) 296–305.

[112] Wanmaker WL, Vrugt JW, Verlijsdonk JG. Synthesis of new compounds with apatite structure. Philips Research Reports 1971;26 373–381.

[113] van't Hoen C, Rheinberger V, Höland W, Apel E. Crystallization of oxyapatite in glass-ceramics. Journal of the European Ceramic Society 2007;27(2-3) 1579–1584.

[114] Gunawardane RP, Howie RA, Glasser FP. Structure of the oxyapatite $NaY_9(SiO_4)_6O_2$. Acta Crystallographica Section B 1982;38(5) 1564–1566.

[115] Redhammer GR, Roth G. Lithium and sodium yttrium orthosilicate oxyapatite, $LiY_9(SiO_4)_6O_2$ and $NaY_9(SiO_4)_6O_2$, at both 100 K and near room temperature. Acta Crystallographica Section C 2003;59(12) 120–124.

[116] Ptáček P, Bartoníčková E, Švec J, Opravil T, Šoukal F, Wasserbauer J, Másilko J. Preparation, kinetics of sinter-crystallization and properties of hexagonal strontium-yttrate-silicate apatite phase: $SrY_4[SiO_4]_3O$. Ceramics International 2015;41(1) 1779–1795.

[117] Heimann RB. Classic and Advanced Ceramics: From Fundamentals to Applications. Vydavatel John Wiley & Sons, 2010. ISBN: 978-3527630189

[118] Fang CM, de Wijs GA, de Groot RA, Metselaar R, Hintzen HT, de With G. O/N ordering in $Y_2Si_3O_3N_4$ with the melilite-type structure from first-principles calculations. Chemistry of Materials 2000;12(4) 1071–1075.

[119] Liddell K, Mandal H, Thompson DP. X-ray data for new Y-Si-Al-O-N glass ceramics. Journal of the European Ceramic Society 1997;17(6) 781–877.

[120] Ahmad S, Ludwig T, Herrmann M, Mahmoud MM, Lippmann W, Seifert HJ. Crystallisation studies of Si_3N_4-Al_2O_3-SiO_2-Y_2O_3 glass-ceramics under different heat-treatment conditions. Journal of the European Ceramic Society 2015;35(8) 2261–2268.

[121] Ching Wai-Y, Xu Yong-Ni, Ouyang L. Electronic structure and bonding in crystalline $Y_{10}[SiO_4]_6N_2$. Journal of the American Ceramic Society 2004;86(8) 1424–1426.

[122] Riedel R, Chen I-W. Ceramics Science and Technology, Structures. Ceramics Science and Technology — Volume 1. John Wiley & Sons, 2011. ISBN: 978-3527631933.

[123] Alper AM (Ed.), Kostorz G, Herman H. Phase Diagrams in Advanced Ceramics. Treatise on Materials Science and Technology. Elsevier, 1995. ISBN: 978-0080538723

[124] Jansen M (Ed.), Haubner R, Herrmann M, Lux B, Petzow G, Weissenbacher R, Wilhelm M. High Performance Non-Oxide Ceramics II. Structure and Bonding — Volume 102. Springer, 2003. ISBN: 978-3540456230

[125] Alper AM. Phase Diagrams: Materials Science and Technology, In: Alper AM (Ed.), Volume 6, Part 5. Academic Press, 1978. ISBN: 978-0120532032

[126] Rae AWJ, Thompson DP, Pikin NJ, Jack KH. pp. 347-360 In: Poper P (Ed.), Special Ceramics. British Ceramic Research Association, Stoke-on Trent, England, D.P. Thompson, p. 358, 1975.

[127] Wills RR, Holmquist S, Wimmer JM, Cunningham JA. Phase relationships in the system SiO_2-Y_2O_3-Si_3N_4. Journal of Materials Science 1976;11(7) 1305–309.

[128] Gauckler LJ, Hohnke H, Tien TY. The system Si_3N_4-SiO_2-Y_2O_3. Journal of the American Ceramic Society 1980;63(1-2) 35–37.

[129] Mitomo M, Izumi F, Haruichi S, Matsui Y. Phase relationships in the system Si_3N_4-SiO_2-La_2O_3. Journal of Materials Science 1982;17(8) 2359–2369.

[130] Buhl H. Advanced Aerospace Materials. Materials Research and Engineering. Springer Science & Business Media, 2012. ISBN: 978-3642501593

[131] Kizilyalli M, Gürbüz G. Lanthanum silicon oxynitrides, synthesis and X-ray diffraction and IR studies. Find out how to access preview-only content. NATO ASI Series 1990;185 45–67.

[132] Lange FF. Si_3N_4-Ce_2O_3-SiO_2 materials: Phase relations and strength. American Ceramic Society Bulletin 1980;59(2) 239–240.

[133] Thompson DP. Special Ceramics 1975;358.

[134] Mitomo N, Kuramoto N, Suzuki H. Journal of Materials Science Letters 1978;13 2523.

[135] Gaude J, Guyader J, Lang J. Comptes Rendus de l'Académie des Sciences, Serie C 1975;280 883–884.

[136] Sakai H, Soma T. Matsui M, Oda I. Oxidation and microstructure of sintered silicon nitride. In: Pask JA, Evans AG (Eds.). Ceramic Microstructures '86: Role of Interfaces, Materials Science Research — Volume 21. Springer Science & Business Media, 2013. ISBN: 978-1461319337

[137] International Aerospace Abstracts, Volume 33, Edition 4-6. American Institute of Aeronautics and Astronautics. Technical Information Service, Cambridge Scientific Abstracts, Inc., United States. National Aeronautics and Space Administration, Institute of the Aerospace Sciences. Technical Information Service, American Institute of Aeronautics and Astronautics, 1993.

[138] Physics Soviet, Crystallography, Volume 14, Pages 1-484. American Institute of Physics, American Crystallographic Association. American Institute of Physics, 1969.

[139] Benmoussa H, Mikou M, Lacout JL. Synthesis and physicochemical study of new rare-earth-containing vanadocalcic oxyapatites. Materials Research Bulletin 1999;34(9) 1429–1434.

[140] Wanmaker WL, ter Vrugt JW, Verlijsdonk JG. Luminescence of alkaline earth yttrium and lanthanum phosphate-silicates with apatite structure. Journal of Solid State Chemistry 1971;3(3) 452–457.

[141] Owada H, Yamashita K, Umegaki T, Kanazawa T, Nagai M. Humidity-sensitivity of yttrium substituted apatite ceramics. Solid State Ionics 1989;35(3-4) 401–404.

[142] Yamashita K, Owada H, Umegaki T, Kanazawa T, Katayama K. Protonic conduction in yttrium-substituted hydroxyapatite ceramics and their applicability to H_2-O_2 fuel cell. Solid State Ionics 1990;40-41(2) 918-921.

Chapter 6

Substituents and Dopants in the Structure of Apatite

Petr Ptáček

Additional information is available at the end of the chapter

http://dx.doi.org/10.5772/62213

Abstract

The structure of apatites allows large variations of composition given by the generic formula ($M_{10}(XO_4)_6Z_2$) including partial or complete substitution of both the cationic as well as the anionic sites, formation of nonstoichiometric forms and solid solutions. More than half of naturally occurring elements can be accommodated by apatite structure in a significant extend. The sixth chapter of this book is divided into five sections. The first, second and the third part deals with many examples of substitution including cationic substitution of M sites, anionic substitution of X-site and anionic substitution of Z-site, respectively. The remaining two sections continue with solid solution of apatites and ends with description of trace elements and their isotopes in the structure of apatite, respectively.

Keywords: Apatite, Substitution, Solid solution, Trace elements, Isotopes, Complexation of Metal Cations, Diffusion

As was already said in **Chapter 1**, the generic formula of apatite ($M_{10}(XO_4)_6Z_2$) enables partial or complete substitution for cationic as well as anionic sites [1],[2]:

1. Cation substitution, where M = Ca^{2+}, Pb^{2+}, Sr^{2+}, Mg^{2+}, Fe^{2+}, Mn^{2+}, Cd^{2+}, Ba^{2+}, Co^{2+}, Ni^{2+}, Cu^{2+}, Zn^{2+}, Sn^{2+}, Eu^{2+}, Na^{+}, K^{+}, Li^{+}, Rb^{+}, NH_4^{+}, La^{3+}, Ce^{3+}, Sm^{3+}, Eu^{3+}, Y^{3+}, Cr^{3+}, Th^{4+}, U^{4+}, U^{6+} and □.

2. Substitution for phosphorus by one or two cations, where X = PO_4^{3-}, AsO_4^{3-}, SiO_4^{3-}, VO_4^{3-}, CrO_4^{3-}, CrO_4^{2-}, MnO_4^{3-}, SO_4^{2-}, SeO_4^{2-}, BeF_4^{2-}, GeO_4^{4-}, ReO_5^{3-}, SbO_3F^{4-}, SiO_3N^{5-}, BO_4^{5-}, BO_3^{3-} and CO_3^{2-}.

3. Z-site substation, where Z = F^{-}, OH^{-}, Cl^{-}, O_2^{-}, O_3^{-}, NCO^{-}, BO_2^{-}, Br^{-}, I^{-}, NO_2^{-}, CO_3^{2-}, O_2^{2-}, O^{2-}, S^{2-}, NCN^{2-}, NO_2^{2-} and □.

where □ represents the vacancy cluster [1].

Besides the monoionic substitution, the co-substitution and mutual combinations of substitutions in anionic and cationic sites (multi-ionic substitution) were also often reported [3],[4],[5],[6]. Mutual substitutions of trace elements into apatite structure brought new physicochemical, mechanical and biological properties in comparison with pure apatite or monoionic substituted apatite materials, e.g. hydroxylapatite [3].

Some substitutions can proceed only at the synthesis stage, while a limited ion exchange between solid apatite and surrounding solution can also occur. Due to their high chemical diversity and ion-exchange capabilities, apatites are considered as materials for toxic waste storage and for wastewater purification. The ion exchange in apatitic structures in human organism also presents an interest for medicine [7].

Recent studies have shown that a number of alkaline-earth-rare-earth silicates and germanates also have the apatite structure, and these have the cell sizes which span the division between the "apatites" and the "pyromorphites". Some, particularly barium and lanthanum apatites, have the lattice parameters comparable with the members of the pyromorphite group. Thus, $Ba_2La_8(SiO_4)_6O_2$ has the cell parameters $\boldsymbol{a}$ = 9.76 Å and $\boldsymbol{c}$ = 7.30 Å and $Pb_{10}(PO_4)_6F_2$ shows $\boldsymbol{a}$ = 9.76 and $\boldsymbol{c}$ = 7.29 Å, while $Ba_3La_7(GeO_4)_6O_{1.5}$ has $\boldsymbol{a}$ = 9.99 Å and $\boldsymbol{c}$ = 7.39 Å and $Pb_{10}(AsO_4)_6F_2$ has $\boldsymbol{a}$ = 10.07 Å and $\boldsymbol{c}$ = 7.42 Å. During synthetic studies, however, it became apparent that the prediction of the composition of compounds with apatite-type structures could not be made solely on the basis of satisfying the valence considerations, since the occurrence of the apatite-type structure also appears to be determined by the ratio of the mean size of "A" ions (i.e. Ca ions in fluorapatite) to the mean size of "X" ions in XO_4 [8],[9].

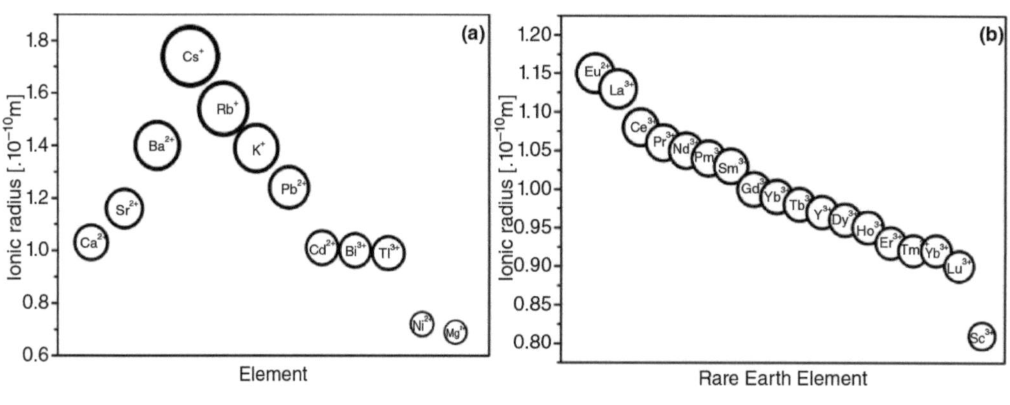

Fig. 1. The ionic radius of elements that can be accommodated instead of M in the lattice of apatite ($M_5(XO_4)_3Z_q$).

The structure of hydroxyapatite allows large variations from its theoretical composition as well as the formation of nonstoichiometric forms and ionic substitutions. More than half of naturally occurring elements are known to be accommodated in the apatite lattice to significant extent. Ca^{2+} cation can be substituted by Na^+, K^+, Mg^{2+}, Sr^{2+}, Pb^{2+}, Mn^{2+} (**Fig. 1(a)**) or rare-earth elements[1] (REE, **Fig. 1(b)**) and PO_4^{3-} anions by AsO_4^{3-}, SO_4^{2-} or CO_3^{2-} without destroying the apatite structure. The changes in lattice parameters must be indicative of the type of substitution occurring. For example, Cl^- interchange for OH^- ions causes a change to lattice parameters from $a = 9.4214$ Å and $c = 6.8814$ Å to 6.628 Å and 6.764 Å, respectively. Another example is the Sr^{2+} substitution for Ca^{2+}, which causes lengthening of a- and c-axes from 9.418 Å and 6.884 Å to 9.76 Å and 7.27 Å, respectively [8],[12],[13],[14],[15].

The substitutions at Z site play a very important role in the crystallography of specific species. The Z site lies in the channel formed by the X sites in fluorapatite and is of just the right size to fit between X atoms, and it lies on (001) mirror planes to yield the space group $P6_3/M$. When Cl substitutes for F, Cl is too large to fit on the mirror plane, so it is displaced along the c-axis and the space group becomes $P6_3$. The OH substitution is even more complex. OH anions are not spherically symmetric due to H^+ (proton) present in the charge cloud of O atoms. H causes a displacement of O off the mirror plane, the O-H orientation tends to align in a given channel, but adjacent channels may have different displacements and orientations. The result is that well-crystallized hydroxylapatites are usually monoclinic with the space group $P2_1/M$ or $P2_1$ [12].

Some of the various families of substitutions that were experimentally established in apatites are summarized in **Table 1**. In general, the ions that substitute for Ca in the A position have the valences from 1 to 3 and the coordination numbers of VII at Ca(2) (6*h*) site and IX at Ca(1) (4*f*) site. **Table 2** introduces the cation radii of possible apatite substituents at M-site [2].

M_{10}	X_6	O_{24}	Z_2	Designation
$M(1)_4M(2)_6$		$O_{12}O_6O_6$		
Ca_{10}	P_6	O_{24}	$(OH)_2$	HAP
C_4Ln_6	Si_6			
Sr_{10}	S_3Si_3		F_2	FAP
Ca_2Ln_8	Si_6		O_2	Oxyapatite
$SrCa_9$	P_6		□	Z-site vacancies
Nd_4Ca_6	Ge_6		$□_2$	
Sr_{10}	P_4Si_2		$□_2$	
Na_2Ca_8	P_6		$□_2$	
$□_2La_2Ca_4La_2$			$□_2$	M-, Z-site vacancy
$□_2La_2La_2$	Ge_6		$(OH)_2$	M-site vacancies (?)

Table 1. Structural formulas of apatites: $M(1)_4M(2)_6(XO_4)_6Z_2$ [2].

[1] The minerals with essential rare-earth elements (REE or rare-earth metals, REM) or chemically related elements Y or Sc are termed as rare-earth minerals. They must be named with suffix (Levinson modifier [10],[11]), indicating the dominant rare-earth element (some examples can be found in **Chapter 1** (**Table 3**)). Please see also note **2**.

Substituents		Coordination number				
		Arens	Shannon and Prewitt			
		VI	VI	VII	VIII	IX
M^{2+}	Ba	1.34	1.36	1.39	—	1.47
	Pb	1.20	1.18	—	1.29	1.33
	Eu	—	1.17	—	1.25	—
	Sn	0.93	—	—	1.22	—
	Sr	1.12	1.16	1.21	1.25	—
	Ca	0.99	1.00	1.07	1.12	1.18
	Cd	0.97	0.95	1.00	1.07	—
	Mn	—	0.83	—	0.93	—
	Zn	0.74	0.745	—	—	0.90 (E)
	Co	0.72	0.735	—	—	—
	Cu	0.72	0.73	—	—	—
	Mg	0.66	0.72	—	0.89	—
	Ni	0.69	0.69	—	—	—
M^{+}	K	1.33	1.38	1.46 (?)	1.51 (?)	1.55 (?, E)
	Na	0.97	1.02	1.13 (?)	1.16	1.32 (?, E)

(E) and (?) denote interpolated and doubtful values, respectively.

Table 2. Cation radii of possible apatite substituents at M_{10}-site of $M_{10}(XO_4)_6Z_2$ unit [2].

An example of charge compensating substitution for phosphorus by two cations is the substitution during the synthesis of apatite species of the composition of $Ca_{10}(SiO_4)_3(SO_4)_3F_2$ (CSSF, fluorellestadite [16]) [2],[17],[18]:

$$2\ P^{5+} \Longleftrightarrow Si^{4+} + S^{6+} \tag{1}$$

These synthetic phases have mineral equivalents in the minerals from the ellestadite group, which are listed in **Table 3**. Since the mineral with ideal end-member formula $Ca_5(SiO_4)_{1.5}(SO_4)_{1.5}Cl$ is assumed not to exist, the name ellestadite-(Cl) is discredited [19].

Descriptor	Brief description
a [Å]	Lattice constant of hexagonal unit cell
b [Å]	
c:*a*	Variable axial ratio
r_{MI} [Å]	Shannon's ionic radii of M(I)-site ion (nine-coordination)

Descriptor	Brief description
r_X [Å]	Shannon's ionic radii of X-site ion
r_{MII} [Å]	Shannon's ionic radii of M(II)-site ion (seven-coordination for X = F^-, eight-coordination for X = Cl^- and Br^-)
R_Z [Å]	Shannon's ionic radii of Z-site ion
Av CR [Å]	Average crystal radius = $[(r_{M(I)}x4) + (r_{M(II)}x6) + (r_Xx6) + (r_Ox24) + (r_Zx2))]/42$
M_{EN} - O_{EN}	Electronegativity difference between M and O atom
X_{EN} - O_{EN}	Electronegativity difference between X and O atom
M_{EN} - Z_{EN}	Electronegativity difference between M atom at M(II) site and Z atom
M_{EN} - X_{EN}	Electronegativity difference between M atom at M(I) site and X atom
M(I) – O(1) [Å]	Distance between M(I) atom and O(1) atom
M(I) – $O(1)^{M(I)z=0}$ [Å]	Distance between M(I) atom and O(1) atom with the constraint $z = 0$ at M(I)
$\Delta_{M(I)-O}$ [Å]	Difference in the lengths M(I) – O(1) and M(I) – O(2).
$\Delta_{M(I)-O}^{M(I)z=0}$	Difference in the lengths M(I) – O(1) and M(I) – O(2) with the constraint $z = 0$ at M(I)
$\Psi_{M(I)-O}$ [°]	Angle that M(I)-O(1) bond makes with respect to c
$\Psi_{M(I)-O}^{M(I)z=0}$ [°]	Angle that M(I)-O(1) bond makes with respect to c with the constraint $z = 0$ at M(I)
$\delta_{M(I)}$ [°]	Counter-rotation angle of $M(I)O_6$ structural unit
$\phi_{M(I)}$ [°]	Metaprism twist angle ($\pi/3 - 2\delta_{M(I)}$)
$\alpha_{M(I)}$ [°]	Orientation of $M(I)_6$ unit with respect to a
<X-O> [Å]	Average X-O bond length
$<\tau_{O-X-O}>$ [°]	Average O-X-O bond-bending angle
$\rho_{M(II)}$ [Å]	M(II)-M(II) triangular side length
M(II) – Z [Å]	Orientation of M(II)-M(II)-M(II) triangles with respect to a
M(II) – O(3) [Å]	Distance between M(II) and O3 atoms
$\varphi_{O(3)-M(II)-O(3)}$ [°]	O(3) – M(II) – O(3) angle
E_{total} [eV]	Total energy calculated from ab initio calculations

Table 3. The list of 29 discrete descriptors of electronic and crystal structure parameters [23].

The fluorellestadite apatite and its solid solutions are minor components of many fluorine-mineralized clinkers. It is stable to liquidus temperature of 1240°C at which it incongruently melts to dicalcium silicate ($2CaO{\cdot}SiO_2$) and liquid [16]. The solid-state synthesis and the luminescence properties of europium-doped fluorellestadite (CSSF:Eu^{2+}) cyan-emitting phosphor were described by Que et al [20]. Ellestadite apatites and their solid solutions are promising materials for the immobilization of toxic metals or hazardous fly ash [21],[22].

The general composition of silico-sulfate apatites, i.e. **ellestadites,** is $Ca_{10}(SiO_4)_{3-x/2}(PO_4)_x(SO_4)_{3-x/2}(F,Cl,OH)_2$ and their structures conform to $P6_3/M$ hexagonal symmetry where F^- is located at the 2a (0, 0, ¼) position, while Cl^- is displaced out of the 6h Ca(2) triangle plane and occupies 4e (0, 0, z) split positions with z ranging from 0.336(3) to 0.4315(3). Si/S randomly occupies the 6h tetrahedral site [19],[21].

The syntheses of Sr and Pb analogues of CSSF are also reported [18]. Strontium silico-sulfate apatite is not stable and decomposes to the mixture of strontium silicate and sulfate when heated to 1130°C for 30 min. Since high temperatures must be avoided, several attempts to prepare cadmium and barium silico-sulfate apatites were unsuccessful and the silicocarnotite-like phase was obtained from a mixture of the composition of $Ca_{10}(GeO_4)_3(SO_4)_8F_2$ rather than apatite [17].

Since there is a huge potential for the substitution in apatite structure $(M(1)_4M(2)_6(XO_4)_6Z_2$ and for the formation of solid solution as well, the classification method enables to identify the key crystallographic parameters which can serve as strong classifiers of crystal chemistries. The structure maps for apatite compounds via data mining were reported by Balachandran and Rajan [23]. The selection of the pair of key parameters from a large set of potential classifiers is accomplished through the linear data dimensionality reduction method. This structure can be represented as a 29-dimensional vector, where the vector components are discrete scalar descriptors (**Table 3**) of electronic and crystal structure attributes utilized for the construction of the map of apatite compounds.

Basically, the structure map approach involves the visualization of the data of known compounds with known crystal structures in a two-dimensional space using two scalar descriptors (normally heuristically chosen), which are associated with physical/chemical properties, crystal chemistry or electronic structure. The objective is to map out the relative geometric position of each structure type from which one tries to discern qualitatively if there are strong associations of certain structure types to certain bivariate combinations of parameters [23].

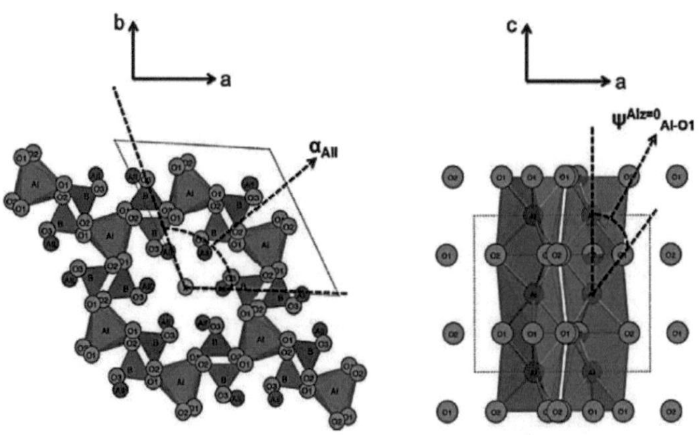

Fig. 2. Bond-distortion angle applied for the construction of structure map [23].

A new structure map, defined using the two distortion angles (**Fig. 3**) [23]:

1. 1.The α_{MII} (rotation angle of M^{II}-M^{II}-M^{II} triangular units);
2. The $\Psi_{MI-O1}^{MIz=0}$ (angle that M(I) – O1 bond makes with the *c* axis when *z* = 0 for the M(I) site).

That enables to classify the apatite crystal chemistries based on the site occupancy at M, X and Z sites and this classification is accomplished using the K-means clustering analysis (**Fig. 3**).

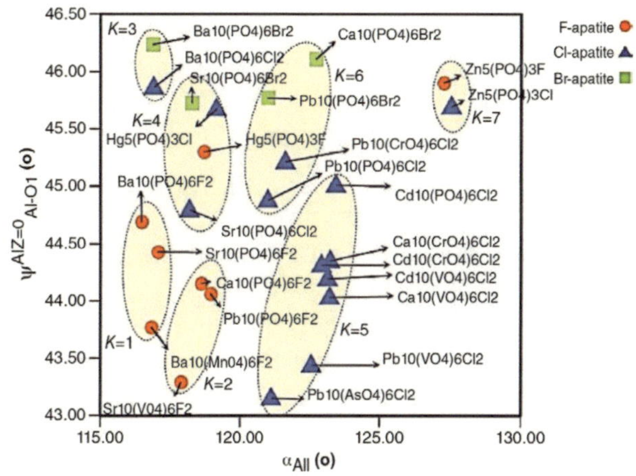

Fig. 3. Structure map for the classification of apatite chemistries based on the site occupancy (**Table 4**) at M, X and Z sites [23].

For example, clusters 1 and 2 (*k* = 1 and *k* = 2) correspond to F-apatites (fluorapatites). They are well localized in the structure map and are characterized by relatively low α_{MII} and $\Psi_{MI-O1}^{MIz=0}$. Two F-apatites which do not belong to the clusters *k* = 1 and *k* = 2 are $Hg_5(PO_4)_3F$ (in *k* = 4) and $Zn_5(PO_4)_3F$ (in *k* = 7). While the existence of fully stoichiometric $Zn_5(PO_4)_3F$ apatite compound is uncertain due to relatively small ionic size of Zn^{2+} cations, the relative position of $Hg_5(PO_4)_3F$ suggests some peculiar characteristics [23].

Cluster	Site occupancy		
k = 1 and *k* = 2	Site	M	Ba, Pb, Sr, Ca
		X	P, V, Mn
		Z	F
k = 3	Site	M	Ba
		X	P
		Z	Cl, Br

Cluster	Site occupancy		
$k = 4$	Site	M	Sr, Hg
		X	P
		Z	Cl, Br
$k = 5$	Site	M	Ca, Cd, Pb
		X	V, Cr, As
		Z	Cl
$k = 6$	Site	M	Ca, Pb
		X	P
		Z	Cl, Br
$k = 7$	Site	M	Zn
		X	P
Z	Z	Z	F, Cl

Table 4. The relationship linking various clusters shown in **Fig. 3** with the site occupancy in the apatite unit [23].

Even though Ca^{2+} and Hg^{2+} cations have roughly the same ionic size (1.18 and 1.23 Å at M(I) site), their electronegativity data indicates that Hg atoms (electronegativity value of 2 in Pauling scale) are relatively highly covalent compared to Ca atoms (electronegativity value of 1 in Pauling scale). In the structure map, this covalent character is predicted to be manifested in the bond distortion angle $\Psi_{MI-O1}^{MIz=0}$ [23].

6.1. Cationic substitution at M(1) and M(2) sites

6.1.1. Strontium-substituted apatites

Sr^{2+} ion, which is larger than Ca^{2+}, is ordered almost completely into the smaller Ca(2) site in the apatite structure (**Fig. 4**). The bond valence sums of Sr ions at two sites demonstrate that Sr is severely overbonded at apatite Ca site but less at Ca(2) site. Complete ordering of Sr into Ca(2) sites has important implications for the diffusion of that element in the apatite structure. It is the subject of several recent studies. The diffusion of Sr in (001) was shown to be as rapid or even more rapid than the diffusion parallel to [001]. As there are neither sites available for Sr, which are linked in (001), nor any interstitial sites, which can contain Sr^{2+} ion, the diffusion mechanism involving the vacancies or defects or both is indicated [24].

A series of Sr-substituted hydroxyapatites, $(Sr_xCa_{1-x})_5(PO_4)_3OH$, where $x = 0.00, 0.25, 0.50, 0.75$ and 1.00, was investigated by O'Donnell et al [25]. The lattice parameters (a and c), the unit cell volume and the density were shown to increase linearly with strontium addition and were consistent with the addition of slightly larger and heavier ion (Sr) instead of Ca. There was a slight preference for strontium to enter Ca(2) site in mixed apatites.

http://dx.doi.org/10.5772/62213

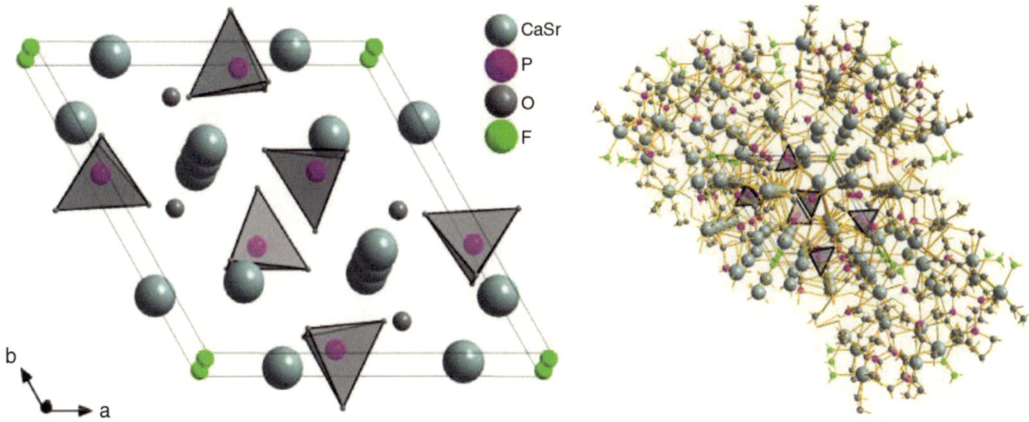

Fig. 4. The structure of natural Sr-bearing apatite refined by Hughes et al [24] and viewed along the c-axis.

Strontium is often substituted for calcium in order to confer the radio-opacity in glasses used for dental cements, biocomposites and bioglass-ceramics. It can be concluded that strontium substitutes for calcium with little change in the glass structure as a result of their similar charge to size ratio. Glasses with low content of strontium nucleate in the bulk to form calcium apatite phase. Glasses with medium strontium content nucleate to mixed calcium-strontium apatite at the surface and glass fully substituted by strontium to strontium fluorapatite [26].

6.1.2. Magnesium-substituted apatite

Magnesium-substituted hydroxyapatite (MgHAP) powders with different crystallinity levels, prepared at room temperature via a heterogeneous reaction between $Mg(OH)_2/Ca(OH)_2$ powders and $(NH_4)_2HPO_4$ solution using the mechanochemical- hydrothermal route, were reported by Suchanek et al [27]. The as-prepared products contained unreacted $Mg(OH)_2$ and therefore had to undergo the purification in ammonium citrate aqueous solutions at room temperature. MgHAP contained 0.24 - 28.4 wt.% of Mg and the concentration of Mg was slightly lower near the surface than that in the bulk.

Two effects of different magnesium sources (magnesium nitrate and magnesium stearate) on the synthesis of Mg-substituted hydroxyapatite (Mg-n-HAP) nanoparticles by the co-precipitation method were investigated by Lijuan et al [28]. There was no obvious difference of morphology, nanoparticle size and thermal stability between those two Mg-n-HAPs. However, Mg-n-HAP synthesized by magnesium stearate had lower crystallinity and better dispersibility, suggesting that magnesium stearate was a novel magnesium source to synthesize Mg-n-HAP, which can effectively reduce the powder crystallinity and prevent the aggregation of Mg-n-HAP nanoparticles, owing to the introduction of organic magnesium source, so as to obtain a promising candidate material to prepare Mg-n-HAP/polymer composite used in a variety of bone applications.

6.1.3. Copper-substituted apatites

Copper-substituted hydroxyapatite ($Ca_{10-x}Cu_x(PO_4)_6(OH)_2$ (where $x = 0.05 – 2.0$) and fluorapatite $Ca_{10-x}Cu_x(PO_4)_6F_2$ ($x = 0.05 – 2.0$) were synthesized by Shanmugam and Gopal [29] via the co-precipitation method and subsequent thermal treatment to 700°C for 30 min. Due to its antimicrobial activity, the copper-substituted fluorapatite could be applied as an antimicrobial biomaterial for various purposes like orthopedic and dental implantations.

6.1.4. Nickel-substituted apatites

According to Mobasherpou et al [30], the reaction mechanism corresponding to equimolar exchange of nickel and calcium and yielding to $Ca_{10-x}Ni_x(PO_4)_6(OH)_2$, where x varies from 0 to 10, could be described by the following equation:

$$Ca_{10}(PO_4)_6(OH)_2 + x\ Ni \rightarrow Ca_{10-x}Ni_x(PO_4)_6(OH)_2 + x\ Ca^{2+} \tag{2}$$

In this process, Ni^{2+} ions are first adsorbed onto the surface of hydroxyapatite (surface complexation, **Section 6.5.2**) and then the substitution of Ca^{2+} for Ni^{2+} ions takes place.

6.1.5. Zinc-substituted apatites

Zinc is a common bioelement. The zinc content in human bones ranges from 0.0126% to 0.0217% by weight [7]. Zinc as a cationic substituent in hydroxyapatite provides the option to counteract the effects of osteoporosis [31]. The incorporation of zinc into the HAP structure (Zn-HAP) was abundantly studied, owing to the key effect of Zn^{2+} cations in several metabolic processes that makes zinc eligible for use in many biomedical applications and to its possible antimicrobial activity [3].

The results of structure analysis indicated that Zn ions substituted partially for Ca ions in the apatite structure and the upper limit of Zn substitution for Ca in HA was about 20 mol.%. In general, the HAP lattice parameters, ***a*** and ***c***, decreased with Zn addition [32].

Zn-substituted apatite was synthesized by the precipitation method as follows [33]:

$$\begin{aligned}&10x\ Ca^{2+} + 10(1-x)\ Zn^{2+} + 6\ PO_4^{3-} + 2\ OH^- \rightarrow \\ &Ca_{10x}Zn_{10(1-x)}(PO_4)_6(OH)_2\end{aligned} \tag{3}$$

where $0 \leq x \leq 1$. The pH of the solution was adjusted to 8 by aqueous solution of NH_3, and the reaction mixture was kept at 90°C for 5 h with stirring. The resulting suspension was then subjected to suction filtration, and the powdery product was dried at 100°C for 10 h. It is known that the usage of chloride or nitrate of calcium as a starting reagent may cause the incorporation of Cl^- or NO_3^- into the structure of apatite. This can be avoided by the utilization of acetate salts, because acetate ions are not incorporated into the apatite, i.e. they would not affect the apatite structure.

http://dx.doi.org/10.5772/62213

6.1.6. Iron-substituted apatites

The synthesis and the characterization of iron-substituted hydroxyapatite via a simple ion-exchange procedure were described by KRAMER et al [34]. Using a ferric chloride solution and a simple soaking procedure, FeHAP can be prepared with no apparent formation of the second phase. The substitution of Fe^{3+} into the HAP lattice results in FeHAP powders with magnetic properties. This novel simplified room temperature soaking procedure can be applied in the future to synthesize magnetic apatite-based nanoparticles for biomedical applications.

6.1.7. Cobalt-substituted apatites

The synthesis and the characterization of cobalt-substituted hydroxyapatite (Co-HAP) powders via the precipitation method were described by KRAMER et al [35]. Using a cobalt chloride solution and a simple soaking procedure, it is possible to prepare CoHAP with no apparent formation of second phases. Cobalt-substituted samples displayed paramagnetic properties as opposed to the diamagnetism of pure HA. The degradation studies showed that Co-HAP did not display markedly different degradation behavior from pure HAP, and the amount of cobalt released over the course of a month was extremely low, alleviating the toxicity concerns. Cobalt-substituted hydroxyapatite nanoparticles, the biomaterial with magnetic properties, could be a promising material to be used in a variety of biomedical applications, including the magnetic imaging, drug delivery or hyperthermia-based cancer treatments.

Single crystals of chlorapatite [$Ca_5(PO_4)_3Cl$] with the substitution of approximately 20% of Ca^{2+} by Co^{2+} (space group $P6_3/M$, $\boldsymbol{a}$ = 9.625(3) and $\boldsymbol{c}$ = 6.747(1) Å, $\boldsymbol{V}$ = 541.3 Å^3) were prepared by ANDERSON and KOSTINER [36] via the flux techniques. Co ion is present at available six-coordinated cation site.

6.1.8. Manganese-substituted apatites

Naturally occurring manganese-substituted apatite is known as manganese-bearing apatite ($Mn,Ca_5(PO_4)_3F$, **Section 1.1**). According to the findings of HUGHES et al [24], the symmetry of Mn-bearing apatite does not degenerate from $P6_3/M$ to P6 or P3 with Mn substitution, nor degenerates the symmetry limit of Mn substitution to one atom/unit cell. Mn^{2+} substituent, which is smaller than Ca^{2+}, preferentially occupies larger apatite Ca(l) site although not completely. Mn atom is underbonded at Ca site, less so at larger Ca(l) site; nine O atoms coordinating that site satisfy more effectively the Mn bond valence than seven ligands coordinating Ca(2) site. The M-O bond lengths of Mn-substituted sites reflect the substitution of the smaller Mn ion.

It is interesting to note that apatite acts effectively as a geochemical sieve that traps Mn^{2+} and excludes Fe^{2+} elements, which are virtually inseparable in most geochemical systems. The bond valence sums for Fe^{2+} at apatite Ca sites yield 1.26 and 1.19 valence units for Ca(1) and Ca(2) sites, respectively; large discrepancy from the formal valence prohibits extensive substitution of Fe^{2+} in the apatite structure (**Fig. 5**) [24].

The crystal structure of pale blue transparent Mn-rich fluorapatite (9.79 wt.% of MnO) with optimized formula $(Ca_{8.56}M^{2+}{}_{1.41}Fe^{2+}{}_{0.01})P_6O_{24}F_2$ was resolved by HUGHES et al [24] to be of the

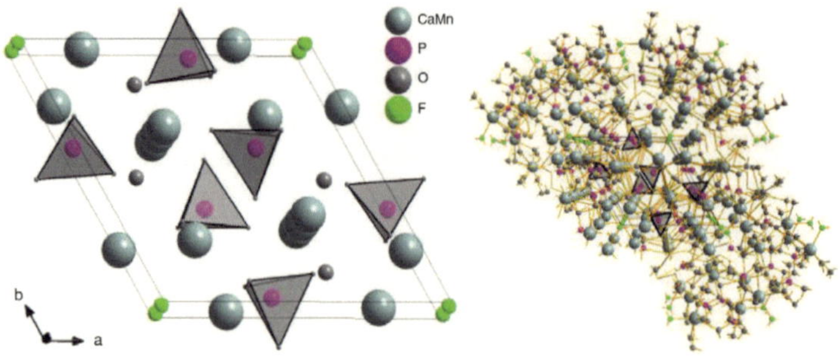

Fig. 5. The structure of natural Mn-bearing apatite refined by HUGHES et al [24] and viewed along the c-axis.

space group of $P6_3/M$ with the cell parameters $a = 9.3429$ Å, $c = 6.8110$ Å and $Z = 2$. Mn is strongly ordered at Ca(1) site ($Ca(1)_{0.72}Mn_{0.28}$, $Ca(2)_{0.96}Mn_{0.04}$). The apatite structure also contains Mn^{5+} at X-site ($P_{5.96}Mn^{5+}{}_{0.04}$).

6.1.9. Substitution of REEs in apatite

Crystals of La-, Gd- and Dy-bearing fluorapatite [La-FAP, Gd-FAP, Dy-FAP, $Ca_{10-x-2y}Na_yREE_{x+y}(P_{1-y}Si_xO_4)Z_2$, where $x = 0.24 - 0.29$ and $y = 0.32 - 0.36$] were synthesized by hydrothermal route by FLEET and PAN [37]. The substitution of trivalent REE in apatite is principally compensated as follows:

$$REE^{3+} + Si^{4+} \Leftrightarrow Ca^{2} + P^{5+} \quad \Rightarrow \quad Ca_{10-x}REE_x\left(P_{1-x}Si_xO_4\right)_6Z_2 \tag{4}$$

$$REE^{3+} + Na^{+} \Leftrightarrow 2\ Ca^{2} \quad \Rightarrow \quad Ca_{10-2y}Na_yREE_y\left(PO_4\right)_6Z_2 \tag{5}$$

$$REE^{3+} + O^{2-} \Leftrightarrow Ca^{2} + Z^{-} \quad \Rightarrow \quad Ca_{10-2z}REE_z\left(PO_4\right)_6Z_{2-z}O_z \tag{6}$$

$$2\ REE^{3+} + [V] \Leftrightarrow 3\ Ca^{2+} \quad \Rightarrow \quad Ca_{10-3w}REE_{2w}[V]_w\left(PO_4\right)_6Z_2 \tag{7}$$

The structure of some REE-bearing apatites [37],[38] is shown in **Fig. 6**.

The partitioning of REE between two Ca positions in apatite contradicts usual first-order dependence on spatial accommodation, with LREE[2] [39],[40],[41],[42],[43], in particular, favoring the smaller Ca(2) position. This behavior was variously ascribed to the control via [37]:

---

[2] Rare-earth elements or metals (REE or REM) are Sc, Y and lanthanoids [40]. Light rare-earth elements (LREE) are Sc, La, Ce, Pr, Nd, Pm, Sm, Eu and Gd (7 elements from La to Eu are known as the cerium group or cerium-group lanthanides). Heavy rare-earth elements (HREE) are Y, Tb, Dy, Ho, Er, Tm, Yb and Lu. The definition of LREE and HREE is based on the electron configuration. LREEs possess unpaired 4*f* electron from 0 to 7 (half-filled 4*f* electron shell). HREEs have paired electron (the clockwise and counterclockwise spinning election) [39],[41]. The element with half-filled *f*-electron shell (Eu) shows enhanced stability of its particular electron configuration [43]. In some cases, REEs are divided into three groups including (1) LREE (La – Pm), (2) MREE (middle rare-earth element, Sm – Dy) and (3) HREE (Ho – Lu) [42].

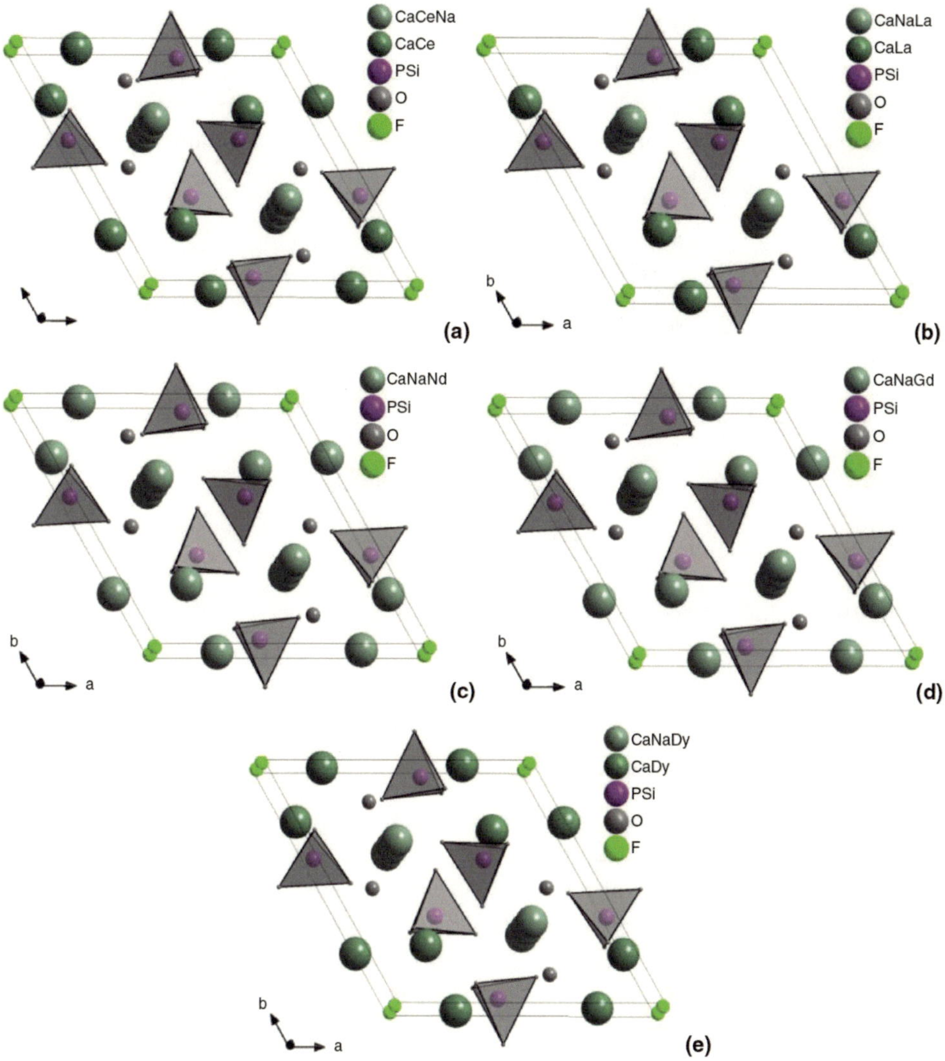

Fig. 6. Rare-earth-element ordering and structural variations in natural rare-earth-bearing fluorapatites (a), LaFAP (b), NdFAP (c), GdFAP (d) and DyFAP (e) [37],[38].

i. Substitution mechanism;

ii. Electronegativity difference;

iii. Bond valence.

The preference of individual REE among multiple Ca positions in minerals (the site occupancy of individual REE) was not extensively studied because of the inability of conventional diffraction methods to distinguish among individual elements at multiply-occupied sites. The site preference for individual LREE from theoretical bond-valence sums was estimated by

HUGHES et al [38], reasoning that La → Pr should preferentially substitute into Ca(2), whereas Pm → Sm should selectively substitute into Ca(1).

The isomorphic substitutions of neodymium for strontium in the structure of synthetic $Sr_5(VO_4)_3OH$ apatite structure type ($P6_3/M$) were reported by GET'MAN et al [44]. The synthesis of apatite specimen was performed via the solution thermolysis on the assumption of the following reaction:

$$(5-x)\ Sr(NO_3)_2 + \frac{x}{2} Nd_2O_3 + 3\ NH_4VO_3 \rightarrow \\ Sr_{5-x}Nd_x(VO_4)_3(OH)_{1-x}O_x + \ldots \tag{8}$$

where x = 0, 0.02, 0.08, 0.10, 0.12, 0.14, 0.16, 0.18 and 0.20. The substitution scheme can be expressed as:

$$Sr^{2+} + OH^- \Longleftrightarrow Nd^{3+} + O^{2-} \tag{9}$$

The procedure includes three stages:

a. Preparation of solution;

b. Thermolysis;

c. Treatment of the dry residue.

The solutions for the thermolysis were prepared by dissolving $Sr(NO_3)_2$ in water; Nd_2O_3 was dissolved in water with nitric acid added; NH_4VO_3 was dissolved in water with hydrogen peroxide added. Dry residues after concentrating the solutions were pestled in an agate mortar and calcined with the temperature steadily raised from 600 to 800°C and intermittent grindings [44].

6.1.10. Actinides-bearing apatites

Thorium and uranium (actinides [45],[46][3])-bearing apatites were synthesized by LUO et al [47] from doped phosphate-halide-rich melts. The structure refinements (**Fig. 7**) of U-doped fluorapatites indicate that U substitutes almost exclusively into Ca(2) site with the site occupancy ratios UCa(2)/UCa(1), which range from 5.00 to 9.33. Similarly, the structure refinements of Th-doped fluorapatites indicate that Th substitutes dominantly into Ca(2) site with ThCa(2)/ThCa(1) values, which range from 4.33 to 8.67.

---

[3] The actinides occupy the second row of the f-block in the periodic table. The actinide group or actinoids (An) include 14 elements with atomic numbers from 90 (Th) to 103 (Lr) [43]. The elements with atomic numbers greater than 92 (U) are termed as transuranes. The elements with atomic numbers greater than 100 are named as the super-heavy elements (SHE). There is also the concept of the periodic table developed by G.T. SEABORG predicting a new inner transition series of 32 elements (from 122 to 153 element), called the superactinite series [45],[46]. Only actinium, thorium, uranium and (in trace quantities) protactinium and plutonium are primordial, while the elements from neptunium onwards are present on Earth solely through artificial generation [46].

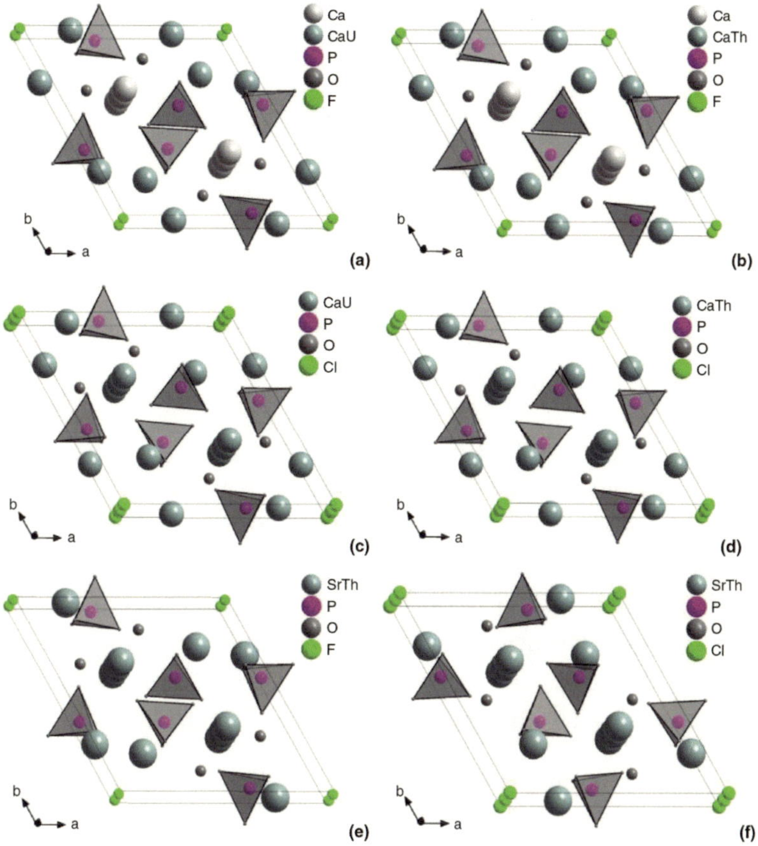

Fig. 7. The structure of UFAP (a), ThFAP (b), UClAP (c), ThClAP (d), ThSrFAP (e) and ThSrClAP (f) [47] viewed along the *c*-axis.

The structure refinements of U-doped chlorapatites show that U is essentially distributed equally between the two Ca sites with UCa(2)/UCa(1) values, which range from 0.89 to 1.17. The results of Th-doped chlorapatites show that Th substitutes into both Ca(1) and Ca(2) sites with ThCa(2)/ThCa(1) values, which range from 0.61 to 0.67. In Th-doped strontium apatites with F and Cl end-members, Th is incorporated into both Ca(1) and Ca(2) sites. The range of ThCa(2)/ThCa(1) values is 0.56 to 1.00 for the F end-member and 0.39 to 0.94 for the Cl end-member. U-doped samples indicate that U in fluorapatite is tetravalent, whereas, in chlorapatite, it is heterovalent but dominantly hexavalent [47].

Based on the chemical analyses of U-, Th-doped fluor-, chlor- and strontiumapatite specimens in this study, local charge compensation may be maintained by the following coupled substitutions (M represents U or Th and [] represents the vacancy) [47]:

$$M^{4+} + [\ ] \rightarrow 2\ Ca^{2+} \tag{10}$$

$$M^{6+} + 2\,[\ ] \rightarrow 3\ Ca^{2+} \tag{11}$$

$$M^{4+} + 2\ Na^{+} \rightarrow 3\ Ca^{2+} \tag{12}$$

$$M^{6+} + 4\ Na^{+} \rightarrow 5\ Ca^{2+} \tag{13}$$

$$M^{4+} + 2\ Si^{4+} \rightarrow Ca^{2+} + 2\ P^{5+} \tag{14}$$

$$M^{6+} + 2\ Si^{4+} \rightarrow Ca^{2+} + 4\ P^{5+} \tag{15}$$

The incorporation of U and Th into fluorapatite results in a decrease of the size of both Ca polyhedra, but the incorporation of U and Th into chlorapatite gives rise to an increase in the volume of both Ca polyhedra. The decrease of both Ca polyhedral volumes in fluorapatite caused by the substitution of U and Th can be explained by the decrease of ionic radius from Ca to U and Th. However, the increase in the volume of both Ca polyhedra in chlorapatite is hard to understand. Because of the effect on Ca(2) polyhedron caused by the replacement of F^- by Cl^-, it can be explained by the structural distortion of Ca(2) polyhedron [47].

Uranium-doped oxy-silicophosphates (britholites) of the composition of $Ca_xLa_y(SiO_4)_{6-u}(PO_4)_uO_t{:}U^{4+}$ were synthesized by Riadh et al [48] via the high-temperature solid-state reaction. The uranium solubility limit was found to be comprised between 4.6 and 4.8 mol.%. The investigation of uranium heated to 1200°C led to the uranium diffusion coefficient of $2.14\cdot10^{-14}$ $m^2\cdot s^{-1}$. The synthesis and the characterization of uranium- ($Ca_9Nd_{1-x}U_x(PO_4)_{5-x}(SiO_4)_{1+x}F_2$) and thorium-bearing britholites ($Ca_9Nd_{1-x}Th_x(PO_4)_{5-x}(SiO_4)_{1+x}F_2$) were also reported by Terra et al [49],[50]:

$$\begin{gathered}\frac{1}{4}Nd_2O_3 + \frac{9}{4}Ca_2P_2O_7 + \frac{7}{2}CaCO_3 + CaF_2 + \frac{3}{2}SiO_2 + \frac{1}{2}An^{4+}O_2 \rightarrow \\ Ca_9Nd_{0.5}An^{4+}_{0.5}(PO_4)_{4.5}(SiO_4)_{1.5}F_2 + \frac{7}{2}CO_2(g)\end{gathered} \tag{16}$$

or

$$\begin{gathered}\frac{1}{4}Nd_2O_3 + \frac{7}{4}Ca_2P_2O_7 + \frac{9}{2}CaCO_3 + CaF_2 + \frac{3}{2}SiO_2 + \frac{1}{2}An^{4+}P_2O_7(\alpha) \rightarrow \\ Ca_9Nd_{0.5}An^{4+}_{0.5}(PO_4)_{4.5}(SiO_4)_{1.5}F_2 + \frac{9}{2}CO_2(g)\end{gathered} \tag{17}$$

where An^{4+} substitutes for tetravalent U^{4+} and Th^{4+}.

The incorporation of thorium in the structure is probably possible due to small differences of ionic radius between calcium (1.06 Å), neodymium (1.05 Å) and thorium (1.00 Å). In order to ensure the quantitative incorporation of thorium, it appeared necessary to consider the coupled substitution [50].

$$Nd^{3+} + PO_4^{3-} \Leftrightarrow Th^{4+} + SiO_4^{4-} \quad (18)$$

instead of the substitution scheme:

$$Nd^{3+} + F^{-} \Leftrightarrow Th^{4+} + O^{2-} \quad (19)$$

Indeed, in the first way, homogeneous and single-phase solid solutions were prepared from $Ca_9Nd(PO_4)_5(SiO_4)F_2$ to $Ca_9Th(PO_4)_4(SiO_4)_2F_2$ leading to full neodymium substitution. Associated small increase of the unit cell parameters results from the simultaneous replacement of phosphate groups by bigger silicate. It was accompanied by a significant change in the grain morphology. These results contrast with those obtained when the coupled substitution according to **Eq. 19** was performed, which confirmed the limitation of about 10 wt.% in the Th substitution [50]. Good resistance of the materials to influence of aqueous solutions enables their utilization for the immobilization of tetravalent actinides in phosphate ceramics [49].

6.1.11. Silver-substituted apatite

The favorable biocompatibility of hydroxyapatite (HA) makes it a popular bone graft material as well as a coating layer on metallic implant. One common and accepted strategy to prevent the implant-related infections is to create antibacterial properties for the implant. Silver ions can be either incorporated into the apatite during the co-precipitation process (AgHAP-CP) or subjected to the ion exchange with calcium ions in apatite (AgHAP-IE). The incorporation of silver ions into apatite is based on the equation [51]:

$$\begin{aligned} &(10-y)\ Ca^{2+} + y\ Ag^{+} + 6\ PO_4^{3+} + (2-y)OH^{-} \rightarrow \\ &Ca_{10-y}Ag_y(PO_4)_6(OH)_{2-y} \end{aligned} \quad (20)$$

where y is the molar amount of silver to be incorporated. However, the distribution of silver ions in AgHAP-CP and AgHAP-IE was different, thus affecting the antibacterial action.

6.1.12. Cadmium-substituted apatites

The absorption of cadmium cations in apatites is relevant both from the medical standpoint of cadmium uptake by human bones, as well as since cadmium migration in nature involves the absorption and desorption equilibria with natural minerals, including apatites. Cadmium has a slight preference for Ca(I) site in fluorapatite and for Ca(II) site in hydroxyapatite [7], [52]. The interactions between these two ions (Cd and Ca) during absorption and ionic change processes in apatites present therefore considerable practical and theoretical interest. Cadmium is also a frequent heavy toxic pollutant element in water [7].

Calculated energy differences (*E*) between these sites are of 12 and 8 $kJ \cdot mol^{-1}$ for fluorapatite and hydroxylapatite, respectively. The preference is not strong, and however, a part of the sites of the other type is also occupied by cadmium ions. The relative site occupation can be expressed by the equation [7]:

$$P = \frac{\text{probability of substitution at } Ca(1) \text{ site}}{\text{probability of substitution at } Ca(2) \text{ site}} = \frac{4}{6} \exp\left(-\frac{E}{kT}\right) \tag{21}$$

where $E = E(Cd^{2+} \text{ or } Zn^{2+} \text{ on } Ca(1)) - E(Cd^{2+} \text{ or } Zn^{2+} \text{ on } Ca(2))$. At $T = 298$ K, $P = 85$ and 17 Cd^{2+} in fluorapatite and hydroxylapatite, respectively. From the value of *P* and from the fact that the sum of the two probabilities is 1, one can calculate that the probability of the lower-energy site occupancy is of 99% and 94%, respectively.

Badraoui et al [53] reported that the maximum amount of cadmium substitution for strontium in the system $Sr_{10-x}Cd_x(PO_4)_6Z_2$ (Z = OH and F) accounts for about 40 at% in HAP and for 60 at% in FAP. The increase of cadmium content induces stronger decrease of the c-axis with respect to the a-axis. The structure refinements evidence found a statistical distribution of Cd atoms in $Sr_{10-x}Cd_x(PO_4)_6(OH)_2$ and a light preference for M(1) site in $Sr_{10-x}Cd_x(PO_4)_6(F)_2$. The stability of the system $M_{10-x}M'_x(PO_4)_6Z_2$ (M and M′ = Ca, Pb, Cd, Sr and Z = OH and F) is strongly affected by the polarizability. As a matter of fact, complete miscibility is possible even when the cations exhibit great size differences, provided they are not both soft acids. Otherwise, the presence of two cations with quite different radii and relevant polarizabilities induces important distortions of the apatite unit cell and PO_4 tetrahedra and consequently limits the possibility of mutual substitution.

6.2. Anionic substitution at X-site

Pentavalent arsenic, vanadium and chromium substitution can completely replace phosphorus in calcium, strontium and barium fluor- and chlorapatites. Calcium fluor-vanadate, -arsenate and -chromate structures were distorted compared to normal hexagonal apatite. Manganese completely replaced phosphorus only in barium apatites, while chromium and manganese could not be incorporated into lead apatites. Excluding these exceptions, continuous solid solutions were formed between the phosphate and/or vanadate and the chromate or manganese analogues for given divalent and halide ions [54]. The substitution of CO_3^{2-} ions at X- (carbonate-apatite of A-type) and Z-site (carbonate-apatite of B-type) was already described in **Section 4.6.**

6.2.1. Arsenate substitution in hydroxylapatite

The arsenate (As^{5+}) substitution in the hydroxylapatite structure was examined by LEE et al [55]. The investigation with samples of hydroxylapatite, the As^{5+}-substituted analogue (synthetic analogue of mineral johnbaumite, **Section 1.6.3**) and of intermediate compositions does not provide any evidence of lowering the symmetry below $P6_3/M$. A series of arsenate-substituted hydroxyapatite was also prepared through aqueous precipitation method by ZHU et al [56]. Prepared solid solutions ($Ca_5(P_xAs_{1-x}O_4)_3(OH)$) showed apatite structure for the whole arsenate/phosphate series. With decreasing arsenate content, the particles changed from smaller needle-like to large tabular crystals and the unit cell dimensions *a* and *c* increased but not in fair agreement with Vegard's law[4] [57]. In FT-IR spectra, the area of phosphate peak was gradually suppressed and the area of arsenate peak increased as the proportion of arsenate increased.

Complete $PO_4^{3-} \leftrightarrow AsO_4^{3-}$ substitution was also recognized in experimental studies of apatite analogues, such as in the system $Sr_5(PO_4)_3OH$-$Sr_5(AsO_4)_3OH$ [58]. The Rietveld refinement of $Sr_5(AsO_4)_3Cl$ (pentastrontium tris[arsenate(V)] chloride, 890.31 g·mol^{-1}) from high-resolution synchrotron data was performed by BELL et al [59]. The hexagonal compound crystallizes in the same structure (**Fig. 8**) as other halogenoapatites in the space group $P6_3/M$ with the cell parameters: $\boldsymbol{a}$ = 10.1969 Å, $\boldsymbol{c}$ = 7.28108 Å, $\boldsymbol{V}$ = 655.63 Å^3, $\boldsymbol{c}:\boldsymbol{a}$ = 0.7140 and $\boldsymbol{Z}$ = 2. The structure consists of isolated tetrahedral AsO_4^{3-} anions (As atom and two O atoms have m-symmetry), separated by two crystallographically independent Sr^{2+} cations, which are located on mirror planes and threefold rotation axes, respectively. One Sr atom is coordinated by nine O atoms and the other one by six. Chloride anions (site symmetry $\bar{3}$) are at 2*a* sites and are located in the channels of the structure.

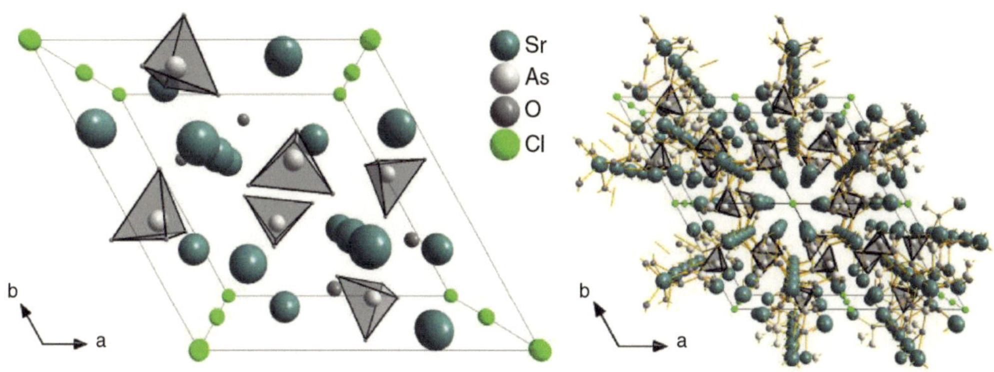

Fig. 8. The structure of $Sr_5(AsO_4)_3Cl$ apatite (perspective view along the c-axis).

[4] Vegard's law, first pronounced in 1921, states that the lattice parameter of a solid solution of two phases with similar structures is a linear function of lattice parameters of the two end-members [57].

6.2.2. Vanadate substitution in hydroxylapatite

The synthesis of synthetic alkaline-earth vanadate hydroxylapatites from hydroxide fluxes was performed by MUGAVERO et al [60]. The hexagonal $Sr_5(VO_4)_3OH$ apatite (pentastrontium tris[vanadate(V)] hydroxide, **Fig. 9(a)**) possesses the $P6_3/M$ space group with the cell parameters: $\boldsymbol{a}$ = 10.0570 Å, $\boldsymbol{c}$ = 7.4349 Å, $\boldsymbol{c}$:$\boldsymbol{a}$ = 0.7393 and $\boldsymbol{V}$ = 651.24 Å^3. The structure of $Ba_5(VO_4)_3OH$ (pentabarium tris[vanadate(V)] hydroxide) apatite is shown in **Fig. 9(b)**. It crystallizes in hexagonal system with the space group $P6_3/M$ and the cell parameters: $\boldsymbol{a}$ = 10.4589 Å, $\boldsymbol{c}$ = 7.8476 Å, $\boldsymbol{c}$:$\boldsymbol{a}$ = 0.7503 and $\boldsymbol{V}$ = 743.43 Å^3.

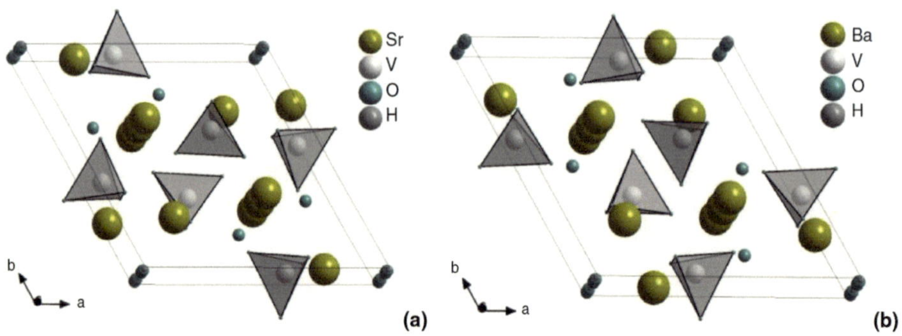

Fig. 9. The structure of $Sr_5(VO_4)_3OH$ (a) and $Ba_5(VO_4)_3OH$ (b) apatite viewed along the c-axis.

The compounds (solid solution) of the composition of $Pb_5(P_xV_{1-x}O_4)_3Cl$ ($0 \le x \le 1$), which are synthetic analogues of minerals pyromorphite, vanadinite and endlichite, were synthesized for the first time by CHERNORUKOV et al [61] via high-temperature solid-phase reactions:

$$
\begin{aligned}
&4.5\ Pb(NO_3)_2 + 0.5\ PbCl_2 + 3x\ (NH_4)_2HPO_4 + 1.5(1-x)\ V_2O_5 \rightarrow \\
&Pb_5(P_xV_{1-x}O_4)_3Cl + 9\ NO_2 + 2.25\ O_2 + 6x\ NH_3 + 4.5x\ H_2O
\end{aligned}
\qquad (22)
$$

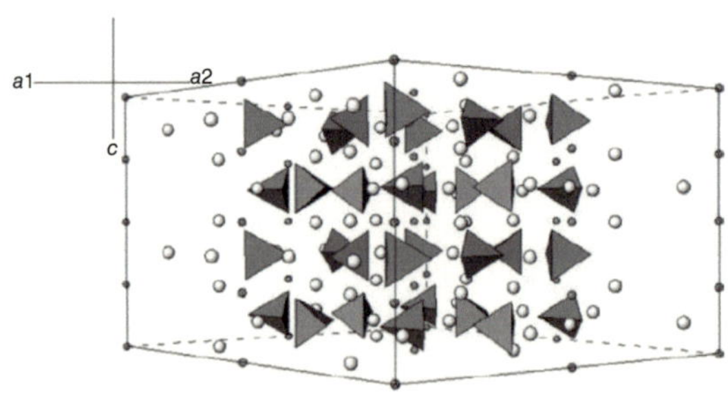

Fig. 10. Fragment of $Pb_5(VO_4)_3Cl$ crystal structure [61].

The variations of unit cell parameters as a function of composition respect Vegard's law. These compounds are structurally built of discrete phosphate or vanadate tetrahedra linked to one another by lead polyhedra, which form joint layers (**Fig. 10**). Apatite-type structures offer typically two crystallographic positions for cations differing in the coordination number and local symmetry. Lead atoms occupying the first positions form polyhedra shaped as three-capped trigonal prisms of PbO_9 having the symmetry C_3, the columns of which run along the threefold axis. Distorted two-capped trigonal prisms of PbO_6Cl_2 residing in the second positions have local symmetry C_1.

6.2.3. Chromium analogues of apatite

The syntheses of chromium (Cr(V) [62]) analogues of apatite were described in literature including the following compounds [63],[64],[65],[66],[67],[67],[69],[70],[71],[72]:

- $Ca_5(CrO_4)_3OH$, which is isomorphous to hydroxyapatite (**Section 1.5.2**): space group $P6_3/M$, $\boldsymbol{a}$ = 9.683 Å and $\boldsymbol{c}$ = 7.010 Å, $\boldsymbol{a}$:$\boldsymbol{c}$ = 1:0.7239, $\boldsymbol{V}$ = 569.20 Å^3, $\boldsymbol{Z}$ = 2.
- $Sr_5(CrO_4)_3OH$ with predicted lattice constants $\boldsymbol{a}$ = 9.9561 Å and $\boldsymbol{c}$ = 7.488 Å.
- $Ba_5(CrO_4)_3OH$.
- $Ca_5(CrO_4)_3F$ with predicted lattice constants $\boldsymbol{a}$ = 9.733 Å and $\boldsymbol{c}$ = 7.0065 Å
- $Ca_5(CrO_4)_3Cl$ with predicted lattice constants $\boldsymbol{a}$ = 10.1288 Å and $\boldsymbol{c}$ = 6.7797 Å.
- $Sr_5(CrO_4)_3F$ with predicted lattice constants $\boldsymbol{a}$ = 9.9349 Å and $\boldsymbol{c}$ = 7.5037Å.
- $Sr_5(CrO_4)_3Cl$ with lattice constant $\boldsymbol{a}$ = 10.125 Å and $\boldsymbol{c}$ = 7.328Å.
- $Sr_5(CrO_4)_3Br$ with predicted lattice constants $\boldsymbol{a}$ = 10.2895 Å and $\boldsymbol{c}$ = 7.2712 Å.

These compounds are in general prepared by the ignition of mixture of alkaline-earth carbonates, hydroxides or oxides with Cr_2O_3 in the presence of water vapor. $Ca_3(CrO_4)_2$ compound (orthochromate), which is isomorphous with $Ca_3(PO_4)_2$, is formed as an intermediate by carrying out the synthesis in dry atmosphere; this compound is often identified as $9CaO{\cdot}4CrO_3{\cdot}Cr_2O_3$ [63],[73],[74],[75].

Theoretical compositions and formula weights of chromium apatite analogues are given in **Table 5**.

Compound	Composition [wt.%]					M
	M	Cr	O	Z	H	[g.mol^{-1}]
$Ca_5(CrO_4)_3OH$	35.44	27.59	36.79	—	0.18	565.39
$Sr_5(CrO_4)_3OH$	54.55	19.42	25.90	—	0.13	803.09
$Ba_5(CrO_4)_3OH$	65.29	14.83	19.78	—	0.10	1051.64
$Ca_5(CrO_4)_3F$	35.32	27.49	33.84	3.35	—	567.38
$Ca_5(CrO_4)_3Cl$	34.33	26.72	32.88	6.07	—	583.33
$Sr_5(CrO_4)_3F$	54.42	19.37	23.85	2.36	—	805.08

Compound	Composition [wt.%]					M
	M	Cr	O	Z	H	[g.mol^{-1}]
$Sr_5(CrO_4)_3Cl$	53.33	18.99	23.37	4.31	—	821.53
$Sr_5(CrO_4)_3Br$	50.59	18.01	22.17	9.23	—	865.98

Table 5. Theoretical compositions of chromium apatite analogues ($M(CrO_3)_4Z$).

$Sr_{10}(CrO_4)_6F_2$ possesses typical hexagonal structure of apatite with the space group $P6_3/M$, which was refined using the powder neutron diffraction (**Fig. 11**) for the first time by Baikie et al [71]. As other chromium analogues of apatite, the material contains chromium in +5 (pentavalent) oxidation state. The material shows the paramagnetic behavior.

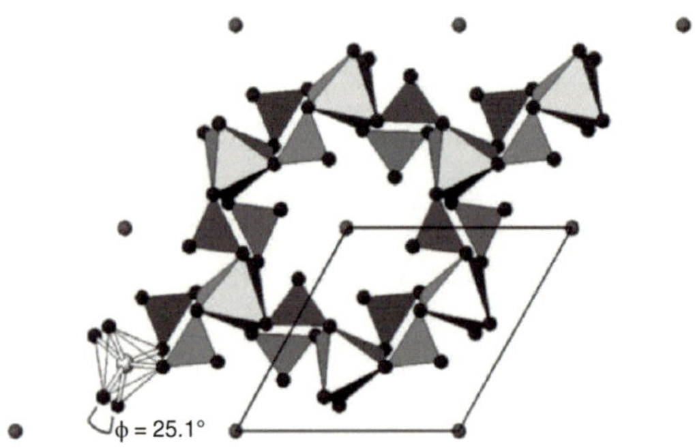

Fig. 11. Structural representation of $Sr_{10}(CrO_4)_6F_2$ apatite with SrO_6 octahedra and CrO_4 tetrahedra: larger and smaller spheres mark F and O atoms, respectively. The unit cell is indicated by black lines [71].

The crystal structure (**Fig. 12**) and the magnetic properties of strontium chromate phase ($Sr_5(CrO_4)_3(Cu_{0.586}O)$) with apatite-like structure were determined by Tyutyunnik and Bazuev [76]. The sample was prepared by solid-state synthesis via the thermal treatment of the mixture of stoichiometric amount of $SrCO_3$, Cr_2O_3 and CuO at the temperature of 1200°C in air for 36 h.

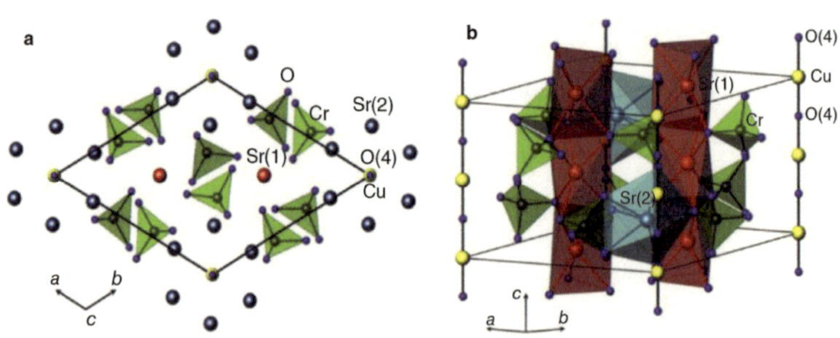

Fig. 12. Crystal structure of $Sr_5(CrO_4)_3(Cu_{0.586}O)$: (a) projection along the c-axis and (b) side view showing the infinite $[CuO]^-$ chains and the coordination polyhedra of Cr and Sr atoms [76].

This phase crystallizes in the space group $P6_3/M$ with hexagonal unit cell parameters: ***a*** = 10.0292 Å and ***c*** = 7.4623 Å, ***V*** = 650.033 Å and ***Z*** = 2. The compound is stable up to 1200°C in air. It was found that copper in the form of Cu^+ cations is located in infinite linear $(CuO)^-$ chains inserted into the tunnels parallel to hexagonal c-axis. The chains contain about 40% vacancies in copper positions. The valence states of Cr and Cu may be mainly +5 and +1, respectively [76].

6.2.4. Selenium analogues of apatite

Selenium oxyanion-substituted hydroxyapatite (SeHAP) was synthesized as a promising material for the treatment of bone cancer to reduce the probability of recurrence, because selenium plays an important role in protein functions and it has significant effect on the induction of cancer cell apoptosis [77]. Another study indicated that selenite (SeO_4^{2-}) or selenite (SeO_3^{2-}) oxyanions exert their cancer chemopreventive effects by direct oxidation of critical thiol-containing cellular substrates and that they are more efficacious anticarcinogenic agents than selenomethionine or selenomethylselenocysteine with a lack of oxidation capability [3],[78].

Selenium was incorporated into the hydroxyapatite lattice by replacing some of the phosphate groups with selenite groups. SeO_4^{2-} (selenate) ion has tetrahedral structure like PO_4^{3-} ion (**Fig. 5** and **Table 4** in **Section 1.2**), but it is slightly larger (2.49 Å) in diameter than phosphate ion, which is 2.38 Å in diameter. By contrast, SeO_3^{2-} (selenite) ion has very similar diameter (2.39 Å), but it has a quite different flat trigonal pyramid geometry. The substitution of bivalent selenium oxyanions forms positively charged vacancy compensated by simultaneous decalcification and dehydroxylation according to the reaction [3],[79]:

$$\begin{aligned} &Ca_{10}(PO_4)_6(OH)_2 + x\ (SeO_n)^{2-} \rightarrow Ca_{10-x}(PO_4)_{6-x}(SeO_n)_x(OH)_{2-x} \\ &+x\ PO_4^{3-} + Ca^{2+} + x\ OH^- \end{aligned} \qquad (23)$$

The SeHAP crystal lattice parameters increased slightly as the Se concentration increased when the Se/P ratios were less than 0.5 [80]. All samples prepared via the precipitation method from aqueous solution by Kolmas et al [79] contained significant amount of carbonates, especially of B-type. Thus, for these samples, the formula $Ca_{10-x}(PO_4)_{6-x}(SeO_n)_x(OH)_{2-x}$ (**Eq. 23**) should be written as $Ca_{10-x-y}(PO_4)_{6-x-y}(SeO_n)_x(CO_3)_y(OH)_{2-x-y}$. Hydroxyapatites containing selenate ions are non-toxic, whereas hydroxyapatite with the highest concentration of selenites is toxic.

6.3. Substitution at Z-site

6.3.1. Nitrogen-containing apatites

Nitrogen was incorporated into hydroxyapatite by dry ammonia treatments at temperatures between 900 and 1200°C in the presence of graphite. The process of synthesis of cyanamidapatite ($Ca_{10}(PO_4)_6CN_2$, $Ca_{10}(PO_4)_6NCN$) can be described by the following chemical equations [13]:

$$2\ NH_3 + C \rightarrow H_2CN_2 + 2\ H_2 \tag{24}$$

$$Ca_{10}(PO_4)_6(OH)_2 + H_2CN_2 \rightarrow Ca_{10}(PO_4)_6CN_2 + 2\ H_2O \tag{25}$$

Ammonia reacts with graphite during the thermal treatment forming $[CN_2]^{2-}$ ions (**Eq. 24**). These cyanamide ions interchange with moveable OH^- ions situated on the sixfold screw axis of apatite to form cyanamidapatite (**Eq. 25**). A similar reaction is known for the synthesis of calcium cyamide from calcium oxide:

$$CaO + 2\ NH_3 + 2\ C \rightarrow CaCN_2 + CO + 3\ H_2 \tag{26}$$

The treatments at temperatures above 1200°C or long-term treatments destroy the apatite lattice completely through the phosphate reduction. Cyanimide ions lose their sites in the apatite lattice and the nitrogen content decreases [13]. The synthesis of $Ca_{10}(PO_4)_6CN_2$ apatite provides the evidence that the hydroxylapatite structure is able to incorporate larger organic molecules [81].

Direct transformation of TCP ($Ca_3(PO_4)_2$) into cyanamidapatite according to the reaction:

$$3\ Ca_3(PO_4)_2 + H_2CN_2 \rightarrow Ca_9(PO_4)_5(HPO_4)HCN_2 \tag{27}$$

was also proposed by Habelitz et al [82].

6.3.2. Peroxide-doped apatites

Although "oxygenated" apatites were not much investigated compared to other substituted apatites, some past studies have, however, reported the possibility of apatitic channels to incorporate oxygenated species such as H_2O_2 or O_2 or molecular ions including O_2^{2-} (the peroxide ion) and superoxide O^{2-}. They are single-phase nanocrystalline apatites, where part of apatitic OH^- ions are replaced by oxygenated species. Typically by peroxide ions (quantified) and at least the traces of superoxide ions can be prepared by the precipitation from aqueous calcium and phosphate solutions in the presence of H_2O_2 under medium room temperature [83],[84].

The local structure of hydroxyl-peroxy apatite was described by Yu et al [85]. Hydroxyl-peroxy apatite contains a small amount of partially dehydroxylated hydroxyapatite phase and calcium hydroxide. The incorporation of peroxide ions into the lattice of HAP causes the perturbations of hydrogen environments and slight changes in its crystal morphology. The distance between H in some structural OH- and adjacent O along the c-axis becomes longer instead of forming the hydrogen bond after the incorporation of peroxide ions.

According to the concentration of peroxide ions in hydroxyl-peroxy apatite and the theoretical value, the corresponding formula for the hydroxyl-peroxy apatite is proposed as follows [85]:

$$Ca_{10}(PO_4)_6(OH)_{1.34-2x}(O_2)_{0.33}(O)_x\square_{0.33+x}.$$

Fig. 13. Possible configuration of hydroxyl ions, peroxide or oxide ions and vacancies in the channel along the crystallographic c-axis in hydroxyl-peroxy apatite. O, H atoms and vacancies are presented by large gray circles, small open circles and gray squares, respectively. Filled small circles represent H atoms perturbed by the incorporation of peroxide ions [85].

A scheme of possible configurations of hydroxyl ions, peroxide or oxide ions and vacancies in the channel along the crystallographic c-axis in hydroxyl-peroxy apatite is illustrated in **Fig. 13**. Peroxide ions incorporated into HAP are located in the channel of apatite structure through the substitution of a portion of OH^- radicals, and the material is a solid solution of hydroxyl- and peroxide apatite.

Zhao et al [86] reported that a new hydrogen bond was formed between peroxide ions and adjacent OH^- radicals in hydroxyl-peroxy apatite. According to the literature [430,446], the formation of hydrogen bond would induce a downfield shift of corresponding proton resonance. Some authors reported a linear correlation of the isotopic proton chemical shift with the O-H…O distance, which was a measure of the hydrogen bond strength. Zhao et al [86] suggested the following mechanism for the incorporation of O_2^{2-}:

$$2\ OH^{-}O^{2-}+[\ ]+H_2O\uparrow \tag{28}$$

where [] was the vacancy. O^{2-} ion was active and could react with O_2 to produce O_2^{2-}.

$$O^{2-}+\frac{1}{2}O_2\rightarrow O_2^{2-} \tag{29}$$

Peroxide ions associated with the vacancies were situated placed in the channel of HA lattice along the c-axis through the substitution of a portion of OH radicals. The molecular ions constituted a symmetric vibrator with a stretching vibration active in Raman spectrometry. This vibration was recorded at 750 cm^{-1} in the Raman spectra of O_2^{2-}-containing HA samples. The final product was a solid solution of hydroxyl- and peroxide-apatite. However, the existence of peroxide ions in the HA lattice caused the contraction of the unit-cell dimensions of HA materials. In addition, a new hydrogen bond was formed between peroxide ions and adjacent OH radicals, which was determined by using molecular spectroscopy analysis. During annealing treatment in air, peroxide ions decomposed and the substituted OH radicals re-entered the HA lattice, resulting in the elimination of the structural aberrations caused by the incorporation of peroxide ions. The concentration of peroxide ions present in HA samples was measured by chemical analysis [86].

6.3.3. Chalcogenide phosphate apatites

The synthesis and the structure of four new chalcogenide[5] [87] phosphate apatitic phases of the composition given by the formula:

1. $Ca_{10}(PO_4)_6S$: a = 9.4619 Å, c = 9.8342 Å, c:a = 0.7223 and V = 529.88 Å^3 (**Fig. 14(a)**);
2. $Sr_{10}(PO_4)_6S$: a = 9.8077 Å, c = 9.2089 Å, c:a = 0.7350 and V = 600.53 Å^3 (**Fig. 14 (b)**);
3. $Ba_{10}(PO_4)_6S$: a = 10.2520 Å, c = 7.6590 Å, c:a = 0.7471 and V = 697.14 Å^3 (**Fig. 14(c)**);
4. $Ca_{10}(PO_4)_6Se$: a = 9.5007 Å, c = 9.8406 Å, c:a = 0.7200 and V = 534.73 Å^3 (**Fig. 14 (d)**).

were reported by Henning et al [88].

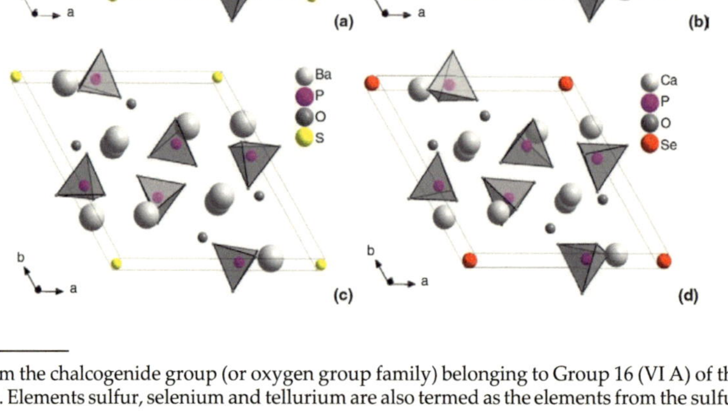

[5] The elements from the chalcogenide group (or oxygen group family) belonging to Group 16 (VI A) of the periodic table: O, S, Se, Te and Po. Elements sulfur, selenium and tellurium are also termed as the elements from the sulfur subgroup [87].

http://dx.doi.org/10.5772/62213

Fig. 14. The structure of $Ca(PO_4)_6S$ (a), $Sr(PO_4)_6S$ (b), $Ba(PO_4)_6S$ (c) and $Ca(PO_4)_6Se$ (d) viewed along the c-axis.

These four apatites are isostructural and crystallize in the trigonal space group P3 over bar with the chalcogenide ion positioned at (001/2). Sulfoapatites show no ability to absorb H_2S in the way that oxyapatite absorbs H_2O at elevated temperatures. This can be attributed to the position of sulfide ion and the way it influences the crystal structure around vacant chalcogenide position at (000) [88].

6.3.4. Metaborate ion-containing apatite phase

Strontium borate-phosphate $Sr_{10}(PO_4)_{5.5}(BO_4)_{0.5}(BO_2)$[6] was prepared from $SrCO_3$, $NH_4H_2PO_4$ and H_3BO_3 at high temperature (from 1150 to 1550°C) and was found to be free of alkali metal compounds. $Sr_{10}(PO_4)_{5.5}(BO_4)_{0.5}(BO_2)$ phase is a derivative of the apatite crystal structure with metaborate ion at Z-site: space group P3, $a = 9.7973$ Å, $c = 7.3056$ Å, $V = 607.29$ Å^3, $Z = 1$ [89],[90],[91].

The strontium sites are found to be fully occupied, while $[PO_4]^{3-}$ tetrahedra are partially replaced by $[BO_4]^{5-}$ groups. The crystal structure contains Sr cations occupying the 6*g* (Sr(1)) and 2*d* (Sr(2), Sr(3)) sites, isolated tetrahedral $[PO_4]^{3-}/[BO_4]^{5-}$ groups and linear $[BO_2]^-$ groups located in hexagonally shaped (trigonal antiprismatic) channels formed by Sr(1) atoms and running along [001] (**Fig. 15**). The space group of the present compound is reduced to P3, because the orientation of the $[PO_4]/[BO_4]$ tetrahedra destroys the mirror plane characteristic for the apatite crystal structure ($P6_3/M$) [89],[91].

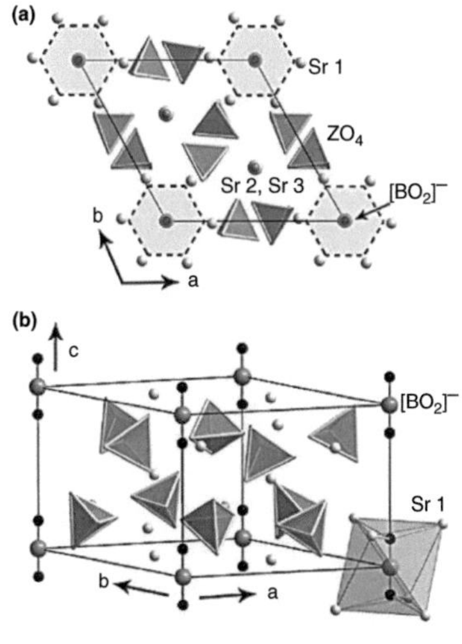

Fig. 15. Crystal structure of $Sr_{10}(PO_4)_{5.5}(BO_4)_{0.5}(BO_2)$: projection along [001] showing the hexagonally shaped channels formed by Sr(1) around the threefold inversion axis ($Z = (P_{0.95}B_{0.05})$ (a) and side view emphasizing the linear $[BO_2]^-$ groups and the corresponding trigonal antiprisma formed by Sr(1) (b) [89].

[6] See also **Section 5.3.**

6.4. Solid solutions of apatites

Crystalline solid solutions[7] [92] of apatites are frequently encountered where the possibility and type depend on the condition of formation or preparation, thermal history after formation and the end-members of the series [2]. The structure of ternary solid solution of hexagonal ($P6_3/M$, $Ca_5(PO_4)_3(F_{0.39}Cl_{0.33}OH_{0.28})$) and monoclinic ($P2_1/B$, $Ca_5(PO_4)_3(F_{0.29}Cl_{0.47}OH_{0.24})$) F-OH-Cl apatite was resolved by HUGHES et al [93]. Phosphate tetrahedra and Ca(1) polyhedra of both structures are generally very similar to analogous polyhedra in the end-member fluor-, chlor- and hydroxylapatite structures. Ca(2) polyhedron, which includes the column anions among its ligands, exhibits significant but regular variations in interatomic distances that can be directly correlated to Cl content.

The solid solution in hexagonal ternary apatite is achieved by a 0.4 Å shift along the c-axis of Cl atom relating to its position in end-member chlorapatite. This adjustment affects the Markovian sequence[8] [94] of anions in the (0,0,z) anion columns by providing a structural environment that includes column OH species at the distance of 2.96 Å from Cl. The shift of Cl atom is accompanied by splitting of Ca(2) atoms into two distinct positions as a function of the kind of anion neighbor (F or OH vs. Cl). Additional nonequivalent Cl site, similar to that in end-member chlorapatite, is also present. Those Cl atoms with adjacent OH occupy a site different from Cl atoms adjacent to vacancies in the anion column [93].

Reduction of symmetry in monoclinic ternary apatite results from the ordering of Cl and OH within the anion columns. The atomic positions of Cl and OH in the anion column are equivalent to those in hexagonal ternary apatite, but each is ordered into only one of the two hexagonal symmetry-equivalent sites [93].

The apatite supergroup minerals of the solid solution [95]:

---

[7] For the purposes of nomenclature, a complete solid-solution series without structural order of ions defining the end-members is arbitrarily divided at 50 mol.% ("50% rule") to two portions with different names. Analogously, the 50% rule applied to members of ternary solid-solution series implies that the mineral names should be given only to the three end-members. Each name should be applied to the compositional range from the end-member to the nearest right bisector of the sides of the composition triangle. For example, the apatite series $Ca_5(PO_4)_3(F,OH,Cl)$ is represented by three compositional fields of fluorapatite (A), hydroxylapatite (B) and chlorapatite (C) [92].

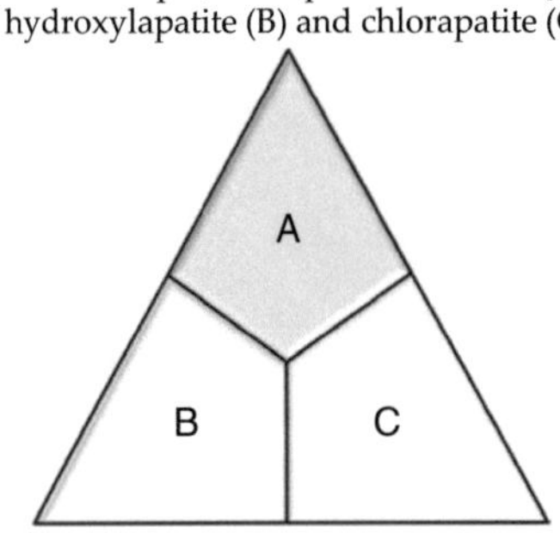

[8] Statistical model where a random sequence k, the probability distribution $p(k)$ of which satisfies the equation [94]:

(a) $p(\bar{k}) = p(k_1)\, p(k_0^{i-1} | k_i)\;\; p(k_{i+1}^{n} | k_i)$;

is referred to as the Markovian sequence or the Markovian chain.

$$Ca_5\left[\left(PO_4\right)_{3-x}\left(AsO_4\right)_x\right]_{3-2y}\left(SO_4\right)_y\left(SiO_4\right)_y]_{2y}\left(OH,F,Cl\right)$$

where $x = 0 - 3$ and $y = 0 - 1.5$ were found in altered calcareous xenoliths within the ignimbrite of the Upper Chegem caldera, Northern Caucasus, Russia. These minerals belonging to the apatite supergroup occur in all zones of skarn from the core to the contact with ignimbrite as follows: brucite-marble, spurrite ($Ca_5(SiO_4)_2CO_3$ [96]), humite ($Mg_7(SiO_4)_3(F,OH)_2$ [97]) and larnite[9] (Ca_2SiO_4 [98],[99],[100],[101]) zones. They are associated with both high-temperature minerals: reinhardbraunsite ($Ca_5(SiO_4)_2(OH)_2$ [102]), chegemite ($Ca_7(SiO_4)_3(OH)_2$ [103]), wadalite ($Ca_6Al_5Si_2O_{16}Cl_3$ [104]), rondorfite ($Ca_8Mg(SiO_4)_4Cl_2$ [105]), cuspidine ($Ca_4(Si_2O_7)F_2$ [106]), lakargiite ($CaZrO_3$ [107]) and srebrodolskite ($Ca_2Fe^{3+}{}_2O_5$ [108]), corresponding to the sanidinite metamorphic facies,[10] and secondary low-temperature minerals: calcium hydrosilicates (hillebrandite [113], awfillite [114], bultfonteinite [115]), hydrogarnets [116] and minerals of the ettringite group [117].

The minerals of the apatite supergroup often form elongated cracked hexagonal or pseudo-hexagonal crystals up to 250 μm in size as well as grain aggregates. A new solid-solution series was found between ellestadite and svabite-johnbaumite (±apatite) with the ellestadite type isomorphic substitution according the following scheme [95]:

$$2\left[RO_4\right]^{3-} \rightarrow \left[SO_4\right]^{2-} + \left[SiO_4\right]^{4-} \tag{30}$$

where $\boldsymbol{R}$ = As^{5+} and P^{5+}. The As content in investigated minerals decreases from the contact skarn zone with the ignimbrite towards the core of altered xenoliths (from 2.11 As pfu[11] to 0), for example [95]:

i. Svabite:

$Ca_5[(AsO_4)_{2.01}(PO_4)_{0.33}(SiO_4)_{0.33}(SO_4)_{0.33}]\cdot{}_3[F_{0.58}(OH)_{0.30}Cl_{0.12}]_{\bullet 1}$;

ii. As-bearing fluorapatite:

$Ca_5[(PO_4)_{1.56}(AsO_4)_{1.06}(SiO_4)_{0.19}(SO_4)_{0.19}]\cdot{}_3[F0_{.59}(OH)_{0.35}Cl_{0.06}]_{\bullet 1}$;

iii. As-bearing hydroxylellestadite:

$Ca_5[(SiO_4)_{1.25}(SO_4)_{1.25}(AsO_4)_{0.43}(PO_4)_{0.07}]\cdot{}_3[(OH)_{0.70}Cl_{0.20}F_{0.10}]_{\bullet 1}$;

iv. Si, S-bearing hydroxylapatite:

$Ca_5[(PO_4)_{0.95}(SO_4)_{0.93}(SiO_4)_{0.93}(AsO_4)_{0.19}]\cdot{}_3[(OH)_{0.67}Cl_{0.18}F_{0.15}]_{\bullet 1}$;

---

[9] The name is also often used for synthetic phase of the same composition (dicalcium silicate, Ca_2SiO_4, C_2S), which is the main component of belite in Portland cement [99],[100],[101].

[10] Since sanidinite facies are formed under conditions of intensive contact metamorphosis at high temperatures and low pressure, volatiles such as carbon dioxide and water are removed from the rock. The components characteristic for the sanidinite facies are sanidine (feldspar, $KAlSi_3O_8$ [109]), corundum (oxide, Al_2O_3 [110]), cordierite (cyclosilicate, $Mg_2Al_4Si_5O_{18}$ [111]), sillimanite (nesosilicate, Al_2OSiO_4 [112]) and glass formed as the product of partial fusion.

[11] The abbreviation for per formula unit (pfu), see also **Footnote 36** in **Chapter 1**.

v. Hydroxylellestadite:

$Ca_5[(SO_4)_{1.49}(SiO_4)_{1.49}(PO_4)_{0.02}] \cdot {}_3[(OH)_{0.74}F_{0.13}Cl_{0.13}]_{\bullet 1}$.

The crystals of As-bearing phases belonging to the investigated solid solution are heterogeneous and small in size. Therefore, X-ray single-crystal data were obtained for only Si, S, As-bearing hydroxylapatite (see the formula above): $P6_3/M$, $\boldsymbol{a}$ = 9.5193 Å, $\boldsymbol{c}$ = 6.9052 Å, $\boldsymbol{V}$ = 541.90 Å^3 and Z = 4. The Raman spectroscopy also confirms that the investigated samples belong to the arsenate phosphate-silicate-sulfate multiple solid solution [95].

The hydrothermal synthesis of vanadate/phosphate hydroxyapatite solid solutions of the composition of $Ca_{10}(VO_4)_x(PO_4)_{6-x}(OH)_2$, where x = 0, 1, 2, 3, 4, 5 and 6, was firstly reported by Onda et al [118]. The lattice parameters increased linearly with increasing content of vanadium according to Vegard's law. The apatite crystals were precipitated in the form of column crystals with the length of about 40 – 100 nm and the width of about 25 – 40 nm. The sizes of the nanoparticle solid solutions increased with increasing vanadium content, whereas the morphology was independent of the vanadate/phosphate ratio. Calcium hydroxyapatites substituted with vanadate were also prepared by Sugiyama et al [119] and used as catalysts in oxidative dehydrogenation of propane. The catalytic activity[12] of vanadate-substituted calcium hydroxyapatites was evidently greater than that of magnesium pyrovanadate, which is one of the most active catalysts for this oxidation.

The crystal structure of 11 samples of synthetic Na-Ca-sulfate apatite systems of the composition of $Na_{6.45}Ca_{3.55}(SO_4)_6(F_xCl_{1-x})_{1.55}$, where $\boldsymbol{x}$ = 0 – 1, was refined by Piotrowski et al [120] in the $P6_3/M$ space group ($\boldsymbol{Z}$ = 1). The sulfate tetrahedra and the two symmetrically independent cation polyhedra around M(1) and M(2) (occupied by Na and Ca, respectively) are generally very similar to analogous polyhedra in phosphate apatites. A common structural feature of all members of the solid-solution series is the deficiency in total Cl^- and F^- content compared to phosphate apatites. The mean value of (Cl + F) for the solid solution equals 1.55(6) atoms per unit cell compared to the ideal value of 2 atoms per unit cell observed in phosphate apatites. The solid-solution series $Na_{6.45}Ca_{3.55}(SO_4)_6Cl_{1.55}$ + $Na_{6.45}Ca_{3.55}(SO_4)_6F_{1.55}$ shows a gap towards the side of fluoride-rich compounds. Under ambient pressure, the gap exists between $0 < \boldsymbol{n}_{Cl}/\boldsymbol{n}_{Cl} + \boldsymbol{n}_F < 0.33$, where $\boldsymbol{n}_{Cl}$ and $\boldsymbol{n}_F$ represent the numbers of Cl- and F-atoms per unit cell, respectively.

Lead apatites form a family of isomorphous compounds, and well-known members of the group are mimetite ($Pb_5(AsO_4)_3Cl$, **Section 1.6.7**) and pyromorphite ($Pb_5(PO_4)_3Cl$, **Section 1.6.4**). Isostructural with vanadinite $Pb_5(VO_4)_3Cl$, these three constituents form a ternary system within the apatite group of $P6_3/M$ symmetry (hexagonal bipyramid). The mimetite and pyromorphite structures can incorporate numerous admixtures, mainly Ca, Ba, As, V, P and others. The most common substitution is the isovalent replacement of part of Pb with Ca and As with P and V. Extensive substitution of $(AsO_4)^{3-}$ group by tetrahedrally coordinated and isovalent $(PO_4)^{3-}$ ion is well established by the existence of a complete solid solution between mimetite and pyromorphite [121].

[12] The utilization of apatites as catalysts is described in **Section 10.7.**

A number of compounds of the mimetite $Pb_5(AsO_4)_3Cl$-pyromorphite $Pb_5(PO_4)_3Cl$ solid-solution series were synthesized at room temperature by Bajda et al [121] and investigated with Raman and infrared spectroscopy. The peak positions of the dominant antisymmetric stretching (ν_3) and bending (ν_4) vibrations in the 720 – 1040 cm^{-1} and 400 – 580 cm^{-1} regions of the Raman and IR spectra of minerals from the mimetite-pyromorphite series vary primarily as a function of the As/(As + P) ratio in the solids' structure. It is due to the effect of the atomic mass and bond forces on the banding energies of the substituting tetrahedra. The observed correlation between the band positions and the extent of the anionic substitution among the series can be used to estimate the As and P content in mimetite-pyromorphite solid solutions [121].

Solid solutions of $Pb_8M_2(XO_4)_6$ lead alkali apatites were studied by Mayer et al [122]. The $Pb_8Na_{2-x}K_x(PO_4)_6$, $Pb_8Na_{2-x}K_x(AsO_4)_6$, $Pb_8Na_{2-x}Rb_x(PO_4)_6$ and $Pb_8K_{2-x}Rb_x(PO_4)_6$ compounds crystallize at all compositions in the $P6_3/m$ hexagonal apatite structure and form true solid solutions.

Some other examples of apatite solid solutions are listed below [2]:

- ▪ $Ca_2Y_8(SiO_4)_6O_2 - Ca_8Y_2(PO_4)_6O_2$;
- $Ca_2La_8(SiO_4)_6O_2 - Ca_8La_2(PO_4)_6O_2$;
- $Ca_2Y_8(SiO_4)_6O_2 - Y_{10}(SiO_4)_4(BO_4)_2O_2$;
- $Mg_2Y_8(SiO_4)_6O_2 - Y_{10}(SiO_4)_4(BO_4)_2O_2$;
- $Pb^{4+}{}_3Pb^{2+}{}_5Y_2(SiO_4)_6O_2 - Pb^{2+}{}_2Y_8(SiO_4)_6O_2$;
- $Ca_{10}(PO_4)_6(OH)_2 - Ca_4Y_6(SiO_4)_6(OH)_2$;
- $M_{10}(PO_4)_6F_2 - M_{10}(PO_4)_6F_2$ (M = Sr, Ba, Pb);
- $M_{10}(PO_4)_6F_2 - M_{10}(MnO_4)_6F_2$ (M = Sr, Ba, Pb).

6.5. Trace elements and their isotopes

Since apatite is an important accessory mineral in most common rock types, it is often used in trace element and isotope investigations of igneous and metamorphic rocks [123]. Stable isotope compositions of biologically precipitated apatite in bone, teeth and scales are widely used to obtain the information on the diet, behavior and physiology of extinct organisms and to reconstruct past climate in terrestrial and marine conditions [124].

Broad spectrum of substitutions in the apatite lattice allows the incorporation of various isotopes, which offer a number of instruments for the interpretation of paleoenvironment and diagenesis. The relative stability of francolite compared to other sedimentary minerals led to an enormous number of studies and applications. Various isotopes occupy the Ca^{2+} and PO_4^{3-} sites in the lattice of apatite (**Fig. 16**).

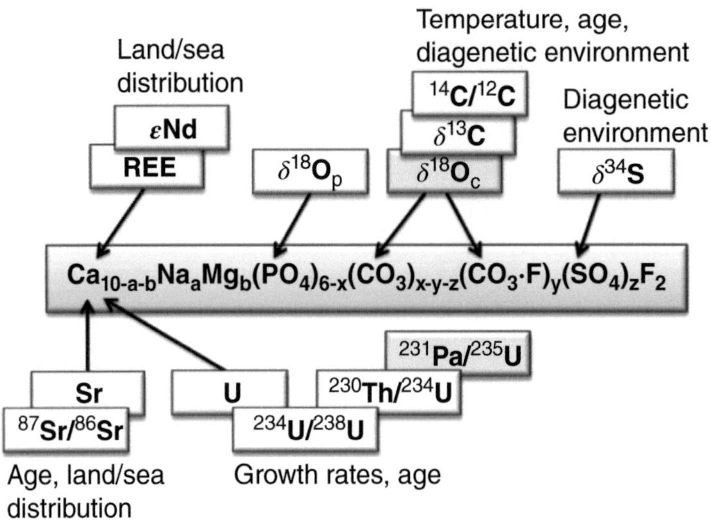

Fig. 16. Possible isotopic substitutions in the structure of francolite [125].

Since the earliest application in deep-time study of Late Cretaceous paleotemperatures in 1950, the oxygen isotope paleothermometry is based on the temperature dependence of oxygen isotope fractionation between authigenic minerals[13] [126] and ambient waters. Under the equilibrium conditions, the $^{18}O/^{16}O$ ratio of sedimentary carbonate and phosphate minerals depends only on the temperature of precipitation and on the $^{18}O/^{16}O$ ratio of ambient water. Thermodynamic relationships and bond vibrational frequencies can be used to determine the mineral-water isotopic fractionation relations but not with the precision and accuracy necessary for the paleothermometry. Such an application requires the calibrations based on mineral-water oxygen exchange experiments at high temperatures, mineral precipitation experiments at low temperatures and/or natural experiments using minerals grown under known conditions [127],[128].

Carbon, oxygen and sulfur isotopes are used to reconstruct the oxygenation stages of the sediments during organic matter degradation and precipitation of apatite. The application of this method gives good results for modern and Neogene deposits. In older occurrences, the signature of carbon and oxygen composition is commonly overprinted by diagenetic and burial diagenesis [125],[129]. The carbon isotope ratios of apatite can be used to interpret the source of carbon in magmas and metamorphic fluids using the assumption that the carbon isotope fractionations between phases are small in igneous and metamorphic systems [123].

The carbon isotope analysis of bioapatites was first applied to terrestrial mammals in early 1980s [130],[131],[132]. While it is now known that some bones do undergo the C-isotope exchange extremely readily, collagen, bone and enamel record different periods of time during

[13] The minerals of sedimentary rocks are subdivided into two main groups [126]: authigenic (formed on their present position) and allogenic (transported to its current position from elsewhere). Both groups will be further described in **Chapter 7**.

the life of a single individual, and the diet may change. That means, there is a fundamental ambiguity (preservation vs. normal intraindividual differences) in interpreting the isotopic differences among different tissues. Unfortunately, early results were taken to imply that all bioapatites are unreliable, and it was not until the 1990s that it became accepted that the tooth enamel, at least, is a robust recorder of diet. Thus, the early work of LEE-THORP and VAN DER MERWE [133],[134] was a struggle against the tide of misplaced opinions [129].

Fossil biogenic apatites display the trace element compositions that can record environmental and biological signals, give insights into past water compositions or be used for dating paleontological and archeological bones and teeth. The mechanisms of the process of trace element and their isotopes incorporation into apatites of skeletal phosphatic tissues are described by REYNARD and BALTER [135]:

1. Partitioning between aqueous fluids and crystals;
2. Surface adsorption, complexation and chelation;
3. Diffusion processes.

6.5.1. Partitioning of divalent cations

Partitioning of divalent cations is defined by the chemical equilibrium expressing the divalent cation (Me^{2+}) exchange between apatite and aqueous solutions [135]:

$$Me^{2+}(aq) + Ca - apatite(ss) \rightarrow Ca^{2+}(aq) + Me - apatite(ss) \tag{31}$$

where (*aq*) and (*ss*) refer to the aqueous solution and to the solid solution, respectively. The equilibrium constant of **Eq. 31** can be written as

$$K_D\left(\frac{Me^{2+}}{Ca}\right) = \frac{\dfrac{x_{Me-apatite}}{x_{Ca-apatite}}}{\dfrac{m_{Me^{2+}}}{m_{Ca^{2+}}}} = \frac{K(T)_{Ca-apatite}\,\lambda_{Ca-apatite}\,\gamma_{Me^{2+}}}{K(T)_{Me-apatite}\,\lambda_{Me-apatite}\,\gamma_{Ca^{2+}}} \tag{32}$$

where $\boldsymbol{x}$ is the molar fraction in apatite solid solution, $\boldsymbol{m}$ is the molality in water, $\boldsymbol{\lambda}$ is the activity coefficient of the component in the solid solution, $\boldsymbol{K}(T)$ is the solubility product of the end-member at temperature $\boldsymbol{T}$ and γ is the ion activity in the aqueous solution, the ratio of which in water is assumed to be equal to one. The activity coefficients in regular solid-solution model are described by Margules parameters[14] [136] and can be approximated by the elastic energy due to the deformation of the host crystal lattice around the substituted cation [135],[137]:

$$W_{G_{ij}} = 4\pi N_A E\left[\frac{r_i}{2}\left(r_j - r_i\right)^2 + \frac{1}{3}\left(r_j - r_i\right)^3\right] \quad (33)$$

where N_A is the Avogadro number, E is the Young's modulus of the crystal, r_i is the ionic radius of cation normally occupying the site in the *i*-compound (Ca in apatite) and r_j is the ionic radius of the substituted cation in compound *j*. The elasticity of hydroxyapatite gives $E = 114 \pm 2$ GPa.

At low concentrations ($X_{Me\text{-}apatite} << 1$) like those of trace elements in biogenic apatites, **Eq. 32** is reduced to the relationships [135]:

$$K_D\left(\frac{Me^{2+}}{Ca}\right) = \frac{K(T)_{Ca-apatite}}{K(T)_{Me-apatite}} \exp\left(\frac{-W_{G_{MeCa}}}{RT}\right) = \exp\left(-\frac{G_{ideal} + W_{MeCa}}{RT}\right) \quad (34)$$

where the term *exp* ($-\Delta G_{ideal}/RT$) is the free enthalpy change of the **reaction 31**, equivalent to the ratio of end-member solubility products. Unlike carbonates, the data of solubility products and thermodynamics for end-member apatites are scarce. When no data are available for the solubility and enthalpy of formation of the end-members, it is assumed that the elastic energy term dominates over the partitioning, i.e. $\Delta G_{ideal} \ll WG_{MeCa}$. Promising ways for obtaining the enthalpies of formation and the substitution energies are the first-principle calculations [138] and the atomistic modeling [135],[139].

For heterovalent substitutions, the equilibrium reaction becomes complex since complementary substitutions are necessary to maintain the charge balance in the crystal. Typically, the substitution of trivalent elements of important rare-earth series requires the compensation by Na^+ for Ca^{2+} at an adjacent site or even more complex substitution scheme involving carbonate groups or fluorine. In that case, most thermodynamic data required for the calculation of the equilibrium constant are not available. Among the series of elements with the same charge and substitution scheme, the pattern of equilibrium constants, or of distribution coefficients, can be approximated by combining **Eqs. 33** and **34** [135]:

---

[14] The Margules equation expresses the Gibbs free energy (G^E) of binary liquid mixture ($x_1 + x_2 = 1$) in the symmetric form [136]:

$\frac{G^E}{RT} = A_1x_1 + A_2x_2$; where $A_1 = A + B$ and $A_2 = A - B$. Applying the partial molar derivative produces the expression for the activity coefficients:

$\ln \gamma_1 = x_2^2[A_2 + 2x_1(A_1 - A_2)]$ and $\ln \gamma_2 = x_1^2[A_1 + 2x_2(A_2 - A_1)]$.

The Margules expressions for activity coefficients are based on the Lewis-Randall standard state (pure substance in the same phase and the same temperature and pressure as the mixture); therefore, they must obey the pure-component limit ($\lim_{x_i \to 1} \gamma_i = 1$). The parameters A_1 and A_2 are simply related to the activity coefficients at infinite dilution: $\lim \gamma_1^\infty = A_1$ and $\lim \gamma_2^\infty = A_2$. For the binary mixture, where $\gamma_1^\infty = \gamma_2^\infty \Rightarrow A_1 = A_2$, the Margules equation collapses to Porter equation. The multicomponent version of the Margules equation is [136]:

$\frac{G^E}{RT} = \sum\sum_{i<j} x_i x_j[A_{ij} + B_{ij}(x_i - x_j)]$.

http://dx.doi.org/10.5772/62213

$$K_D = K_D^0 \exp\left(-4\pi N_A E_{eff}\left[\frac{r_0}{2}\left(r_j - r_0\right)^2 + \frac{1}{3}\left(r_j - r_0\right)^3\right] / RT\right) \tag{35}$$

where E_{eff} is the effective Young's modulus and r_0 is the optimum radius for maximum equilibrium constant K_D^0, all of which will depend on the charge of the considered series of elements. These parameters can be adjusted to experimental data such as partition coefficients between minerals and liquids and lead to parabola-like curves, the position and curvature of which depend on the charge of the element. This approach was applied so far only to rare-earth elements in apatite, where the relative partition coefficients were extrapolated from magmatic temperatures around 800°C to low temperatures appropriate for fossil diagenesis [135].

6.5.2. Complexation of metal cations

The complexation of metal cations in aqueous fluids involves binding with a broad range of molecules from simple inorganic ones (e.g. carbonates, phosphates and sulfates) to complex organic ones (humic acids, amino acids, proteins, enzymes, etc.). For molecules with several bonding sites and structural flexibility (e.g. multidentate or chelator), the complexation is thermodynamically favored with respect to the complexation with several monodentates having one bonding site; the process is named chelation. Chelators can be adsorbed on mineral surfaces while remaining complexed to metallic cations. The pattern of the partition coefficients associated with this process was measured for rare-earth elements complexed with humic acids and manganese oxides. It shows null fractionation along the whole series; the effect of chelation is therefore to screen the trace element in the crystal or ligand field and to suppress the fractionation associated with ionic radius variations and tetrad effects, and most of the anomalies associated with redox of Ce [140]. Similar effects might occur for the adsorption of chelated metals on other mineral surfaces and in particular phosphates. In addition to chelators, the transition metals also form complexes with proteins and enzymes that interact with bones and teeth in living organisms and may influence their incorporation in bioapatite [135].

6.5.3. Diffusion processes

Solid-state diffusion in crystals is a thermally activated process governed by the enthalpy of formation and of migration of defects and usually well described by the Arrhenius relation [135]:

$$\mathrm{D} = \mathrm{D}_0 \exp\left(-\frac{\Delta \mathrm{H_a}}{\mathrm{RT}}\right) \tag{36}$$

where D_0 is the pre-exponential factor corresponding to the diffusion coefficient at infinite temperature and ΔH_a is the activation enthalpy (or energy) of the diffusion process. The extrapolation of high-temperature diffusion data of trace elements in apatite shows that these processes are inefficient at temperatures below 300°C, which cover the conditions of diagenetic alteration up to low-grade metamorphism [135],[141].

The differing initial and boundary conditions imposed in three sets of diffusion experiments:

1. Ion implantation;

2. In-diffusion with powder sources experiment;

3. Out-diffusion.

consequently resulting in different solutions to the diffusion equation. However, in all cases, the process can be described as one-dimensional, concentration-independent diffusion [141].

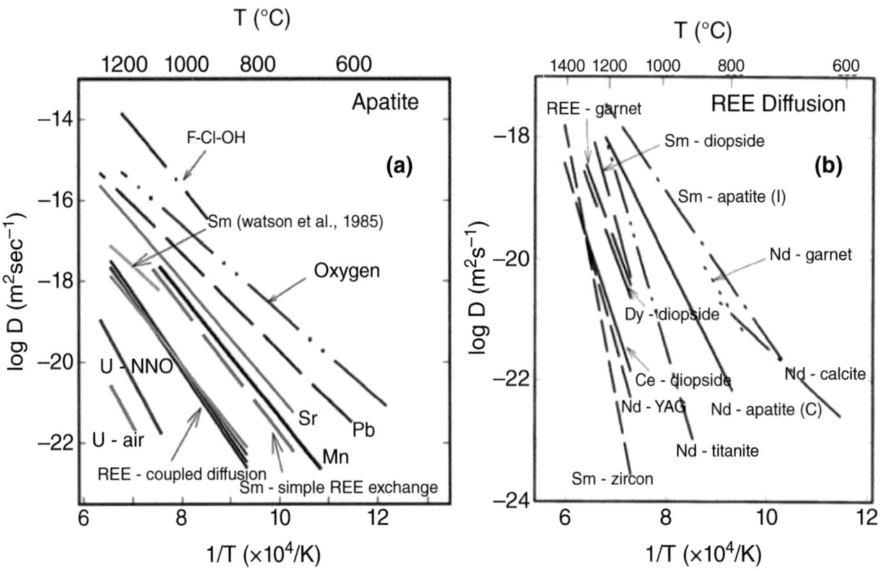

Fig. 17. The summary of data diffusion for cations and anions in apatite (a) [142] and the diffusion of Sm and Nd for various minerals and oxides (a) [141].

A plot of diffusivities of various cations and anions in apatite is shown in **Fig. 17(a)**. The diffusivities of Mn are similar to those of Sr and about an order of magnitude slower than those of Pb. On the other hand, the diffusion of Mn^{2+} in apatite is about two orders of magnitude faster than the diffusion of (trivalent) REE when coupled substitutions according to **Eqs. 4** and **5** are involved [141],[142]. The diffusion coefficients of Nd and Sm in various minerals and related oxides are plotted in **Fig. 17(b)**. The diffusion of REE in apatite is relatively fast; when only simple REE exchange is involved, it is among the fastest in rock-forming minerals for which the data exist. Even when the chemical diffusion involving coupled exchange is considered, REE transport in apatite is considerably faster than the REE diffusion in other accessory minerals [141].

Author details

Petr Ptáček

Brno University of Technology, Czech Republic

References

[1] Valsami-Jones E. Phosphorus in Environmental Technologies: Principles and Applications: Principles and Applications (Integrated Environmental Technology). 1st ed., London: IWA Publishing, 2004. ISBN 1-84339-001-9

[2] Alper A. Phase Diagrams 6-V: Materials Science and Technology. Elsevier, 2012. ISBN 978-0323154895

[3] Šupová M. Substituted hydroxyapatites for biomedical applications: A review. Ceramics International 2015;41(8) 9203–9231.

[4] Cox SC, Jamshidi P, Grover LM, Mallick KK. Preparation and characterisation of nanophase Sr, Mg, and Zn substituted hydroxyapatite by aqueous precipitation. Materials Science and Engineering: C 2014;35 106–114.

[5] Iafisco M, Ruffini A, Adamiano A, Sprio S, Tampieri A. Biomimetic magnesium–carbonate-apatite nanocrystals endowed with strontium ions as anti-osteoporotic trigger. Materials Science and Engineering: C 2014;35 212–219.

[6] Landi E, Tampieri A, Mattioli-Belmonte M, Celotti G, Sandri M, Gigante A, Fava P, Biagini G. Biomimetic Mg- and Mg,CO_3-substituted hydroxyapatites: synthesis characterization and in vitro behavior. Journal of the European Ceramic Society 2006;26(13) 2593–2601.

[7] Tamm T, Peld M. Computational study of cation substitutions in apatite. Journal of Solid State Chemistry 2006;179(5) 1581–1587.

[8] Cockbain AG. The crystal chemistry of the apatites. Mineralogical Magazine 1968;36(281) 654–660.

[9] Cockbain AG, Smith GV. Alkaline-earth-rare-earth silicate and germanate apatites. Mineralogical Magazine, 1967;36 411–421.

[10] Nickel EH, Grice JD. The IMA commission on new minerals and mineral names: procedures and guidelines on mineral nomenclature. Canadian Mineralogist 1998;36 3–16.

[11] Levinson AA. A system of nomenclature for rare-earth minerals. American Mineralogist 1966;51 152–158.

[12] Brown PW, Constanz B. Hydroxyapatite and Related Materials. CRC Press, 1994. ISBN: 978-0849347504

[13] Habelitz S, Pascual L, Durán A. Nitrogen-containing apatite. Journal of the European Ceramic Society 1999;19(15) 2685–2694.

[14] Ding T, Ma D, Lu J, Zhang R. Apatite in granitoids related to polymetallic mineral deposits in southeastern Hunan Province, Shi-Hang zone, China: Implications for petrogenesis and metallogenesis. Ore Geology Reviews 2015;69 104–117.

[15] Deer WA. Rock-forming minerals: Non-silicates, volume 5B, second edition. Geological Society of London, 1998. ISBN: 978-1897799901

[16] Hewlett P. Lea's Chemistry of Cement and Concrete. 4th ed., Butterworth-Heinemann, 2003. ISBN: 978-0080535418

[17] Kreidler ER, Hummel FA. The crystal chemistry of apatite: Structure fields of fluor- and chlorapatite. Journal of Physics and Chemistry of Solids 2007;68 1863–1871

[18] Klement R. Isomorphic replacement of phosphorus in apatites by silicon and sulfur. Naturwissenschaften 1939;27 57–58.

[19] Pasero M, Kampf AR, Ferraris C, Pekov IV, Rakovan JR, White TJ. Nomenclature of the apatite supergroup minerals. European Journal of Mineralogy 2010;22 163–179.

[20] Que M, Ci Z, Wang Y, Zhu G, Xin S, Shi Y, Wang Q. Crystal structure and luminescence properties of a cyan emitting $Ca_{10}(SiO_4)_3(SO_4)_3F_2:Eu^{2+}$ phosphor. Crystal Engineering Communications 2013;15 6389–6394. DOI: 10.1039/C3CE40482J

[21] Fang YN, Ritter C, White TJ. Crystal chemical characteristics of ellestadite-type apatite: implications for toxic metal immobilization. Dalton Transactions 2014;43(42) 16031–16043. DOI: 10.1039/c4dt02088j

[22] Fang Y, Ritter C, White T. The crystal chemistry of $Ca_{10-y}(SiO_4)_3(SO_4)_3Cl_{2-x-2y}F_x$ ellestadite. Inorganic Chemistry 2011;50(24) 12641–12650. DOI: 10.1021/ic201673r.

[23] Balachandran PV, Rajan K. Structure maps for $AI_4AII_6(BO_4)_6X_2$ apatite compounds via data mining. Acta Crystallographica Section B 2012;68(1) 24–33.

[24] Hughes JM, Cameron M, Crowley KD. Ordering of divalent cations in the apatite structure; crystal structure refinements of natural Mn- and Sr-bearing apatite. American Mineralogist 1991;76 1857–1862.

[25] O'Donnell MD, Fredholm Y, de Rouffignac A, Hill RG. Structural analysis of a series of strontium-substituted apatites. Acta Biomaterialia 2008;4(5) 1455–1464.

[26] Hill RG, Stamboulis A, Law RV, Clifford A, Towler MR, Crowley C. The influence of strontium substitution in fluorapatite glasses and glass-ceramics. Journal of Non-Crystalline Solids 2004;336(3) 223–229.

[27] Suchanek WL, Byrappa K, Shuk P, Riman RE, Janas VF, Ten Huisen KS. Preparation of magnesium-substituted hydroxyapatite powders by the mechanochemical-hydrothermal method. Biomaterials 2004;25(19) 4647–4657.

[28] Lijuan X, Liuyun J, Lixin J, Chengdong X. Synthesis of Mg-substituted hydroxyapatite nanopowders: Effect of two different magnesium sources. Materials Letters 2013;106 246–249.

[29] Shanmugam S, Gopal B. Copper substituted hydroxyapatite and fluorapatite: Synthesis, characterization and antimicrobial properties. Ceramics International 2014;40(10) 15655–15662.

[30] Mobasherpour I, Salahi E, Pazouki M. Removal of nickel (II) from aqueous solutions by using nano-crystalline calcium hydroxyapatite. Journal of Saudi Chemical Society 2011;15(2) 105–112.

[31] Shepherd D. Zinc-Substituted Hydroxyapatite for the Inhibition of Osteoporosis Hydroxyapatite (Hap) for Biomedical Applications. Woodhead Publishing, 2015. ISBN: 978-1-78242-033-0

[32] Ren F, Xin R, Ge X, Leng Y. Characterization and structural analysis of zinc-substituted hydroxyapatites. Acta Biomaterialia 2009;5(8) 3141–3149.

[33] Miyaji F, Kono Y, Suyama Y. Formation and structure of zinc-substituted calcium hydroxyapatite. Materials Research Bulletin 2005;40(2) 209–220.

[34] Kramer ER, Morey AM, Staruch M, Suib SL, Jain M, Budnick JI, Wei M. Synthesis and characterization of iron-substituted hydroxyapatite via a simple ion-exchange procedure. Journal of Materials Science 2013;48(2) 665–673.

[35] Kramer E, Itzkowitz E, Wei M. Synthesis and characterization of cobalt-substituted hydroxyapatite powders. Ceramics International 2014;40(8) 13471–13480.

[36] Anderson JB, Kostiner E. The crystal structure of cobalt-substituted calcium chlorapatite. Journal of Solid State Chemistry 1987;66(2) 343–349.

[37] Fleet ME, Pan Y. Site preference of rare earth elements in fluorapatite. American Mineralogist 1995;80 329–335.

[38] Hughes JM, Cameron M, Mariano AN. Rare-earth-element ordering and structural variations in natural rare-earth-bearing apatites. American Mineralogist 1991;76 1165–1173.

[39] Cornwall HR. B065: Mineral and Water Resources of Nevada. Vydavatel NV Bureau of Mines & Geology, 1964.

[40] Connelly NG. IUPAC Recommendations 2005. Royal Society of Chemistry (Great Britain), International Union of Pure and Applied Chemistry, 2005. ISBN: 978-0854044382

[41] Letcher TM, Scott JL, Peter L. Materials for a Sustainable Future. Royal Society of Chemistry, 2012. ISBN: 978-1849734073

[42] Kothe E, Varma A. Bio-Geo Interactions in Metal-Contaminated Soils. Soil Biology — Volume 31. Springer Science & Business Media, 2012. ISBN: 978-3642233272

[43] Seaborg GT, Katz JJ, Morss LR. The Chemistry of the Actinide Elements, Volume 2. 2nd ed., Springer Science & Business Media, 2012. ISBN: 978-9400931558

[44] Get'man EI, Marchenko VI, Loboda SN, Yablochkova NV. Substitutions of neodymium for strontium in the $Sr_5(VO_4)_3OH$ structure. Russian Journal of Inorganic Chemistry 2007;52(2) 147–149.

[45] Report of the Twelfth Annual Meeting of the National Research Council, Washington D.C. 1969. National Research Council (U.S.), National Academies, 1969.

[46] Natrajan LS, Swinburne AN, Andrews MB, Randall S, Heath SL. Redox and environmentally relevant aspects of actinide(IV) coordination chemistry. Coordination Chemistry Reviews 2014;266-267 171-193.

[47] Luo Y, Hughes JM, Rakovan J, Pan YM. Site preference of U and Th in Cl, F, and Sr apatites. American Mineralogist 2009;94 345–351.

[48] Riadh QE, Cohen-Adad M, Goutaudier C, Panczer G. Uranium-doped britholites $Ca_xLa_y(SiO_4)_{6-u}(PO_4)_uO_t$:U synthesis, characterization and preliminary study of uranium diffusion. Solid State Ionics 2005;176(1) 225–231.

[49] Terra O, Audubert F, Dacheux N, Guy C, Podor R. Synthesis and characterization of uranium-bearing britholites. Journal of Nuclear Materials 2007;366(1-2) 70–86.

[50] Terra O, Audubert F, Dacheux N, Guy C, Podor R. Synthesis and characterization of thorium-bearing britholites. Journal of Nuclear Materials 2006;354(1-3) 49–65.

[51] Lim PN, Chang L, Thian ES. Development of nanosized silver-substituted apatite for biomedical applications: A review. Nanomedicine: Nanotechnology, Biology and Medicine 2015;11(6) 1331–1344.

[52] Nounah A, Lacout JL, Savariault JM. Localization of cadmium in cadmium-containing hydroxy- and fluorapatites. Journal of Alloys and Compounds 1992;188 141–146.

[53] Badraoui B, Aissa A, Bigi A, Debbabi M, Gazzano M. Synthesis and characterization of $Sr_{(10-x)}Cd_x(PO_4)_6Y_2$ (Y = OH and F): A comparison of apatites containing two divalent cations. Materials Research Bulletin 2009;44(3) 522–530.

[54] Grisafe DA, Hummel FA. Pentavalent ion substitutions in the apatite structure part A. Crystal chemistry. Journal of Solid State Chemistry 1970;2(2) 160–166.

[55] Lee YK, Stephens PW, Tang Y, Li W, Phillips BL, Parise JB, Reeder RJ. Arsenate substitution in hydroxylapatite: Structural characterization of the $Ca_5(P_xAs_{1-x}O_4)_3OH$ solid solution. American Mineralogist 2009;94 666–675

[56] Zhu Y, Zhang X, Long F, Liu H, Qian M, He N. Synthesis and characterization of arsenate/phosphate hydroxyapatite solid solution. Materials Letters 2009;63(13-14) 1185–1188.

[57] Tilley RJD. Understanding Solids: The Science of Materials. 2nd ed., New York: John Wiley & Sons, 2013. ISBN: 978-1-118-42328-8

[58] George G, Gupta SK, Rao PVR, Narasaraju TSB. Preparation and characterization of phosphate and arsenate apatites of strontium and their solid solutions. Journal of Materials Science 1987;22 2274–2276.

[59] Bell AMT, Henderson CMB, Wendlandt RF, Harrisonc WJ. Rietveld refinement of $Sr_5(AsO_4)_3Cl$ from high-resolution synchrotron data. Acta Crystallographica Section E 2009;65 i16–i17.

[60] Mugavero SJ, Bharathy M, McAlum J, Loye HC. Crystal growth of alkaline earth vanadates from hydroxide fluxes. Solid State Sciences 2008;10 370–376.

[61] Chernorukov NG, Knyazev AV, Bulanov EN. Isomorphism and phase diagram of the $Pb_5(PO_4)_3Cl$-$Pb_5(VO_4)_3Cl$ system. Russian Journal of Inorganic Chemistry 2010;55(9) 1463–1470.

[62] Glasser FP, Osborn EF. Phase equilibrium studies in the system CaO-Cr_2O_3-SiO_2. Journal of the American Ceramic Society 1958;41 358–367.

[63] Klemm W. Anomale Wertigkeiten. Angewandte Chemie 1951;63(17-18) 396–402.

[64] Scholder R, Klemm W. Über neue Metallate mit Sauerstoff und Fluor als Liganden. Angewandte Chemie 1954;66 461–474.

[65] Scholder R, Suchy H. Zeitschrift für anorganische und allgemeine Chemie 1961;308 295.

[66] Scholder R, Schwarz H. Z. Über Alkalichromate(V). Zeitschrift für anorganische und allgemeine Chemie 1963;326(1-2) 1–10.

[67] Banks E, Jaunarajs KL. Chromium analogs of apatite and spodiosite. Inorganic Chemistry 1965;1(4) 78–83.

[68] Scholder R, Schwochow F, Schwarz H. Über Alkalichromate(V). Zeitschrift für anorganische und allgemeine Chemie 1968;363(1-2) 10–23.

[69] Banks E, Greenblatt M, McGarvey BR. Electron spin resonance of $CrO^{3-}{}_4$ in chloroapatite $Ca_5(PO_4)_3Cl$. Journal of Solid State Chemistry 1971;3(2) 308–313.

[70] Guldotti RA, Roth EP. Specific heats of calcium chromate chlorides, calcium chromate hydroxide and calcium chromate ($Ca_5(CrO_4)_3Cl$, Ca_2CrO_4Cl, $Ca_5(CrO_4)_3OH$, and $Ca(CrO_4)$ at elevated temperatures. Journal of Chemical & Engineering Data 1985;30(3) 328–329.

[71] Baikie T, Ahmad Z, Srinivasan M, Maignan A, Pramana SS, White TJ. The crystallographic and magnetic characteristics of Sr_2CrO_4 (K_2NiF_4-type) and $Sr_{10}(CrO_4)_6F_2$ (apatite-type). Journal of Solid State Chemistry 2007;180(5) 1538–1546.

[72] Wu P, Zeng YZ, Wang CM. Prediction of apatite lattice constants from their constituent elemental radii and artificial intelligence methods. Biomaterials 2004;25(6) 1123–1130.

[73] Ford WF, Rees WJ. The CaO-MgO-Cr_2O_3 ternary system: Part I. Partial investigation of the CaO-Cr_2O_3 system. Transactions of the British Ceramic Society 1948;47(6) 207–231.

[74] Ford WF, Rees WJ. The CaO-MgO-Cr_2O_3 ternary system: Part II. Further experiments. Transactions of the British Ceramic Society 1949;48(8) 291–312.

[75] Ford WF, White J. The CaO-MgO-Cr_2O_3 ternary system. Part III. CaO-CrO_3-Cr_2O, ternary and the MgO-Cr_2O_3 and CaO-MgO binary systems. Transactions of the British Ceramic Society 1949;48 417–428.

[76] Tyutyunnik AP, Bazuev GV. Synthesis, crystal structure and magnetic properties of $Sr_5(CrO_4)_3(Cu_{0.586}O)$ with apatite-like structure. Journal of Alloys and Compounds 2012;522 141–143.

[77] Wang Y, Ma J, Zhou L, Chen J, Liu Y, Qiu Z, Zhang S. Dual functional selenium-substituted hydroxyapatite, Interface Focus 2012;2(3) 378–386.

[78] Monteil-Rivera F, Masset S, Dumonceau J, Fedoroff M, Jeanjean J. Sorption of selenite ions on hydroxyapatite. Journal of Materials Science Letters 1999;18(14) 1143–1145.

[79] Kolmas J, Oledzka E, Sobczak M, Nałęcz-Jawecki G. Nanocrystalline hydroxyapatite doped with selenium oxyanions: A new material for potential biomedical applications. Materials Science and Engineering C 2014;39 134–142.

[80] Zhang W, Chai Y, Cao N, Wang Y. Synthesis and characterization of selenium substituted hydroxyapatite via a hydrothermal procedure. Materials Letters 2014;134 123–125.

[81] Derek EC. Chemistry, Biochemistry and Technology. 6th ed., CRC Press, 2013. ISBN: 978-1439840887

[82] Habelitz S, Pascual L, Durán A. Transformation of tricalcium phosphate into apatite by ammonia treatment. Journal of Materials Science 2001;36(17) 4131–4135.

[83] Vandecandelaere N, Bosc F, Rey C, Drouet C. Peroxide-doped apatites: Preparation and effect of synthesis parameters. Powder Technology 2014;255 3–9.

[84] Dugas J, Rey C. Electron spin resonance characterization of superoxide ions in some oxygenated apatites. The Journal of Physical Chemistry 1977;81(14) 1417–1419.

[85] Yu H, Zhang H, Wang X, Gu Z, Li X, Deng F. Local structure of hydroxy–peroxy apatite: A combined XRD, FT-IR, Raman, SEM, and solid-state NMR study. Journal of Physics and Chemistry of Solids 2007;68(10) 1863–1871.

[86] Zhao H, Li X, Wang J, Qu S, Weng J, Zhang X. Characterization of peroxide ions in hydroxyapatite lattice. Journal of Biomedical Materials Research 2000;52(1) 157–163.

[87] Bouroushian M. Electrochemistry of Metal Chalcogenides. Monographs in Electrochemistry. Springer Science & Business Media, 2010. ISBN: 978-3642039676

[88] Henning PA, Adolfsson E, Grins J. The chalcogenide phosphate apatites $Ca_{10}(PO_4)_6S$, $Sr_{10}(PO_4)_6S$, $Ba_{10}(PO_4)_6S$ and $Ca_{10}(PO_4)_6Se$. Zeitschrift fur Kristallographie 2000;215 226–230.

[89] Chen S, Hoffmann S, Carrillo-Cabrera W, Akselrud LG, Prots Y, Schwarz U, Zhao Jing-T, Kniep R. $Sr_{10}[(PO_4)_{5.5}(BO_4)_{0.5}](BO_2)$: Growth and crystal structure of a strontium phosphate orthoborate metaborate closely related to the apatite-type crystal structure. Journal of Solid State Chemistry 2010;183(3) 658–661.

[90] Chen S, Hoffmann S, Carrillo-Cabrera W, Akselrud LG, Prots Y, Zhao Jing-T, Kniep R. FA2-MS01-P20: $Sr_{10}(PO_4)_{5.5}(BO_4)_{0.5}(BO_2)$: A strontium borate-phosphate closely related to the apatite crystal structure. Acta Crystallographica 2009;A65 182.

[91] Calvo C, Faggiani R, Krishnamurthy N. The crystal structure of $Sr_{9.402}Na_{0.209}(PO_4)_6B_{0.996}O_2$ — A deviant apatite. Acta Crystallographica 1975;31 B 188-192.

[92] Nickel EH. Solid solution in mineral nomenclature. Canadian Mineralogist 1992;30 231–234.

[93] Hughes JM, Cameron M, Crowley KD. Crystal structures of natural ternary apatites: Solid solution in the $Ca_5(PO_4)_3X$ (X = F, OH, Cl) system. American Mineralogist 1990;75 295–304.

[94] Schlesinger MI, Hlavác V. Ten Lectures on Statistical and Structural Pattern Recognition. Computational Imaging and Vision — Volume 24. Springer Science & Business Media, 2013. ISBN: 978-9401732178

[95] Banasik K, Galuskin EV, Armbruster T, Lazic B, Galuskina IO, Gazeev VM. Apatite supergroup minerals of the solid solution $Ca_5[(PO_4)_{3-x}(AsO_4)_x]_{\bullet 3-2y}[(SO_4)_y(SiO_4)_y]_{\bullet 2y}$ (OH, F, Cl), x = 0-3, y = 0-1.5, from high temperature skarn of the Upper Chegem caldera, Northern Caucasus, Russia. European Mineralogical Conference 2012, Vol. 1, EMC2012-298, 2012.

[96] Wright FE. On three contact minerals from Valardeña, Durango, Mexico (Gehlenite, spurrite and hillebrandite). American Journal of Science 1908;176 545–554.

[97] Bournon JL. Humite, in Catalogue de la Collection Minéralogique du Compte de Bournon, De l'imprimetie de R. Juigné (London) 1813; 52-54.

[98] Tilley CE. On larnite (calcium orthosilicate, a new mineral) and its associated minerals from the limestone-contact zone of Scawt Hill, Co. Antrim. Mineralogical Magazine 1929;22 77–86.

[99] Li X, Xu W, Wang S, Tang M, Shen X. Effect of SO_3 and MgO on Portland cement clinker: Formation of clinker phases and alite polymorphism. Construction and Building Materials 58;2014 182–192.

[100] Fukuda K, Ito S, Taguchi H. Thermoelasticity of belite in Portland cement clinker. Cement and Concrete Research 1998;28(8) 1141–1145.

[101] Hjorth L, Laurén K-G. Belite in Portland cement. Cement and Concrete Research 1971;1(1) 27–40.

[102] Hamm HM, Hentschel G. Reinhardbraunsite, $Ca_5(SiO_4)_2(OH,F)_2$, a new mineral — The natural equivalent of synthetic "calcio-chondrodite". Neues Jahrbuch für Mineralogie, Monatshefte 1983; 119-129.

[103] Galuskin EV, Gazeev VM, Lazic B, Armbruster T, Galuskina IO, Zadov AE, Pertsev NN, Wrzalik R, Dzierzanowski P, Gurbanov AG, Bzowska G. Chegemite $Ca_7(SiO_4)_3(OH)_2$ — A new humite group calcium mineral from the northern Caucasus, Kabardino-Balkaria, Russia. European Journal of Mineralogy 2009;21 1045–1059.

[104] Tsukimura K, Kanazawa Y, Aoki M, Bunno M. Structure of wadalite $Ca_6Al_5Si_2O_{16}Cl_3$. Acta Crystallographica 1993;C49 205–207.

[105] Mihajlovic T, Lengauer CL, Ntaflos T, Lolitsch U, Tillmanns E. Two new minerals, rondorfite, $Ca_8Mg[SiO_4]_4Cl_2$, and almarudite, $K(\square,Na)_2(Mn,Fe,Mg)_2(Be,Al)_3[Si_{12}O_{30}]$, and a study of iron-rich wadalite, $Ca_{12}[(Al_8Si_4Fe_2)O_{32}]Cl_6$, from the Bellerberg (Bellberg) volcano, Eifel, Germany. Neues Jahrbuch für Mineralogie, Abhandlungen 2004;179 265–294.

[106] Scacchi A. Della cuspidina e del neocrisolito, nuovi minerali vesuviani, Rendiconto dell'Accademia delle Scienze Fisiche e Matematiche (sezione della Societa reale di Napoli) 1876;15, 208-209.

[107] Galuskin EV, Gazeev VM, Armbruster T, Zadov AE, Galuskina IO, Pertsev NN, Dzierzanowski P, Kadiyski M, Gurbanov AG, Wrzalik R, Winiarski A. Lakargiite $CaZrO_3$: A new mineral of the perovskite group from the North Caucasus, Kabardino-Balkaria, Russia. American Mineralogist 2008;93 1903–1910.

[108] Chesnokov BV, Bazhenova LF. Srebrodolskite $Ca_2Fe_2O_5$ — A new mineral, Zapiski Vsesoyuznogo Mineralogicheskogo Obshchestva 1985;114(2) 195–199.

[109] Nöggerath JJ. Sanidin, in Mineralogische Studien über die Gebirge am Niederrhein. Johann Christian Hermann (Frankfurt) 1808; 24-39.

[110] Newnham RE, de Haan YM. Refinement of the α-Al_2O_3, Ti_2O_3, V_2O_3 and Cr_2O_3 structures. Zeitschrift für Kristallographie 1962;117 235–237.

[111] Gibbs GV. The polymorphism of cordierite I: The crystal structure of low cordierite. American Mineralogist 1966;51 1068–1087.

[112] Winter JK, Subrata Ghose. Thermal expansion and high-temperature crystal chemistry of the Al_2SiO_5 polymorphs. American Mineralogist 1979;64 573–586.

[113] Wright FE. On three contact minerals from Valardeña, Durango, Mexico. (gehlenite, spurrite and hillebrandite). American Journal of Science 1908;176, 545–554.

[114] Parry J, Wright FE. Afwillite, a new hydrous calcium silicate, from Dutoitspan mine, Kimberley, South Africa. Mineralogical Magazine 1925;20 277–285.

[115] Parry J, Williams AF, Wright FE. On bultfonteinite, a new fluorine-bearing hydrous calcium silicate from South Africa. Mineralogical Magazine 1932;23 145–162.

[116] Grew ES, Locock A, Mills SJ, Galuskina IO, Galuskin EV, Hälenius U. Nomenclature of the garnet supergroup. American Mineralogist 2013;98(4) 785–811.

[117] Lehmann J. Über den ettringit, ein neues mineral, in Kalkeinschlüssen der Lava von Ettringen (Laacher Gebiet). Neues Jahrbuch für Mineralogie, Geologie und Palaontologie 1874; 273-275.

[118] Onda A, Ogo S, Kajiyoshi K, Yanagisawa K. Hydrothermal synthesis of vanadate/phosphate hydroxyapatite solid solutions. Materials Letters 2008;62(8-9) 1406–1409.

[119] Sugiyama S, Osaka T, Hirata Y, Sotowa Ken-I. Enhancement of the activity for oxidative dehydrogenation of propane on calcium hydroxyapatite substituted with vanadate. Applied Catalysis A: General 2006;312 52–58.

[120] Piotrowski A, Kahlenberg V, Fischer RX. The solid solution series of the sulfate apatite system $Na_{6.45}Ca_{3.55}(SO_4)_6(F_xCl_{1-x})_{1.55}$. Journal of Solid State Chemistry 2002;163(2) 398–405.

[121] Bajda T, Mozgawa W, Manecki M, Flis J. Vibrational spectroscopic study of mimetite–pyromorphite solid solutions. Polyhedron 2011;30(15) 2479–2485.

[122] Mayer I, Cohen S, Matalon JR. Solid solutions of $Pb_8M_2(XO_4)_6$ lead alkali apatites. Journal of Solid State Chemistry 1981;36(3) 271–274.

[123] Peck WH, Tumpane KP. Low carbon isotope ratios in apatite: An unreliable biomarker in igneous and metamorphic rocks. Chemical Geology 2007;245(3-4) 305–314.

[124] Eagle R, Schauble EA, Tripati AK, Fricke HC, Tuetken T, Eiler JM. A paleothermometer based on abundances of ^{13}C-^{18}O bonds in bioapatite: Calibration and reconstruction of the body temperatures of extinct Cenozoic mammals and Mesozoic dinosaurs. American Geophysical Union, Fall Meeting 2009, abstract #PP31B-1342.

[125] Trappe J. Phanerozoic Phosphorite Depositional Systems: A Dynamic Model for a Sedimentary Resource System. Lecture Notes in Earth Sciences — Volume 76, ISSN 0930-0317. Springer, 1998. ISBN: 978-3540696049

[126] Teodorovich GI. Authigenic Minerals in Sedimentary Rocks. Springer Science & Business Media, 2012. ISBN: 978-1468406528

[127] Grossan EL. Applying oxygen isotope paleothermometry in deep time. The Paleontological Society Papers 2012;18 39–67.

[128] LaPorte DF, Holmden C, Patterson WP, Prokopiuk T, Eglington BM. Oxygen isotope analysis of phosphate: improved precision using TC/EA CF-IRMS. Journal of Mass Spectrometry 2009;44(6) 879–890.

[129] Kohn MJ, Cerling TE. Stable isotope compositions of biological apatite. Reviews in Mineralogy and Geochemistry 2002;48 455–488.

[130] Land LS, Lundelius EL Jr., Valastro S. Isotopic ecology of deer bones. Palaeogeography, Palaeoclimatology, Palaeoecology 1980;32 143–151.

[131] Ericson JE, Sullivan CH, Boaz NT. Diets of Pliocene mammals from Omo, Ethiopia, deduced from carbon isotopic ratios in tooth apatite. Palaeogeography, Palaeoclimatology, Palaeoecology 1981;36 69–73.

[132] Sullivan CH, Krueger HW. Carbon isotope analysis of separate chemical phases in modern and fossil bone. Nature 1981;292 333–335.

[133] Lee-Thorp JA, van der Merwe NJ. Carbon isotope analysis of fossil bone apatite. South African Journal of Science 1987;83 712–715.

[134] Lee-Thorp JA, Sealy JC, van der Merwe NJ. Stable carbon isotope ratio differences between bone collagen and bone apatite, and their relationship to diet. Journal of Archaeological Science 1989;16 585–599.

[135] Reynard B, Balter V. Trace elements and their isotopes in bones and teeth: Diet, environments, diagenesis, and dating of archeological and paleontological samples. Palaeogeography, Palaeoclimatology, Palaeoecology 2014;416 4–16.

[136] O'Connell JP, Haile JM. Thermodynamics: Fundamentals for Applications. Cambridge University Press, 2005. ISBN: 978-1139443173

[137] Brice JC. Some thermodynamic aspects of the growth of strained crystals. Journal of Crystal Growth 1975;28 249–253.

[138] Almora-Barrios N, Grau-Crespo R, de Leeuw NH. A computational study of magnesium incorporation in the bulk and surfaces of hydroxyapatite. Langmuir 29;2013 5851–5856.

[139] Rabone JAL, De Leeuw NH. Interatomic potential models for natural apatite crystals: Incorporating strontium and the lanthanides. Journal of Computational Chemistry 2006;27 253–266.

[140] Davranche M, Pourret O, Gruau G, Dia A, Le Coz-Bouhnik M. Adsorption of REE(III)-humate complexes onto MnO_2: experimental evidence for cerium anomaly and lanthanide tetrad effect suppression. Geochimica et Cosmochimica Acta 2005;69, 4825–4835.

[141] Cherniak DJ. Rare earth element diffusion in apatite. Geochimica et Cosmochimica Acta 2000;64 3871–3885.

[142] Cherniak DJ. Uranium and manganese diffusion in apatite. Chemical Geology 2005;219(1-4) 297–308.

Chapter 7

Phosphate Rocks

Petr Ptáček

Additional information is available at the end of the chapter

http://dx.doi.org/10.5772/62214

Abstract

Apatite is the most abundant phosphate mineral which include more than 95% of al phosphorus in the Earth´s crust. The seventh chapter of this book provides brief description of sedimentary and igneous phosphate rocks and introduces basic ideas for characterization and classification of phosphate rocks. The chapter continues with description of biogenic apatites, description of phosphate rocks deposits and introduces other sources of phosphorus. Furthermore, geological role of apatite, cycle of phosphorus, weathering of apatite, fission track analysis and extraterrestrial apatites were described. The last section is dedicated to structure and properties important of non-apatitic phosphate minerals, such as atuanite, crandallite, lazulite, millisite, monazite, tobernite, xenotime etc.

Keywords: Apatite, Phosphate Rock, Biogenic Apatites, Fission Track, Extraterrestrial Epatite, Non-Apatitic Phosphate Minerals

Apatite [$Ca_5(PO_4,CO_3)_3(OH,F,Cl)$] is the most abundant phosphate mineral, which accounts for more than 95% of all phosphorus in the Earth's crust and is found as an accessory mineral in most rock types[1] on the Earth's surface, primarily because it is stable in a wide variety of geological conditions and over a range of different geological processes [1],[2],[3],[4],[5],[6],[7],[8]. According to the list of symbols for rock- and ore-forming minerals, the abbreviated symbol used for apatite is **Ap** [9].

However, exploitable deposits of apatite are mainly found in igneous rocks and also in sedimentary and metamorphic rocks. The former comprises the stratiform phosphorite deposits in shelf-type shale-carbonate sequences, which contain high phosphorus ores of microcrystal-

[1] A rock may be best defined as any mineral or aggregate of minerals that forms an essential part of the Earth [10]. Apatite is found in all classes of rock: igneous, metamorphic and sedimentary [2]. Also, much of phosphorus in coal is present in the form of apatite [7]. Phosphorus is the 10th most abundant element on Earth, with an average crustal abundance of 0.1% [8].

line CO_2-rich fluorapatite (francolite) and cryptocrystalline collophane. The igneous deposits comprise fluorapatite ores, which mostly accommodate carbonatites and other types of alkaline intrusions. The magmatic ores are generally of lower grade but give higher-quality beneficiation products with low contents of unwanted contaminants (Cd, Pb, As, U, Th, Mg and Al) [11], [12].

The beneficiation products of apatite ores as a commodity are traded as phosphate rock. It is the only significant global resource of phosphorus used dominantly in the manufacturing of nitrogen-phosphorus-potassium (NPK) fertilizers for food-crop nutrition and in the production of animal feed supplements. Only 10 – 15% of the world's production of phosphate rock has other applications (e.g. pharmaceuticals, ceramics, textiles and explosives) and represents an important alternative source of rare-earth elements (REE) [12],[13]. The REE contents in apatites are useful in paleoceanographic studies to identify the seawater masses and circulation patterns or to quantify the redox state of the ocean [14].

The composition of phosphate rocks varies from one deposit to another. Therefore, phosphate rocks from different sources may be expected to behave differently in beneficiation and acidulation processes. Phosphate rocks are primarily composed of the apatite group in association with a wide assortment of accessory minerals, mainly fluorides, carbonates, clays, quartz, silicates and metal oxides [13],[15],[16].

Si, Ca, Fe and Al are the most common companion elements in phosphate rocks, with the median abundances of 53.3 wt.%, 30.0 wt.%, 13.6 wt.% and 8.0 wt.%, respectively, compared

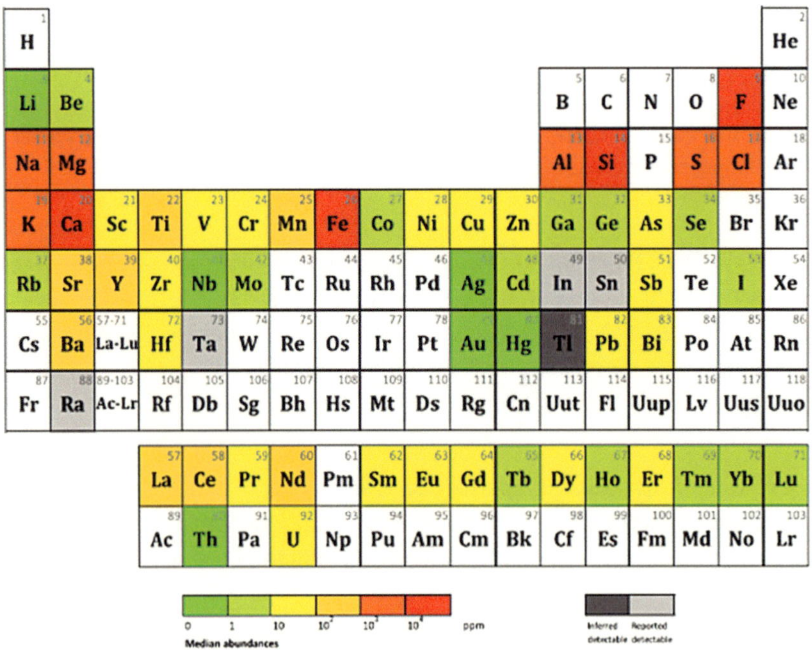

Fig. 1. The average distribution of trace elements in phosphate rock [17].

with P_2O_5. In some low-grade phosphate rock mines, the content of Fe and Al is even higher than that of P_2O_5, and it is usual that the P_2O_5 content in phosphate rock is less than the Si and Ca content. In addition, some elements that are very rare in the Earth's crust are found to be relatively abundant (**Fig. 1**) in phosphate rocks [17].

7.1. Characterization and classification of phosphate rocks

When rock contains phosphate components between 5 and 50% (by volume), then it is phosphatic, and the name of the main lithology is used as a suffix (phosphatic limestone, phosphatic claystone, etc.). In addition, the dominant textural form of the phosphate components in a phosphorite can be used in defining the rock name (e.g. peloidal[2]) (phosphorite, coprolitic phosphorite, etc.) [10],[18].

There are two main kinds of phosphate rocks deposits in the world [10],[20],[21],[22],[17],[23]:

1. **Sedimentary phosphate rocks**: marine phosphate deposit, metamorphic deposit, biogenic deposit (bird and bat guano accumulation) and phosphate deposit as the result of weathering. The sedimentary deposits contain the varieties of carbonate-fluorapatite that are collectively called as francolite (**Section 2.6**). The most common non-phosphatic accessory minerals associated with sedimentary phosphate rocks are quartz, clay and carbonates (calcite and dolomite). Phosphate rocks of high concentration of phosphates (10 – 15% of P_2O_5) are called phosphorites.

2. **Igneous phosphate rocks**: apatite is a common accessory mineral occurring in practically all types of igneous rocks (acid, basic or ultrabasic).

Depending on their origin (igneous or sedimentary), phosphate rocks have widely varying mineralogical, textural and chemical characteristics [23]. The locations of the major phosphate rocks deposit and producers are shown in **Fig. 2** [20],[24].

Sedimentary phosphate deposits are exploited to produce about 80% of the total world production of phosphate rocks. Igneous phosphate deposits are often associated with carbonatites[3] and/or alkalic (silica-deficient) intrusions. Igneous phosphate rock concentrates are produced from the deposits mainly exploited in Russia, the Republic of South Africa, Brazil, Finland and Zimbabwe. Igneous phosphate ores are often low in grade (less than 5% P_2O_5) but can be upgraded to high-grade products (from about 35% to over 40% P_2O_5) [22],[23].

The atom ratio of P:N = 1:15(16) in the oceans is not greatly different from that found in living organisms. The availability of soluble phosphate from weathering of apatite-containing rocks may initially has been the rate-determining factor in early live development. In most ecolog-

---

[2] Peloid is a comprehensive descriptive term for polygenetic grains composed of micro- and polycrystalline carbonate. The term was proposed in order to replace the widely used name "pellet," which for many authors had become a synonym for pelletal coprolites (fossilized faces). Peloids differ from ooids and oncoids by the absence of centrosymmetric or radial internal structures [19].

[3] Igneous rocks (intrusive or extrusive) that contain carbonates in the amount higher than 50%.

ical systems, the phosphate content is the limiting factor for growth. Nearly all igneous rocks contain some phosphate, even if it is only ~0.1% (0.2% P_2O_5 on average in lithosphere), with nearly all of it in the form of apatite. Sedimentary rocks generally contain rather less (~0.1% P_2O_5 on average). Sedimentary phosphorite is believed to originate from widely dispersed apatite mainly in igneous rocks [25].

7.1.1. Sedimentary phosphate rocks

Most marine sediments and rocks contain less than 0.3% of P_2O_5. However, periodically through geological time, phosphorites (with the content of P_2O_5 of 5% or greater) formed on the seafloor in response to specialized oceanic conditions and accumulated in sufficient concentrations to produce major deposits of regional extent[4] [26].

Marine phosphate formation and deposition represent the periods of low rates of sedimentation in combination with large supplies of nutrients. Phosphorus is then concentrated by various mechanisms, possibly bacterial (refer to discussion of **Fig. 7**), at either the sediment-water interface or within interstitial pore waters. This process leads to primary formation and growth of phosphate grains, which remain where they were formed or are transported as clastic particles within the environment of formation. During subsequent periods of time, some primary phosphate grains may be physically reworked into another sediment unit in response to either changing or different environmental processes [26].

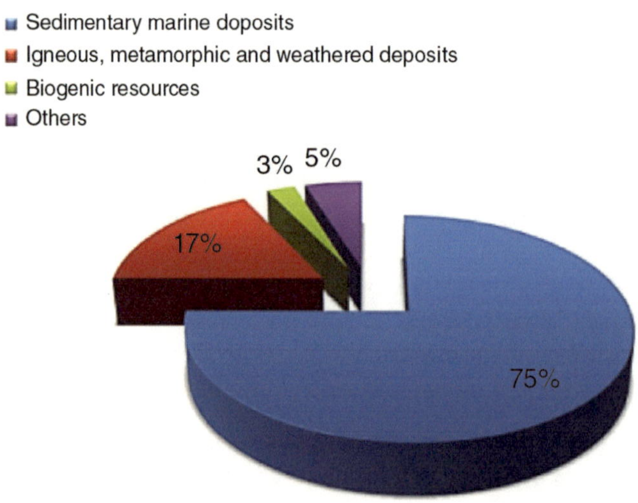

Fig. 2. The distribution of the world's phosphate resources [20].

[4] Most of the world's phosphate production comes from marine phosphorites [27].

http://dx.doi.org/10.5772/62214

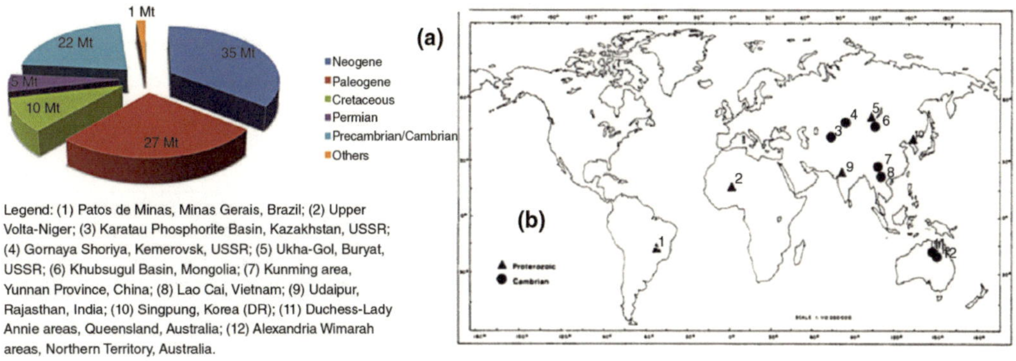

Fig. 3. Stratigraphic distribution of phosphorites based on 1982 production data (a) and distribution of major Proterozoic-Cambrian phosphorites (b) [28].

The stratigraphic distribution[5] [29] of phosphorites [28] is shown in **Fig. 3(a)**. The major discoveries in Proterozoic and Cambrian rocks were not made until late 1930s, initially in the USSR (Kazakhstan), Poland, Korea, China and northern Vietnam. Early discoveries were made in the course of regional geological mapping and exploration for metal deposits, but later deposits were found mainly using more direct exploration techniques. It is possible that the greatest global phosphogenic episode in geological history took place in Late Proterozoic and Cambrian. While pelletal phosphorites are common in most Cambrian deposits, the Proterozoic age phosphorites contain mudstone (**Fig. 3(b)**, microphosphorite) and stromatolitic phosphorite[6] [28],[30].

Some phosphates were formed during all major sea-level transgressions during 67 million years of Cenozoic history; however, some periods were more important than others with respect to producing large volumes of phosphorites and preserving them in the geologic column. During the Paleocene and Eocene, several major episodes of phosphogenesis occurred within the major episodes of phosphogenesis in the major east-west ocean, which included Tethys [31],[7] producing extensive amounts of phosphorites throughout the Middle East, Mediterranean and northern South American regions. By the Neogene, this circum-global ocean had been destroyed by the plate tectonic processes, and the north-south Pacific and

[5] Since the Earth is stratified, in a broad sense, all rocks and classes of rocks (sedimentary, igneous and metamorphic) fall within the scope of stratigraphy and stratigraphic classification. Rocks can be classified according to lithology, fossil content, magnetic polarity, electrical properties, seismic response, chemical or mineralogical composition, etc. Rocks can be also classified according to time of their origin or environment of genesis. Rock bodies can be classified into many different categories of stratigraphic classification, including lithostratigraphic units (1), biostratigraphic units (2), chronostratigraphic units (3), unconformity-bounded units (4) and magnetostratigraphic units (5). Please see work [29] for further details.

[6] Phosphate occurs as concentrated in stromatolite columns, laminar algal (stromatolitic) phosphorite, reworked fragments of stromatolites forming silicified conglomeratic or brecciated phosphorite, massive-bedded phosphorite with sandy and clayey laminate and disseminate pellets and nodules in dolomite [30].

[7] The Tethys Ocean divided the continental masses into northern and southern groups. It merged at both ends with the Panthalassa or proto-Pacific Ocean. The remnants of Tethys are now located within Alpine mountain belts from the Caribbean in the west to recent collision zone between Australia and Eurasia in the east and within still-growing Central Atlantic Ocean between Africa and North America [31].

Atlantic oceans dominated global circulation patterns. Upper Cenozoic phosphogenesis (**Table 1**) occurred along the north-south ocean-ways, which now contain modern continental margins. On the basis of the extent of known phosphate deposits, the Miocene was by far the most important episode of phosphate formation in Upper Cenozoic [26].

Geological period			**Age**	**Duration**	
Phanerozoic (541 *ma* to present)	**Quaternary**	**Holocene**	<3 *ma*	10 *ta*	
Cenozoic (66 *ma* to present)	258 *ma* to present	0.0117 *ma* to present			
		Pleistocene			
		2.58 – 0.0117 *ma*			
	Neogene	**Pliocene**	5 – 4 *ma*	1 *ma*	
	23.03 – 2.58 *ma*	5.333 – 2.58 *ma*			
		Miocene	19 – 13 *ma*	6 *ma*	
		23.03 – 5.333 *ma*			
		Paleogene	**Oligocene**	29 – 25 *ma*	4 *ma*
		66 – 23.03 *ma*	33.9 – 23.03 *ma*		

ma – million years ago, *ta* – thousand years

Table 1. Formation of phosphorites during Upper Cenozoic phosphogenesis [26].

Phosphorites and phosphatic sediments are known on the floor of the Pacific, Indian and Atlantic oceans. They occur in a number of inshore areas (the shelves and upper part of the continental slopes) and in pelagic zones, chiefly on seamounts. Most of the shelf phosphorites are localized in four very large oceanic phosphorite provinces [32],[26],[33]:

a. **East Atlantic**: Portugal, northwest Africa throughout South Africa and Agulhas Bank;

b. **West Atlantic**: North Carolina throughout Florida, Cuba, Venezuela and Argentina;

c. **East Pacific**: California throughout Baja California, Mexico and Peru throughout Chile;

d. **West Pacific**: Sakalin Island, Sea of Japan, Indonesia, Chathman Rise east of New Zealand and East Australian shelf.

Sedimentary rocks with the content of 18 – 20 wt.% of P_2O_5 are termed as phosphorites. The main phosphate mineral in phosphorites is carbonate-fluorapatite (CAF, francolite): $Ca_{10-a-b-c}Na_aMg_b(PO_4)_{6-x}(CO_3)_{x-y-z}(CO_3,F)_y(SO_4)_zF_2$, where $\boldsymbol{x} = \boldsymbol{y} + \boldsymbol{a} + 2\boldsymbol{c}$ and $\boldsymbol{c}$ denotes the number of Ca vacancies, present as grain or mud. Most phosphorites are of marine origin [34],[35],[36]. Phosphorites on the sea floor occur in two types of environments on [37]:

i. Continental margins in association with terrigenous;

ii. Submerged mountains in association with calcareous and volcanogenic rocks.

http://dx.doi.org/10.5772/62214

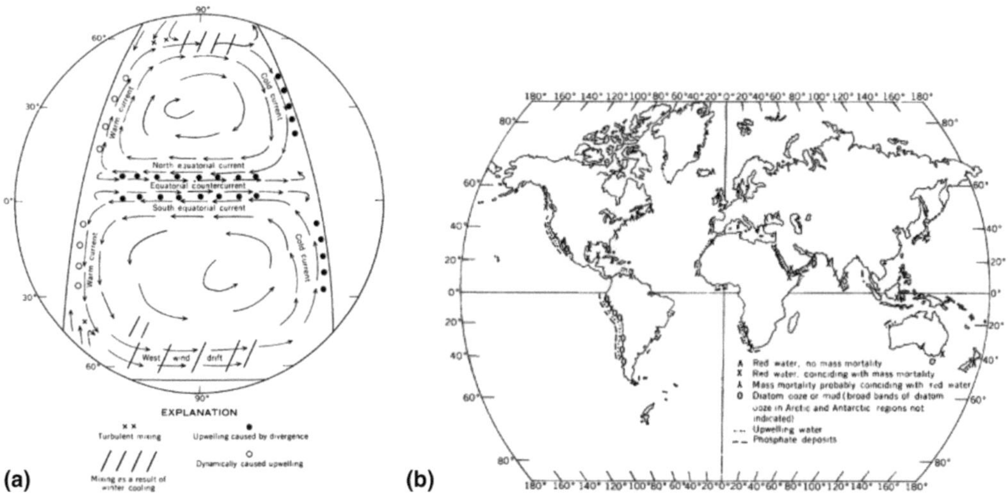

Fig. 4. The surface currents in an idealized ocean, showing the areas of ascending nutrient-rich water (a) and the distribution of upwelling water and related phenomena in modern oceans (b) [27].

Phosphorites consist mainly of phosphate cement enveloping small grains of phosphatic and non-phosphatic materials. Phosphate in the cavities of foraminifera is purer than that enveloping the grains [37].

The largest phosphorite-bearing regions are situated along the west coasts of Africa and America, at the east coast of the USA, off New Zealand and in the central part of the northern Pacific. The phosphatic matter of phosphorites consists of carbonate-fluorapatite and is intermixed with variable amounts of terrigenous, biogenic and diagenetic non-phosphatic components, which are the cause of a wide range of fluctuations in their chemical compositions. The age of sea-floor phosphorites varies from Cretaceous to Recent. Recent phosphorites are localized in the south west of Africa and at Peru-Chile shelves, which are the areas influenced by strong upwelling of nutrient-rich waters (**Fig. 4(a)**), resulting in high biological productivity, intensive biogenic sedimentation and diagenetic redistribution of geochemically active, mobile, organic-derived phosphorus in sediments. This phosphorus is accreted in the form of initially soft and friable nodules undergoing gradual lithification [27],[37].

Pronounced climatic, biological and geologic effects accompany upwelling, especially where it is produced by the divergence in coastal areas (**Fig. 4(b)**). The presence of cold waters along the coasts produces the coastal fogs and humid-air deserts, such as those of northern Chile and southwest Africa. The nutrient-rich waters that lie alongside these deserts are the lushest gardens of the sea, as the upwelling cold waters there support tremendous quantities of organisms. Most of large accumulations of guano (**Section 7.2.2**) are formed by the seafowl colonies feeding in these waters, and it is the extremely dry climate created by upwelling that makes the preservation of guano possible [27].

Sedimentary deposits usually contain varieties of carbonate-fluorapatite called francolite (described in **Section 2.6**). Francolite is defined as apatite that contains significant amount of

CO_2 with less than 1% of fluorine. Apatite associated with igneous source rocks may be of primary magmatic, hydrothermal or secondary origin. Primary apatite from igneous sources may be of fluorapatite, hydroxylapatite or chlorapatite varieties. Pure apatites from igneous deposits contain slightly over 42% of P_2O_5 [23].

Since a wide range of very different particles and processes of formation complicates the simple classification of phosphorites, there is not any unified phosphorite classification[8] [18],[36],[38]. The proposed classification schemes for phosphorites describe their constituent particles, such as pelletal phosphorite. Nonetheless, the descriptor pelletal indicates nothing more than rounded phosphate particles of any origin. A widely recognized distinction in phosphorites is based on the grain size and holds specifically among phosphorites where the phosphate particles are of sand- or coarse silt-size and those that are of clay- and fine silt-size [28]. The phosphorites can be classified as follows [37]:

1. **Non-conglomeratic** (also termed as **nodular**) phosphorites: consist of phosphatized limestones and gluconate-quartz sandstones. Two varieties of nodular phosphorites can be recognized:

    a. **Ferruginized** with glazed surface. The cement of ferruginized phosphorites is much richer in finely dispersed goethite than that of non-ferruginized phosphorites. Furthermore, ferruginized phosphorites do not contain the fragments of macrofauna.

    b. **Non-ferruginized** with rough surface. In non-ferruginized phosphorites, the cement is micrite-collophane and its color varies form yellow (collophane, described in **Section 2.6**) to gray (micrite[9] [19]). The chambers of foraminifera are filled with phosphate-carbonate cement, less often with glauconite or goethite (FeO(OH) [33]).

    Some nodules of phosphatized limestone are coated with a discontinuous layer of secondary phosphate with the thickness up to 1 cm.

2. **Conglomeratic phosphorites** consist of pebbles of phosphatized limestone (up to 50% of the rock) held together by the cement similar in composition to the phosphatized glauconite-quartz sandstones described above. In many samples of this type, two or three conglomerate layers are clearly visible, differing in size of pebbles and in content of glauconite. The bedding planes separating the layers with denser or less dense packing of grains are also distinguished in the cement. The surface of these planes is glazed and brown due to higher content of iron hydroxides and organic matter. Upon impact, the rock breaks along the planes. In addition, irregular microerosion surfaces are observed in

[8] The nature and origin of phosphorites have been a matter of much speculation since they were first discovered more than 150 years ago [38], and there is not any commonly accepted nomenclature [39].

[9] Micrite (the abbreviation for microcrystalline calcite) is characterized by crypto- to microcrystalline crystal texture. As the synonym of micrite, the names as lime mud, lime ooze, lime mudstone, calcimudstone and calcilutite are also used. The original definition sets a grain-size limit to < 4 μm, but current terminology distinguishes between minimicrite (<1 μm), micrite I (1 – 4 μm) and micrite II (4 – 30 μm). Furthermore, primary micrites (orthomicrites and nannoagorites), secondary micrites and pseudomicrites are recognized. Orthomicrites consist of subhedral polygonal calcite grains meeting at the interfaces. Nannoagorites are composed of calcareous pelagic biota. Secondary and pseudo micrites result from the diagenetic processes [19].

the conglomeratic phosphorites, which run across the grains of glauconite, shells and bedding planes [33].

The investigation of the microstructure of phosphorites by electron microscopy enables to recognize the following varieties [33]:

i. **Gel-like,** phosphatized diatomaceous oozes, analogous (except phosphate content) to the enclosing diatomaceous oozes.

ii. **Microgranular**, forming solid masses and globules from 1 to 3 mm in diameter. The rough surface of phosphate is caused by the fact that it consists of granules less than 0.1 mm in size.

iii. **Fibrous**, constituting inner parts of globules.

iv. **Ultramicrocrystalline phosphate**, forming a "jacket" on the surface of globules of amorphous phosphate. The size of crystal of apatite is 0.1 – 0.3 μm.

v. **Microcrystalline phosphate,** consisting of crystal of 1 – 3 μm in size. Euhedral crystals are often formed in the cavities within the carbonate grains.

vi. **Multiphase microgranular cement** consisting of carbonate, phosphate, quartz and layered silicates.

With regard to their texture and petrographic character, phosphorites can be classified according to the predominant size of the phosphorite component into four types [40]:

a. **Microgranular (oolitic microgranular) phosphorites**, conditionally including aphanite: 0.01 – 0.1 mm;

b. **Granular phosphorites**: 0.1 – 1 mm;

c. **Nodular phosphorites**: 1 – 5 mm;

d. **Shelly phosphorites**: 5 – 100 mm.

Although the types are named on the structural basis, the phosphate grains do not always have the dimensions given above. At the same time, the classified types fairly differ in many features, such as the association with various geological formations, the phosphate mineralogy and the stratigraphic sequence, thus being of lithologic character [40].

Most attempts to classify the phosphorite rocks adopt and modify the classification scheme for carbonates [36]. In 1962, Dunham [41] published the classification scheme for limestone. This scheme for carbonate rocks was modified for non-genetic classification of phosphorites (**Fig. 5**) [28].

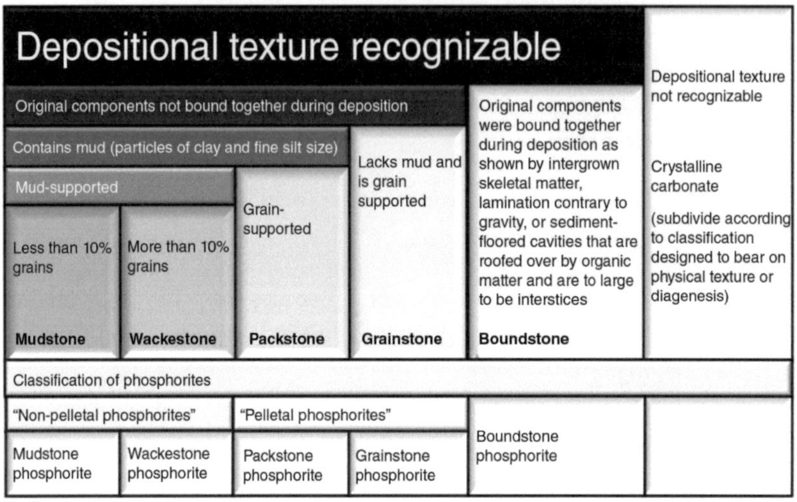

Fig. 5. The classification scheme of carbonate rocks modified for phosphorites [28].

The macroscopic classification scheme for phosphate sediments suggested by RIGGS [42] is shown in **Fig. 6(a)**. The ancient deposits are better characterized via the scheme in **Fig. 6(b)**, which was proposed by KASTERN and GARRISON [43]. In this model, three types of phosphates were recognized [39],[43]:

i. F-phosphates are friable, light-colored micronodules and peloids of carbonate-fluorapatite (CFAP); they were formed by the precipitation of CFAP in laminated diatom muds deposited within the oxygen-minimum zone.

ii. Phosphatic sands, termed as P-phosphates, consist of phosphatic peloids, coated grains and fish debris, often having an admixture of fine siliciclastic grains. These sands occur in thin layers and burrowed beds up to 2 m thick.

iii. Dark and dense phosphates, herein called D-phosphates, are the most abundant. They occur as nodules, gravels and hard grounds. These phosphates were formed through complicated cycles of CFAP precipitation during early diagenesis, erosion and exhumation and reburial and rephosphatization processes associated with changing energy conditions, which may reflect the effects of changes in the sea level.

CFAP cements in P- and D-phosphates are often replaced microbial structures, but our data do not reveal whether this microbial involvement was passive or active. F-phosphates are most common in deeper water, outer-shelf/upper-slope sites, whereas D- and P-phosphates tend to predominate at shallower shelf sites more subjected to episodic high-energy conditions, especially during the low stands of sea level. This concept reveals the paleoenvironmental and time relationships of various phosphate sediments [39],[43].

http://dx.doi.org/10.5772/62214

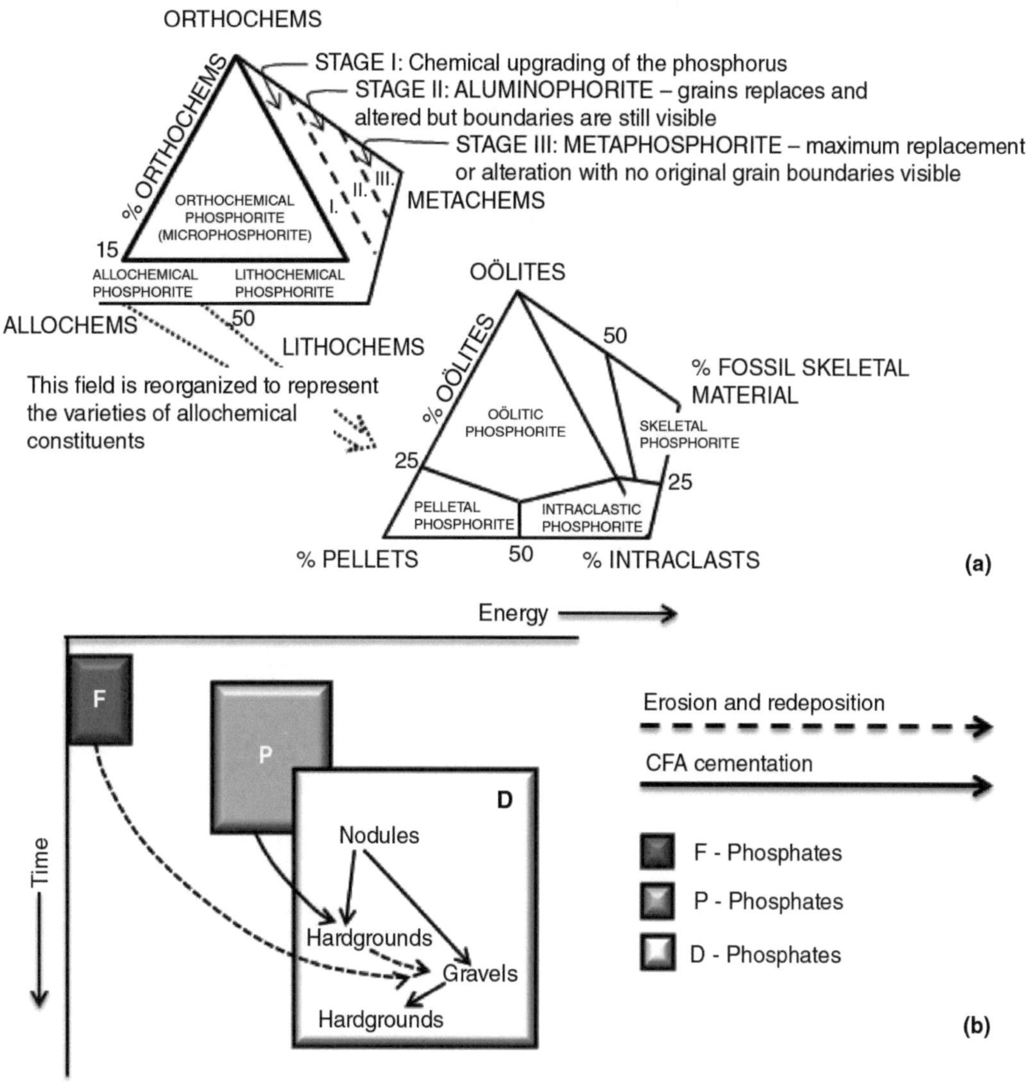

Fig. 6. The classification scheme of phosphorites proposed by RIGGS (a) and GARRISON and KASTNER [43].

Phosphorites can be formed in nature **authigenically** or **diagenically**. In authigenesis, phosphorite forms as a result of the reaction of soluble phosphate with calcium ions forming corresponding insoluble phosphate compound. The role of microbes in these processes may be one or more of the following [44]:

a. Making reactive phosphate available;

b. Making reactive calcium available;

c. Generating or maintaining the pH and redox conditions, which favor the precipitation of phosphate.

The models of authigenic phosphorite genesis (**Fig.** 7) assume the occurrence of mineralization of organic phosphorus in biologically productive waters, such as at ocean margins, that is, at shallow depths on continental slopes, shelf areas or plateaus [44].

Here, detrital accumulations may be mineralized at the sediment-water interface and in interstitial pore waters, liberating phosphate, some of which may then interact chemically with calcium in seawater to form phosphorite grains. These grains may be subsequently redistributed within the sediments units. The dissolution of fish debris (bones) is also considered an important source of phosphate in authigenic phosphorite genesis. The upwelling probably also plays an important role in many cases of authigenic formation of phosphorite. During non-upwelling period in winter, the phosphate-sequestering bacteria of oxidative genera *Pseudomonas* and *Acinetobacter* become dominant in the water column. Fermentative Vibrios and Enterobacteriaceae are dominant during upwelling in summer. It was suggested that *Pseudomonas* and *Acinetobacter*, which sequester phosphate as polyphosphate under aerobic conditions and hydrolyze polyphosphate under anaerobic conditions to obtain the energy of maintenance and to sequester volatile fatty acid from polyhydroxybutyrate formation,

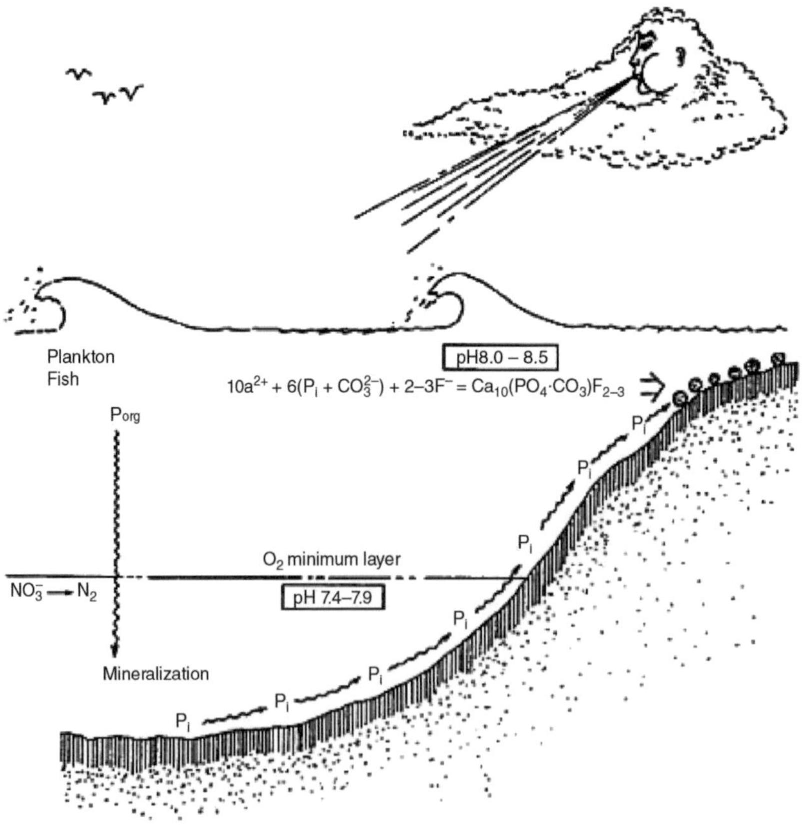

Fig. 7. The schematic presentation of formation of phosphorite in marine environment [44].

http://dx.doi.org/10.5772/62214

contribute to the phosphorite formation. Locally elevated, excreted orthophosphate becomes available for the precipitation as phosphorite by reacting with seawater calcium [44].

Authigenic phosphorite formation at some eastern continental margins, where upwelling, if occurred at all, was a weak and intermittent process that may have been formed more directly as a result of intracellular bacterial phosphate accumulation, which became transformed into carbonate-fluorapatite upon the death of cells accumulated in sediments in areas where the sedimentation rate was very low [44].

The model of diagenetic formation of phosphorite generally assumes the exchange of phosphate for carbonate in accretions that have the form of calcite and aragonite. The role of bacteria in this process is to mobilize phosphate by mineralizing detrital organic matter. The demonstration of this process in marine and freshwater environment under laboratory conditions leads to the hypothesis that the diagenesis of calcite to form apatite explains the origin of some deposits in the North Atlantic. The phosphorite deposits of Baja California and in the core of eastern Pacific Ocean seem to have formed as a result of partial diagenesis [44].

7.1.2. Igneous rocks

Apatites of igneous origin include hydrothermal veins and disseminated replacements, marginal differentiations near the boundaries of intrusions and pegmatites, but the largest deposits are intrusive masses or sheets associated with carbonatite, nepheline-syenite and other alkalic rocks [27]. Igneous rocks are classified on the basis of their [21],[45]:

1. **Color index**: means the volume percentage of dark-colored mineral and divides the rocks to **leucocratic** (color index varies from 0 to 30%), **mesocratic** (from 30 to 60%) and **melanocratic** (from 60 to 100%).

2. **Texture**: coarse-grained rocks, which crystallize at depth, are termed as **plutonic rocks** (as the rate of crystallization is slow, such rocks are holocrystalline[10]). Fine-grained rocks containing minerals embedded in glassy matrix[11] (often devitrified) are named as **volcanic** or **effusive rocks**. The crystallization of volcanic rocks takes place on the surface and is associated with rapid cooling and loss of volatile constituents of the lava. Their texture is therefore vesicular [46][12] (vesicle-rich) and hypocrystalline.[13]

   Rocks that crystallize partly at depth and partly near the surface are called **hypabyssal** [14] (**subvolcanic**). The term **porphyry** is also related to hypabyssal rocks, which are characterized by one or more than one minerals present as phenocrysts in fine-grained groundmass.

---

[10] The name for fully (100%) crystallized igneous rock [45].

[11] The matrix is defined as interstitial material between larger (skeletal) grains [46].

[12] The texture is characterized by many cavities (vesicles), which were formed by bubbles of volatile gasses during the decrease of pressure at extrusion of magma to the surface. Lava solidifies before bubbles of gases can escape to the atmosphere [46].

[13] The name for igneous rocks where the ratio of crystals to glassy phase is higher than 3:5. Rocks containing higher amount of glass are termed as hypohyaline or holohyaline [45].

[14] Denotes the intrusions of magma at shallow depths in the crust, often directly related to overlying volcanic edifices [45].

3. **Chemistry and mineralogy**: the rocks comprising more than 90 vol.% of ferromagnesian minerals, such as olivine, pyroxene, amphibole and biotite, are called **ultramafic (ultrabasic) rocks**. The rocks composed from essentially one or more ferromagnesian minerals are termed as **mafic** [15] **(basic) rocks**. In **mafelsic rocks**, the mafic and felsic minerals are present in approximately equal amounts. **Felsic rocks** [16] contain predominantly light-colored minerals, such as quartz, feldspar, feldspathoid and muscovite.

   An acidic rock contains > 60% SiO_2, whereas a basic rock is characterized by silica content ranging from 44 to 52% of SiO_2. Many of ultramafic rocks are ultrabasic with the content of silica < 44%, but such ultramafic rocks as pyroxenites and amphibolites are not ultrabasic, but they are rather basic [21].

Igneous rocks are formed by the solidification of silicate melt from high temperatures. Since the sequence of crystallization follows the liquidus-solidus phase relationships, the minerals of low content will normally crystallize the least, but diorite and granodiorite melts may have enough phosphorus present for the FAP phase field to intersect the liquidus and to allow early formation of fluorapatite. Later-crystallizing phases should form in the interstices between early-crystallizing phases of alkali-rich igneous rocks and should form an immiscible phosphate-rich liquid phase, which leads to large late-stage segregations of FAP, some of which are associated with magnetite [47].

Where the content of phosphorus is very low, phosphorus may remain in the fluid phase, and apatite will form during the time at which the rock re-reacts with this fluid. This reaction is termed as pneumatolitic, and formed crystals will be small and often euhedral (with crystal facets). They may be included inside preexisting mineral grains. This situation is often encountered in granites and other related siliceous igneous rocks. The concentration of apatite minerals in igneous rocks is rarely sufficient to yield the source for mining the deposits for the phosphorus content [47].

7.1.3. Biogenic apatites

The biogenic (endogenous) mineral deposits form in surface environments as the transformation of primary organic aggregates or as a result of biochemical processes. Since the organism produces many of the same substances that form inorganically in rocks, the biogenic minerals are not minerals in the conventional sense[1]. Biogenic minerals originate from living organisms or with their assistance (**Table 2**). These compounds are crystallized within living organisms as a result of cell activity and are surrounded by organic matter. Classical examples are the bones of vertebrates. The bones and teeth consist of fine fibers or platy crystals (**Section 10.9.2**) of a mineral closely related to carbonate-hydroxylapatite (**Section 4.6**). These crystals are suspended in organic collagen. The crystals of apatite, which often do not exceed 10 nm in length, comprise up to 70% of weight of dried bone. The proteins make up the remaining 30% [48].

---

[15] Term mafic is the abbreviation of names of elements **ma**gnesium and iron (Latin word **f**errum) [45].

[16] Term felsic is the abbreviator of names of minerals **fel**dspar and **si**lica [45].

Composition		Plant or animal examples
Silica (opal, chalcedony, quartz)		Radiolaria, siliceous sponges, diatomic algae
Carbonate	Calcite	Archeocyatha, foraminifera, stromatoporoids, carbonate sponges, echinoderms, brachiopods, belemnites, ostracods, coccolithophora, cyanophycerae, purple algae, some mollusk shells, eggshells or birds and reptiles
	Calcite crystals	Eyes of trilobites and brittle star
	Aragonite	Corals, shells of mollusks and cephalopods
	Aragonite transforming to calcite	Corals, bryozoa, gastropods, pelecypods
Phosphate	Apatite	Bones, teeth, scales of vertebrates, brachiopods
	Barite, gypsum	Ear stones of animals
	Struvite	Kidney and gall stones
Oxalates	Whewellite, weddellite	Kidney and gall stones
Phosphate-bearing carbonates		Brachiopods
Magnetite		In brain tissue of birds and insects (carrier pigeons, bees, etc.), bacteria (*Magnetospirillum magnetotacticum*). Magnetite is used for navigation and orientation.
Fe-hydroxides		Shells of diatoms, pediculates of Protozoa

Table 2. Mineralogical composition of solid plant and animal tissues [48].

In addition to the occurrence within the bones and teeth of vertebrates, mineral-organic aggregates are also found in mollusk shells, solid tissues of foraminifera corals, trilobites and other arthropods, echinoderms, some algae, etc. Some other biogenic processes involve bacteria. Large deposits of native sulfur, manganese oxides and hydroxides and iron are attributed to bacterial activity. Bacterial activity is also involved in the weathering processes of sulfide oxidation and transformation of kaolinite into bauxites [48].

Biogenic apatite is one of the most promising authigenic phases in this respect, as it is present in most siliciclastic deposits and is strongly enriched in a large suite of trace elements. Biogenic (fish teeth and bones) and diagenetic apatites are essential repositories of sedimentary phosphorus. They occasionally form huge deposits, as in West Africa, which are actively mined to provide agricultural fertilizers. Some of these deposits, found in particular in the Late Precambrian of China, after chemical precipitates seem to be associated with the episodes of global glaciation. In low-temperature waters, phosphates form numerous complexes. The concentration of phosphorus in sea and river water is limited by very low solubility of apatite. Phosphate radicals often attach to the surface of iron oxyhydroxide colloids when they precipitate in estuaries [49].

7.1.4. Genetic classification of phosphate rocks

The genetic classification proposed by Egorov [50] includes three types of apatite ores according to the mineral assemblages [6],[51],[52],[53],[54],[55]:

i. **Silicate-apatite** (**ijolite**[17]): ijolite is a medium- to coarse-grained equigranular rock composed of nephelite and aegirite-augite, or other pyroxene, in nearly equal proportions. There are varieties richer in nephelite, approaching the composition of urtite.[18] Nephelite is an equant anhedrons, pyroxene forms euhedral crystals with zonal structure and margin of more sodic pyroxenes than that in the center, which is commonly titaniferous. Some varieties contain titaniferous melanite and iivaarite. Apatite is a prominent constituent and titanate is generally present in small amounts.

ii. **Silicate-magnetite-apatite** (**phoscorite**): phoscorites[19] are spatially and temporally associated with carbonatites, often forming multiphase phoscorite-carbonatite series [56]. The term "phoscorite" was originally used to describe the magnetite-olivine-apatite rock with a carbonate core by Russell et al [57],[58],[59].

iii. **Carbonate-apatite** (**carbonatite**): the crystalline products of low-volume and high-temperature carbonate melts that have been evolving from the upper mantle (**Fig. 8**) for at least the past 2 Ga. Many are associated with crustal complexes of alkali-rich silicate rocks from which they may have evolved by liquid immiscibility.

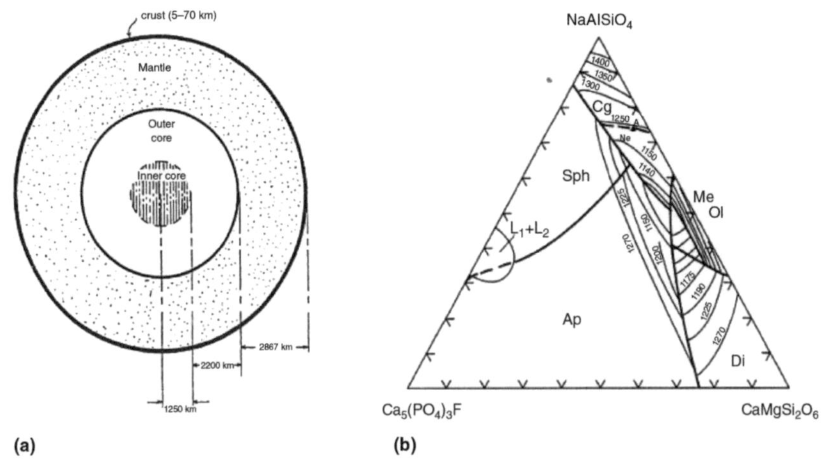

Fig. 8. The location of upper mantle on the cross-section of the Earth, core and mantle drawn to scale [44] (a). Ternary equilibrium phase diagram of the apatite bearing ijolite-urtite rock (b) approximated by the system $NaAlSiO_4$-$CaMgSi_2O_6$-$Ca_5(PO_4)_3F$ [6]: olivine (Ol), melilite (Me), silicophosphate (Sph), apatite (Ap) and immiscibility field (L_1+L_2).

---

[17] Ijolite, theralite and teschenite belong to phanerites (prelenic rocks characterized by equal or nearly equal amounts of femic components) [55].

[18] Urtite is medium-grained and light gray rock composed of about 70% nephelite, 25% aegirite or aegirite-augite [55].

[19] In the work [58], the name camaforite, i.e. Russian neologism [59], denotes the calcite-magnetite-forsterite assemblage, used as a synonym with phoscorite.

The endogenous apatite deposits can be classified[20] into the following types [60],[61]:

1. **Crystallized and concentrated in the late magmatic stage**, which can be further divided to (a) apatite-magnetite deposits (e.g. Kiruna deposit in Sweden) and (b) nepheline-apatite (e.g. Khibiny in Kola Peninsula in Russia).

2. **Skarns** (e.g. Ontario and Quebec in Canada).

3. **Carbonatites** (e.g. Sukulu in Uganda and Dorowa in Zimbabwe).

4. **Hypothermal veins with phlogopite** (e.g. Kaceres in Spain).

5. **Mesothermal type** (e.g. Toledo in Spain).

6. **Metamorphosed sedimentary phosphates** (e.g. Southern Peribaikalia, the Aldan massif in Russia (regional-metamorphosed subtype) and Karatau in Russia (contact-metamorphosis subtype)).

7. **Sedimentary phosphates**, which yield 85% of the world production, form in the sea by biochemical processes, in either:

   - **Geosynclinal sea**: e.g. in the Late Cretaceous and in Paleogene on the shelf of the Tethys geosyncline with the deposits in Morocco (Khouribga and Youssoufia), Algeria (Djebel Onk, El Kouit), Tunis (Gafsa), of the Permian geosyncline of the Rocky Mountains, with the deposits in the Phosphoria Formation in Idaho, Wyoming, Utah and Montana (USA), of the Caledonian geosyncline Karatau (Russia), large deposits are in the Upper Cretaceous in Kazakhstan, in the neighborhood of Aktyubinsk.

   - **Epicontinental sea**: e.g. in the Cenomanian of the south Russian digression and in the Jurassic of the Moscow region, at the margin of the African Shield in the Eocene complex near Hahotoe (Togo) and Taiba (Senegal), in the upper Cretaceous of Egypt [60].

7.2 Phosphate rock reserves

The Earth's crust contains about 0.27% of P_2O_5. About 200 minerals are known, which contain 1% or more P_2O_5. Minable concentrations of phosphate, containing from 5 to 35% of P_2O_5, are formed in all phases of the phosphate cycle (**Section 7.3.1**). The primary deposits include igneous apatites, sedimentary phosphorites and guano. The secondary deposits form from each of these as the result of weathering. Apatite is the principal primary mineral, but a number of others (**Section 7.5**) are common in the deposits formed during weathering of phosphate rocks and guano, e.g. brushite ($CaHPO_4{\cdot}2H_2O$), monetite ($CaHPO_4$), whitlockite (β-$Ca_3(PO_4)_2$), crandallite ($CaAl_3(PO_4)_2(OH)_5{\cdot}H_2O$), wavellite ($Al_3(OH)_3(PO_4)_2{\cdot}5H_2O$), taranakite ($K_2Al_6(PO_4)_6(OH)_2{\cdot}18H_2O$), millisite ($(Na,K)CaAl_6(PO_4)_4(OH)_9{\cdot}3H_2O$), variscite ($AlPO_4{\cdot}2H_2O$) and strengite ($FePO_4{\cdot}2H_2O$) [27].

[20] The classification according to Dybkov and Karyakin.

7.2.1. Phosphate rock deposits

In general, phosphate rock reserves are non-metallic ores[21] that can be economically produced at the present time using existing technology. Phosphate rock resources include reserves and any other materials of interest that are not reserves. A reserve base is a portion of the resource from which future reserves may be developed. The classification applying to phosphate rocks include [23].

- **Resource** is defined as a concentration of naturally occurring phosphate material in such a form or amount for which the economic extraction of a product is currently or potentially feasible. The resources are divided into many categories depending on the amount of pertinent information available to define the amount of material potentially available and if it is economic, marginally economic or sub-economic to exploit these resources.
- **Reserve base** is the part of an identified resource that meets the minimum criteria related to current mining and production practices including grade, quality, thickness and depth.
- **Reserves** are the part of the reserve base that can be economically extracted or produced at the time of the determination. They may be termed as marginal, inferred or inferred marginal reserves. This does not signify that the extraction facilities are in place or functional.
- The grades of apatite deposits from the economic point of view are introduced in **Table 3**. The locations of the world's phosphate deposits are shown in **Fig. 9** and the average content of P_2O_5 is listed in **Table 4**.

The most known Miocene phosphate deposits (**Table 1**) are in North Carolina, Florida, Venezuela, California, Baja California and Peru. In several cases, these emerged deposits are only the up dip limit of a larger Miocene section that extends seaward beyond the coastal plain and constitute large portions of the upper sediment regime that built the modern continental shelves [26].

Grade	P_2O_5 [wt.%]	BLP [%]	Location
1 Economic	20	40.70	Florida and Moroccan sedimentary phosphorites, Kola and Palabora crystalline igneous apatites
2 Sub-economic	5 – 20	10.93 – 40.70	Western USA phosphoria, Russia nepheline-apatites
3 Non-economic	1 – 5	2.19 – 10.93	Low-grade ores, phosphatic limestones
4 Non-phosphatic	0.1 – 1.0	0.22 – 2.19	Widely distributed apatite in almost all igneous rocks

Table 3. Grades of apatite deposits [25].

[21] Ores of economic value can be classified as metallic or non-metallic according to the use of the mineral. Certain minerals may be mined and processed for more than one purpose. In one category, the mineral may be metal and non-metallic ore, e.g. bauxite used for the production of aluminum and ceramics, respectively [62].

http://dx.doi.org/10.5772/62214

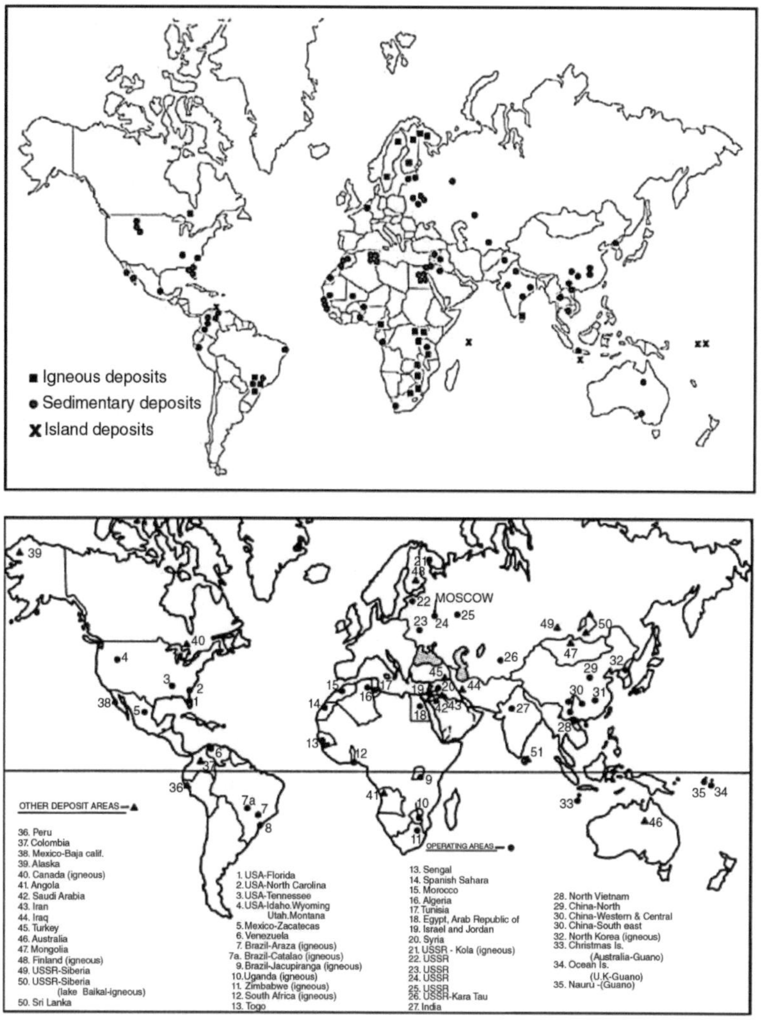

Fig. 9. The overview of the world's phosphate deposits (a) [20] and the location of the major phosphate rock-producing areas of some known deposits (b) [26],[63].

Locally, these Miocene sediments are exposed on the seafloor; however, generally, they are buried below thin covers of Plio-Pleistocene and Holocene surface sediments. Thus, there is high potential for discovering new phosphate deposits within the Miocene sediments on the world's continental shelves, because [26]:

a. Phosphate genesis is known to occur throughout the shelf in upper slope environments.

b. Thicker and more extensive sequences of Miocene sediments occur on the shelves than on adjacent coastal plains.

c. The shallow subsurface Neogene geology of most of the world's shelves is poorly known.

Source	%	Source	%	Source	%
Fluorapatite	42	Tunisia (sedimentary)	28	Basic slag	10 – 20
Kola (igneous)	40	West USA (phosphoria)	18 – 30	Bone meal	20
Nauru (phosphorite)	39	Queensland	16 – 30	Guano	12 – 15
Florida (sedimentary)	35	Venezuela	20	California (seabed)	30
Kazakhstan	23	China (Yunnan)	32 – 36	Australia (Queensland)	24
Morocco (sedimentary)	35	Kola (nepheline)	12 – 20	–	–

Table 4. Grades of apatite deposits [25].

Two important sediment relationships were developed concerning reworked phosphate in surficial sediments on the North Carolina continental shelf [26],[63]:

i. The distribution of phosphate in surface sediments closely reflects the distribution within underlying Miocene sediments units.

ii. The process of reworking significantly dilutes the concentration of phosphate in the surficial sediments.

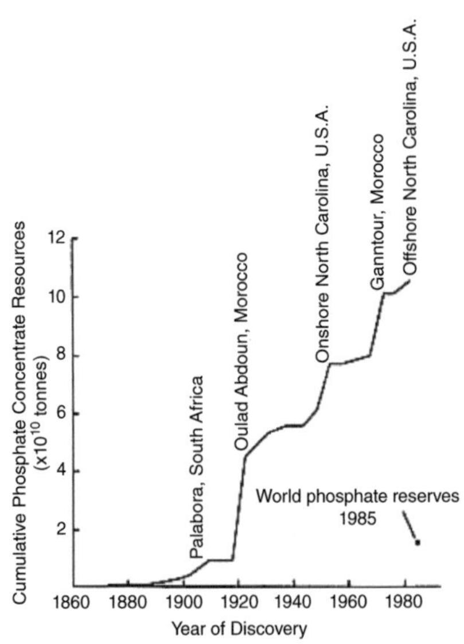

Fig. 10. The history of the discovery of the world's phosphate resources [22].

These relationships were also recognized on the shelves of northwest and southwest Africa and could represent important exploration tools for richer Tertiary[22] phosphorites occurring within the shallow subsurface on many continental shelves in the world [26].

[22] The term Tertiary (geologic period from 66 to 2.58 ma) is no longer recognized by the International Commission on Stratigraphy.

http://dx.doi.org/10.5772/62214

The deposits of apatite of igneous origin occur as intrusive masses or sheets, as hydrothermal veins or disseminated replacements, as marginal differentiations along or near the boundaries of intrusions or as pegmatites. Intrusive masses are the largest of these deposits. They are commonly associated with alkalic igneous rock complexes, many of which such as those in Africa, Brazil and Sweden are associated with the rift valley structures. Carbonatite, ijolite, nepheline-syenite and pyroxenite are common members of the rock assemblage. Many of these complexes have a ring like structure, with carbonatite as the central core [27].

Phosphate deposits have been discovered within the past 100 years at the rate far greater than the rate of consumption (**Fig. 10**). Since new phosphate deposits are expected to be discovered in the future, the oil exploration programs have probed most of the coastal sedimentary basins of the world during the past 20 – 30 years, and any large-scale discoveries of phosphate rock would probably have occurred in conjunction with these activities [22].

7.2.2. Other sources of phosphorus

Other important commercial sources of phosphorus (**Table 4**) include [25]:

i. **Guano** [25],[27],[64],[65],[66]: natural deposit (accumulation) formed from decaying bones and excreta from fish-eating birds.[23] Fresh seafowl droppings contain about 22% N and 4% P_2O_5. Bat guanos are the most abundant in the cave areas of temperate and tropical regions. Although many bat guano deposits were found and mined, most of them are measured in hundreds or thousands of tons, and only sporadic production is obtained from them now. Seafowl deposits are mainly confined to islands and coastal regions at low latitudes. The largest lie along the west coasts of lower California, South America and Africa and on islands near the equatorial currents.

   It was known that bird dung was utilized by the Carthaginians as early as 200 BC in order to improve crop yields. The content of P_2O_5 in guano can vary in dependence on the age of the deposit,[24] the local climate and the bird kind. Guano deposits are found in Chile, Peru, Mexico, Seychelles, Philippines, the Arabian Gulf and elsewhere, but they account for less than 2% of the world's phosphate production.

   Guano is used almost exclusively as fertilizers. The Nauru and Christmas Island phosphorite deposits may be guano in origin, but they are of very limited extent. It

---

[23] It is estimated that marine birds may take out as much as 5.10^{10} g of phosphorus from the ocean each year [67]. The Spanish name "guano" has an origin in Quechua word "huanu" (i.e. dung).

Guano is composed of bird droppings, and although birds existed as early as the mid-Mesozoic, their major development did not occur until the Cenozoic era, probably in the Eocene period. It follows that some seamount deposits are too old to have originated from bird droppings [64].

Most of large accumulations of guano are formed on the surface by seafowl, but smaller quantities are formed by bats and to a lesser extent by other cave-dwelling mammals and birds [27].

[24] Recent guano contains 10 – 12% of P_2O_5, but leached guano contains 20 – 32%. The mineralogy of guano is complex. Slightly decomposed deposits contain soluble ammonium and alkali oxalates, sulfates and nitrates and a variety of magnesium and ammonium-magnesium phosphates. Deeply decomposed guano consists chiefly of calcium phosphates (for example, monetite or whitlockite) [27].

is believed that rainwater can carry soluble phosphate from guano and trickle over rocks, where phosphate interacts to form phosphatic layers, e.g. phosphatized coral rock. Bird guano, mainly from Peru, achieved the greatest importance in about the middle of the 19th century, shortly before the phosphate rock industry began to establish itself.

ii. **Basic slag** [25]: is a minor source of phosphorus. This waste is the product from blast furnaces operating on iron ores with significant content of phosphorus. Basic slag contains tetracalcium phosphate ($Ca_3(PO_4)_2 \cdot CaO$) and silicocarnotite ($Ca_3(PO_4)_2 \cdot Ca_2SiO_4$), which are applied directly as fertilizers. Recorded world production is mainly from France, Germany and Luxembourg.

iii. **Meat and bone meal (MBM) or bone ash** [25],[68]: as an animal byproduct, MBM contains not only substantial amounts of phosphorus in soluble organic form but also calcium and microelements. Ground (bone meal) or calcined and ground (bone ash) bones were recognized as a source of phosphorus at an early date.

Other important commercial sources of phosphorus are casein and lecithin. Casein is obtained from bovine milk. Lecithin was extracted from soy bean oil [25].

7.2.3. Phosphorite weathering derivates

Weathering (**Section 7.3.2**) leads to the formation of enriched residual and replacing deposits from phosphatic deposits not otherwise minable. The Tennessee "brown rock" phosphate deposits consist of nowadays residuum developed through the decomposition of phosphatic limestones of Ordovician age. The "river pebble" deposits prominent in early history of phosphate mining in Florida and South Carolina are mostly placers formed by alluvial concentration of phosphatic pebbles eroded from the phosphatic formations of adjacent terrain [27].

The Tennessee "white rock" and Florida "hard rock" deposits were formed by the redeposition of phosphate derived from the decomposition of apatite under more advanced weathering. The same decomposition-phosphatization process accounts for the formation of calcium aluminum phosphate and aluminum phosphate in the "leached zone" of the Bone Valley field and deeply weathered Cretaceous and Eocene deposits of west Africa [27].

7.3. Geological role of apatite

As was described above, the number of different elements can substitute into the structure of apatite,[25] and this mineral can contain a number of trace elements by the substitution in both anion and cation sites. This means that apatite can be used as an indicator of planetary halogen compositions. The quantitative ion microprobe measurements of apatite from lunar basalts showed that portions of the lunar mantle and/or crust are richer in volatile species than previously thought [4].

[25] The tools based on the isotope composition of apatite were already described in **Section 6.5**.

http://dx.doi.org/10.5772/62214

Apatite was also used to determine the characteristics of metamorphic fluids within the mantle. For example, O'Reilly and Griffin [69] and Douce et al [70] classified apatite within Phanerozoic mantle material into two geochemically distinct types:

1. **Apatite A**: is inferred to result from the metasomatism by CO_2- and H_2O-rich fluids derived from a primitive mantle source region;

2. **Apatite B**: compositions are consistent with the crystallization from magmas within the carbonate-silicate compositional spectrum.

This classification is based on halogen content, presence or absence of structural CO_2, Sr and trace elements (especially U, Th and light rare-earth) and association with either metasomatized mantle wall-rock peridotites (Apatite A) or high-pressure magmatic crystallization products (Apatite B) [4],[69],[70].

In addition, apatite can be used as a probe to determine the petrogenetic evolution of granites, and significant amounts of research were devoted to the use of apatite in granitic rocks to distinguish between S- and I-type granites [4],[71].

The chemical composition of apatite is also a useful guideline for the petrogenetic and metallogenic history of magmas for the following reasons [4]:

- Apatite Eu and Ce anomalies provide the evidence of the redox state of the magmas that formed the host granitic rocks, with Eu enrichment and Ce depletion being indicative of oxidized magma and Eu depletion and Ce enrichment being indicative of reduced magma.

- Apatite $^{87}Sr/^{86}Sr$ ratios reflect the Sr isotopic composition of the host granitic rocks.

- Apatite F and Cl concentrations can reflect the enrichment or depletion of halogens within the host granitoids, with apatite associated with slab dehydration containing more Cl and less F, whereas apatites related to magmas formed by partial melting of the crust contain less Cl and more F.

In recent sedimentary systems, the major phosphorus deposition occurs within upwelling zones at continental margins. Upwelling of deep ocean waters rich in phosphorus triggers high biological production in the photic zone and eventually high concentration of phosphorus in organic-rich sediments, as in recent Namibian and Peruan shelves [72],[73], [74],[75].

7.3.1. Cycle of phosphorus

In the Earth's crust, phosphorus takes the second place after carbon, and in comparison with all known elements, it takes about 12th place in natural abundance [25]. The phosphorus cycle is quite different from the nitrogen and sulfur cycles in which phosphorus is present in only one oxidation state and it forms no gases stable in biosphere or atmosphere. Also, in contrast to nitrogen and sulfur, substantial proportions of phosphorus in soil appear in inorganic form [76],[77].

About 10 Mt of phosphorus are released by weathering of apatite annually. In soil, monobasic ($H_2PO_4^{\ -}$) and dibasic ($HPO_4^{\ 2-}$) phosphates are generally available to plants. Phosphates are precipitated by calcium in alkaline soils and most of phosphate is adsorbed on aluminum and iron oxides in acidic soils. Phosphates are most readily available in slightly acidic to neutral soil. Much of such phosphorus in surface soils appears in organic matter. This phosphorus is used repeatedly by recycling in plants and organisms that decompose the plant detritus. Little amount of phosphorus is lost by leaching through soils, but the erosion losses of soil particles and the plant detritus carried off to aquatic systems may be substantial [76].

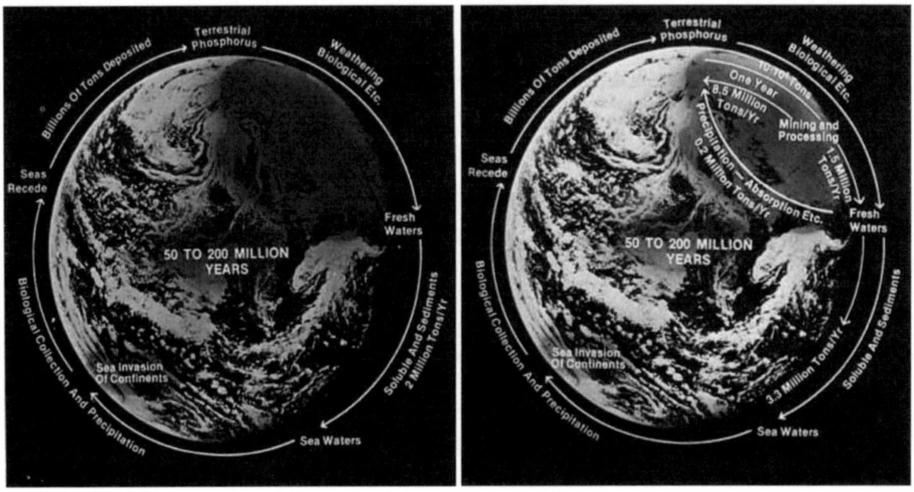

Fig. 11. Major natural cycle of phosphorus (a) and the contribution of the man to the cycle (b) [78].

The availability of phosphorus is a major factor limiting the biomass production in both terrestrial and aquatic ecosystems. Mycorrhizas are efficient scavengers of phosphorus for plants growing in soils with limited availability of this element. The phosphorus fertilization in agricultural lands can have detrimental effect, as it increases phosphate amounts in the runoff soil resulting in the accumulation of phosphate in aquatic plants and algal growth. If the decomposers of plants and algae use practically all oxygen from water, the habitat becomes unsuitable for fish and other aquatic animals. The process of abundant nutrient-induced biomass production in lakes and rivers and its decay to deplete the water oxygen is called the eutrophication [25],[76],[78].

Despite the advantage for which phosphorus is used, it is doubtful that man has significant contribution to the Earth's cycle of phosphorus (**Fig. 11**). The $2{\cdot}10^9$ tons of mined phosphate rock is less than 0.15% of known reserves of phosphorus ore and less than $1{\cdot}10^{-5}$% of the Earth's cycle of phosphorus [78].

At least 100 millions years before humankind exerted any influence on the cycles of phosphorus, the pattern had already been established (**Fig. 12**). Phosphorus was continuously leached from igneous rocks as the rocks were weathered to sedimentary deposits and this released phosphorus flowed to the seas, which had long since become saturated with phosphorus. Each new addition causes a similar quantity of phosphorus to precipitate as sediment. If the

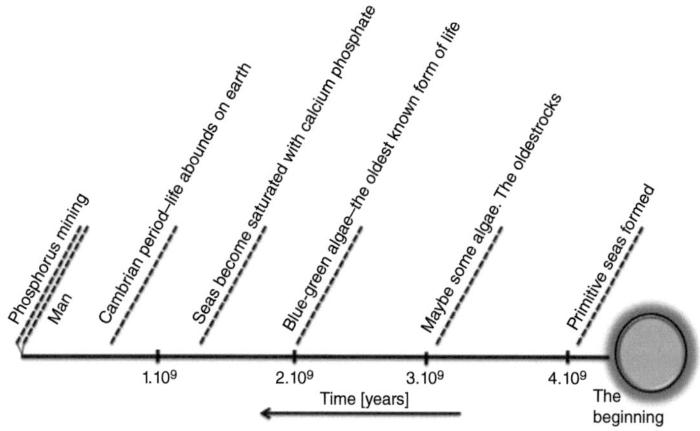

Fig. 12. The establishment of the Earth's cycle pattern of phosphorus [78].

precipitate formed when an island sea had invaded a land area, the new sediment became landlocked. The new landlocked sedimentary deposits are more easily leached than igneous rocks from which they are derived. When the seas recede sufficiently to expose the new sediments to the greater solvent action of fresh water, the sediments begin to weather and the cycle is complete. Best estimates of the cycle time of phosphorus in the oceans today are in the range of 50,000 years. This is a short period compared to $3{\cdot}10^9$ years, which were required for the saturation of oceans for the first time [78].

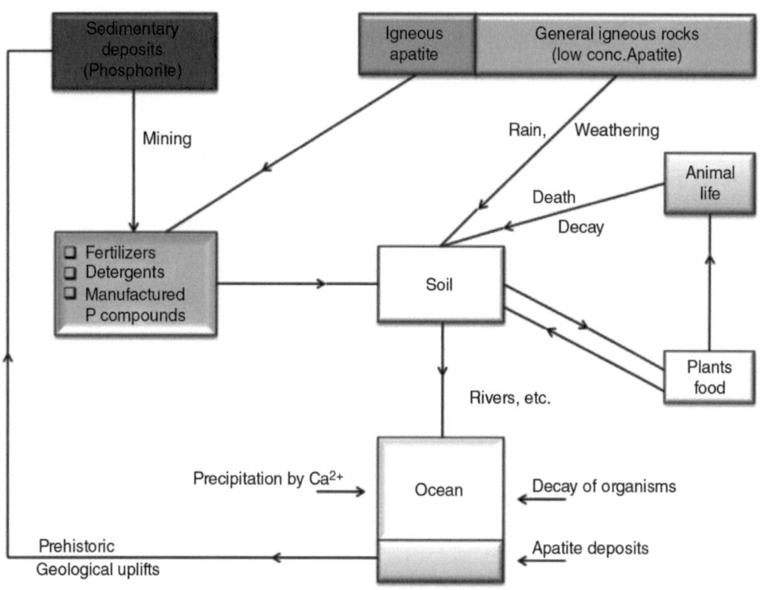

Fig. 13. Natural and artificial cycles of phosphorus [25].

Recently, man has only a slightly stronger influence on the total amount of the Earth's phosphorus than his prehistoric ancestors. If man made a significant alteration in the cycles of phosphorus, it had an impact on the cycles of fresh surface waters. The detergent phosphates have been blamed for degrading freshwater lakes and there is no doubt that several lakes have been overabundant with phosphates and sewage. Sewage treatment will alleviate most of the problems associated with point-source loading of lakes [78].

The overall natural and artificial cycles involving phosphorus are introduced in **Fig. 13**.

7.3.2. Weathering of apatite

Weathering and leaching processes from millions of years ago led to the transfer of phosphate to rivers and oceans where it was concentrated in shells, bones and marine organism that were deposited on the sea floor. Subsequent uplift and other geological movements led to these accumulations becoming dry land deposits [25],[78].

Generally, weathering of apatite occurs synergistically through biotic and abiotic processes and leads to the release of mineral phosphate. Inorganic phosphate cannot be assimilated by plants, but it can be converted to the bioavailable form orthophosphate ($HPO_4^{\ 2-}$, $H_2PO_4^{\ -}$) by some species of phosphate-solubilizing fungi and bacteria. The main mechanism underlying the microbial phosphate solubilization is the secretion of organic acids that, by changing the soil pH and acting as chelators, may induce the dissolution of phosphorus from minerals and its release into the pore water of soils [79],[80]. The dissolution of apatite is described in **Section 3.4**.

Apatite represents an important source of inorganic P for natural ecosystems and may favor the establishment of microbial communities able to exploit it [79]. The microorganisms can cause the fixation or immobilization of phosphate, either by promoting the formation of inorganic precipitates or by the assimilation of phosphate into organic cell constituents on intracellular polyphosphate granules. Insoluble forms of inorganic phosphorus, e.g. calcium, aluminum and iron phosphates, may be solubilized through the microbial action. The mechanisms by which the microbes accomplish this solubilization vary [44]:

i. The first mechanism may be the production of inorganic or organic acids that attack the insoluble phosphates.

ii. The second mechanism may be the production of chelators such as gluconate and 2-ketogluconate, citrate, oxalate and lactate. All these chelators can complex the cation portion of insoluble phosphate salts and thus force their dissociation.

iii. The third mechanism of phosphate solubilization may be the reduction of iron in ferric phosphate, e.g. strengite ($Fe^{3+}PO_4{\cdot}2H_2O$ [81]), to ferrous iron by enzymes and metabolic products of nitrate reducers such as *Pseudomonas fluorescens* and *Alcaligenes* spp. in sediments.

iv. The fourth mechanism is the production of hydrogen sulfide (H_2S), which can react with iron phosphate and precipitate it as iron sulfide, thereby mobilizing phosphate, as in the reaction [44]:

http://dx.doi.org/10.5772/62214

$$2\ FePO_4 + 3\ H_2S \rightarrow 2\ FeS + 2\ H_3PO_4 + S^0 \quad (1)$$

The phosphate-solubilizing ability is a feature of many free-living and plant-symbiotic bacterial taxa, such as [79]:

- ***Pseudomonas*** [82]: the genus *Pseudomonas* sensu stricto comprises many species characterized by their metabolic diversity and by a wide range of niches that they can colonize.
- ***Rhizobium*** [83]: specific group of bacteria that have the capability of symbiotic nitrogen fixation.
- ***Burkholderia*** [84]: aerobic, non-spore-forming bacteria. *Burkholderia* is very versatile and occupies a wide range of ecological niches.

Microbial rock weathering is common in all climate zones and usually acts very slowly [80].

7.3.3. Fission track and apatite fission-track analysis

The fission-track (FT) dating is a radiometric dating method[26] based on the analysis of radiation damage trails (fission tracks) in uranium-bearing, non-conductive minerals and glasses. It is routinely applied to the minerals apatite, zircon and titanite. Fission tracks are produced continuously through geological time as a result of spontaneous fission track of ^{238}U atoms[27] that undergo spontaneous fission.[28] The atom splits into two parts that move rapidly in opposite directions, creating a long thin region of damage. The submicroscopic features with an initial width of approximately 10 nm and the length of up to 20 μm can be revealed by chemical etching.[29] Crucially, fission tracks are semi-stable features that can self-repair (shorten and eventually disappear) by the process known as annealing at a rate that is a function of both time and temperature. The extent of any track shortening (exposure to elevated temperatures) in a sample can be quantified by examining the distribution of fission-track lengths [6], [85],[86],[87],[88].

This unique sensitivity of the apatite fission-track system is now of considerable economic importance due to the coincidence between the temperature range over which annealing occurs and that over which liquid hydrocarbons are generated. Other applications include the determination of timing of emplacement and the thermal history of ore deposits. There is

[26] Radiometric dating techniques are, in general, complementary to one another, and each method produces an age with the special meaning, such as the last outgassing, the last melting accompanied by mixing with isotopically separate material and the last heating to remove the track. Fission-track dating is conceptually the simplest of several dating techniques that provide absolute measures of time from slow but statistically steady decay of radioactive nuclides [86].

[27] Most radiometric dating processes are based on the statistical regularity of the decay of one parent radionuclide into a daughter nuclide, for example, ^{40}K into ^{40}Ar, ^{87}Rb into ^{87} Sr or ^{238}U ,^{235}U and ^{232}Th into ^{206} Pb, ^{207}Pb and ^{208}Pb, respectively. The age of sample can be then determined by measuring relative abundance of parent and daughter in any pair. The major isotope of uranium (^{238}U) decays at a spontaneous rate of ~10^{-16} per year [86].

[28] Fission track can also be created artificially (induced track) by irradiating the mineral specimen with thermal neutrons in a nuclear [88].

[29] The mineral grain is ground and polished to expose a flat surface inside the crystal. It is then immersed in a chemical etchant that preferentially attacks the regions of damage, widening them and making them visible under optical microscope. The track appears in the apatite torch readily in 20 to 30 s when immersed in diluted nitric acid [88].

abundant literature on both fission-track dating and its use in evaluating the tectonic and thermal history of rocks [6],[89],[90],[91].

Apatite is the most frequently used material for fission-track dating [92]. Apatite fission-track (AFT) analysis serves as a thermochronological tool to investigate the low-temperature thermal history of rocks below ~120°C [93],[94]. The estimates of closure temperatures for fission-track retention in apatite are usually in the range from 75 to 120°C at cooling rates between 1 and 100°C/m.y. [6].

Thermochronology may be described as the quantitative study of the thermal histories of rocks using temperature-sensitive radiometric dating methods such as $^{40}Ar/^{39}Ar$ and K-Ar, fission track and (U-Th)/He. Among these different methods, apatite fission track and apatite (U-Th-Sm)/He (AHe) are now, perhaps, the most widely used thermochronometers, as they are the most sensitive to low temperatures (typically between 40 and 125°C for the durations of heating and cooling in the extent of 10^6 years). They are ideal for investigating the tectonic and climate-driven surficial interactions that take place within the top few (<5 km) kilometers of the Earth's crust. These processes govern the landscape evolution, influence the climate and generate the natural resources essential to the well-being of mankind [85],[95].

7.3.4. Extraterrestrial apatite

On Earth, magmatic volatiles (i.e. H_2O, F, Cl, C-species and S-species) play an important role in the physicochemical processes that control thermal stabilities of minerals and melts, in magma eruptive processes and in the transportation of economically important metals. On the Moon, magmatic volatiles in igneous systems are poorly understood, and the magmatic volatile inventory of lunar interior, aside from being very low, is not well constrained. Although the Moon is a volatile-depleted planetary body, there is evidence indicating that magmatic volatiles have played a role in igneous processes on the Moon. Specifically, magmatic volatiles were implicated as the propellants that drove fire-fountain eruptions, which produced the pyroclastic glass deposits encountered at the Apollo 15 and 17 sites [96]. That is supported by recent discoveries of water-rich apatite from lunar mare basalts [97],[98].

Apatite was found in a large number of samples of igneous lunar rocks, although it typically occurred in only trace amounts and is typically reported as coexisting with REE-merrillite [$(Mg,Fe)_2REE_2Ca_{16}P_{14}O_{56}$], and those two minerals make up the primary mineralogical budget for P on the Moon [96]. Merrillite, also known as the mineral whitlockite (or more precious and dehydrogenated whitlockite) [99], is one of the main phosphate minerals, along with apatite, which occur in lunar rocks, in Martian meteorites and in many other groups of meteorites. Significant structural differences between terrestrial whitlockite and lunar (and meteoritic) varieties require the use of "merrillite" name for the H-free extraterrestrial material, and the systematic enrichment of REE in lunar merrillite requires the use of "REE-merrillite". Lunar merrillite, ideally $(Mg,Fe^{2+},Mn^{2+})_2[Ca_{18-x}(Y,REE)_x]$ $(Na_{2-x})(P,Si)_{14}O_{56}$, contains high concentrations of Y + REE [100],[101].

http://dx.doi.org/10.5772/62214

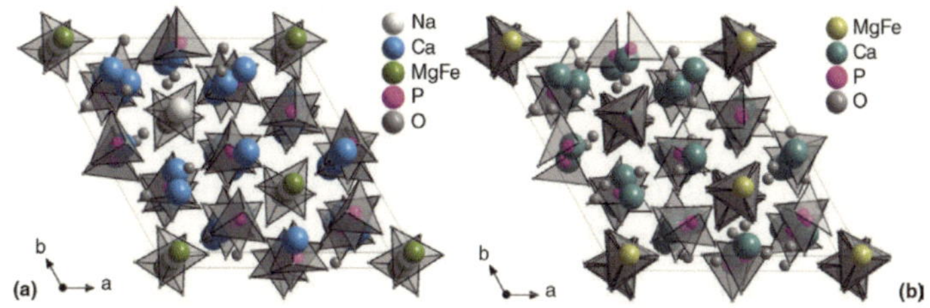

Fig. 14. The structure of lunar merrillite (a) [102] and terrestrial whitlockite (b).

Lunar merrillite (**Fig. 14(a)**, trigonal, the space group R3c with the cell parameters $\boldsymbol{a}$ = 10.2909 Å, $\boldsymbol{c}$ = 36.8746 Å, $\boldsymbol{c}$:$\boldsymbol{a}$ = 3.5832 and $\boldsymbol{V}$ = 3381.93 Å^3) and terrestrial whitlockite (**Fig. 14(b)**, trigonal, the space group R3c with the cell parameters $\boldsymbol{a}$ = 10.3300 Å, $\boldsymbol{c}$ = 37.1030 Å, $\boldsymbol{c}$:$\boldsymbol{a}$ = 3.5918 and $\boldsymbol{V}$ = 3428.79 Å^3) have largely similar atomic arrangements, but the phases differ due to the presence or absence of hydrogen. In whitlockite, H is an essential element and allows the charge balance. Hydrogen is incorporated into the whitlockite atomic arrangement by disordering one of the phosphate tetrahedra and forming the $PO_3(OH)$ group. Lunar merrillite is devoid of hydrogen; thus, no disordered tetrahedral groups exist. The charge balance for the substituents Y and REE (for Ca) is maintained by $Si^{4+} \leftrightarrow P^{5+}$ tetrahedral substitution and $\square \leftrightarrow Na^+$ substitution at Na site [99],[102].

A number of sources potentially contributed to the overall inventory of lunar water, including primary indigenous water acquired during lunar accretion, late addition of water through asteroidal and cometary impacts and solar wind implanting H into lunar soils. The average D/H ratios of apatite in norite (Apollo sample 78235) and in the granite clast (14303) are consistent with the estimates for the H isotopic composition of recent bulk-Earth and terrestrial mantle. By contrast, the average H isotopic composition of apatites in norite 77215 is lower. The content of water in norite parental melts provides strong evidence that the magmas involved in secondary crust production on the Moon were hydrated, in agreement with recent findings of water in lunar ferroan anorthosites. Water they contain, locked in the crystalline structure of apatite, is characterized by an H isotopic composition similar to that on Earth and in some carbonaceous chondrites [103].

Apatite preserves a record of halogen and water fugacities that existed during the waning stages of crystallization of planetary magmas, when they became saturated in phosphates. The thermodynamic formalism based on apatite-merrillite equilibria that makes it possible to compare the relative values of halogen and water fugacities in Martian, lunar and terrestrial basalts, accounting for possible differences in pressure, temperature and oxygen fugacities among the planets, was described by Douce and Roden [104].

They showed that planetary bodies have distinctive ratios among volatile fugacities at apatite saturation and that these fugacities are, in some cases, related in a consistent way to volatile fugacities in the mantle magma sources. Their analysis shows that the Martian mantle parental to basaltic SNC meteorites was dry and poor in both fluorine and chlorine compared to the terrestrial mantle. Limited data available from Mars show no secular variations in mantle halogen and water fugacities from ~4 Ga to ~180 Ma. Water and halogens found in recent

Martian surface rocks have thus resided in the planet's surficial systems since at least 4 Ga and may have been degassed from the planet's interior during the primordial crust-forming event. In comparison to the Earth and Mars, the Moon, and possibly the eucrite parent body too, appear to be strongly depleted not only in H_2O but also in Cl_2 relative to H_2O. The chlorine depletion is the strongest in mare basalts, perhaps reflecting an eruptive process characteristic with large-scale lunar magmatism [104].

Mars does not recycle crustal materials via the plate tectonics. For this reason, the magmatic water reservoir of the Martian mantle has not been affected by the surface processes, and the deuterium/hydrogen (D/H) ratio of this water should represent the original primordial Martian value. Following this logic, hydrous primary igneous minerals on the Martian surface should also carry this primordial D/H ratio, assuming no assimilation of Martian atmospheric water during the crystallization and no major hydrogen fractionation during the melt degassing. Hydrous primary igneous minerals, such as apatite and amphibole, are present in Martian meteorites here on Earth. Provided these minerals have not been affected by terrestrial weathering, Martian atmospheric water or shock processes after the crystallization, they should contain a good approximation of the primordial Martian D/H ratio. As Nakhla was seen to fall on the Egyptian desert in 1911, the terrestrial contamination is minimized in this meteorite. The nakhlites are also among the least shocked Martian meteorites. Therefore, apatite within Nakhla could contain primordial Martian hydrogen isotope ratios. The similar D/H ratios indicate that the Earth and Mars, and possibly the other terrestrial planets, accreted water from the same source [105].

Vesta, as the second most massive asteroid, has long been perceived as anhydrous. Recent studies suggesting the presence of hydrated minerals and past subsurface water have challenged this long-standing perception. The volatile components indicate the presence of apatite in eucrites. Eucritic apatite is fluorine rich with minor chlorine and hydroxyl (calculated by difference) [106],[107],[108].

7.4. The future of phosphate rocks

The biochemist and sci-fi author Izaak Asimov said [60]: "In the future coal will be probably substituted by nuclear energy, wood by plastics, meat by yeast and suspecting solitude by friendship, but there is no substitute for phosphorus."

The search for phosphate rock deposits became a global effort in the 20th century as the demand for phosphate rocks increased. The development of deposits further intensified in the 1950s and 1960s. The world production reached its peak in 1987 – 1988 and then again in 2008 with over 160 million metric tons (mmt) of the product. Phosphate rock mining has evolved over time, and worldwide, it relies on high volume and advanced technology using mainly open-pit mining methods and advanced transportation systems to move hundreds of millions of tons of overburden to produce hundreds of millions of tons of ore, which are beneficiated to produce approximately 160 mmt of phosphate rock concentrate per year. The concentrate of suitable grade and chemical quality is then used to produce phosphoric acid,

http://dx.doi.org/10.5772/62214

the basis of many fertilizer and non-fertilizer products [23]. The world phosphate production rate since 1850 according to Jasinski [109] and Abouzeid et al [20] is shown in **Fig. 15**.

The estimates of the world's phosphate reserves and availability of exploitable deposits vary greatly and the assessments of how long it will take until these reserves are exhausted vary also considerably. Furthermore, it is commonly recognized that the high-quality reserves are being depleted expeditiously and that the prevailing management of phosphate, a finite non-renewable source, is not fully in accordance with the principles of sustainability. The depletion of current economically exploitable reserves is estimated to be completed in some 60 to 130 years. Using the median reserve estimates and under reasonable predictions, it appears that phosphate reserves would last for at least 100+ years [20].

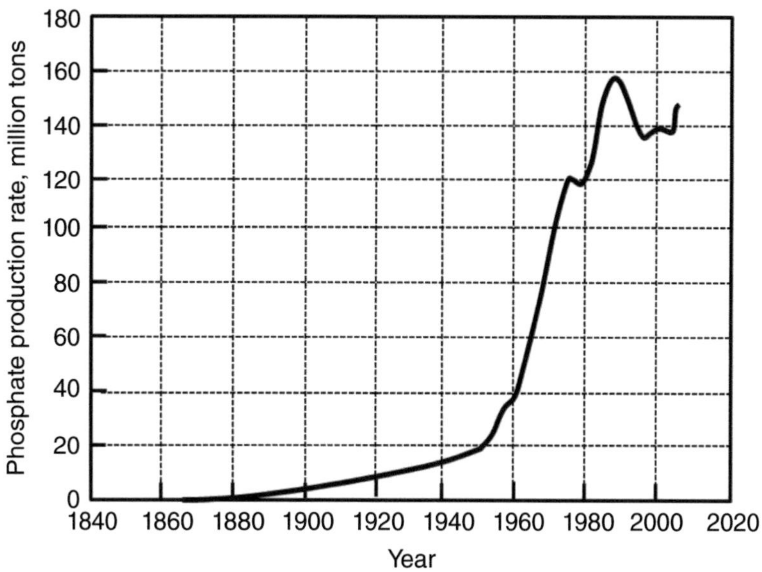

Fig. 15. World phosphate production rate [20].

Preliminary estimates of phosphate rock reserves range from 15,000 mmt to over 1,000,000 mmt, while the estimates of phosphate rock resources range from about 91,000 mmt to over 1,000,000 mmt. Using available literature, the reserves of various countries were assessed in the terms of reserves of concentrate. The IFDC[30] estimate of worldwide reserve is approximately 60,000 mmt of concentrate. Based on the data gathered, collated and analyzed for the IFDC report, there is no indication that a "peak phosphorus" event will occur in 20 – 25 years. Based on the data reviewed, and assuming current rates of production, phosphate rock concentrate reserves to produce fertilizers will be available for the next 300 – 400 years [23].

Phosphate rock prices will increase when the demand approaches the limits of supply. When the phosphate rock prices increase, some resources will become reserves, marginal mining

[30] International Fertilizer Development Center.

projects will become viable and the production will be stimulated. In the future, fuel and fuel-related transportation costs may become even more important components in the world phosphate rock production scenario. The political disruption is always an unknown factor, and it can profoundly influence the supply and demand for fertilizer raw materials on a worldwide basis [22].

7.5. Non-apatitic phosphate minerals

Apart from those in the supergroup of apatite minerals, the well-known phosphate minerals include [25]:

1. **Autunite** [110],[111]: is orthorhombic mineral (space group I4/MMM) of the composition of ($Ca[(UO_2)(PO_4)]_2(H_2O)_{11}$, **Fig. 16**), which crystallizes in the space group Pnma with the cell parameters: $\boldsymbol{a}$ = 14.0135 Å, $\boldsymbol{b}$ = 20.7121 Å, c = 6.9959 Å, $\boldsymbol{V}$ = 2030.55 Å^3 and $\boldsymbol{Z}$ = 4. It belongs to the most abundant and widely distributed uranyl phosphate minerals. The structure contains the well-known autunite-type sheet with the composition $[(UO_2)(PO_4)]^-$ resulting from the sharing of equatorial vertices of uranyl square bipyramids with phosphate tetrahedra. Calcium atom in the interlayer is coordinated by seven H_2O groups and two longer distances to uranyl apical O atoms. Two symmetrically independent H_2O groups are held in the structure only by hydrogen bonding. The bond-length-constrained refinement provides a crystal-chemically reasonable description of the hydrogen bonding.

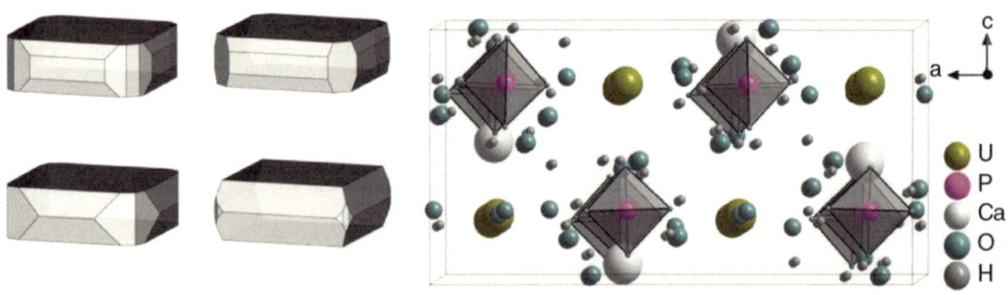

Fig. 16. The examples of forms and the structure of mineral atunite [110] viewed along the b-axis.

The mineral was named by HENRY J. BROOKE and WILLIAM H. MILLER in 1854 after the typical locality at Saint Symphorien, Autun District, Saône-et-Loire, France. Autunite dehydrates rapidly in air (except for high relative humidity) to tetragonal meta-autunite (P4/NMM, $\boldsymbol{a}$ = 6.96 Å, c = 8.40 Å and c:$\boldsymbol{a}$ = 1.21) [112]. The loss of O12 and O13 from autunite results in the formula $Ca[(UO_2)(PO_4)]_2(H_2O)_7$ (**Fig. 17**) [110].

http://dx.doi.org/10.5772/62214

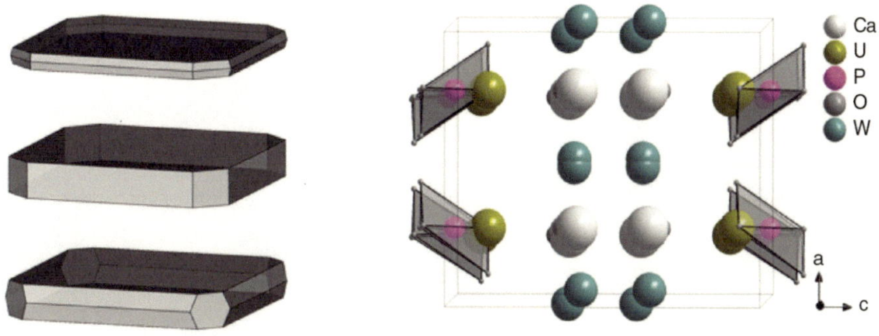

Fig. 17. The examples of forms and the structure of mineral meta-atunite [112] viewed along the b-axis.

2. **Crandallite** [113]: $CaAl_3(OH)_6[PO_3(O_{1/2}(OH)_{1/2}]_2$, has hexagonal structure ($\boldsymbol{a}$ = 7.005 Å and c = 16.192 Å), which is analogous with alunite. The structure of mineral (**Fig. 18**) consists of corner-sharing Al octahedra, which are linked into trigonal and hexagonal rings to form the sheets perpendicular to the c-axis. Ca ions, surrounded by 12 oxygen and hydroxyl ions, lie in large cavities between the sheets. Each phosphate tetrahedron shares three corners with three Al octahedra from a trigonal ring in the sheet. The unshared corner is turned away from the trigonal hole towards the adjacent sheet to which it is hydrogen bonded. The mineral "deltaite" was found to be identical to crandallite, within the accuracy of the structural results.

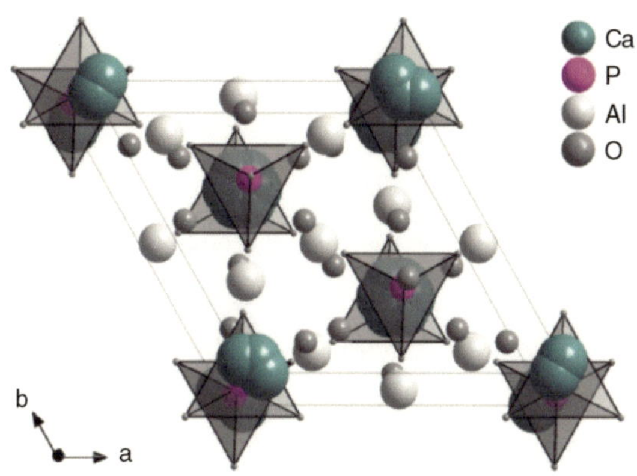

Fig. 18. The structure of mineral crandallite [113] viewed along the c-axis.

3. **Lazulite** [114],[115]: monoclinic mineral of the composition of $MgAl_2(PO_4)_2(OH)_2$, which crystallizes in the $P2_1/c$ space group and has the cell parameters: $\boldsymbol{a}$ = 7.16 Å, $\boldsymbol{b}$ = 7.26 Å and c = 7.24 Å and β = 120.67°. The mineral is the magnesium analogue of scorzalite ($Fe^{2+}Al_2(PO_4)_2(OH)_2$). Named in 1795 by Marten H. Klaproth from the Arabic word meaning "heaven," in allusion to its color.

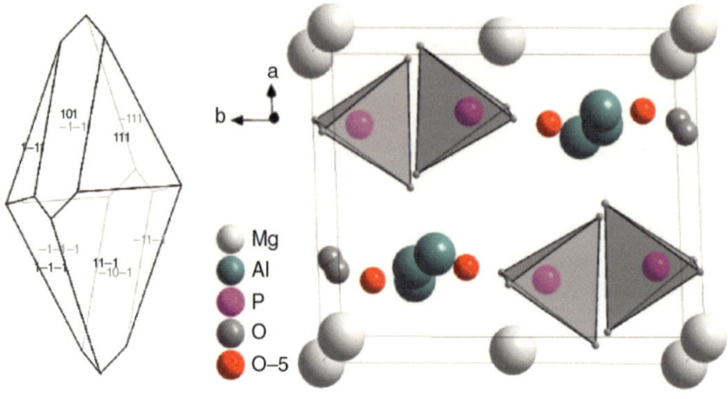

Fig. 19. The form and the structure of lazulite [114] viewed along the c-axis.

4. **Millisite** [116]: tetragonal mineral of the composition of $NaCaAl_6(PO_4)_4(OH)_9{\cdot}3H_2O$ (the space group $P4_12_12$, $\boldsymbol{a}$ = 7.00 Å and $\boldsymbol{c}$ = 18.99Å, $\boldsymbol{a{:}c}$ = 1:2.713 and $\boldsymbol{Z}$ = 2), which was named in 1942 by Esper Signius Larsen III in the honor of Mr. F.T. Millis of Lehi, Utah, who had collected the first specimens.

5. **Monazite** [117],[118]: is natural light rare-earth element phosphate that generally contains large amounts of uranium and thorium. The monazite-type compounds (AXO_4, **Fig. 21**) form an extended family that is described in this review in the terms of field of stability versus composition. They crystallize in a monoclinic lattice with the space group $P2_1/N$ ($\boldsymbol{Z}$ = 4).

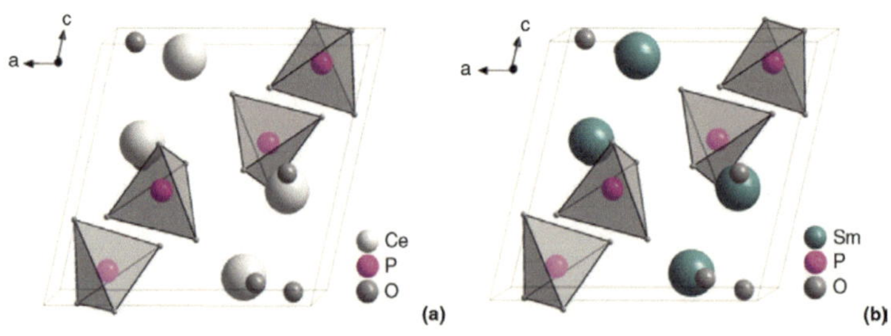

Fig. 20. The structure of monazite-(Ce): $\boldsymbol{a}$ = 6.7902 Å, $\boldsymbol{b}$ = 7.0203 Å, $\boldsymbol{c}$ = 6.4674 Å and $\boldsymbol{\beta}$ = 103.38° (a) and monazite-(Sm) (b): $\boldsymbol{a}$ = 6.6818 Å, $\boldsymbol{b}$ = 6.8877 Å, $\boldsymbol{c}$ = 6.3653 Å and $\boldsymbol{\beta}$ = 103.386° (b) [118] viewed along the b-axis.

Monazite (**Fig. 20**) and xenotime dimorphs[31] [119] are the most ubiquitous rare-earth (REE) minerals. Monazite incorporates preferentially larger, light rare-earth elements (LREEs, here, La-Gd), whereas xenotime tends to incorporate smaller, heavy rare-earth elements (HREEs, here, Tb-Lu, and Y).

---

[31] One chemical compound is capable of crystallizing in two different systems, e.g. $CaCO_3$ can occur as calcite or aragonite [119].

http://dx.doi.org/10.5772/62214

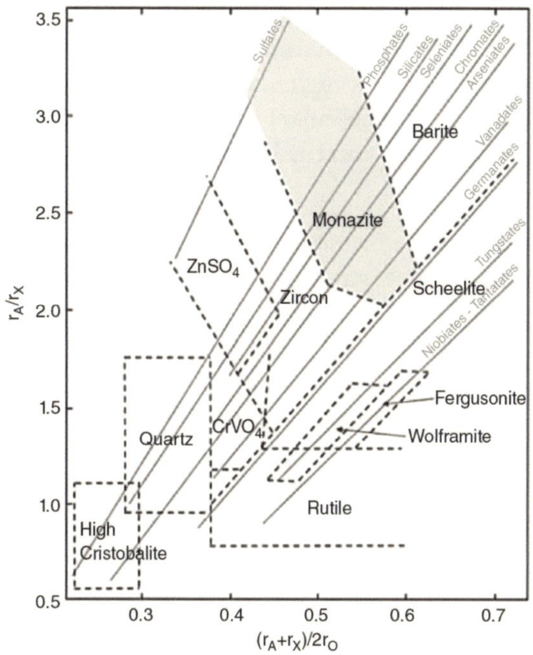

Fig. 21. The classification diagram of AXO_4-type compounds. The monazite stability domain is colored by gray [117].

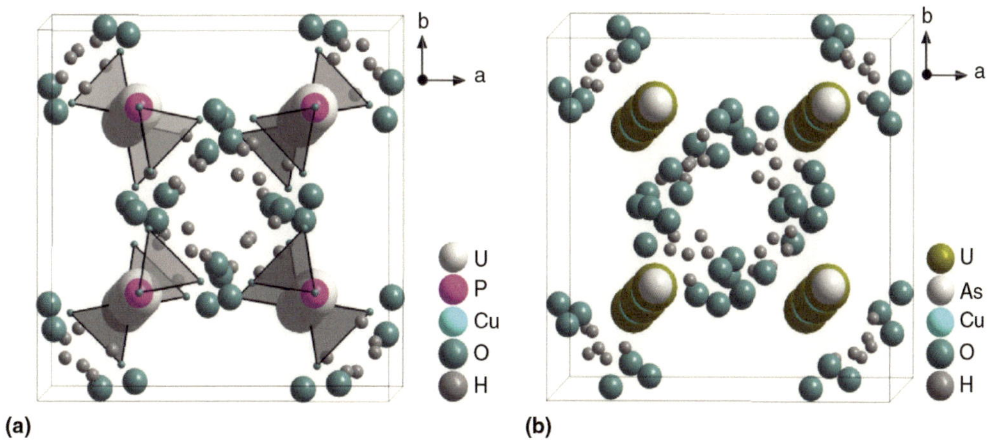

Fig. 22. The structure of tobernite and isostructural zeunerite [120] viewed along the c-axis.

6. **Tobernite** [120],[121]: tetragonal torbernite (P4/NNC, $\boldsymbol{a}$ = 7.0267 Å, $\boldsymbol{c}$ = 20.8070 Å, $\boldsymbol{c}$:$\boldsymbol{a}$ = 2.9611, $Cu[(UO_2)(PO_4)]_2(H_2O)_{12}$, **Fig. 22(a)**) and zeunerite ($Cu[(UO_2)(AsO_4)]_2(H_2O)_{12}$, **Fig. 22(b)**), as well as metatorbernite ($Cu[(UO_2)(PO_4)]_2(H_2O)_8$) and metazeunerite ($Cu[(UO_2)(AsO_4)]_2(H_2O)_8$), belong to the autunite and meta-autunite groups, respectively, which

make up together approximately 40 mineral species of hydrated uranyl phosphates and arsenates. The structures, compositions and stabilities of minerals of the autunite and meta-autunite groups are of high interest because of their environmental significance. They are widespread and abundant and exert an impact on the mobility of uranium in phosphate-bearing systems and soils contaminated by actinides.

These minerals contain the autunite-type sheet, of the composition of $[(UO_2)(PO_4)]^-$, which involves the sharing of equatorial vertices of uranyl square bipyramids with phosphate tetrahedra. In each of these structures, Cu^{2+} cations are located between the sheets in Jahn-Teller[32] [122] distorted (4 + 2) octahedra, with short bonds to four H_2O groups in a square-planar arrangement and two longer distances to oxygen atoms of uranyl ions. A symmetrically independent H_2O group is held in each structure only by H-bonding, and in torbernite (and in zeunerite), it forms the square-planar sets of interstitial H_2O groups both above and below the planes of Cu^{2+} cations. In metatorbernite (and in metazeunerite), the square-planar sets of interstitial H_2O groups are either above or below the planes of Cu^{2+} cations. The bond-length-constrained refinement provides the crystal-chemically reasonable descriptions of H-bonding in those four structures [120].

7. **Turquoise** [123],[124]: $CuAl_6(PO_4)_4(OH)_8{\cdot}4H_2O$ is a copper analogue of triclinic mineral fausite $ZnAl_6(PO_4)_4(OH)_8{\cdot}4H_2O$ (the space group P1 with the cell parameters $\boldsymbol{a}$ = 7.410 Å, $\boldsymbol{b}$ = 7.633 Å, $\boldsymbol{c}$ = 9.904 Å, $\boldsymbol{\alpha}$ = 68.4°, $\boldsymbol{\beta}$ = 69.65° and γ = 65.05°). The structure (**Fig. 23**) consists of distorted MO_6 polyhedra (M= Zn, Cu), AlO_6 octahedra and PO_4 tetrahedra. By the edge- and corner-sharing of these polyhedra, a fairly dense three-dimensional framework is formed, which is further strengthened by a system of hydrogen bonds. The metal atoms in the unique MO_6 (M = Zn or Cu) polyhedron show a distorted [2 + 2 + 2] coordination, the distortion being more pronounced in turquoise. About 10% of the M site is vacant in both minerals. In turquoise, a previously undetected structural site with very low occupancy of (possibly) Cu is present at the position (1/2,0,1/2).

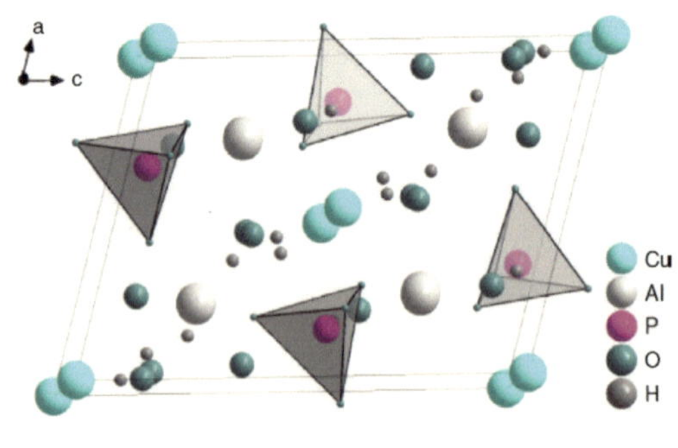

Fig. 23. The structure of turquoise viewed along the b-axis [124].

[32] According to the Jahn-Teller theorem, any nonlinear molecule in a degenerate electronic state will be unstable and will undergo some kind of distortion that will lower its symmetry so as to remove the degeneracy of the electronic state and also to attain lower energy. The Jahn-Teller effect is termed as static when there is permanent distortion in the structure of molecule [122].

http://dx.doi.org/10.5772/62214

8. **Vivianite** [125],[126],[127]: is monoclinic mineral of the composition of $Fe_3(PO_4)_3{\cdot}8H_2O$, which crystallizes in the space group of C1 $_2$ /M with the cell parameters $\boldsymbol{a}$ = 10.08 Å, $\boldsymbol{b}$ = 13.43 Å, $\boldsymbol{c}$ = 4.70 Å and $\boldsymbol{\beta}$ = 104.50° (**Fig. 24**). The mineral was named by ABRAHAM GOTTLOB WERNER in 1817 after JOHN HENRY VIVIAN. Vivianite belongs to the simplest group of minerals with the composition given by general formula: $A_3(XO_4)_2{\cdot}8H_2O$, where A = Mg, Zn, Ni, Co or Fe and X is P or As [128].

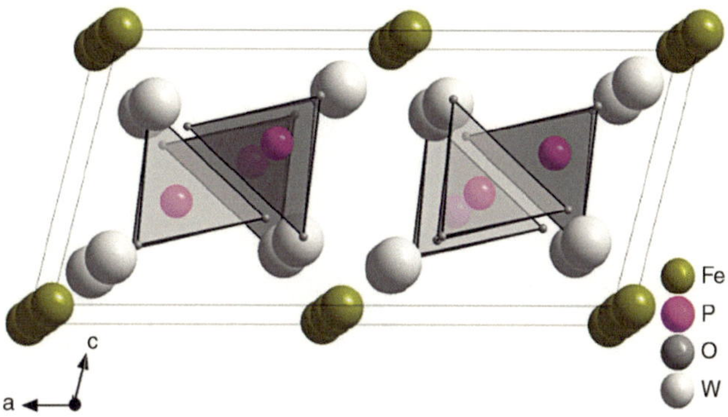

Fig. 24. The structure of vivanite viewed along the b-axis [125].

9. **Wavellite** [129]: the mineral of the composition of $Al_3(PO_4)_2(OH)_3{\cdot}4.5\text{-}5H_2O$ (the space group *Pcmn*, $\boldsymbol{a}$ = 9.62 Å, $\boldsymbol{b}$ = 17.36 Å and $\boldsymbol{c}$ = 6.99 Å). The two aluminum atoms in the structure are octahedrally coordinated (**Fig. 25**): one is bonded to two O atoms, two –OH groups and two H_2O molecules and the other to three O, two (–OH) and one H_2O. Phosphorus is in tetrahedral coordination with oxygen. Al octahedra, linked through (OH) corners, form chains parallel to the c-axis, and P tetrahedra are attached to this chain by sharing O atoms of subsequent octahedra. An extra H_2O molecule occupies the large cavity between the chains, and as indicated by a high temperature factor, it has a statistical distribution within this cavity.

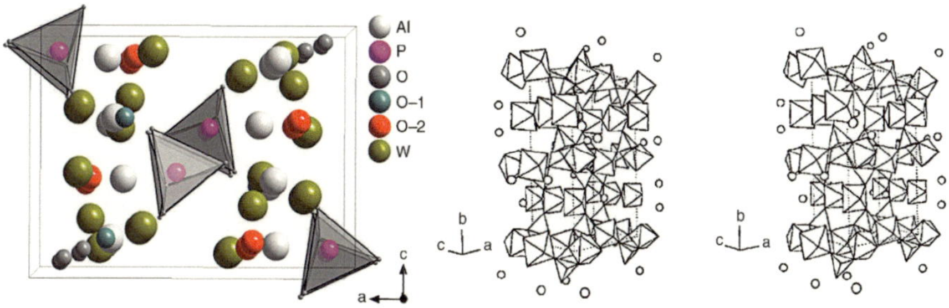

Fig. 25. The structure of wavellite viewed along the b-axis (a) and stereoscopic view of the wavellite structure (b) [129].

10. **Xenotime** [118]: monazite (**Fig. 26**) is isostructural with zircon ($I4_1/AMD$). Monazite atomic arrangement as well as that of xenotime is based on [001] chains of intervening phosphate tetrahedra and RE polyhedra, with REO, polyhedron in xenotime that accommodates heavy lanthanides (Tb-Lu in the synthetic phases) and REO polyhedron in monazite that preferentially incorporates larger light rare-earth elements (La-Gd). As the structure "transforms" from xenotime to monazite, the crystallographic properties are comparable along the [001] chains, with the structural adjustments to the different sizes of REE atoms occurring principally in (001).

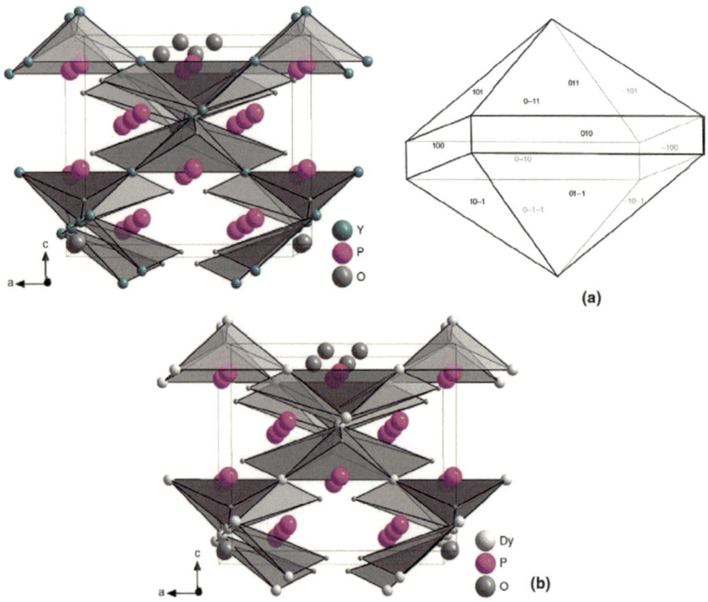

Fig. 26. The structure of xenotime-(Y): $\boldsymbol{a}$ = 6.8947 Å, $\boldsymbol{b}$ = 6.8947 Å and $\boldsymbol{c}$ = 6.0276 Å (a) and xenotime-(Dy) (b) $\boldsymbol{a}$ = 6.9052 Å, $\boldsymbol{b}$ = 6.9053 Å and $\boldsymbol{c}$ = 6.0384 Å (b) [118] viewed along the b-axis.

Isostructural arsenate analogues of many phosphate minerals are known, and in some cases, vanadates too. Some orthophosphates capable of forming complete ranges of solid solutions with the corresponding orthoarsenates are [25]:

- **Variscite group**: $MXO_4 \cdot 2H_2O$, where M = Fe, Al and X = P or As;
- **Fairfieldite group**: $Ca_2M(XO_4)_2 \cdot 2H_2O$, where M = Mn, Fe, Mg, Ni, Zn, Co and X = P, As;
- **Vivianite group**: $M_3(XO_4)_2 \cdot 8H_2O$, where M = Fe, Mn, Mg, Zn, Co, Ni and X = P, As;
- **Monazite group**: MXO_4, where M = Ce, La, Nd, Th, Bi and X = P, As;
- **Rhabdophane group**: $MXO_4 \cdot H_2O$, where M = Ce, La, Nd, Th and X = P, As;
- **Xenotime group**: MXO_4, where M = Y, Ce, Bi and X = P, V;

http://dx.doi.org/10.5772/62214

- **Autunite group**: $M[(UO_2)_2(XO_4)]{\cdot}nH_2O$, where M = Ca, Ce, Ba, K, NH_4^+, Sr, Pb, Mg, Na, Zn and X = P, As, V;
- **Crandallite group**: $MM'_3(XO_4)_2(OH)_6{\cdot}H_2O$, where M = Ca, Sr, Ba, M'= Al, Fe and X = P, As.

Phosphate minerals, like silicate minerals, are found with a great variety of cations. Unlike the latter group that contains numerous types of condensed silicate anions, almost all phosphate minerals are orthophosphates that contain PO_4^{3-} anion. Non-phosphorus anions, such as O^{2-}, OH^-, F^-, Cl^-, SO_4^{2-}, SiO_4^{4-} and AsO_4^{3-}, may also be present in these stoichiometric (or as occluded) materials [25].

Author details

Petr Ptáček

Brno University of Technology, Czech Republic

References

[1] Bishop AC, Woolley AR, Woolley WRH. Cambridge Guide to Minerals, Rocks and Fossils. 2nd ed., Cambridge: Cambridge University Press, 1999. ISBN: 0-521-77881-6

[2] Clarke FW. The Data of Geochemistry. Bulletin (Geological Survey (U.S.)), U.S. Government Printing Office, 1920.

[3] McClellan GH, Lehr JR. Crystal chemical investigation of natural apatites. American Mineralogist 1969;54 1374–1391.

[4] Ding T, Ma D, Lu J, Zhang R. Apatite in granitoids related to polymetallic mineral deposits in southeastern Hunan Province, Shi-Hang Zone, China: Implications for petrogenesis and metallogenesis. Ore Geology Reviews 2015;69 104–117.

[5] Watson EB. Apatite and phosphorus in mantle source regions: an experimental study of apatite/melt equilibria at pressures to 25 kbar. Earth and Planetary Science Letters 1980;51(2) 322–335.

[6] Deer WA. Rock-Forming Minerals: Non-silicates. Volume 5B, 2nd ed., Geological Society of London, 1998. ISBN: 978-1897799901

[7] Raask E. Mineral Impurities in Coal Combustion: Behavior, Problems, and Remedial Measures. Taylor & Francis, 1985. ISBN: 978-3540138174

[8] Jacobson M, Charlson RJ, Rodhe H, Orians GH. Earth System Science: From Biogeochemical Cycles to Global Changes. International Geophysics — Volume 72. Academic Press, 2000. ISBN: 978-0080530642

[9] Spear FS. Metamorphic Phase Equilibria and Pressure-Temperature-Time Paths. Mineralogical Society of America short course notes. 2nd ed., Mineralogical Society of America, 1993. ISBN: 978-0939950348

[10] Kemp JF. A Handbook of Rocks for Use Without the Microscope. BiblioBazaar, 2009. ISBN: 978-1110466528

[11] Drouet Ch. A comprehensive guide to experimental and predicted thermodynamic properties of phosphate apatite minerals in view of applicative purposes. Journal of Chemical Thermodynamics 2015;81 143–159.

[12] Ihlen PM, Schiellerup H, Gautneb H, Skår Ø. Characterization of apatite resources in Norway and their REE potential — A review. Ore Geology Reviews 2014;58 126–147.

[13] Santana RC, Farnese ACC, Fortes MCB, Ataíde CH, Barrozo MAS. Influence of particle size and reagent dosage on the performance of apatite flotation. Separation and Purification Technology 2008;64(1) 8–15.

[14] Pucéat E, Reynard B, Lécuyer C. Can crystallinity be used to determine the degree of chemical alteration of biogenic apatites? Chemical Geology 2004;205(1-2) 83–97.

[15] Zafar ZI, Anwar MM, Pritchard DW. Optimization of thermal beneficiation of a low grade dolomitic phosphate rock. International Journal of Mineral Processing 1995;43(1-2) 123–131.

[16] dos Santos MA, Santana RC, Capponi F, Ataíde CH, Barrozo MAS. Effect of ionic species on the performance of apatite flotation. Separation and Purification Technology 2010;76(1) 15–20.

[17] Chen M, Graedel TE. The potential for mining trace elements from phosphate rock. Journal of Cleaner Production 2015;91 337–346.

[18] Al-Bassam KS, Aba-Hussain AA, Mohamed IQ, Al-Rawi Yehya T. Petrographic classification of phosphate components of East Mediterranean phosphorite deposits. Iraqi Bulletin of Geology and Mining 2010;6(1) 59–79.

[19] Flügel E. Microfacies of Carbonate Rocks: Analysis, Interpretation and Application. Springer Science & Business Media, 2013. ISBN: 978-3662087268

[20] Abouzeid A-ZM. Physical and thermal treatment of phosphate ores — An overview. International Journal of Mineral Processing 2008;85(4) 59–84.

[21] Gupta AK. Igneous Rocks. Allied Publishers, 1998. ISBN: 978-8170237846

[22] Manual Fertilizer, UN Industrial Development Organization, International Fertilizer Development Center. Springer Science & Business Media, 1998. ISBN: 978-0792350323

[23] World Phosphate Rock Reserves and Resources. International Center for Soil Fertility and Agricultural Development, 2010. ISBN: 978-0880901673

[24] Emich GD. Phosphate rock. Industrial Minerals and Rocks 1984;2 1017–1047.

[25] Derek EC. Chemistry, Biochemistry and Technology. 6th ed., CRC Press, 2013. ISBN: 978-1439840887

[26] Teleki PG, Dobson MR, Moore JR, von Stackelberg U. Marine Minerals: Advances in Research and Resource Assessment. Nato Science Series C: Volume 194. Springer Science & Business Media, 2012. ISBN: 978-9400938038

[27] McKelvey VE. Phosphate Deposits. Geological Survey Bulletin 1252-D. A Summary of Salient Features of the Geology of Phosphate Deposits, Their Origin, and Distribution. Washington: United States Government Printing Office, 1967.

[28] Cook PJ, Shergold JH. Phosphate Deposits of the World: Volume 1: Proterozoic and Cambrian Phosphorites. Cambridge Earth Science Series. Cambridge University Press, 2005. ISBN: 978-0521619219

[29] Salvador A. International Stratigraphic Guide: A Guide to Stratigraphic Classification, Terminology, and Procedure. Publication (International Union of Geological Sciences). 30th ed., Geological Society of America, 1994. ISBN: 978-0813774015

[30] Walter MR. Stromatolites. Developments in Sedimentology. Elsevier, 1976. ISBN: 978-0080869322

[31] Nairn AEM, Ricou Luc-E, Vrielynck B, Dercourt J. The Tethys Ocean. Ocean Basins and Margins — Volume 8. Springer Science & Business Media, 2013. ISBN: 978-1489915580

[32] Notholt AJG, Sheldon RP, Davidson DF. Phosphate Deposits of the World: Volume 2, Phosphate Rock Resources. Cambridge Earth Science Series. Revised edition, Cambridge University Press, 2005. ISBN: 978-0521673334

[33] Gualtieri A, Venturelli P. In situ study of the goethite-hematite phase transformation by real time, synchrotron powder diffraction, sample at T = 25 C. American Mineralogist 1999;84 895–904.

[34] Nathan Y. Phosphate minerals. In: Nriagu JO, Moore PB (Eds.), Chapter 8. The Mineralogy and Geochemistry of Phosphorites. Springer Science & Business Media, 2012. ISBN: 978-3642617362

[35] Rao VP. Phosphorites from the Error Seamount, northern Arabian Sea. Marine Geology 1986;71(1-2) 177–186.

[36] Glenn CR, Föllmi KB, Riggs SR, Baturin GN, Grimm KA, Trape J. Phosphorus and phosphorites: Sedimentology and environments of formation. Eclogae Geologicae Helvetiae 1994;87(3) 747–788.

[37] Baturin GN, Bezrukov PL. Phosphorites on the sea floor and their origin. Marine Geology 1979;31(3-4) 317–332.

[38] Cook PJ, Shergold JH, Burnett WC, Riggs SR. Phosphorite research: A historical overview. Geological Society, London, Special Publications 1990;52 1–22.

[39] Trappe J. Phanerozoic Phosphorite Depositional Systems: A Dynamic Model for a Sedimentary Resource System. Lecture Notes in Earth Sciences — Volume 76, ISSN 0930-0317. Springer, 1998. ISBN: 978-3540696049

[40] Bogdanov NA. Non-metallic Mineral Ores. Proceedings of the 27th International Geological Congress — Volume 15. VSP, 1984. ISBN: 978-9067640244

[41] Dunham RJ. Classification of carbonate rocks according to depositional texture, in: Ham WE. Classification of carbonate rocks. American Association of Petroleum Geologists Memoir 1962;1 108–121.

[42] Riggs SR. A petrographic classification of sedimentary phosphorites. In: Cook PJ, Shergold JH. (Eds.), Proterozoic–Cambrian Phosphorites. IGCP, 156, 1st Field Workshop and Seminar. Canberra: National Library of Australia, 1979.

[43] Suess E, von Huene R. Proceedings of the Ocean Drilling Program, Scientific Results, Volume 112. Phosphatic Sediments and Rocks Recovered from the Peru Margin during ODP LEG 1121. The Program, 1986.

[44] Ehrlich HL, Newman DK. Geomicrobiology. 5th ed., CRC Press, 2008. ISBN: 978-0849379079

[45] Tyrrell GW. The Principles of Petrology: An Introduction to the Science of Rocks. Springer Science & Business Media, 2012. ISBN: 978-9401160261

[46] Gill R. Igneous Rocks and Processes: A Practical Guide. John Wiley & Sons, 2011. ISBN: 978-1444362435

[47] Brown PW, Constanz B. Hydroxyapatite and Related Materials. CRC Press, 1994. ISBN: 978-0849347504

[48] Wenk Hans-R, Bulakh A. Minerals: Their Constitution and Origin. Cambridge University Press, 2004. ISBN: 978-0521529587

[49] Laenen B, Hertogen J, Vandenberghe N. The variation of the trace-element content of fossil biogenic apatite through eustatic sea-level cycles. Palaeogeography, Palaeoclimatology, Palaeoecology 1997;132(1-4) 325–342.

[50] Egorov LS. Ijolite-Carbonatite Plutonism (in Russian). Nedra, Leningrad, 1991.

[51] Isakova AT, Panina LI, Rokosova EY. Carbonatite melts and genesis of apatite mineralization in the Guli pluton (northern East Siberia). Russian Geology and Geophysics 2015;56(3) 466–475.

[52] Dawson JB, Steele IM, Smith JV, Rivers ML. Minor and trace element chemistry of carbonates, apatites and magnetites in some African carbonatites. Mineralogical Magazine 1996;60 415–425.

[53] Aspden JA. The mineralogy of primary inclusions in apatite crystals extracted from Alnö ijolite. Lithos 1980;13(3) 263–268.

[54] Cordeiro PFO, Brod JA, Dantas EL, Barbosa ESR. Mineral chemistry, isotope geochemistry and petrogenesis of niobium-rich rocks from the Catalão I carbonatite-phoscorite complex, Central Brazil. Lithos 2010;118(3-4) 223–237.

[55] Iddings JP. Igneous Rocks: Composition, Texture and Classification, Description and Occurrence, Volume 1. Рипол Классик, 1920. ISBN: 978-5883586766

[56] Wall F, Zaitsev AN. Phoscorites and Carbonatites from Mantle to Mine: The Key Example of the Kola Alkaline Province. Mineralogical Society of Great Britain & Ireland, 2004. ISBN: 978-0903056229

[57] Russell HD, Hiemstra SA, Groeneveld D. New aspects of the Lower Katanga rocks of the Kafue Gorge. Transactions of the Geological Society of South Africa 1955;58 197–208.

[58] Le Maitre RW, Streckeisen A, Zanettin B, Le Bas MJ, Bonin B, Bateman P. Igneous Rocks: A Classification and Glossary of Terms: Recommendations of the International Union of Geological Sciences Subcommission on the Systematics of Igneous Rocks. 2nd ed., Cambridge University Press, 2002. ISBN: 978-1139439398

[59] Reviews Book: Igneous Rocks: A Classification and Glossary of Terms (Recommendations of the IUGS Subcommission on the Systematics of Igneous Rocks). 2nd ed., LeMaitre RW (ed.). Cambridge University Press, New York, N.Y., 2002. ISBN: 0-521-66215-X. The Canadian Mineralogist 2002;40 1737-1742.

[60] Kucera M. Industrial Minerals and Rocks: Developments in Economic Geology. Elsevier, 2013. ISBN: 978-0444597502

[61] Tatarinov PM. Nonmetallic mineral deposits course (in Russian). Moscow: Nedra, 1969.

[62] Wills BA. Mineral Processing Technology: An Introduction to the Practical Aspects of Ore Treatment and Mineral Recovery (in SI/Metric Units). International Series on Materials Science and Technology. 4th ed., Elsevier, 2013. ISBN: 978-483286679

[63] Iijima A, Abed AM. Siliceous, Phosphatic and Clauconitic Sediments of the Tertiary and Mesozoic. Proceedings of the 29th International Geological Congress: Kyoto, Japan, 24 August-3 September 1992 — Part 3. VSP, 1994. ISBN: 978-9067641753

[64] Cronan DS. Marine Minerals in Exclusive Economic Zones. Topics in the Earth Sciences — Volume 5. Springer Science & Business Media, 1992. ISBN: 978-0412292705

[65] Cushman GT. Guano and the Opening of the Pacific World: A Global Ecological History. Studies in Environment and History. Cambridge University Press, 2013. ISBN: 978-1107004139

[66] Hollett D. More Precious Than Gold: The Story of the Peruvian Guano Trade. Associated University Press, 2008. ISBN: 978-0838641316

[67] Chang Jen-S, Kelly A, Crowley JM. Handbook of Electrostatic Processes. CRC Press, 1995. ISBN: 978-1420066166

[68] Nogalska A, Zalewska M. The effect of meat and bone meal on phosphorus concentrations in soil and crop plants. Plant Soil Environment 2013;59(12) 575–580.

[69] O'Reilly SY, Griffin WL. Apatite in the mantle: implications for metasomatic processes and high heat production in Phanerozoic mantle. Lithos 2000;53 217–232.

[70] Douce AEP, Roden MF, Chaumba J, Fleisher C, Yogodzinski G. Compositional variability of terrestrial mantle apatites, thermodynamic modeling of apatite volatile contents, and the halogen and water budgets of planetary mantles. Chemical Geology 2011;288 14–31.

[71] Sha LK, Chappell BW. Apatite chemical composition, determined by electron microprobe and laser-ablation inductively coupled plasma mass spectrometry, as a probe into granite petrogenesis. Geochimica et Cosmochimica Acta 1999;63(22) 3861–3881.

[72] Joosu L, Lepland A, Kirsimäe K, Romashkin AE, Roberts NMW, Martin AP, Črne AE. The REE-composition and petrography of apatite in 2 Ga Zaonega Formation, Russia: The environmental setting for phosphogenesis. Chemical Geology 2015;395 88–107.

[73] Follmi KB. The phosphorus cycle, phosphogenesis and marine phosphate-rich deposits. Earth-Science Reviews 1996;40(1-2) 55–124.

[74] Glenn CR, Föllmi KB, Riggs Sr, Baturin GN, Grimm KA, Trappe J. Phosphorus and phosphorites: sedimentology and environments of formation. Eclogae Geologicae Helvetiae 1994;87(3) 747–788.

[75] Jarvis I, Burnett WC, Nathan Y, Almbaydin FSM, Attia AKM, Castro LN, Flicoteaux R, Hilmy ME, Husain V, Qutawnah AA, Serjani A, Zanin YN. Phosphorite geochemistry — State-of-the-art and environmental concerns. Eclogae Geologicae Helvetiae 1994;87(3) 643–700.

[76] Alexander EB. Soils in Natural Landscapes. CRC Press, 2013. ISBN: 978-1466594364

[77] Jørgensen SE. Global Ecology. Academic Press, 2010. ISBN: 978-0444536273

[78] CIBA Foundation Symposium. Phosphorus in the Environment: Its Chemistry and Biochemistry. Novartis Foundation Symposia — Volume 919. John Wiley & Sons, 2009. ISBN: 978-0470718070

[79] Ragot S, Zeyer J, Zehnder L, Reusser E, Brandl H, Lazzaro A. Bacterial community structures of an alpine apatite deposit. Geoderma 2013;202-203 30-37.

[80] Puente ME, Rodriguez-Jaramillo MC, Li CY, Bashan Y. Image analysis for quantification of bacterial rock weathering. Journal of Microbiological Methods 2006;64(2) 275–286.

[81] Nies A. Strengit, ein neues Mineral. Neues Jahrbuch für Mineralogie, Geologie und Palaontologie 1877; 8-16.

[82] Cornelis P. *Pseudomonas*: Genomics and Molecular Biology. Horizon Scientific Press, 2008. ISBN: 978-1904455196

[83] Pongslip N. Phenotypic and Genotypic Diversity of Rhizobia. Bentham Science Publishers, 2012. ISBN: 978-1608054619

[84] Vanlaere E. *Burkholderia cepacia* complex. DCL Print & Sign, 2008. ISBN: 978-9087560379

[85] Lisker F, Ventura B, Glasmacher UA. Apatite thermochronology in modern geology. Geological Society, London, Special Publications 2009;324 1–23. DOI: 10.1144/SP324.1.

[86] Fleischer RL, Price PB, Walker RM. Nuclear Tracks in Solids: Principles and Applications. University of California Press, 1975. ISBN: 978-0520026650

[87] Fleisher RL, Price PB. Techniques for geological dating of minerals by chemical etching of fission fragment tracks. Geochimica et Cosmochimica Acta 1964;28(11) 1705–1714.

[88] Galbraith RF. Statistics for Fission Track Analysis. CRC Press, 2005. ISBN: 978-1420034929

[89] Li W, Wang L, Lang M, Trautmann C, Ewing RC. Thermal annealing mechanisms of latent fission tracks: Apatite vs. zircon. Earth and Planetary Science Letters 2011;302(1-2) 227–235.

[90] Li W, Lang M, Gleadow AJW, Zdorovets MV, Ewing RC. Thermal annealing of unetched fission tracks in apatite. Earth and Planetary Science Letters 2012;321-322 121-127.

[91] Chen H, Hu J, Wu G, Shi W, Geng Y, Qu H. Apatite fission-track thermochronological constraints on the pattern of late Mesozoic-Cenozoic uplift and exhumation of the Qinling Orogen, central China. Journal of Asian Earth Sciences 2014. In press. DOI: 10.1016/j.jseaes.2014.10.004.

[92] Hejl E. Evidence for unetchable gaps in apatite fission tracks. Chemical Geology 1995;122(1-4) 259–269.

[93] Peternell M, Kohlmann F, Wilson CJL, Seiler C, Gleadow AJW. A new approach to crystallographic orientation measurement for apatite fission track analysis: Effects of crystal morphology and implications for automation. Chemical Geology 2009;265(3-4) 527–539.

[94] Enkelmann E, Ehlers TA, Buck G, Schatz Ann-K. Advantages and challenges of automated apatite fission track counting. Chemical Geology 2012;322-323 278-289.

[95] van den Haute P, De Corte F. Advances in Fission-Track Geochronology. Solid Earth Sciences Library — Volume 10. Springer Science & Business Media, 2013. ISBN: 978-9401591331

[96] McCubbin FM, Jolliff BL, Nekvasil H, Carpenter PK, Zeigler RA, Steele A, Elardo SM, Lindsley DH. Fluorine and chlorine abundances in lunar apatite: Implications for heterogeneous distributions of magmatic volatiles in the lunar interior. Geochimica et Cosmochimica Acta 2011;75(17) 5073–5093.

[97] Boyce JW, Tomlinson SM, McCubbin FM, Greenwood JP, Treiman AH. The lunar apatite paradox. Science 25;344(6182) 400-402.

[98] Crotts APS. Lunar outgassing, transient phenomena, and the return to the Moon. Part 1: Existing data. The Astrophysical Journal 2008;687 692–705.

[99] Calvo C, Gopal R. The crystal structure of whitlockite from the Palermo quarry. American Mineralogist 1975;60 120–133.

[100] McConnell D. Apatite: Its Crystal Chemistry, Mineralogy, Utilization, and Geologic and Biologic Occurrences: Applied Mineralogy (Volume 5). Springer Science & Business Media, 2012. ISBN: 978-3709183144

[101] Jolliff BL, Hughes JM, Freeman JJ. Zeigler RA. Crystal chemistry of lunar merrillite and comparison to other meteoritic and planetary suites of whitlockite and merrillite. American Mineralogist 2006;91 1583–1595. DOI: 10.2138/am.2006.2185 1583.

[102] Hughes JM, Jolliff BL, Gunter ME. The atomic arrangement of merrillite from the Fra Mauro Formation, Apollo 14 lunar mission: The first structure of merrillite from the Moon. American Mineralogist 2006;91 1547–1552.

[103] Barnes JJ, Tartèse R, Anand M, McCubbin FM, Franchi IA, Starkey NA, Russell SS. The origin of water in the primitive Moon as revealed by the lunar highlands samples. Earth and Planetary Science Letters 2014;390 244–252.

[104] Douce AEP, Roden M. Apatite as a probe of halogen and water fugacities in the terrestrial planets. Geochimica et Cosmochimica Acta 2006;70(12) 3173–3196.

[105] Hallis LJ, Taylor GJ, Nagashima K, Huss GR. Magmatic water in the Martian meteorite Nakhla. Earth and Planetary Science Letters 2012;359-360 84-92.

[106] Sarafian AR, Roden MF, Patiño-Douce AE. The volatile content of Vesta: Clues from apatite in eucrites. Meteoritics & Planetary Science 2013,48(11) 2135–2154.

[107] Scully JEC, Russell CT, Yin A, Jaumann R, Carey E, Castillo-Rogez J, McSween HY, Raymond CA, Reddy V, Le Corre L. Geomorphological evidence for transient water flow on Vesta. Earth and Planetary Science Letters 2015;411 151–163.

[108] Treiman AH, Lanzirotti A, Xirouchakis D. Ancient water on asteroid 4 Vesta: evidence from a quartz veinlet in the Serra de Magé eucrite meteorite. Earth and Planetary Science Letters 2004;219(3-4) 189–199.

[109] Jasinski SM. Phosphate rock. U.S. Geological Survey Mineral Commodity Summaries 2007; 120-121.

[110] Locock AJ, Burns PC. The crystal structure of synthetic autunite, $Ca[(UO_2)(PO_4)]_2(H_2O)_{11}$. American Mineralogist 2003;88 240–244.

[111] Phillips W, Brooke HJ, Miller WH. Autunite. In: An Elementary Introduction to Mineralogy. London: Longman, Brown, Green, and Longman, 1852; 519-519.

[112] Makarov ES, Ivanov VI. The crystal structure of meta-autunite, $Ca(UO_2)_2(PO_4)_2{\cdot}6H_2O$. Doklady Akademii Nauk SSSR 1960;132 601–603.

[113] Blount AM. The crystal structure of crandallite. American Mineralogist 1974;59 41–47.

[114] Lindberg ML, Christ CL. Crystal structures of the isostructural minerals lazulite, scorzalite and barbosalite. Acta Crystallographica 1959;12 695–697.

[115] Giuseppetti G, Tadini C. Lazulite, $(Mg,Fe)Al_2(OH)_2(PO_4)_2$: structure refinement and hydrogen bonding. Neues Jahrbuch fur Mineralogie, Monatshefte 1983; 410-416.

[116] Larsen ES, Shannon EV. The minerals of the phosphate nodules from near Fairfield, Utah. American Mineralogist 1930;15 307–337.

[117] Clavier N, Podor R, Dacheux N. Crystal chemistry of the monazite structure. Journal of the European Ceramic Society 2011;31(6) 941–976.

[118] Ni Y, Hughes JM, Mariano AN. Crystal chemistry of the monazite and xenotime structures. American Mineralogist 1995;80 21–26.

[119] Rutley F. Mineralogy Murby's "Science and Art Department" series of textbooks. Thomas Murby & Company, 1802.

[120] Locock AJ, Burns PC. Crystal structures and synthesis of the copper-dominant members of the autunite and meta-autunite groups: Torbernite, zeunerite, metatorbernite and metazeunerite. The Canadian Mineralogist 2003;41 489–502.

[121] Hennig C, Reck G, Reich T, Robberg A, Kraus W, Sieler J. EXAFS and XRD investigations of zeunerite and meta-zeunerite. Zeitschrift fur Kristallographie 2003;218 37–45.

[122] Sathyanarayana DN. Electronic Absorption Spectroscopy and Related Techniques. Universities Press, 2001. ISBN: 978-8173713712

[123] Cid-Dresdner H. Determination and refinement of the crystal structure of turquoise, $CuAl_6(PO_4)_4(OH)_8{\cdot}4H_2O$. Zeitschrift fur Kristallographie 1965;121 87–113.

[124] Kolitsch U, Giester G. The crystal structure of faustite and its copper analogue turquoise. Mineralogical Magazine 2000;64 905–913.

[125] Werner AG. Vivianit, in Letztes Mineral-System, bey Craz und Gerlach, und bey Carl Gerold (Freiberg und Wien) 1817; 41-42.

[126] Fejdi P, Poullen JF, Gasperin M. Affinement de la structure de la vivianite $Fe_3(PO_4)_2{\cdot}8H_2O$. Bulletin de Mineralogie 1980;103 135–138.

[127] Bartl H. Water of crystallization and its hydrogen-bonded crosslinking in vivianite $Fe_3(PO_4)_2{\cdot}8H_2O$; a neutron diffraction investigation. Zeitschrift fur Analytische Chemie 1989;333 401–403.

[128] Mori H, Ito T. The structure of vivianite and symplesite. Acta Crystallographica 1950;3 1–6.

[129] Araki T, Zoltai T. The crystal structure of wavellite. Zeitschrift fur Kristallographie 1968;127 21–33.

Chapter 8

Mining and Beneficiation of Phosphate Ore

Petr Ptáček

Additional information is available at the end of the chapter

http://dx.doi.org/10.5772/62215

Abstract

The first commercial production of phosphate rock began in England in 1847. A wide variety of techniques and equipment is used to mine and process phosphate rocks in order to beneficiate low-grade ores and remove impurities. The eighth chapter of this book deals with mining and beneficiation of phosphate ore. The principle and operating conditions of important parts of manufacturing process including separation, classification, removing of carbonates, calcination and flotation was described. The chapter ends with description of techniques used for extraction of rare earth element.

Keywords: Apatite, Phosphate Rock, Mining, Benefication, Separation, Calcination, Flotation, Extraction of Rare-Earth Element

Phosphate rock is an important mineral commodity used in the chemical industry and production of food. Phosphate ores show a wide diversity in the composition of their gangue materials but generally fall into one of the following categories based on major associated gangue materials [1],[2],[3],[4]:

1. **Siliceous ores**: these ores contain quartz, chalcedony or different forms of silica. Such ores could be upgraded economically by techniques such as flotation or gravity separation methods.

2. **Clayey ores**: contain mainly clays and hydrous iron and aluminum silicates or oxides as gangue materials. These impurities could be removed by simple beneficiation techniques such as scrubbing and washing. In some cases, dispersing agents would be necessary.

3. **Calcareous ores of sedimentary origin**: contain calcite and/or dolomite as the major impurities with small amounts of silica. It is usually difficult to remove the carbonate minerals efficiently from such ores by conventional techniques such as flotation or by physical separation methods. This is because the physical properties of carbonates and

phosphates are very similar. The separation by physical means becomes even impossible when the carbonate minerals are finely disseminated into the phosphate particles.

An alternative technique for the beneficiation of these ores is the calcination. Calcination is the process of heating the ore to a high temperature ranging from 800 to 1000°C to decompose $CaCO_3$ and $MgCO_3$ to CaO, MgO and gaseous CO_2. The CaO and MgO formed are then removed as hydroxides by quenching the calcined product in water and washing. The most common chemical reagent used to enhance the removal of calcium and magnesium hydroxides is ammonium chloride.

Chemical dissolution of carbonate minerals (calcite and dolomite) from calcareous phosphate ores, without the calcination, using organic acids also proved to be capable of beneficiating the calcareous phosphate ores on the laboratory scale.

4. **Phosphate ores associated with organic matter** (**black or brown phosphates**): ores of this type are generally beneficiated by heating the ore up to about 800°C. This type of calcination burns organic material and residual organic carbon without significantly affecting the superior qualities of sedimentary phosphates such as the solubility and reactivity.

   Furthermore, as a result of low calcination temperature, the reduction of calcium sulfate, present in ore, to corrosive calcium sulfide by the organic matter is minimized. During the burning of organic matter, the following two conditions must be kept: organic carbon must be decreased to less than 0.3% to minimize the gassing in the wet phosphoric acid processing, and apatite CO_2 must be maintained at a level close to 2% to allow good reactivity of calcined product.

5. **Phosphate ores containing more than one type of gangue minerals**: many sedimentary phosphate deposits contain mixtures of undesired constituents. These ores require a series of beneficiating operations during their processing depending on the type of gangue minerals present in each ore. This may include, after the size reduction, the combination of attrition scrubbing, desliming, flotation, gravity separation and/or calcination. Each flow sheet is to be designed after thorough characterization and testing of a representative sample of the exploited ore.

   Igneous and metamorphic phosphate ores: the main gangue materials in these ores are sulfides, magnetite, carbonates (calcite, dolomite, siderite and ankerite), nepheline syenite, pyroxenite, foskorite, etc. The processing of these ores may include, after crushing and grinding, washing, desliming, magnetic separation and flotation depending on the types of present gangue minerals. However, the flotation is a common step in all of them [1].

The quality factors of commercial phosphate rocks include [5]:

i. **Physical factors**, which include the parameters such as the texture (hardness, porosity, cementing of coating phases), the particle size (coarse or cryptocrystalline), the degree of crystallinity of apatite and the effect of physical treatments (natural or calcined state).

ii. **Chemical factors**, which include the parameters such as the content of phosphorus (BLP grade, **Section 9.3**), fluorine, carbonate and free carbonates in apatite. The

http://dx.doi.org/10.5772/62215

content of Fe and Al (expressed as R_2O_3[1]), the content of Mg in phosphate and accessory minerals, the content of inert gangue mineral (insoluble oxides and silicates), the content of Na and K (phosphate and accessory minerals), organic matter (native and beneficiation reagents), chlorides (from evaporite salts), heavy metals (Cd, Pb, Zn, Hg), potentially toxic elements (Se, As, Cr, V) and radionuclides (U, Th, Ra, Rn).

Increasing world demand on fertilizer in the 1960s and 1970s and the need for phosphate feedstock stimulated the efforts to develop the techniques to beneficiate low-grade ores and remove impurities. It is highly desirable, for both economic and technical reasons, to remove as much of these impurities as possible, thus to increase the apatite content and the grade of phosphate feedstock and to improve the chemical quality. Phosphate ores can be beneficiated by many methods, and usually a combination of more methods is used [6],[7], [8].

The phosphate rock concentrate must meet the following conditions to be salable [4]:

a. High P_2O_5 content (>30%);

b. Low CaO/P_2O_5 ratio (<1.6);

c. Low MgO content (<1%).

As will be mentioned in **Section 9.3**, the content of P_2O_5 is usually expressed as bone phosphate of lime. The treatment and utilization of phosphate ore is shown in **Fig. 1**.

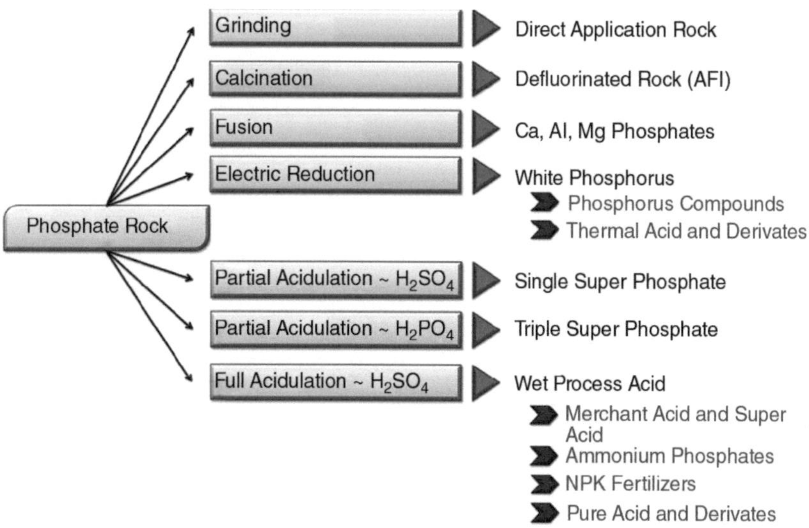

Fig. 1. The treatment of phosphate rock and end-product [8].

[1] Sesquioxides (R_2O_3) that consist of three atoms of oxygen and two atoms or radicals of other elements, e.g. Al_2O_3, Fe_2O_3 and La_2O_3.

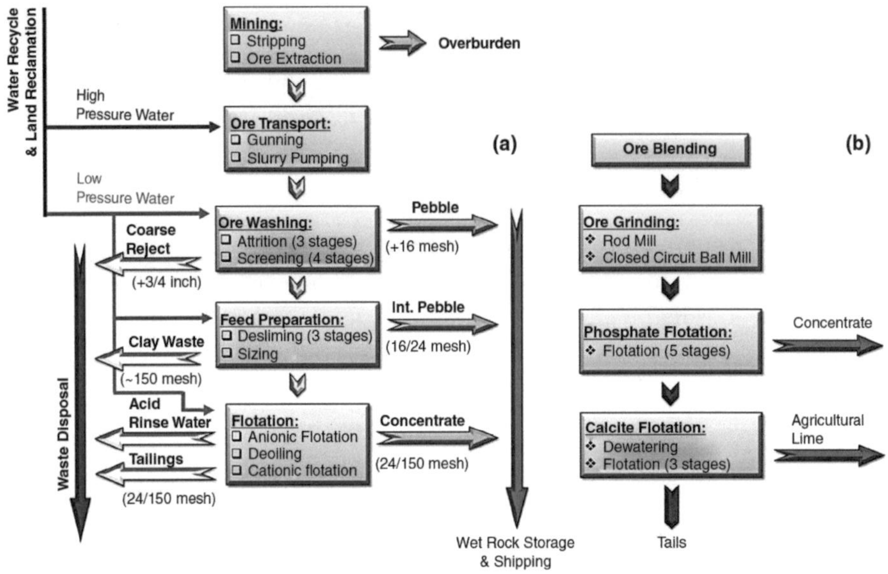

Fig. 2. Mining and beneficiation of sedimentary (a) and igneous (b) phosphate ore [8].

The example of a generic scheme for mining and beneficiation of sedimentary and igneous phosphate ore is shown in **Fig. 2** [8].

8.1. Mining of phosphate ore

The first commercial production of phosphate rock began in England in 1847 and mining was undoubtedly by hand methods. Phosphate mining began in the United States in South Carolina in 1867. Platy phosphate rock beds were mined by hand and later by dredges; sorting was mainly by hand. Phosphate rock deposits were discovered in north and west Africa in the late 1800s. The exploitation of deposits in Algeria and Tunisia began prior to 1900. The production of phosphate rock began at many deposits in the north and west African region in the early to mid-20th century [5],[6],[9]. Currently, there are about 1635 operating world phosphate mines or occurrence worldwide [10].

A wide variety of techniques and many types of equipment are used to mine and process phosphate rock. The methods and the equipment used are very similar to methods and equipment used for coal mining. Phosphate rock is mined by both surface (open-cast, open-pit or strip mining) and underground methods. The surface mining can take many forms — from manual methods employing picks and shovels to highly mechanized operations. Surface mining is the most utilized method by far for mining phosphate deposits. In high-volume applications, the surface mining methods are typically less costly and are generally the preferred methods when the deposit geometry and other factors are favorable [5],[6],[11].

http://dx.doi.org/10.5772/62215

Open-cast mining has developed into a versatile method with plenty of variations to match the mining depth, the slope of the original topography and the types of equipment available. The two major variations of open-cast mining are [12]:

1. **Area mining**, which is carried out on relatively flat terrain with flat-lying seams. Mining cuts are made in straight, parallel panels, running across the property.

2. **Contour mining** is conducted in hilly or mountains terrain, with cuts placed on the contours of the topography.

3. **Other** variations of the open-cast mining, including the box-cut and block-cut mining, are often utilized as well.

The open-cast mining cycle consist mainly of [11],[12]:

i. **Cleaning**: rotary brush of dozer cleans the top of the seam;

ii. **Drilling**: small auger or percussion drill used where needed;

iii. **Blasting**: ANFO or alternatively ripping with dozer;

iv. **Excavation**: front-end loader, power shovel, continuous miner (designed for surface mines);

v. **Haulage**: truck, tractor-trailer, belt conveyor, hydraulic conveyor, rail;

vi. **Auxiliary operations**: reclamation, slope stability, haul road construction and maintenance, equipment maintenance, drainage and pumping, communications, power distribution, dust control and safety.

Because the profile of the deposit layer is different, the mining process selection is needed to be adapted in order to reach the best economic effect [11]:

The open-cast mining has numerous advantages and disadvantages as well. The main advantages are [12]:

a. Higher productivity, efficiency and safety than for underground mining;

b. The utilization of large equipment reduces the unit cost;

c. Modest development requirements allow rapid exploitation;

d. Relatively flexible, can increase the production by expanding operations.

The disadvantages of open-cast mining process are:

a. The economic limits of the method and the technological limit of the equipment impose the depth limit, which is generally about 90 m;

b. The economics imposes the limits of stripping ratios;

c. Extensive environmental reclamation of the surface is required.

8.2. Separation and classification

Separation and classification[2] are very important elemental manufacturing processes in many industries such as the mining and chemical industries. The equipment using many different methods of separation is applied in these processes [13],[14]. Used concentrating devices depend on the fluid, the force field and the specific properties of particles, such as the density, size, shape, chemistry, surface chemistry, magnetism, conductivity, color and porosity. Various concentrating devices are applicable to particles according to their size ranges (**Fig. 3**), and for any given size range, several processes or devices might be used. The gravity concentration works the best in the range from 130 mm to 74 μm [15].

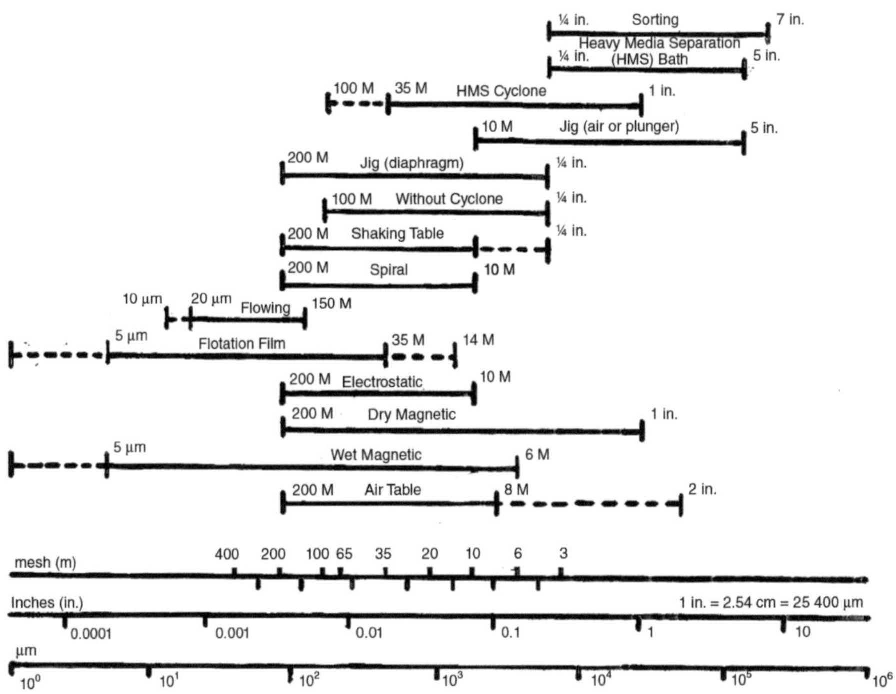

Fig. 3. Approximate range of applicability of various concentrating devices (M = mesh, Tyler standard) [15].

In phosphate rock beneficiation, the availability of water is of primary importance and may determine the process or processes used. Fine-grained impurities can often be removed from phosphate ores by using the combinations of comminution, scrubbing, water washing, screening and/or hydrocyclones. The disposal of tiny ore constituents (slimes) can be problematic. The beneficiation technique of froth flotation (described in **Section 8.7**) is widely used within the world phosphate rock industry [4],[6].

[2] Classification is defined as the separation of a mixture of solid particles into various fractions according to their sizes or densities [14].

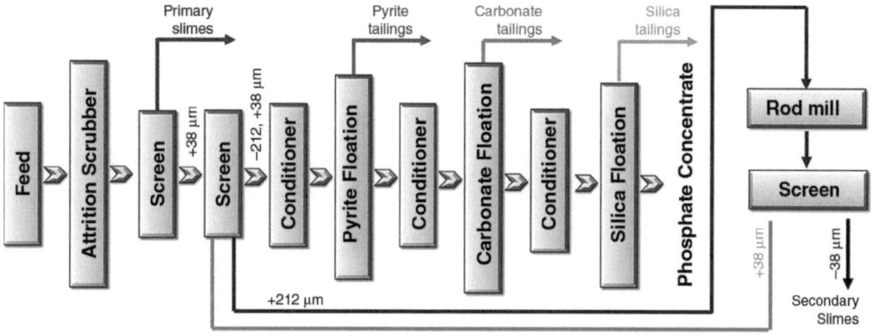

Fig. 4. Beneficiation of phosphate rock containing carbonates, silica and pyrite as the main gangue materials [1],[7].

Ore type	**Beneficiation method**		
	Flotation	**Calcination**	**Organic acid leaching**
Igneous	The best approach	Not applicable	Not applicable
Sedimentary with siliceous gangue	The best approach	Not applicable	Not applicable
Sedimentary with calcareous gangue	**Advantages** –	**Advantages** Low water consumption. Complete elimination of the carbonate gangue.	**Advantages** Very selective. Acid plants have low capital cost. Method has few environmental hazards. Leaching does not affect phosphate minerals. Organic acid can be recycled. Water consumption equal to convectional beneficiation. Final product has good quality and purity. Organic acid salts are soluble in water and easily filtered from solid product.
	Disadvantages Method has not given satisfactory results. Not applicable in many cases.	**Disadvantages** Calcination. Plants high capital cost. Needs high thermal energy. Calcined product has no desirable quality. Calcination decreases the product solubility. Process is time consuming.	**Disadvantages** Economic aspects are not well established. Organic acid price is high.

Table 1. The comparison of phosphate ore beneficiation methods [16].

Effective beneficiation can be achieved by various processes depending on the liberation size of phosphate and gangue minerals and other ore specifications. Different processes like screening, scrubbing, heavy media separation, washing, roasting, calcinations, leaching and flotation may be used. For example, crushing and screening are used to remove coarse hard siliceous material, and attrition scrubbing and desliming are used to remove clayey fine fractions. If silica is the main gangue material, flotation is the conventional mineral processing technique used. Igneous-type ores are also amenable to flotation, which is the best approach for the processing of this type of phosphate ore [16].

8.3. Electrostatic separation

Almost all minerals show some degree of conductivity. The electronic separation process uses the difference in electrical conductivity or surface charge of the mineral species of interest. The electrostatic separation process is generally confined to recovering valuable heavy minerals from beach-sand deposits. However, the growing interest in plastic and meta recycling has opened up new applications in secondary materials recovery [15].

When particles come under the influence of electrical field, depending on their conductivity, they accumulate charge that depends directly on the maximum achievable charge density on the particle surface. These charged particles can be separated by differential attraction or repulsion. Therefore, the first important step in electrostatic separation is to impose an electrostatic charge onto particles. Three main types of charging mechanism are the contact electrification or triboelectrification, the conductive induction and the ion bombardment (**Fig. 5**). Once the particles are charged, the separation can be achieved by the equipment with various electrode configurations [13],[15].

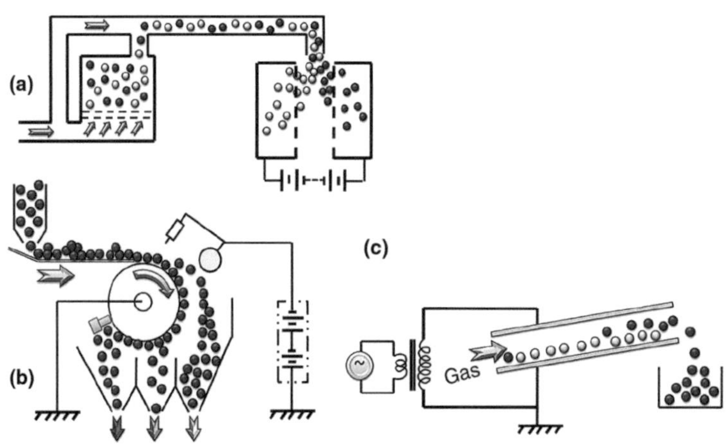

Fig. 5. Representative methods of electrostatic separation: contact charge (a), ion attachment (b) and induced charge (c) [13].

http://dx.doi.org/10.5772/62215

In the combination with attrition, desliming and gravity separation, the electrostatic separation technique is successful in the beneficiation of phosphate ores by removing silica and/or carbonates, mostly on laboratory scale. However, low capacity of electrostatic separators limits their use in large-scale production. This technique is used to concentrate the phosphate ores of different types [1].

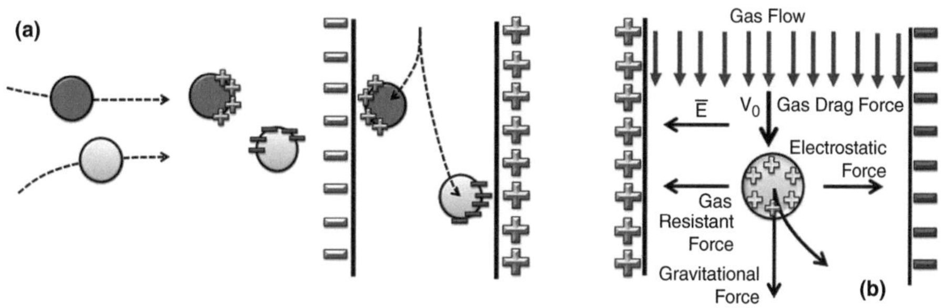

Fig. 6. Particle charging mechanism: the particle charged positively has lower work function and the particle charged negatively has higher work function (a) [1] and the illustration of the separator chamber (b) [17].

The triboelectrification is a type of electrostatic separation in which two nonconductive mineral species acquire opposite charges by contact with each other. The particle charging process is the critical step for the triboelectrostatic separation since the separation efficiency is a function of the difference in charge polarity and the magnitude of different particles.[3] Oppositely charged particles can be separated under the influence of electric fields. This process uses the difference in the electronic surface structure of the particles involved. A good example is the strong negative surface charge the silica acquires when it touches carbonates and phosphates. The surface phenomenon that comes into play is the work function, which may be defined as the energy required to remove electrons from any surface (**Fig. 6(a)**). The work function is defined as the minimum energy that must be supplied to extract an electron from a solid. The particle that is charged positively after particle-particle charging has lower work function than the particle that is charged negatively [15],[17].

The particle residence time, i.e. the time for the particle traveling through the separation chamber (**Fig. 6(b)**), is controlled by the particle vertical motion. However, the horizontal particle motion (*y*) is controlled by electric field deflection. The relation governing the horizontal displacement (*x*) of moving particle is [17]:

$$\frac{d^2\overline{x}}{dt^2} = \overline{E}\frac{q}{m} \tag{1}$$

[3] The charge density achieved with conventional pneumatic chargers (including tubing charger, cyclone, honeycomb, static mixer, etc.) and belt charger is about $5-8{\cdot}10^{-6}$ C/m^2. Since the theoretical limit for the charge is $2.63{\cdot}10^{-5}$ C/m^2, clearly, there is a huge potential in improving the charging efficiency [17].

where ***m*** is the mass of particle, ***x*** is the horizontal displacement vector, ***t*** is the time, ***E*** is the electric field intensity and ***q*** is the charge of particle. The charge-to-mass ratio ***q*/*m*** is referred to as the particle specific charge. If the resistance of air with the viscosity η is also considered, the horizontal motion of moving spherical particle of radius ***r*** is given by the equation:

$$\frac{d^2\bar{x}}{dt^2} + 6\pi\frac{\eta}{m}r\frac{d\bar{x}}{dt} = \bar{E}\frac{q}{m} \tag{2}$$

From **Eq. 2**, the speed of the particle as a function of time can be derived:

$$\frac{d\bar{x}}{dt} = \bar{E}\frac{q}{6\pi\eta r}\left[1 - \exp\left(-\frac{t}{m/6\pi\eta r}\right)\right] \tag{3}$$

where $t >> \frac{m}{6\pi\eta r}$ or $t \to \infty$. The terminal horizontal speed of particle is:

$$\left(\frac{d\bar{x}}{dt}\right)_{horizontal} = \bar{E}\frac{q}{6\pi\eta r} \tag{4}$$

Under these conditions, the terminal horizontal speed is independent of the mass. However, since the time ***t*** is in the range of milliseconds, the mass does play an important role in determining the horizontal motion of the particle as well as the resultant trajectory that affects the separation performance [17].

The particle motion in the vertical direction is influenced by the gravitational force and gas drag force. The governing equation is [17]:

$$\frac{d^2y}{dt^2} = 6\pi r\frac{\eta}{m}\frac{dy}{dt} + g \tag{5}$$

where η is the dynamic viscosity of gas and ***g*** is the gravitational acceleration. For the initial conditions of $t = 0$, $y(0) = 0$ and $dy(0)/dt = V_0$, **Eq. 5** can be solved as follows:

$$y(t) = \frac{(g + V_0E)\ \exp\ (Bt) - Bgt - g - V_0B}{B^2} \tag{6}$$

where $B = 6\pi\eta t/m$. The particle trajectories can be obtained from **Eqs. 4** and **6**.

The tube-type separator has the pre-charging zone and the separation zone as the integral parts of the machine (**Fig. 7(a)**). The pre-charging zone, or the triboelectrification process, exploits the difference in the electronic appearance of the particles involved. The particles become charged by the particle-particle contact, particle-wall contact or both. The particle-particle contact between different particles results in the transfer of electrons (charges) from the surface of one particle to the surface of the other one. After this transfer, one of the particles is positively charged and the other one possesses the negative charge. The separation zone consists of two vertical walls of rotating tubes, which oppose each other and which are electrified by opposite potential. As the charged particles enter the separation zone, they become attracted by oppositely charged electrodes. The separated products are collected at the base of separator. This separator removes very effectively silica from other nonconductive minerals, such as calcium carbonate, phosphate and talc [15].

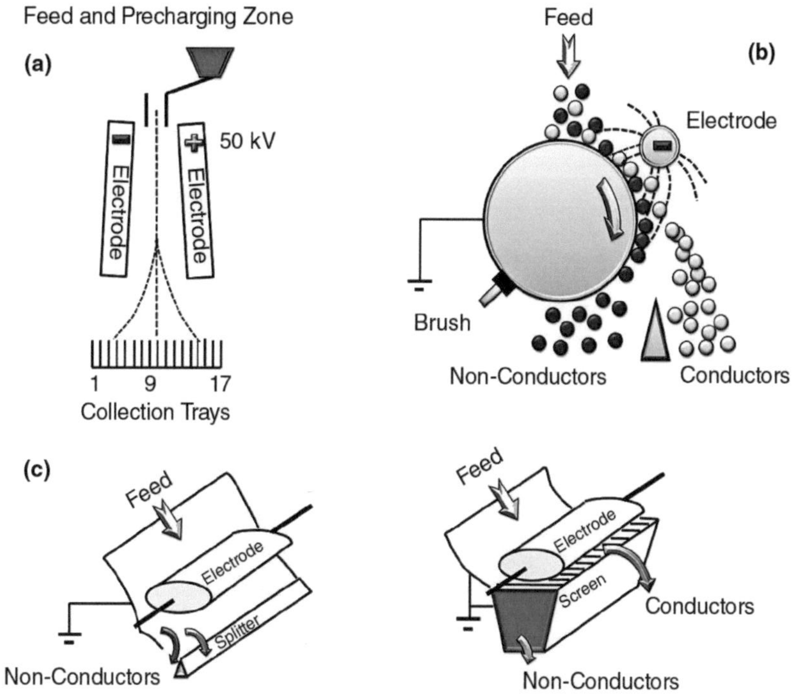

Fig. 7. Operating principle of electrostatic separator: (a) V-stat separator, (b) plate-type separator and (c) roll-type separator [15].

In the horizontal belt-type separator, fast-moving belts travel in opposite directions adjacent to suitably placed plate electrodes of the opposite polarity. The material is fed into a narrow gap between two parallel electrodes. The particles are swept upward by a moving open-mesh belt and conveyed in opposite directions, thus facilitating the particles' charging by contact with other particles. The electric field attracts the particles up or down depending on their charge. The moving belts transport the particles adjacent to each electrode towards opposite ends of the separator [15].

8.4. Magnetic separation

In 1792, a patent was filed by WILLIAM FULLARTON describing the separation of iron minerals with a magnet.[4] The early applications were based on the intrinsic magnetic properties of sediments for the separation. In 1852, magnetite was separated from apatite by a New York company on a conveyor belt separator. Later, a new line of separators was introduced for the separation of iron from brass fillings, turnings, of metallic iron from furnace products and of magnetite from plain gangue. The 1950s were the time of great expansion in the field of magnetic separations as the introduction of high-gradient magnetic separation (HGMS) systems permitted faster and more general magnetic separation processes. More recently, the separations using external magnetic fields have become common processes in biotechnology, where they are used for both protein purification as well as flow cytometry [18],[19].

Electromagnets almost completely replaced permanent magnets as the field-generating elements in drum separators [20]. Recent progress in magnet technology has realized economically and operationally favorable cryocooler-cooled[5] [21] superconducting magnets, which can be used for commercial applications [22]. The first large superconducting[6] [23] separator has been operating successfully in the USA since May 1986 and a larger system was installed with twice the capacity in 1989. A revolutionary design for the superconducting magnetic separator with a reciprocating canister system was installed and successfully operated for clay processing in May 1989. Following this, a number of other reciprocators have been installed for kaolin processing in the places as far apart as Brazil and Germany [24].

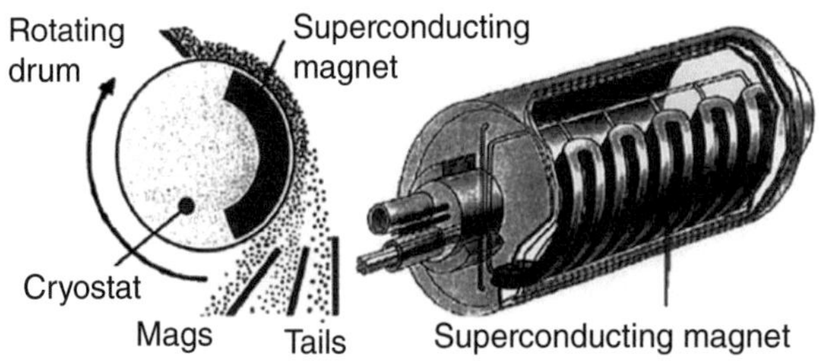

Fig. 8. Drum separator using a multi-pole superconducting magnet (Klochner Humbolt-Deutz Cologne, Germany) [22].

[4] The properties of magnetic materials were identified as early as 6th century BC, but the means by which magnets could move material remained only a curious phenomenon until the late 18th century. The background for electricity and magnetism, the reasons that magnets could move materials, were explained by GAUSS AND HELMHOLTZ [18].

[5] All superconducting devices share the need for sufficient refrigeration to overcome their low-temperature heat loading. This loading comes typically in two forms: (1) heal leaks from the surrounding and (2) internal heat generation in the device. In addition, the refrigeration system needs to bring the superconducting device from ambient temperature to its low operating temperature in reasonable length of time [21].

[6] After Kamerlingh Onnes's pioneering the demonstration in 1908 that the last so-called permanent gas helium could indeed be liquefied. His follow-up discovery of superconductivity in 1911 introduced the zero electrical current resistivity to the world. It was theorized that one could go beyond the resistive limit of a copper wire to develop a superconductor that could carry any amount of current but without the ohmic loss [23].

Despite all this progress, the majority of the commercial magnetic separators fulfill only the simple technological objective of the removal of magnetic substances without the ability to classify them. Only three classical separation products (tails, middlings and mags, **Fig. 9**) are usually obtained [25].

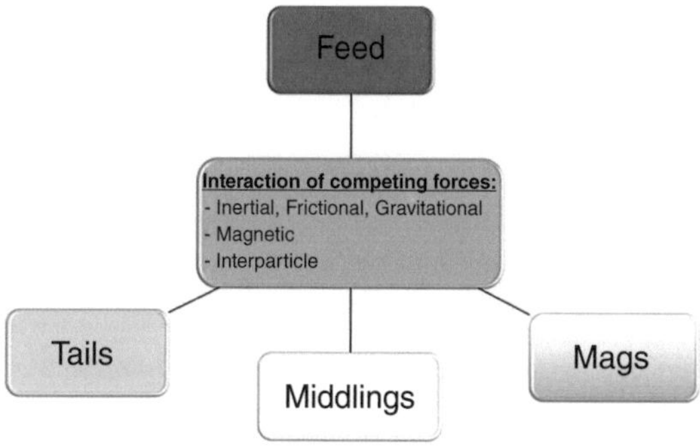

Fig. 9. Classical magnetic separation products [25].

Unlike the conventional filtration methods that use the blocking-type filtration, the secondary waste is not produced in high-gradient magnetic separation (HGMS), which is also known as the magnetic or electromagnetic filtration. Furthermore, because HGMS systems use much higher magnetic forces than conventional magnetic separation techniques, it can also be used to separate rapidly large quantities of diluted suspension [22].

According to the applied separation method, two classes of magnetic separators are recognized [26]:

i. Separators that deflect the magnetic particles from the main stream, e.g. open-gradient magnetic separation (OGMS);

ii. Separators that usually collect the magnetic particles in matrices, e.g. high-gradient magnetic separation (HGMS).

Although current separators usually achieve high grades of separation, they cannot classify the particles[7] as they are being separated. The magnetic separator in which these two steps are performed at the same time and in the same machine was proposed by Augusto and Martins [26].

Magnetic separation has been considered for many years a valuable method to achieve the purification of streams of particles (dry or wet) [26]. Magnetic separators have unrestricted industrial applications and are widely used in mineral beneficiation, food, textiles, plastic and ceramic processing industries. The separation efficiency of magnetic separator depends on the

[7] Differential magnetic classification and the selectivity are different definitions. The selectivity is defined as the ability to separate one certain kind of magnetic particles from all others, independently of how close their magnetic susceptibilities may be [25].

material characteristics and the design features of equipment along with the optimization of process variables [27].

The magnetic force ($\vec{F}_m$) acting on weakly magnetic particle flowing in a fluid is given by the equation [19]:

$$\vec{F}_m = \frac{1}{\mu_0}(\kappa_p - \kappa_f)V_p B \nabla B \quad (7)$$

where $\boldsymbol{B}$ is the magnitude of magnetic flux density at the particle position, $\boldsymbol{\mu}_0$ is the magnetic permeability of vacuum, $\boldsymbol{\kappa}_\mathrm{p}$ is the volume susceptibility of the fluid and $\boldsymbol{V}_\mathrm{p}$ is the volume of the particle. The magnetic force on a particle is then proportional to the magnitude of magnetic flux density and the gradient. The magnetic field can be increased using a stronger magnet having more ampere turns, and the field gradient can be increased by changing the magnetic polarities and using a steel wool matrix. For sufficiently strong magnetic particles such as iron, magnetite and maghemite, it is advantageous, and **Eq. 7** can be written as:

$$\vec{F}_m = \mu_0 V_p M \nabla H \quad (8)$$

where $\boldsymbol{M}$ is the magnetization of the particle and $\boldsymbol{H}$ is the magnitude of magnetic field intensity at the particle position [19].

The basic principle behind magnetic separations is remarkably simple and remains unchanged from these early examples. It is based on a simple fact that materials with differing magnetic moments experience different forces in the presence of magnetic field gradients; thus, externally applied field can hand pick the components with distinctive magnetic characteristics out of physically similar mixtures [18]. When one of the major gangue constituents is magnetic, magnetic separators are used as one of the steps in the flow sheet to remove the magnetic constituents. This is mostly used in the beneficiation of igneous phosphate rocks. However, it was also used for the beneficiation of some sedimentary phosphate ores [1].

Paramagnetic minerals have higher magnetic permeability than the surrounding medium, usually air or water, and they concentrate the lines of source of an external magnetic field. The higher the magnetic susceptibility, the higher the field intensity in the particle and the greater the attraction up the field gradient toward increasing field strength. Diamagnetic minerals, on the other hand, have lower magnetic permeability than the surrounding medium and they repel the lines of force of magnetic field. These characteristics cause the expulsion of diamagnetic minerals down the gradient of the field towards decreasing field strength. This negative diamagnetic effect is usually orders of magnitude smaller than the positive paramagnetic attraction. Thus, a magnetic circuit can be designed to produce higher field intensity or higher field gradient or both to achieve the effective separation [15].

Magnets are used in the mineral industry to remove the tramp iron that might damage the equipment and to separate minerals according to their magnetic susceptibility. According to the intensity of the magnetic field, two types of magnetic separators are recognized [15]:

a. **Low-intensity magnetic separators** have the flux densities up to 2000 G[8] [28]. These separators are mainly used to remove the ferromagnetic material, such as iron, to protect downstream unit operations, such as conveyor belts, or to scalp ferromagnetic materials to improve the performance for permanent or electromagnetic separators used to separate weakly magnetic materials. Low-intensity separators can treat wet slurry or dry solids.

b. **High-intensity magnetic separators** separating paramagnetic or weakly magnetic particles require higher flux density. This higher density is achieved by designing the electromagnetic circuitry that can generate the magnetic force of up to 2 tesla. For example, in a silica sand processing plant, these separators are used to remove weakly magnetic iron-bearing particles.

Rotating-drum magnetic separators (**Fig. 10**) are mainly used in mines. The rotating disc magnetic separator is used in so-called ferritic processes [22].

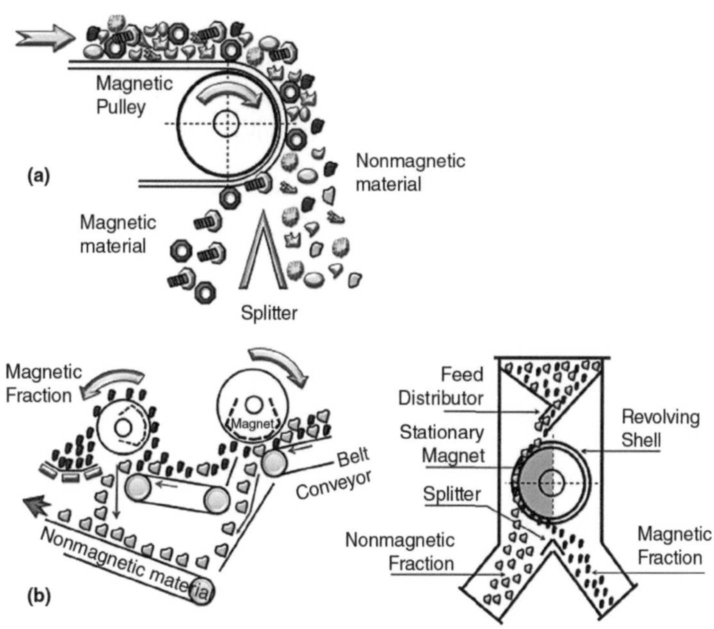

Fig. 10. Typical magnetic pulley (a) and magnetic drum operating as lifting magnet (b) [15].

---

[8] The gauss (G or Gs) is a unit of magnetic field (magnetic flux density) named after Carl Friedrich Gauss ($1\ G = 10^{-4}\ T = 1\ cm^{-1/2}g^{1/2}s^{1}$) [28].

8.5. Removing of carbonates

The removal of carbonates from phosphate rock has been the focus of significant research efforts. Several countries have large deposits of phosphate rock that contain significant amounts of calcite ($CaCO_3$) and dolomite ($CaMg(CO_3)_2$). The calcination of phosphate ores to remove carbonates is expensive because of high costs of energy. Calcination is practiced commercially at several phosphate rock mining operations around the world, mainly to improve final product quality by removing minor amounts of carbonates and organic matter. Calcination is also used to remove carbonates where the cost of natural gas is very low [1],[6].

Calcium and magnesium carbonates are readily dissolvable in both mineral (strong acids) and organic acids (weak acids). In the case of calcareous phosphate ores, although mineral acids dissolve carbonates at high reaction rates, they also attack the phosphorus-bearing minerals and cause losses in the P_2O_5 content of the ore; hence, they are not appropriate if the intention is only to beneficiate the ore not to dissolve phosphates. To avoid this problem, organic acids were studied as carbonate leaching agents, although their reaction rates are low. These organic acids may be expensive and will certainly add to the production cost. On the other hand, they are selective to leaching carbonates, their capital cost is low, they do not cause environmental hazards and they can be recycled [1].

The organic acids most commonly used in carbonate leaching are acetic acid, citric acid and formic acid. They are used for some specific advantages (may be the cost, availability, etc.). Suggested reaction between acetic acid and carbonates is [1],[16],[29]:

$$CaCO_3 + 2\ CH_3COOH \rightarrow Ca\left(CH_3OO\right)_2 + CO_2 + H_2O \tag{9}$$

The dissolution kinetics of calcareous material with acetic acid solution was found to fit the shrinking core model for the reaction-controlled process. The activation energy was determined to be 41.0 kJ·mol^{-1}, which is consistent with a chemically controlled reaction. The process is driven by the surface chemical reaction kinetic model: $(1 - (1 - \alpha)^{1/3})$ [30].

Acetic acid may be recovered by reversing the above reaction at high CO_2 pressure in a separate reactor or by using sulfuric acid to precipitate calcium sulfate and to liberate acetic acid:

$$\begin{aligned} &Ca\left(CH_3OO\right)_2 + H_2SO_4 + 2\ H_2O \rightarrow 2\ CH_3COOH \\ &+CaSO_4 \cdot 2H_2O \end{aligned} \tag{10}$$

It is noted that the by-products such as calcium sulfate (gypsum) could be used and/or sold to lower the costs of acetic acid and its recovery by sulfuric acid (**Eq. 10**). Similarly, formic and lactic acids (**Eq. 11**) can be used to dissolve carbonate minerals [16],[30]:

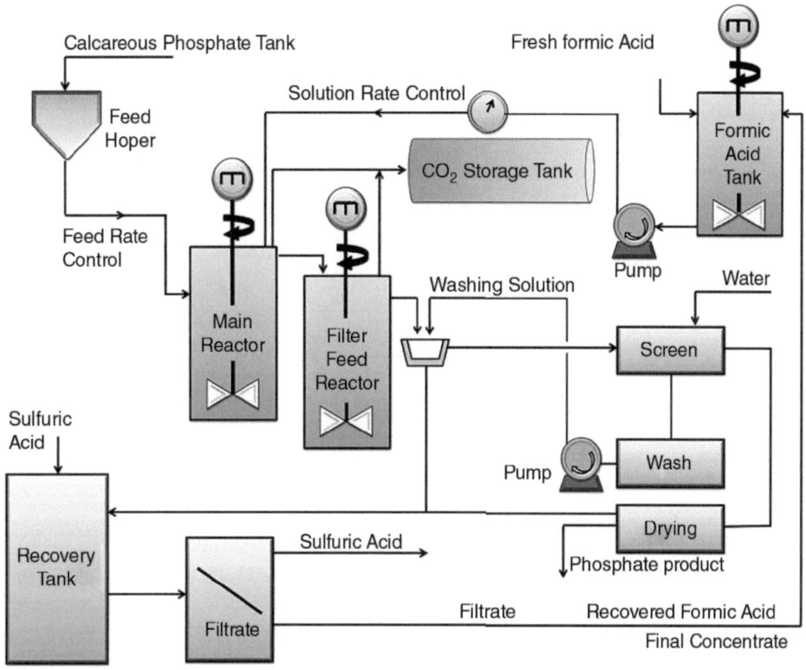

Fig. 11. Proposed flow sheet for leaching of phosphate rock/ores using formic acid [16].

$$CaCO_3 + 2\ CH_3CH(OH)COOH \rightarrow Ca\left(CH_3CH(OH)OO\right)_2 + CO_2 + H_2O \quad (11)$$

The main factors investigated by the researchers were: leaching reagent, acid concentration, reaction time, liquid/solid ratio (pulp solid percent), temperature, particle size distribution, stirring speed and type and nature of ore [16].

8.6. Calcination

More than 10% of the world's marketable phosphates are produced by calcination. Traditionally, the heat treatment of phosphate ores is defined as heating up the ore to a certain temperature to obtain a product with specific properties. The main processes that take place during the thermal treatment of apatite ore are [1],[3],[31],[32]:

1. **Drying**, i.e. the evaporation of water within the temperature range from 120 to 150°C;

2. **Pyrolysis of organic matter** within the temperature range from 650 to 750°C is important for black or brown phosphates;

3. **Thermal decomposition of carbonates**, i.e. the calcination within the temperature range from 850 to 1000°C;

4. **Removal of fluorine**, i.e. the **defluorination** at temperatures higher than 1350°C.

The calcination process of phosphate ore is schematically shown in **Fig. 12**.

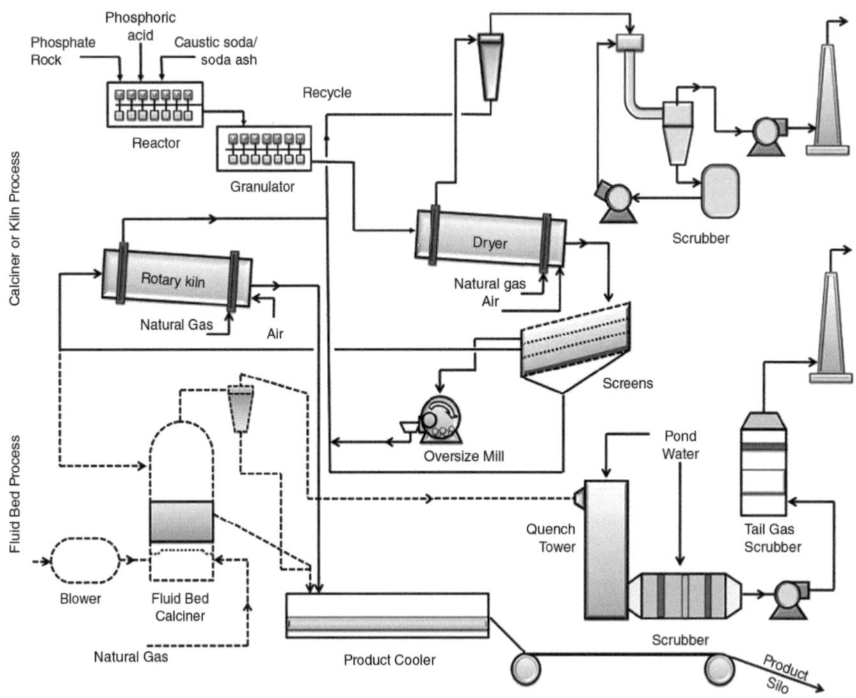

Fig. 12. Illustration of the defluorination process of phosphate rocks [1].

There are various types of units that can be used for the calcination of phosphate ores, such as [1]:

i. **Vertical-shaft kilns** [33],[34]: are the most popular type of kilns, having varying heights, diameters and constructional details. There are two types, namely mixed- (a) and unmixed-fuel type (b). The construction of a vertical shaft may be cylindrical, conical or a combination of both shapes with varying diameters in different zones (**Fig. 13(a)**).

ii. **Fluidized-bed reactors (calciners)** [33],[34]: the hog gases perform two functions: (1) fluidize the particles and (2) transfer the heat to the particles (**Fig. 13(b)**). Since the fluidization is a function of particle size, only fine particles can be introduced as the feed particles.

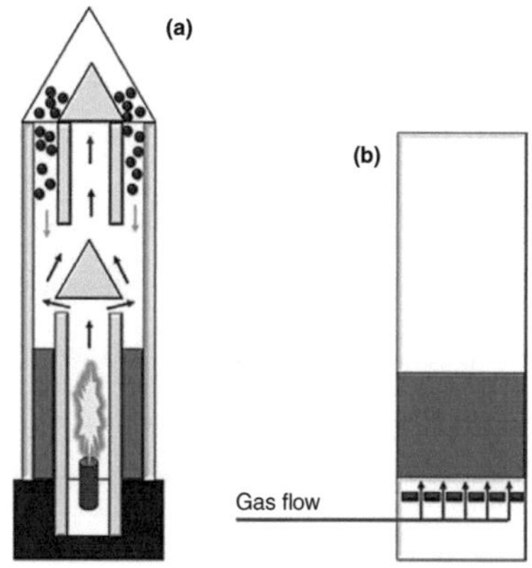

Fig. 13. Schematic diagram of annular-shaft kiln (a) and fluidized-bed calciner (b) [34].

iii. **Rotary kilns**[9] [33],[34],[35]: are extremely versatile incineration systems. They differ greatly in size with respect to their diameter (150 – 390 cm) and length (1800 – 1350 cm). Basic rotary kiln is composed of a cylindrical, refractory-lined steel shell, supported on two or more trunnions. The kiln is gently sloped (usually up to 0.03 m/m) and rotates slowly (1 – 5 rpm, the rotation rate is usually less than 2 rpm). The kiln may be operated in the co-current (parallel) or countercurrent mode (**Fig. 14**) with respect to the relative direction of gas and solid flow.

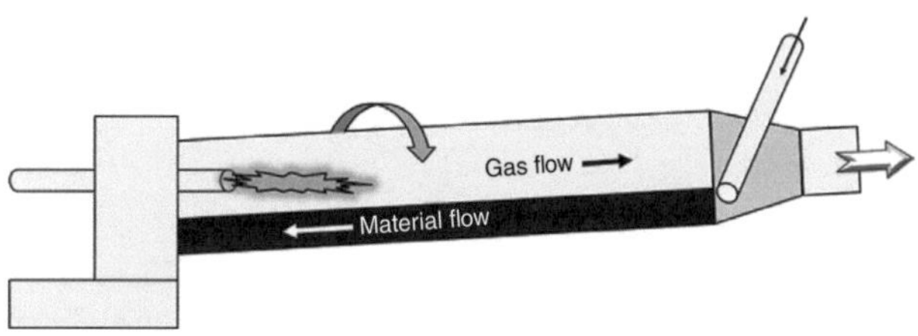

Fig. 14. Schematic representation of countercurrent flow rotary kiln [34].

The rate of movement of the material through the kiln may be estimated using several relationships, e.g.:

$$\Theta = \frac{0.19\ L_T}{NDS} \tag{12}$$

where $\boldsymbol{\Theta}$ is the residence time [min], $\boldsymbol{L_T}$ is the length of the kiln [m], ***N*** is the kiln rotation velocity (rpm), ***D*** is the kiln internal diameter [m] and ***S*** is the slope of the

[9] Rotary kilns are synonymous with cement and lime kilns probably because of the history of their evolution and development.

kiln [m/m]. Since the rotary kiln is divided to zones, the relationships should, more appropriately, be used for several reasonably uniform zones along the kiln and the total residence time can be calculated as the sum of the residence times for the individual zones.

iv. **Traveling grate-kilns, rotary kilns systems** [36],[37]: use low strength, somewhat wet pellets. These pellets are placed in a uniform bed upon a traveling grate, hot air being blown upward from below. The dehydration and partial calcination occur on the grate. Pellets are then fed to a short rotary kiln. The example of grate-kiln technology for the thermal treatment of pellets is shown in **Fig. 15**. The main advantages of this system are controlled feed rate, no flushing of materials into the kiln, no segregation of raw material due to different shapes and densities, avoidance of fluidization of the material bed, minimal dusting, etc.

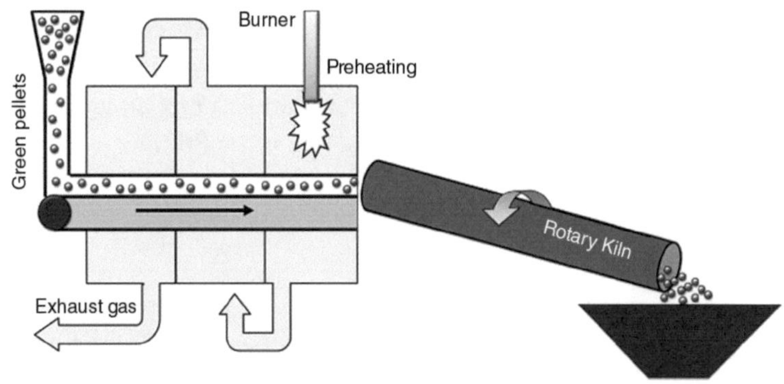

Fig. 15. Thermal treatment of pellet using grate-kiln technology [36].

v. **Flash calciner** [31],[38],[39]: is one of more recent developments in calcination, but it is not really a kiln. There are three main elements including preheater (1), flash calciner (2, **Fig. 16**) and cooler (3).

The unique characteristics of flash calcination are particularly suited to pressing phosphate. Phosphate is a complicated mineral that varies from deposit to deposit with each ore requiring its own special processing consideration. During thermal treatment, it is important not to destroy the delicate crystal structure of phosphate by overheating. Flash calcination rates, very good oxygen contact and rapid cooling, all of these characteristics, are very important in the production of high-quality calcined phosphate. The operating conditions in the range from 800°C to 1000°C are required.

http://dx.doi.org/10.5772/62215

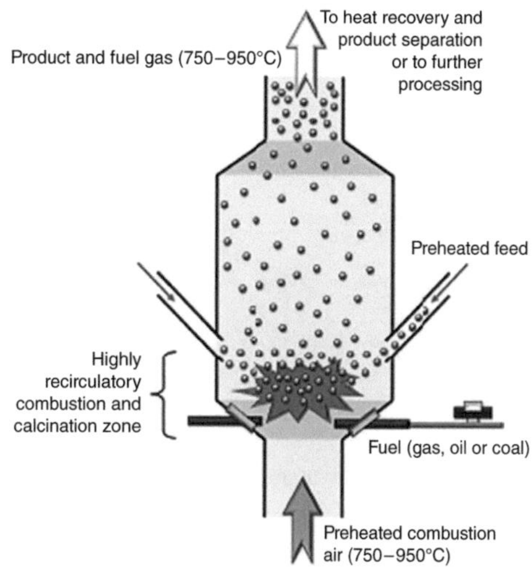

Fig. 16. Schematic representation of flash furnace [38].

8.7. Flotation

Flotation is a selective separation process that consists of attaching hydrophobic particles to rising air bubbles to form a particle-rich froth on the suspension surface, which flows over the lip of the cell. Hydrophilic particles do not attach to the bubbles and settle at the bottom to be discharged. Flotation has been the workhorse of mineral industry for over 100 years and has been expanded into many other areas, including deinking of wastepaper for recycling, water treatment and separation of plastics, crude oils, effluents, microorganisms and proteins [40].

The beneficiation of phosphate ores using froth flotation method has been practiced for at least 65 years. Extensive research work has been carried out in the last 25 years on various phosphate-containing ores. Despite extensive research and industrial experience, there are some challenges remaining in particular in beneficiation of siliceous-, calcite- and heavy mineral-containing phosphate ores [41].

Despite the fact that the flotation of apatite is difficult due to its physicochemical properties being similar to other minerals present in phosphate ores [42],[43], the froth flotation is widely used in mineral processing technologies to separate finely ground valuable minerals from a mixture with gangue minerals initially present in a pulp. The technique involves the contact of air bubbles with the solids [44]. Flotation technology is also used to remove suspended impurities during the treatment of wastewater, water purification, recovery of bacteria, cereal cleaning, recovery of metal and colloidal matters and recovery of ions and surfactants from the solution [45].

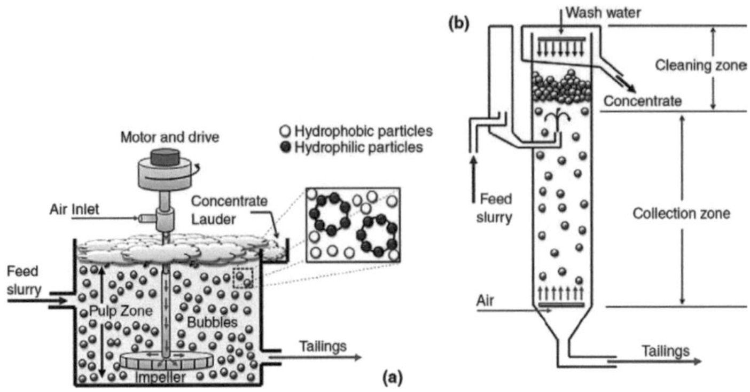

Fig. 17. Mechanical flotation cell (a) and flotation column (b) [4].

Currently, more than half of the world's marketable phosphates are concentrated by the flotation process [46]. Two types of flotation machines are available [4]:

1. Mechanical flotation cell (**Fig. 17(a)**);

2. Colum flotation cell (**Fig. 17(b)**).

The comparison of both systems is shown in **Table 2** [4]. Phosphoric tailings are fine-grained rock produced from the flotation processes [47].

Mechanical Flotation Cells	Column Flotation
Cell sizes ranging from ~0.1 to 350 m^3	Available up to 4 m in diameter
	Typical heights around 9 – 15 m
Air induced or injected through the impeller to generate bubbles	Internal or external spargers generate air bubbles
	Produces smaller bubbles
Bubble-particle interaction through mixing by impeller	Bubble-particle interaction through the countercurrent action-descending slurry and rising bubbles
Less favorable for the bubble-particle attachment	Generally considered more favorable for the bubble-particle attachment
	Better metallurgical performance (grade and recovery)
	Axial mixing can significantly reduce the overall performance (especially in larger-diameter columns)
	No moving parts
Well known to operators and easier to operate	Newer and less known by operators
Convectional plant operation history and knowledge base on mechanical cells	Harder to operate

Table 2. Comparison of mechanical flotation cell and column flotation [4].

Flotation cells are usually designed to perform several functions simultaneously, some of them are [45]:

a. Agitation and circulation of the pulp mixture to keep all particles in suspension;

b. Aeration, which provides further agitation, involves dissemination of fine air bubbles throughout the pulp;

c. Promotion of particle-bubble collision facilitating selective attachment and transport. Such collisions can also be enhanced by the countercurrent flow of discrete particles and bubbles;

d. Maintenance of quiescent pulp conditions immediately under the froth layer.

In most flotation plants, the cells are interconnected in batteries, and the first flotation stage, called roughing, permits quick rejection of most of the gangue and achieves high recoveries with low grades. A schematic circuit, which includes roughing, cleaning, re-cleaning and scavenging stages, is given in **Fig. 18**.

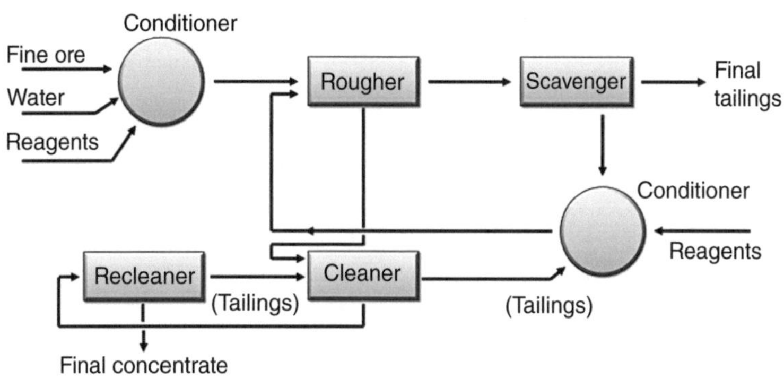

Fig. 18. Schematic representation of a flotation circuit [45]

Flotation is a dynamic process [48]. A whole range of variables can affect the performance of flotation systems (**Fig. 19**), such as their operating variables, particle size, reagents, ore composition and also the presence of ionic species in water [42],[46],[49]. The suspension of soluble minerals such as apatite is bonding large amounts of ions that interact with the mineral surface and affects the flotation performance. The interactions of dissolved anions from minerals in a pulp with a collector can form insoluble surfactant salts, which can precipitate non-selectively on mineral surface. These ions are so called potential-determining ions of fluorapatite such as Ca^{2+}, $CaOH^{+}$, PO_4^{3-}, HPO_4^{2-}, $H_2PO_4^{-}$, F^{-}, H^{+} and OH^{-} [46],[50]. The effect of water quality on the flotation process was described by Liu et al [51].

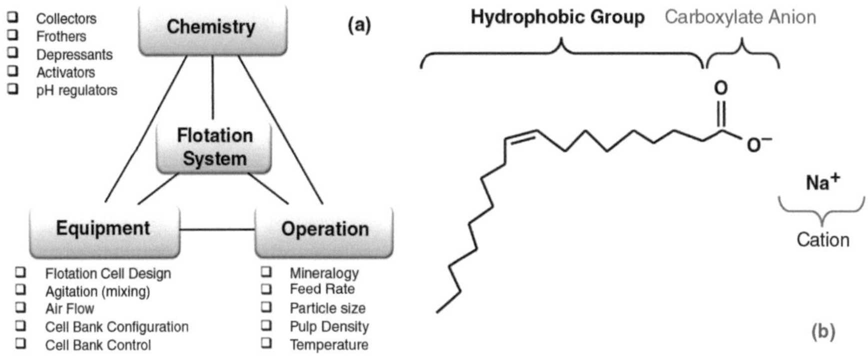

Fig. 19. Flotation system with interrelating subsystems (chemistry, equipment and operating components)(a) and sodium oleate in an anionic collector that can be used to render apatite hydrophobic in alkaline environments (b) [4].

The selectivity of froth flotation processes is highly influenced by the specificity of integrations between minerals and reagents, which are used to control the hydrophobic/hydrophilic character of mineral/water interfaces [52].

The use of additives is a tool for the control of surface tension of the flotation system. Additives (flotation reagents) used in phosphate flotation are synthetic organic species. They are produced via the ethoxylation of fatty alcohols. Alcohols are obtained from vegetable oils or animal fats. Ethylene oxide comes from the petroleum industry. These reagents may exhibit variable molecular composition and number of carbon atoms in the hydrocarbon chain, as well as the presence of double bonds, different stereochemistry (*cis-trans* isomerism) and also several levels of ethoxylation. The additives employed in phosphate ore flotation contain the carbon chains of different lengths, with a predominance of 18 carbon atoms. The ethoxylation level is represented by the average number of ethylene oxide groups in the molecule. Best results were achieved with three or four groups. The dosage of additives is 5% with respect to the collector dosage, reaching 10% under special conditions [53].

The organic reagents, such as guar gum, cashew gum, tannins, dextrin, ethyl cellulose and carboxymethylcellulose, are capable of acting as depressor in the flotation of igneous phosphate ores. The performance of corn starches was consistently superior to that of those reagents [53],[54]. The depressing ability of starch and ethyl cellulose appears to be related to steric compatibility between the positions of cations present on the mineral surface and hydroxyl groups within the molecular structure of reagents [52].

The role of surface and porosity was investigated by Zhong et al [55]. When the samples were not aged prior to the collector (potassium oleate) addition, the floatability was controlled by the dissolution (of calcium) and adsorption (of oleate) behaviors, which, in turn, were governed by the surface area. It appears that the surface constituted by pores had lower influence on the adsorption and dissolution characteristics than the external surface. This was suggested to be due to slow diffusion of calcium through the pores, which resulted in reduced dissolution rate, as well as the non-participation of a substantial portion of pores in the adsorption process. When the samples were aged prior to the oleate addition, the bulk precipitation of calcium oleate complex was found to play a crucial role. Since the bulk precipitation is not an interfacial process, the effect of surface area was slighter with aged samples.

A critical review of reagents used in the flotation of phosphate ores was performed by Sis and Chander [56]. Based on the literature, it was concluded that the usage of surfactant mixtures has certain advantages over single surfactant as the synergistic effects between surfactant mixtures were observed during different experiments such as surface tension, contact angle, adsorption and flotation. The synergism of surfactant mixtures at air/liquid, liquid/oil and liquid/solid interfaces arises from the improvement of froth properties, emulsification of hydrocarbon oil (e.g. fuel oil) and homogenous adsorption of collector on the minerals and protection of the collector from harmful effect of dissolved ions in the presence of auxiliary surfactant.

The activation of apatite particles during dry milling may enhance the adsorption of reagents, which favors the recovery of apatite. However, active defects may serve as the sites for the adsorption of water and some very fine gangue particles on the apatite surfaces, causing apatite particles to be less responsive to flotation. As a result, dry milling did not have much impact on the recovery and flotation kinetics of apatite [42].

The fact that microorganisms, both living and dead, and products derived from the organisms can function as flotation agents and flocculation agents is abundantly clear. They can modify the surfaces of minerals. They can function as flotation collectors and as flotation depressants and activators. In many cases, they or their products can function as specific flocculation agents [57].

Many strains of bacteria are able to adsorb Ca(II) and Mg(II) ions from aqueous solution and, in some cases, the adsorption can be very specific. For example, *Bacillus subtilis* typically binds Mg(II) much more readily than Ca(II). Bacteria can also adhere to the surfaces of minerals containing calcium and magnesium, either enhancing or depressing the flotation of these minerals. Since *B. subtilis* binds Mg(II) preferentially, it was reasoned that the adhesion to a mineral containing magnesium and calcium (dolomite) might be quite different from the adhesion to a mineral containing only calcium (apatite) and this difference could possibly be utilized in mineral processing. The experiments investigating the binding of Ca(II) and Mg(II) to *B. subtilis* cells were initiated, and anionic collector microflotation of pure dolomite and apatite mineral samples in the presence and absence of these bacteria was performed. Since Ca(II) and Mg(II) also bind to dolomite and apatite, the zeta-potential measurements as a function of pH in the presence and absence of these ions were performed in order to better elucidate the effect this binding may have on the attachment of *B. subtilis* to those two minerals [58].

8.8. Extraction of rare-earth elements

The group of rare earths consists of 14 lanthanides or 4f elements in the periodic table along with three more elements: lanthanum, scandium and yttrium. Lanthanides comprise 15 elements with atomic numbers 57 – 71, which include lanthanum (La), cerium (Ce), praseodymium (Pr), neodymium (Nd), promethium (Pm), samarium (Sm), europium (Eu), gadolinium (Gd), terbium (Tb), dysprosium (Dy), holmium (Ho), erbium (Er), thulium (Tm), ytterbium (Yb) and lutetium (Lu). All elements occur in nature, while promethium (Pm) originates as a part of radioactive decay. Elements La, Sc and Y have physiochemical properties similar to rare earths and are associated with the same minerals. Since they have similar chemical properties, the elements in the lanthanide series, yttrium and scandium, are considered as rare-earth elements (REE). Another classification used is light rare-earth elements (LREEs, atomic numbers 57 – 63: La, Ce, Nd, Pr, Pm, Sm and Eu) and heavy rare-earth elements (HREEs, atomic numbers 64 – 71: Gd, Tb, Dy, Ho, Er, Tm and Yb) [59],[60].

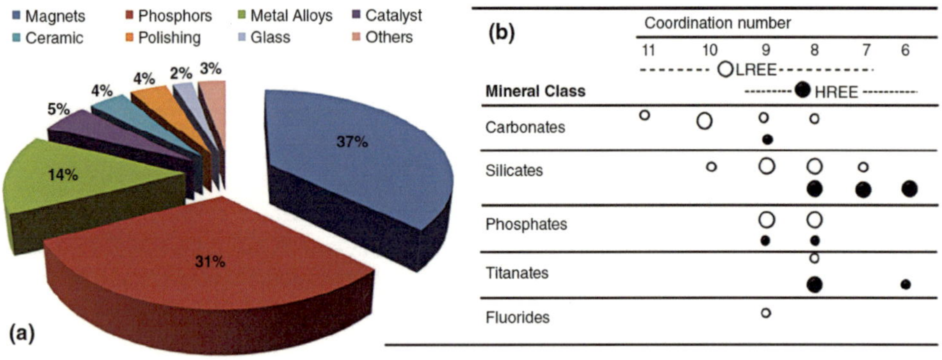

Fig. 20. Applications of REE (a) [62] and the coordination numbers and abundances of LREEs (white cycle)- and HREEs (black cycle)-bearing minerals. The size of circle indicates rough abundance of REEs for each mineral class (b) [64].

Most of the REE deposits exist in China, America, India, Middle Asian nations, South Africa, Australia and Canada. The demand for REEs has increased in recent years due to the uncertainty of the supply and high technological applications associated with their characteristic electronic, optical and magnetic properties (**Fig. 20**). RE phosphate minerals, such as monazite, florencite, xenotime, cheralite and britholite, are the most naturally abundant forms that are associated with fluorapatite [47],[61],[62],[63],[64].

The techniques described in **Chapter 8** are usually used for concentrating REE minerals prior to the extraction of REEs from phosphate rocks (PR).[10] A pre-leaching stage with mineral acid (**Eq. 19** and **Eqs. 13 – 16**) can be useful in order to selectively leach the FAP fraction as well as other impurities such as sodium, potassium, magnesium, aluminum, iron, manganese, uranium and thorium associated with the FAP lattice, resulting in REE-enriched concentrate [47].

[10] The processing chain for PR results in the majority of trace elements being lost either to waste disposal or to the environment (mainly soil and water) through fertilizer consumption and the food chain [47].

http://dx.doi.org/10.5772/62215

$$Ca_{10}(PO_4)_6F_2 + 20\ HCl \rightarrow 10\ CaCl_2 + 6\ H_3PO_4 + 2\ HF \tag{13}$$

$$Ca_{10}(PO_4)_6F_2 + 20\ HNO_3 \rightarrow 10\ Ca(NO_3)_2 + 6\ H_3PO_4 + 2\ HF \tag{14}$$

$$Ca_{10}(PO_4)_6F_2 + 20\ HClO_4 \rightarrow 10\ Ca(ClO_4)_2 + 6\ H_3PO_4 + 2\ HF \tag{15}$$

$$Ca_{10}(PO_4)_6F_2 + 14\ H_3PO_4 \rightarrow 10\ Ca(H_2PO_4)_2 + 2\ HF \tag{16}$$

However, leaching efficiencies can vary significantly depending on the mineralogy of the ore and the type of acid used. H_3PO_4 and HF acids formed during the leaching process of FAP with acids interfere and change the leaching efficiency [61].

The effect of aliphatic and aromatic low molecular weight organic acid on the release of REEs and yttrium from phosphate minerals was investigated by Goyne et al [65]. The performance of acid increases in the following order:

No ligand ≈ salicylic acid < phthalic acid ≈ oxalic acid < citric acid.

The utilization of organophosphorus reagents, such as Talcher organic phosphorus solvent (TOPS 99), an equivalent to di-2-ethylhexyl phosphoric acid, 2-ethylhexylphosphonic acid mono-2-ethylhexyl ester (PC-88A) and bis(2,4,4-trimethylpentyl) phosphinic acid (Cyanex 272), etc., for the extraction of REEs was also reported [66],[67],[68],[69].

Systematic study of the thermal decomposition of monazite to remove phosphate in order to achieve more complete conversion of rare-earth phosphate into its oxides was performed by Kumari et al [62]. The method is based on roasting of monazite with CaO, Na_2CO_3 and NaOH (**Fig. 21**):

$$2\ REEPO_4 + 3\ CaO \xrightarrow{610-700°C} REE_2O_3 + Ca_3(PO_4)_2 \tag{17}$$

$$2\ REEPO_4 + 3\ Na_2CO_3 \xrightarrow{600-900°C} REE_2O_3 + 2\ Na_3PO_4 + 3\ CO_2 \tag{18}$$

$$REEPO_4 + 3\ NaOH \xrightarrow{250-450°C} REE(OH)_3 + Na_3PO_4 \tag{19}$$

Washed monazite concentrate achieved from roasting was dried and leached by diluted HCl:

$$REE_2O_3 + 6\ HCl \rightarrow 2\ REECl_3 + 3\ H_2O \tag{20}$$

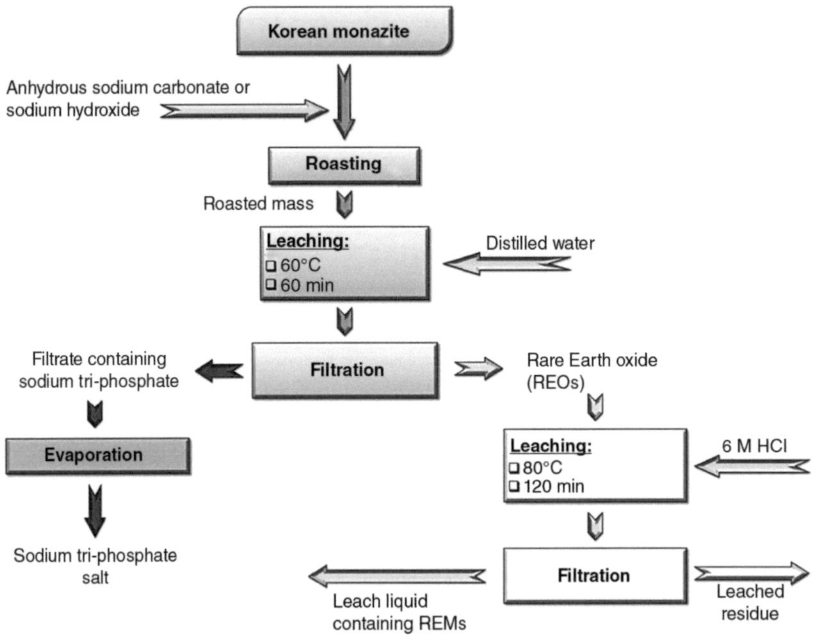

Fig. 21. Process flow sheet for the separation of phosphate and recovery of REMs from monazite [62].

Optimal condition includes 2 h of leaching by 6 M HCl at the temperature of 80°C. The pulp density should be of 30 g·dm^{-3} [62].

The optimization of leaching operation of rare-earth-bearing ores is a complex process since many attributes simultaneously affect the operation, with some of them being conflicting in nature. Therefore, a proper selection of leaching process with pertinent attributes is crucial for the user in order to maximize the percentage recovery at minimal operating costs. The methodology is proposed by Baral et al [70]. The parameters affecting the performance of leaching operation are listed in **Table 3**.

n	Attributes of leaching process		
	Physical	**Chemical**	**General**
1	Leaching temperature	Nature of ore	Source of ore
2	Leaching time	Grade of ore	Cost of ore
3	Agitation time	Choice of leaching agent	Mode and cost of transportation of ore to the plant
4	Pressure	Acidity of leaching agent	Percentage recovery/extraction of REEs
5	Partial of mesh size of the ore sample	Corrosiveness of leaching agent	Pretreatment prior to leaching

n	Attributes of leaching process		
	Physical	Chemical	General
6	Liquid-to-solid ratio	Reducing agents along with leaching agent	Cost of leaching agent
7	Size of the leaching tank	Percentage gangue/impurities in ore	Cost involved in pretreatment of ore
8	Amount of ore to be processed	Roasting of ore before leaching	Cost involved in mechanical agitation
9	Amount of leaching agent to be handled	Dissolution rate of REEs	
10	Number of leaching stages	Solubility of other elements along with REEs	

Table 3. Operating conditions affecting the performance of leaching procedure [70].

Author details

Petr Ptáček

Brno University of Technology, Czech Republic

References

[1] Abouzeid A-ZM. Physical and thermal treatment of phosphate ores — An overview. International Journal of Mineral Processing 2008;85(4) 59–84.

[2] Abouzeid A-ZM, El-Jallad IS, Orphy MK. Calcareous phosphates and their calcined products. Minerals Science and Engineering 1980;12(2) 73–83.

[3] Khoshjavan S, Rezai B. Beneficiation of refractory rock phosphate by calcination and flotation. Minerals & Metallurgical Processing 2011;28(4) 187–192.

[4] Komar KS, Carlson JT. Beneficiation of Phosphate Ore. SME, 2013. ISBN: 978-0873353915

[5] Fertilizer Manual, UN Industrial Development Organization, International Fertilizer Development Center. Springer Science & Business Media, 1998. ISBN: 978-0792350323

[6] World Phosphate Rock Reserves and Resources. International Center for Soil Fertility and Agricultural Development, 2010. ISBN: 978-0880901673

[7] Rule AR, Kirby DE, Dahlin DC. Recent advances in beneficiation of western phosphates. Minerals Engineering 1978;30 37–44.

[8] Kent JA. Kent and Riegel's Handbook of Industrial Chemistry and Biotechnology. Springer Science & Business Media, 2010. ISBN: 978-0387278438

[9] Barker AV, Pilbeam DJ. Titul Handbook of Plant Nutrition. Books in Soils, Plants, and the Environment — Volume 117. CRC Press, 2014. ISBN: 978-1420014877

[10] USGS Open-File Report 2002-156. Data Set of World Phosphate Mines, Deposits, and Occurrences — Part A. Geologic Data; Part B. Location and Mineral Economic Data http://geopubs.wr.usgs.gov/open-file/of02-156/

[11] Sung Wen-P, Kao JCM, Chen R. Environment, Energy and Sustainable Development. CRC Press, 2013. ISBN: 978-0203799017

[12] Hartman HL, Mutmansky JM. Introductory Mining Engineering. John Wiley & Sons, 2002. ISBN: 978-0471348511

[13] Chang Jen-S, Kelly A, Crowley JM. Handbook of Electrostatic Processes. CRC Press, 1995. ISBN: 978-1420066166

[14] Swain AK, Patra H, Roy GK. Mechanical Operations, 1st ed., Tata McGraw-Hill Education, 2011. ISBN: 978-0070700222

[15] Fuerstenau MC, Han KN. Principles of Mineral Processing. SME, 2003. ISBN: 978-0873351676

[16] Gharabaghi M, Irannajad M, Noaparast M. A review of the beneficiation of calcareous phosphate ores using organic acid leaching. Hydrometallurgy 2010;103 96–107.

[17] Sobhy A, Tao D. Innovative RTS technology for dry beneficiation of phosphate. Procedia Engineering 2014;83 111–121.

[18] Yavuz CT, Prakash A, Mayo JT, Colvin VL. Magnetic separations: From steel plants to biotechnology. Chemical Engineering Science 2009;64(10) 2510–2521.

[19] Baik SK, Ha DW, Kwon JM, Lee YJ, Ko RK. Magnetic force on a magnetic particle within a high gradient magnetic separator. Physica C: Superconductivity 2013;484 333–337.

[20] Watson JHP, Younas I. Superconducting discs as permanent magnets for magnetic separation. Materials Science and Engineering: B 1998;53(1-2) 220–224.

[21] Van Sciver SW. Cryogenic systems for superconducting devices. Physica C: Superconductivity 2001;354(1-4) 129–135.

[22] Ohara T, Kumakura H, Wada H. Magnetic separation using superconducting magnets. Physica C: Superconductivity 2001;357-360(2) 1272-1280.

[23] Wang Z, van Oort JM, Zou MX. Development of superconducting magnet for high-field MR systems in China. Physica C: Superconductivity 2012;482 80–86.

[24] Watson JHP. Status of superconducting magnetic separation in the minerals industry. Minerals Engineering 1994;7(5-6) 737–746.

[25] Augusto PA, Augusto P, Castelo-Grande T. Magnetic classification. Minerals Engineering 2002;15(1-2) 35–43.

[26] Augusto PA, Martins JP. A new magnetic separator and classifier: prototype design. Minerals Engineering 1999;12(7) 799–807.

[27] Singh V, Nag S, Tripathy SK. Particle flow modelling of dry induced roll magnetic separator. Powder Technology 2013;244 85–92.

[28] Bureau International des Poids et Mesures, The International System of Units (SI). 8th ed. Retrieved, 2006.

[29] Gu Z. Fine Particle Flotation for Florida Dolomitic Phosphate Pebbles. M.S. Thesis, College of Engineering and Mineral resources, West Virginia University, 2002.

[30] Gharabaghi M, Noaparast M, Irannajad M. Selective leaching kinetics of low-grade calcareous phosphate ore in acetic acid. Hydrometallurgy 2009;95(3-4) 341–345.

[31] Kogel JE. Industrial Minerals & Rocks: Commodities, Markets, and Uses. SME, 2006. ISBN: 978-0873352338

[32] Sobczak-Kupiec A, Wzorek Z. The influence of calcination parameters on free calcium oxide content in natural hydroxyapatite. Ceramics International 2012;38(1) 641–647.

[33] Vertical-shaft Limekiln Technology. Spolupracovník United Nations Centre for Human Settlements. Un-Habitat, 1993. ISBN: 978-9211312256

[34] Boateng AA. Rotary Kilns: Transport Phenomena and Transport Processes. Butterworth-Heinemann, 2011. ISBN: 978-0080557120

[35] Niessen WR. Combustion and Incineration Processes: Applications in Environmental Engineering. Environmental Science and Pollution Control Series — Volume 25. 3rd ed., CRC Press, 2002. ISBN: 978-0203908365

[36] Seetharaman S. Treatise on Process Metallurgy, Volume 3: Industrial Processes. Newnes, 2013. ISBN: 978-0080969893

[37] Sell NJ. Industrial Pollution Control: Issues and Techniques Environmental Engineering Series. John Wiley & Sons, 1992. ISBN: 978-0471284192

[38] Jenkins B, Mullinger P. Industrial and Process Furnaces: Principles, Design and Operation. 2nd ed., Butterworth-Heinemann, 2013. ISBN: 978-0080993782

[39] Plumpton AJ. Production and Processing of Fine Particles: Proceedings of the International Symposium on the Production and Processing of Fine Particles, Montreal, August 28-31, 1988. Elsevier, 2013. ISBN: 978-1483286907

[40] Nguyen AV. Froth Flotation. Reference Module in Chemistry, Molecular Sciences and Chemical Engineering, 2013. DOI: 10.1016/B978-0-12-409547-2.04401-2

[41] Bulatovic SM. Handbook of Flotation Reagents: Chemistry, Theory and Practice. Chapter 26 — Flotation of Phosphate Ore. 1st ed., Elsevier, 2015. ISBN: 978-0444530820

[42] Feng D, Aldrich C. Influence of operating parameters on the flotation of apatite. Minerals Engineering 2004;17(3) 453–455.

[43] Oliveira MS, Queiroz GM, Guimarães RC, Ataíde CH, Barrozo MAS. Selectivity in phosphate column flotation. Minerals Engineering 2007;20(2) 197–199.

[44] Yekeler M, Sönmez I. Effect of the hydrophobic fraction and particle size in the collectorless column flotation kinetics. Colloids and Surfaces A Physicochemical and Engineering Aspects 1997;121(1) 9–13.

[45] McKetta JJ Jr. Unit Operations Handbook: Volume 2 (in Two Volumes). CRC Press, 1992. ISBN: 978-0824786700

[46] dos Santos MA, Santana RC, Capponi F, Ataíde CH, Barrozo MAS. Effect of ionic species on the performance of apatite flotation. Separation and Purification Technology 2010;76(1) 15–20.

[47] Chen M, Graedel TE. The potential for mining trace elements from phosphate rock. Journal of Cleaner Production 2015;91 337–346.

[48] Fuerstenau MC, Jameson GJ, Yoon Roe-H. Froth Flotation: A Century of Innovation. SME, 2007. ISBN: 978-0873352529

[49] Santana RC, Farnese AAC, Fortes MCB, Ataíde CH,. Barrozo MAS. Influence of particle size and reagent dosage on the performance of apatite flotation. Separation and Purification Technology 2008;64(1) 8–15.

[50] Sis H, Chander S. Improving froth characteristics and flotation recovery of phosphate ores with nonionic surfactants. Minerals Engineering 2003;16(7) 587–595.

[51] Liu W, Moran CJ, Vink S. A review of the effect of water quality on flotation. Minerals Engineering 2013;53 91–100.

[52] Filho LSL, Seidl PR, Correia JCG, Cerqueira LCK. Molecular modelling of reagents for flotation processes. Minerals Engineering 2000;13(14-15) 1495–1503.

[53] Guimarães RC, Araujo AC, Peres AEC. Reagents in igneous phosphate ores flotation. Minerals Engineering 2005;18(2) 199–204.

[54] Ribeiro RCC, Correia JCG, Monte MBM, Seidl PR, Mothé CG, Lima CA. Cashew gum: a new depressor for limestone in the phosphate minerals flotation. Minerals Engineering 2003;16(9) 873–875.

[55] Zhong K, Vasudevan TV, Somasundaran P. Floatability of apatites of different type and origin: role of surface area and porosity. International Journal of Mineral Processing 1993;38(3-4) 177–188.

[56] Sis H, Chander S. Reagents used in the flotation of phosphate ores: a critical review. Minerals Engineering 2003;16(7) 577–585.

[57] Smith RW, Miettinen M. Microorganisms in flotation and flocculation: Future technology or laboratory curiosity? Minerals Engineering 2006;19(6-8) 548–553.

[58] Zheng X, Arps PJ, Smith RW. Adsorption of *Bacillus subtilis* to minerals: effect on the flotation of dolomite and apatite. Process Metallurgy 1999;9 127–136.

[59] Balachandran G. Case Study 1 — Extraction of Rare Earths for Advanced Applications Treatise on Process Metallurgy 2014; 1291-1340.

[60] Laveuf C, Cornu S. A review on the potentiality of rare earth elements to trace pedogenetic processes. Geoderma 2009;154(1-2) 1–12.

[61] Bandara AMTS, Senanayake G. Leachability of rare-earth, calcium and minor metal ions from natural fluorapatite in perchloric, hydrochloric, nitric and phosphoric acid solutions: Effect of proton activity and anion participation. Hydrometallurgy 2015;153 179–189.

[62] Kumari A, Panda R, Jha MK, Lee JY, J. Kumar R, Kumar V. Thermal treatment for the separation of phosphate and recovery of rare earth metals (REMs) from Korean monazite. Journal of Industrial and Engineering Chemistry 2015;21 696–703.

[63] Foley NK, De Vivo B, Salminen R. Rare earth elements: The role of geology, exploration, and analytical geochemistry in ensuring diverse sources of supply and a globally sustainable resource. Journal of Geochemical Exploration 2013;133 1–5.

[64] Golev A, Scott M, Erskine PD, Ali SH, Ballantyne GR. Rare earths supply chains: Current status, constraints and opportunities. Resources Policy 2014;41 52–59.

[65] Goyne KW, Brantley SL, Chorover J. Rare earth element release from phosphate minerals in the presence of organic acids. Chemical Geology 2010;278(1-2) 1–14.

[66] Radhika S, Nagaphani Kumar B, Lakshmi Kantam M, Ramachandra Reddy B. Liquid-liquid extraction and separation possibilities of heavy and light rare-earths from phosphoric acid solutions with acidic organophosphorus reagents. Separation and Purification Technology 2010;75(3) 295–302.

[67] Nagaphani Kumar B, Radhika S, Ramachandra Reddy B. Solid-liquid extraction of heavy rare-earths from phosphoric acid solutions using Tulsion CH-96 and T-PAR resins. Chemical Engineering Journal 2010;160(1) 138–144.

[68] Bunuş F, Dumitrescu R. Simultaneous extraction of rare earth elements and uranium from phosphoric acid. Hydrometallurgy 1992;28(3) 331–338.

[69] Wang L, Long Z, Huang X, Yu Y, Cui D, Zhang G. Recovery of rare earths from wet-process phosphoric acid. Hydrometallurgy 2010;101(1-2) 41–47.

[70] Baral SS, Shekar KR, Sharma M, Rao PV. Optimization of leaching parameters for the extraction of rare earth metal using decision making method. Hydrometallurgy 2014;143 60–67.

Chapter 9

Utilization of Apatite Ores

Petr Ptáček

Additional information is available at the end of the chapter

http://dx.doi.org/10.5772/62215

Abstract

Phosphate rock is an important mineral commodity used in the chemical industry and production of food. The first section of ninth chapter of this book introduces utilization of apatite ores for manufacturing of phosphorus. The second part deals with production of phosphoric acid via wet and thermal process and utilization of byproducts such as phosphogypsum, phosphorous slag and ferrophosphorus. The last section of this chapter describes the methods for production of fertilizers, such as supephosphates, Thomas slag, ammonium phosphates, thermophosphates, etc., and the chapter ends with environmental demand of phosphate fertilizers.

Keywords: Apatite, Apatite Ore, Phosphorus, Phosphoric Acid, Fertilizers, Superphosphate, Thomas Slag

The group of apatite minerals was studied by the physiologists, biochemists or soil scientists as well as the geologists or mineralogists. Each group of scientists concerned with the apatite minerals see their own problems and make attempts to solve them in their own way. The soil scientists are mainly interested in the availability of P_2O_5 to plants. The physiologists focus on the prevention of dental caries or on the mechanism of various types of metal poisoning. The geologists study the mode of formation of phosphorite deposits, the chemists carry out the equilibrium studies and analytical methods and the mineralogists investigate the crystal chemistry of entire group [1].

9.1. Manufacturing of phosphorus

The original procedure for the production of elemental phosphorus, as described by R. Boyle in 1680, was based on the method of H. Brandt [2],[3],[4]. The substance glowed in

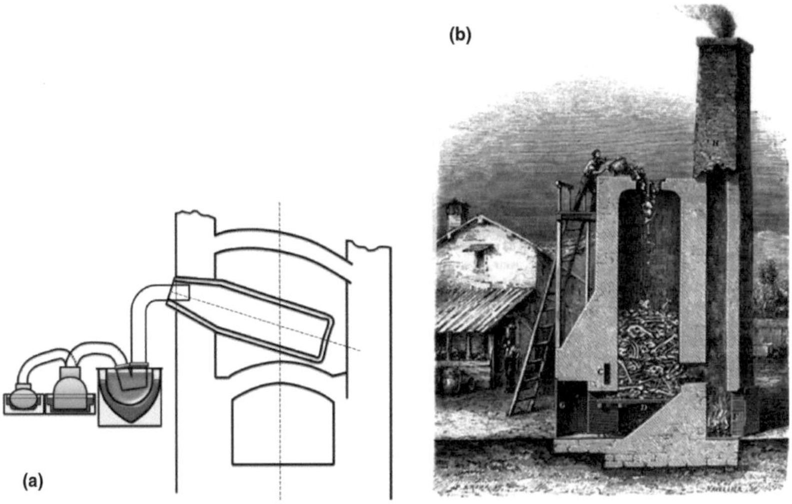

Fig. 1. Phosphorus retort condensation arrangement (Albricht & Wilson) (a) and a French bone furnace (Four pour la calcination des os) (b) [7].

the dark[1] and burst in flame into P_4O_{10} when exposed to air [2],[5]. The procedure involved the distillation of large quantity of partly digested urine to the consistency of thick syrup. Fine white sand was added and the mixture was heated in a retort at first gently to remove the volatiles and then very intensely to produce phosphorus, which was distilled over and cooled under water. When the bone ash (M.M. Coignet) was used as a raw material for the production of phosphorus, the process consisted of the treatment with sulfuric acid to produce phosphoric acid, which was then concentrated and heated with coke in a retort to produce phosphorus according to the following equations [6],[7]:

$$Ca_3(PO_4)_2 + 3\ H_2SO_4 \rightarrow 3\ CaSO_4 + 2\ H_3PO_4 \tag{1}$$

$$H_3PO_4 \rightarrow HPO_3 + H_2O \tag{2}$$

$$4\ HPO_3 + 12\ C \rightarrow P_4 + 2\ H_2 + 12\ CO \tag{3}$$

The drawing of the kiln for the calcination of bones[2] is shown in **Fig. 1(b)**.

[1] The name phosphorus (light-bearing) was often used by alchemists to name various light-bearing materials, which were devoid of the element, e.g. the Bologna phosphorus (luminescent barium sulfide), the Baldwin's phosphorus (luminescent calcium nitrate), etc. Probably the earliest prepared phosphorus salt was sodium ammonium hydrogen phosphate, which has been known since the ancient times [2].

[2] Bones became the source increasingly difficult to achieve. It is stated that the use of bones was so great in England during the eighties of 19th century that the battlefields on the continent of Europe were plundered to supply Great Britain's demand for phosphates [7].

The discovery of elemental (white) phosphorus[3] was soon followed by the characterization of its combustion products, phosphorus pentaoxide,[4] and in 1694, R. Boyle prepared phosphoric acid by dissolving P_2O_5 in water. Phosphorus was found in plants by B. Albino in 1688, and the element was detected in human brain tissue by J.T. Hensing in 1719. In about 1770, phosphorus was recognized as an essential ingredient of animal bones and teeth by C.W. Sheele, when he prepared the element from bone ash, carbon and sand. By 1779, the first phosphorus-containing mineral pyromorphyte (**Section 1.6.4**) was identified by J.T. Hensing [2].

The first organic phosphorus compound to be identified was probably lecithin, isolated from barin fat in 1811 by Vauquelin and characterized as a phosphorus-containing lipid by Gobley in 1850. The earliest laboratory synthesis of an organic phosphorus compound was reported by Lassaigne, who in 1820 obtained crude alkyl phosphates by the reaction of alcohols with phosphoric acid. The discovery of the first metal-phosphine complex by Rose in 1847 was followed by Cahors and Hoffmann who prepared the first organic complex with metal-phosphorus bond. It can be considered as the beginning of the metallophosphorus chemistry [2].

The following major classes of phosphorus compounds (**Fig. 2**) are recognized [2]:

1. **Oxyphosphorus compounds**, which contain covalent P-O linkages.

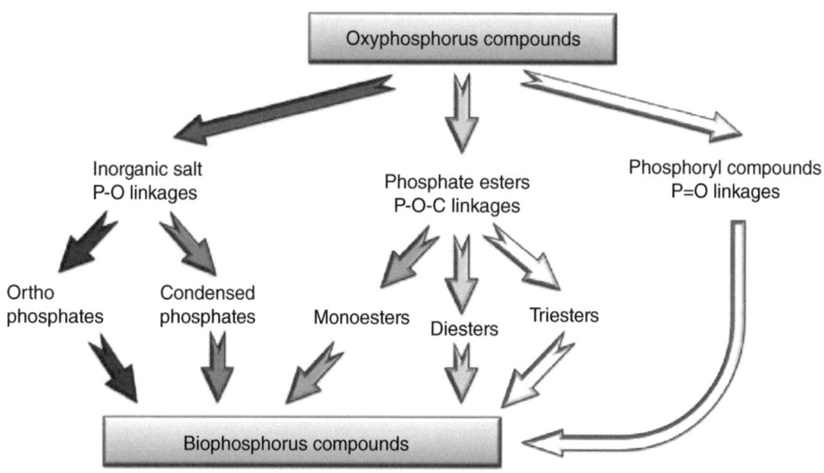

Fig. 2. The major division of oxyphosphorus compounds [2]:

---

[3] Three forms of phosphorus include (a) white phosphorus (the most reactive form) with at least two crystalline forms, (b) red phosphorus (discovered in 1847 by von Schrotter) and (c) black phosphorus (discovered in 1914 by Bridgeman) [5].

[4] Three forms of P_4O_{10} are recognized [5]: (a) **H-form**, (b) **O-form** and (c) **O′-form**. The H-form is rhombohedral (R3c, a = 7.43 Å, α = 87.0° and Z = 2). In the hexagonal setting, Z = 6, a = 10.31 and c = 13.3 Å. The O-form is orthorhombic of the space group Fdd2 and the cell parameters a =13.3, b = 8.14 and c = 5.26 Å and Z = 8. The O′-form is orthorhombic, the space group Pnam, a = 9.23, b = 7.18 and c = 4.94 Å and Z =2.

2. **Carbophosphorus** (**organophosphorus**) **compounds**, which contain the P-C linkages including carbophosphanes (organophosphanes, P-C), carbophosphenes (organophosphens, P=C) and carbophosphynes (organophosphynes, P≡C).

3. **Azophosphorus compounds**, which contain the P-N linkages including azophosphanes (phosphazanes, P-N), azophosphenes (phosphazenes, P=M) and azophosphynes (phosphazynes, P≡N).

4. **Metalophosphorus compounds**, which contain the P-metal linkages including metallophosphanes (P-M), metallophosphenes (P=M) and metallophosphynes (P≡M).

These compounds vary greatly in their abundance and importance. The compounds with the P-O linkages dominate the phosphorus chemistry. The most important types of oxyphosphorus compounds are phosphates (salts with PO_4^{3-} anion), phosphate esters (P-O-C linkage) and phosphoric compounds (P=O linkage) [2].

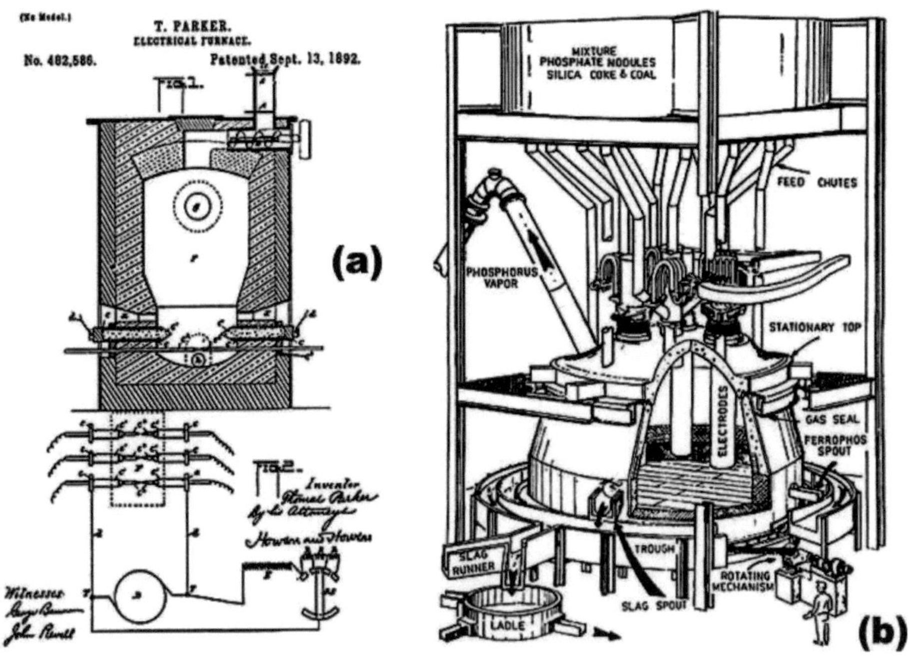

Fig. 3. Parker's electric furnace (a) [7] and the scheme of phosphorus furnace (b) [11].

The production of phosphorus by heating the mixture of silica, coke and phosphate rock was first proposed by AUBERTON and BOBLIQUE in 1867, and the use of electric furnace for the heating of the mixture was proposed in 1888 by the patents by J.B. READMAN [8] and T. PARKER and A.E. ROBINSON [9]. **Fig. 3**(**a**) shows Parker's electric furnace from his later patent in 1892 [10].

The basic method for the production of elemental phosphorus today (**Fig. 4**), except for engineering improvements, is essentially that of the method originated by READMAN. Lower-grade phosphate sand contaminated with clay is concentrated by washing to an average content of P_2O_5 in the range from 28 to 30%. Higher-grade sand with the content of P_2O_5 of 26 – 28% is used directly in the combination with washed sand. These fine phosphatic grains are compacted or "nodulized" and then sintered into fused agglomerates. The nodules are then mixed with silica and coke particles of similar size. Such mixture is called the "furnace burden". A typical furnace burden has the SiO_2/CaO ratio of 0.8 to 1.2 and the P_2O_5/C ratio of 2.3 to 2.6 [6].

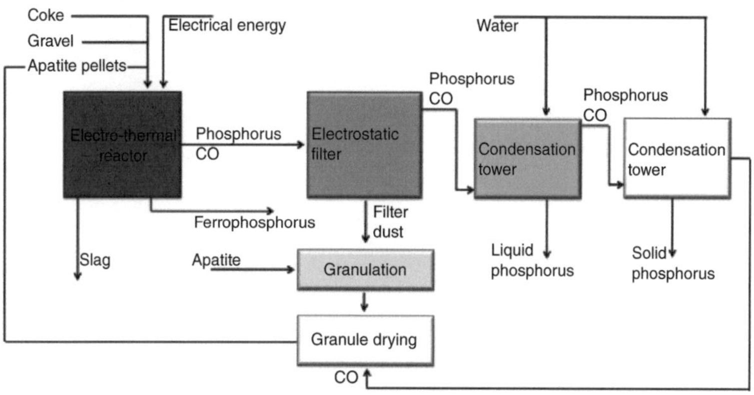

Fig. 4. The flow sheet for the electrothermal manufacture of elemental phosphorus [12].

Modern reduction plants for the manufacture of elemental phosphorus (**Fig. 4**) has three main units [12]:

i. The electrothermal reactor (furnace);

ii. The gas purifier (electrostatic precipitator);

iii. The phosphorus condenser.

The electric furnace for the reaction **Fig. 3(b)** is lined on the bottom and sides with thick carbon blocks, while the top is lined with refractory bricks. The furnace is heated by symmetrically positioned carbon electrodes. They go through almost airtight seal in the furnace roof into the reaction zone and are supported in a manner that enables them to move vertically depending on the power requirements of fluctuating furnace conditions. The temperature in the reaction zone ranges from 1400 to 1500°C. Under this condition, phosphorus vaporizes and rises with carbon monoxide and entrained dust through the space between the furnace burden particles. The mixture next passes through an electrostatic precipitator where the most of the dust is removed. The phosphorus vapor is then cooled, condensed and collected under water [6],[11], [12],[13].

The condensation of white phosphorus[5] is carried out in two stages. In the first stage, the condensation tower water of 50 to 60°C is sprayed from the top to meet the phosphorus vapor

[5] Red phosphorus is produced in much less quantities than white phosphorus. The conversion of white phosphorus to red phosphorus is an exothermic reaction producing red phosphorus as a solid product. Since the conduction of heat from the reaction is difficult, the conversion is carried out semi-continuously in a ball mill at 350°C. White phosphorus is fed into the mill with such rate that the reaction temperature is kept constant [12].

being transported countercurrently from below, whereupon phosphorus condenses as a liquid. Solid phosphorus is formed in the second condensation tower, which uses water with the temperature of 10 to 25°C. CO gas is recovered for the use as a fuel in the sintering operation. The by-product calcium silicate is drawn off from the bottom of the furnace as molten liquid. Iron phosphide, "ferrophos" or ferrophosphorus (**Section 9.2.7**) formed from the iron impurities present in the phosphate ore is also drawn off as a melt [6],[12].

The mechanism of phosphate reduction is complex and there is no complete agreement among the exact path of each step in the reaction sequence. The overall reaction can be presented by the following equation [6]:

$$Ca_3(PO_4)_2 + 3\ SiO_2 + 5\ C \rightarrow 3\ CaSiO_3 + 5\ CO + P_2 \quad (4)$$

Proposed mechanisms for the reactions are:

1. Phosphide mechanism [14]:

$$5\ Ca_3(PO_4)_2 + 40\ C \rightarrow 5\ Ca_3P_2 + 40\ CO \quad (5)$$

$$3\ Ca_3(PO_4)_2 + 5\ Ca_3P_2 \rightarrow 24\ CaO + 8\ P_2 \quad (6)$$

2. Acid displacement mechanism:

$$2\ Ca_3(PO_4)_2 + 6\ SiO_2 \rightarrow 6\ CaSiO_3 + P_4O_{10} \quad (7)$$

$$P_4O_{10} + 10\ C \rightarrow 2\ P_2 + 10\ CO \quad (8)$$

3. CO reduction mechanism:

$$Ca_3(PO_4)_2 + 5\ CO \rightarrow 3\ CaO + 5\ CO_2 + P_2 \quad (9)$$

$$5\ CO_2 + 5\ C \rightarrow 10\ CO \quad (10)$$

The phosphide theory is generally considered as unlikely due to the thermodynamic reasons, but the acid displacement mechanism has considerable experimental support [6].

http://dx.doi.org/10.5772/62215

9.2. Properties and manufacturing of phosphoric acids

9.2.1. Orthophosphoric acid

Orthophosphoric acid[6] [15] is a colorless polyprotic weak acid with stepwise dissociation that involves three equilibrium reactions described by the following equations [2],[16], [17],[18]:

$$\begin{aligned} &H_3PO_4 \overset{k_1}{\leftrightarrow} H^+ + H_2PO_4^- \qquad K_1 = 7.52 \cdot 10^{-3} \\ &2.15 \le pH \le 7.2 \end{aligned} \tag{11}$$

$$\begin{aligned} &H_2PO_4^- \overset{k_2}{\leftrightarrow} H^+ + HPO_4^{2-} \qquad K_2 = 6.23 \cdot 10^{-8} \\ &7.2 \le pH \le 12.37 \end{aligned} \tag{12}$$

$$\begin{aligned} &HPO_4^{2-} \overset{k_3}{\leftrightarrow} H^+ + PO_4^{3-} \qquad K_3 = 2.2 \cdot 10^{-13} \\ &pH \ge 12.37 \end{aligned} \tag{13}$$

However, due to very low values of the equilibrium constants associated with reactions (**Eqs. 12** and **13**), orthophosphoric acid has only one strongly ionizing hydrogen atom and only the first acidic dissociation (**Eq. 11**) has significant effect on the system composition. HPO_4^{2-} and PO_4^{3-} ionic species are present at significant concentrations in very highly diluted solutions only [2].

Orthophosphoric acid gives three series of salts (**Eqs. 11 – 13**):

1. Normal phosphates with trivalent PO_4^{3-} anion, e.g. trisodium phosphate dodecahydrate ($Na_3PO_4 \cdot 12H_2O$);
2. Hydrogen phosphates with divalent HPO_4^{2-} anion, e.g. disodium hydrogen phosphate dedecahydrate ($Na_2HPO_4 \cdot 12H_2O$);
3. Dihydrogen phosphates with monovalent $H_2PO_4^-$ anion, e.g. sodium dihydrogen phosphate monohydrate ($NaH_2PO_4 \cdot H_2O$).

When heated, orthophosphoric acid condenses to pyrophosphoric acid (250°C) and then to metaphosphoric acid (316°C) [19]:

$$2\ H_3PO_4 \overset{\ge 250°C}{\rightarrow} H_4P_2O_7 + H_2O \tag{14}$$

[6] Orthophosphoric acid (H_3PO_4, phosphoric(V) acid) is often termed simply as phosphoric acid, but other phosphoric acids exist: metaphosphoric acid (HPO_3) and pyrophosphoric acid ($H_4P_2O_7$) [15].

$$H_4P_2O_7 \xrightarrow{\geq 316°C} 2\ HPO_3 + H_2O \qquad (15)$$

Hence, from the PO_4 building block, a long series of two- and three-dimensional phosphates originates through P-O-P linkages. There is a continuous series of phosphates from orthophosphate (one P atom) to phosphorus pentaoxide (P_2O_5) followed by homologous series of straight-chained, branched and cyclic phosphates. The members of the series having one atom of phosphorus are called orthophosphates; the dimers (two P atoms) are pyrophosphates followed by the triphosphates (tripolyphosphates, three P atoms) and by the tetraphosphates (four P atoms). The members of homologous series ($H_2P_nO_{3n+1}$) with 5 – 15 P atoms are referred to as oligophosphates. In general, any phosphate having three or more P atoms is considered to be polyphosphate. Metaphosphates are cyclic with general formula $(HPO_3)_n$ or $(P_nO_{3n})^-$. The phosphates with three-dimensional structure are termed as ultraphosphates with the composition given by general formula: P_nO_{3n+x}, where $1 \geq x \leq n/2$ [20].

In the crystalline state, phosphoric acid exists as prismatic crystal of H_3PO_4 and as hemihydrate ($H_3PO_4 \cdot 1/2H_2O$, **Fig. 5**), but $10H_3PO_4 \cdot H_2O$ hydrate was also reported. The crystals of H_3PO_4 have layered structure, where each molecule is connected to six others via hydrogen bonds. Heated to temperatures higher than the melting point (42.35°C), phosphoric acid slowly dehydrates according to the equation:

$$2\ H_3PO_4 \leftrightarrow H_4P_2O_7 + H_2O \qquad (16)$$

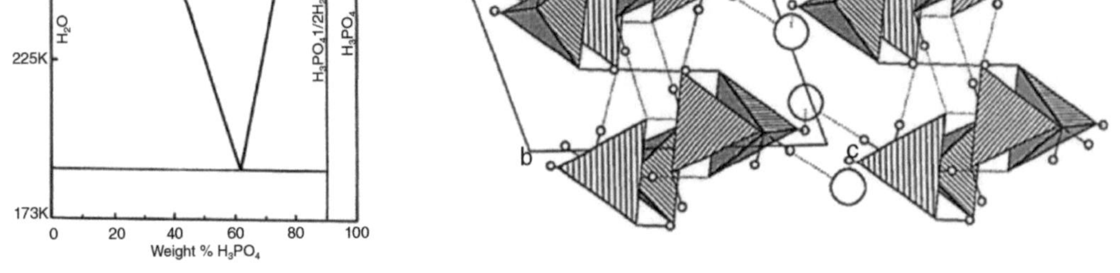

Fig. 5. The H_2O-H_3PO_4 phase diagram (a) and the structure of $H_3PO_4 \cdot 1/2H_2O$ (b) viewed along the b-direction: large and small empty cycles represent water and hydrogen atoms, respectively. PO_4^{3-} tetrahedra are given in polyhedral representation [5].

The crystal structure of H_3PO_4 (**Fig. 6**) is monoclinic with the space group $P2_1/C$ and the cell parameters ***a*** = 5.80 Å, ***b*** = 4.85 Å, ***c*** = 11.62 Å, ***β*** = 95.20° and **Z** = 4. $H_3PO_4 \cdot 1/2H_2O$ is monoclinic (space group $P2_1/A$) with the cell parameters ***a*** = 7.92 Å, ***b*** = 12.99 Å, ***c*** = 7.47 Å, ***β*** =109.9° and **Z** = 8 [5],[21].

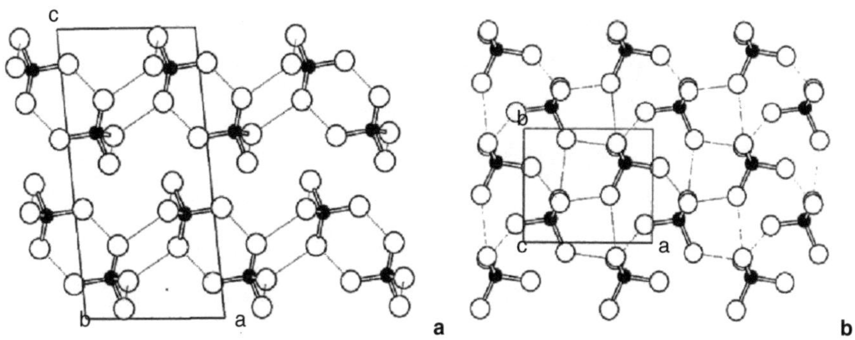

Fig. 6. The projection of the structure of H_3PO_4 along the b-axis (a) and c-axis (b). H-bonds are figured by dotted lines. This view induces the feeling that some H-bonds connect two O atoms of the same phosphoric tetrahedron, but they are, in fact, established between two superimposed tetrahedra [5].

Furthermore, polyphosphoric acids and deutronic phosphoric acid are known. Deutrophosphoric acid (D_3PO_4) can be prepared by dissolving phosphorus pentaoxide in D_2O or by the hydrolysis of $POCl_3$ with D_2O. This acid has slightly higher melting point, density and viscosity but lower electrical conductivity than the hydrogen analogue [2],[5].

9.2.2. Properties of phosphoric acid

The solutions of phosphoric acid show unique phenomena of sonoluminescence (SL) in which the nonlinear oscillations of a single bubble under the influence of sufficiently strong sound field lead to a violent collapse of the bubble and to the production of very brief light pulse with the duration of several picoseconds at the end of the bubble collapse. The standard spectrum of SL radiation is a featureless continuum spectrum, ranging from 200 to 800 nm, which is well fitted by the black-body radiation with the temperature of 6000 – 20,000 K. The temperature inside the bubble has been predicted to be even much higher, as the spectrum of radiation from the regions close to the bubble center is fitted by thermal bremsstrahlung radiation from the plasma with temperature more than 10^6 K. The experiments show that the SL emission from sulfuric acid (**Fig. 7**) is about 2700 times greater than the brightest SL in water. Also, the SL radiation from phosphoric acid can be up to four orders of magnitude brighter than SL from water [22],[23],[24],[25].

Phosphoric acid is a key material for the manufacture of detergents, food products and alimentary supplies for cattle, toothpastes and fertilizers such as monoammonium phosphate (MAP, $(NH_4)H_2PO_4$, ammonium dihydrogen phosphate), diammonium phosphate (DAP, $(NH_4)_2HPO_4$, diammonium hydrogen phosphate) and triple superphosphate (TSP) [26], [27], which are described in **Section 9.3**.

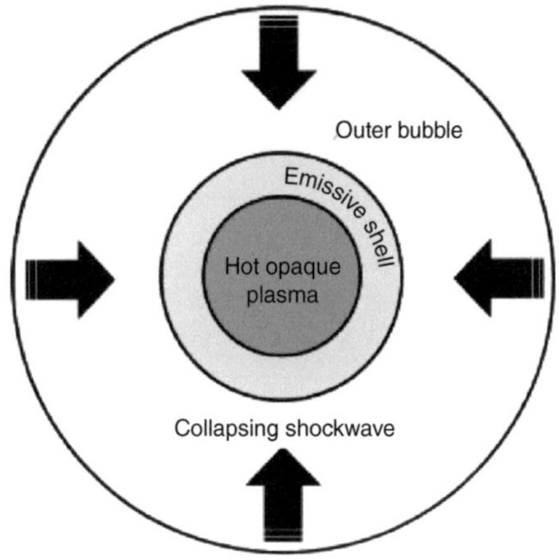

Fig. 7. Sonoluminescence occurs as the bubble collapse under some specific conditions including very low vapor pressure liquids [25].

Another applications of phosphoric acid include the treatment of surface of metals (**Section 10.8**), the utilization in dentistry [28],[29],[30],[31],[37],[38] (described in **Section 10.1.2**), phosphate binders [32], geopolymers [33], phosphoric acid fuel cells [34],[35],[36], gel-based electrolytes [39] and solid membranes [40],[41],[42] for fuel cells, the activation of carbon adsorbents [43],[44] and catalysts [45],[46], the modification of zeolites [47], catalytic decomposition of H_2O_2 [48] and organic syntheses, e.g. esterification [49],[50]. Direct applications of phosphoric acid are shown in **Fig. 8**.

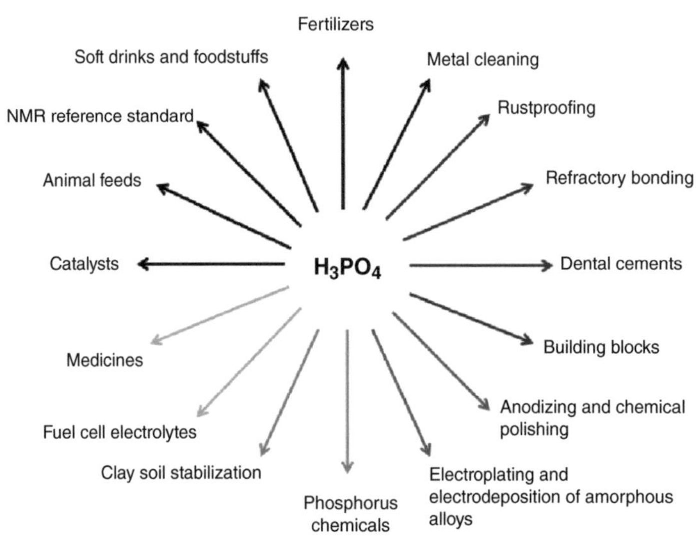

Fig. 8. Direct applications of orthophosphoric acid [2].

9.2.3. Manufacturing of phosphoric acids

Globally, about 81% of phosphate rock is used for the production of phosphoric acid [51],[52]. There are two processes (**Fig. 9**) described below used for the preparation of phosphoric acid [7],[26],[53],[54],[56],[57]:

i. **Wet process of phosphoric acid (WPA);**

ii. **Thermal process phosphoric acid (TPA).**

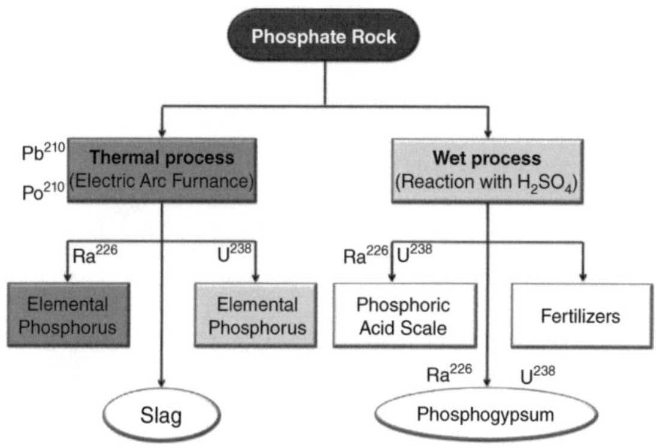

Fig. 9. The scheme of phosphate processes [57].

Since the concentration and the purity grade of phosphoric acid depend on the purpose, the content of waterless H_3PO_4 in the product is usually expressed as P_4O_{10}, where waterless $H_3PO_4 = 1.38 \times P_4O_{10}$. For example, the product of TPA with the content of 66.33% P_4O_{10} contains 66.33 × 1.38 = 91.54% of waterless H_3PO_4. With increasing content of P_4O_{10} in the system, pyrophosphoric acid (diphosphoric acid, $H_4P_2O_4$) becomes the dominant species.

9.2.4. Wet process

Several companies began developing the hemihydrate technologies during the 1960s and full-scale plants began to appear during the 1970s [53]. The wet process (**Fig. 10**) is named as wet because the concentrated solution of sulfuric acid (93%) was used to digest the apatite ore [5], [26],[58]:

$$\begin{aligned} &Ca_{10}(PO_4)_6(OH)_2 + 10\ H_2SO_4 \rightarrow 10\ CaSO_4(s) \\ &+6\ H_3PO_4 + 2\ H_2O \end{aligned} \tag{17}$$

Since the ore is invariably contaminated by fluorapatite ($Ca_{10}(PO_4)_6F_2$) and calcium carbonate ($CaCO_3$), hydrofluoric acid is formed during the dissolution process:

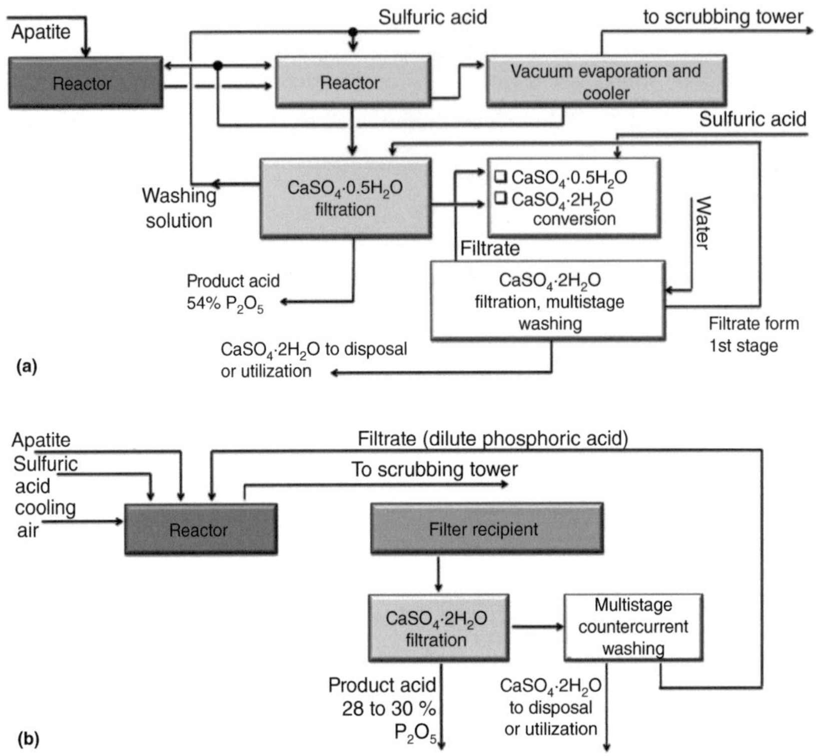

Fig. 10. The flow sheet for the production of orthophosphoric acid via the fission hemihydrate process (a) and the dihydrate process (b) [12].

$$Ca_{10}(PO_4)_6F_2 + 10\ H_2SO_4 \rightarrow 10\ CaSO_4(s) + 6\ H_3PO_4 + 2\ HF \tag{18}$$

or

$$Ca_{10}(PO_4)_6F_2 + 10\ H_2SO_4 + 20\ H_2O \rightarrow 10\ CaSO_4 \cdot 2H_2O(s) + 6\ H_3PO_4 + 2\ HF \tag{19}$$

$$CaCO_3 + H_2SO_4 \rightarrow CaSO_4 + CO_2(g) + H_2O \tag{20}$$

The dissolution of silicate minerals in the ore by HF leads to the formation of fluorosilicates, including volatile silicon tetrafluoride (SiF_4) and hexafluorsilicic acid (H_2SiF_6) or its salts (Na_2SiF_6).

According to the kind of produced hydrate of calcium sulfate (**Fig. 11**), the wet process is further divided to:

http://dx.doi.org/10.5772/62215

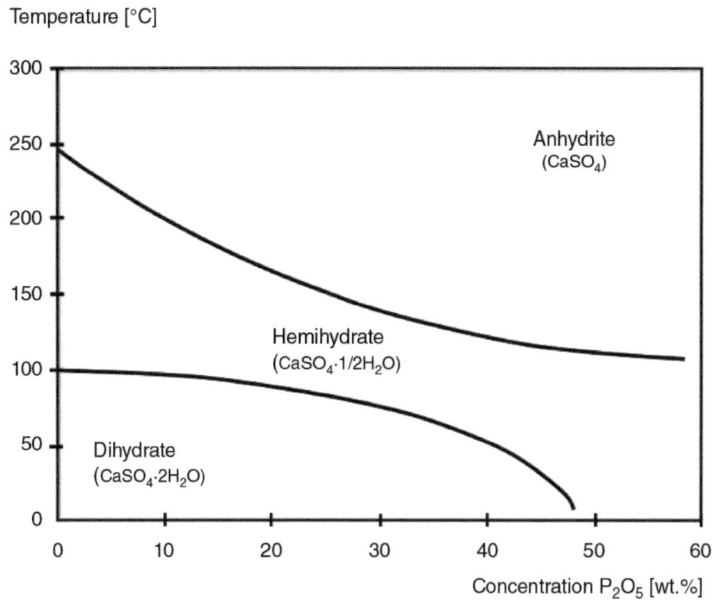

Fig. 11. The influence of reaction conditions on the crystallization of calcium sulfate [53].

a. **Hemihydrate wet process** in which $CaSO_4 \cdot ½H_2O$ (HH) is formed in the temperature range from 360 to 380 K;

b. **Dihydrate wet process** in which $CaSO_4 \cdot 2H_2O$ (DH) is formed in the temperature range from 340 to 350 K;

c. **Hemidihydrate wet process** (HDH) where HH crystals are converted to DH and next filtered out to recover weak phosphoric acid from the filtration cake (**Fig. 12**).

Calcium sulfate and other insoluble impurities such as silica were filtered out and the vacuum distillation process was used to increase the concentration of phosphoric acid[7] to 56%. The major part (about 85%) of phosphorous fertilizers is produced by wet process [26]. Since phosphate ores are a potential resource of rare-earth elements (REE) as well, the recovery of rare earths during the wet processing of phosphoric acid is investigated [59],[60],[61],[62],[63].

The crucial step in the decomposition of apatite is the formation of calcium sulfate. Its properties, in particular, its ability to be filtered, are very important, e.g. for the throughput optimization. The incorporation of phosphate in the crystal lattice of calcium sulfate reduces the phosphate yield and can render the calcium sulfate unusable in the building industry [12].

[7] The Karl Fisher titration was used to determine the amount of water in phosphoric acid [64].

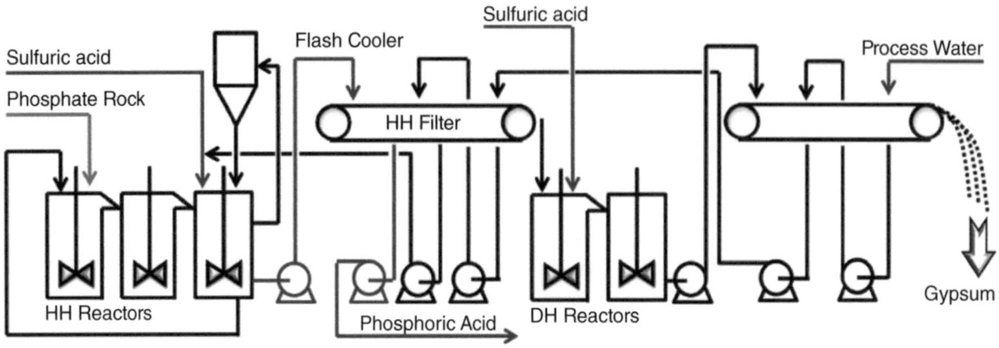

Fig. 12. Yara HDH process [53].

Yara, as Fisons Fertilizers, started its research and development to find an alternative to the traditional dihydrate process route (**Fig. 12**) during the late 1950s and early 1960s at its R&D establishment at Levington, UK. The comprehensive laboratory-scale work established that the hemihydrate process route was a feasible alternative. Subsequent pilot plant testing was employed at the Fisons (James Fison) Kings Lynn factory to develop the process on a larger scale and to confirm earlier findings that the hemihydrate process could be adopted successfully [53].

When the Kings Lynn factory closed, all research and development was concentrated at the R&D center in Levington, UK. A new pilot plant facility was constructed, this time using reactors made from high-grade stainless steel. The size of the equipment was somewhat smaller because all raw materials had to be specially delivered and produced acid and the by-product phosphogypsum dispatched to the nearest production unit for disposal. During the 1980s and 1990s, the research facilities were subsequently based in Sweden, the Netherlands and Norway. Currently, all phosphoric acid research and development is carried out at Yara's phosphate R&D center in Siilinjarvi, Finland [53].

However, the organic matter (OM) contained in acid may interact with organic solvents to form stable foams, preventing the phase settling, or simply by forming cross-layers and organic phases and denaturing part of the solvent. Hence, the removal of these organics seems to be an important step for the production of decontaminated phosphoric acid. The utilization of activated bentonite [65],[66], activated carbon and coal [67] for the adsorption of organic matter was investigated.

9.2.5. Thermal process

The **thermal (dry) process** or the electric furnace process is based on the oxidation of phosphorus in the furnace heated to temperatures from 1800 to 3000 K (**Eq. 21**). If wet (undried) air is used for the oxidation, the gaseous product of this reaction is led directly through the hydration tower, where P_4O_{10} gas is adsorbed and hydrated by water to phosphoric acid (**Eq. 22**) [55]:

http://dx.doi.org/10.5772/62215

$$P_4(\ell) + 5\ O_2(g) \rightarrow P_4O_{10}(g) \tag{21}$$

$$P_4O_{10}(g) + 6\ H_2O(\ell) \rightarrow 4\ H_3PO_4(aq) \tag{22}$$

The product of phosphorus oxidized by dry air, P_2O_5, was condensed as white powder and next hydrated to phosphoric acid.

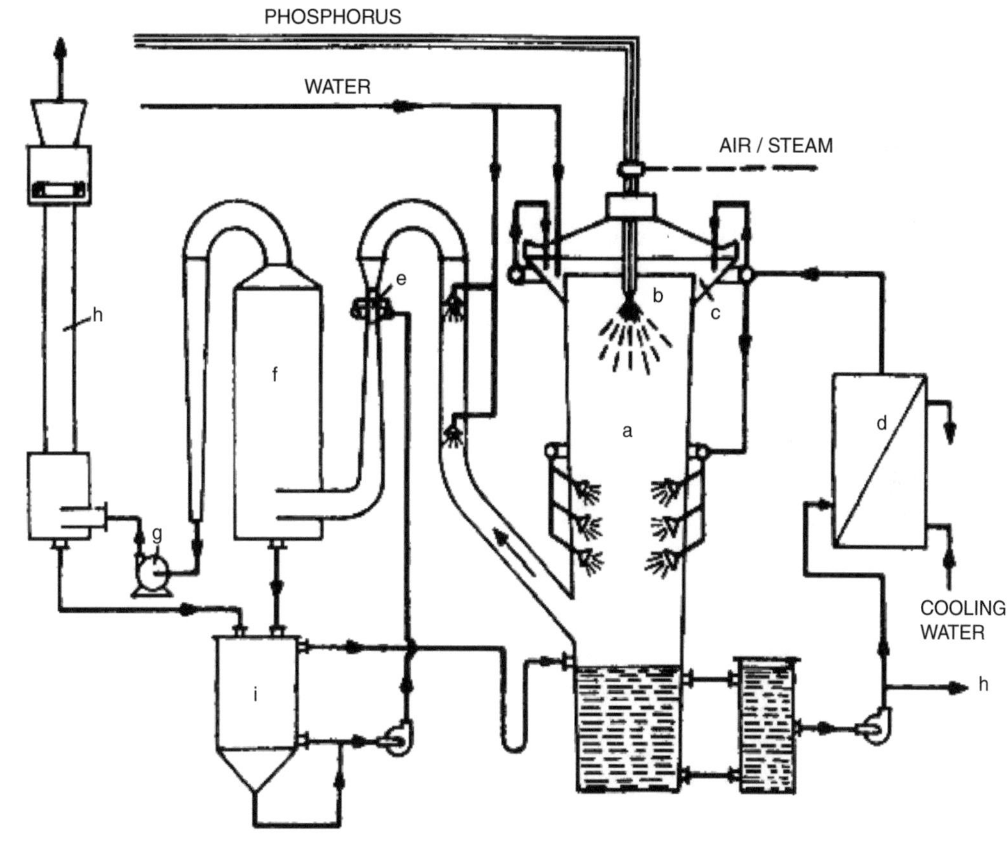

Fig. 13. The scheme of the plant for the production of thermal phosphoric acid [11].

These two operations can be carried out in the apparatus connected in series (**Fig. 13**), but the hydration takes place in large extent directly in the combustion tower. The combustion nozzle is arranged centrally at the head of the tower and directed downwards. It is a nozzle for two components. Phosphorus is atomized by compressed air or vapor into fine particles, which spontaneously ignite immediately upon leaving the nozzle. Phosphorus is fed in the liquid state through pipes by means of pumps or pressure vessels. The combustion aid is either forced through the apparatus by means of suction fan arranged at the outlet of final purification device. If complete oxidation to P_2O_5 is assured, the waste gases must contain 4 to 5% of oxygen. The nebulous P_2O_5 is conducted from the combustion chamber in a tower like so-called hydrator. The gas entering at the bottom is washed with water or diluted phosphoric acid sprayed in through several rows of nozzles [11].

Small amount of pure phosphoric acid can be prepared by heating white phosphor with diluted nitric acid (using concentrated HNO_3 turns the course of reaction to explosive) or by the oxidation of red phosphor by concentrated nitric acid:

$$P + 3HNO_3 \rightarrow H_3PO_4 + NO + 2\ NO_2 \tag{23}$$

The oxidative reaction is catalyzed by the trace of iodine anions (I^-). The evaporation of solution on platinum dish (thickening) leads to the viscous (syrupy) liquid from which concentrated H_3PO_4 precipitates [55].

9.2.6. Phosphogypsum

Phosphogypsum (PG) is a by-product produced by phosphate fertilizer industry during the production of phosphoric acid (**Eq. 19**) from phosphate rocks (**Section 9.2.4**). About 4.5 – 5 kg of phosphogypsum are produced for every kilogram of P_2O_5 manufactured [68]. PG is mainly composed of gypsum, but it also contains high level of impurities, which include naturally occurring radionuclides, metals and other trace elements, the quantity varying with element and the production process. Major PG reuse includes the production of cement, china and crystallite glass as well as soil amendments in agriculture (PG appears to be good source of S and Ca for crops [69]) without a consideration of element recovery, but even these latter reuses are limited due to the radioactivity within PG. Presently, PG is mainly stockpiled without any treatment. It can, however, be discharged into aquatic environments and pose a radioactive threat to ecosystems [51],[57],[70],[71][72],[73],[74],[75], [76],[77].

Potential utilization of phosphogypsum (**Fig. 14**), the by-product from fertilizer industries, as a bitumen modifier for paving industry was reported by Cuadri et al [68]. It was found that when activated with small quantity of sulfuric acid (0.5 wt.%), the addition of 10 wt.% phosphogypsum leads to a notable improvement in the rheological response of resulting material at high temperatures. On the contrary, poor level of modification was noticed when in such formulation phosphogypsum was substituted by the same amount of commercial gypsum.

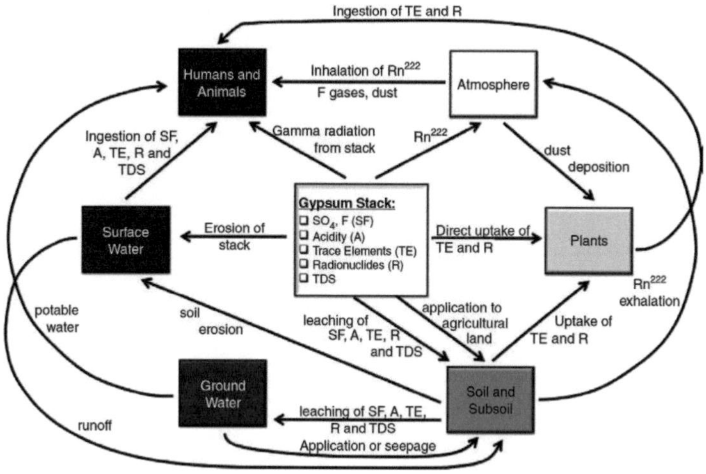

Fig. 14. Main environmental concerns due to the storage and agricultural use of phosphogypsum [77].

The reduction in the concentration of radionuclides content from PG was based on leaching of ^{226}Ra, ^{210}Pb, ^{238}U and ^{40}K using tributyl phosphate (TBP) and trioctyl phosphine oxide (TOPO) in kerosene [70].

9.2.7. Phosphorus slag and ferrophosphorus

Phosphorus slag[8] and ferrophosphorus are the by-products of the production of elemental phosphorus. Phosphorus slag has applications similar to those of PG, but more than 80% of phosphorus slag is stored near the factory instead of being recycled. At present, the recycling of phosphorus slag appears not being considered for its element recovery potential. Ferrophosphorus containing roughly 75% Fe and 25% P is usually sold as an additive for the steel industry after being crushed and screened [51],[78]. Ferrophosphorus is used in the steel industry to prevent the steel plates from sticking together during the pack annealing [79]. Ferrophosphorus contains FeP, Fe_2P and perhaps little amount of Fe_3P [2]. The utilization of ferrophosphorus as a heavyweight additive for API oil well cements was investigated by Bensted [80]. Ferrophosphorus is used for steelmaking, providing a convenient source for phosphorus addition to alloy steels [13].

9.3. Production of phosphatic fertilizers

Phosphorus is essential to plant and animal life.[9] It provides the matter for the skeletal bone structure in animals and for the cell membranes in both plants and animals. Phosphorus is also

---

[8] The by-product known as Thomas (basic or phosphatic) slag is obtained from the iron industry [78].

[9] Three primary (major) plant nutrients are N, P and K. The elements Ca, Mg and S are the secondary nutrients and Fe, Cu, Zn, Mn, B, Mo and Cl are the micronutrients [84]. Primary and secondary nutrients together with C, H and O (available from air and water) are termed as macronutrients [85].

the second most abundant element in the human body after calcium. It is essential for healthy formation of bones and teeth and energy storage system and helps maintain the blood sugar level. The largest and least expensive source of phosphorus is obtained by mining and beneficiation of phosphate rocks from numerous phosphate deposits throughout the world (**Section 7.2**). The principal use of these phosphate rocks is the manufacture of fertilizers (**Fig. 15**) [81],[82]. Phosphorus is important because it is an essential component of energy metabolism of all forms of life. Together with N and K, phosphorus presents the three macronutrients needed by all crops [83].

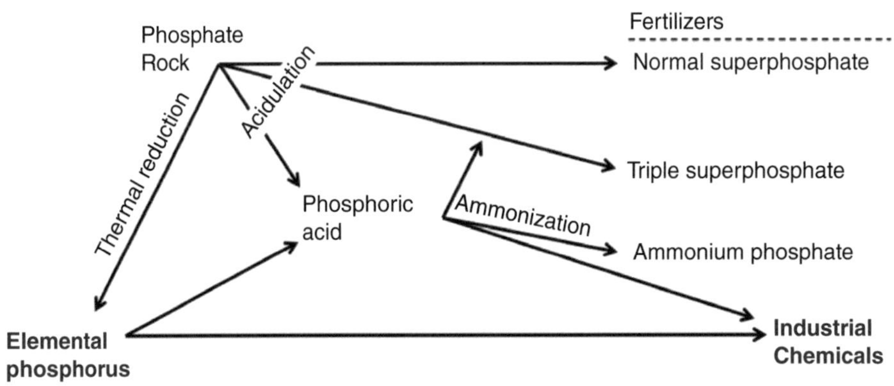

Fig. 15. The utilization of phosphate ore [82].

Most virgin soils in temperate climates contain enough phosphorus to support good crop production, but many soils in tropical climates are naturally deficient in phosphorus because it is leached out during advanced stages of chemical weathering. Phosphorus can be added to deficient soils in the form of natural or artificial fertilizers,[10] and for sustainably good yields, it must be added to all soils when they are cropped heavily for long periods. Phosphate fertilizers are used extensively now in the developed countries, but in many less developed countries their use must be strongly increased to bring the crop production to acceptable levels [86].

Incidental phosphorus fertilization in the form of manures, such as plant and animal biomass, and other natural materials, such as bones, has probably been practiced since the agriculture began. Although the specific nitrification benefits were unknown, ARTHUR YOUNG described in the *Annals of Agriculture* in the mid-19th century the experiment evaluating a wide range of products including the poultry dung, the gunpowder, the charcoal, ashes and various salts. The results showed positive crop responses to certain materials. Benefiting from the achievements in chemistry by ANTOINE LAVOISIER and others, THEODORE DE SAUSSURE was perhaps the first to develop the concept that plants absorb specific mineral elements from the soil. The science of plant nutrition advanced considerably in the 19th century, owing to the contribu-

[10] A fertilizer is any material, organic or inorganic, natural or synthetic, that supplies the plants with one or more chemical elements necessary for normal growth [85].

tion by Carl Sprengel, A.F. Wiegmann, Jean-Baptiste Boussingault and Justus von Liebig. Based on the ubiquitous presence of phosphorus in soil and plant materials and the crop response to phosphorus-containing products, it became apparent that phosphorus is essential for the plant growth [87].

An important concept in the nitrification of plant was developed by J. von Liebig in about 1840. It is known as "The Law of the Minimum" (**Fig. 16**). According to this concept, the plant growth is limited by particular growth factors that are in the shortest supply to the plant. Different factors could potentially control the rate of plant growth at different times during the crop growth. For example, the temperature might limit early spring growth, the moisture affects the growth during a droughty period and nitrogen supply may influence the growth even in cases when the previous factors are inactive [88].

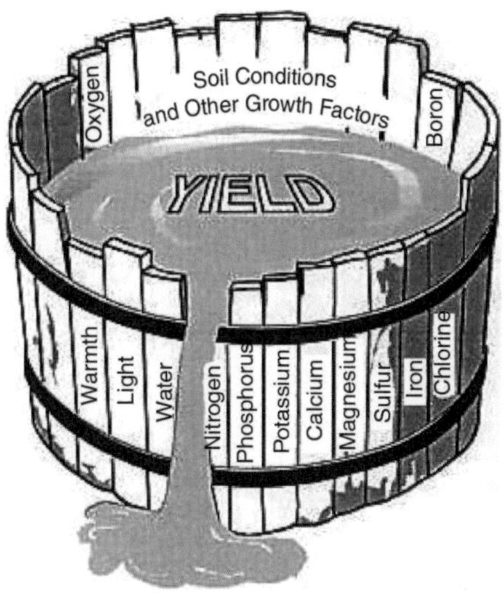

Fig. 16. Illustration of the Liebig's law of the minimum [88].

Liebig observed that dissolving bones in sulfuric acid enhanced the phosphorus availability to plants. Familiar with Liebig's work, John Lawes and coworkers evaluated several apatite-containing products as the phosphorus nutritional source for plants. They performed experiments in which they ultimately became the world's most famous agricultural experiment station – his estate in Rothamsted. Limited supply of bones prompted the developments in the utilization of phosphate rocks where Lawes obtained the first patent concerning the utilization of acid-treated phosphate rock in 1842. The first commercial production of rock phosphate began in Suffolk, England, in 1847. Mining of phosphate in the United States began in 1867. Thus, the phosphorous fertilizer industry began [87].

Non-organic fertilizers contain mainly phosphate, nitrate, ammonium and potassium salts [90]. Phosphorus as an essential element for all life on earth and global food production is highly dependent on the use of fertilizers produced from phosphate rock [91]. Globally, about 150 million tones of phosphate rock are extracted each year for the production of phosphate fertilizers (**Fig. 17**) and the demand for this finite resource is projected to increase to feed growing global population [92]. Due to higher cost per unit of nutrient and availability the bone meal, guano and other natural organic phosphate sources are of only minor commercial importance [89].

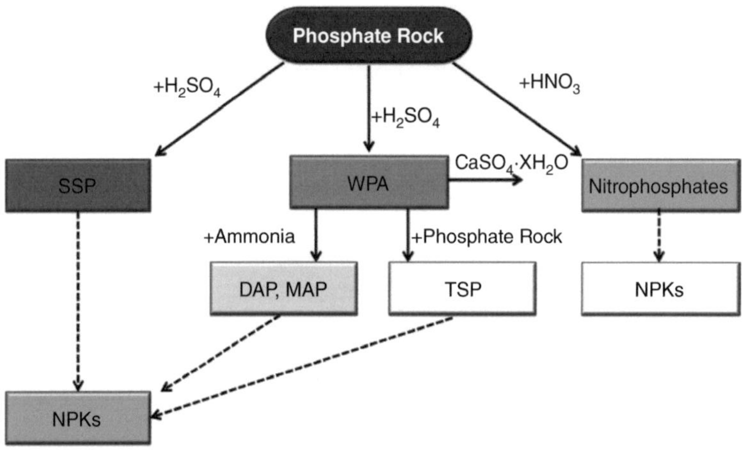

Fig. 17. The relation between phosphate rock and fertilizers [89].

The content of phosphorus in fertilizers (**Table 1**, the grade of fertilizer[11]) is usually expressed as P_2O_5 (or as P_4O_{10}) and the phosphate rock grade is often listed in the trade publications as **BPL,** [12] referring to "**bone phosphate of lime**" [93],[94], the common name for tricalcium phosphate ($Ca_3(PO_4)_2$). Early workers believed that tricalcium phosphate was the chief constituent of phosphate rock. These commercial terms are widely used and the following conversion factors are included for the reference purposes [51],[85],[88],[89],[95].

- $P_2O_5 = 0.4576{\cdot}BLP$;
- $BLP = 2.1852{\cdot}P_2O_5$;
- TPL[13] = BLP;
- $P = 0.1997{\cdot}BLP$;
- $P_2O_5 = P{\cdot}2.2914$ and $P = P_2O_5{\cdot}0.4364$.

[11] The grade of a fertilizer is the nutrient content expressed in weight percentages [85].

[12] 1% P_2O_5 = 2.1852% BLP [89]. The peculiar compound occurs in bones after the calcination, which gives a gelatinous precipitate when pouring calcium chloride into a solution of rhombic phosphate of soda or when adding ammonia to the solution of any phosphate of in acids [93].

[13] Denotes the triphosphate of lime. This salt is commonly called neutral phosphate and appears as a granular precipitate when rhombic phosphate of soda is added drop by drop to calcium chloride in excess [93].

[14] A product made by treating phosphate rock with sulfuric or phosphoric acid or by a mixture of two acids [85].

Name	Chemical composition	Solubility	P_2O_5 [%]
Superphosphate[14]	$Ca(H_2PO_4)_2 + CaSO_4$	Water sol.	18 – 22
Triple superphosphate	$Ca(H_2PO_4)_2$		46 – 47
Monoammonium phosphate	$NH_4H_2PO_4$		48 – 50
Diammonium phosphate	$(NH_4)_2HPO_4$		54
Basic (Thomas) slag	$Ca_3P_2O_8{\cdot}CaO + CaO{\cdot}SiO_2$	Citric acid sol.	10 – 22
Sinter(Thermo)phosphate	$CaNaPO_4{\cdot}Ca_2SiO_4$	NH_4^+ citrate sol.	25 – 29
Ground rock phosphate	Apatite	Soluble in citric acid	29
Fused Mg phosphate	Ca-Mg phosphate		20

Table 1. Direct phosphate fertilizers [96].

Commercial phosphate rocks vary in grade from about 80 to 60 BLP. Most international trade involves higher-grade phosphate rocks and lower-grade rocks are often used at local processing facilities [88].

Phosphorus is utilized in fully oxidized and hydrated forms as orthophosphate. Plants typically absorb either $H_2PO_4^-$ or HPO_4^{2-} depending on the pH of growing medium. However, under certain conditions, plants might absorb soluble organic phosphates, including nucleic acids. A portion of absorbed inorganic phosphorus is quickly combined into organic molecules upon the entry into the roots or after it is transported into the roots. Total phosphorus in plant tissue ranges from 0.1 to 1%. Typical plant might contain approximately 0.004% P as DNA (deoxyribonucleic acid), 0.04% P as RNA (ribonucleic acid), 0.03% as lipid, 0.02% as ester and 0.13% as inorganics [87].

9.3.1. Single superphosphate

Single superphosphate (SSP, ordinary termed as normal or ordinary superphosphate) was the main phosphate fertilizer for more than a century and supplied over 60% of the world's phosphate until as late as 1955. Since then, the importance of SSP has been steadily decreasing with time. In 1975, it supplied only 20% of the phosphate fertilizers and this amount fell to 17% in 1988. Recent decline in actual tonnage has been small, the single superphosphate is still an important phosphate fertilizer and is likely to remain so even though its importance will decrease [97].

The advantages of single superphosphate fertilizer are [97]:

1. The process of manufacturing of SSP is simple, requiring only little technical skill and small capital investment.

2. The economies of scale are minor; thus, small plants can be economical.

3. Since the process is not capital intensive, there is little advantage of high percentage utilization of capacity. In fact, many SSP plants operate on a planned seasonal schedule.

4. The fertilizer effectiveness of SSP is unquestioned. In fact, it is a standard for the comparison of other phosphate fertilizers.

5. SSP supplies two secondary elements, sulfur and calcium, which are sometimes deficient in the soil.

Despite many advantages, the disadvantage of low analysis (from 16 to 22% P_2O_5, i.e. from 35 to 48% BLP) and consequent high distribution costs have caused the declining interest in its production because the delivered cost at the farm level is usually higher per unit of P_2O_5 than that of TSP or ammonium phosphates.

Single superphosphate is an optimal choice in the following situations [97]:

a. Where both P_2O_5 and sulfur are deficient, SSP can be the most economical product to meet these needs.

b. In small countries or remote regions where the demand is insufficient to justify the economic scale of production of concentrated phosphate fertilizers and where the importation is expensive, SSP can be the most economical means of supplying local needs.

c. In many cases, SSP can be an attractive way to use the by-product of sulfuric acid that cannot be used to produce more concentrated product because the quality or the quantity of acid is unsuitable. Likewise, SSP can use the deposits of phosphate rock, which are too small to supply more extensive plant.

Since the grade of phosphate rock determines the grade of the SSP product, high-grade rocks are desirable. Since less reactive rocks must be ground more finely, the reactivity is also important. It is extremely difficult to produce single superphosphate from some igneous rocks (**Section 7.1.2**). Iron and aluminum compounds can be tolerated up to a certain point, although they decrease the P_2O_5 water solubility. Silica has no other adverse effect than the decrease in grade. An increase of $CaO:P_2O_5$ ratio raises the consumption of sulfuric acid per unit of P_2O_5 and decreases the grade. High chloride rocks (with the content of up to 0.5% Cl and perhaps higher) can be used without serious disadvantage, since the corrosion is not a serious problem in the production of SSP [88].

The preparation of single superphosphate (**Fig. 18**) via the treatment of finely ground phosphate rocks by sulfuric acid is based on the conversion of insoluble fluorapatite or natural tricalcium phosphate to monocalcium phosphate [97]:

$$
\begin{aligned}
&2\ Ca_5(PO_4)_3F + 7\ H_2SO_4 + 3\ H_2O \rightarrow 7\ CaSO_4 \\
&+3\ Ca(H_2PO_4)_2 \cdot H_2O + 2\ HF
\end{aligned}
\tag{24}
$$

$$
\begin{aligned}
&Ca_3(PO_4)_2 + 2\ H_2SO_4 + 5\ H_2O \rightarrow 2\ CaSO_4 \cdot 2H_2O \\
&+Ca(H_2PO_4)_2 \cdot H_2O
\end{aligned}
\tag{25}
$$

It has been generally accepted that this process proceeds in two stages [97]:

http://dx.doi.org/10.5772/62215

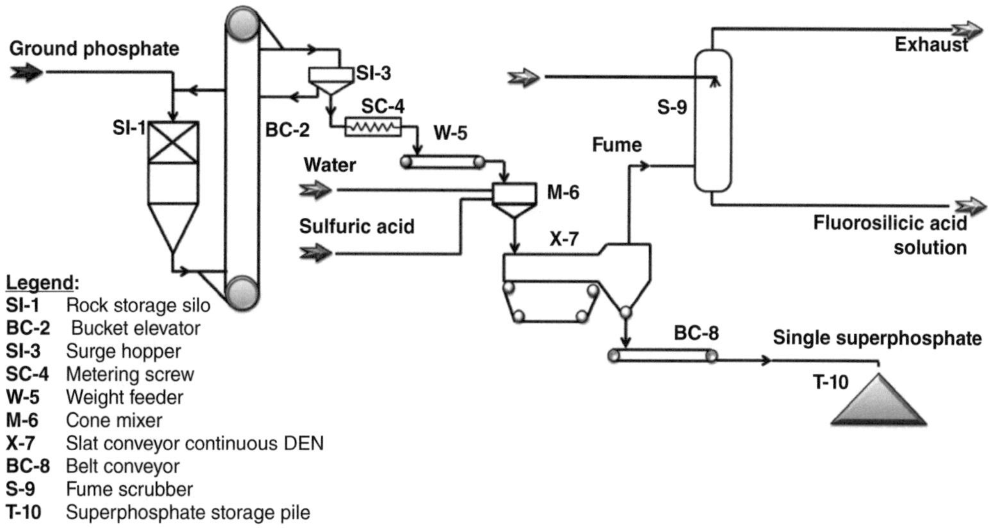

Fig. 18. Manufacturing of superphosphate [88].

i. Phosphoric acid and calcium sulfate are formed by the reaction of sulfuric acid with part of the phosphate rock.

ii. Phosphoric acid formed in the first step of the process reacts with other portion of phosphate rock to form monocalcium phosphate.

These two reactions proceed concurrently, but the first one is fast, while the second one continues for several days or weeks.

9.3.2. Triple superphosphate

Triple superphosphate (TSP) was prepared by the reaction of phosphoric acid with phosphate rock or bone ash [78]:

$$Ca_5(PO_4)_3F + 7\ H_3PO_4 + 5\ H_2O \rightarrow 5\ Ca(H_2PO_4)_2 \cdot H_2O + HF \tag{26}$$

$$Ca_3(PO_4)_2 + 4\ H_3PO_4 \rightarrow 3\ Ca(H_2PO_4)_2 \tag{27}$$

Triple phosphate is a concentrated form of single superphosphate and it contains 46 – 48% P_2O_5, i.e. the content of P_2O_5 is three times higher than that of single superphosphate.

9.3.3. Enriched superphosphate

Enriched superphosphate is essentially a mixture of SSP and TSP, usually made by the acidification of phosphate rock with a mixture of sulfuric and phosphoric acid. Theoretically, any grade between SSP and TSP can be produced, but the usual range is 25 – 35% P_2O_5. The processes and used equipment are the same as for SSP. Enriched superphosphate may be useful product for the applications in sulfur-deficient areas where SSP would supply more sulfur than necessary. One advantage is that mixed acid of proper concentration can be obtained by mixing concentrated sulfuric acid (93 – 98% H_2SO_4) with diluted phosphoric acid (30% P_2O_5), thereby avoiding the need for concentrating the latter [88].

9.3.4. Thomas slag

Thomas, basic or phosphatic slag[15] is actually obtained as a by-product in iron industries. This results from the presence of small amounts of phosphorus in iron ores. When the iron ore is burnt with limestone in the presence of air, calcium phosphate and calcium silicate appear as slag. Thus, the mixture of calcium phosphate and calcium silicate is known as Thomas (basic or phosphatic) slag and contains 14 – 18% of P_2O_5 and 40% of lime. The following reactions accompany the formation of Thomas slag [78],[99],[100]:

$$4\ P + 5\ O_2 \rightarrow 2\ P_2O_5 \tag{28}$$

$$3\ CaO + P_2O_5 \rightarrow Ca_3(PO_4)_2 \tag{29}$$

$$CaO + SiO_2 \rightarrow CaSiO_3 \tag{30}$$

The composition of Thomas slag is shown in **Fig. 20**.

9.3.5. Ammonium phosphate

Ammonium phosphate (ammoniated superphosphate) was prepared from phosphate ore treated by ammonium sulfate and sulfuric acid [78]:

$$Ca_3(PO_4)_2 + (NH_4)_2SO_4 + 2\ H_2SO_4 \rightarrow 2\ NH_4H_2PO_4 + 3\ CaSO_4 \tag{31}$$

This product contains 61 – 73% P_2O_5.

Anhydrous ammonia passes through triple superphosphate in a rotary drum ammoniator where **reaction 32** takes place to evolve considerable amount of heat to granulate the ammoniate superphosphate, which is subsequently dried and bagged [101]:

---

[15] Also known as Thomas phosphate powder. The utilization as fertilizer was firstly tried as a manure on fields in Germany, and in November 1883, HERREN HOYRMANN and MEYER were able to report to the German Royal Agricultural Society the most excellent effect of its use [99].

$$NH_3 + Ca(H_2PO_4)_2 \rightarrow NH_4 \cdot H_2PO_4 + CaHPO_4 \quad (32)$$

Orthophosphoric acid neutralized by ammonia yields to monoammonium phosphate (MAP):

$$NH_3 + H_3PO_4 \rightarrow NH_4H_2PO_4 \quad (33)$$

When ammonia passes through phosphoric acid to maintain the pH of the resulting solution in the range from 5.8 to 6, diammonium phosphate (diammonium hydrogen phosphate, DAP) is formed:

$$2\ NH_3 + H_3PO_4 \rightarrow (NH_4)_2\,HPO_4 \quad (34)$$

Mono- and diammonium phosphates are manufactured by treating orthophosphoric acid in a preneutralizer where the ratio of NH_3:PO_4 is adjusted to 0.6 in the case of monoammonium phosphate and to 1.4 in the case of diammonium phosphate. The heat of reaction raises the slurry temperature to the boiling point and some moisture evaporates. Further addition of ammonia, so that the NH_3:PO_4 ratio increases to 1.0 (monoammonium phosphate) and to 2.0 (diammonium phosphate), generates additional heat to evaporate water during the granulation [84],[101].

Diammonium phosphate provides the following additional favorable factors [84]:

i. Decreases the amount of reactants otherwise required in ammoniators.

ii. Lowers the moisture content in formulation, which decreases the requirement for drying.

iii. May contribute to higher production rates.

iv. Formulations using DAP are often less expensive than other convectional formulations.

v. Improves physical stability.

vi. Less tendency of product caking in storage.

DAP can favorably be used in irrigation systems and in the production of liquid suspension fertilizers, because it is completely soluble in water [84].

9.3.6. Nitrophosphate and urea nitrate phosphate

Nitrophosphate (calcium superphosphate nitrate) is prepared by the same way as superphosphate using nitric acid instead of sulfuric acid [78]:

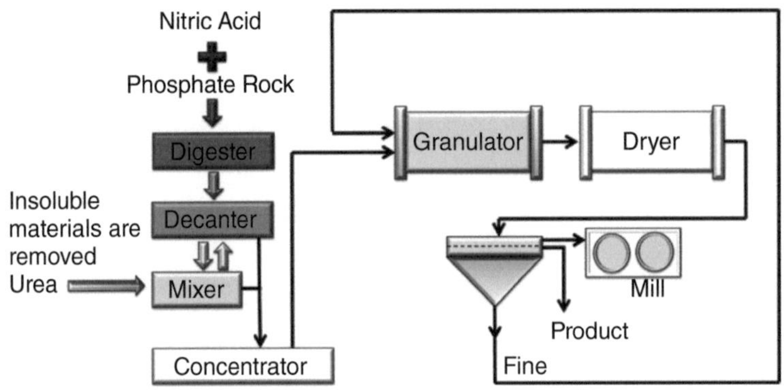

Fig. 19. The scheme of a plant for the production of thermal phosphoric acid [101].

$$Ca_3(PO_4)_2 + 4\ HNO_3 \rightarrow Ca(H_2PO_4)_2 + 2\ Ca(NO_3)_2 \tag{35}$$

The amount of nitric acid applied for the treatment of phosphate rock affects the ratio between formed monocalcium phosphate and calcium nitrate, as the following equations reveal [101]:

$$Ca_5(PO_4)_3F + 10\ HNO_3 \rightarrow 5\ Ca(NO_3)_2 + 3\ H_3PO_4 + HF \tag{36}$$

$$\begin{aligned} &Ca_5(PO_4)_3F + 9\ HNO_3 \rightarrow 0.5\ Ca(H_2PO_4)_2 \\ &+4.5\ Ca(NO_3)_2 + 2\ H_3PO_4 + HF \end{aligned} \tag{37}$$

$$\begin{aligned} &Ca_5(PO_4)_3F + 8\ HNO_3 \rightarrow Ca(H_2PO_4)_2 \\ &+4\ Ca(NO_3)_2 + H_3PO_4 + HF \end{aligned} \tag{38}$$

$$\begin{aligned} &Ca_5(PO_4)_3F + 7\ HNO_3 \rightarrow 1.5\ Ca(H_2PO_4)_2 \\ &+3.5\ Ca(NO_3)_2 + HF \end{aligned} \tag{39}$$

These fertilizers are also the source of nitrogen.[16] For the preparation of urea nitrate phosphate, nitrophosphate contains the product of **Eqs. 35 – 39** conveyed to the decanter (**Fig. 19**) to get rid of the insoluble impurities. Urea is added to the solution in the mixer [78],[101].

The following reactions accompany the formation of urea nitrate phosphate [101]:

[16] Fertilizers containing two or more nutrients are termed as compound nutrients (not in the United States). The names complex fertilizers and chemically mixed fertilizers are used in the same meaning in some countries [84].

http://dx.doi.org/10.5772/62215

$$Ca(NO_3)_2 + H_3PO_4 + 2\ CO(NH_2)_2 \rightarrow Ca(H_2PO_4)(NO_3)CO(NH_2)_2 + CO(NH_2)_2 HNO_3 \quad (40)$$

$$Ca(NO_3)_2 + 4\ CO(NH_2)_2 \rightarrow Ca(NO_3)_2 \cdot 4CO(NH_2)_2 \quad (41)$$

$$Ca(NO_3)_2 + H_3PO_4 \rightarrow Ca(NO_3)_2 \cdot H_3PO_4 \quad (42)$$

$$Ca(H_2PO_4)_2 + Ca(NO_3)_2 + 2\ CO(NH_2)_2 \rightarrow Ca_2(H_2PO_4)_2(NO_3)_2 CO(NH_2)_2 \quad (43)$$

$$2\ HF + Ca(NO_3)_2 + 2\ CO(NH_2)_2 \rightarrow CaF_2 + 2CO(NH_2)_2 \cdot HNO_3 \quad (44)$$

After removing insoluble impurities in the decanter, the reaction product, the slurry is dehydrated and concentrated, granulated and dried.

9.3.7. Dicalcium phosphate

Dicalcium phosphate is a citrate soluble fertilizer. Phosphate rock is first converted to orthophosphoric acid via the treatment by HCl. Orthophosphoric acid reacting with lime gives dicalcium phosphate [78]:

$$H_3PO_4 + CaO \rightarrow CaHPO_4 + H_2O \quad (45)$$

9.3.8. Thermophosphates

Thermophosphates (rhenania phosphates, thermal phosphate) are manufactured by the reaction of rock phosphates with soda and quartz [96],[102]:

$$Ca_5(PO_4)_3F + 2\ Na_2CO_3 + SiO_2 \rightarrow 3\ NaCaPO_4 + Ca_2SiO_4 + NaF + 2\ CO_2(g) \quad (46)$$

Rhenania phosphate is prepared by the calcination of mixture of phosphate rock, sodium carbonate and silica in a rotary kiln at 1250°C. The fertilizer contains 28 – 30% P_2O_5 and has an alkaline effect and hence is more efficient in acid soils. In neutral or basic soils, it reacts more slowly.

A somewhat similar product, Roechling phosphate uses soda slag, which is a by-product from the steel industry, as the source of sodium. Also, naturally occurring source of minerals such as trona ($Na_3(HCO_3)(CO_3)\cdot 2H_2O$ [103]) or natron ($Na_2CO_3\cdot 10H_2O$ [104]) can be applied. A similar product can be also prepared by sintering potassium carbonate with phosphate rocks and silica where the formation of $CaKPO_4$ is supposed [85]:

$$Ca_5(PO_4)_3F + 2\ K_2CO_3 + SiO_2 \rightarrow 3\ KCaPO_4 + Ca_2SiO_4 + KF + 2\ CO_2(g) \tag{47}$$

The origin of thermal phosphate can be derived from Thomas slag (**Fig. 20**). Since Thomas slag became popular, numerous attempts have been made to produce fertilizers by thermal treatment of phosphate rock with additives. The most of these attempts are not successful, except for a few that attained commercial production of fertilizers such as rhenania phosphate, fused magnesium phosphate (FMP) and calcined defluorinated phosphate [84].

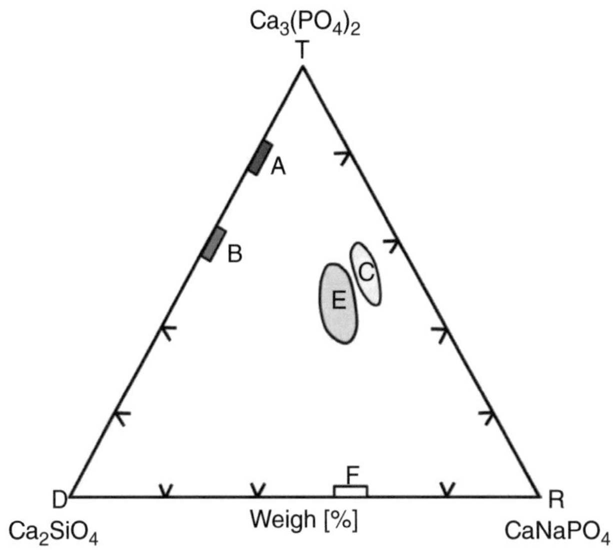

Fig. 20. Ternary diagram of thermal phosphates: calcined defluorinated phosphate tested in the United States (A), Thomas slag (B), calcined defluorinated phosphates produced in Japan, United States, etc. (C), silicophosphate tested in England (E) and Rhenania phosphate in Germany (F) [84].

The composition of some systems is shown in the ternary diagram in **Fig. 20**. Rhenanite (R, $CaNaPO_4$) has two forms (β- and α-rhenanite) with $\beta \rightarrow \alpha$ transition temperature of 670°C. Both phases are highly soluble in 2% citric acid and ammonium citrate. β-Rhenanite is the major constituent of Rhenania phosphate [84].

9.3.9. Environmental demand on phosphate fertilizers

Fertilizers are essential to provide adequate nutrients for the crop growth and to ensure successful harvests. Continuing exponential growth in human population and increasing demand for biofuels point to ever-increasing demand for fertilizers. Despite the apparent success of current agricultural production systems, the overuse of fertilizers has caused severe environmental problems and increasing number of health concerns. Overall, environmental and human health concerns associated with the overuse of fertilizer result in two main problems [90],[105],[106],[107],[108]:

i. Disruption of natural nitrogen and phosphorus nutrient cycles because of massive infusion of nitrogen and phosphorus from fertilizers.

ii. Inadvertent release of heavy metals and radionuclides from mineral fertilizers and pathogens, veterinary pharmaceuticals and endocrine disrupters from organic fertilizers.

Moreover, fertilizers can be adulterated products containing the raw material sometimes from unknown and/or questionable sources. Besides certified nutritional ingredients for plants, they may contain, most notably, the trace element contaminants that can be inadvertently introduced into soils. The fertilizer applications are by far one of the most consistent sources of trace elements to accumulate in cropland soils. Based on the analysis of existing data from literature and of results from model simulations, we have concluded that a long-term use of phosphorus fertilizers and micronutrients could cause the As, Cd and Pb contents of the cropland soils to rise if the products used contained high levels of these elements [92],[105],[106].

Since naturally occurring nuclides ^{238}U, ^{232}Th and ^{40}K have strong association with phosphate ore, which is the major raw material for the production of phosphate fertilizers, radon 222 (^{222}Rn) (the most significant natural isotope of radioactive element radon) was formed as a decay product of ^{238}U. Radon is colorless, odorless poisonous gas, and sustained exposure of humans to its increased level can lead to lung cancer. Radon is a noble gas and does not undergo chemical reaction. When concentrated in enclosed environment, it can only diminish by diffusion, advection and radioactive decay [70],[109],[110],[111],[112].

Phosphate fertilizers are being enriched with ^{238}U during their production from phosphate rocks. The activity of ^{238}U is higher in phosphate-rich fertilizers like TSP and SSP. The application of phosphate fertilizer significantly increases the radioactivity level of cultivated soil as compared to soil from barren land [113],[114].

Current waste of phosphorus fertilizers causes a great deal of environmental problems, and it is questionable if it is a good idea to extract all the phosphate rock reserves if it would still end up in lakes, streams and sea. As more and more phosphorus has been added to the ecosystem, many lakes and coasts have seen an increased algae growth, which in some cases have led to serious eutrophication and dead zones due to lack of oxygen [91]. Carpenter and Bennett [115] even consider that the planetary limits for the eutrophication of freshwater due to phosphorus have already been exceeded. It is possible to recycle phosphorus from different sources (human excreta, manure, different types of waste products, etc.) and improve the efficiency in the production and usage in order to postpone the potential production peak.

Author details

Petr Ptáček

Brno University of Technology, Czech Republic

References

[1] Jaffe EB. Abstract of the literature on synthesis of apatites and some related phosphates. Geological Survey Circular 135, 1951.

[2] Derek EC. Chemistry, Biochemistry and Technology. 6th ed., CRC Press, 2013. ISBN: 978-1439840887

[3] Emsley J. The Shocking History of Phosphorus: A Biography of the Devil's Element. Pan, 2001. ISBN: 978-0330390057

[4] Valsami-Jones E. Phosphorus in Environmental Technologies: Principles and Applications. Integrated Environmental Technology Series. IWA Publishing, 2004. ISBN: 978-1843390015

[5] Averbuch-Pouchot Marie-T, Durif A. Topics in Phosphate Chemistry. World Scientific, 1996. ISBN: 978-9810226343

[6] Toy ADF. The Chemistry of Phosphorus: Pergamon Texts in Inorganic Chemistry, Volume 3. Elsevier, 2013. ISBN: 9781483147413

[7] Gilmour R. Phosphoric Acid: Purification, Uses, Technology, and Economics. CRC Press, 2013. ISBN: 978-1439895160

[8] Patent GB1888/14962

[9] Patent GB1888/17719

[10] Patent US482586

[11] Fauser G. Chemical Fertilizers: Proceedings of the XVII International Congress Chemistry Days 1966 On. Elsevier, 2013. ISBN: 978-1483184159

[12] Büchel KH, Moretto Hans-H, Werner D. Industrial Inorganic Chemistry. 2nd ed., John Wiley & Sons, 2008. ISBN: 978-3527613335

[13] Hocking MB. Modern Chemical Technology and Emission Control. Springer Science & Business Media, 2012. ISBN: 978-3642697739

[14] Franck HH, Füldner H. Beiträge zur Kenntnis der Phosphatreduktion. Zeitschrift für anorganische und allgemeine Chemie 1932;204(1-2) 97–139.

[15] Gerrans GC, Hartmann-Petersen P, Hartmann-Petersen R. Sasol Encyclopaedia of Science and Technology. New Africa Books, 2004. ISBN: 978-1869283841

[16] Cherif M, Mgaidi A, Ammar MN, Abderrabba M, Fürst W. Modelling of the equilibrium properties of the system H_3PO_4-H_2O: Representation of VLE and liquid phase composition. Fluid Phase Equilibria 2000;175(1-2) 197–212.

[17] Cherif M, Mgaidi A, Ammar MN, Abderrabba M, Fürst W. Representation of VLE and liquid phase composition with an electrolyte model: Application to H_3PO_4-H_2O and H_2SO_4-H_2O. Fluid Phase Equilibria 2002;194-197 729-738.

[18] Wagh AS. Chemically Bonded Phosphate Ceramics: Twenty-First Century Materials with Diverse Applications. Elsevier, 2004. ISBN: 978-0080445052

[19] Prakash S. Advanced Inorganic Chemistry, Volume 1. S. Chand, 2000. ISBN: 978-8121902632

[20] Molins RA. Phosphates in Food. CRC Press, 1990. ISBN: 978-0849345883

[21] Atul S. The Pearson Guide to Objective Chemistry for the AIEEE. Pearson Education India, 2010. ISBN: 978-8131733813

[22] Moshaii A, Faraji M, Tajik-Nezhad S. Study of single bubble Sonoluminescence in phosphoric acid. Ultrasonics Sonochemistry 2011;18(5) 1148–1152.

[23] Faraji M, Moshaii A. Noble gas dependence of single-bubble sonoluminescence in phosphoric acid. Physics Letters A 2012;376(42-43) 2703–2706.

[24] Troia A, Ripa DM, Spagnolo R. Moving single bubble sonoluminescence in phosphoric acid and sulphuric acid solutions. Ultrasonics Sonochemistry 2006;13(3) 278–282.

[25] Suslick KS, Eddingsaas NC, Flannigan DJ, Hopkins SD, Xu H. Characterization of stable and transient cavitation bubbles in a milliflow reactor using a multibubble sonoluminescence quenching technique. Ultrasonics Sonochemistry 2011;18(4) 842–846.

[26] Awwad NS, El-Nadi YA, Hamed MM. Successive processes for purification and extraction of phosphoric acid produced by wet process. Chemical Engineering and Processing: Process Intensification 2013;74 69–74.

[27] Spainhour CB. Phosphoric Acid. Reference Module in Biomedical Sciences, From Encyclopedia of Toxicology (Third Edition), 2014; 916-919. DOI: 10.1016/B978-0-12-386454-3.00904-0

[28] Scheffel DLS, Sacono NT, Ribeiro APD, Soares DG, Basso FG, Pashley D, de Souza Costa CA, Hebling J. Immediate human pulp response to ethanol-wet bonding technique. Journal of Dentistry 2015;43(5) 537–545.

[29] Legler LR, Retief DH, Bradley EL. Effects of phosphoric acid concentration and etch duration on enamel depth of etch: An in vitro study. American Journal of Orthodontics and Dentofacial Orthopedics 1990;98(2) 154–160.

[30] Sauro S, Mannocci F, Toledano M, Osorio R, Pashley DH, Watson TF. EDTA or H_3PO_4/NaOCl dentine treatments may increase hybrid layers' resistance to degradation: A

microtensile bond strength and confocal-micropermeability study. Journal of Dentistry 2009;34(4) 279–288.

[31] Gross JD, Retief DH, Bradley EL. An optimal concentration of phosphoric acid as an etching agent. Part II: Microleakage studies. The Journal of Prosthetic Dentistry 1984;52(6) 786–789.

[32] Sahnoun RD, Bouaziz J. Sintering characteristics of kaolin in the presence of phosphoric acid binder. Ceramics International 2012;38(1) 1–7.

[33] Douiri H, Louati S, Baklouti S, Arous M, Fakhfakh Z. Structural, thermal and dielectric properties of phosphoric acid-based geopolymers with different amounts of H_3PO_4. Materials Letters 2014;116 9–12.

[34] Appleby AJ. Fuel Cells — Phosphoric Acid Fuel Cells, Overview. Reference Module in Chemistry, Molecular Sciences and Chemical Engineering, From Encyclopedia of Electrochemical Power Sources 2009; 533-547.

[35] Oka Y. Fuel Cells — Phosphoric Acid Fuel Cells, Cells and Stacks. Reference Module in Chemistry, Molecular Sciences and Chemical Engineering, From Encyclopedia of Electrochemical Power Sources 2009; 568-578.

[36] Kuwabara T. Fuel Cells — Phosphoric Acid Fuel Cells, Cathodes. Reference Module in Chemistry, Molecular Sciences and Chemical Engineering, From Encyclopedia of Electrochemical Power Sources 2009, 557–5.63.

[37] Kuwabara T, Kurzweil P. Fuel Cells — Phosphoric Acid Fuel Cells, Anodes. Reference Module in Chemistry, Molecular Sciences and Chemical Engineering, From Encyclopedia of Electrochemical Power Sources 2009, 548–556.

[38] Sammes N, Bove R, Stahl K. Phosphoric acid fuel cells: Fundamentals and applications. Current Opinion in Solid State and Materials Science 2004;8(5) 372–378.

[39] Kargupta K, Saha S, Banerjee D, Seal M, Ganguly S. Performance enhancement of phosphoric acid fuel cell by using phosphosilicate gel based electrolyte. Journal of Fuel Chemistry and Technology 2012;40(6) 707–713.

[40] Ahmad F, Sheha E. Preparation and physical properties of $(PVA)_{0.7}(NaBr)_{0.3}(H_3PO_4)_xM$ solid acid membrane for phosphoric acid — Fuel cells. Journal of Advanced Research 2013;4(2) 155–161.

[41] Wannek C, Glüsen A, Stolten D. Materials, manufacturing technology and costs of fuel cell membranes. Desalination 2010;250(3) 1038–1041.

[42] Wang X, Xu C, Golding BT, Sadeghi M, Cao Y, Scott K. A novel phosphoric acid loaded quaternary 1,4-diazabicyclo-[2.2.2]-octane polysulfone membrane for intermediate temperature fuel cells. International Journal of Hydrogen Energy 2011;36(14) 8550–8556.

[43] Myglovets M, Poddubnaya OI, Sevastyanova O, Lindström ME, Gawdzik B, Sobiesiak M, Tsyba MM, Sapsay VI, Klymchuk DO, Puziy AM. Preparation of carbon adsorbents

from lignosulfonate by phosphoric acid activation for the adsorption of metal ions. Carbon 2014;80 771–783 .

[44] Liou Tzong-H. Development of mesoporous structure and high adsorption capacity of biomass-based activated carbon by phosphoric acid and zinc chloride activation. Chemical Engineering Journal 2010;158(2) 129–142.

[45] Cordero RL, Llambías FJG, Palacios JM, Fierro JLG, Agudo AL. Surface changes of alumina induced by phosphoric acid impregnation. Applied Catalysis 1989;56(1) 197–206.

[46] Shaterian HR, Azizi K, Fahimi N. Phosphoric acid supported on alumina: A useful and effective heterogeneous catalyst in the preparation of α-amidoalkyl-β-naphthols, α-carbamato-alkyl-β-naphthols, and 2-arylbenzothiazoles. Arabian Journal of Chemistry 2012. In press. DOI: 10.1016/j.arabjc.2012.07.006

[47] Panneerselvam P, Thinakaran N, Thiruvenkataravi KV, Palanichamy M, Sivanesan S. Phosphoric acid modified-Y zeolites: A novel, efficient and versatile ion exchanger. Journal of Hazardous Materials 2008;159(2-3) 427–434.

[48] Zeineb O, Hedi BA, Jeday MR, Cheker C. Kinetic study of the catalytic decomposition of H_2O_2 in phosphoric acid medium. International Journal of Hydrogen Energy 2015;40(2) 1278–1282.

[49] Shen L, Leng Y, Wang J, Ren X, Wu Y, Zhang M, Xu Y. Esterification of phosphoric acid with equimolar lauryl alcohol catalyzed by a Schiff base manganese complex anchored to a Keggin heteropolyacid. Chinese Journal of Catalysis 2010;31(2) 156–162.

[50] Saravanan K, Tyagi B, Shukla RS, Bajaj HC. Esterification of palmitic acid with methanol over template-assisted mesoporous sulfated zirconia solid acid catalyst. Applied Catalysis B: Environmental 2015;172-173 108-115.

[51] Chen M, Graedel TE. The potential for mining trace elements from phosphate rock. Journal of Cleaner Production 2015;91 337–346.

[52] Zhang P, Miller J, Hassan E. Beneficiation of Phosphates: New Thought, New Technology, New Development. SME, 2012. ISBN: 978-0873353588

[53] Gobbitt JM. Yara Hemihydrate (HH) and hemidihydrate (HDH) processes for phosphoric acid production. Procedia Engineering 2012;46 143–153.

[54] Amin MI, Ali MM, Kamal HM, Youssef AM, Akl MA. Recovery of high grade phosphoric acid from wet process acid by solvent extraction with aliphatic alcohols. Hydrometallurgy 2010;105(1-2) 115–119.

[55] Zan C, Shi L, Song YZ, Zhu MS. Evaluation method for thermal processing of phosphoric acid with waste heat recovery. Energy 2006;31(14) 2791–2804.

[56] Abu-Eishah SI, Abu-Jabal NM. Parametric study on the production of phosphoric acid by the dihydrate process. Chemical Engineering Journal 2001;81(1-3) 231–250.

[57] Tayibi H, Choura M, López FA, Alguacil FJ, López-Delgado A. Environmental impact and management of phosphogypsum. Journal of Environmental Management 2009;90(8) 2377–2386.

[58] Papadopoulos AI, Seferlis P. Generic modelling, design and optimization of industrial phosphoric acid production processes. Chemical Engineering and Processing: Process Intensification 2009;48(1) 493–506.

[59] Wang L, Long Z, Huang X, Yu Y, Cui D, Zhang G. Recovery of rare earths from wet-process phosphoric acid. Hydrometallurgy 2010;101(1-2) 41–47.

[60] Preston JS, Cole PM, Craig WM, Feather AM. The recovery of rare earth oxides from a phosphoric acid by-product. Part 1: Leaching of rare earth values and recovery of a mixed rare earth oxide by solvent extraction. Hydrometallurgy 1996;41(1) 1–19.

[61] Preston JS, Cole PM, du Preez AC, Fox MH, Fleming AM. The recovery of rare earth oxides from a phosphoric acid by-product. Part 2: The preparation of high-purity cerium dioxide and recovery of a heavy rare earth oxide concentrate. Hydrometallurgy 1996;41(1) 21–44.

[62] Preston JS, du Preez AC, Cole PM, Fox MH. The recovery of rare earth oxides from a phosphoric acid by-product. Part 3: The separation of the middle and light rare earth fractions and the preparation of pure europium oxide. Hydrometallurgy 1996;42(2) 131–149.

[63] Preston JS. The recovery of rare earth oxides from a phosphoric acid byproduct. Part 4: The preparation of magnet-grade neodymium oxide from the light rare earth fraction. Hydrometallurgy 1996;42(2) 151–167.

[64] Fisher K. Neues Verfahren zur maßanalytischen Bestimmung des Wassergehaltes von Flüssigkeiten und festen Körpern. Angewandte Chemie 1935;48(26) 394–396.

[65] Khoualdia B, Loungou M, Elaloui E. Adsorption of organic matter from industrial phosphoric acid (H_3PO_4) onto activated bentonite. Arabian Journal of Chemistry 2013. In press. ISBN: 10.1016/j.arabjc.2013.01.014

[66] Boualia A, Mellah A, Aissaoui TT, Menacer K, Silem A. Adsorption of organic matter contained in industrial H_3PO_4 onto bentonite: Batch-contact time and kinetic study. Applied Clay Science 1993;7(6) 431–445.

[67] Zermane S, Meniai AH. Experimental study of competitive adsorption of heavy metals and organic matter for the phosphoric acid purification. Energy Procedia 2012;18 888–895.

[68] Cuadri AA, Navarro FJ, García-Morales M, Bolívar JP. Valorization of phosphogypsum waste as asphaltic bitumen modifier. Journal of Hazardous Materials 2014;279 11–16.

[69] Alcordo IS, Rechcigl JE. Phosphogypsum in agriculture: A review. Advances in Agronomy 1993;49 55–118.

[70] El-Didamony H, Gado HS, Awwad NS, Fawzy MM, Attallah MF. Treatment of phosphogypsum waste produced from phosphate ore processing. Journal of Hazardous Materials 2013;244-245 596-602.

[71] Borges RC, Ribeiro FCA, da Costa Lauria D, Bernedo AVB. Radioactive characterization of phosphogypsum from Imbituba, Brazil. Journal of Environmental Radioactivity 2013;126 188–195.

[72] Mousa S, Hanna A. Synthesis of nano-crystalline hydroxyapatite and ammonium sulfate from phosphogypsum waste. Materials Research Bulletin 2013;48(2) 823–828.

[73] Shweikani R, Kousa M, Mizban F. The use of phosphogypsum in Syrian cement industry: Radiation dose to public. Annals of Nuclear Energy 2013;54 197–201.

[74] Fuding M, Jiaojiao H, Zude L. Research on activity characteristics on composite cementitious materials based on phosphogypsum. Procedia Engineering 2012;43 9–15.

[75] Kuryatnyk T, da Luz CA, Ambroise J, Pera J. Valorization of phosphogypsum as hydraulic binder. Journal of Hazardous Materials 2008;160(2-3) 681–687.

[76] Radwan MM, Heikal M. Hydration characteristics of tricalcium aluminate phase in mixes containing β-hemihydate and phosphogypsum. Cement and Concrete Research 2005;35(8) 1601–1608.

[77] Rutherford PM, Dudas MJ, Samek RA. Environmental impacts of phosphogypsum. Science of The Total Environment 1994;149(1-2) 1–38.

[78] Darpan P. Competition Science Vision 2008;125.

[79] Singh R. Inorganic Chemistry. Mittal Publications, 2002. ISBN: 978-8170998266

[80] Bensted J. Use of ferrophosphorus as a heavyweight additive for API oil well cements. Cement and Concrete Research 1993;23(4) 988–990.

[81] Kogel JE. Industrial Minerals & Rocks: Commodities, Markets, and Uses. SME, 2006. ISBN: 978-0873352338

[82] Rose KE, Hardy RG. Recovery of Phosphate from the Cabaniss and Pleasanton Shales of Kansas. Volume 187, Part 4. Kansas Geological Survey, 1967.

[83] Jørgensen SE. Global Ecology. Academic Press, 2010. ISBN: 978-0444536273.

[84] Nielsson FT. Manual of Fertilizer Processing. Fertilizer Science and Technology — Volume 5. CRC Press, 1986. ISBN: 978-0824775223

[85] Hignett TP. Fertilizer Manual. Developments in Plant and Soil Sciences — Volume 15. Springer Science & Business Media, 2013. ISBN: 978-9401715386

[86] McKelvey VE. Phosphate Deposits. Geological Survey Bulletin 1252-D. A Summary of Salient Features of the Geology of Phosphate Deposits, Their Origin, and Distribution. United States Government Printing Office, Washington, 1967.

[87] Barker AV, Pilbeam DJ. Titul Handbook of Plant Nutrition. Books in Soils, Plants, and the Environment — Volume 117. CRC Press, 2014. ISBN: 978-1420014877

[88] Fertilizer Manual, UN Industrial Development Organization, International Fertilizer Development Center. Springer Science & Business Media, 1998. ISBN: 978-079235032

[89] World Phosphate Rock Reserves and Resources. International Center for Soil Fertility and Agricultural Development, 2010. ISBN: 978-0880901673

[90] Savci S. Investigation of effect of chemical fertilizers on environment. APCBEE Procedia 2012;1 287–292.

[91] Walan P, Davidsson S, Johansson S, Höök M. Phosphate rock production and depletion: Regional disaggregated modeling and global implications. Resources, Conservation and Recycling 2014;93 178–187.

[92] Hartley TN, Macdonald AJ, McGrath SP, Zhao Fang-J. Historical arsenic contamination of soil due to long-term phosphate fertiliser applications. Environmental Pollution 2013;180 259–264.

[93] Webster JW. A Manual of Chemistry: Containing the Principal Facts of the Science, in the Order in Which They Are Discussed and Illustrated in the Lectures at Harvard University, N.E. and Several Other Colleges and Medical Schools in the United States: Compiled and Arranged as a Text Book for the Use of Students and Persons Attending Lectures on Chemistry. 3rd ed., Marsh, Capen, Lyon and Webb, 1839.

[94] von Liebig JF, Gregory W. Edward Turner, Elements of Chemistry. 7th ed., Taylor and Walton, and sold, 1841.

[95] Gharabaghi M, Irannajad M, Noaparast M. A review of the beneficiation of calcareous phosphate ores using organic acid leaching. Hydrometallurgy 2010;103 96–107.

[96] Mengel K, Kosegarten H, Kirkby EA, Appel T. Principles of Plant Nutrition. Springer Science & Business Media, 2001. ISBN: 978-9401010092

[97] Abouzeid A.-ZM, El-Jallad IS, Orphy MK. Calcareous phosphates and their calcined products. Minerals Science and Engineering 1980;12(2) 73–83.

[98] Manual Fertilizer. UN Organization, International Fertilizer Development Center. Springer Science & Business Media, 1998. ISBN: 978-0792350323

[99] Thomas SG, Burnie RW. Memoirs and Letters of Sidney Gilchrist Thomas, Inventor Cambridge Library Collection — Technology. Cambridge University Press, 2011. ISBN: 978-1108026918

[100] Selim HM. Phosphate in Soils: Interaction with Micronutrients, Radionuclides and Heavy Metals. Advances in Trace Elements in the Environment. CRC Press, 2015. ISBN: 978-1482236804

[101] Kolay AK. Manures and Fertilizers. Atlantic Publishers & Dist, 2007. ISBN: 978-8126908103

[102] Chesworth W. Encyclopedia of Soil Science. Encyclopedia of Earth Sciences Series. Springer Science & Business Media, 2008. ISBN: 978-1402039942

[103] Choi CS, Mighell AD. Neutron diffraction study of sodium sesquicarbonate dihydrate. Acta Crystallographica Section B 1982;38(11) 2874–2876.

[104] Taga T. Crystal structure of $Na_2CO_3{\cdot}10H_2O$. Acta Crystallographica Section B 1969;25 2656.

[105] Jiao W, Chen W, Chang AC, Page AL. Environmental risks of trace elements associated with long-term phosphate fertilizers applications: A review. Environmental Pollution 2012;168 44–53.

[106] Kilmer VJ. Fertilizers of the future and factors affecting their role in crop production. Agrochemicals in Soils 1980; 3-10.

[107] Chaney RL. Chapter Two — Food Safety Issues for Mineral and Organic Fertilizers. Advances in Agronomy 2012;117 51–116.

[108] Morari F, Vellidis G, Gay P. Reference Module in Earth Systems and Environmental Sciences, From Encyclopedia of Environmental Health 2011, 727–737.

[109] Okeji MC, Agwu KK. Assessment of indoor radon concentration in phosphate fertilizer warehouses in Nigeria. Radiation Physics and Chemistry 2012;81(3) 253–255.

[110] Ibrahim N. Determination of natural radioactivity in fertilizers by gamma ray spectroscopy. Radiation Physics and Chemistry 1998;51(4-6) 621.

[111] Mustonen R. Radioactivity of fertilizers in Finland. Science of The Total Environment 1985;45 127–134.

[112] Makweba MM, Holm E. The natural radioactivity of the rock phosphates, phosphatic products and their environmental implications. Science of The Total Environment 1993;133(1-2) 99–110.

[113] Khalifa NA, El-Arabi AM. Natural radioactivity in farm soil and phosphate fertilizer and its environmental implications in Qena governorate, Upper Egypt. Journal of Environmental Radioactivity 2005;84(1) 51–64.

[114] Hameed PS, Pillai GS, Mathiyarasu R. A study on the impact of phosphate fertilizers on the radioactivity profile of cultivated soils in Srirangam (Tamil Nadu, India). Journal of Radiation Research and Applied Sciences 2014;7(4) 463–471.

[115] Carpenter SR, Bennett EM. Reconsideration of the planetary boundary for phosphorus. Environmental Research Letters 2011;6(1) 014009.

Chapter 10

Utilization of Compounds of Phosphorus

Petr Ptáček

Additional information is available at the end of the chapter

http://dx.doi.org/10.5772/62217

Abstract

The last chapter of this book provides brief description of utilization of apatites and phosphorus-bearing compounds in industry and material science research. Since the chemistry of phosphorus is quite complicated and a quickly developing field of science, the topics described below are only limited insights to chemically bonded ceramics and refractories, dental phosphate cements, oil-well cements, phosphate glasses and glass ceramics. Chapter continues with description of functional phosphate materials applied as solid oxide fuel cells electrolytes, sensors, phosphors, catalysts and coatings. The chapter ends with introduction of basic ideas for biological apatite in bone tissue engineering, collagen apatite composites, apatite layers and biocoatings.

Keywords: Apatite, Chemically Bonded Ceramics, Phosphate Cement, SFOC Electrolytes, Sensors, Biological Apatite, Collagen, Biocoatings

In the previous chapter, the utilization of phosphate ore was described. Elemental phosphorus (**Section 9.1**) and phosphoric acid (**Section 9.2**) are used in the manufacturing of other phosphorus compounds (as illustrated in **Fig. 1**) and materials, which are briefly reported in this chapter. Since the chemistry of phosphorus is quite complicated and a quickly developing field of science, the topics described below are only limited insights to phosphorus-bearing compounds utilized in the industry and material science research.

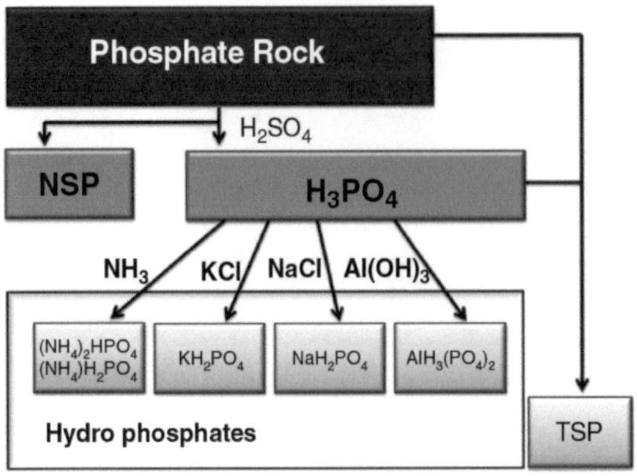

Fig. 1. Flow chart of production of acid phosphates from phosphate ores [1].

10.1. Chemically bonded phosphate ceramics

Ceramics are formed by the compaction of powders and their subsequent fusion at high to very high temperatures, ranging somewhere from ~700° to 2000°C. Once fused, the resulting ceramics are hard and dense and exhibit very good corrosion resistance [2]. Among the conventional ceramic bonds, the chemical (organic and inorganic) and hydraulic bonds were used during the preparation of ceramic materials.

Phosphate bond can be utilized for the preparation of hard and quick-setting ceramic materials, chemically bonded phosphate ceramics (PCBC) and ceramic composites [3],[4] via the reaction of metal cation with phosphate anions (**Fig. 2**). The reaction is attained by mixing a cation donor, generally an oxide (CaO [5],[6],[7],[8], MgO [2],[6],[9], ZnO [2],[8], Al_2O_3 [2], Fe_2O_3 [2],

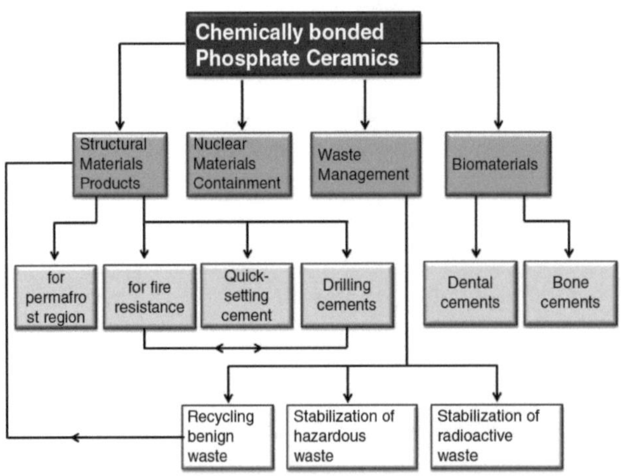

Fig. 2. Potential application of chemically bonded phosphate ceramics [1].

etc.), raw materials [10],or secondary raw materials (fly ash [11] and blast furnace slag [12]) with either phosphoric acid or acid phosphate such as ammonium phosphate solution or magnesium dihydrogen phosphate ($Mg(H_2PO_4)_2{\cdot}2H_2O$) and aluminum hydrogen phosphate ($AlH_3(PO_4)_2{\cdot}H_2O$). Typical acid-base reaction between a metal oxide and phosphoric acid can be written as [2],[13]:

$$MO_x + n\ H_3PO_4 + m\ H_2O \rightarrow MH_{3n-2x}\left(PO_4\right)_n + \left(m + x\right)\ H_2O \tag{1}$$

where x denotes half of valence of **M**, $n \geq 2x/3$, and m is an arbitrary integer that decides the amount of water to be added in the reaction.

In the case of magnesium phosphate chemically bonded ceramics (MPCBC), the acido-basic **reaction 1** of orthophosporic acid with MgO yields the following products [2]:

$$MgO + 2\ H_3PO_4 \rightarrow Mg\left(H_2PO_4\right)_2 \cdot H_2O \quad \Rightarrow x = 1,\ n = 2,\ \text{and} m = 0 \tag{2}$$

$$MgO + H_3PO_4 + 2\ H_2O \rightarrow Mg\left(H_2PO_4\right)_2 \cdot 3H_2O \quad \Rightarrow x = 1,\ n = 1,\ \text{and} m = 2 \tag{3}$$

$$3\ MgO + 2\ H_3PO_4 \rightarrow Mg3\left(PO_4\right)_2 + 3\ H_2O \quad \Rightarrow x = 1,\ n = \frac{2}{3},\ \text{and} m = 0 \tag{4}$$

If acidic phosphate is used, the reaction proceeds as follows [2]:

$$MgO + Mg\left(H_2PO_4\right)_2 \rightarrow 2\ MgHPO_4 + H_2O \tag{5}$$

and similar reaction can be also written for the reaction of $Al(H_2PO_4)_3$ with $AlO_{3/2}$:

$$2\ AlO_{3/2} + Al\left(H_2PO_4\right)_3 \rightarrow 3\ AlPO_4 + 3\ H_2O \tag{6}$$

Reactions **5** and **6** indicate that partially acidic phosphate salt is only an intermediate phase. The dissociation of orthophosporic acid and the pH stability range of formed ionic species are given by **Eqs. 11–13** in **Chapter 9**. General dissociation reaction for acidic dihydrogen phosphate and its dissociation constant in the form of pK_{dis} ($pK_{dis} = -log\ K_{dis}$) can be written as follows:

$$M(H_2PO_4)_n \leftrightarrow M^{n+} + n\ (H_2PO_4)^{n-} \quad (7)$$

and

$$pK_{dis} = -\log\left(\left[M^{n+}\right]\left[H_2PO_4\right]^{n-}\right) \quad (8)$$

The values of pK_{dis} for KH_2PO_4, $(NH_4)H_2PO_4$, $Mg(H_2PO)_2{\cdot}2H_2O$ and $Ca(H_2PO_4)_2{\cdot}H_2O$ are 0.15, -0.69, 2.97 and 1.15, respectively [2].

The preparation scheme for the synthesis of aluminum phosphate binder is shown in **Fig. 3**. The evolution of chemical composition of phosphate binder with Al:P ratio of 1.4:3 with temperature is introduced by **Table 1**.

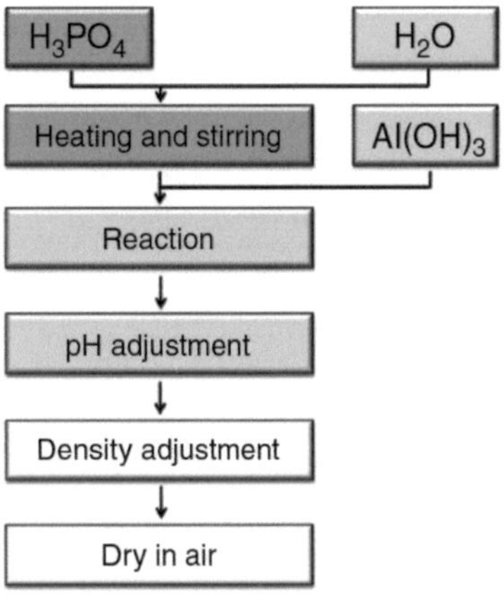

Fig. 3. Schematic procedure for the synthesis of aluminum phosphate binder [14].

Initially, the motivation for the development of these ceramic materials was the preparation of dental cements. Phosphate chemically bonded[1] ceramics (PCBC) find recently the application in diverse fields (**Fig. 2**), which include structural ceramics, refractory materials [15], toxic, radioactive [16],[17],[18],[19] or asbestos-containing[2] [20],[21],[22] waste management [23], oil drilling and bioceramics, pigments [1], etc. [2].

10.1.1. Ceramics and refractories

The broad term "ceramics" refers to any of a large family of materials that are usually inorganic and require high temperatures in their processing or manufacture. They are generally classified into glass, whitewares, including artware and structural ceramics, and refractories. In general,

[1] Chemical bonding as a means of solidification is very widely observed in nature. The formation of sedimentary rocks, such as carbonate rocks, lateritic soils and solidification of desert soils, are examples of this process [1].

Temperature [°C]	Dominant phase	Secondary phase
60	$AlH_3(PO_4)_2{\cdot}3H_2O$	—
105	$AlH_3(PO_4)_2{\cdot}3H_2O$ (monoclinic)	$Al(H_2PO_4)_3$ (hexagonal)
200	$Al(H_2PO_4)_3$ (hexagonal)	—
220	$AlPO_4$ (trigonal)	$AlH_2P_3O_{10}{\cdot}2H_2O$
250		
300		
400	$Al(PO_3)_3$	$AlH_2P_3O_{10}{\cdot}2.5H_2O$
500	$Al_2P_6O_{18}$	$Al(PO_3)_3$ (cubic)
600		
700		
800	$Al(PO_3)_3$ (cubic)	$AlPO_4$ (rhombic system)
900		
1000		

Table 1. The alteration of composition of phosphate binder with increasing temperature [14].

the term "glass" refers to an amorphous solid with non-directional properties, characterized by its transparency, hardness, rigidity at ordinary temperatures and capacity for plastic working at elevated temperatures. Major commercial uses of glass include plate or "float" glass, as used for windows and windshields; glass tubing, or formed shapes, used for electric lighting envelopes; and glass containers such as tumblers and bottles. Whiteware is characterized by a crystalline matrix held together by a glassy phase and usually covered by a glazed coating. Major classifications of whiteware are ceramic tiles, glazed and unglazed, for floors, walls and external use; sanitaryware in the form of toilets and lavatories; and tableware, from earthenware to fine china [24].

Kaolin is major raw material used for the fabrication of conventional ceramics. It is obtained from the alteration of granitoid rocks[3] [25]. It consists mostly of kaolinite and a small amount of impurities such as quartz, micas and other phyllosilicates. Firing of kaolinite induces numerous complex structural and microstructural transformations leading to the formation

[2] SEM picture of thin long fibrous structure of chrysotile ($H_4Mg_3Si_2O_9$ [20],[21],[22]) asbestos under the magnification of 98× (a), 5000× (b) and 10,000× (c).

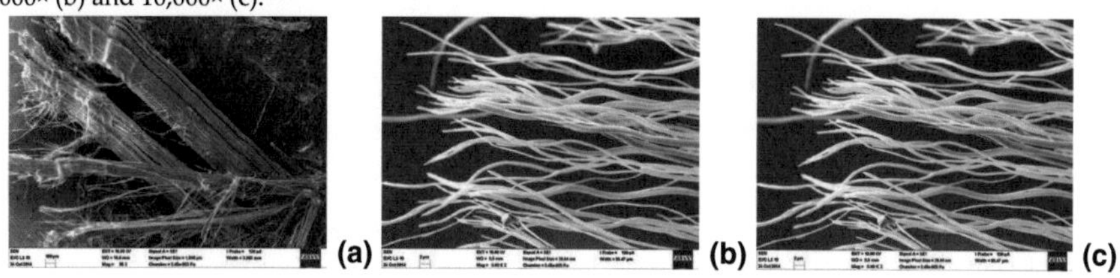

of mullite and silica (cristobalite) phase. Mullite phase is characterized by some advantageous properties such as good corrosion resistance, low dilatation coefficient, good creep and thermal shock resistances, thermal stability and high strength. These advantages make this phase favorable for different applications. Orthophosphoric acid reacts with aluminum from kaolin to provide new compounds, which are $Al(H_2PO_4)_3$ at room temperature and $AlPO_4$ when heated to temperatures above 800°C [10].

Approximately 70% of all refractories used in industry are in the form of bricks, which are cast in the shapes such as straights, soaps, splits, arches, wedges, keys, skews, jambs or other special and frequently patented shapes. Most industrial refractories are composed of metal oxides or of carbon, graphite or silicon carbide. Most refractory bricks are shaped by combining the size-graded refractory aggregate with a small amount of moisture and casting in a dry press. The important properties of any refractory, including its high-temperature strength, depend on its mineral makeup, the particle-size distribution of minerals and the way these materials react at high temperatures and in furnace environments. When a refractory is chosen for a particular service, the service conditions must be considered in the design. In proper selection of a refractory, these factors, together with an economic balance, must be considered so that the refractory ultimately produces the lowest cost per unit weight of product per unit weight of refractory consumed [26].

Fig. 4. Cutting plane throughout the block of chemically phosphate bonded refractory material containing large grains of calcined bauxite (a) and electron microscopy (SEM) picture of ceramic body (b).

The materials must be chosen with generous safety margins in the temperature capability and the refractory construction system, whether brick, monolithic or fiber, should be suitable not only for the operating conditions but also for the type of equipment concerned and the construction conditions. Refractory concretes and castables (monolithic refractories) are especially suitable for these small burner quarls. Fibers and refractory ceramic fibers (glass,

[3] Granular crystalline rock consisting essentially from quartz, orthoclase-feldspar and mica. Usually is light gray, white or light in color [25].

mineral, ceramic fibers and whiskers) become an important construction as well as insulation material, although the brickwork has the longest history of the development [27].

Aluminosilicate refractories are manufactured using refractory clays, sillimanite minerals, bauxite (**Fig. 4**) and mixtures of alumina and silica sand. They will refer, somewhat arbitrarily, to common crystalline compounds with melting temperatures of at least 1500°C. The major categories of traditional refractories are fire clays, high alumina and silica. The choice of material for traditional refractory applications, as well as for advanced material applications, was and is based on balancing the cost and the performance lifetime [28].

Industries involved in steel melting and casting of special alloys are always interested in using the maintenance-free, functional refractories to achieve energy-efficient metal processing. They seek for the development of new and advanced thermal ceramics for protective thermal insulation linings and molten metal handling crucibles. Conventionally, porous to dense Al_2O_3, SiO_2, MgO, ZrO_2, SiC[4] [29] and fireclay ceramic bodies were used as thermal insulation refractory liners [30]. Phosphate bond can also be utilized in manufacturing of refractory and wear-resistant and of protective coatings on metal [14] or ceramic surfaces [31]. Intensive development of ceramic materials has increased the availability of well-characterized engineering ceramics capable of the utilization over the range of temperatures and atmospheres [32].

Refractory castables can be classified according to different aspects including the content of calcium, binder source, overall chemical composition, bulk density, application method and others. Binders commonly used in monolithic products may be of various classes [33]:

1. Hydraulic (cement and hydratable alumina) [34]
2. Chemical (phosphoric acid, silicates, geopolymers, etc.) [33],[35]
3. Sol-gel (colloidal silica or alumina) [33]

Regarding the chemical ones, the bond strength can be provided by the addition of phosphates (dry or solution) or by in situ generation of phosphates (formed via the reaction with added H_3PO_4) in the refractory structure. The reaction of Al_2O_3 or $Al(OH)_3$ with orthophosphoric acid can be described by reactions [33]:

$$6\ H_3PO_4 + Al_2O_3 \rightarrow 2\ Al(H_2PO_4)_3 + 3\ H_2O \tag{9}$$

$$3\ H_3PO_4 + Al(OH)_3 \rightarrow Al(H_2PO_4)_3 + 3\ H_2O \tag{10}$$

A cold-setting refractory material was developed by Hipedinger et al [36] via the magnesia-phosphate reaction. A cement paste based on alumina, silica fume, magnesia and orthophosphoric acid or monoaluminum phosphate was designed to form cordierite-mullite during

[4] It was proved that H_3PO_4 is effective binder for SiC [29].

heating. This cement paste set at room temperature and $MgHPO_4 \cdot 3H_2O$ phase (newberyite) was observed, but amorphous phases were predominant. Two exothermic effects were detected during the setting process corresponding to the acido-basic reaction of magnesia with phosphates and to the formation of bonding hydrates. At 1100°C, *c*-$AlPO_4$ was formed by the reaction of alumina with orthophosphoric acid or monoaluminum phosphate. At 1350°C, the dominant crystalline phases were cordierite and mullite. A refractory concrete with obtained cement paste and a cordierite-mullite aggregate (scrap refractory material) was prepared.

The acid phosphate impregnation, with ozone pretreatment, improves the oxidation resistance of carbon materials (polycrystalline graphite and pitch-based carbon fiber), as shown by the weight measurement in air up to 1500°C. The impregnation involves using phosphoric acid and dissolved aluminum hydroxide in the molar ratio of 12:1 and results in rough, white and hard aluminum metaphosphate coating of the weight of about 20% of that of the carbon before the treatment. Without ozone pretreatment, the impregnation is not effective. Without aluminum hydroxide, the impregnation even degrades the oxidation resistance of carbon [37].

10.1.2. Dental phosphate cement

A variety of cements are used in modern clinical dentistry, such as glass ionomers, zinc phosphate and zinc polycarboxylate [38],[39]. Dental zinc phosphate cement is primarily used for the cementation of indirect restorations, such as crown and bridges. It has the longest record of any cement, approximate 100 years, and has remained popular throughout this time. Zinc phosphate cements are also considered the strongest among the dental cements. However, it is also applied for temporary fillings, cavity bases and buildings of teeth beneath crowns. Zinc phosphate cement is primarily in contact with the pulp-dentin system and in certain cases (e.g. temporary fillings) with the gingiva. A variety of cementing materials are currently used as the bases and luting[5] agents, but zinc phosphate cement has been used for many decades. Phosphoric acid-based cements originated from Ostermann's formula from 1832, which was composed of calcium oxide and anhydrous phosphoric acid. In 1902, Fleck established a formula that is similar to that being in use today [40],[41],[42],[43],[44].

The powder is mainly a mixture of zinc oxide and up to 13% magnesium oxide. The liquid is an aqueous solution of phosphoric acid containing 38 – 59% H_3PO_4, 30 – 55% water and 0 – 10% zinc. Aluminum is essential to the cement-forming reaction, and zinc moderates the reaction between powder and liquid, allowing adequate working time and sufficient quantity of powder to be added for optimum properties of cement. When the powder is mixed with liquid, phosphoric acid attacks the surface of particles, dissolving zinc oxide, which releases zinc ions into the solution. Aluminum in the liquid reacts with phosphoric acid to form zinc aluminophosphate gel on remaining portion of particles. Thus, the cement reveals a cored structure consisting primarily of non-reacted zinc oxide particle core embedded in a cohesive amorphous matrix of zinc aluminophosphate (glasslike phosphate). Aluminum phos-

[5] The word 'luting' implies the use of a molded or moldable article to seal a space or to cement two components together [44].

phate, in addition to its role as a retarder, contributes to the increase in mechanical hardness of cement [40],[45].

Setting reactions and resultant structure of zinc phosphate cements are largely based on the formation of hopeite ($Zn_3(PO_4)_2{\cdot}4H_2O$) and/or zinc phosphate hydrate ($Zn_2P_2O_7{\cdot}3H_2O$) when using orthophosphoric acid (OPA) cement-forming liquids. OPA solutions buffered with aluminum and zinc ion produced better mechanical properties than non-buffered OPA solutions because of the formation of hopeite and amorphous phase. The development of crystalline forms of phosphate hydrates of zinc and magnesium was retarded and/or prevented by the incorporation of aluminum and zinc ion in the cement-forming liquid [43].

This mechanism is similar to the cement-forming reaction described by Wilson [43],[46] in a dental silicate cement (**Fig. 5**). When H^+ ions attack the glass powder, Al^{3+}, Ca^{2+}, Na^+ and F^- are liberated from the glass, leaving behind an ion-depleted layer of silicate gel at the surface of glass particles. Liberated ions migrate and react with $H_2PO_4^-$, and salts precipitate. The principal reaction is the formation of an insoluble aluminum phosphate, the gel matrix. Associated side reactions are the precipitation of calcium fluoride and the formation of soluble sodium dihydrogen phosphate.

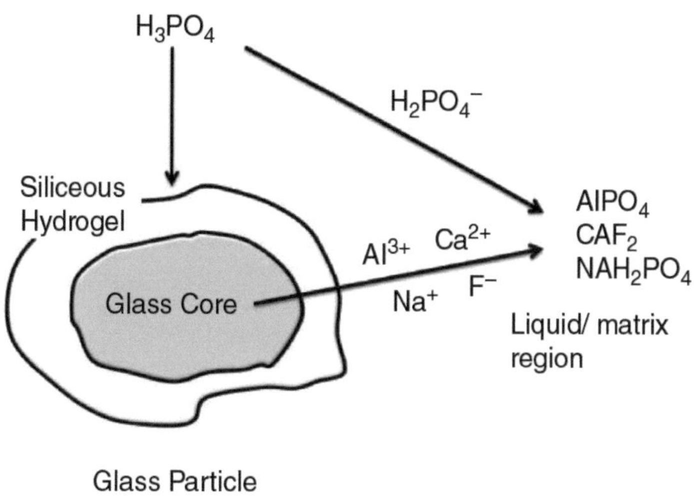

Fig. 5. Setting reaction of dental silicate cement [43].

10.2. Oil-well cement

The main application of the cement in an oil well is to stabilize the steel casing in the borehole and to protect it from corrosion. The cement is pumped through the borehole and is pushed upwards through the annulus between the casing and the formation. The cement is exposed to the temperature and pressure gradients of the borehole [47].

Various types of oil-well cements are distinguished. The most important is the phase composition of cement, primarily the C_3A phase content, which causes quick paste thickening. The main feature of oil-well cement is that it must remain sufficiently fluid for a long period, required for its pumping to deep well. Simultaneously, the temperature in the borehole increases with depth[6] [48],[49]. The types and properties of oil-well cements specified by the American Petroleum Institute (API) are introduced by Clinkers containing limestone loess, diatomite, pyritic ash and sand modified with gypsum and apatite were used for the manufacture of heat resistant oil-well cements. Apatite is also a good stabilizer for high belite cements [50].

API class	Special properties	Intended use
A	Same as ASTM Type I	Well depths up to 1800 m and temperatures of 27 – 77°C.
B	Similar to ASTM Type II, low C_3A, high sulfate resistance	
C	Similar to ASTM Type III, low C_3A, high sulfate resistance	Well depths up to 1800 m and temperatures of 80 – 170°C.
D	Low C_3A with set retarder	Depths up to 1800 – 3600 m and temperatures of 77 – 138°C.
E		Depths up to 1800 – 4200 m and temperatures of 77 – 138°C.
F		Depths up to 3000 – 4800 m and temperatures of 127 – 160°C.
G and H	Coarse-ground ASTM Types II and IV	Temperatures of 27 – 93°C.
J	Essentially β-C_2S and pulverized silica sand	Depths below 6000 m and temperatures > 177°C

Table 2 Types and properties of oil-well cements [49].

10.3. Phosphate glasses and glass-ceramics

In the past decades, optical waveguides have raised great interest, as they are the most fundamental and integral part of integrated optic circuits. Glass-based integrated optical devices have several obvious advantages over other technologies such as low intrinsic absorption in near-infrared region of the spectrum, minimized coupling losses to optical fibers and no intrinsic material birefringence compared to crystalline semiconductors. Phosphate glasses are regarded as excellent glass host for the waveguide laser fabrication mainly because of high solubility of rare-earth ions compared to other oxide glasses, which allows for high doping concentrations without significant lifetime reduction, resulting in high gain in short waveguides or cavities and a desirable feature in single-frequency lasers. High-performance

[6] The rate depends on the geothermal degree, which in Europe is about 33 $m \cdot °C^{-1}$ [48].

waveguide amplifiers or lasers are fabricated in various earth-ion-doped phosphate glasses and commercial phosphate glasses such as Kigre Q89 and Schott-IOG 1 [51].

Glass-ceramics are polycrystalline materials with an inorganic–inorganic microstructure, which are prepared from the base glass by controlled crystallization. This can be achieved by subjecting glasses to regulated heat treatment, which results in the nucleation and growth of one or more crystal phases within the glass [52]. Once a stable crystal nucleus has formed and begun to grow, there are a number of possible crystal growth mechanisms and these determine the final crystal morphology [53],[54]:

a. **Faceted crystal growth**: the favored growth sites on an atomic scale are the steps in rows of atoms. However, these sites can be easily eliminated by the very growth that they promote. Despite this, several types of lattice or crystal defects were shown to provide the growth sites that are impossible to be eliminated by the growth.

b. **Dendritic crystal growth**: formed dendritic crystals are characterized by their tree-like appearance. This type of solidification usually takes place in metals and sometimes occurs naturally in silicate minerals, especially in olivine during rapid cooling of lavas.

c. **Spherulitic crystal growth**: is most commonly associated with organic polymeric materials. The geological definition of a spherulite is: "A crystalline spherical body built of exceedingly thin fibers radiating outwards from a center and terminating on the surface of the sphere…".

Bioactive glass-ceramics are an alternative to synthetic HAP (**Section 10.9**) for the use in vivo both in restorative dental applications and in bone implantation [54],[55]. Artificial materials implanted into the bone defects are generally encapsulated by fibrous tissue isolating them from the surrounding bone. This is the normal response of the body towards inert artificial materials. However, some ceramics, such as bioglass, A-W glass-ceramics and sintered hydroxyapatite form a bone-like apatite on their surfaces in the living body and bond to living bone through this apatite layer. This bone-bonding ability is called the bioactivity [56].

These bioactive ceramics are already used clinically as important bone-repairing materials. Their bone-bonding ability is achieved by the formation of a biologically active apatite layer after the reaction of the ceramics with surrounding body fluid. Controlled surface reaction of the ceramics is an important factor governing its bioactivity as well as its biodegradability [56]. The preparation of glass-ceramics with high CaO/P_2O_5 ratio, containing large amounts of calcium phosphate crystals, is believed to be one of the best approaches to obtain the bioceramic implants suitable for bone replacement/regeneration [57].

10.3.1. Alkaline aluminum phosphate glasses

Alkaline aluminum phosphate glasses (NMAP) with excellent chemical durability for thermal ion-exchanged optical waveguide were investigated by Wang et al [51]. The transition temperature (T_g = 470°C) is higher than the ion-exchange temperature (390°C), which is favorable for sustaining the stability of the glass structure for planar waveguide fabrication. The glass-forming region in the Na_2O-Al_2O_3-P_2O_5 system is shown in **Fig. 6** [51],[58].

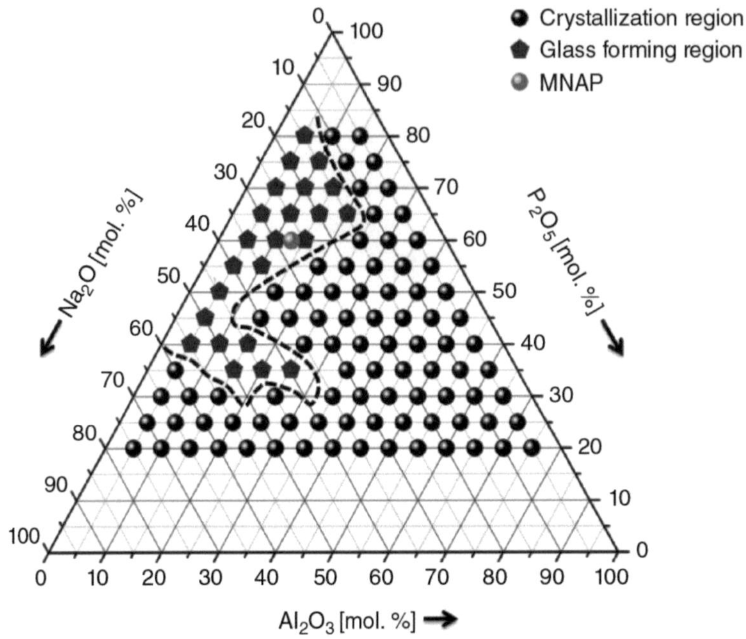

Fig. 6. Ternary phase diagram of Na_2O-Al_2O_3-P_2O_5 system [51],[58].

10.3.2. Iron phosphate glasses

Iron phosphate glass is a versatile matrix for the immobilization of various radioactive elements found in high-level nuclear waste (HLW). Among various compositions of iron phosphate glass, the one with 40 mol.% Fe_2O_3-60 mol.% P_2O_5 was found to be chemically durable. It also has the ability to accommodate large amounts of certain nuclear wastes, especially those that are not well suited for borosilicate glasses. Better chemical durability of iron phosphate glass is attributed to the presence of more hydration-resistant Fe-O-P bonds compared to P-O-P bonds available in other phosphate glasses [59],[60],[61].

10.3.3. Lithium vanado-phosphate glasses

Lithium vanado-phosphate (LiVP) glasses have been largely studied due to their potential application as cathode materials as a result of mixed electronic-ionic conductivity character.[7] Furthermore, lithium and vanadium structural rearrangements in the glass matrix could modify the transport properties of the systems. Interestingly, the modifier ions depolymerize the glass network, creating useful channels that enhance the ionic conductivity, but they can also break some $V^{4+}/O/V^{5+}$ linkages that are essential to the electronic conductivity because they are supposed to be the preferential path for small polaron hopping. The population of

---

[7] Fast ion conducting (FIC) phosphate glasses have become very important due to a wide range of applications in solid-state devices [63].

$V^{4+}/O/V^{5+}$ paths depends even on the V/P ratio and it can modify the main structure role of vanadium ions in glasses [62],[63],[64].

10.3.4. Apatite-wollastonite glass-ceramics

Wollastonite ($CaSiO_3$) is white glassy silicate mineral that occurs as masses or tabular crystals of metamorphosed limestone. A silica chain GC that contained crystalline apatite and wollastonite (AW) was introduced in MgO-CaO-SiO_2-P_2O_5 glassy matrix and it showed excellent bioactivity, biocompatibility, machinability and adequate mechanical properties such as Young's modulus (117 GPa), compressive strength (1080 MPa) and bending strength (215 MPa) [52],[65],[66],[67].

Wollastonite-2M and pseudowollastonite (low- and high-temperature forms of wollastonite, respectively) are the most common calcium silicate biomaterials proposed for bone tissue regeneration [68]. The major drawback of the $CaSiO_3$ bioceramics is their relatively fast dissolution rate that could reduce their mechanical strength. In addition, the pH of surrounding medium significantly increases, which could affect the osseointegration of the substitute material within the natural bone. The problem can potentially be solved by the development of multiphase materials containing highly dissolvable phases such as wollastonite, on one hand, and stable phases such as HAP, on the other hand [69].

From the bioactive ceramics, the A/W glass-ceramics show high bioactivity and high mechanical strength [56]. The A/W glass-ceramics composed of apatite and wollastonite crystalline phases in a glassy matrix was developed by Kokubo [70]. This bioceramics is highly bioactive and also mechanically strong in comparison with other glasses and glass-ceramics because of wollastonite and apatite crystals' presence.

The A/W glass-ceramics is used in some medical applications, either in powder form as a bone filler or as a bulk material. These materials are currently manufactured by the powder processing methods, providing uniform crystallization of apatite and wollastonite phases in the glassy matrix, as the crystallization of parent glass in a bulk form leads to the appearance of large cracks [69],[71]. The glass-ceramics exposed to the SBF releases predominantly Ca and Si ions, due to the dissolution of amorphous phases and wollastonite-2M, leading to the formation of an apatite-like layer on the surface of the material [69].

The addition of ZnO increased the chemical durability of A-W glass-ceramics, resulting in a decrease in the rate of apatite formation in simulated body fluid. On the other hand, the release of zinc from the glass-ceramics increased with increasing ZnO content. The addition of ZnO may provide bioactive CaO-SiO_2-P_2O_5-CaF_2 glass-ceramics with the capacity for appropriate biodegradation as well as the enhancement of bone formation [56]. The effect of MgO was investigated by Ma et al [72]. As the MgO content increased, the glass crystallization temperature increased and the crystallization of the glass-ceramics was changed from the bulk crystallization to the surface crystallization. The addition of MgO slowed down the rate of dissolution and retarded the formation of apatite layer.

10.3.5. Apatite-mullite glass-ceramics

Apatite-mullite glass-ceramics crystallize from the glass of generic composition SiO_2-Al_2O_3-P_2O_5-CaO-CaF_2 to form an osseoconductive apatite phase existing as spherulites within a mullite matrix. The formation of apatite spherulites is accompanied by a depletion zone, from which calcium, phosphate and fluorine is taken and added to the growing crystal. The depletion zone also inhibits the formation of further apatite crystals in the immediate vicinity due to the glass' compositional similarity with a mullite composition. As mullite begins to crystallize, there is interdependence between the growth of apatite and mullite. Spherulitic grain boundaries in partially devitrified apatite-mullite ceramics act as crack promoters, offering preferential paths to propagation due to the grain boundary interfacial surface energy [54],[73].

10.3.6. FAP-anorthite-diopside glass-ceramics

There is a considerable interest in oxyfluoride glasses and glass-ceramics for laser amplifiers and up-conversion processors. Fluoride-containing crystals have low phonon energies and apatite crystals, in particular, are good host phases to adsorb rare-earth ions. Fluoride-containing glass-ceramics also often crystallize on a nanoscale, which is an added advantage, since optically transparent materials are required for the applications such as fiber amplifiers. The evidence of the nanoscale crystallization in an FAP-anorthite-diopside-based glass-ceramics was found by Hill et al [74].

10.3.7. Apatite-wollastonite ceramics

Wollastonite-hydroxyapatite ceramics was successfully prepared by a novel method, corresponding to the thermal treatment of a silicone embedding micro- and nanosized fillers in air. $CaCO_3$ nanosized particles, providing CaO upon the decomposition, acted as "active" filler, whereas different commercially available or synthesized hydroxyapatite particles were used as "passive" filler. The homogeneous distribution of CaO, at a quasi-molecular level, favored the reaction with silica derived from the polymer, at only 900°C, preventing extensive decomposition of hydroxyapatite. Open-celled porous ceramics suitable for scaffolds for bone-tissue engineering applications were easily prepared from the filler-containing silicone resin mixed with sacrificial PMMA microbeads as the templates [75].

10.3.8. Oxyapatite glass-ceramics

The crystallization of oxyapatite of the composition $NaY_9(SiO_4)_6O_2$ ($P6_3/m$, $\boldsymbol{a}$ = 9.334 Å and $\boldsymbol{c}$ = 6.759 Å, $c{:}a$ = 1:0.7241, V = 509.97 Å, **Fig.** 7) from three different glasses from the SiO_2-B_2O_3-Al_2O_3-Y_2O_3-CaO-Na_2O-K_2O-F glass-ceramics system with different F and B_2O_3 content after the

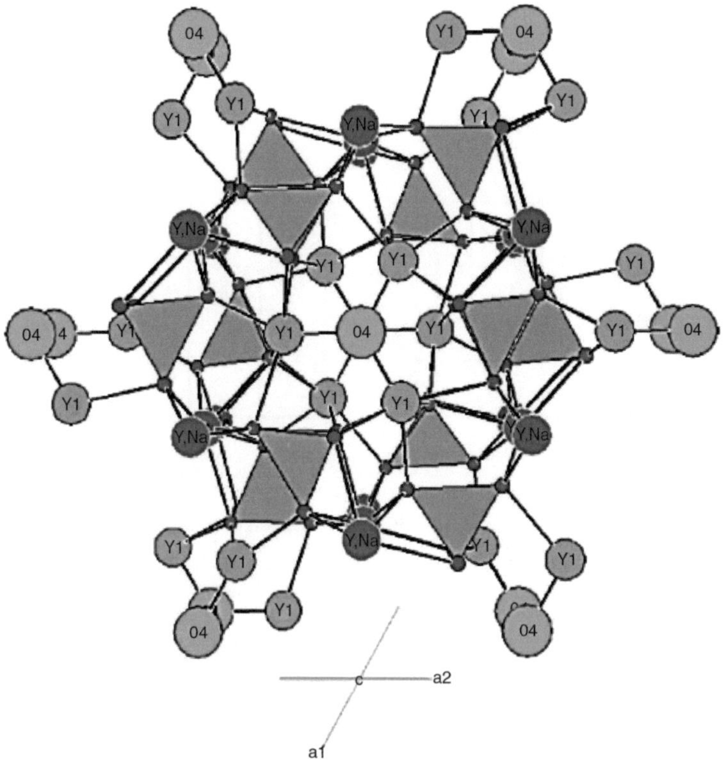

Fig. 7. Crystal structure of oxyapatite $NaY_9(SiO_4)_6O_2$ (perspective view along the *c*-axis) where the tetrahedra represents $[SiO_4]^{4-}$ structural units [76].

heat treatment was observed by VAN'T HOEN et al [76]. The formation of oxyapatite proceeds according to the mechanism of controlled surface nucleation and crystallization. Therefore, applying suitable chemical composition and heat treatment, it is possible to produce glass-ceramics containing oxyapatite crystals. This type of crystal exhibits similar crystal structure as that of fluorapatite. Therefore, there is an isotype relationship between those two phases: the differences in the crystal structures are found in the structural units, with fluorapatite containing $[PO_4]^{3-}$ and $[F]^-$, while oxyapatite containing $[SiO_4]^{4-}$ and $[O]^{2-}$. Because of its specific optical properties (high opacity), this new glass-ceramic material may be used as layering material for dental restoration [76].

10.4. Solid oxide fuel cells

Solid oxide fuel cell (SOFC) is an electrochemical energy-conversion device, which offers tremendous promise for delivering high electrical efficiency and significant environmental benefits in the terms of fuel flexibility (hydrocarbons and municipal waste) as well as clean and efficient (>70% with fuel regeneration) electric power generation. SOFC produces useful electricity by the reaction of fuel with an oxidant via the diffusion of oxide ions (or protons) through an ion-conducting solid-electrolyte layer [77].

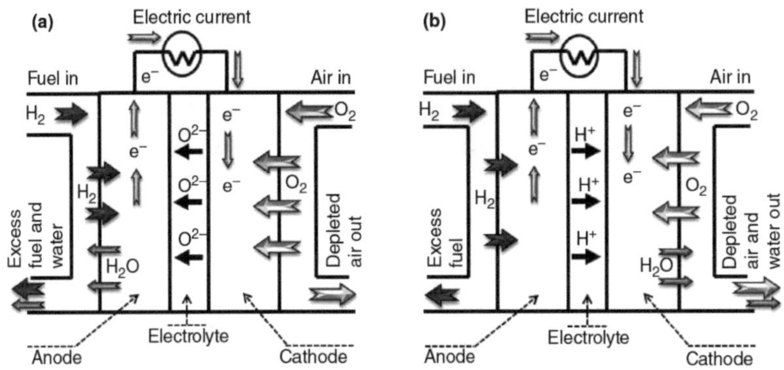

Fig. 8. Schematic diagram of solid oxide fuel cell (SOFC) showing non-ion-conducting electrolyte (a) and proton-conducting electrolyte during its operation (b) [77].

SOFC is composed of a dense electrolyte layer that is sandwiched between two porous electrodes (i.e. cathode and anode) as shown in **Fig. 8**. SOFC can use either oxide ion (a) and/or proton conduction through the electrolyte (b). Electrons generated through the oxidation of fuel on anode are accepted for the oxygen reduction on cathode, which completes the external circuit. The electricity is, thus, produced by the flow of electrons in the external circuit (from the anode to the cathode). Since the current is obtained via the diffusion of oxide ions (or protons) through a solid electrolyte, it becomes imperative to use high operating temperatures (~800 – 1000°C) for achieving high ionic conductivity (of ~0.1 $S{\cdot}cm^{-1}$) [77],[78].

The first fuel cell was invented in 1838 by an English scientist, William Grove. He named it "wet cell battery" or "Grove cell", which operated by reversing the electrolysis phenomena of water [77],[79]. The fuel cell, invented in 1839 by Grove, is an electrochemical device that converts the chemical energy of fuels directly into electricity and heat by electrochemically combining H_2, CO/H_2 or reformed hydrocarbons in fuel and an oxidant gas transported via an ion-conducting electrolyte. Direct combustion of fuels is eliminated here, which renders the fuel cells much higher conversion efficiencies compared to other conventional thermomechanical methods. Moreover, with fuel cells, the power generation is virtually noise-free and can produce 0.9 times lower emissions of NO_x and SO_x per unit of power output compared to that of conventional technologies. Additionally, it is possible to use the fuel cells for combined heat and power (CHP or cogeneration) generation [77],[80].

The development of high-performance SOFC involves the material selection and operation-related issues (of anode, cathode, electrolyte, sealant and interconnects). These challenges open up the myriad research opportunities for researchers in the field of SOFC. A list of various materials used in SOFC is presented in **Fig. 9**.

http://dx.doi.org/10.5772/62217

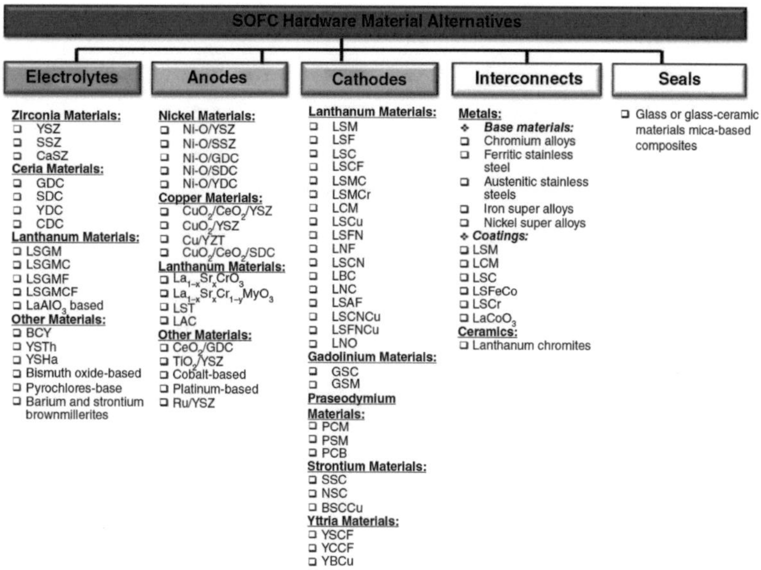

Fig. 9. Comprehensive list of various materials used in SOFC [77].

Apatite-type silicates are considered as promising electrolytes for solid oxide fuel cells. Lanthanum silicate with the composition of $La_{10}Si_{5.5}Al_{0.5}O_{26.75}$ was evaluated as an electrolyte with the electrode materials commonly used in SOFC, i.e. manganite, ferrite and cobaltite as cathode materials and NiO-CGO composite, chromium-manganite and Sr_2MgMoO_6 as anode materials by Marrero-López et al [81]. This electrolyte has conductivity values higher than those of YSZ and comparable to most important solid electrolytes proposed for the intermediate-temperature range, such as doped ceria and lanthanum gallate-based electrolytes. The chemical compatibility did not reveal appreciable bulk reactivity between silicate and many electrodes up to 1300°C [81]. Among several reported rare earth apatites, lanthanum silicates exhibit the ionic conductivity higher than their germinate counterparts [77].

On examining and modeling the probable conduction mechanism in apatite silicates, the atomistic simulation results suggest that the conduction in $La_{9.33}(SiO_4)_6O_2$ and $La_8Sr_2(SiO_4)_6O_2$ takes place via interstitial and vacancy mechanism, respectively [82],[83]. The predicted pathway appears to be a complex nonlinear "sinusoidal-like" process for the interstitial oxygen migration along the c-axis (**Fig. 10**), while a direct linear pathway is predicted for oxygen migration via the vacancy mechanism [77].

Zeng et al [84] developed a model capable of prediction of oxygen ion conduction from relative Coulomb electronic interactions in oxyapatites and rationalized observed experimental trends reported in the literature. Two types of fundamental chemical property, i.e. the electronegativities and ionic radii of the constituents, control oxygen ionic conduction in oxyapatites. Those two properties were used to represent the relative charge densities and the distances between charged units, respectively, and then to formulate the relative Coulomb energy. It was found that this relative Coulomb energy is linearly correlated to the oxygen ionic

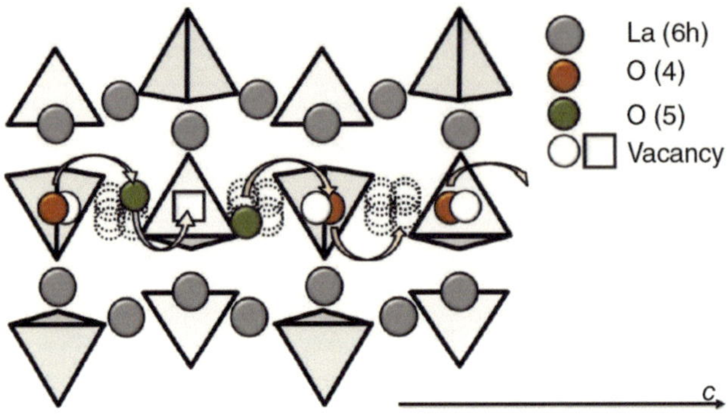

Fig. 10. Structural defect position and possible conduction mechanism along the c-axis representation of two adjacent unit-cells [77].

conductivity (in logarithmic form) in the oxyapatite systems. Doping a cation with large ionic radius and low electronegativity tends to increase the ionic conductivity of oxyapatite.

The c-axis-oriented apatite-type lanthanum silicate ($La_{10}Si_6O_{27}$) ceramics was prepared by NAKAYAMA et al [85] via the sintering process under high magnetic field. The degree of orientation in the $La_{10}Si_6O_{27}$ ceramics sintered at 1700°C was 48.1%. The conductivity of the c-axis-oriented ceramics is about 0.5 orders of magnitude higher than that of non-oriented ceramics. Higher conductivity is caused by the orientation of oxide ions in the grains composing the ceramics, which are located along the c-axis and are responsible for the ionic conduction.

10.5. Sensors

Compacted sinters of $Ln_{9.33+x/3}Si_{6-x}Al_xO_{26}$ ($0 \le x \le 2.0$, Ln = La, Nd and Sm) are composed of an apatite-like phase with a hexagonal structure. Compacted sinters were used as potentiometric oxygen sensors (**Fig. 11**). The concentration dependence of EMF was well expressed by the Nernst equation:

$$E_{obs} = \frac{RT}{nF} \ln \frac{p_{O_2}(I)}{p_{O_2}(II)} \tag{11}$$

Furthermore, the electron number $\boldsymbol{n}$ is comparable to the theoretical value of 4. The sensing characteristics of the sinters are comparable to those of the sensors with 3 and 8 mol.% YSZ [86].

http://dx.doi.org/10.5772/62217

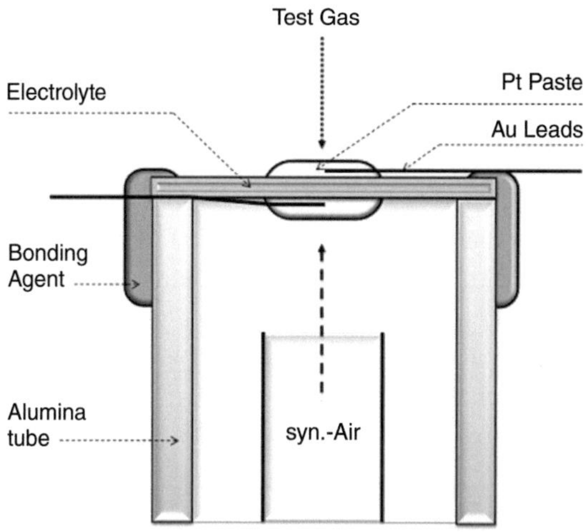

Fig. 11. Schematic picture of the O_2 concentration cell with the electrolyte of $Ln_{9.83}Si_{4.5}Al_{1.5}O_{26}$ where Ln = La, Nd and Sm [86].

Electrical properties and humidity sensor characteristics of lead hydroxyapatite material were reported by Tudorache et al [87]. The electrical characteristics of lead hydroxyapatite material treated at different temperatures made us focus on the analysis of the influence of water vapors upon the electrical characteristics. Thus, the electrical response to humidity adsorptive processes of lead hydroxyapatite material suggested that we analyze the material characteristics in terms of its use as a humidity sensor.

The humidity-sensitivity of yttrium-substituted calcium oxyhydroxyapatites was studied by Owada et al [88]. The logarithm of electrical resistance of the present sensors decreased linearly with increasing relative humidity (RH) from 30 to 65%. The resistance of $[Ca_{9.0}Y_{1.0}](PO_4)_6[O_{1.5}\square_{0.5}]$ with the highest OH vacancy content was about one order of magnitude lower than that of calcium hydroxyapatite. It was found that the larger ratio of surface hydroxyl groups per unit surface area in the sample, the lower the resistance and the higher the amount of OH vacancies.

Using the hydroxylapatite ceramics as CO_2 sensor, which is based on electrical conductivity changes, was investigated by Nagai et al [89]. Starting powders prepared by usual wet process were cast in film with an organic vehicle and fired on alumina substrates after the electrodes had been arranged. It was necessary to soak the samples in a $CaCl_2$ solution in order to make them reactive with CO_2. Both D.C. and A.C. measurements were carried out in various atmospheres including air, CO_2 and air containing different amounts of CO_2.

10.6. Phosphors

Fluorescent lamps typically have transparent glass envelope enclosing sealed discharge space containing an inert gas and mercury vapor. When subjected to a current provided by electrodes, mercury ionizes to produce the radiation having the primary wavelengths of 185 nm and 254 nm. This ultraviolet radiation, in turn, excites phosphors on the inside surface of the envelope to produce visible light that is emitted through the glass. Generally, the fluorescent lamp for illumination uses a phosphor that absorbs the 254 nm Hg-resonance wave; the phosphor is activated so as to convert the ultraviolet light into the visible light. In order to improve the color-rendering properties and the emission output of fluorescent lamps, efficient illumination of a white color has been recently provided using a three-band-type fluorescent lamp, which employs a mixture of red, green and blue-emitting phosphors. In such three-band-type phosphor lamp, the emitting colors of the respective phosphors are considerably different from one another. Therefore, if the emitting intensity of any of the three corresponding phosphors is decreased, the color deviation occurs, degrading the color-rendering properties of the lamp [90]. The literature dedicated to the preparation of apatite-type light-emitting phosphors is really abundant.

A series of orange-red-emitting $Ba_2Y_3(SiO_4)_3F$:xSm^{3+} ($0.003 \le x \le 0.08$) fluorosilicate apatite phosphors were synthesized via the conventional solid-state reaction by Yu et al [91]. The emission spectra of the $Ba_2Y_3(SiO_4)_3F:Sm^{3+}$ phosphors contained some sharp emission peaks of Sm^{3+} ions centered at 564, 601, 648 and 710 nm. The strongest one is located at 601 nm. The optimum dopant concentration of Sm^{3+} ions in $Ba_2Y_3(SiO_4)_3F$:xSm^{3+} is around 3 mol.% and the critical transfer distance of Sm^{3+} was calculated to be 26 Å. The quenching temperature is above 500 K.

Red-emitting phosphors $Ba_2Gd_8(SiO_4)_6O_2:Eu^{3+}$ (BGS:Eu^{3+}) with silicate apatite structure were prepared by Liu et al [92] via the conventional high-temperature solid-state reaction method. There are two different sites (4*f* and 6*h* [93]) for Eu^{3+} occupying the host. It was found that the phosphors BGS:Eu^{3+} exhibit red emission with high quenching concentration at ~70.75 at.%, and the critical transfer distance of Eu^{3+} in BGS:Eu^{3+} was calculated to be ~12.3 Å. More importantly, it has better CIE chromaticity coordinate for white light-emitting diode (w-LED) application in comparison with commercial phosphor $(Y,Gd)BO_3:Eu^{3+}$ (YGB:Eu^{3+}) under near-ultraviolet (n-UV) 393 nm excitation [92],[94]. White Tb^{3+}/Sm^{3+} ions co-doped $Ca_2La_8(GeO4)_6O_2$ (CLGO) phosphors prepared by Jeon et al [95] show observable emission spectra under 374 nm excitation.

A novel blue-emitting phosphor $Sr_8La_2(PO_4)_6O_2:Eu^{2+}$ was synthesized by Liu et al [96] via conventional high-temperature solid-state method and its photoluminescence (PL) properties were investigated for the applications in white light-emitting diodes. The phosphor exhibited strong broad absorption band in the near-ultraviolet (n-UV) range and generated bright-blue emission centered at 442 nm upon 365 nm excitation light. The critical Eu^{2+} quenching concentration (QC) mechanism was verified to be the dipole-dipole interaction.

A green-emitting phosphor of Eu^{2+}-doped $Ca_5(PO_4)_2SiO_4$ was prepared via a solid-state reaction by Roh et al [97]. The phosphor was excited at the wavelengths of 220 – 450 nm, which was suitable for the emission band of near-ultraviolet (n-UV) light-emitting diode (LED) (350 – 430 nm). In $Ca_5(PO_4)_2SiO_4:Eu^{2+}$ phosphor, there were three distinguishable Eu^{2+} sites, which resulted in a strong green emission peaking at 530 nm and broad bands up to 700 nm.

10.7. Catalysts

Catalysts are usually defined as the substances that increase the rate at which a chemical reaction approaches the equilibrium without becoming permanently involved in this reaction. Basically, the catalysis can be divided to [98]:

i. Homogeneous catalysis,

ii. Heterogeneous catalysis.

Heterogeneously catalyzed process is more complex because the catalyst is not uniformly distributed throughout the reaction medium. Considering a two-phase system, either vapor/solid or liquid/solid, with the catalyst in the solid phase, the several steps need to be realized to complete the catalytic cycle [98]:

1. Transport of the reactant to the catalyst;
2. Adsorption, i.e. the interaction of the reactant with the catalyst;
3. Reaction of adsorbed species to the products;
4. Desorption of the products from the catalyst;
5. Transport of the product away from the catalyst surface.

Due to their versatility in anionic and cationic composition and their ability to adsorb organic and organometallic molecules as well as metallic salts, the surface properties of apatites can be tuned and they can behave as powerful catalysts in a wide range of organic reactions. In many cases, the apatite-based catalysts can be used without a solvent and show good recycling capacity. The catalytic properties mainly arise from the acid-base character of the apatite's surface. In some cases, adsorbed moieties are responsible for the catalytic properties, apatites playing the role of a solid support. Finally, the catalytic activities can result from the combination of properties of the apatite's surface and of the adsorbed or anchored moieties [99].

The oxidative Glaser-Hay coupling reaction of terminal alkynes is a very important reaction in organic chemistry to achieve the synthesis of diyne compounds. In general, the reaction is performed under homogeneous conditions using Cu(I) or Cu(II) salts in the presence of a reagent such as tetramethylethylenediamine (TMEDA), which can bind to copper ions, an organic base and dioxygen. Although this reaction is known for a long time, the mechanism is still under the discussion. It is possible to catalyze the Glaser-Hay reaction under heterogeneous conditions using Cu-modified hydroxyapatite (Cu-HAp). With several para-substituted phenyl-acetylenes and alkynols, we can show that Cu-HAp acts as a catalyst for single-bond coupling reactions leading to diyne derivatives in high yields without using auxiliary chelating molecules and organic bases. These heterogeneous conditions allow easy recovery of the catalyst and simplify the purification work-up.

The oxidative Glaser-Hay coupling reaction of terminal alkynes (acetylenes) is a very important reaction in organic chemistry to achieve the synthesis of diyne compounds. In general, the reaction is performed under homogeneous conditions using Cu(I) or Cu(II) salts in the presence of base (ethanolic ammonia solution, tetramethylethylenediamine, pyridine, ...), which can bind to copper ions, an organic base and dioxygen [100],[101]:

$$R-C\equiv CH \xrightarrow[\text{base}]{\text{CuCl}} R-C\equiv C-C\equiv -R \quad (12)$$

Although this reaction is known for a long time, the mechanism is still under the discussion. It is possible to catalyze the Glaser-Hay reaction under heterogeneous conditions using Cu-modified hydroxyapatite (Cu-HAp). With several para-substituted phenyl-acetylenes and alkynols, where Cu-HAp acts as a catalyst for single-bond coupling reactions leading to diyne derivatives in high yields without using auxiliary chelating molecules and organic bases. These heterogeneous conditions allow easy recovery of the catalyst and simplify the purification work-up [100].

The apatite catalyst was utilized for the catalysis of the synthesis of n-butanol, 1,3-butadiene and high octane fuel from bioethanol. The process requires relatively low temperature. The synthesis shows significantly lower cost compared to n-butanol derived from petroleum-based processes. The technology offers a closed-loop system with no waste or emissions [102].

10.8. Phosphate conversion coatings

Conversion coatings provide the resistance to corrosive environments. Phosphate conversion coatings (PCC) bring about the transformation of metal substrates into new surfaces having non-metallic and non-conducting properties. The transformations occur in phosphating solution containing divalent metal phosphates and, in some cases, in solutions containing monovalent metal phosphates. Generally, the solutions are prepared from liquid concentrates containing one or more divalent metals (zinc, magnesium, calcium, etc., phosphates), free phosphoric acid and an accelerator. Three types of phosphate conversion coatings are currently being used [103],[104]:

1. **Zinc phosphate coatings** are often used as a pretreatment for painted parts. They are also used to impart the corrosion resistance and to aid in cold-forming operations.

2. **Iron phosphate coatings** are primarily used to form a passive substrate under paints.

3. **Manganese phosphate coatings** are primarily on machined parts such as gears and internal combustion engine components as an anti-scuff film for the break-in wear.

The addition of metal ions, such as cupric ions, to a conversion bath greatly reduces the formation time and the size and nonuniformity of coating crystals. Copper, which is cathodic to dissolving metals, deposits on the base metal to form many local cells and thus to increase the potential difference between local anode and cathode site. Nickel ions behave differently than cupric ions, and their benefit results from the catalytic action associated with the release of molecular hydrogen. Furthermore, the addition of Ni^{2+} and Mn^{2+} into the treating solution refines the grain size and reduces the porosity of phosphate conversion coatings on electro-galvanized steels. While Ni exists in both the zero-valance state and the two-valance state, Mn is mainly present in the two-valance state in the phosphate conversion coating [103],[105].

In fact, all of the chemical reactions of phosphating process are based on mutual interactions of metal immersed in phosphate bath and redox of the accelerators. Generally, phosphating proceeds in the acidic solution containing Zn^{2+}, Mn^{2+}, Ca^{2+}, Na^{+}, Fe^{2+} and Mg^{2+}. Phosphating with different metal substrates and types is not of the same reaction mechanism. For example, when a pure iron is immersed in the phosphating solution, iron dissolves on the micro-anodes through the following reaction [106]:

$$Fe + 2\ H^{+} \rightarrow Fe^{2+} + H_2 \tag{13}$$

The hydrogen evolution occurs at the micro-cathodic sites resulting in an increase of pH value at the metal-solution interface. This change in pH alters the dissociation equilibrium, which leads to the formation of PO_4^{3-}:

$$H_3PO_4 \rightarrow H_2PO_4^{-} + H^{+} \rightarrow HPO_4^{2-} + 2\ H^{+} \rightarrow PO_4^{3-} + 3\ H^{+} \tag{14}$$

When PO_4^{3-} and Me^{2+} (metal ion, e.g. Zn^{2+}, Mn^{2+}, Ca^{2+} and Fe^{2+}) in the solution reach the saturation, the deposition of insoluble phosphate will be achieved. Then, it can crystallize in PCC coating, as shown in **Fig. 12**.

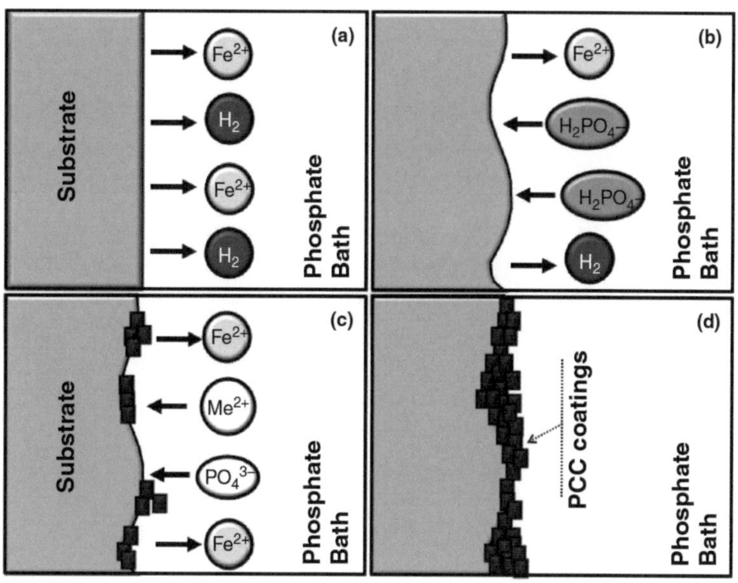

Fig. 12. Schematic representation of the deposition process of PCC coating on the surface of pure iron [106].

In the zinc phosphating solution, the addition of Zn^{2+} supports the formation of crystals of zinc phosphate. Zinc ions combine with phosphate ions to form an insoluble film. The formation of hopeite is described by the reaction [106]:

$$3\ Zn^{2+} + 2\ H_2PO_4^- + 2\ H^+ + 4\ H_2O + 6\ e^- \rightarrow$$
$$Zn_3(PO_4)_2 \cdot 4H_2O + 3\ H_2 \tag{15}$$

In some cases, Zn and ZnO were found in the phosphating process in reactions:

$$Zn + 2\ e^- \rightarrow Zn \tag{16}$$

$$Zn + 2\ H_2O \rightarrow Zn(OH)_2 + H_2 \tag{17}$$

$$Zn(OH)_2 \rightarrow ZnO + H_2O \tag{18}$$

In the phosphate solution of coexisting Zn^{2+} and Ca^{2+}, zinc calcium phosphate can be formed by the reactions:

$$Ca^{2+} + HPO_4^{2-} \rightarrow CaHPO_4 \tag{19}$$

$$Ca^{2+} + 2\ Zn^{2+} + 2\ H_2PO_4^- + 2\ H_2O \rightarrow CaZn_2(PO_4)_2 \cdot 2H_2O + 4\ H^+ \tag{20}$$

It is notable that the phosphating mechanism varies in different phosphating systems and materials [106].

The formation of conversion coating on zinc-coated samples under cathodic conditions was studied by Perrin et al [107] in a chromating bath containing phosphate (phosphate-chromate solution). Thick chromium phosphate (Cr-P) coating has two distinct layers: an outer porous layer and an inner and thinner pore-free adherent one. Both layers contain chromium phosphate as the main constituent and, to a lower extent, zinc phosphate species, the concentrations of which decrease from the metal-coating interface outwards. Formed zinc phosphates have a general formula of $xCrPO_4 \cdot yZn_3(PO_4)_2 \cdot z\ H_2O$ where $x > (y, z)$.

10.9. Synthetic apatite analogues in tissue engineering

Biomaterials in general are based on the materials such as metals, polymers, and ceramics. Typical metallic biomaterials are based on stainless steel, cobalt-based alloys, titanium or titanium alloys and amalgam alloys. Polymeric biomaterial composites from monomers are based on amides, ethylene, propylene, styrene, methacrylates and/or methyl methacrylates. Biomaterials based on ceramics are found within all classical ceramic families (**Table 3**) including traditional ceramics, special ceramics, glasses, glass-ceramics, coatings and chemically bonded ceramics (CBCs) [13].

Ceramic material classification	Example
Traditional ceramics	Dental porcelain, leucite-based ceramics
Special ceramics	Al, Zr and Ti oxides
Glass	Bioglass (Na_2O-CaO-P_2O_5-SiO_2)
Glass-ceramics	Apatite-wollastonite, Li-silicate-based
CBCs	Phosphates, aluminates, silicates and sulfates

Table 3 Examples of biomaterials based on ceramics [13].

Whereas many chemists and materials scientists consider the biomaterial to be synthetically produced material, most biologists, geologists and mineralogists consider the materials such as bone and tooth, which are biologically produced, to be the biomaterials. Also a common reference to Ca:P (1.67, **Table 7** in **Chapter 1**) ratio is usually used in the biomaterials literature, which disregards the fact that different calcium phosphate phases have different crystalline structures.

There are many phosphate minerals and salts (**Table 4**) that do not have the crystalline structure of apatite [108],[109]. Apatite-based materials have attracted a considerable interest for orthopedic and dental applications because of their biocompatibility and tight bonding to bone, resulting in the growth of healthy tissue directly onto their surface. Several combinations of apatite and other phases were proposed in order to improve poor mechanical properties of apatite [110].

Typical acronym	Chemical name	Chemical formula	Mineral name	Structure	Ca/P ratio
HAP, HA	Tribasic calcium phosphate	$Ca_5(PO_4)_3OH$	Hydroxylapatite	Apatitic	1.67
ACP	Amorphous calcium phosphate	?	–	–	?
PCHA, PCA	Poorly crystalline hydroxylapatite	$Ca_5(PO_4)_3OH$	Hydroxylapatite	Apatitic	1.67
CAP	Carbonated apatite	Refer to **Sections 2.6** and **4.6**			
TCP	Tricalcium phosphate	$Ca_9(PO_4)_6$	Whitlockite	Non-apatitic	1.5
β-TCMP	Magnesium-substituted TCP	$Ca_9(PO_4)_6$	Whitlockite	Non-apatitic	≤1.5

Typical acronym	Chemical name	Chemical formula	Mineral name	Structure	Ca/P ratio
?	"Tricalcium phosphate"	$Ca_9(Mg,Fe^{2+})(PO_4)_6(HPO_4)$	Geologically occurring whitlockite	Non-apatitic	1.28
CPPD	Calcium pyrophosphate dihydrate	$Ca_2P_2O_7{\cdot}2H_2O$	–	Non-apatitic	1.0
γ-CCP	γ-Calcium pyrophosphate	$Ca_2P_2O_7$	–	Non-apatitic	1.0
OCP	Octacalcium phosphate	$Ca_8H_2(PO_4)_6{\cdot}5H_2O$	–	Non-apatitic	1.33
MON	Dibasic calcium phosphate	$Ca(HPO_4)$	Monetite	Non-apatitic	1.0
DCPD	Dicalcium phosphate dihydrate	$Ca(HPO_4){\cdot}2H_2O$	Brushite	Non-apatitic	1.0

Table 4 Different apatitic and non-apatitic calcium phosphates [108].

Small organic molecules incorporated in apatite crystals act as porogens that control the porous structure of apatite single crystal. The presence of amino acid under the apatite synthesis conditions leads to firm bindings and encapsulation of amino acid within apatite single crystals. The amino acid elimination by heating or electron beam irradiation enhances the pore formation in apatite single crystals. Moreover, the incorporation of acidic amino acid into apatite induces the peapod-like nanotubes in apatite single crystals. That suggests the potential of using small organics for nanostructural control of apatite single crystals, which would be valuable for enhancing the drug loadings or for modulating the material digestion in vivo [111].

10.9.1. Biological apatite in bone tissue engineering

Tissue engineering (TE) techniques were developed to recover or enhance lost tissue function and structure. In biological hard tissues, for example, lost portions can be effectively reconstructed by the control of environmental factors, physical stimulation, addition of growth factors and by the use of degradable materials. These factors strongly facilitate the regeneration of macroscopic shape of defected hard tissues. Nevertheless, the differences in microstructure, and also in mechanical and physical properties, between regenerated and original hard tissues must be examined prior to clinical application [112],[113],[114].

For in vitro engineering of living tissues, cultured cells are grown on bioactive degradable substrates (scaffolds[8]) that provide the physical and chemical cues to guide their differentiation and assembly into three-dimensional structures. One of the most critical issues in TE is the realization of scaffolds with specific physical, mechanical and biological properties. Scaffolds act as substrate for cellular growth, proliferation and the support for new tissue

[8] Scaffolds might be defined as artificial structure capable of supporting the three-dimensional tissue formation, which allows the cell attachment and migration, the delivery and retaining of cells and biochemical factors and enables the diffusion of vital cell nutrients and expressed products. In the case of bone, scaffolds should replicate its architecture and three-dimensional structure with predetermined density, hierarchical pore distribution and interconnected pathways [109],[115].

formation. Biomaterials and fabrication technologies play a key role in tissue engineering. Materials used for tissue engineering applications must be designed to stimulate specific cell response on the molecular level. They should elicit specific interactions with the cell and thereby direct cell attachment, proliferation, differentiation and extracellular matrix production and organization [109],[115]:

- **Biocompatibility** is defined as the ability to perform its function in the host tissue without eliciting any immune response.
- **Biodegradability** denotes tunable rate of degeneration to match the growth of new bone tissue as scaffold gets replaced by new bone.
- **Mechanical properties** include the properties such as sufficient mechanical strength to provide temporary support to the defect region and withstand in vivo loading forces.
- **Microarchitecture** is an interconnected scaffold structure that uniformly distributes stresses throughout the scaffold.
- **Osteoinductivity** includes osteoinductive properties that enable to recruit and differentiate the osteoprogenitors to the defect region.
- **Porosity** is large surface area, where the volume and the pore size allow the tissue in-growth, neovascularization, mass transport and the osteogenesis.
- **Surface properties** include appropriate chemical and topographical properties for influencing cellular adhesion, proliferation and differentiation.

Inorganic-organic composites aiming at mimicking the composite nature of real bone combine the toughness of the polymer phase with the compressive strength of an inorganic one to generate bioactive materials with improved mechanical properties and degradation profiles. Hydroxylapatite (HAP) is widely used as a biocompatible ceramic material in many areas of medicine, but mainly for the contact with bone tissue, due to its resemblance to mineral bone [115].

Under normal conditions, human body fluid is supersaturated with respect to apatite, so that, when apatite nuclei form, crystals can grow spontaneously. Chemical species that are capable of supporting the nucleation of apatite are calcium and silicate ions, while, in contrast, phosphate ion does not affect this process. Hence, ceramics that release the former mentioned ions are capable of developing the surface apatite layer when exposed to human body fluids. This explains why glasses that do not contain P_2O_5 are able to develop this surface layer and show greater bioactivity than glasses containing P_2O_5, even if the latter compositions approximate to that of hydroxyapatite [116].

10.9.2. Preparation of nanocrystalline apatites

Nanocrystalline calcium phosphate apatites play an important role in biomineralization[9] and in the field of biomaterials. Biological nanocrystalline apatites are the main inorganic compo-

[9] The enamel of vertebrate teeth, vertebrate bone and tooth-like microfossils of conodonts.

nents of hard tissues in mammals (bone[10] and tooth[11] (**Fig. 13**)) with the exception of enamel (which is closer to stoichiometric hydroxylapatite) and are involved in several pathological calcifications such as dental calculi, salivary stones and blood vessel calcification. In comparison with hydroxylapatite, which is a stoichiometric apatitic phase and is the most stable and the least soluble calcium phosphate at ambient conditions, nanocrystalline apatites are nonstoichiometric and calcium (and OH-) deficient and may incorporate substituted ions in their nanosized crystals. Their calcium and hydroxide deficiencies are responsible for higher solubility than HA. Besides, they have the ability to mature when submitted to humid environments [108],[112],[117],[118].

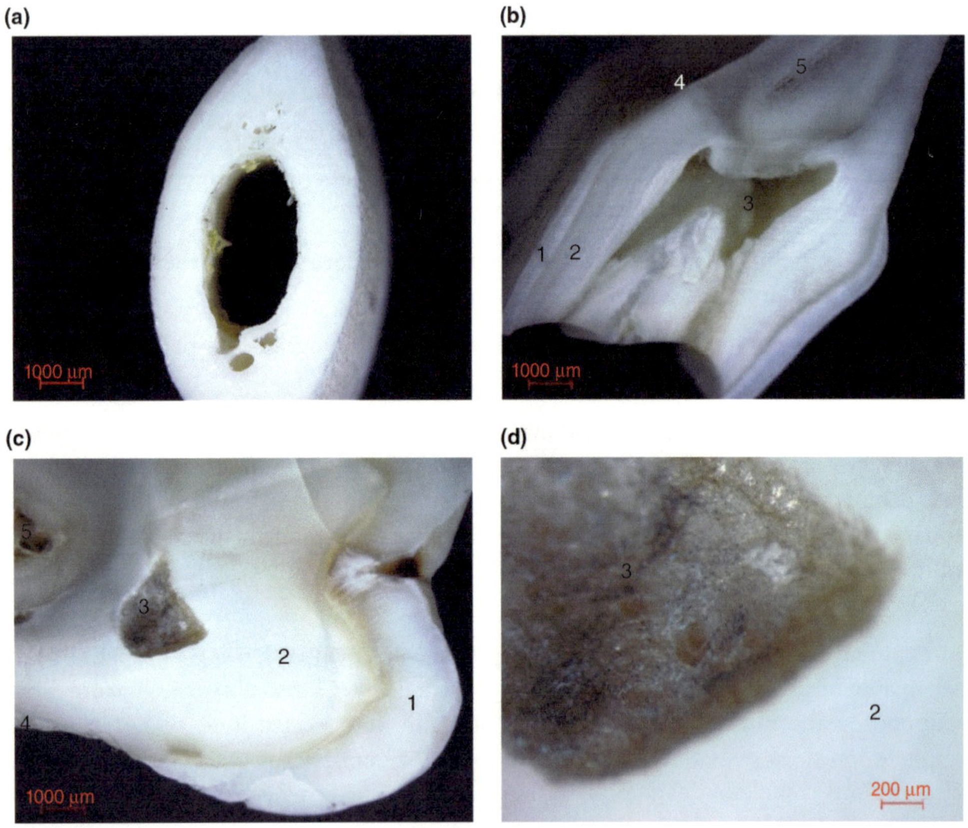

Fig. 13. The cross-section of lower jaw bone (a) and tooth (b) of European roe deer (*Capreolus capreolus*) and human tooth (c, d): enamel (1), dentin (2), pulp chamber (3), cementum (4) and root canal (5).

[10] Bones are rigid organs that form a part of the endoskeleton of vertebrates. Their function is to move, support and protect various organs of the body, produce red and white blood cells and store minerals. Bones appear in a variety of shapes and have a complex internal and external structure described by various hierarchical models [117]. Bone is a complex and hierarchical tissue consisting of nano-hydroxylapatite (70 wt.%) and collagen (30%) as major portions [113].

[11] Teeth consist of a bulk of dentin covered with (inorganic) enamel on the crown and cementum on the root surface. Thick collagen bundles, called periodontal ligaments (PDL), attach to cementum at one end and to the alveolar bone at the other end. The alveolar bone is supported by the jaw [117].

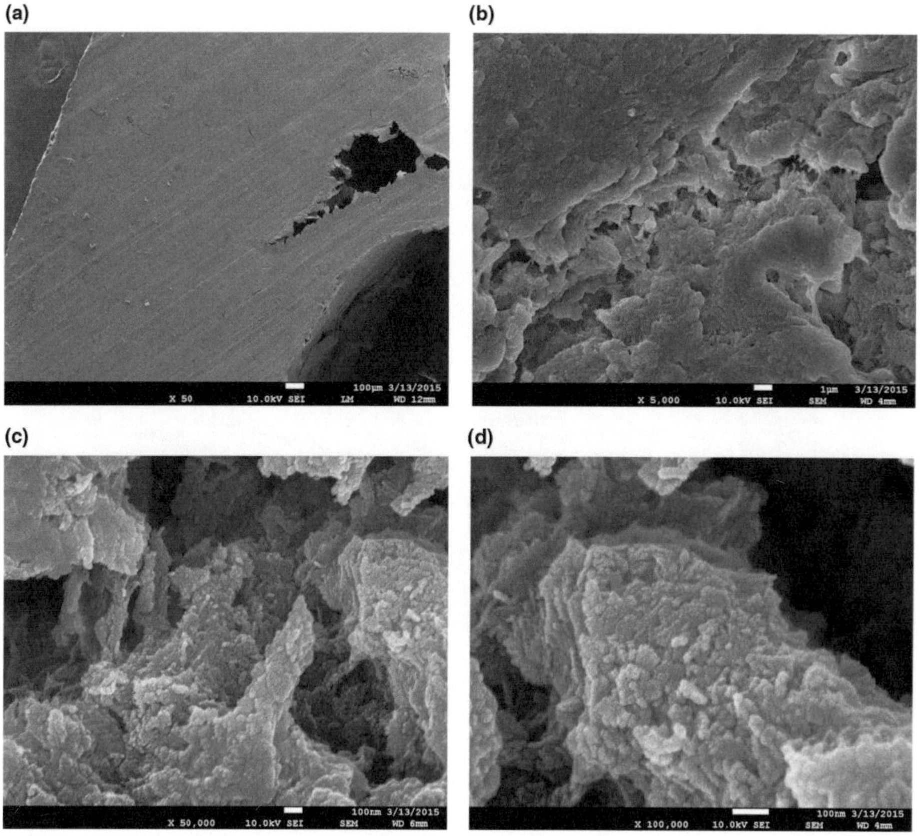

Fig. 14. Electron microscopy picture (SEM) of lower jawbone of European roe deer from **Fig. 13** under magnification of 50× (a), 5000× (b), 50,000× (c) and 100,000× (d).

Scanning electron microscopy of deer bone (**Fig. 14**) shows a spongiform texture (*a, b*) formed by crystals of carbonated hydroxylapatite with the size below 100 nm (*e, f*).

The inorganic portion of bone contains two major mineral phases

i. Non-crystalline calcium phosphate;

ii. Crystalline bone apatite.

In amorphous calcium phosphate, the Ca:P ratio is about 1.33. The non-crystalline or amorphous bone mineral is metastable with respect to bone apatite. Bone apatite crystals are calcium deficient due to the defects in the crystalline lattice and isomorphous substitutions, that is, the replacement of some ions by others in the crystal without disrupting the general symmetry. Young bone tissue was found to be richer in amorphous mineral than crystalline apatite [119].

In the body, only the bone collagen has the property of inducing the mineralization through in vitro; collagens of other tissues do not possess this property. One of the differences between nucleating and non-nucleating collagens might be the presence of a sort of inhibitor bound to the latter. This inhibiting substance was shown to be pyrophosphate. Calcium is accreted in bone tissue in the process of new bone formation or remodeling and resorbed from the bone tissue in the process of bone destruction. The loss of endogenous calcium in urine and feces is compensated for by an equivalent intake of this element (**Fig. 15**) [119].

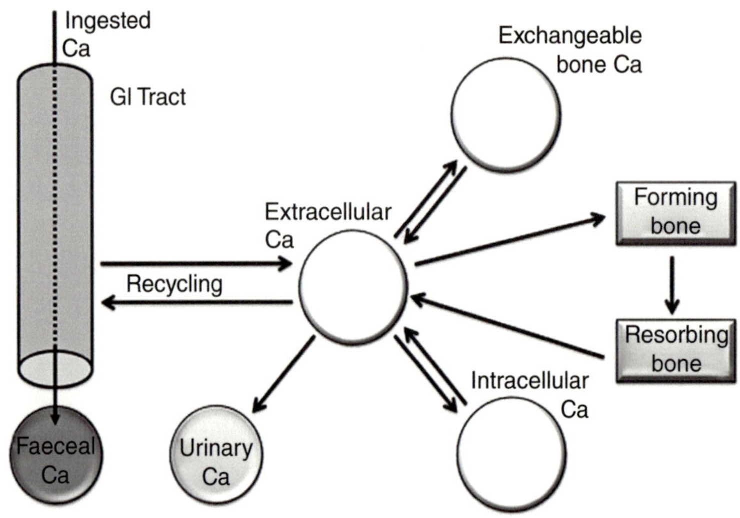

Fig. 15. General scheme of metabolism of calcium [119].

Enamel (**Fig. 17**) is normally the best preserved from hard tissues. It is almost completely a mineral, so the decomposition of organic matter has little effect on it. Archeological enamel nearly always yields good microscope sections, often indistinguishable from fresh enamel [120]. The structure of carbonate dental enamel refined by WILSON et al [121] is shown in **Fig. 16**.

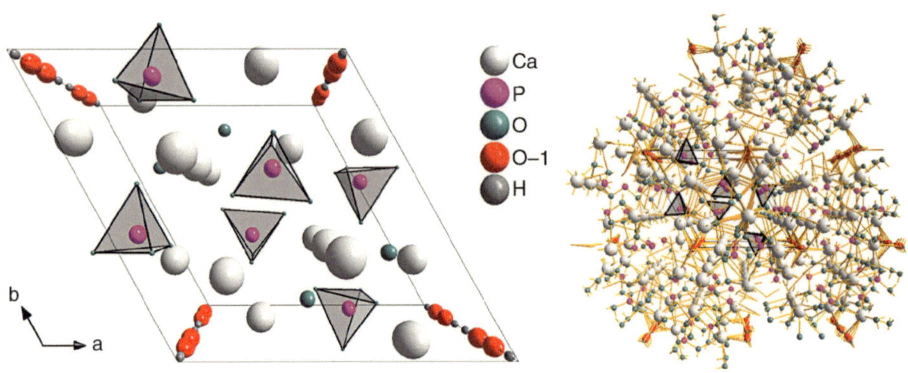

Fig. 16. Crystallographic structure of human dental enamel apatite (perspective view along the c-axis): $P6_3/m$, $\boldsymbol{a}$ = 9.4081 Å, $\boldsymbol{b}$ = 6.8887 Å, $\boldsymbol{c}$:$\boldsymbol{a}$ = 0.7322 and $\boldsymbol{V}$ = 528.05 Å^3 [121].

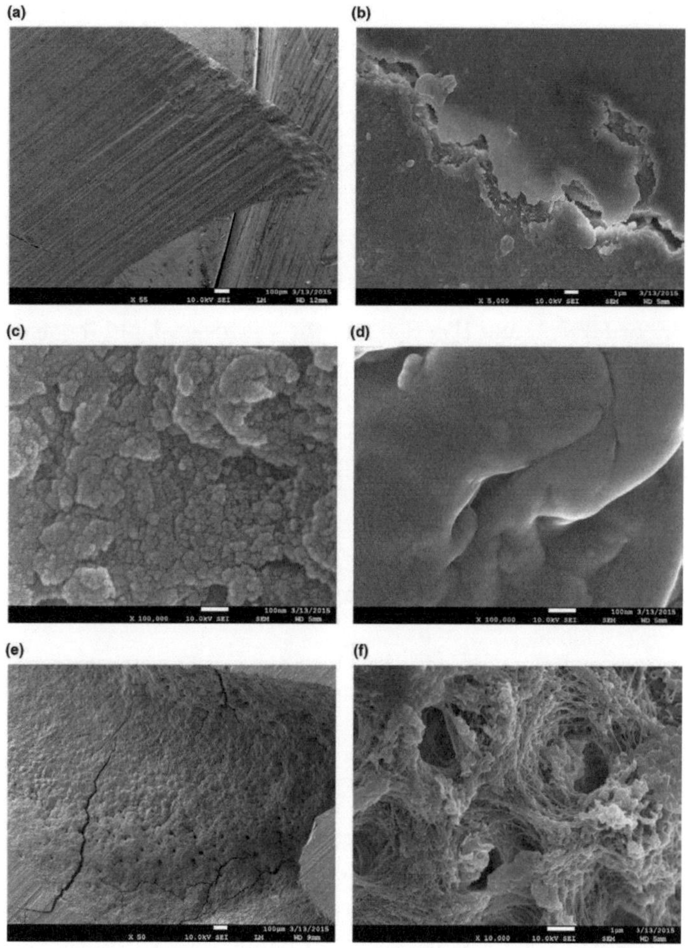

Fig. 17. Electron microscopy picture (SEM) of tooth from **Fig. 13**: the crown with enamel-dentin interface (*a, b*), nanocrystals of carbonated hydroxylapatite in the texture of dentin (*c*), glass-like texture of enamel (*d*) and surface of pulp chamber (*e, f*).

Among carbonate apatites, which are formed in mouth cavity, only dental enamel mine belongs to the B-type of carbonate-apatite (**Section 4.6**) and the others (of dentin, salivary and dental stones) belong to the AB-type (B > A). According to the variations in unit-cell parameters, isomorphic replacements in crystal structures of apatites of pathogenic origin (renal, salivary and dental stones) are more intensive in comparison with physiogenic dental enamel apatites. Among pathogenic apatites, the most considerable compositional variations are observed in renal stone apatites. That indicates strong variability of conditions of their formation. The changes in unit-cell parameters of bone apatites are not completely interpretable, because these apatites consist of nanosized crystals that are smaller than those of other biological apatites [122].

The age variations of the crystal lattice parameters of human enamel apatites are related to complicated processes of de- and remineralization, which result in the increase or reduction of vacancies in Ca positions and in the respective changes of CO_3^{2-}, H_2O and HPO_4^{2-} contents in the unit cell:

$$\begin{aligned} &(n + m/2)\ Ca^{2+} n\ PO_4^{3-} + m\ OH^- \rightarrow \\ &(n + m/2)\ [\ ]_{Ca} + n\ A^{2-} + m\left([\ ]_{OH^-}, H_2O\right) \end{aligned} \tag{21}$$

where $A^{2-} = CO_3^{2-}$ or HPO_4^{2-}, and $[]_{Ca}$, $[]_{OH-}$ are the vacancies. Until the age of 50 years, the values of ***a*** and ***c***-parameter of enamel apatites change considerably without any dependence of particular age that may be explained by essential fluctuations of the content of Ca in human organism. After 50 years of age, significant direct correlation between the age and the ***a***-parameter appears [122].

The surface of apatite nanocrystals is possibly doped with foreign elements or functionalized with organic molecules [117],[123],[124]. The course of facile synthesis of B-type carbonated nanoapatite with tailored microstructure is described by GUALTIERI et al [125].

10.10. Collagen apatite composites

Tissue engineering techniques have been developed to recover or enhance lost tissue function and structure.[12] Collagen-apatite[13] (or collagen/apatite, **Col-AP**) composite resembling the composition of natural bone[10] has been studied extensively and considered as a promising bone tissue engineering material, which can be used to replace or regenerate damaged tissue, resulting from an accident, trauma or cancer. Such synthetic or hybrid biomaterials must have high porosity with interconnected pores to allow the vascularization as well as the nutrients and gases diffusion. Moreover, they should be biodegradable to act as temporary cellular support [114],[126],[127],[128].

The defining feature of collagen is a structural motif in which three parallel polypeptide strands in a left-handed, polyproline II-type (**PPII**) helical conformation coil around each other with a one-residue stagger to form a right-handed triple helix. Tight packing of **PPII** helices within the triple helix mandates that every third residue be **GLY**, resulting in a repeating **XAAYAAGLY** sequence, where **XAA** and **YAA** can be any amino acid. This repeating occurs in all types of collagen, although it is disrupted at certain locations within the triple-helix domain of nonfibrillar collagens. The amino acids in the **XAA** and **YAA** positions of collagen are often (2S)-

[12] One of the first polymers used for bone tissue engineering was based on a hydrolytically copolymer of polylactic-co-glycolic acid (PLGA) but its use for large bone defect regeneration was controversial as inflammatory events were observe. The utilization of chitosan and alginate was also investigated [128].

[13] Names such as collagen-hydroxylapatite or collagen-hydroxyapatite composite were also often applied in published literature.

proline (**Pro**, 28%) and (2S,4R)-4-hydroxyproline (**Hyp**, 38%), respectively. The **ProHypGly** is the most common triplet (10.5%) in collagen (**Fig. 18**) [129].

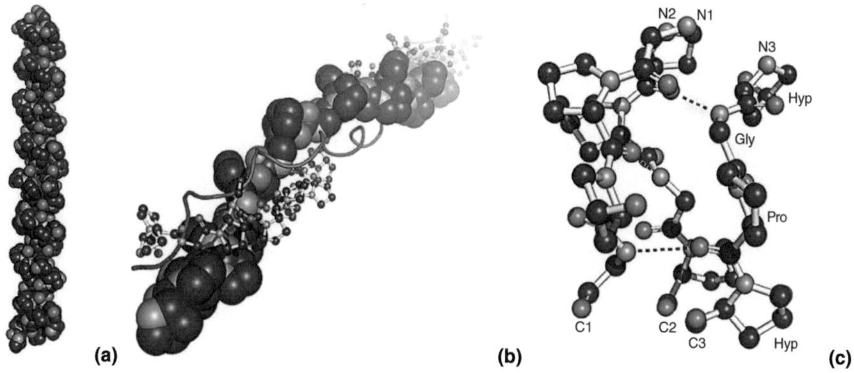

Fig. 18. The triple helix of collagen formed from $(ProHypGly)_4$-(ProHypAla)-$(ProHypGly)_5$ (a), the view down the axis of a $(ProProGly)_{10}$ triple helix (b) and the segment of triple helix (c) with hydrogen bonds (—) [129].

The categories of collagen[14] include the classical fibrillar and network-forming collagens, the FACITs (fibril-associated collagens with interrupted triple helices), MACITs (membrane-associated collagens with interrupted triple helices) and MULTIPLEXINs (multiple triple-helix domains and interruptions). The collagen of type I is the most abundant type used in tissue engineering. Natural polymer collagen that represents the matrix material of bone, teeth and connective tissue can be extracted from animal or human sources (skin, bones, tendons, ligaments and cornea). The treatment includes the separation and isolation (in soluble or insoluble form), decalcification, purification (purification is required to eliminate the antigenic component of protein), sterilization and chemical modification process to achieve polyanionic or purified protein. Type I polyanionic collagen was found to improve the cell adhesion [129],[130].

Collagen, the most abundant protein in extracellular matrix, is chemotactic to fibroblasts. It shows high affinity to cells and good resorbability in vivo. Nevertheless, its poor mechanical properties have restricted its usage in load-bearing applications. Carbonated apatite and collagen interact to form a composite material, the mechanical, physicochemical and biological properties of which differ considerably from those of either constituent considered separately. Collagen and non-collagenous proteins (**NCPs**) are thought to control the crystal deposition, size, crystallization and multiplication/maturation. On the other hand, the crystal deposition in intrafibrillary spaces is likely to modify the three-dimensional conformation of collagen [114],[131],[132],[133]. The solubility of apatite-collagen composites is significantly reduced by UV radiation [134],[135].

[14] There are 27 types of collagen described in literature composed of at least 46 distinct polypeptide (CRP, collagen-related peptide) chains [129], but the types I – V are the most common. More than 90% of collagen in human body is the fibrillar type I.

Calcium phosphates are available commercially, as hydroxylapatites are extracted from bones or they can be produced wet by direct precipitation from pH-adjusted solutions of calcium and phosphate salts [130]. The crystallographic c-axes of the plate-shaped apatite crystals are well aligned with long axes of collagen fibrils (**Fig. 19**), and this preferred orientation between the mineral and the organic framework is assumed to be the general feature of the calcium phosphate biomineralization process. Several attempts were made to mimic this lowest level of hierarchical organization of bone by using proteins as a site for the heterogeneous nucleation and subsequent growth of stoichiometric hydroxylapatite crystals. In special approaches, the biomimetic apatite coatings on surfaces were prepared by soaking the materials in simulated body fluid (SBF) solutions, which contained ions in the concentrations similar to those in inorganic part of human blood plasma. It is generally accepted that the in vitro apatite growth during the exposure to SBF is an indicator for the in vivo bioactivity of materials surface [136],[137].

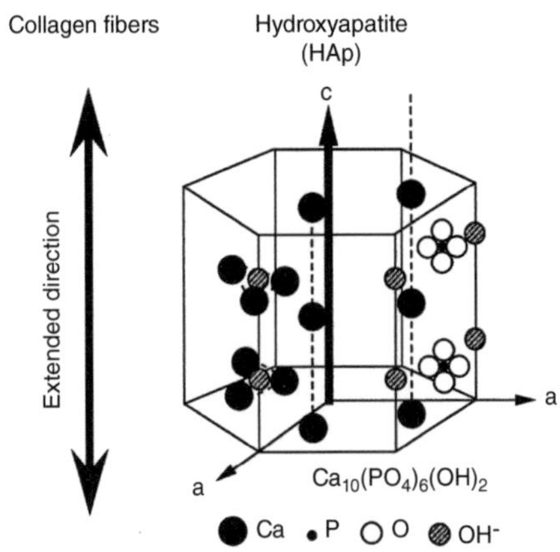

Fig. 19. The crystal structure of hydroxylapatite is oriented according to the c-axis with the extended direction of collagen fibrils [112].

Biomimetic materials are able to mimic the morphological and physicochemical features of biological apatite compounds, i.e. they are synthetic analogues of inorganic part of hard tissues [112],[117]. The biomimetic deposition is considered as an ideal method to produce calcium phosphate ceramics such as apatite coatings on titanium and its alloys for medical applications. It has also been proved that the chemical pretreatment in alkali solution can improve the bonding between the titanium substrate[15] and calcium phosphate coatings fabricated by subsequent biomimetic deposition in simulated body fluid (SBF) [138],[139],[140],[141]. The biomimetic growth of apatite was also described on hydrogen-implanted silicon [142], polyvinyl alcohol (PVA) [143], TiO_2 nanotubes [144], alumina [145], zirconia ceramics (Y-TZP) [146], forsterite [147], akermanite [148], magnetite [149], glasses [150] and

[15] Titanium and its alloys are widely used as orthopedic and dental implant materials because of their high mechanical strength, low modulus and good corrosion resistance [139].

http://dx.doi.org/10.5772/62217

bioglasses [151],[152], cements [153], geopolymers [154], carbon nanotubes [155] and microspheres [156].

The ability to form apatite [150],[153] from supersaturated solution has been widely used to imply the bioactivity of an implant in vivo. However, the method itself may provide at best incomplete information, primarily because it is determined only by the solution supersaturation, irrespective of biological processes. The bone regeneration is triggered mainly by the vitality of osteoblasts and regulated by the expression of growth factors such as estrogen, parathyroid hormone and bone morphogenetic proteins, while ions or other species released from an implant may affect the expression of such growth factors and so the bone resorption or formation. The misinterpretation of the outcome of such tests must result in the misunderstanding of the true effects and behavior of materials intended for use in embedded biological applications. Moreover, SBF may not be able to mimic properly the physiological conditions because it is based on analytical concentrations and not on the activity of key components [157].

The methods of the preparation of collagen-hydroxylapatite composites include the production of composite gels, films, collagen-coated ceramics, ceramic-coated collagen matrices and composite scaffolds for bone substitutes and hard tissue repair via the following techniques [130],[158],[159]:

- **In vitro collagen mineralization**: the method is based on direct mineralization of a collagen substrate (film) through which calcium and phosphate ions diffuse into the fibrils or as phosphate-containing collagen solution[16] (in situ precipitation [159]). **Fig. 20** introduces the experimental set-up for direct mineralization of a collagen sheet. The growth of HAP crystals with c-axis oriented along the collagen fibrils requires the pH in the range from 8 to 9 and the temperature of 40°C. These conditions promote the accumulation of calcium ions on the carboxyl groups of collagen molecules, leading to the nucleation of hydroxylapatite.

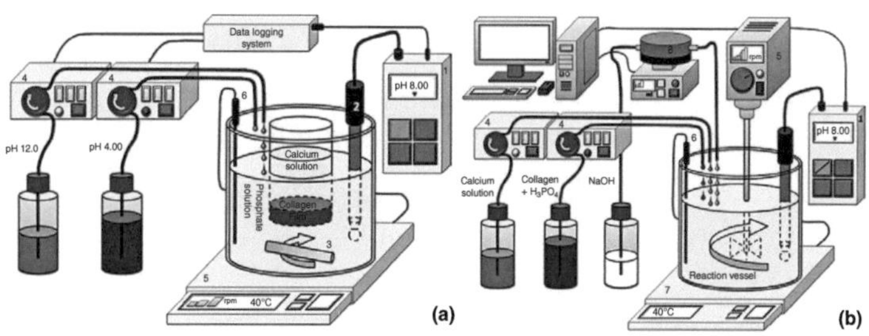

Fig. 20. Simplified scheme of apparatus for the preparation of collagen-hydroxylapatite composite by in vitro collagen mineralization method: modified scheme according to Wahl and Czernuszka [130] for in situ collagen/HAP precipitation *(a)*; modified scheme according to Wang and Liu [159]: pH meter (1), pH electrode (2), stirrer (3), peristaltic pump (4), hot-plate magnetic stirrer (5a) or stirrer (5b), thermocouple (6), hot plate (7) and peristaltic pump driven by pH controller (8) (b).

[16] These techniques usually continue with freeze-drying.

- **Thermally triggered assembly of HA/collagen gels**: the solution of calcium and phosphate ions is encapsulated within the liposomes and next inserted into the acidic suspension of collagen. After the injection into a skeletal defect, the increasing temperature due to body heat initiates the gelation process, which leads to fibrous network. The mineralization occurs after reaching the liposome's transition temperature of 37°C.
- **Vacuum infiltration of collagen into a ceramic matrix**: ceramic scaffold is prepared by heating aqueous hydroxylapatite slurry containing poly(butyl methacrylate) (PBMA) spheres to high temperatures. The pyrolysis of PBMA particles leads to porous HAP green body. Pores in this matrix are then filled with collagen suspension under vacuum. The final composite was then freeze-dried to produce the microsponges within.
- **Enzymatic mineralization of collagen sheets**: is a method based on the cycle (**Fig. 21**) where collagen-containing alkaline (basic) phosphatase (ALP) is treated by aqueous solution of calcium ions and phosphate ester. The enzyme provides a reservoir for PO_4^{3-} ions for calcium phosphate to crystallize and the mineralization occurs on coated area. The sample is then coated again with collagen suspension, air-dried and cross-linked with UV irradiation. Repeating this cycle results in multilayered composite sheets of calcium/phosphate and collagen.

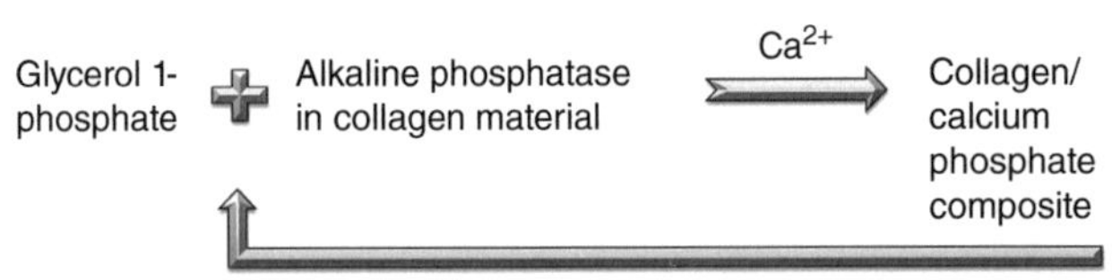

Fig. 21. The cycle of enzymatic mineralization of collagen sheets [130].

- **Water-in-oil emulsion system**: purified collagen suspension mixed with HAP powders at the temperature of 4°C is next dispersed in olive oil and stirred at 37°C. The collagen aggregates and reconstitutes into the aqueous droplets. The addition of phosphate-buffered saline[17] (PBS) leads to the gelation of Col-HAP microspheres (gel beads) of bone filler. The main disadvantages of this method are the problem with complete removing of the oil content from the composite and too low viscosity of the mixture.
- **Freeze-drying and critical point drying (CPD) scaffolds**: the ice crystals with collagen fibers at the interstices can be formed by freezing the suspension of collagen and HAP in water under controlled conditions. In the case when the freeze-drying is applied (temperature and pressure corresponding to CPD), ice crystals sublimate to water vapor.

  Under the condition of critical point,[18] the density of liquid and gas phase converge and become identical (supercritical fluid) as well as the surface tension is negligible.

---

[17] Aqueous solution of sodium phosphate and chloride, where the ion concentrations and osmolarity (osmotic concentration) correspond to human body. In some cases, potassium phosphate and chloride were used.

[18] Critical point is defined by the value of three parameters including critical temperature, critical pressure and critical volume. Critical temperature is the highest temperature at which pure matter can exist as liquid.

- Col-HAP was cast into the mould and frozen, ice crystals replaced with ethanol, ethanol-liquid CO_2 exchanged and critical point dried to finally arrive with an exact porous replica of the original bone. Solid freeform fabrication techniques have recently been developed with artificial polymers and ceramic materials. These have the ability to change the pore interconnectivity, pore size and pore shape but have the disadvantage of not having the affinity of collagen to cell attachment. Another major advantage of Col-HAP scaffolds produced through the SFF method is the ability to control variables such as the control of external and internal structure, porosity and cross-linking [130].

The collagen-hydroxyapatite/pectin (Col-HA/pectin) composite was prepared in situ by the introduction of pectin, a kind of plant polysaccharide, into the collagen-hydroxyapatite composite. The structure of composite consisted of hydroxylapatite of low crystallinity with particles uniformly dispersed in organic materials. There is strong bonding interaction between HAP, collagen and pectin. The mechanical properties, water absorption, enzyme degradation and cytotoxicity indicate a potential use in bone replacement for the new composite [160].

Most mammalian biofluids are supersaturated with respect to the bone and tooth mineral hydroxylapatite. Nevertheless, the biofluids, which are in contact with soft tissues, especially those like milk that must be stored for any length of time, should be highly stable with respect to calcium phosphate precipitation. In contrast, saliva and the extracellular matrix of hard tissues, especially near to the sites of mineralization or remineralization, are required not only to maintain the mineral phase with which they are in contact but also to deposit calcium phosphate in a highly controlled manner [161].

Collagen composites were also used as a scaffold for the repair of soft tissues. These materials provide analogous environment to extracellular matrix (ECM) and induced rate of synthesis or growth of new tissues. Several natural polymers as collagen, chitosan,[19] gelatin[20] and keratin[21] possess the ability to induce the proliferation of cells and hence find their use as biomaterials for a wide range of biomedical applications. Among all biopolymers, collagen is a widely accepted material for the tissue engineering applications in the view of its low antigenicity, excellent biocompatibility and biodegradability [162].

The preparation of collagen-based biocomposite constructed from micro-crimped long collagen fiber bundles extracted from a soft coral embedded in alginate hydrogel matrix was described by Sharabi et al [163]. This biocomposite demonstrated the hyperelastic behavior similar to human native tissues.

---

[19] Chitosan is (1–4)-linked 2-amino-2-deoxy-b-glucan, a byproduct of N-deacetylation of chitin. It is a major constituent of crab and shrimp shells and of cuticles of insects.

[20] Gelatin is a denatured form of collagen. Gelatin has low antigenicity and promotes the cell adhesion, differentiation and proliferation. Gelatin also possesses high cytocompatibility, which makes it a potential candidate as a biomaterial for various tissue engineering applications [162].

[21] Keratin is a family of fibrous proteins, which is found abundantly in nature. It is the main constituent of hair, wool, nails, horns and hooves of mammals, birds and reptiles [162].

10.11. Layers and biocoatings

The nanostructured coatings are expected to enhance the mechanical properties and improve the strength of bonding between coating and implant. These materials can also promote the deposition of calcium-containing minerals on their surface (bioactivity). Some typical examples of the research carried out in recent years, in the field of fabrication of nanostructured glass-ceramic coatings, utilize various techniques such as conventional enameling, sputtering, sol-gel processing, ion beam deposition, plasma spraying, electrophoretic deposition and pulsed laser deposition [164].

As was mentioned earlier, among the bioactive ceramics, the apatite/wollastonite (A/W) glass-ceramics, containing apatite and wollastonite crystals in the glassy matrix, has been largely studied because of good bioactivity and was used in some fields of medicine, especially in orthopedics and dentistry. However, medical applications of bioceramics are limited to non-load-bearing applications because of their poor mechanical properties. The solution of this problem can be the preparation of layers on the titanium alloy base material. Thermally sprayed layers by APS (atmospheric plasma spraying) on Ti-6Al-4V substrates combine good bioactivity of the bioceramics and good mechanical strength. The microstructure and the resulting properties were evaluated depending on the processing parameters and postprocessing thermal treatments. Thermal treatments decreased the bioactivity of the coatings, and after specific treatments, some bioactive materials were transformed into inert materials [67].

The biological performance of a porous apatite-mullite glass-ceramics, manufactured via the selective laser sintering (SLS) method, was investigated by Goodridge et al [165]. Laser-sintered A-M has shown similar cytotoxicity, in vitro bioactivity and in vivo results to cast A-M, indicating that the laser-sintering processing route does not alter the behavior of the material. The porous structure produced by SLS was seen by in vivo testing to be excellent for bone in-growth. However, the inability of the material to form apatite in vitro raised the concerns over the material's ability to prove bioactive in vivo, and further assessment is required to confirm whether this material is not bioactive or whether it is just not particularly suited for the characterization through SBF testing [165].

Author details

Petr Ptáček

Brno University of Technology, Czech Republic

http://dx.doi.org/10.5772/62217

References

[1] Meseguer S, Tena MA, Gargori C, Badenes JA, Llusar M, Monrós G. Structure and colour of cobalt ceramic pigments from phosphates. Ceramics International 2007;33(5) 843–849.

[2] Wagh AS. Introduction to Chemically Bonded Ceramics. Chemically Bonded Phosphate Ceramics, 2004. ISBN: 978-0-08-044505-2.

[3] Colorado HA, Hiel C, Hahn HT. Chemically bonded phosphate ceramics composites reinforced with graphite nanoplatelets. Composites Part A: Applied Science and Manufacturing 2011;42(4) 376–384.

[4] Dimitry SA. Characterization of reinforced chemically bonded ceramics. Cement and Concrete Composites 1991;13(4) 257–263.

[5] Yang L, Harink B, Habibovic P. Calcium phosphate ceramics with inorganic additives. Comprehensive Biomaterials 2011;1 299–312.

[6] Klammert U, Ignatius A, Wolfram U, Reuther T, Gbureck U. In vivo degradation of low temperature calcium and magnesium phosphate ceramics in a heterotopic model. Acta Biomaterialia 2011;7(9) 3469–3475.

[7] Chanda A, Dasgupta S, Bose S, Bandyopadhyay A. Microwave sintering of calcium phosphate ceramics. Materials Science and Engineering: C 2009;29(4) 1144–1149.

[8] Bandyopadhyay A, Withey EA, Moore J, Bose S. Influence of ZnO doping in calcium phosphate ceramics. Materials Science and Engineering: C 2007;27(1) 14–17.

[9] Formosa J, Chimenos JM, Lacasta AM, Niubó M. Interaction between low-grade magnesium oxide and boric acid in chemically bonded phosphate ceramics formulation. Ceramics International 2012;38(3) 2483–2493.

[10] Sahnoun RD, Bouaziz J. Sintering characteristics of kaolin in the presence of phosphoric acid binder. Ceramics International 2012;38(1) 1–7.

[11] Gardner LJ, Bernal SA, Walling SA, Corkhill CL, Provis JL, Hyatt NC. Characterisation of magnesium potassium phosphate cements blended with fly ash and ground granulated blast furnace slag. Cement and Concrete Research 2015;74 78–87.

[12] Liu Z, Qian G, Zhou J, Li C, Xu Y, Qin Z. Improvement of ground granulated blast furnace slag on stabilization/solidification of simulated mercury-doped wastes in chemically bonded phosphate ceramics. Journal of Hazardous Materials 2008;157(1) 146–153.

[13] Hermansson L. Nanostructural Bioceramics: Advances in Chemically Bonded Ceramics. CRC Press, 2014. ISBN: 978-9814463430

[14] Chen D, He L, Shang S. Study on aluminum phosphate binder and related Al_2O_3-SiC ceramic coating. Materials Science and Engineering: A 2003;348(1-2) 29–35.

[15] Hipedinger NE, Scian AN, Aglietti EF. Magnesia-ammonium phosphate-bonded cordierite refractory castables: Phase evolution on heating and mechanical properties. Cement and Concrete Research 2004;34(1) 157–164.

[16] Choi J, Um W, Choung S. Development of iron phosphate ceramic waste form to immobilize radioactive waste solution. Journal of Nuclear Materials 2014;452(1-3) 16–23.

[17] Metcalfe BL, Donald IW, Fong SK, Gerrard LA, Strachan DM, Scheele RD. Ageing of a phosphate ceramic used to immobilize chloride contaminated actinide waste. Journal of Nuclear Materials 2009;385(2) 485–488.

[18] Singh D, Mandalika VR, Parulekar SJ, Wagh AS. Magnesium potassium phosphate ceramic for ^{99}Tc immobilization. Journal of Nuclear Materials 2006;348(3) 272–282.

[19] Scheetz BE, Agrawal DK, Breval E, Roy R. Sodium zirconium phosphate (NZP) as a host structure for nuclear waste immobilization: A review. Waste Management 1994;14(6) 489–505.

[20] von Kobell F. Über den schillernden asbest von Reichenstein in Schlesien. Journal für Praktische Chemie 1834;2 297–298.

[21] Warren BE, Bragg WL. The structure of the chrysotile $H_4Mg_3Si_2O_9$. Zeitschrift für Kristallographie 1931;76 201–210.

[22] Falini G, Foresti E, Gazzano M, Gualtieri AF, Leoni M, Lesci IG, Roveri N. Tubular-shaped stoichiometric chrysotile nanocrystals. Chemistry — A European Journal 2004;10 3043–3049.

[23] Viani A, Gualtieri AF. Preparation of magnesium phosphate cement by recycling the product of thermal transformation of asbestos containing wastes. Cement and Concrete Research 2014;58 56–66.

[24] Ciullo PA. Industrial Minerals and Their Uses: A Handbook and Formulary. William Andrew, 1996. ISBN: 978-0815518082

[25] Clarke DB. Granitoid Rocks. International Studies in Economic Modelling — Volume 7. Topics in the Earth Sciences. Springer Science & Business Media, 1992. ISBN: 978-0412291708

[26] Ruh E. Concise Encyclopedia of Advanced Ceramic Materials. Advances in Materials Sciences and Engineering. Elsevier, 2012. ISBN: 978-0080983707

[27] Huggett LG. Refractories in the chemical industries. International Journal of Materials in Engineering Applications 1979;1(5) 280–294.

[28] Sadik C, El Amrani Iz-E, Albizane A. Recent advances in silica-alumina refractory: A review. Journal of Asian Ceramic Societies 2014;2(2) 83–96.

[29] Marghussian VK, Naghizadeh R. Chemical bonding of silicon carbide. Journal of the European Ceramic Society 1999;19(16) 2815–2821.

[30] Sujith SS, Kumar ASL, Mangalaraja RV, Mohamed AP, Ananthakumar S. Porous to dense $LaPO_4$ sintered ceramics for advanced refractories. Ceramics International 2014;40(9) 15121–15129.

[31] Souza TM, Luz AP, Santos T Jr, Gimenes DC, Miglioli MM, Correa AM, Pandolfelli VC. Phosphate chemical binder as an anti-hydration additive for Al_2O_3-MgO refractory castables. Ceramics International 2014;40(1) 1503–1512.

[32] Gale WF, Totemeier TC. Smithells Metals Reference Book. 8th ed., Butterworth-Heinemann, 2003. ISBN: 978-0080480961

[33] Luz AP, Gomes DT, Pandolfelli VC. High-alumina phosphate-bonded refractory castables: $Al(OH)_3$ sources and their effects. Ceramics International 2015;41(7) 9041–9050.

[34] García-Prieto A, Ramos-Lotito MD, Gutiérrez-Campos D, Pena P, Baudín C. Influence of microstructural characteristics on fracture toughness of refractory materials. Journal of the European Ceramic Society 2015;35(6) 1955–1970.

[35] Toy C, Whittemore OJ. Phosphate bonding with several calcined aluminas. Ceramics International 1989;15(3) 167–171.

[36] Hipedinger NE, Scian AN, Aglietti EF. Magnesia-phosphate bond for cold-setting cordierite-based refractories. Cement and Concrete Research 2002;32(5) 675–682.

[37] Lu W, Chung DDL. Oxidation protection of carbon materials by acid phosphate impregnation. Carbon 2002;40(8) 1249–1254.

[38] Smith BGN, Wright PS, Brown D. The Clinical Handling of Materials. 2nd ed., Oxford: Butterworth-Heinemann, 1994. ISBN: 978-0723610038

[39] Nicholson JW, Czarnecka B, Limanowska-Shaw H. A preliminary study of the effect of glass-ionomer and related dental cements on the pH of lactic acid storage solutions. Biomaterials 1999;20(2) 155–158.

[40] Liu Y, Yu Hai-Y. Does dental zinc phosphate cement really shrink in clinical applications? Medical Hypotheses 2009;73(2) 257–258.

[41] Schmalz G, Bindslev DA. Titul Biocompatibility of Dental Materials. Springer Science & Business Media, 2008. ISBN: 978-3540777823

[42] Paffenbarger GC, Sweeney WT, Isaacs A. Zinc phosphate cements: Physical properties and a specification. The Journal of the American Dental Association 1922;21(11) 1907–1924.

[43] Park C.-K, Silsbee MR, Roy DM. Setting reaction and resultant structure of zinc phosphate cement in various orthophosphoric acid cement-forming liquids. Cement and Concrete Research 1998;28(1) 141–150.

[44] Mitra SB. Dental cements: formulations and handling techniques. In: Curtis RV, Watson TF (Eds.), Dental Biomaterials. Elsevier, 2014. ISBN: 978-1845694241

[45] Jabri M, Mejdoubi E, El Gadi M, Hammouti B. Optimisation of hardness and setting time of dental zinc phosphate cement using a design of experiments. Arabian Journal of Chemistry 2012;5(3) 347–351.

[46] Wilson AD. The chemistry of dental cements. Chemical Society Reviews 1978;7(2) 265–296.

[47] Ramkumar N. Phosphate based oil well cements. ProQuest Dissertations and Theses; Thesis (Ph.D.)-University of Illinois at Chicago, 2005; Publication Number: AAI3199873; ISBN: 9780542467271; Source: Dissertation Abstracts International, Volume: 66-12, Section: B, page: 6863; 131 p.

[48] Kurdowski W. Cement and Concrete Chemistry. Springer Science & Business, 2014. ISBN: 978-9400779457

[49] Warner J. Practical Handbook of Grouting: Soil, Rock, and Structures. John Wiley & Sons, 2004. ISBN: 978-0471463030

[50] Ghosh SN. Cement and Concrete Science and Technology. Cement and Concrete Science & Technology — Part 1. Progress in Cement and Concrete. Thomas Telford, 1991. ISBN: 978-8185522005

[51] Wang F, Chen B, Pun EYB, Lin H. Alkaline aluminum phosphate glasses for thermal ion-exchanged optical waveguide. Optical Materials 2015;42 484–490.

[52] Saadaldin SA, Rizkalla AS. Synthesis and characterization of wollastonite glass-ceramics for dental implant applications. Dental Materials 2014;30(3) 364–371.

[53] Lewis M, Metcalf-Johansen J, Bell P. Crystallisation mechanisms in glassceramics. Journal of Non-Crystalline Solids 1979;62 5–6.

[54] Stanton KT, O'Flynn KP, Kiernan S, Menuge J, Hill R. Spherulitic crystallization of apatite-mullite glass-ceramics: Mechanisms of formation and implications for fracture properties. Journal of Non-Crystalline Solids 2010;356(35-36) 1802–1813.

[55] Hill RG. Bioactive Glass-Ceramics, 181-186. Comprehensive Biomaterials, Volume 1. In: Ducheyne P (Ed.). Elsevier, 2011. ISBN: 978-0-08-055294-1

[56] Kamitakahara M, Ohtsuki C, Inada H, Tanihara M, Miyazaki T. Effect of ZnO addition on bioactive $CaO-SiO_2-P_2O_5-CaF_2$ glass-ceramics containing apatite and wollastonite. Acta Biomaterialia 2006;2(4) 467–471.

[57] Zhang Y, Santos JD. Microstructural characterization and in vitro apatite formation in $CaO-P_2O_5-TiO_2-MgO-Na_2O$ glass-ceramics. Journal of the European Ceramic Society 2001;21(2) 169–175.

[58] Kishioka A, Hayashi M, Kinoshita M. Glass formation and crystallization in ternary phosphate systems containing Al_2O_3. Bulletin of the Chemical Society of Japan 1976;49(11) 3032–3036.

[59] Joseph K, Jolley K, Smith R. Iron phosphate glasses: Structure determination and displacement energy thresholds, using a fixed charge potential model. Journal of Non-Crystalline Solids 2015;411 137–144.

[60] Stefanovsky SV, Stefanovsky OI, Kadyko MI, Presniakov IA, Myasoedov BF. The effect of Fe_2O_3 substitution for Al_2O_3 on the phase composition and structure of sodium-aluminum-iron phosphate glasses. Journal of Non-Crystalline Solids 2015;425 138–145.

[61] Premila M, Rajaraman R, Abhaya S, Ravindran TR, Kamali K, Amarendra G, Sundar CS. Spectroscopic evidence of irreversible changes in cesium loaded iron phosphate glasses under pressure. Journal of Non-Crystalline Solids 2014;406 111–118.

[62] Broglia G, Mugoni C, Du J, Siligardi C, Montorsi M. Lithium vanado-phosphate glasses: Structure and dynamics properties studied by molecular dynamics simulations. Journal of Non-Crystalline Solids 2014;403 53–61.

[63] Das SS, Singh NP, Srivastava PK. Ion conducting phosphate glassy materials. Progress in Crystal Growth and Characterization of Materials 2009;55(3-4) 47–62.

[64] Kishore MS, Pralong V, Caignaert V, Varadaraju UV, Raveau B. Synthesis and electrochemical properties of a new vanadyl phosphate: $Li_4VO(PO_4)_2$. Electrochemistry Communications 2006;8(10) 1558–1562.

[65] Kokubo T. Bioactive glass ceramics: properties and applications. Biomaterials 1991;12(2) 155–163.

[66] Cannillo V, Pierli F, Sampath S, Siligardi C. Thermal and physical characterisation of apatite/wollastonite bioactive glass-ceramics. Journal of the European Ceramic Society 2009;29(4) 611–619.

[67] Cannillo V, Colmenares-Angulo J, Lusvarghi L, Pierli F, Sampath S. In vitro characterisation of plasma-sprayed apatite/wollastonite glass-ceramic biocoatings on titanium alloys. Journal of the European Ceramic Society 2009;29(9) 1665–1677.

[68] Liu X, Ding C, Chu PK. Mechanism of apatite formation on wollastonite coatings in simulated body fluids. Biomaterials 2004;25(10) 1755–1761.

[69] Magallanes-Perdomo M, Luklinska ZB, De Aza AH, Carrodeguas RG, De Aza S, Pena P. Bone-like forming ability of apatite-wollastonite glass ceramic. Journal of the European Ceramic Society 2011;31(9) 1549–1561.

[70] Kokubo T. A/W glass-ceramic: Processing and properties. Chapter 5. In: Hench LL (Ed.). An Introduction to Bioceramics. Advanced Series in Ceramics — Volume 1. World Scientific, 1993. ISBN: 978-9810214005.

[71] Kokubo T, Shigematsu M, Nagashima Y, Tashiro M, Nakamura T, Yamamuro T, Higash S. Apatite- and wollastonite-containing glass ceramics for prosthetic applications. Bulletin of the Institute of Chemistry Research of Kyoto University 1982;60 260–268.

[72] Ma J, Chen CZ, Wang DG, Shao X, Wang CZ, Zhang HM. Effect of MgO addition on the crystallization and in vitro bioactivity of glass ceramics in the CaO-MgO-SiO_2-P_2O_5 system. Ceramics International 2012;38(8) 6677–6684.

[73] Clifford A., Hill R. Apatite-mullite glass-ceramics. Journal of Non-Crystalline Solids 1996;196 346–351.

[74] Hill RG, O'Donnell MD, Law RV, Karpukhina N, Cochrane B, Tulyaganov DU. The early stages of nucleation and crystallisation of an apatite glass-ceramic: Evidence for nano-scale crystallization. Journal of Non-Crystalline Solids 2010;356(52-54) 2935–2941.

[75] Bernardo E, Colombo P, Cacciotti I, Bianco A, Bedini R, Pecci R, Pardun K, Treccani L, Rezwan K. Porous wollastonite-hydroxyapatite bioceramics from a preceramic polymer and micro- or nano-sized fillers. Journal of the European Ceramic Society 2012;32(2) 399–408.

[76] van't Hoen C, Rheinberger V, Höland W, Apel E. Crystallization of oxyapatite in glass-ceramics. Journal of the European Ceramic Society 2007;27(2-3) 1579–1584.

[77] Mahato N, Banerjee A, Gupta A, Omar S, Balani K. Progress in material selection for solid oxide fuel cell technology: A review. Progress in Materials Science 2015;72 141–337.

[78] Fergus J, Hui R, Li X, Wilkinson DP, Zhang J. Solid Oxide Fuel Cells: Materials Properties and Performance. Chemistry and Chemical Engineering. CRC Press, 2008. ISBN: 978-1420088847.

[79] Grove WR. On voltaic series and combination of gases by platinum. Philosophical Magazine Series 3 1839;14(86) 127–130.

[80] Camaratta M. Microstructural engineering of composite cathode systems for intermediate and low temperature solid oxide fuel cells, Ph.D. Thesis. Gainesville, Florida: University of Florida, 2007.

[81] Marrero-López D, Martín-Sedeño MC, Peña-Martínez J, Ruiz-Morales JC, Núñez P, Aranda MAG, Ramos-Barrado JR. Evaluation of apatite silicates as solid oxide fuel cell electrolytes. Journal of Power Sources 2010;195(9) 2496–2506.

[82] Islam MS, Tolchard JR, Slater PR. An apatite for fast oxide-ion conduction. Chemical Communications 2003;(13) 1486–1487.

[83] Tolchard JR, Islam MS, Slater PR. Defect chemistry and oxygen ion migration in the apatite-type materials $La_{9.33}Si_6O_{26}$ and $La_8Sr_2Si_6O_{26}$. Journal of Materials Chemistry 2003;13(8) 1956–1961.

[84] Zeng Y, Mao Pei-L, Jiang SP, Wu P, Zhang L, Wu P. Prediction of oxygen ion conduction from relative Coulomb electronic interactions in oxyapatites. Journal of Power Sources 2011;196(10) 4524–4532.

[85] Nakayama S, Higuchi Y, Sugawara M, Makiya A, Uematsu K, Sakamoto M. Fabrication of c-axis-oriented apatite-type polycrystalline $La_{10}Si_6O_{27}$ ceramic and its anisotropic oxide ionic conductivity. Ceramics International 2014;40(1) 1221–1224.

[86] Takeda N, Itagaki Y, Aono H, Sadaoka Y. Preparation and characterization of $Ln_{9.33+x/3}Si_{6-x}Al_xO_{26}$ (Ln = La, Nd and Sm) with apatite-type structure and its application to a potentiometric O_2 gas sensor. Sensors and Actuators B: Chemical 2006;115(1) 455–459.

[87] Tudorache F, Petrila I, Popa K, Catargiu AM. Electrical properties and humidity sensor characteristics of lead hydroxyapatite material. Applied Surface Science 2014;303 175–179.

[88] Owada H, Yamashita K, Umegaki T, Kanazawa T, Nagai M. Humidity-sensitivity of yttrium substituted apatite ceramics. Solid State Ionics 1989;35(3-4) 401–404.

[89] Nagai M, Nishino T, Saeki T. A new type of CO_2 gas sensor comprising porous hydroxyapatite ceramics. Sensors and Actuators 1988;15(2) 145–151.

[90] Phosphor system for improved efficacy lighting sources. Patent US 8866372 B2, 2011.

[91] Yu R, Xue N, Wang T, Zhao Z, Wang J, Hei Z, Li M, Noh HM, Jeong JH. Photoluminescence characteristics of high thermal stable fluorosilicate apatite $Ba_2Y_3(SiO_4)_3F$: Sm^{3+} orange-red emitting phosphor. Ceramics International 2015;41(4) 6030–6036.

[92] Liu Y, Wang Z, Zhong J, Pan F, Liang H, Xiao Z. Synthesis and photoluminescence properties of red-emitting phosphors $Ba_2Gd_8(SiO_4)_6O_2:Eu^{3+}$. Materials Letters 2014;129 130–133.

[93] Sokolnicki J, Zych E. Synthesis and spectroscopic investigations of $Sr_2Y_8(SiO_4)_6O_2:Eu^{2+},Eu^{3+}$ phosphor for white LEDs. Journal of Luminescence 2015;158 65–69.

[94] Zhang F, Liu B. An intense red-emitting phosphor $Ca_3Gd_7(SiO_4)_5(PO_4)O_2$: Eu^{3+} for NUV-LEDs application. Journal of Alloys and Compounds 2012;542 276–279.

[95] Jeon YI, Bharat LK, Yu JS. White-light emission of $Ca_2La_8(GeO_4)_6O_2$: Tb^{3+}/Sm^{3+} nanocrystalline phosphors for solid-state lighting applications. Journal of Luminescence 2015;166 93–100.

[96] Liu H, Zhang Y, Liao L, Guo Q, Mei L. Synthesis, broad-band absorption and luminescence properties of blue-emitting phosphor $Sr_8La_2(PO_4)_6O_2:Eu^{2+}$ for n-UV white-light-emitting diodes. Ceramics International 2014;40(8) 13709–13713.

[97] Roh Hee-S, Hur S, Song HJ, Park IJ, Yim DK, Kim Dong-W, Hong KS. Luminescence properties of $Ca_5(PO_4)_2SiO_4:Eu^{2+}$ green phosphor for near UV-based white LED. Materials Letters 2012;70 37–39.

[98] Augustine RL. Heterogeneous Catalysis for the Synthetic Chemist. CRC Press, 1995. ISBN: 978-0824790219

[99] Gruselle M. Apatites: A new family of catalysts in organic synthesis. Journal of Organometallic Chemistry 2015. In press. DOI: 10.1016/j.jorganchem.2015.01.018

[100] Maaten B, Moussa J, Desmarets C, Gredin P, Beaunier P, Kanger T, Tõnsuaadu K, Villemin D, Gruselle M. Cu-modified hydroxy-apatite as catalyst for Glaser-Hay Csingle bondC homo-coupling reaction of terminal alkynes. Journal of Molecular Catalysis A: Chemical 2014;393 112–116.

[101] Li JJ. Name Reactions for Homologation, Part I. John Wiley & Sons, 2009. ISBN: 978-0470487013.

[102] Edition D. Apatite catalyst used to synthesise n-butanol from ethanol. Focus on Catalysts 2005;4 5–6.

[103] Cubberly WH, Bakerjian R. Tool and Manufacturing Engineers Handbook Desk Edition. Society of Manufacturing Engineers, 1989. ISBN: 978-0872633513

[104] Wang LK, Hung Yung-T, Shammas NK. Handbook of Advanced Industrial and Hazardous Wastes Treatment. Advances in Industrial and Hazardous Wastes Treatment. CRC Press, 2009. ISBN: 978-1420072228

[105] Su Hsiang-Y, Lin Chao-S. Effect of additives on the properties of phosphate conversion coating on electrogalvanized steel sheet. Corrosion Science 2014;83 137–146.

[106] Liu B, Zhang X, Gui-y Xiao, Lu Yu-p. Phosphate chemical conversion coatings on metallic substrates for biomedical application: A review. Materials Science and Engineering: C 2015;47 97–104.

[107] Perrin FX, Gigandet MP, Wery M, Pagetti J. Chromium phosphate conversion coatings on zinc electroplates: cathodic formation and characterization. Surface and Coatings Technology 1998;105(1-2) 135–140.

[108] Wopenka B, Pasteris JD. A mineralogical perspective on the apatite in bone. Materials Science and Engineering: C 2005;25(2) 131–143.

[109] López-Álvarez M, Rodríguez-Valencia C, Serra J, González P. Bio-inspired ceramics: promising scaffolds for bone tissue engineering. Procedia Engineering 2013;59 51–58.

[110] Abdel-Hameed SAM, Hessien MM, Azooz MA. Preparation and characterization of some ferromagnetic glass-ceramics contains high quantity of magnetite. Ceramics International 2009;35(4) 1539–1544.

[111] Matsumotoa T, Uddina MH, Ana SH, Arakawab K, Taguchib E, Nakahirac A, Okazakid M. Modulation of nanotube formation in apatite single crystal via organic molecule incorporation. Materials Chemistry and Physics 2011;128(3) 495–499.

[112] Nakano T, Kaibara K, Ishimoto T, Tabata Y, Umakoshi Y. Biological apatite (BAp) crystallographic orientation and texture as a new index for assessing the microstructure and function of bone regenerated by tissue engineering. Bone 2012;51(4) 741–747.

[113] Venkatesan J, Bhatnagar I, Manivasagan P, Kang Kyong-H, Kim Se-K. Alginate composites for bone tissue engineering: A review. International Journal of Biological Macromolecules 2015;72 269–281.

[114] Liu Y, Lim J, Teoh Swee-H. Review: Development of clinically relevant scaffolds for vascularised bone tissue engineering. Biotechnology Advances 2013;31(5) 688–705.

[115] Armentano I, Dottori M, Fortunati E, Mattioli S, Kenny JM. Biodegradable polymer matrix nanocomposites for tissue engineering: A review. Polymer Degradation and Stability 2010;95(11) 2126–2146.

[116] Nicholson JW. The Chemistry of Medical and Dental Materials. RSC Materials Monographs — Volume 3. Royal Society of Chemistry, 2002. ISBN: 978-0854045723

[117] Gómez-Morales J, Iafisco Mi, Delgado-López JM, Sarda S, Drouet C. Progress on the preparation of nanocrystalline apatites and surface characterization: Overview of fundamental and applied aspects. Progress in Crystal Growth and Characterization of Materials 2013;59(1) 1–46.

[118] Rey C, Combes C, Drouet C, Sfihi H, Barroug A. Physico-chemical properties of nanocrystalline apatites: Implications for biominerals and biomaterials. Materials Science and Engineering: C 2007;27(2) 198–205.

[119] Szymendera J. Bone Mineral Metabolism in Cancer. Recent Results in Cancer Research — Volume 27. Springer Science & Business Media, 2012. ISBN: 978-3642999789

[120] Hillson S. Teeth. Cambridge Manuals in Archaeology. 2nd ed., Cambridge University Press, 2005. ISBN: 978-1139444057

[121] Wilson RM, Elliot JC, Dowker SEP. Rietveld refinement of the crystallographic structure of human dental enamel apatites Sample: RHB-enamel. American Mineralogist 1999;84 1406–1414.

[122] Krivovichev SV. Titul Minerals as Advanced Materials I. Springer Science & Business Media, 2008. ISBN: 978-3540771234

[123] Jamuna-Thevi K, Daud NM, Kadir MRA, Hermawan H. The influence of new wet synthesis route on the morphology, crystallinity and thermal stability of multiple ions doped nanoapatite. Ceramics International 2014;40(1) 1001–1012.

[124] Delgado-López JM, Iafisco M, Rodríguez I, Tampieri A, Prat M, Gómez-Morales J. Crystallization of bioinspired citrate-functionalized nanoapatite with tailored carbonate content. Acta Biomaterialia 2012;8(9) 3491–3499.

[125] Gualtieri ML, Ma Romagnoli, Hanuskova M, Fabbri E, Gualtieri AF. Facile synthesis of B-type carbonated nanoapatite with tailored microstructure. Journal of Solid State Chemistry 2014;220 60–69.

[126] Xia Z, Yu X, Jiang X, Brody HD, Rowe DW, Wei M. Fabrication and characterization of biomimetic collagen-apatite scaffolds with tunable structures for bone tissue engineering. Acta Biomaterialia 2013;9(7) 7308–7319.

[127] Sader MS, Martins VCA, Gomez S, LeGeros RZ, Soares GA. Production and in vitro characterization of 3D porous scaffolds made of magnesium carbonate apatite (MCA)/ anionic collagen using a biomimetic approach. Materials Science and Engineering: C 2013;33(7) 4188–4196.

[128] Fricain JC, Schlaubitz S, Le Visage C, Arnault I, Derkaoui SM, Siadous R, Catros S, Lalande C, Bareille R, Renard M, Fabre T, Cornet S, Durand M, Léonard A, Sahraoui N, Letourneur D, Amédée J. A nano-hydroxyapatite — Pullulan/dextran polysaccharide composite macroporous material for bone tissue engineering. Biomaterials 2013;34(12) 2947–2959.

[129] Shoulders MD, Raines RT: Collagen structure and stability. Annual Review of Biochemistry 2009 ;78 929–958.

[130] Wahl DA, Czernuszka JT. Collagen-hydroxyapatite composites for hard tissue repair. European Cells and Materials 2006;11 43–56.

[131] Magne D, Pilet P, Weiss P, Daculsi G. Fourier transform infrared microspectroscopic investigation of the maturation of nonstoichiometric apatites in mineralized tissues: a horse dentin study. Bone 2001;29(6) 547–552.

[132] Li H, Guo Z, Xue B, Zhang Y, Huang W. Collagen modulating crystallization of apatite in a biomimetic gel system. Ceramics International 2011;37(7) 2305–2310.

[133] Su X, Sun K, Cui FZ, Landis WJ. Organization of apatite crystals in human woven bone. Bone 2003;32(2) 150–162.

[134] Okazaki M, Ohmae H, Hino T. Insolubilization of apatite-collagen composites by UV irradiation. Biomaterials 1989;10(8) 564–568.

[135] Okazakl M, Ohmae H, Takahashi J, Kimura H, Sakuda M. Insolubilized properties of UV-irradiated C03 apatite-collagen composites. Biomaterials 1990;11(8) 568–572.

[136] Müller FA, Müller L, Caillard D, Conforto E. Preferred growth orientation of biomimetic apatite crystals. Journal of Crystal Growth 2007;304(2) 464–471.

[137] Fernandez-Yague MA, Abbah SA, McNamara L, Zeugolis DI, Pandit A, Biggs MJ. Biomimetic approaches in bone tissue engineering: Integrating biological and physicomechanical strategies. Advanced Drug Delivery Reviews 2015;84 1–29.

[138] Chu CL, Pu YP, Yin LH, Chung CY, Yeung KWK, Chu PK. Biomimetic deposition process of an apatite coating on NiTi shape memory alloy. Materials Letters 2006;60(24) 3002–3006.

[139] Shi J, Ding C, Wu Y. Biomimetic apatite layers on plasma-sprayed titanium coatings after surface modification. Surface and Coatings Technology 2001;137(1) 97–103.

[140] Song Won-H, Jun Youn-K, Han Y, Hong Seong-H. Biomimetic apatite coatings on micro-arc oxidized titania. Biomaterials 2004;25(17) 3341–3349.

[141] Faure J, Balamurugan A, Benhayoune H, Torres P, Balossier G, Ferreira JMF. Morphological and chemical characterisation of biomimetic bone like apatite formation

on alkali treated Ti_6Al_4V titanium alloy. Materials Science and Engineering: C 2009;29(4) 1252–1257.

[142] Liu X, Fu RKY, Poon RWY, Chen P, Chu PK, Ding C. Biomimetic growth of apatite on hydrogen-implanted silicon. Biomaterials 2004;25(25) 5575–5581.

[143] Ye M, Mohanty P, Ghosh G. Biomimetic apatite-coated porous PVA scaffolds promote the growth of breast cancer cells. Materials Science and Engineering: C 2014;44 310–316.

[144] Kunze J, Müller L, Macak JM, Greil P, Schmuki P, Müller FA. Time-dependent growth of biomimetic apatite on anodic TiO_2 nanotubes. Electrochimica Acta 2008;53(23) 6995–7003.

[145] Rambo CR, Müller FA, Müller L, Sieber H, Hofmann I, Greil P. Biomimetic apatite coating on biomorphous alumina scaffolds. Materials Science and Engineering: C 2006;26(1) 92–99.

[146] Stefanic M, Krnel K, Pribosic I, Kosmac T. Rapid biomimetic deposition of octacalcium phosphate coatings on zirconia ceramics (Y-TZP) for dental implant applications. Applied Surface Science 2012;258(10) 4649–4656.

[147] Tavangarian F, Emadi R. Improving degradation rate and apatite formation ability of nanostructure forsterite. Ceramics International 2011;37(7) 2275–2280.

[148] Wu C, Chang J. Synthesis and apatite-formation ability of akermanite. Materials Letters 2004;58(19) 2415–2417.

[149] Múzquiz-Ramos EM, Cortés-Hernández DA, Escobedo-Bocardo J. Biomimetic apatite coating on magnetite particles. Materials Letters 2010;64(9) 1117–1119.

[150] Fujibayashi S, Neo M, Kim Hyun-M, Kokubo T, Nakamura T. A comparative study between in vivo bone ingrowth and in vitro apatite formation on Na_2O-CaO-SiO_2 glasses. Biomaterials 2003;24(8) 1349–1356.

[151] Berbecaru C, Alexandru HV, Ianculescu Ad, Popescu A, Socol G, Sima F, Mihailescu I. Bioglass thin films for biomimetic implants. Applied Surface Science 2009;255(10) 5476–5479.

[152] Bingel L, Groh D, Karpukhina N, Brauer DS. Influence of dissolution medium pH on ion release and apatite formation of Bioglass® 45S5. Materials Letters 2015;143 279–282.

[153] Gandolfi MG, Taddei P, Tinti A, De Stefano Dorigo E, Prati C. Alpha-TCP improves the apatite-formation ability of calcium-silicate hydraulic cement soaked in phosphate solutions. Materials Science and Engineering: C 2011;31(7) 1412–1422.

[154] Pangdaeng S, Sata V, Aguiar JB, Pacheco-Torgal F, Chindaprasirt P. Apatite formation on calcined kaolin-white Portland cement geopolymer. Materials Science and Engineering: C 2015;51 1–6.

[155] Akasaka T, Watari F, Sato Y, Tohji K. Apatite formation on carbon nanotubes. Materials Science and Engineering: C 2006;26(4) 675–678.

[156] Wu C, Chan J. Bonelike apatite formation on carbon microspheres. Materials Letters 2007;61(11-12) 2502–2505.

[157] Pan H, Zhao X, Darvell BW, Lu WW. Apatite-formation ability — Predictor of "bioactivity"? Acta Biomaterialia 2010;6(11) 4181–4188.

[158] Tien Wen-B, Chen Ming-T, Yao Pin-C. Effects of pH and temperature on microstructure and morphology of hydroxyapatite/collagen composites synthesized in vitro. Materials Science and Engineering: C 2012;32(7) 2096–2102.

[159] Wang J, Liu C. Biomimetic collagen/hydroxyapatite composite scaffolds: Fabrication and characterizations. Journal of Bionic Engineering 2014;11(4) 600–609.

[160] Wenpo F, Gaofeng L, Shuying F, Yuanming Q, Keyong T. Preparation and characterization of collagen-hydroxyapatite/pectin composite. International Journal of Biological Macromolecules 2015;74 218–223.

[161] Holt C, Lenton S, Nylander T, Sørensen ES, Teixeira SCM. Mineralisation of soft and hard tissues and the stability of biofluids. Journal of Structural Biology 2014;185(3) 383–396.

[162] Kakkar P, Verma S, Manjubala I, Madhan B. Development of keratin-chitosan-gelatin composite scaffold for soft tissue engineering. Materials Science and Engineering: C 2014;45 343–347.

[163] Sharabi M, Mandelberg Y, Benayahu D, Benayahu Y, Azem A, Haj-Ali R. A new class of bio-composite materials of unique collagen fibers. Journal of the Mechanical Behavior of Biomedical Materials 2014;36 71–81.

[164] Marghussian V. Nano-Glass Ceramics: Processing, Properties and Applications. Elsevier Science, 2015. ISBN: 9780323354325

[165] Goodridge RD, Wood DJ, Ohtsuki C, Dalgarno KW. Biological evaluation of an apatite-mullite glass-ceramic produced via selective laser sintering. Acta Biomaterialia 2007;3(2) 221–231.